Human Embryology and Developmental Biology

WITHDRAWN

Health Library
Clinical Education Centre
ST4 6QG

Human Embryology and Developmental Biology

Fifth Edition

Bruce M. Carlson, MD, PhD
Professor Emeritus
Department of Cell and Developmental Biology
University of Michigan
Ann Arbor, Michigan

Contributor:
Piranit Nik Kantaputra, DDS, MS
Division of Pediatric Dentistry
Department of Orthodontics and Pediatric Dentistry
Faculty of Dentistry
Chiang Mai University
Chiang Mai, Thailand

SAUNDERS

ELSEVIER

1600 John F. Kennedy Blvd.
Ste 1800
Philadelphia, PA 19103-2899

Library of Congress Cataloging-in-Publication Data

Carlson, Bruce M.
    Human embryology and developmental biology / Bruce M. Carlson.—5th ed.
        p. ; cm.
    Includes bibliographical references and index.
    ISBN 978-1-4557-2794-0 (pbk.)
    I. Title.
    [DNLM: 1. Embryonic Development—physiology.   2. Fetal Development—physiology. WQ
210.5]   612.6'4—dc23

                                                                            2012036372

Content Strategist: Meghan Ziegler
Content Development Specialist: Andrea Vosburgh
Publishing Services Manager: Hemamalini Rajendrababu
Project Manager: Saravanan Thavamani
Design Direction: Louis Forgione
Illustrator: Alex Baker, DNA Illustrations, Inc.
Marketing Manager: Abigail Swartz

Printed in China

Last digit is the print number: 9 8 7 6 5 4 3 2 1

To Jean, for many wonderful years together.

Preface to the Fifth Edition

As was the case in the preparation of the fourth edition (and for that matter, also the previous editions), the conundrum facing me was what to include and what not to include in the text, given the continuing explosion of new information on almost every aspect of embryonic development. This question always leads me back to the fundamental question of what kind of book I am writing and what are my goals in writing it. As a starting point, I would go back to first principles and the reason why I wrote the first edition of this text. In the early 1990s, medical embryology was confronted with the issue of integrating traditional developmental anatomy with the newly burgeoning field of molecular embryology and introducing those already past their formal learning years to the fact that genes in organisms as foreign as *Drosophila* could have relevance in understanding the cause of human pathology or even normal development. This is no longer the case, and the issue today is how to place reasonable limits on coverage for an embryology text that is not designed to be encyclopedic.

For this text, my intention is to remain focused both on structure and on developmental mechanisms leading to structural and functional outcomes during embryogenesis. A good example is mention of the many hundreds of genes, mutations of which are known to produce abnormal developmental outcomes. If the mutation can be tied to a known mechanism that can illuminate how an organ develops, it would be a candidate for inclusion, whereas without that I feel that at present it is normally more appropriate to leave its inclusion in comprehensive human genetic compendia. Similarly, the issue of the level of detail of intracellular pathways to include often arises. Other than a few illustrative examples, I have chosen not to emphasize these pathways.

The enormous amount of new information on molecular networks and interacting pathways is accumulating to the point where new texts stressing these above other aspects of development could be profitably written. Often, where many molecules, whether transcription factors or signaling molecules, are involved in a developmental process, I have tried to choose what I feel are the most important and most distinctive, rather than to strive for completeness. Especially because so many major molecules or pathways are reused at different stages in the development of a single structure, my sense is that by including everything, the distinctiveness of the development of the different parts of the body would be blurred for the beginning student. As usual, I welcome feedback (brcarl@ umich.edu) and would be particularly interested to learn whether students or instructors believe that there is too much or too little molecular detail either overall or in specific areas.

In this edition, almost every chapter has been extensively revised, and more than 50 new figures have been added. Major additions of relevant knowledge of early development, especially related to the endoderm, have led to significant changes in Chapters 3, 5, 6, 14, and 15. Chapter 12 on the neural crest has been completely reorganized and was largely rewritten. Chapter 9 (on skin, skeleton, and muscle) has also seen major changes. Much new information on germ cells and early development of the gonads has been added to Chapter 16, and in Chapter 17 new information on the development of blood vessels and lymphatics has resulted in major changes.

For this edition, I have been fortunate in being allowed to use photographs from several important sources. From the late Professor Gerd Steding's *The Anatomy of the Human Embryo* (Karger) I have taken eight scanning electron micrographs of human embryos that illustrate better than drawings the external features of aspects of human development. I was also able to borrow six photographs of important congenital malformations from the extensive collection of the late Dr. Robert Gorlin, one of the fathers of syndromology. This inclusion is particularly poignant to me because while we were students at the University of Minnesota in the early 1960s, both my wife and I got to know him before he became famous. This edition includes a new Clinical Correlation on dental anomalies written by Dr. Pranit N. Kantaputra from the Department of Orthodontics and Pediatric Dentistry at Chiang Mai University in Chiang Mai, Thailand. He has assembled a wonderful collection of dental anomalies that have a genetic basis, and I am delighted to share his text and photos with the readers. Finally, I was able to include one digitized photograph of a sectioned human embryo from the Carnegie Collection. For this I thank Dr. Raymond Gasser for his herculean efforts in digitizing important specimens from that collection and making them available to the public. All these sections (labeled) are now available online through the Endowment for Human Development (www.ehd.ord), which is without question the best source of information on human embryology on the Internet. I would recommend this source to any student or instructor.

In producing this edition, I have been fortunate to be able to work with much of the team that was involved on the last edition. Alexandra Baker of DNA Illustrations, Inc. has successfully transformed my sketches into wonderful artwork for the past three editions. I thank her for her patience and her care. Similarly, Andrea Vosburgh and her colleagues at Elsevier have cheerfully succeeded in transforming a manuscript and all the trimmings into a recognizable book. Madelene Hyde efficiently guided the initial stages of contracts through the corporate labyrinth. Thanks, as always, to Jean, who provided a home environment compatible with the job of putting together a book and for putting up with me during the process.

Bruce M. Carlson

Contents

Developmental Tables

Carnegie Stages of Early Human Embryonic Development (Weeks 1 to 8)

Age (days)*	External Features	Carnegie Stage	Crown-Rump Length (mm)	Pairs of Somites
1	Fertilized oocyte	1	0.1	
2-3	Morula (4-16 cells)	2	0.1	
4-5	Free blastocyst	3	0.1	
6	Attachment of blastocyst to endometrium	4	0.1	
7-12	Implantation, bilaminar embryo with primary yolk sac	5	0.1-0.2	
17	Trilaminar embryo with primitive streak, chorionic villi	6	0.2-0.3	
19	Gastrulation, formation of notochordal process	7	0.4	
23	Hensen's node and primitive pit, notochord and neurenteric canal, appearance of neural plate, neural folds, and blood islands	8	1-1.5	
25	Appearance of first somites, deep neural groove, elevation of cranial neural folds, early heart tubes	9	1.5-2.5	1-3
28	Beginning of fusion of neural folds, formation of optic sulci, presence of first two pharyngeal arches, beginning heart beat, curving of embryo	10	2-3.5	4-12
29	Closure of cranial neuropore, formation of optic vesicles, rupture of oropharyngeal membrane	11	2.5-4.5	13-20
30	Closure of caudal neuropore, formation of pharyngeal arches 3 and 4, appearance of upper limb buds and tail bud, formation of otic vesicle	12	3-5	21-29
32	Appearance of lower limb buds, lens placode, separation of otic vesicle from surface ectoderm	13	4-6	30-31
33	Formation of lens vesicle, optic cup, and nasal pits	14	5-7	
36	Development of hand plates, primary urogenital sinus, prominent nasal pits, evidence of cerebral hemispheres	15	7-9	
38	Development of foot plates, visible retinal pigment, development of auricular hillocks, formation of upper lip	16	8-11	
41	Appearance of finger rays, rapid head enlargement, six auricular hillocks, formation of nasolacrimal groove	17	11-14	
44	Appearance of toe rays and elbow regions, beginning of formation of eyelids, tip of nose distinct, presence of nipples	18	13-17	
46	Elongation and straightening of trunk, beginning of herniation of midgut into umbilical cord	19	16-18	
49	Bending of arms at elbows, distinct but webbed fingers, appearance of scalp vascular plexus, degeneration of anal and urogenital membranes	20	18-22	
51	Longer and free fingers, distinct but webbed toes, indifferent external genitalia	21	22-24	
53	Longer and free toes, better development of eyelids and external ear	22	23-28	
56	More rounded head, fusion of eyelids	23	27-31	

*Based on additional specimen information, the ages of the embryos at specific stages have been updated from those listed in O'Rahilly and Müller in 1987. See O'Rahilly R, Müller F: *Human embryology and teratology,* ed 3, New York, 2001, Wiley-Liss, p 490.
Data from O'Rahilly R, Müller F: *Developmental stages in human embryos,* Publication 637, Washington, DC, 1987, Carnegie Institution of Washington.

Major Developmental Events during the Fetal Period

External Features	Internal Features
8 WEEKS	
Head is almost half the total length of fetus	Midgut herniation into umbilical cord occurs
Cervical flexure is about 30 degrees	Extraembryonic portion of allantois has degenerated
Indifferent external genitalia are present	Ducts and alveoli of lacrimal glands form
Eyes are converging	Paramesonephric ducts begin to regress in males
Eyelids are unfused	Recanalization of lumen of gut tube occurs
Tail disappears	Lungs are becoming glandlike
Nostrils are closed by epithelial plugs	Diaphragm is completed
Eyebrows appear	First ossification begins in skeleton
Urine is released into amniotic fluid	Definitive aortic arch system takes shape
9 WEEKS	
Neck develops and chin rises from thorax	Intestines are herniated into umbilical cord
Cranial flexure is about 22 degrees	Early muscular movements occur
Chorion is divided into chorion laeve and chorion frondosum	Adrenocorticotropic hormone and gonadotropins are produced by pituitary
Eyelids meet and fuse	Corticosteroids are produced by adrenal cortex
External genitalia begin to become gender specific	Semilunar valves in heart are completed
Amniotic fluid is swallowed	Fused paramesonephric ducts join vaginal plate
Thumb sucking and grasping begin	Urethral folds begin to fuse in males
10 WEEKS	
Cervical flexure is about 15 degrees	Intestines return into body cavity from umbilical cord
Gender differences are apparent in external genitalia	Bile is secreted
Fingernails appear	Blood islands are established in spleen
Eyelids are fused	Thymus is infiltrated by lymphoid stem cells
Fetal yawning occurs	Prolactin production by pituitary occurs First permanent tooth buds form Deciduous teeth are in early bell stage Epidermis has three layers
11 WEEKS	
Cervical flexure is about 8 degrees	Stomach musculature can contract
Nose begins to develop bridge	T lymphocytes emigrate into bloodstream
Taste buds cover inside of mouth	Colloid appears in thyroid follicles Intestinal absorption begins
12 WEEKS	
Head is erect	Ovaries descend below pelvic rim
Neck is almost straight and well defined	Parathyroid hormone is produced
External ear is taking form and has moved close to its definitive position in the head	Blood can coagulate
Yolk sac has shrunk	
Fetus can respond to skin stimulation	
Bowel movements begin (meconium expelled)	
4 MONTHS	
Skin is thin; blood vessels can easily be seen through it	Seminal vesicle forms
Nostrils are almost formed	Transverse grooves appear on dorsal surface of cerebellum
Eyes have moved to front of face	Bile is produced by liver and stains meconium green
Legs are longer than arms	Gastric glands bud off from gastric pits
Fine lanugo hairs appear on head	Brown fat begins to form

Major Developmental Events during the Fetal Period—cont'd

External Features	Internal Features
Fingernails are well formed; toenails are forming	Pyramidal tracts begin to form in brain
Epidermal ridges appear on fingers and palms of hand	Hematopoiesis begins in bone marrow
Enough amniotic fluid is present to permit amniocentesis	Ovaries contain primordial follicles
Mother can feel fetal movements	
5 MONTHS	
Epidermal ridges form on toes and soles of feet	Myelination of spinal cord begins
Vernix caseosa begins to be deposited on skin	Sebaceous glands begin to function
Abdomen begins to fill out	Thyroid-stimulating hormone is released by pituitary
Eyelids and eyebrows develop	Testes begin to descend
Lanugo hairs cover most of body	
6 MONTHS	
Skin is wrinkled and red	Surfactant begins to be secreted
Decidua capsularis degenerates because of reduced blood supply	Tip of spinal cord is at S1 level
Lanugo hairs darken	
Odor detection and taste occur	
7 MONTHS	
Eyelids begin to open	Sulci and gyri begin to appear on brain
Eyelashes are well developed	Subcutaneous fat storage begins
Scalp hairs are lengthening (longer than lanugo)	Testes are descending into scrotum
Skin is slightly wrinkled	Termination of splenic erythropoiesis occurs
Breathing movements are common	
8 MONTHS	
Skin is pink and smooth	Regression of hyaloid vessels from lens occurs
Eyes are capable of pupillary light reflex	Testes enter scrotum
Fingernails have reached tips of fingers	
9 MONTHS	
Toenails have reached tips of toes	Larger amounts of pulmonary surfactant are secreted
Most lanugo hairs are shed	Ovaries are still above brim of pelvis
Skin is covered with vernix caseosa	Testes have descended into scrotum
Attachment of umbilical cord becomes central in abdomen	Tip of spinal cord is at L3
About 1 L of amniotic fluid is present	Myelination of brain begins
Placenta weighs about 500 g	
Fingernails extend beyond fingertips	
Breasts protrude and secrete "witch's milk"	

Part I

Early Development and the Fetal-Maternal Relationship

Chapter 1

Getting Ready for Pregnancy

Human pregnancy begins with the fusion of an egg and a sperm within the female reproductive tract, but extensive preparation precedes this event. First, both male and female sex cells must pass through a long series of changes (**gametogenesis**) that convert them genetically and phenotypically into mature **gametes**, which are capable of participating in the process of fertilization. Next, the gametes must be released from the gonads and make their way to the upper part of the uterine tube, where fertilization normally takes place. Finally, the fertilized egg, now properly called an **embryo**, must enter the uterus, where it sinks into the uterine lining (**implantation**) to be nourished by the mother. All these events involve interactions between the gametes or embryo and the adult body in which they are housed, and most of them are mediated or influenced by parental hormones. This chapter focuses on gametogenesis and the hormonal modifications of the body that enable reproduction to occur.

Gametogenesis

Gametogenesis is typically divided into four phases: (1) the extraembryonic origin of the germ cells and their migration into the gonads, (2) an increase in the number of germ cells by mitosis, (3) a reduction in chromosomal number by meiosis, and (4) structural and functional maturation of the eggs and spermatozoa. The first phase of gametogenesis is identical in males and females, whereas distinct differences exist between the male and female patterns in the last three phases.

Phase 1: Origin and Migration of Germ Cells

Primordial germ cells, the earliest recognizable precursors of gametes, arise outside the gonads and migrate into the gonads during early embryonic development. Human primordial germ cells first become readily recognizable at 24 days after fertilization in the endodermal layer of the yolk sac (**Fig. 1.1A**) by their large size and high content of the enzyme alkaline phosphatase. In the mouse, their origin has been traced even earlier in development (see p. 390). Germ cells exit from the

yolk sac into the hindgut epithelium and then migrate* through the dorsal mesentery until they reach the primordia of the gonads (Fig. 1.1B). In the mouse, an estimated 100 cells leave the yolk sac, and through mitotic multiplication (6 to 7 rounds of cell division), about 4000 primordial germ cells enter the primitive gonads.

Misdirected primordial germ cells that lodge in extragonadal sites usually die, but if such cells survive, they may develop into **teratomas**. Teratomas are bizarre growths that contain scrambled mixtures of highly differentiated tissues, such as skin, hair, cartilage, and even teeth (**Fig. 1.2**). They are found in the mediastinum, the sacrococcygeal region, and the oral region.

Phase 2: Increase in the Number of Germ Cells by Mitosis

After they arrive in the gonads, the primordial germ cells begin a phase of rapid mitotic proliferation. In a mitotic division, each germ cell produces two **diploid** progeny that are genetically equal. Through several series of mitotic divisions, the number of primordial germ cells increases exponentially from hundreds to millions. The pattern of mitotic proliferation differs markedly between male and female germ cells. **Oogonia**, as mitotically active germ cells in the female are called, go through a period of intense mitotic activity in the embryonic ovary from the second through the fifth month of pregnancy in the human. During this period, the population of germ cells increases from only a few thousand to nearly 7 million (**Fig. 1.3**). This number represents the maximum number of germ cells that is ever found in the ovaries. Shortly thereafter, numerous oogonia undergo a natural degeneration called **atresia**. Atresia of germ cells is a continuing feature of the histological landscape of the human ovary until menopause.

---

*Considerable controversy surrounds the use of the term "migration" with respect to embryonic development. On the one hand, some believe that displacements of cells relative to other structural landmarks in the embryo are due to active migration (often through ameboid motion). On the other hand, others emphasize the importance of directed cell proliferation and growth forces in causing what is interpreted as apparent migration of cells. As is often true in scientific controversies, both active migration and displacement as a result of growth seem to operate in many cases where cells in the growing embryo appear to shift with respect to other structural landmarks.

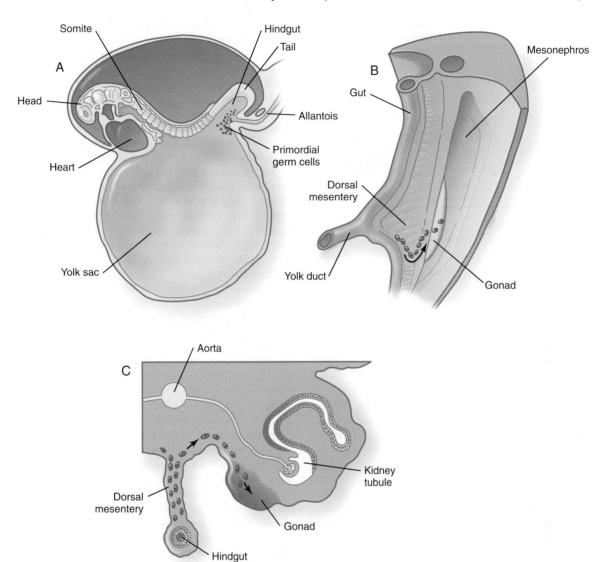

Fig. 1.1 **Origin and migration of primordial germ cells in the human embryo.** A, Location of primordial germ cells in the 16-somite human embryo (midsagittal view). B, Pathway of migration *(arrow)* through the dorsal mesentery. C, Cross section showing the pathway of migration *(arrows)* through the dorsal mesentery and into the gonad.

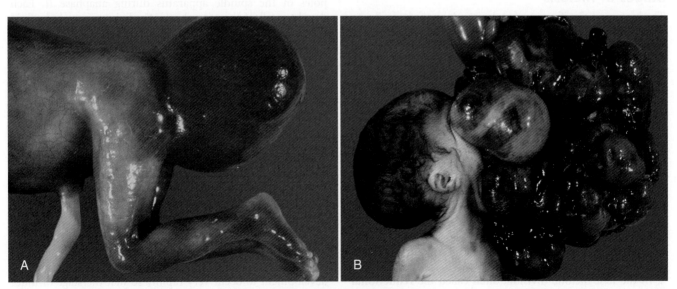

Fig. 1.2 A, Sacrococcygeal teratoma in a fetus. B, Massive oropharyngeal teratoma. *(Courtesy of M. Barr, Ann Arbor, Mich.)*

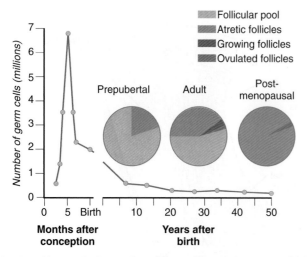

Fig. 1.3 Changes in the number of germ cells and proportions of follicle types in the human ovary with increasing age. *(Based on Baker TG: In Austin CR, Short RV: Germ cells and fertilization (reproduction in mammals), vol 1, Cambridge, 1970, Cambridge University Press, p 20; and Goodman AL, Hodgen GD: The ovarian triad of the primate menstrual cycle, Recent Prog Horm Res 39:1-73, 1983.)*

Spermatogonia, which are the male counterparts of oogonia, follow a pattern of mitotic proliferation that differs greatly from that in the female. Mitosis also begins early in the embryonic testes, but in contrast to female germ cells, male germ cells maintain the ability to divide throughout postnatal life. The seminiferous tubules of the testes are lined with a germinative population of spermatogonia. Beginning at puberty, subpopulations of spermatogonia undergo periodic waves of mitosis. The progeny of these divisions enter meiosis as synchronous groups. This pattern of spermatogonial mitosis continues throughout life.

Phase 3: Reduction in Chromosomal Number by Meiosis

Stages of Meiosis

The biological significance of meiosis in humans is similar to that in other species. Of primary importance are (1) reduction of the number of chromosomes from the diploid (2n) to the **haploid** (1n) number so that the species number of chromosomes can be maintained from generation to generation, (2) independent reassortment of maternal and paternal chromosomes for better mixing of genetic characteristics, and (3) further redistribution of maternal and paternal genetic information through the process of crossing-over during the first meiotic division.

Meiosis involves two sets of divisions (**Fig. 1.4**). Before the first meiotic division, deoxyribonucleic acid (DNA) replication has already occurred, so at the beginning of meiosis, the cell is 2n, 4c. (In this designation, **n** is the species number of chromosomes, and **c** is the amount of DNA in a single set [n] of chromosomes.) The cell contains the normal number (2n) of chromosomes, but as a result of replication, its DNA content (4c) is double the normal amount (2c).

In the first meiotic division, often called the **reductional division**, a prolonged prophase (see Fig. 1.4) leads to the

pairing of homologous chromosomes and frequent **crossing-over**, resulting in the exchange of segments between members of the paired chromosomes. Crossing-over even occurs in the sex chromosomes. This takes place in a small region of homology between the X and Y chromosomes. Crossing-over is not a purely random process. Rather, it occurs at sites along the chromosomes known as **hot spots**. Their location is based on configurations of proteins that organize the chromosomes early in meiosis. One such protein is **cohesin**, which helps to hold sister chromatids together during division. Hypermethylation of histone proteins in the chromatin indicates specific sites where the DNA strands break and are later repaired after crossing-over is completed. Another protein, **condensin**, is important in compaction of the chromosomes, which is necessary for both mitotic and meiotic divisions to occur.

During metaphase of the first meiotic division, the chromosome pairs (**tetrads**) line up at the metaphase (equatorial) plate so that at anaphase I, one chromosome of a homologous pair moves toward one pole of the spindle, and the other chromosome moves toward the opposite pole. This represents one of the principal differences between a meiotic and a mitotic division. In a mitotic anaphase, the centromere between the sister chromatids of each chromosome splits after the chromosomes have lined up at the metaphase plate, and one chromatid from each chromosome migrates to each pole of the mitotic spindle. This activity results in genetically equal daughter cells after a mitotic division, whereas the daughter cells are genetically unequal after the first meiotic division. Each daughter cell of the first meiotic division contains the haploid (1n) number of chromosomes, but each chromosome still consists of two chromatids (2c) connected by a centromere. No new duplication of chromosomal DNA is required between the first and second meiotic divisions because each haploid daughter cell resulting from the first meiotic division already contains chromosomes in the replicated state.

The second meiotic division, called the **equational division**, is similar to an ordinary mitotic division except that before division the cell is haploid (1n, 2c). When the chromosomes line up along the equatorial plate at metaphase II, the centromeres between sister chromatids divide, allowing the sister chromatids of each chromosome to migrate to opposite poles of the spindle apparatus during anaphase II. Each daughter cell of the second meiotic division is truly haploid (1n, 1c).

Meiosis in Females

The period of meiosis involves other cellular activities in addition to the redistribution of chromosomal material. As the oogonia enter the first meiotic division late in the fetal period, they are called **primary oocytes**.

Meiosis in the human female is a very leisurely process. As the primary oocytes enter the diplotene stage of the first meiotic division in the early months after birth, the first of two blocks in the meiotic process occurs (**Fig. 1.5**). The suspended diplotene phase of meiosis is the period when the primary oocyte prepares for the needs of the embryo. In oocytes of amphibians and other lower vertebrates, which must develop outside the mother's body and often in a hostile environment, it is highly advantageous for the early stages of development to occur very rapidly so that the stage of independent locomotion and feeding is attained as soon as pos-

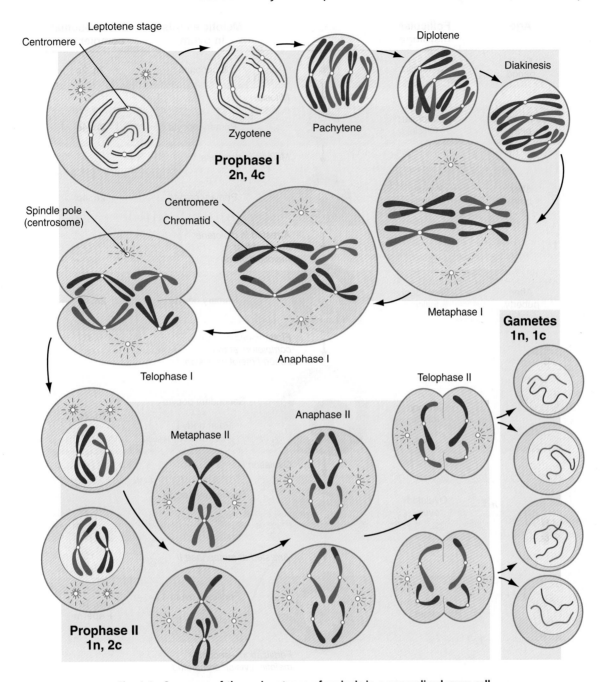

Fig. 1.4 **Summary of the major stages of meiosis in a generalized germ cell.**

sible. These conditions necessitate a strategy of storing up the materials needed for early development well in advance of ovulation and fertilization because normal synthetic processes would not be rapid enough to produce the materials required for the rapidly cleaving embryo. In such species, yolk is accumulated, the genes for producing ribosomal ribonucleic acid (rRNA) are amplified, and many types of RNA molecules are synthesized and stored in an inactive form for later use.

RNA synthesis in the amphibian oocyte occurs on the lampbrush chromosomes, which are characterized by many prominent loops of spread-out DNA on which messenger RNA (mRNA) molecules are synthesized. The amplified genes for producing rRNA are manifested by the presence of 600 to 1000 nucleoli within the nucleus. Primary oocytes also prepare

for fertilization by producing several thousand cortical granules, which are of great importance during the fertilization process (see Chapter 2).

The mammalian oocyte prepares for an early embryonic period that is more prolonged than that of amphibians and that occurs in the nutritive environment of the maternal reproductive tract. Therefore, it is not faced with the need to store as great a quantity of materials as are the eggs of lower vertebrates. As a consequence, the buildup of yolk is negligible. Evidence indicates, however, a low level of ribosomal DNA (rDNA) amplification (two to three times) in diplotene human oocytes, a finding suggesting that some degree of molecular advance planning is also required to support early cleavage in the human. The presence of 2 to 40 small (2-μm) RNA-

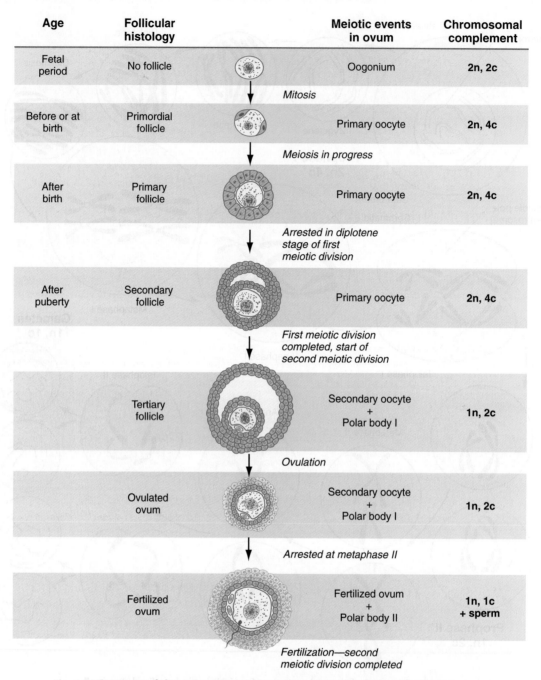

Age	Follicular histology	Meiotic events in ovum	Chromosomal complement
Fetal period	No follicle	Oogonium	2n, 2c
		Mitosis	
Before or at birth	Primordial follicle	Primary oocyte	2n, 4c
		Meiosis in progress	
After birth	Primary follicle	Primary oocyte	2n, 4c
		Arrested in diplotene stage of first meiotic division	
After puberty	Secondary follicle	Primary oocyte	2n, 4c
		First meiotic division completed, start of second meiotic division	
	Tertiary follicle	Secondary oocyte + Polar body I	1n, 2c
		Ovulation	
	Ovulated ovum	Secondary oocyte + Polar body I	1n, 2c
		Arrested at metaphase II	
	Fertilized ovum	Fertilized ovum + Polar body II	1n, 1c + sperm
		Fertilization—second meiotic division completed	

Fig. 1.5 **Summary of the major events in human oogenesis and follicular development.**

containing micronuclei (miniature nucleoli) per oocyte nucleus correlates with the molecular data.

Human diplotene chromosomes do not appear to be arranged in a true lampbrush configuration, and massive amounts of RNA synthesis seem unlikely. The developing mammalian (mouse) oocyte produces 10,000 times less rRNA and 1000 times less mRNA than its amphibian counterpart. Nevertheless, there is a steady accumulation of mRNA and a proportional accumulation of rRNA. These amounts of maternally derived RNA seem to be enough to take the fertilized egg through the first couple of cleavage divisions, after which the embryonic genome takes control of macromolecular synthetic processes.

Because cortical granules play an important role in preventing the entry of excess spermatozoa during fertilization in human eggs (see p. 31), the formation of cortical granules (mainly from the Golgi apparatus) continues to be one of the functions of the diplotene stage that is preserved in humans. Roughly 4500 cortical granules are produced in the mouse oocyte. A higher number is likely in the human oocyte.

Unless they degenerate, all primary oocytes remain arrested in the diplotene stage of meiosis until puberty. During the reproductive years, small numbers (10 to 30) of primary oocytes complete the first meiotic division with each menstrual cycle and begin to develop further. The other primary oocytes remain arrested in the diplotene stage, some for 50 years.

With the completion of the first meiotic division shortly before ovulation, two unequal cellular progeny result. One is a large cell, called the **secondary oocyte**. The other is a small cell called the **first polar body** (see Fig. 1.5). The secondary oocytes begin the second meiotic division, but again the meiotic process is arrested, this time at metaphase. The stimulus for the release from this meiotic block is fertilization by a spermatozoon. Unfertilized secondary oocytes fail to complete the second meiotic division. The second meiotic division is also unequal; one of the daughter cells is relegated to becoming a second polar body. The first polar body may also divide during the second meiotic division. Formation of both the first and second polar bodies involves highly asymmetric cell divisions. To a large extent, this is accomplished by displacement of the mitotic spindle apparatus toward the periphery of the oocyte through the actions of the cytoskeletal protein **actin** (see Fig. 2.7).

Meiosis in Males

Meiosis in the male does not begin until after puberty. In contrast to the primary oocytes in the female, not all spermatogonia enter meiosis at the same time. Large numbers of spermatogonia remain in the mitotic cycle throughout much of the reproductive lifetime of males. When the progeny of a spermatogonium have entered the meiotic cycle as **primary spermatocytes**, they spend several weeks passing through the first meiotic division (**Fig. 1.6**). The result of the first meiotic division is the formation of two **secondary spermatocytes**, which immediately enter the second meiotic division. About 8 hours later, the second meiotic division is completed, and four haploid (1n, 1c) **spermatids** remain as progeny of the single primary spermatocyte. The total length of human spermatogenesis is 64 days.

Disturbances that can occur during meiosis and result in chromosomal aberrations are discussed in **Clinical Correlation 1.1** and Figure 1.7.

Phase 4: Final Structural and Functional Maturation of Eggs and Sperm
Oogenesis

Of the roughly 2 million primary oocytes present in the ovaries at birth, only about 40,000—all of which are arrested in the diplotene stage of the first meiotic division—survive until puberty. From this number, approximately 400 (1 per menstrual cycle) are actually ovulated. The rest of the primary oocytes degenerate without leaving the ovary, but many of them undergo some further development before becoming atretic. Although some studies suggested that adult mammalian ovaries contain primitive cells that can give rise to new oocytes, such reports remain controversial.

The egg, along with its surrounding cells, is called a **follicle**. Maturation of the egg is intimately bound with the development of its cellular covering. Because of this, considering the

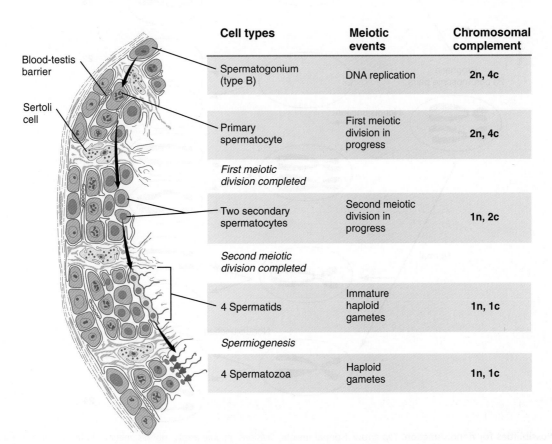

Cell types	Meiotic events	Chromosomal complement
Spermatogonium (type B)	DNA replication	**2n, 4c**
Primary spermatocyte	First meiotic division in progress	**2n, 4c**
First meiotic division completed		
Two secondary spermatocytes	Second meiotic division in progress	**1n, 2c**
Second meiotic division completed		
4 Spermatids	Immature haploid gametes	**1n, 1c**
Spermiogenesis		
4 Spermatozoa	Haploid gametes	**1n, 1c**

Blood-testis barrier

Sertoli cell

Fig. 1.6   **Summary of the major events in human spermatogenesis.**

CLINICAL CORRELATION 1.1
Meiotic Disturbances Resulting in Chromosomal Aberrations

Chromosomes sometimes fail to separate during meiosis, a phenomenon known as **nondisjunction**. As a result, one haploid daughter gamete contains both members of a chromosomal pair for a total of 24 chromosomes, whereas the other haploid gamete contains only 22 chromosomes (**Fig. 1.7**). When such gametes combine with normal gametes of the opposite sex (with 23 chromosomes), the resulting embryos contain 47 chromosomes (with a **trisomy** of 1 chromosome) or 45 chromosomes (**monosomy** of 1 chromosome). (Specific syndromes associated with the nondisjunction of chromosomes are summarized in Chapter 8.) The generic term given to a condition characterized by an abnormal number of chromosomes is **aneuploidy**.

In other cases, part of a chromosome can be **translocated** to another chromosome during meiosis, or part of a chromosome can be **deleted**. Similarly, duplications or inversions of parts of chromosomes occasionally occur during meiosis. These conditions may result in syndromes similar to those seen after the nondisjunction of entire chromosomes. Under some circumstances (e.g., simultaneous fertilization by two spermatozoa, failure of the second polar body to separate from the oocyte during the second meiotic division), the cells of the embryo contain more than two multiples of the haploid number of chromosomes (**polyploidy**).

Chromosomal abnormalities are the underlying cause of a high percentage of spontaneous abortions during the early

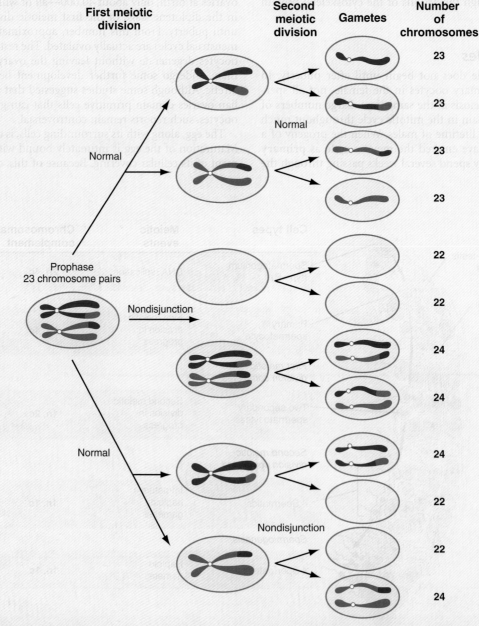

Fig. 1.7 **Possibilities for nondisjunction.** *Top arrow*, Normal meiotic divisions; *middle arrow*, nondisjunction during the first meiotic division; *bottom arrow*, nondisjunction during the second meiotic division.

CLINICAL CORRELATION 1.1
Meiotic Disturbances Resulting in Chromosomal Aberrations—cont'd

weeks of pregnancy. More than 75% of spontaneous abortions occurring before the second week and more than 60% of those occurring during the first half of pregnancy contain chromosomal abnormalities ranging from trisomies of individual chromosomes to overall polyploidy. Although the incidence of chromosomal anomalies declines with stillbirths occurring after the fifth month of pregnancy, it is close to 6%, a 10-fold higher incidence over the 0.5% of living infants who are born with chromosomal anomalies. In counseling patients who have had a stillbirth or a spontaneous abortion, it can be useful to mention that this is often nature's way of handling an embryo destined to be highly abnormal.

development of the egg and its surrounding follicular cells as an integrated unit is a useful approach in the study of oogenesis.

In the embryo, oogonia are naked, but after meiosis begins, cells from the ovary partially surround the primary oocytes to form **primordial follicles** (see Fig. 1.5). By birth, the primary oocytes are invested with a complete layer of follicular cells, and the complex of primary oocyte and the follicular (granulosa) cells is called a **primary follicle** (**Fig. 1.8**). Both the oocyte and the surrounding follicular cells develop prominent microvilli and gap junctions that connect the two cell types.

The meiotic arrest at the diplotene stage of the first meiotic division is the result of a complex set of interactions between the oocyte and its surrounding layer of follicular (granulosa) cells. The principal factor in maintaining meiotic arrest is a high concentration of cyclic adenosine monophosphate (cAMP) in the cytoplasm of the oocyte (**Fig. 1.9**). This is accomplished by both the intrinsic production of cAMP by the oocyte and the production of cAMP by the follicular cells and its transport into the oocyte through gap junctions connecting the follicular cells to the oocyte. In addition, the follicular cells produce and transport into the oocyte cyclic guanosine monophosphate (cGMP), which inactivates **phosphodiesterase 3A** (**PDE3A**), an enzyme that converts cAMP to 5′AMP. The high cAMP within the oocyte inactivates **maturation promoting factor** (**MPF**), which at a later time functions to lead the oocyte out of meiotic arrest and to complete the first meiotic division.

As the primary follicle takes shape, a prominent, translucent, noncellular membrane called the **zona pellucida** forms between the primary oocyte and its enveloping follicular cells (**Fig. 1.10**). The microvillous connections between the oocyte and follicular cells are maintained through the zona pellucida. In rodents, the components of the zona pellucida (four glycoproteins and glycosaminoglycans) are synthesized almost entirely by the egg, but in other mammals, follicular cells also contribute materials to the zona. The zona pellucida contains sperm receptors and other components that are important in fertilization and early postfertilization development. (The functions of these molecules are discussed more fully in Chapter 2.)

In the prepubertal years, many of the primary follicles enlarge, mainly because of an increase in the size of the oocyte (up to 300-fold) and the number of follicular cells. An oocyte

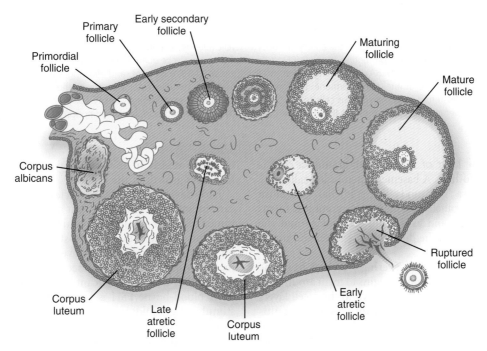

Fig. 1.8 The sequence of maturation of follicles within the ovary, starting with the primordial follicle and ending with the formation of a corpus albicans.

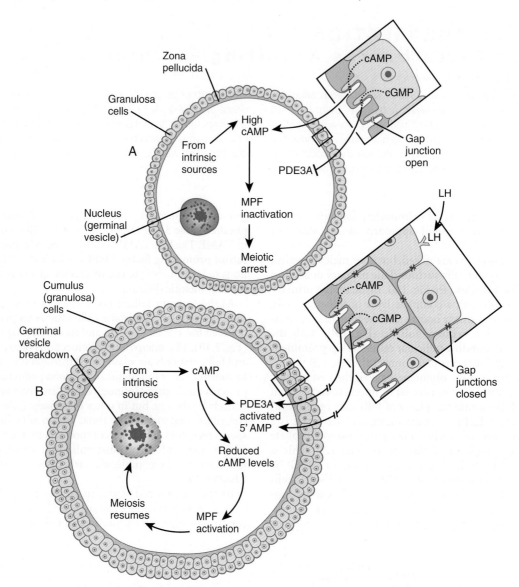

Fig. 1.9 **A,** Major steps leading to meiotic arrest in the embryonic oocyte. Cyclic adenosine monophosphate (cAMP), contributed by both the oocyte and follicular cells, inactivates maturation promoting factor (MPF), a driver of meiosis. Cyclic guanosine monophosphate (cGMP) from the follicular cells inactivates phosphodiesterase 3A (PDE3A), preventing it from breaking down cAMP molecules, and allowing a high concentration of cAMP in the oocyte. **B,** Under the influence of luteinizing hormone (LH), gap junctions of the cumulus cells close down, thus reducing the amount of both cAMP and cGMP that is transferred from the cumulus cells to the oocyte. The reduction in cGMP activates PDE3A, whose action breaks down cAMP within the oocyte. The lowered concentration of cAMP within the oocyte activates MPF and stimulates resumption of meiosis.

with more than one layer of surrounding granulosa cells is called a **secondary follicle**. A basement membrane called the **membrana granulosa** surrounds the epithelial **granulosa cells** of the secondary follicle. The membrana granulosa forms a barrier to capillaries, and as a result, the oocyte and the granulosa cells depend on the diffusion of oxygen and nutrients for their survival.

An additional set of cellular coverings, derived from the ovarian connective tissue (**stroma**), begins to form around the developing follicle after it has become two to three cell layers thick. Known initially as the **theca folliculi**, this covering ultimately differentiates into two layers: a highly vascularized and glandular **theca interna** and a more connective tissue–like outer capsule called the **theca externa**. The early thecal cells secrete an **angiogenesis factor**, which stimulates the growth

of blood vessels in the thecal layer. This nutritive support facilitates growth of the follicle.

Early development of the follicle occurs without the significant influence of hormones, but as puberty approaches, continued follicular maturation requires the action of the pituitary gonadotropic hormone **follicle-stimulating hormone** (**FSH**) on the granulosa cells, which have by this time developed FSH receptors on their surfaces (see Fig. 1.10). In addition, the oocyte itself exerts a significant influence on follicular growth. After blood-borne FSH is bound to the FSH receptors, the stimulated granulosa cells produce small amounts of **estrogens**. The most obvious indication of the further development of some of the follicles is the formation of an **antrum**, a cavity filled with a fluid called **liquor folliculi**. Initially formed by secretions of the follicular cells, the antral fluid is later formed

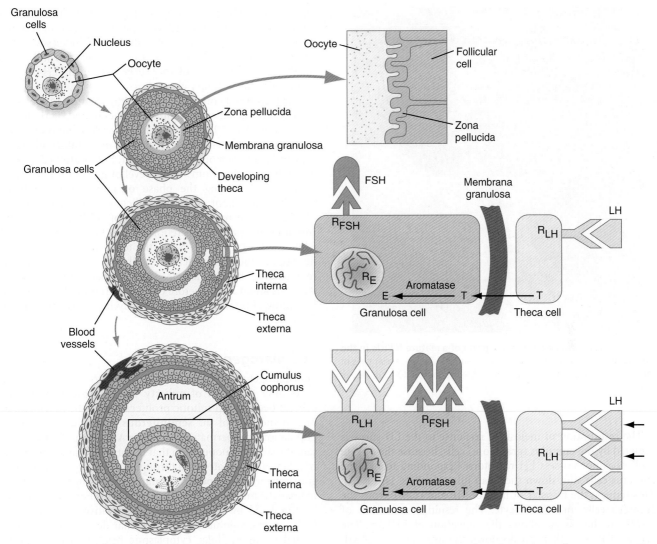

Fig. 1.10 Growth and maturation of a follicle along with major endocrine interactions in the theca cells and granulosa cells. E, estrogen; FSH, follicle-stimulating hormone; LH, luteinizing hormone; R, receptor; T, testosterone.

mostly as a transudate from the capillaries on the outer side of the membrana granulosa.

Formation of the antrum divides the follicular cells into two groups. The cells immediately surrounding the oocyte are called **cumulus cells**, and the cells between the antrum and the membrana granulosa become the **mural granulosa cells**. Factors secreted by the oocyte confer different properties on the cumulus cells from the mural granulosa cells. In the absence of a direct stimulus from the oocyte, the granulosa cells follow a default pathway and begin to assemble hormone receptors on their surface (see Fig. 1.10). In contrast, the cumulus cells do not express hormone receptors, but under the influence of the oocyte, they undergo changes that facilitate the release of the ovum at the time of ovulation.

Enlargement of the follicle results largely from the proliferation of granulosa cells. The direct stimulus for granulosa cell proliferation is a locally produced signaling protein, **activin**, a member of the **transforming growth factor-β** family of signaling molecules (see Table 4.1). The local action of activin is enhanced by the actions of FSH.

Responding to the stimulus of pituitary hormones, secondary follicles produce significant amounts of steroid hormones. The cells of the theca interna possess receptors for **luteinizing hormone** (**LH**), also secreted by the anterior pituitary (see Fig. 1.15). The theca interna cells produce **androgens** (e.g., testosterone), which pass through the membrana granulosa to the granulosa cells. The influence of FSH induces the granulosa cells to synthesize the enzyme (**aromatase**), which converts the theca-derived androgens into estrogens (mainly 17β-estradiol). Not only does the estradiol leave the follicle to exert important effects on other parts of the body, but also it stimulates the formation of LH receptors on the granulosa cells. Through this mechanism, the follicular cells are able to respond to the large LH surge that immediately precedes ovulation (see Fig. 1.16).

Under multiple hormonal influences, the follicle enlarges rapidly (**Fig. 1.11**; see Fig. 1.10) and presses against the surface of the ovary. At this point, it is called a **tertiary** (**graafian**) **follicle**. About 10 to 12 hours before ovulation, meiosis resumes.

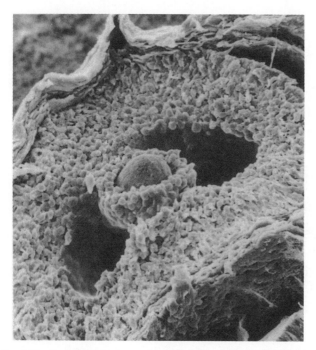

Fig. 1.11 **Scanning electron micrograph of a mature follicle in the rat ovary.** The spherical oocyte *(center)* is surrounded by smaller cells of the corona radiata, which projects into the antrum. (×840.) *(Courtesy of P. Bagavandoss, Ann Arbor, Mich.)*

glycans. The strong negative charge of the proteoglycans attracts water molecules, and with greater amounts of secreted proteoglycans, the volume of antral fluid increases correspondingly. The follicle is now poised for ovulation and awaits the stimulus of the preovulatory surge of FSH and LH released by the anterior pituitary gland.

The reason only one follicle normally matures to the point of ovulation is still not completely understood. Early in the cycle, as many as 50 follicles begin to develop, but only about 3 attain a diameter of as great as 8 mm. Initial follicular growth is gonadotropin independent, but continued growth depends on a minimum "tonic" level of gonadotropins, principally FSH. During the phase of gonadotropin-induced growth, a dominant enlarging follicle becomes independent of FSH and secretes large amounts of **inhibin** (see p. 19). Inhibin suppresses the secretion of FSH by the pituitary, and when the FSH levels fall below the tonic threshold, the other developing follicles, which are still dependent on FSH for maintenance, become atretic. The dominant follicle acquires its status about 7 days before ovulation. It may also secrete an inhibiting substance that acts directly on the other growing follicles.

Spermatogenesis

Spermatogenesis begins in the seminiferous tubules of the testes after the onset of puberty. In the broadest sense, the process begins with mitotic proliferation of the spermatogonia. At the base of the **seminiferous epithelium** are several populations of spermatogonia. **Type A spermatogonia** represent the stem cell population that mitotically maintains proper numbers of spermatogonia throughout life. Type A spermatogonia give rise to **type B spermatogonia**, which are destined to leave the mitotic cycle and enter meiosis. Entry into meiosis is stimulated by **retinoic acid** (a derivative of vitamin A). Many spermatogonia and their cellular descendants are connected by intercellular cytoplasmic bridges, which may be instrumental in maintaining the synchronous development of large clusters of sperm cells.

All spermatogonia are sequestered at the base of the seminiferous epithelium by interlocking processes of **Sertoli cells**, which are complex cells that are regularly distributed throughout the periphery of the seminiferous epithelium and that occupy about 30% of its volume (see Fig. 1.6). As the progeny of the type B spermatogonia (called **primary spermatocytes**) complete the leptotene stage of the first meiotic division, they pass through the Sertoli cell barrier to the interior of the seminiferous tubule. This translocation is accomplished by the formation of a new layer of Sertoli cell processes beneath these cells and, slightly later, the dissolution of the original layer that was between them and the interior of the seminiferous tubule. The Sertoli cell processes are very tightly joined and form an immunological barrier (**blood-testis barrier** [see Fig. 1.6]) between the forming sperm cells and the rest of the body, including the spermatogonia. When they have begun meiosis, developing sperm cells are immunologically different from the rest of the body. Autoimmune infertility can arise if the blood-testis barrier is broken down.

The progeny of the type B spermatogonia, which have entered the first meiotic division, are the **primary spermatocytes** (see Fig. 1.6). Located in a characteristic position just inside the layer of spermatogonia and still deeply embedded

The resumption of meiosis in response to the LH surge is initiated by the cumulus (granulosa) cells, because the oocyte itself does not possess LH receptors. Responding to LH, the cumulus cells shut down their gap junctions (see Fig. 1.9B). This reduces the transfer of both cAMP and cGMP from the cumulus cells into the oocyte. The resulting reduction of cGMP in the oocyte allows the activation of PDE3A. The activated PDE3A then breaks down the intra-oocytic cAMP into 5′AMP. The decline in the concentration of cAMP sets off a signaling pathway leading to the activation of MPF and the subsequent resumption of meiosis.

The egg, now a secondary oocyte, is located in a small mound of cells known as the **cumulus oophorus**, which lies on one side of the greatly enlarged antrum. In response to the preovulatory surge of gonadotropic hormones, factors secreted by the oocyte pass through gap junctions into the surrounding cumulus cells and stimulate the cumulus cells to secrete hyaluronic acid into the intercellular spaces. The hyaluronic acid binds water molecules and enlarges the intercellular spaces, thus expanding the cumulus oophorus. In keeping with the hormonally induced internal changes, the diameter of the follicle increases from about 6 mm early in the second week to almost 2 cm at ovulation.

The tertiary follicle protrudes from the surface of the ovary like a blister. The granulosa cells contain numerous FSH and LH receptors, and LH receptors are abundant in the cells of the theca interna. The follicular cells secrete large amounts of estradiol (see Fig. 1.16), which prepares many other components of the female reproductive tract for gamete transport. Within the antrum, the follicular fluid contains the following: (1) a complement of proteins similar to that seen in serum, but in a lower concentration; (2) 20 enzymes; (3) dissolved hormones, including FSH, LH, and steroids; and (4) proteo-

in Sertoli cell cytoplasm, primary spermatocytes spend 24 days passing through the first meiotic division. During this time, the developing sperm cells use a strategy similar to that of the egg—producing in advance molecules that are needed at later periods when changes occur very rapidly. Such preparation involves the production of mRNA molecules and their storage in an inactive form until they are needed to produce the necessary proteins.

A well-known example of preparatory mRNA synthesis involves the formation of **protamines**, which are small, arginine-rich, and cysteine-rich proteins that displace the lysine-rich nuclear histones and allow the high degree of compaction of nuclear chromatin required during the final stages of sperm formation. Protamine mRNAs are first synthesized in primary spermatocytes but are not translated into proteins until the spermatid stage. In the meantime, the protamine mRNAs are complexed with proteins and are inaccessible to the translational machinery. If protamine mRNAs are translated before the spermatid stage, the chromosomes condense prematurely, and sterility results.

After completion of the first meiotic division, the primary spermatocyte gives rise to 2 **secondary spermatocytes**, which remain connected by a cytoplasmic bridge. The secondary spermatocytes enter the second meiotic division without delay. This phase of meiosis is very rapid, typically completed in approximately 8 hours. Each secondary spermatocyte produces 2 immature haploid gametes, the **spermatids**. The 4 spermatids produced from a primary spermatocyte progenitor are still connected to one another and typically to as many as 100 other spermatids as well. In mice, some genes are transcribed as late as the spermatid stage.

Spermatids do not divide further, but they undergo a series of profound changes that transform them from ordinary-looking cells to highly specialized **spermatozoa** (singular, **spermatozoon**). The process of transformation from spermatids to spermatozoa is called **spermiogenesis** or **spermatid metamorphosis**.

Several major categories of change occur during spermiogenesis (**Fig. 1.12**). One is the progressive reduction in the size of the nucleus and tremendous condensation of the chromosomal material, which is associated with the replacement of histones by protamines. Along with the changes in the nucleus, a profound reorganization of the cytoplasm occurs. Cytoplasm streams away from the nucleus, but a condensation of

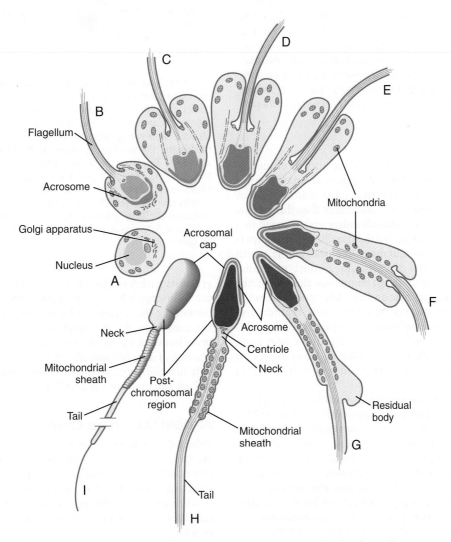

Fig. 1.12   Summary of the major stages in spermiogenesis, starting with a spermatid (A) and ending with a mature spermatozoon (I).

Box 1.1 **Passage of Sperm Precursors through the Blood-Testis Barrier**

During spermatogenesis, developing sperm cells are closely linked with Sertoli cells, and the topography of maturation occurs in regular but complex patterns. A striking example involves the coordinated detachment of mature spermatids from the apical surface of Sertoli cells and the remodeling of the inter-Sertoli cell tight-junction complex that constitutes the blood-testis barrier (see Fig. 1.13). Type B spermatogonia, which are just entering the preleptotene stage of the first meiotic division and becoming primary spermatocytes, are located outside (basal to) the blood-testis barrier. Late-stage spermatids are attached to the apical surface of Sertoli cells by aggregates of tight-junction proteins, called **surface adhesion complexes**.

At a specific stage in spermatid development, the surface adhesion complexes break down, and the mature spermatids are released into the lumen of the seminiferous tubule. Biologically active laminin fragments, originating from the degenerating surface adhesion complexes, make their way to the tight-junction complex that constitutes the blood-testis barrier. These fragments, along with certain cytokines and proteinases, degrade the tight-junctional proteins of the blood-testis barrier, and the blood-testis barrier, located apically to the preleptotene primary spermatocyte, breaks down. Then testosterone, which is 50 to 100 times more concentrated in the seminiferous tubule than in the general circulation, stimulates the synthesis of new tight-junction proteins on the basal side of that preleptotene spermatocyte, thus reestablishing the integrity of the blood-testis barrier. In parallel, a new set of spermatids becomes adherent to the apical surface of the Sertoli cells through the formation of new surface adhesion complexes.

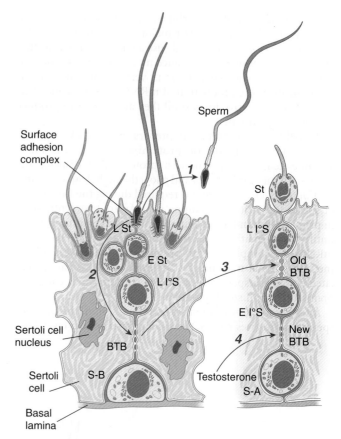

Fig. 1.13 Diagram showing coordination between release of mature spermatids and the dissolution and reconstruction of the blood-testis barrier; (1) with degradation of the surface adhesion complex, mature spermatids are released into the lumen of the seminiferous tubule; (2) active laminin fragments join with cytokines and proteinases to begin to degrade the junctional proteins at the blood-testis barrier located apically to the late type B spermatogonium; (3) the old blood-testis barrier breaks down; (4) under the influence of testosterone, a new blood-testis barrier forms basal to what is now a preleptotene primary spermatocyte. BTB, blood-testis barrier; EI°S, early primary spermatocyte; ESt, early spermatid; LI°S, late primary spermatocyte; LSt, late spermatid; S-A, type A spermatogonium; S-B, type B spermatogonium; St, spermatid.

the Golgi apparatus at the apical end of the nucleus ultimately gives rise to the **acrosome**. The acrosome is an enzyme-filled structure that plays a crucial role in the fertilization process. At the other end of the nucleus, a prominent **flagellum** grows out of the centriolar region. **Mitochondria** are arranged in a spiral around the proximal part of the flagellum. During spermiogenesis, the plasma membrane of the head of the sperm is partitioned into several antigenically distinct molecular domains. These domains undergo numerous changes as the sperm cells mature in the male and at a later point when the spermatozoa are traveling through the female reproductive tract. As spermiogenesis continues, the remainder of the cytoplasm (**residual body** [see G in Fig. 1.12]) moves away from the nucleus and is shed along the developing tail of the sperm cell. The residual bodies are phagocytized by Sertoli cells (**Box 1.1** and **Fig. 1.13**).

For many years, gene expression in postmeiotic (haploid) spermatids was considered to be impossible. Molecular biological research on mice has shown, however, that gene expression in postmeiotic spermatids is not only possible but also common. Nearly 100 proteins are produced only after the completion of the second meiotic division, and many additional proteins are synthesized during and after meiosis.

On completion of spermiogenesis (approximately 64 days after the start of spermatogenesis), the **spermatozoon** is a highly specialized cell well adapted for motion and the deliv-

ery of its packet of DNA to the egg. The sperm cell consists of the following: a head (2 to 3 µm wide and 4 to 5 µm long) containing the nucleus and acrosome; a midpiece containing the centrioles, the proximal part of the flagellum, and the mitochondrial helix; and the tail (about 50 µm long), which consists of a highly specialized flagellum (see Fig. 1.12). (Specific functional properties of these components of the sperm cell are discussed in Chapter 2.)

ABNORMAL SPERMATOZOA

Substantial numbers (up to 10%) of mature spermatozoa are grossly abnormal. The spectrum of anomalies ranges from double heads or tails to defective flagella or variability in head size. Such defective sperm cells are highly unlikely to fertilize an egg. If the percentage of defective spermatozoa increases to greater than 20% of the total, reduced fertility may result.

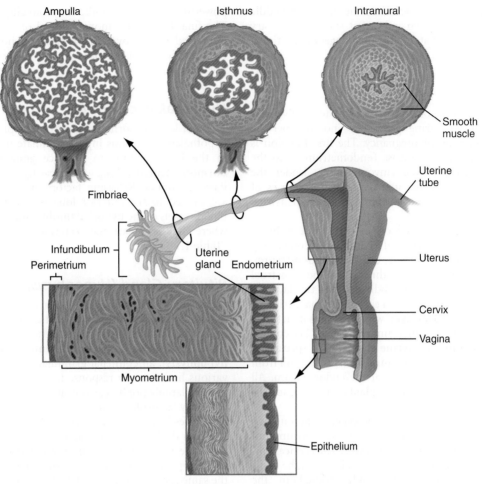

Fig. 1.14 **Structure of the female reproductive tract.**

Preparation of the Female Reproductive Tract for Pregnancy

Structure

The structure and function of the female reproductive tract are well adapted for the transport of gametes and maintenance of the embryo. Many of the subtler features of this adaptation are under hormonal control and are cyclic. This section briefly reviews the aspects of female reproductive structure that are of greatest importance in understanding gamete transport and embryonic development.

Ovaries and Uterine Tubes

The **ovaries** and **uterine** (or **fallopian**) tubes form a functional complex devoted to the production and transport of eggs. In addition, the uterine tubes play an important role as a conduit for spermatozoa and in preparing them to be fully functional during the fertilization process. The uterine tube consists of three anatomically and functionally recognizable segments: the **ampulla**, the **isthmus**, and the **intramural** segments.

The almond-shaped ovaries, located on either side of the uterus, are positioned very near the open, funnel-shaped ends

of the ampullary segments of the uterine tubes. Numerous fingerlike projections, called **fimbriae** (**Fig. 1.14**), project toward the ovary from the open **infundibulum** of the uterine tube and are involved in directing the ovulated egg into the tube. The uterine tube is characterized by a complex internal lining, with a high density of prominent longitudinal folds in the upper ampulla. These folds become progressively simpler in parts of the tube closer to the uterus. The lining epithelium of the uterine tubes contains a mixture of ciliated cells that assist in gamete transport and secretory cells that produce a fluid supporting the early development of the embryo. Layers of smooth muscle cells throughout the uterine tubes provide the basis for peristaltic contractions. The amount and function of many of these components are under cyclic hormonal control, and the overall effect of these changes is to facilitate the transport of gametes and the fertilized egg.

The two segments of the uterine tubes closest to the uterus play particularly important roles as pathways for sperm transport toward the ovulated egg. The intramural segment, which is embedded in the uterine wall, has a very thin lumen containing mucus, the composition of which varies with phases in the menstrual cycle. This segment serves as a gateway regulating the passage of spermatozoa into the uterine tube, but it also restricts the entry of bacteria into the tube. The middle isthmus segment serves as an important site of temporary sperm storage and participates in the final stages of functional maturation of sperm cells (see Chapter 2).

Uterus

The principal functions of the uterus are to receive and maintain the embryo during pregnancy and to expel the fetus at the termination of pregnancy. The first function is carried out by the uterine mucosa (endometrium) and the second by the muscular wall (myometrium). Under the cyclic effect of hormones, the uterus undergoes a series of prominent changes throughout the course of each menstrual cycle.

The **uterus** is a pear-shaped organ with thick walls of smooth muscle (**myometrium**) and a complex mucosal lining (see Fig. 1.14). The mucosal lining, called the **endometrium**, has a structure that changes daily throughout the menstrual cycle. The endometrium can be subdivided into two layers: a **functional layer**, which is shed with each menstrual period or after parturition, and a **basal layer**, which remains intact. The general structure of the endometrium consists of (1) a columnar **surface epithelium**, (2) **uterine glands**, (3) a specialized connective tissue stroma, and (4) **spiral arteries** that coil from the basal layer toward the surface of the endometrium. All these structures participate in the implantation and nourishment of the embryo.

The lower outlet of the uterus is the **cervix**. The mucosal surface of the cervix is not typical uterine endometrium, but is studded with a variety of irregular crypts. The cervical epithelium produces glycoprotein-rich cervical mucus, the composition of which varies considerably throughout the menstrual cycle. The differing physical properties of cervical mucus make it easier or more difficult for spermatozoa to penetrate the cervix and find their way into the uterus.

Vagina

The **vagina** is a channel for sexual intercourse and serves as the birth canal. It is lined with a stratified squamous epithelium, but the epithelial cells contain deposits of **glycogen**, which vary in amount throughout the menstrual cycle. Glycogen breakdown products contribute to the acidity (pH 4.3) of the vaginal fluids. The low pH of the upper vagina serves a bacteriostatic function and prevents infectious agents from entering the upper genital tract through the cervix and ultimately spreading to the peritoneal cavity through the open ends of the uterine tubes.

Hormonal Control of the Female Reproductive Cycle

Reproduction in women is governed by a complex series of interactions between hormones and the tissues that they influence. The hierarchy of cyclic control begins with input to the **hypothalamus** of the brain (**Fig. 1.15**). The hypothalamus influences hormone production by the anterior lobe of the pituitary gland. The pituitary hormones are spread via the blood throughout the entire body and act on the ovaries, which are stimulated to produce their own sex steroid hormones. During pregnancy, the placenta exerts a powerful effect on the mother by producing several hormones. The final level of hormonal control of female reproduction is exerted by the ovarian or placental hormones on other reproductive target organs (e.g., uterus, uterine tubes, vagina, breasts).

Hypothalamic Control

The first level of hormonal control of reproduction is in the hypothalamus. Various inputs stimulate neurosecretory cells in the hypothalamus to produce **gonadotropin-releasing hormone** (**GnRH**), along with releasing factors for other pituitary hormones. Releasing factors and an inhibiting factor are carried to the anterior lobe of the pituitary gland by blood vessels of the **hypothalamohypophyseal portal** system, where they stimulate the secretion of pituitary hormones (**Table 1.1**).

Pituitary Gland (Hypophysis)

Producing its hormones in response to stimulation by the hypothalamus, the **pituitary gland** constitutes a second level of hormonal control of reproduction. The pituitary gland consists of two components: the **anterior pituitary** (**adenohypophysis**), an epithelial glandular structure that produces various hormones in response to factors carried to it by the hypothalamohypophyseal portal system; and the **posterior pituitary** (**neurohypophysis**), a neural structure that releases hormones by a neurosecretory mechanism.

Under the influence of GnRH and direct feedback by steroid hormone levels in the blood, the anterior pituitary secretes two polypeptide **gonadotropic hormones**, FSH and LH, from the same cell type (see Table 1.1). In the absence of an inhibiting factor (**dopamine**) from the hypothalamus, the anterior pituitary also produces **prolactin**, which acts on the mammary glands.

The only hormone from the posterior pituitary that is directly involved in reproduction is **oxytocin**, an oligopeptide involved in childbirth and the stimulus for milk let-down from the mammary glands in lactating women.

Ovaries and Placenta

The ovaries and, during pregnancy, the placenta constitute a third level of hormonal control. Responding to blood levels of the anterior pituitary hormones, the granulosa cells of the ovarian follicles convert androgens (**androstenedione** and **testosterone**) synthesized by the theca interna into estrogens (mainly **estrone** and the 10-fold more powerful **17β-estradiol**), which then pass into the bloodstream. After ovulation, **progesterone** is the principal secretory product of the follicle after its conversion into the corpus luteum (see Chapter 2). During later pregnancy, the placenta supplements the production of ovarian steroid hormone by synthesizing its own estrogens and progesterone. It also produces two polypeptide hormones (see Table 1.1). **Human chorionic gonadotropin** (**HCG**) acts on the ovary to maintain the activity of the corpus luteum during pregnancy. **Human placental lactogen** (**somatomammotropin**) acts on the corpus luteum; it also promotes breast development by enhancing the effects of estrogens and progesterone and stimulates the synthesis of milk constituents.

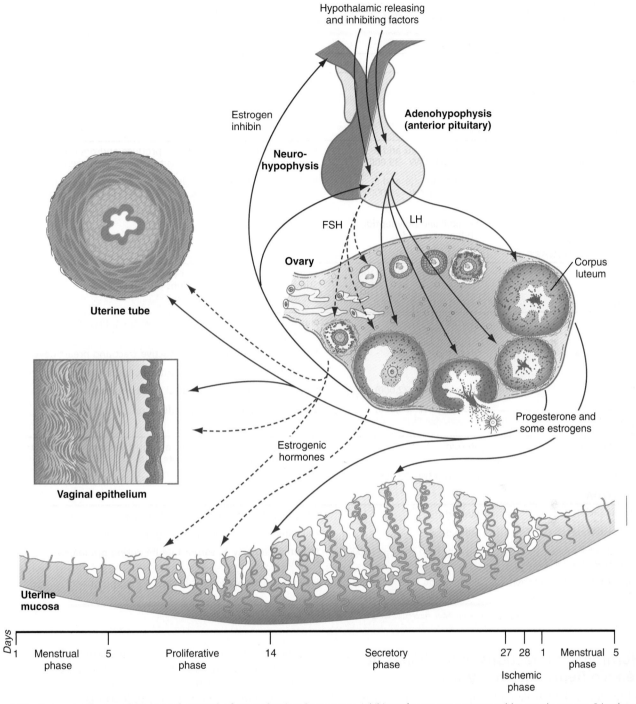

Fig. 1.15 **General scheme of hormonal control of reproduction in women.** Inhibitory factors are represented by *purple arrows*. Stimulatory factors are represented by *red arrows*. Hormones involved principally in the proliferative phase of the menstrual cycle are represented by *dashed arrows*; those involved principally in the secretory phase are represented by *solid arrows*. FSH, follicle-stimulating hormone; LH, luteinizing hormone.

Reproductive Target Tissues

The last level in the hierarchy of reproductive hormonal control constitutes the target tissues, which ready themselves structurally and functionally for gamete transport or pregnancy in response to ovarian and placental hormones binding to specific cellular receptors. Changes in the number of ciliated cells and in smooth muscle activity in the uterine tubes, the profound changes in the endometrial lining of the uterus,

and the cyclic changes in the glandular tissues of the breasts are some of the more prominent examples of hormonal effects on target tissues. These changes are described more fully later.

A general principle recognized some time ago is the efficacy of first priming reproductive target tissues with estrogen so that progesterone can exert its full effects. Estrogen induces the target cells to produce large quantities of progesterone receptors, which must be in place for progesterone to act on these same cells.

Table 1.1　Major Hormones Involved in Mammalian Reproduction

Hormone	Chemical Nature	Function
HYPOTHALAMUS		
Gonadotropin-releasing hormone (GnRH, LHRH)	Decapeptide	Stimulates release of LH and FSH by anterior pituitary
Prolactin-inhibiting factor	Dopamine	Inhibits release of prolactin by anterior pituitary
ANTERIOR PITUITARY		
Follicle-stimulating hormone (FSH)	Glycoprotein (α and β subunits) (MW $\approx$35,000)	Male: Stimulates Sertoli cells to produce androgen-binding protein Female: Stimulates follicle cells to produce estrogen
Luteinizing hormone (LH)	Glycoprotein (α and β subunits) (MW $\approx$28,000)	Male: Stimulates Leydig cells to secrete testosterone Female: Stimulates follicle cells and corpus luteum to produce progesterone
Prolactin	Single-chain polypeptide (198 amino acids)	Promotes lactation
POSTERIOR PITUITARY		
Oxytocin	Oligopeptide (MW $\approx$1100)	Stimulates ejection of milk by mammary gland Stimulates uterine contractions during labor
OVARY		
Estrogens	Steroid	Has multiple effects on reproductive tract, breasts, body fat, and bone growth
Progesterone	Steroid	Has multiple effects on reproductive tract and breast development
Testosterone	Steroid	Is precursor for estrogen biosynthesis, induces follicular atresia
Inhibin	Protein (MW $\approx$32,000)	Inhibits FSH secretion, has local effects on ovaries
Activin	Protein (MW $\approx$28,000)	Stimulates granulosa cell proliferation
TESTIS		
Testosterone	Steroid	Has multiple effects on male reproductive tract, hair growth, and other secondary sexual characteristics
Inhibin	Protein (MW $\approx$32,000)	Inhibits FSH secretion, has local effects on testis
PLACENTA		
Estrogens	Steroid	Has same functions as ovarian estrogens
Progesterone	Steroid	Has same functions as ovarian progesterone
Human chorionic gonadotropin (HCG)	Glycoprotein (MW $\approx$30,000)	Maintains activity of corpus luteum during pregnancy
Human placental lactogen (somatomammotropin)	Polypeptide (MW $\approx$20,000)	Promotes development of breasts during pregnancy

LHRH, luteinizing hormone–releasing hormone; MW, molecular weight.

Hormonal Interactions with Tissues during Female Reproductive Cycles

All tissues of the female reproductive tract are influenced by the reproductive hormones. In response to the hormonal environment of the body, these tissues undergo cyclic modifications that improve the chances for successful reproduction.

Knowledge of the changes the ovaries undergo is necessary to understand hormonal interactions and tissue responses during the female reproductive cycle. Responding to both FSH and LH secreted by the pituitary just before and during a menstrual period, a set of secondary ovarian follicles begins to mature and secrete 17β-estradiol. By ovulation, all of these follicles except one have undergone atresia, their main contribution having been to produce part of the supply of estrogens needed to prepare the body for ovulation and gamete transport.

During the preovulatory, or **proliferative**, **phase** (days 5 to 14) of the menstrual cycle, estrogens produced by the ovary act on the female reproductive tissues (see Fig. 1.15). The uterine lining becomes re-epithelialized from the just-completed menstrual period. Then, under the influence of estrogens, the endometrial stroma progressively thickens, the uterine glands elongate, and the spiral arteries begin to grow toward the surface of the endometrium. The mucous glands of the cervix secrete glycoprotein-rich but relatively watery mucus, which facilitates the passage of spermatozoa through the cervical canal. As the proliferative phase progresses, a higher percentage of the epithelial cells lining the uterine tubes becomes ciliated, and smooth muscle activity in the tubes increases. In the days preceding ovulation, the fimbriated ends of the uterine tubes move closer to the ovaries.

Toward the end of the proliferative period, a pronounced increase in the levels of estradiol secreted by the developing

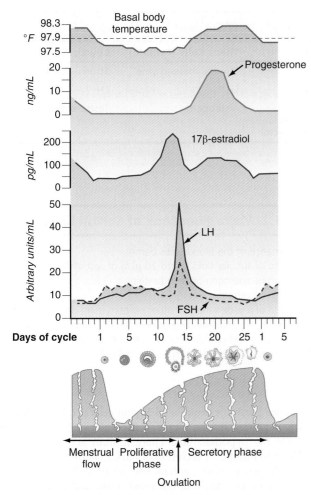

Fig. 1.16 Comparison of curves representing daily serum concentrations of gonadotropins and sex steroids and basal body temperature in relation to events in the human menstrual cycle. FSH, follicle-stimulating hormone; LH, luteinizing hormone. *(Redrawn from Midgley AR and others: In Hafez ES, Evans TN, eds: Human reproduction, New York, 1973, Harper & Row.)*

After the LH surge and with the increasing concentration of progesterone in the blood, the basal body temperature increases (see Fig. 1.16). Because of the link between an increase in basal body temperature and the time of ovulation, accurate temperature records are the basis of the **rhythm method** of birth control.

Around the time of ovulation, the combined presence of estrogen and progesterone in the blood causes the uterine tube to engage in a rhythmic series of muscular contractions designed to promote transport of the ovulated egg. Progesterone prompts epithelial cells of the uterine tube to secrete fluids that provide nutrition for the cleaving embryo. Later during the secretory phase, high levels of progesterone induce regression of some of the ciliated cells in the tubal epithelium.

In the uterus, progesterone prepares the estrogen-primed endometrium for implantation of the embryo. The endometrium, which has thickened under the influence of estrogen during the proliferative phase, undergoes further changes. The straight uterine glands begin to coil and accumulate glycogen and other secretory products in the epithelium. The spiral arteries grow farther toward the endometrial surface, but mitosis in the endometrial epithelial cells decreases. Through the action of progesterone, the cervical mucus becomes highly viscous and acts as a protective block, inhibiting the passage of materials into or out of the uterus. During the secretory period, the vaginal epithelium becomes thinner.

In the mammary glands, progesterone furthers the estrogen-primed development of the secretory components and causes water retention in the tissues. More extensive development of the lactational apparatus awaits its stimulation by placental hormones.

Midway through the secretory phase of the menstrual cycle, the epithelium of the uterine tubes has already undergone considerable regression from its midcycle peak, whereas the uterine endometrium is at full readiness to receive a cleaving embryo. If pregnancy does not occur, a series of hormonal interactions brings the menstrual cycle to a close. One of the early feedback mechanisms is the production of the protein **inhibin** by the granulosa cells. Inhibin is carried by the bloodstream to the anterior pituitary, where it directly inhibits the secretion of gonadotropins, especially FSH. Through mechanisms that are unclear, the secretion of LH is also reduced. This inhibition results in regression of the corpus luteum and marked reduction in the secretion of progesterone by the ovary.

Some of the main consequences of the regression of the corpus luteum are the infiltration of the endometrial stroma with leukocytes, the loss of interstitial fluid, and the spasmodic constriction and breakdown of the spiral arteries that cause local ischemia. The ischemia results in local hemorrhage and the loss of integrity of areas of the endometrium. These changes initiate menstruation (by convention, constituting days 1 to 5 of the menstrual cycle). Over the next few days, the entire functional layer of the endometrium is shed in small bits, along with the attendant loss of about 30 mL of blood. By the time the menstrual period is over, only a raw endometrial base interspersed with the basal epithelium of the uterine glands remains as the basis for the healing and reconstitution of the endometrium during the next proliferative period.

ovarian follicle acts on the hypothalamohypophyseal system, thus causing increased responsiveness of the anterior pituitary to GnRH and a surge in the hypothalamic secretion of GnRH. Approximately 24 hours after the level of 17β-estradiol reaches its peak in the blood, a preovulatory surge of LH and FSH is sent into the bloodstream by the pituitary gland (**Fig. 1.16**). The **LH surge** is not a steady increase in gonadotropin secretion; rather, it constitutes a series of sharp pulses of secretion that appear to be responding to a hypothalamic timing mechanism.

The LH surge leads to ovulation, and the graafian follicle becomes transformed into a **corpus luteum** (yellow body). The basal lamina surrounding the granulosa of the follicle breaks down and allows blood vessels to grow into the layer of granulosa cells. Through proliferation and hypertrophy, the granulosa cells undergo major structural and biochemical changes and now produce progesterone as their primary secretory product. Some estrogen is still secreted by the corpus luteum. After ovulation, the menstrual cycle, which is now dominated by the secretion of progesterone, is said to be in the **secretory phase** (days 14 to 28 of the menstrual cycle).

Table 1.2	Homologies between Hormone-Producing Cells in Male and Female Gonads			
Parameter	Granulosa Cells (Female)	Sertoli Cells (Male)	Theca Cells (Female)	Leydig Cells (Male)
Origin	Rete ovarii	Rete testis	Stromal mesenchyme	Stromal mesenchyme
Major receptors	FSH	FSH	LH	LH
Major secretory products	Estrogens, progesterone, inhibin	Estrogen, inhibin, androgen-binding protein, Leydig cell stimulatory factor	Androgens	Testosterone

FSH, follicle-stimulating hormone; LH, luteinizing hormone.

Clinical Vignette

A 33-year-old woman has had both ovaries removed because of large bilateral ovarian cysts. The next year she is on an extended expedition in northern Canada, and her canoe tips, sending her replacement hormonal medication to the bottom of the lake. More than 6 weeks elapse before she is able to obtain a new supply of medication.

Which of the following would be least affected by the loss of the woman's medication?

A. Blood levels of follicle-stimulating hormone and luteinizing hormone
B. Ciliated cells of the uterine tube
C. Mass of the heart
D. Glandular tissue of the breasts
E. Thickness of the endometrium

Box 1.2 Major Functions of Sertoli Cells

Maintenance of the blood-testis barrier
Secretion of tubular fluid (10 to 20 µL/g of testis/hr)
Secretion of androgen-binding protein
Secretion of estrogen and inhibin
Secretion of a wide variety of other proteins (e.g., growth factors, transferrin, retinal-binding protein, metal-binding proteins)
Maintenance and coordination of spermatogenesis
Phagocytosis of residual bodies of sperm cells

Hormonal Interactions Involved with Reproduction in Males

Along with the homologies of certain structures between the testis and ovary, some strong parallels exist between the hormonal interactions involved in reproduction in males and females. The most important homologies are between granulosa cells in the ovarian follicle and Sertoli cells in the seminiferous tubule of the testis and between theca cells of the ovary and Leydig cells in the testis (**Table 1.2**).

The hypothalamic secretion of GnRH stimulates the anterior pituitary to secrete FSH and LH. The LH binds to the nearly 20,000 LH receptors on the surface of each Leydig (interstitial) cell, and through a cascade of second messengers, LH stimulates the synthesis of testosterone from cholesterol. Testosterone is released into the blood and is taken to the Sertoli cells and throughout the body, where it affects a variety of secondary sexual tissues, often after it has been locally converted to dihydrotestosterone.

Sertoli cells are stimulated by pituitary FSH via surface FSH receptors and by testosterone from the Leydig cells via cytoplasmic receptors. After FSH stimulation, the Sertoli cells convert some of the testosterone to estrogens (as the granulosa cells in the ovary do). Some of the estrogen diffuses back to the Leydig cells along with a **Leydig cell stimulatory factor**, which is produced by the Sertoli cells and reaches the Leydig cells by a **paracrine** (non–blood-borne) mode of secretion (**Fig. 1.17**). The FSH-stimulated Sertoli cell produces **androgen-binding protein**, which binds testosterone and is carried into the fluid compartment of the seminiferous tubule, where it exerts a strong influence on the course of spermatogenesis. Similarly, their granulosa cell counterparts in the

ovary, the hormone-stimulated Sertoli cells produce inhibin, which is carried by the blood to the anterior pituitary and possibly the hypothalamus. There inhibin acts by negative feedback to inhibit the secretion of FSH. In addition to inhibin and androgen-binding protein, the Sertoli cells have a wide variety of other functions, the most important of which are summarized in **Box 1.2** and **Clinical Correlation 1.2**.

Summary

- Gametogenesis is divided into four phases:
  1. Extraembryonic origin of germ cells and their migration into the gonads
  2. An increase in the number of germ cells by mitosis
  3. A reduction in chromosomal material by meiosis
  4. Structural and functional maturation
- Primordial germ cells are first readily recognizable in the yolk sac endoderm. They then migrate through the dorsal mesentery to the primordia of the gonads.
- In the female, oogonia undergo intense mitotic activity in the embryo only. In the male, spermatogonia are capable of mitosis throughout life.
- Meiosis involves a reduction in chromosome number from diploid to haploid, independent reassortment of paternal and maternal chromosomes, and further redistribution of genetic material through the process of crossing-over.
- In the oocyte, there are two meiotic blocks—in diplotene of prophase I and in metaphase II. In the female, meiosis begins in the 5-month embryo; in the male, meiosis begins at puberty.
- Failure of chromosomes to separate properly during meiosis results in nondisjunction, which is associated with multiple anomalies, depending on which chromosome is affected.

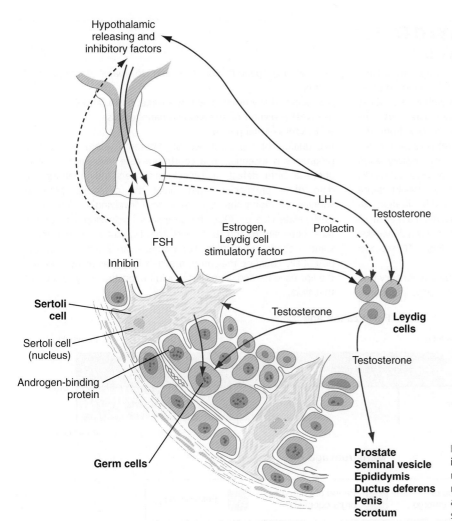

Hypothalamic releasing and inhibitory factors

LH

Testosterone

Estrogen, Leydig cell stimulatory factor

Prolactin

FSH

Inhibin

Sertoli cell

Sertoli cell (nucleus)

Androgen-binding protein

Testosterone

Leydig cells

Testosterone

Germ cells

**Prostate
Seminal vesicle
Epididymis
Ductus deferens
Penis
Scrotum**

Fig. 1.17 **General scheme of hormonal control in the male reproductive system.** *Red arrows* represent stimulatory influences. *Purple arrows* represent inhibitory influences. Suspected interactions are represented by *dashed arrows*. FSH, follicle-stimulating hormone; LH, luteinizing hormone.

- Developing oocytes are surrounded by layers of follicular cells and interact with them through gap junctions. When stimulated by pituitary hormones (e.g., FSH, LH), the follicular cells produce steroid hormones (estrogens and progesterone). The combination of oocyte and follicular (granulosa) cells is called a *follicle*. Under hormonal stimulation, certain follicles greatly increase in size, and each month, one of these follicles undergoes ovulation.
- Spermatogenesis occurs in the testis and involves successive waves of mitosis of spermatogonia, meiosis of primary and secondary spermatocytes, and final maturation (spermiogenesis) of postmeiotic spermatids into spermatozoa. Functional maturation of spermatozoa occurs in the epididymis.
- Female reproductive tissues undergo cyclic, hormonally induced preparatory changes for pregnancy. In the uterine tubes, this involves the degree of ciliation of the epithelium and smooth muscle activity of the wall. Under the influence of estrogens and then progesterone, the endometrium of the uterus builds up in preparation to receive the embryo. In the absence of fertilization and with the subsequent withdrawal of hormonal support, the endometrium breaks down and is shed (menstruation). Cyclic

changes in the cervix involve thinning of the cervical mucus at the time of ovulation.
- Hormonal control of the female reproductive cycle is hierarchical, with releasing or inhibiting factors from the hypothalamus acting on the adenohypophysis and causing the release of pituitary hormones (e.g., FSH, LH). The pituitary hormones sequentially stimulate the ovarian follicles to produce estrogens and progesterone, which act on the female reproductive tissues. In pregnancy, the remains of the follicle (corpus luteum) continue to produce progesterone, which maintains the early embryo until the placenta begins to produce sufficient hormones to maintain pregnancy.
- In the male, LH stimulates the Leydig cells to produce testosterone, and FSH acts on the Sertoli cells, which support spermatogenesis. In males and females, feedback inhibition decreases the production of pituitary hormones.
- There are two systems for dating pregnancy:
  1. Fertilization age: dates the age of the embryo from the time of fertilization.
  2. Menstrual age: dates the age of the embryo from the start of the mother's last menstrual period. The menstrual age is 2 weeks greater than the fertilization age.

CLINICAL CORRELATION 1.2
Dating of Pregnancy

Two different systems for dating pregnancies have evolved. One, used by embryologists, dates pregnancy from the time of fertilization (**fertilization age**), so that a 6-week-old embryo is 6 weeks (42 days) from the day of fertilization. The other system, used by obstetricians and many clinicians, dates pregnancy from the woman's last menstrual period (**menstrual age**) because this is a convenient reference point from the standpoint of a history taken from a patient. The menstrual age of a human embryo is 2 weeks greater than the fertilization age because usually 2 weeks elapse between the start of the last menstrual period and fertilization. An embryo with a fertilization age of 6 weeks is assigned a menstrual age of 8 weeks, and the typical duration of pregnancy is 38 weeks' fertilization age and 40 weeks' menstrual age (**Fig. 1.18**; see also Fig. 18.16).

For valid clinical reasons, obstetricians subdivide pregnancy into three equal **trimesters**, whereas embryologists divide pregnancy

into unequal periods corresponding to major developmental events.
0-3 weeks—Early development (cleavage, gastrulation)
4-8 weeks—Period of embryonic organogenesis
9-38 weeks—Fetal period
Recognition of the existence of different systems for dating pregnancy is essential. In a courtroom case involving a lawsuit about a birth defect, a 2-week misunderstanding about the date of a pregnancy could make the difference between winning or losing the case. In a case involving a cleft lip or cleft palate (see p. 300), the difference in development of the face between 6 and 8 weeks (see Fig. 14.6) would make some scenarios impossible. For example, an insult at 6 weeks potentially could be the cause of a cleft lip, whereas by 8 weeks, the lips have formed, so a cleft would be most unlikely to form at that time.

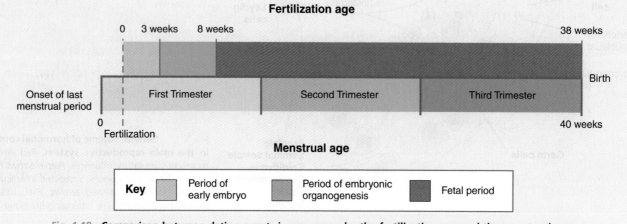

Fig. 1.18 **Comparison between dating events in pregnancy by the fertilization age and the menstrual age.**

Review Questions

1. During spermatogenesis, histone is replaced by which of the following, to allow better packing of the condensed chromatin in the head of the spermatozoon?
A. Inhibin
B. Prostaglandin E
C. Testosterone
D. Protamine
E. Androgen-binding protein

2. Which cell type is located outside the blood-testis barrier?
A. Spermatozoon
B. Secondary spermatocyte
C. Spermatid
D. Primary spermatocyte
E. Spermatogonium

3. Which of the following cells normally participates in mitotic divisions?
A. Primary oocyte
B. Oogonium

C. Primary spermatocyte
D. Spermatid
E. Secondary spermatocyte

4. In a routine chest x-ray examination, the radiologist sees what appear to be teeth in a mediastinal mass. What is the likely diagnosis, and what is a probable embryological explanation for its appearance?

5. When does meiosis begin in the female and in the male?

6. At what stages of oogenesis is meiosis arrested in the female?

7. What is the underlying cause of most spontaneous abortions during the early weeks of pregnancy?

8. What is the difference between spermatogenesis and spermiogenesis?

9. **The actions of what hormones are responsible for the changes in the endometrium during the menstrual cycle?**

10. **Sertoli cells in the testis are stimulated by what two major reproductive hormones?**

References

Abou-Haila A, Tulsiani DRP: Mammalian sperm acrosome: formation, contents, and function, *Arch Biochem Biophys* 379:173-182, 2000.

Bardhan A: Many functions of the meiotic cohesin, *Chromosome Res* 18:909-924, 2010.

Bellve AR, ed: The male germ cell: migration to fertilization, *Semin Cell Dev Biol* 9:379-489, 1998.

Bowles J, Koopman P: Retinoic acid, meiosis and germ cell fate in mammals, *Development* 134:3401-3411, 2007.

Cheng CY and others: Regulation of spermatogenesis in the microenvironment of the seminiferous epithelium: new insights and advances, *Mol Cell Endocrinol* 315:49-56, 2010.

Clermont Y: The cycle of the seminiferous epithelium in man, *Am J Anat* 112:35-51, 1963.

Dym M: Spermatogonial stem cells of the testis, *Proc Natl Acad Sci U S A* 91:11287-11289, 1994.

Eppig JJ: Oocyte control of ovarian follicular development and function in mammals, *Reproduction* 122:829-838, 2001.

Erickson RP: Post-meiotic gene expression, *Trends Genet* 6:264-269, 1990.

Ewen KA, Koopman P: Mouse germ cell development: from specification to sex determination, *Mol Cell Endocrinol* 323:76-93, 2010.

Filicori M: The role of luteinizing hormone in folliculogenesis and ovulation induction, *Fertil Steril* 71:405-414, 1999.

Freeman B: The active migration of germ cells in the embryos of mice and men is a myth, *Reproduction* 125:635-643, 2003.

Gosden R, Lee B: Portrait of an oocyte: our obscure origin, *J Clin Invest* 120:973-983, 2010.

Gougeon A: Regulation of ovarian follicular development in primates: facts and hypotheses, *Endocr Rev* 17:121-155, 1996.

Grootegoed JA, Siep M, Baarends WM: Molecular and cellular mechanisms in spermatogenesis, *Bailleres Clin Endocrinol Metab* 14:331-343, 2000.

Halvorson LM, DeCherney AH: Inhibin, activin, and follistatin in reproductive medicine, *Fertil Steril* 65:459-469, 1996.

Handel MA: The XY body: a specialized meiotic chromatin domain, *Exp Cell Res* 296:57-63, 2004.

Hogarth CA, Griswold MD: The key role of vitamin A in spermatogenesis, *J Clin Invest* 120:956-962, 2010.

Kauppi L and others: Distinct properties of the XY pseudoautosomal region crucial for male meiosis, *Science* 331:916-920, 2011.

Kehler J and others: Oct4 is required for primordial germ cell survival, *EMBO Rep* 5:1078-1083, 2004.

Kota SK, Feil R: Epigenetic transitions in germ cell development and meiosis, *Dev Cell* 19:675-686, 2010.

Kurahashi H and others: Recent advance in our understanding of the molecular nature of chromosomal abnormalities, *J Hum Genet* 54:253-260, 2009.

Lie PPY and others: Coordinating cellular events during spermatogenesis: a biochemical model, *Trends Biochem Sci* 334:366-373, 2009.

Lin Y and others: Germ cell–intrinsic and –extrinsic factors govern meiotic initiation in mouse embryos, *Science* 322:1685-1687, 2008.

Liu K and others: Control of mammalian oocyte growth and early follicular development by the oocyte PI3 kinase pathway: new roles for an old timer, *Dev Biol* 299:1-11, 2006.

Mather JP, Moore A, Li R-H: Activins, inhibins, and follistatins: further thoughts on a growing family of regulators, *Proc Soc Exp Biol Med* 215:209-222, 1997.

Matzuk MM and others: Intercellular communication in the mammalian ovary: oocytes carry the conversation, *Science* 296:2178-2180, 2002.

Mehlmann LM, Jones TLZ, Jaffe LA: Meiotic arrest in the mouse follicle maintained by a G_s protein in the oocyte, *Science* 297:1343-1345, 2002.

Neill JD, ed: *The physiology of reproduction*, ed 3, Amsterdam, 2006, Academic Press.

Page SL, Hawley RS: Chromosome choreography: the meiotic ballet, *Science* 301:785-789, 2003.

Richards JS: Ovulation: new factors that prepare the oocyte for fertilization, *Mol Cell Endocrinol* 234:75-79, 2005.

Richards JS, Pangas SA: The ovary: basic biology and clinical implications, *J Clin Invest* 120:963-972, 2010.

Salustri A and others: Oocyte-granulosa cell interactions. In Adashi EY, Leung PCK, eds: *The ovary*, New York, 1993, Raven, pp 209-225.

Shoham Z and others: The luteinizing hormone surge: the final stage in ovulation induction: modern aspects of ovulation triggering, *Fertil Steril* 64:237-251, 1995.

Taieb F, Thibier C, Jessus C: On cyclins, oocytes and eggs, *Mol Reprod Dev* 48:397-411, 1997.

Tripathi A and others: Meiotic cell cycle arrest in mammalian oocytes, *J Cell Physiol* 223:592-600, 2010.

Turner JMA: Meiotic sex chromosome inactivation, *Development* 134:1823-1831, 2007.

Turner TT: De Graaf's thread: The human epididymis, *J Androl* 29:237-250, 2008.

Visser JA, Themmen APN: Anti-Müllerian hormone and folliculogenesis, *Mol Cell Endocrinol* 234:81-86, 2005.

Willard HF: Centromeres of mammalian chromosomes, *Trends Genet* 6:410-416, 1990.

Wood AJ and others: Condensin and cohesin complexity: the expanding repertoire of functions, *Nat Rev Genet* 11:391-404, 2010.

Wynn RM: *Biology of the uterus*, New York, 1977, Plenum.

Yanowitz J: Meiosis: making a break for it, *Curr Opin Cell Biol* 22:744-751, 2010.

Zamboni L: Physiology and pathophysiology of the human spermatozoon: the role of electron microscopy, *J Electron Microsc Tech* 17:412-436, 1991.

Zhang M and others: Granulosa cell ligand NPPC and its receptor NPR2 maintain meiotic arrest in mouse oocytes, *Science* 330:366-369, 2010.

Transport of Gametes and Fertilization

Chapter 1 describes the origins and maturation of male and female gametes and the hormonal conditions that make such maturation possible. It also describes the cyclic, hormonally controlled changes in the female reproductive tract that ready it for fertilization and the support of embryonic development. This chapter first explains the way the egg and sperm cells come together in the female reproductive tract so that fertilization can occur. It then outlines the complex set of interactions involved in fertilization of the egg by a sperm.

Ovulation and Egg and Sperm Transport

Ovulation

Toward the midpoint of the menstrual cycle, the mature graafian follicle, containing the egg that has been arrested in prophase of the first meiotic division, has moved to the surface of the ovary. Under the influence of follicle-stimulating hormone (FSH) and luteinizing hormone (LH), the follicle expands dramatically. The first meiotic division is completed, and the second meiotic division proceeds until the metaphase stage, at which the second meiotic arrest occurs. After the first meiotic division, the first polar body is expelled. By this point, the follicle bulges from the surface of the ovary. The apex of the protrusion is the **stigma**.

The stimulus for ovulation is the surge of LH secreted by the anterior pituitary at the midpoint of the menstrual cycle (see Fig. 1.16). Within hours of exposure to the LH surge, the follicle reorganizes its program of gene expression from one directed toward development of the follicle to one producing molecules that set into gear the processes of follicular rupture and ovulation. Shortly after the LH peak, local blood flow increases in the outer layers of the follicular wall. Along with the increased blood flow, plasma proteins leak into the tissues through the postcapillary venules, with resulting local edema. The edema and the release of certain pharmacologically active compounds, such as prostaglandins, histamine, vasopressin, and plasminogen activator, provide the starting point for a series of reactions that result in the local production of **matrix metalloproteinases**—a family of lytic enzymes that degrade components of the extracellular matrix. At the same time, the secretion of hyaluronic acid by cells of the cumulus results in a loosening of the cells surrounding the egg. The lytic action

of the matrix metalloproteinases produces an inflammatory-like reaction that ultimately results in rupture of the outer follicular wall about 28 to 36 hours after the LH surge (**Fig. 2.1**). Within minutes after rupture of the follicular wall, the cumulus oophorus detaches from the granulosa, and the egg is released from the ovary.

Ovulation results in the expulsion of both antral fluid and the ovum from the ovary into the peritoneal cavity. The ovum is not ovulated as a single naked cell, but as a complex consisting of (1) the ovum, (2) the zona pellucida, (3) the two- to three-cell-thick corona radiata, and (4) a sticky matrix containing surrounding cells of the cumulus oophorus. By convention, the adhering cumulus cells are designated the **corona radiata** after ovulation has occurred. Normally, one egg is released at ovulation. The release and fertilization of two eggs can result in fraternal twinning.

Some women experience mild to pronounced pain at the time of ovulation. Often called **mittelschmerz** (German for "middle pain"), this pain may accompany slight bleeding from the ruptured follicle.

Egg Transport

The first step in egg transport is capture of the ovulated egg by the uterine tube. Shortly before ovulation, the epithelial cells of the uterine tube become more highly ciliated, and smooth muscle activity in the tube and its suspensory ligament increases as the result of hormonal influences. By ovulation, the fimbriae of the uterine tube move closer to the ovary and seem to sweep rhythmically over its surface. This action, in addition to the currents set up by the cilia, efficiently captures the ovulated egg complex. Experimental studies on rabbits have shown that the bulk provided by the cellular coverings of the ovulated egg is important in facilitating the egg's capture and transport by the uterine tube. Denuded ova or inert objects of that size are not so readily transported. Capture of the egg by the uterine tube also involves an adhesive interaction between the egg complex and the ciliary surface of the tube.

Even without these types of natural adaptations, the ability of the uterine tubes to capture eggs is remarkable. If the fimbriated end of the tube has been removed, egg capture occurs remarkably often, and pregnancies have even occurred in women who have had one ovary and the contralateral uterine tube removed. In such cases, the ovulated egg would have to

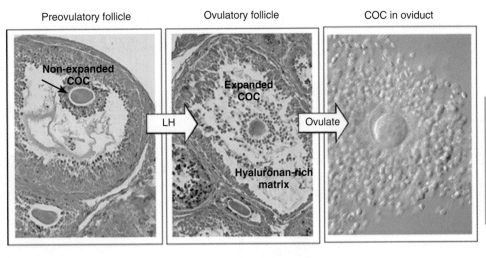

Fig. 2.1 Changes in the cumulus-oocyte complex (COC) of rabbits during follicular maturation and ovulation. In the preovulatory follicle, the cumulus cells (arrow) are tightly packed around the oocyte. Because the follicle is stimulated by luteinizing hormone (LH) before ovulation, the cumulus cells produce extracellular matrix and become much less tightly packed by the time of ovulation. The ovulated oocyte is still surrounded by cumulus cells. *(From Espey LL, Richards JS: Ovulation. In Neill JD, ed:* Physiology of reproduction, *ed 3, Amsterdam, 2006, Elsevier.)*

Preovulatory follicle

Ovulatory follicle

COC in oviduct

Non-expanded COC

LH

Expanded COC

Hyaluronan-rich matrix

Ovulate

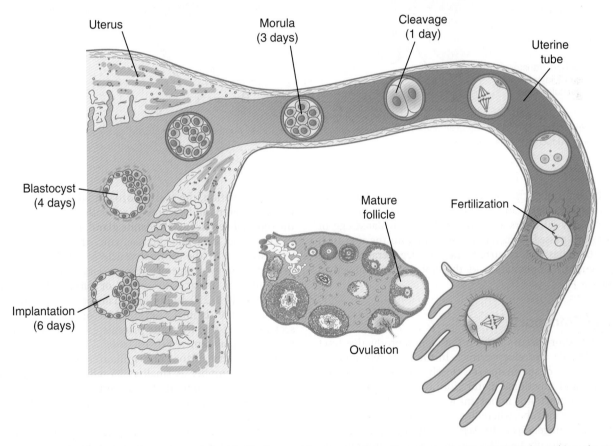

Uterus

Morula (3 days)

Cleavage (1 day)

Uterine tube

Blastocyst (4 days)

Implantation (6 days)

Mature follicle

Fertilization

Ovulation

Fig. 2.2 Follicular development in the ovary, ovulation, fertilization, and transport of the early embryo down the uterine tube and into the uterus.

travel free in the pelvic cavity for a considerable distance before entering the ostium of the uterine tube on the other side.

When inside the uterine tube, the egg is transported toward the uterus, mainly as the result of contractions of the smooth musculature of the tubal wall. Although the cilia lining the tubal mucosa may also play a role in egg transport, their action is not obligatory because women with **immotile cilia syndrome** are often fertile.

While in the uterine tube, the egg is bathed in **tubal fluid**, which is a combination of secretion by the tubal epithelial cells and transudate from capillaries just below the epithelium. In

some mammals, exposure to oviductal secretions is important to the survival of the ovum and for modifying the composition of the zona pellucida, but the role of tubal fluid in humans is less clear.

Tubal transport of the egg usually takes 3 to 4 days, whether or not fertilization occurs (**Fig. 2.2**). Egg transport typically occurs in two phases: slow transport in the ampulla (approximately 72 hours) and a more rapid phase (8 hours) during which the egg or embryo passes through the isthmus and into the uterus (see p. 51). By a poorly understood mechanism, possibly local edema or reduced muscular activity, the egg is temporarily prevented from entering the isthmic portion of

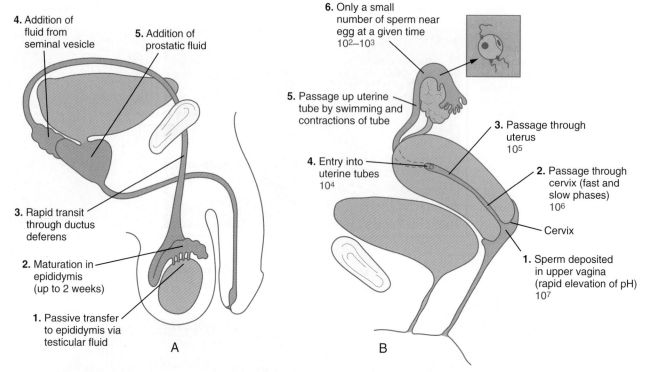

Fig. 2.3 Sperm transport in (**A**) the male and (**B**) the female reproductive tract. In **B,** numbers of spermatozoa typically found in various parts of the female reproductive tract are indicated in red.

the tube, but under the influence of progesterone, the uterotubal junction relaxes and permits entry of the ovum.

By roughly 80 hours after ovulation, the ovulated egg or embryo has passed from the uterine tube into the uterus. If fertilization has not occurred, the egg degenerates and is phagocytized. (Implantation of the embryo is discussed in Chapter 3.)

Sperm Transport

Sperm transport occurs in both the male reproductive tract and the female reproductive tract. In the male reproductive tract, transport of spermatozoa is closely connected with their structural and functional maturation, whereas in the female reproductive tract, it is important for spermatozoa to pass to the upper uterine tube, where they can meet the ovulated egg.

After spermiogenesis in the seminiferous tubules, the spermatozoa are morphologically mature but are nonmotile and incapable of fertilizing an egg (**Fig. 2.3**). Spermatozoa are passively transported via testicular fluid from the seminiferous tubules to the caput (head) of the epididymis through the rete testis and the efferent ductules. They are propelled by fluid pressure generated in the seminiferous tubules and are assisted by smooth muscle contractions and ciliary currents in the efferent ductules. Spermatozoa spend about 12 days in the highly convoluted duct of the epididymis, which measures 6 m in the human, during which time they undergo biochemical maturation. This period of maturation is associated with changes in the glycoproteins in the plasma membrane of the sperm head. By the time the spermatozoa have reached the cauda (tail) of the epididymis, they are capable of fertilizing an egg.

On ejaculation, the spermatozoa rapidly pass through the **ductus deferens** and become mixed with fluid secretions from the **seminal vesicles** and **prostate gland**. Prostatic fluid is rich in citric acid, acid phosphatase, zinc, and magnesium ions, whereas fluid of the seminal vesicle is rich in fructose (the principal energy source of spermatozoa) and prostaglandins. The 2 to 6 mL of ejaculate (**semen,** or **seminal fluid**) typically consists of 40 to 250 million spermatozoa mixed with alkaline fluid from the seminal vesicles (60% of the total) and acid secretion (pH 6.5) from the prostate (30% of the total). The pH of normal semen ranges from 7.2 to 7.8. Despite the numerous spermatozoa (>100 million) normally present in an ejaculate, a number as small as 25 million spermatozoa per ejaculate may be compatible with fertility.

In the female reproductive tract, sperm transport begins in the upper vagina and ends in the ampulla of the uterine tube, where the spermatozoa make contact with the ovulated egg. During copulation, the seminal fluid is normally deposited in the upper vagina (see Fig. 2.3), where its composition and buffering capacity immediately protect the spermatozoa from the acid fluid found in the upper vaginal area. The acidic vaginal fluid normally serves a bactericidal function in protecting the cervical canal from pathogenic organisms. Within about 10 seconds, the pH of the upper vagina is increased from 4.3 to as much as 7.2. The buffering effect lasts only a few minutes in humans, but it provides enough time for the spermatozoa to approach the cervix in an environment (pH 6.0 to 6.5) optimal for sperm motility.

The next barriers that the sperm cells must overcome are the cervical canal and the cervical mucus that blocks it. Changes in intravaginal pressure may suck spermatozoa into the cervical os, but swimming movements also seem to be important for most spermatozoa in penetrating the cervical mucus.

The composition and viscosity of cervical mucus vary considerably throughout the menstrual cycle. Composed of **cervical mucin** (a glycoprotein with a high carbohydrate composition) and soluble components, cervical mucus is not readily penetrable. Between days 9 and 16 of the cycle, however, its water content increases, and this change facilitates the passage of sperm through the cervix around the time of ovulation; such mucus is sometimes called **E mucus**. After ovulation, under the influence of progesterone, the production of watery cervical mucus ceases, and a new type of sticky mucus, which has a much decreased water content, is produced. This progestational mucus, sometimes called **G mucus**, is almost completely resistant to sperm penetration. A highly effective method of natural family planning makes use of the properties of cervical mucus.

There are two main modes of sperm transport through the cervix. One is a phase of initial rapid transport, by which some spermatozoa can reach the uterine tubes within 5 to 20 minutes of ejaculation. Such rapid transport relies more on muscular movements of the female reproductive tract than on the motility of the spermatozoa themselves. These early-arriving sperm, however, appear not to be as capable of fertilizing an egg as do those that have spent more time in the female reproductive tract. The second, slow phase of sperm transport involves the swimming of spermatozoa through the cervical mucus (traveling at a rate of 2 to 3 mm/hour), their storage in cervical crypts, and their final passage through the cervical canal as much as 2 to 4 days later.

Relatively little is known about the passage of spermatozoa through the uterine cavity, but the contraction of uterine smooth muscle, rather than sperm motility, seems to be the main intrauterine transport mechanism. At this point, the spermatozoa enter one of the uterine tubes. According to some more recent estimates, only several hundred spermatozoa enter the uterine tubes, and most enter the tube containing the ovulated egg.

Once inside the uterine tube, the spermatozoa collect in the isthmus and bind to the epithelium for about 24 hours. During this time, they are influenced by secretions of the tube to undergo the **capacitation** reaction. One phase of capacitation is removal of **cholesterol** from the surface of the sperm. Cholesterol is a component of semen and acts to inhibit premature capacitation. The next phase of capacitation consists of removal of many of the glycoproteins that were deposited on the surface of the spermatozoa during their tenure in the epididymis. Capacitation is required for spermatozoa to be able to fertilize an egg (specifically, to undergo the acrosome reaction; see p. 29). After the capacitation reaction, the spermatozoa undergo a period of hyperactivity and detach from the tubal epithelium. Hyperactivation helps the spermatozoa to break free of the bonds that held them to the tubal epithelium. It also assists the sperm in penetrating isthmic mucus, as well as the corona radiata and the zona pellucida, which surround the ovum. Only small numbers of sperm are released at a given time. This may reduce the chances of polyspermy (see p. 31).

On their release from the isthmus, the spermatozoa make their way up the tube through a combination of muscular movements of the tube and some swimming movements. The simultaneous transport of an egg down and spermatozoa up the tube is currently explained on the basis of peristaltic contractions of the uterine tube muscles. These contractions subdivide the tube into compartments. Within a given compartment, the gametes are caught up in churning movements that over 1 or 2 days bring the egg and spermatozoa together. Fertilization of the egg normally occurs in the ampullary portion (upper third) of the uterine tube. Estimates suggest that spermatozoa retain their function in the female reproductive tract for about 80 hours.

After years of debate concerning the possibility that mammalian spermatozoa may be guided to the egg through attractants, more recent research suggests that this could be the case. Mammalian spermatozoa have been found to possess odorant receptors of the same family as olfactory receptors in the nose, and they can respond behaviorally to chemically defined odorants. Human spermatozoa also respond to cumulus-derived progesterone and to yet undefined chemoattractants emanating from follicular fluid and cumulus cells. Human spermatozoa are also known to respond to a temperature gradient, and studies on rabbits have shown that the site of sperm storage in the oviduct is cooler than that farther up the tube where fertilization occurs. It seems that only capacitated spermatozoa have the capability of responding to chemical or thermal stimuli. Because many of the sperm cells that enter the uterine tube fail to become capacitated, these spermatozoa are less likely to find their way to the egg.

Formation and Function of the Corpus Luteum of Ovulation and Pregnancy

While the ovulated egg is passing through the uterine tubes, the ruptured follicle from which it arose undergoes a series of striking changes that are essential for the progression of events leading to and supporting pregnancy (see Fig. 1.8). Soon after ovulation, the basement membrane that separates the granulosa cells from the theca interna breaks down, thus allowing thecal blood vessels to grow into the cavity of the ruptured follicle. The granulosa cells simultaneously undergo a series of major changes in form and function (**luteinization**). Within 30 to 40 hours of the LH surge, these cells, now called **granulosa lutein cells**, begin secreting increasing amounts of progesterone along with some estrogen. This pattern of secretion provides the hormonal basis for the changes in the female reproductive tissues during the last half of the menstrual cycle. During this period, the follicle continues to enlarge. Because of its yellow color, it is known as the **corpus luteum**. The granulosa lutein cells are terminally differentiated. They have stopped dividing, but they continue to secrete progesterone for 10 days.

In the absence of fertilization and a hormonal stimulus provided by the early embryo, the corpus luteum begins to deteriorate (**luteolysis**) late in the menstrual cycle. Luteolysis seems to involve both the preprogramming of the luteal cells to **apoptosis** (cell death) and **uterine luteolytic factors**, such as **prostaglandin F_2**. Regression of the corpus luteum and the accompanying reduction in progesterone production cause the hormonal withdrawal that results in the degenerative changes of the endometrial tissue during the last days of the menstrual cycle.

During the regression of the corpus luteum, the granulosa lutein cells degenerate and are replaced with collagenous scar tissue. Because of its white color, the former corpus

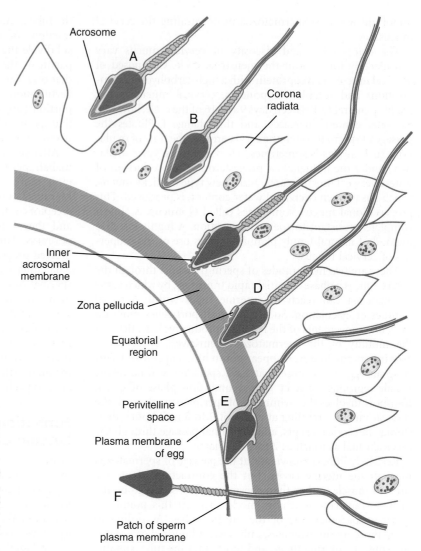

Fig. 2.4   The sequence of events in penetration of the coverings and plasma membrane of the egg. **A** and **B,** Penetration of the corona radiata. **C** and **D,** Attachment to the zona pellucida, the acrosomal reaction, and penetration of the zona. **E** and **F,** Binding to plasma membrane and entry into the egg.

luteum now becomes known as the **corpus albicans** ("white body").

If fertilization occurs, the production of the protein hormone **chorionic gonadotropin** by the future placental tissues maintains the corpus luteum in a functional condition and causes an increase in its size and hormone production. Because the granulosa lutein cells are unable to divide and cease producing progesterone after 10 days, the large **corpus luteum of pregnancy** is composed principally of theca lutein cells. The corpus luteum of pregnancy remains functional for the first few months of pregnancy. After the second month, the placenta produces enough estrogens and progesterone to maintain pregnancy on its own. At this point, the ovaries can be removed, and pregnancy would continue.

Fertilization

Fertilization is a series of processes rather than a single event. Viewed in the broadest sense, these processes begin when spermatozoa start to penetrate the corona radiata that surrounds the egg and end with the intermingling of the maternal and paternal chromosomes after the spermatozoon has entered the egg.

Penetration of the Corona Radiata

When the spermatozoa first encounter the ovulated egg in the ampullary part of the uterine tube, they are confronted by the corona radiata and some remnants of the cumulus oophorus, which represents the outer layer of the egg complex (**Fig. 2.4**). The corona radiata is a highly cellular layer with an intercellular matrix consisting of proteins and a high concentration of carbohydrates, especially hyaluronic acid. It is widely believed that hyaluronidase emanating from the sperm head plays a major role in penetration of the corona radiata, but the active swimming movements of the spermatozoa are also important.

Attachment to and Penetration of the Zona Pellucida

The **zona pellucida**, which is 13 μm thick in humans, consists principally of four glycoproteins—ZP_1 to ZP_4. ZP_2 and ZP_3 combine to form basic units that polymerize into long filaments. These filaments are periodically linked by cross-bridges of ZP_1 and ZP_4 molecules (**Fig. 2.5**). The zona pellucida of an unfertilized mouse egg is estimated to contain more than 1 billion copies of the ZP_3 protein.

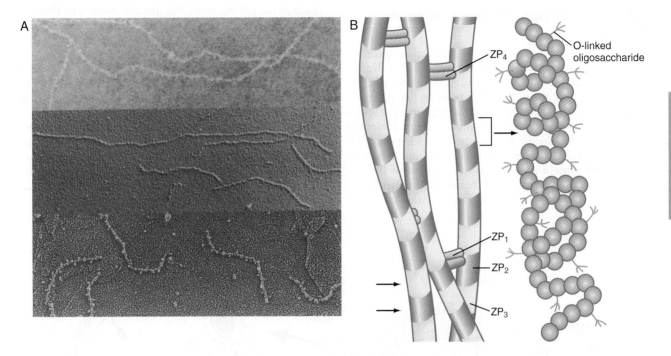

Fig. 2.5　**A,** Filamentous components of the mammalian (mouse) zona pellucida. **B,** Molecular organization of the filaments in the zona pellucida (ZP). *Far right,* Structure of the ZP_3 glycoprotein. *(From Wassarman PM: Sci Am 259(6):82, 1988.)*

After they have penetrated the corona radiata, spermatozoa bind tightly to the zona pellucida by means of the plasma membrane of the sperm head (see Fig. 2.4). Spermatozoa bind specifically to a sialic acid molecule, which is the terminal part of a sequence of four sugars at the end of O-linked oligosaccharides that are attached to the polypeptide core of the ZP_3 molecule. Molecules on the surface of the sperm head are specific binding sites for the ZP_3 sperm receptors on the zona pellucida. More than 24 molecules have been proposed, but the identity of the zona-binding molecules remains unknown. Interspecies molecular differences in the sperm-binding regions of the ZP_3 molecule may serve as the basis for the inability of spermatozoa of one species to fertilize an egg of another species. In mammals, there is less species variation in the composition of ZP_3; this may explain why penetration of the zona pellucida by spermatozoa of closely related mammalian species is sometimes possible, whereas it is rare among lower animals.

On binding to the zona pellucida, mammalian spermatozoa undergo the **acrosomal reaction**. The essence of the acrosomal reaction is the fusion of parts of the outer acrosomal membrane with the overlying plasma membrane and the pinching off of fused parts as small vesicles. This results in the liberation of the multitude of enzymes that are stored in the acrosome (**Box 2.1**).

The acrosomal reaction in mammals is stimulated by the ZP_3 molecule acting through G proteins in the plasma membrane on the sperm head. In contrast to the sperm receptor function of ZP_3, a large segment of the polypeptide chain of the ZP_3 molecule must be present to induce the acrosomal reaction. An initiating event of the acrosomal reaction is a massive influx of calcium (Ca^{++}) through the plasma membrane of the sperm head. This process, accompanied by an influx of sodium (Na^+) and an efflux of hydrogen (H^+),

---

Box 2.1　Some Major Mammalian Acrosomal Enzymes

Acid proteinase	β-Galactosidase
Acrosin	β-Glucuronidase
Arylaminidase	Hyaluronidase
Arylsulfatase	Neuraminidase
Collagenase	Phospholipase C
Esterase	Proacrosin

---

increases the intracellular pH. Fusion of the outer acrosomal membrane with the overlying plasma membrane soon follows. As the vesicles of the fused membranes are shed, the enzymatic contents of the acrosome are freed and can assist the spermatozoa in making their way through the zona pellucida.

After the acrosomal reaction, the inner acrosomal membrane forms the outer surface covering of most of the sperm head (see Fig. 2.4D). Toward the base of the sperm head (in the equatorial region), the inner acrosomal membrane fuses with the remaining **postacrosomal plasma membrane** to maintain membrane continuity around the sperm head.

Only after completing the acrosomal reaction can the spermatozoon successfully begin to penetrate the zona pellucida. Penetration of the zona is accomplished by a combination of mechanical propulsion by movements of the sperm's tail and digestion of a pathway through the action of acrosomal enzymes. The most important enzyme is **acrosin**, a serine proteinase that is bound to the inner acrosomal membrane. When the sperm has made its way through the zona and into the **perivitelline space** (the space between the egg's plasma membrane and the zona pellucida), it can make direct contact with the plasma membrane of the egg.

Binding and Fusion of Spermatozoon and Egg

After a brief transit period through the perivitelline space, the spermatozoon makes contact with the egg. In two distinct steps, the spermatozoon first binds to and then fuses with the plasma membrane of the egg. Binding between the spermatozoon and egg occurs when the **equatorial region** of the sperm head contacts the microvilli surrounding the egg. Molecules on the plasma membrane of the sperm head, principally sperm proteins called **fertilins** and **cyritestin**, bind to α_6 **integrin** and **CD9 protein** molecules on the surface of the egg. The acrosomal reaction causes a change in the membrane properties of the spermatozoon because, if the acrosomal reaction has not occurred, the spermatozoon is unable to fuse with the egg. Actual fusion between spermatozoon and egg, mediated by integrin on the membrane of the oocyte, brings their plasma membranes into continuity.

After initial fusion, the contents of the spermatozoon (the head, the midpiece, and usually the tail) sink into the egg (**Fig. 2.6**), whereas the sperm's plasma membrane, which is antigenically distinct from that of the egg, becomes incorporated into the egg's plasma membrane and remains recognizable at least until the start of cleavage. Although mitochondria located in the sperm neck enter the egg, they do not contribute to the functional mitochondrial complement of the zygote. In humans, the sperm contributes the centrosome, which is required for cell cleavage.

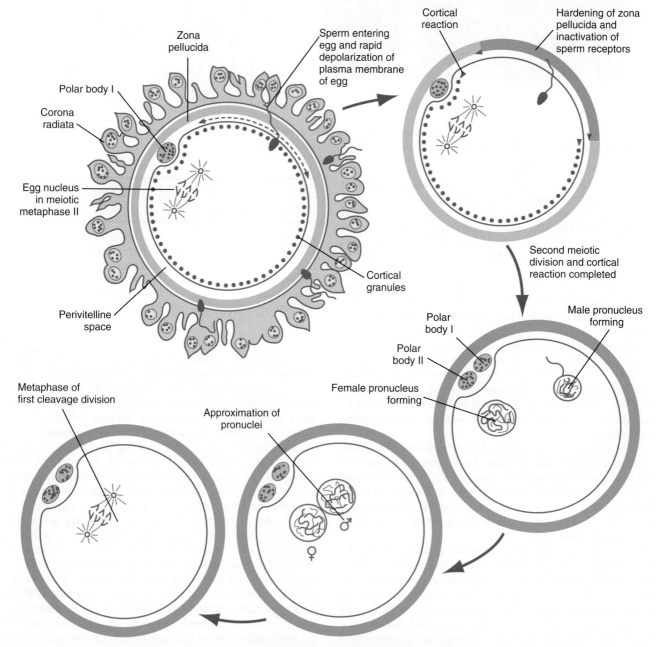

Fig. 2.6   Summary of the main events involved in fertilization.

Prevention of Polyspermy

When a spermatozoon has fused with an egg, the entry of other spermatozoa into the egg (**polyspermy**) must be prevented, or abnormal development is likely to result. Two blocks to polyspermy, fast and slow, are typically present in vertebrate fertilization.

The **fast block to polyspermy**, which has been best studied in sea urchins, consists of a rapid electrical depolarization of the plasma membrane of the egg. The resting membrane potential of the egg changes from about −70 to +10 mV within 2 to 3 seconds after fusion of the spermatozoon with the egg. This change in membrane potential prevents other spermatozoa from adhering to the egg's plasma membrane. The fast block in mammals is short-lived, lasting only several minutes, and may not be as heavily based on membrane depolarization as that in sea urchins. This time is sufficient for the egg to mount a permanent slow block. The exact nature of the fast block in the human egg is still not well defined.

Very soon after sperm entry, successive waves of Ca^{++} pass through the cytoplasm of the egg. The first set of waves, spreading from the site of sperm-egg fusion, is involved in stimulating completion of the second meiotic division of the egg. Later waves of Ca^{++} initiate recruitment of maternal RNAs in the egg and act on the cortical granules as they pass by them. Exposure to Ca^{++} causes the cortical granules to fuse with the plasma membrane and to release their contents (hydrolytic enzymes and polysaccharides) into the perivitelline space. The polysaccharides released into the perivitelline space become hydrated and swell, thus causing the zona pellucida to rise from the surface of the egg.

The secretory products of the cortical granules diffuse into the porous zona pellucida and hydrolyze the sperm receptor molecules (ZP_3 in the mouse) in the zona. This reaction, called the **zona reaction**, essentially eliminates the ability of spermatozoa to adhere to and penetrate the zona. The zona reaction has been observed in human eggs that have undergone in vitro fertilization. In addition to changes in the zona pellucida, alterations in sperm receptor molecules on the plasma membrane of the human egg cause the egg itself to become refractory to penetration by other spermatozoa.

Metabolic Activation of the Egg

The entry of the spermatozoon into the egg initiates several significant changes within the egg, including the aforementioned fast and slow blocks to polyspermy. In effect, the sperm introduces into the egg a soluble factor (currently thought to be a phospholipase [phospholipase C zeta]), which stimulates a pathway leading to the release of pulses of Ca^{++} within the cytoplasm of the egg. In addition to initiating the blocks to polyspermy, the released Ca^{++} stimulates a rapid intensification of the egg's respiration and metabolism through an exchange of extracellular Na^+ for intracellular H^+. This exchange results in a rise in intracellular pH and an increase in oxidative metabolism.

Decondensation of the Sperm Nucleus

In the mature spermatozoon, the nuclear chromatin is very tightly packed, in large part because of the —SS— (disulfide) cross-linking that occurs among the protamine molecules complexed with the DNA during spermatogenesis. Shortly after the head of the sperm enters the cytoplasm of the egg, the permeability of its nuclear membrane begins to increase, thereby allowing cytoplasmic factors within the egg to affect the nuclear contents of the sperm. After reduction of the —SS— cross-links of the protamines to sulfhydryl (—SH) groups by reduced glutathione in the ooplasm, the protamines are rapidly lost from the chromatin of the spermatozoon, and the chromatin begins to spread out within the nucleus (now called a **pronucleus**) as it moves closer to the nuclear material of the egg.

Remodeling of the sperm head takes about 6 to 8 hours. After a short period during which the male chromosomes are naked, histones begin to associate with the chromosomes. During the period of pronuclear formation, the genetic material of the male pronucleus becomes demethylated, whereas methylation in the female genome is maintained.

Completion of Meiosis and the Development of Pronuclei in the Egg

After penetration of the egg by the spermatozoon, the nucleus of the egg, which had been arrested in metaphase of the second meiotic division, completes the last division and releases a **second polar body** into the perivitelline space (see Fig. 2.6).

The nucleus of the oocyte moves toward the cortex as the result of the action of myosin molecules acting on a network of actin filaments that connect one pole of the mitotic spindle to the cortex. The resulting contraction draws the entire mitotic apparatus toward the surface of the cell (**Fig. 2.7**). This determines the location at which both the first and second polar bodies are extruded.

A pronuclear membrane, derived largely from the endoplasmic reticulum of the egg, forms around the female

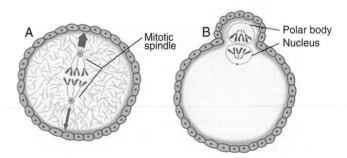

Fig. 2.7 Schematic representation showing how the dividing nucleus of the oocyte becomes translocated to the cortex of the egg and how that determines where the polar body forms. **A,** The mitotic spindle is situated within a meshwork of cytoplasmic actin filaments (*green*). Powered by myosin molecules (*blue*), contractions of the actin-myosin complex pull on either end of the mitotic spindle (*red arrow*). On the end of the spindle closest to the cell surface, the intensity of the pull is greater (*thick red arrow*), and the entire spindle apparatus moves toward that surface. **B,** As the mitotic process nears completion, one daughter nucleus buds off from the oocyte as a polar body. The nucleus remaining in the oocyte then divides again after fertilization and produces a second polar body in the same location as the first, because the nucleus of the oocyte is already near the cortex in that area. (*Based on Schuh M, Ellenberg J: Curr Biol 18:1986-1992, 2008.*)

chromosomal material. Cytoplasmic factors seem to control the growth of the female and the male pronuclei. Pronuclei appear 6 to 8 hours after sperm penetration, and they persist for about 10 to 12 hours. DNA replication occurs in the developing haploid pronuclei, and each chromosome forms two chromatids as the pronuclei approach each other. When the male and female pronuclei come into contact, their membranes break down, and the chromosomes intermingle. The maternal and paternal chromosomes quickly become organized around a mitotic spindle, derived from the centrosome of the sperm, in preparation for an ordinary mitotic division. At this point, the process of fertilization can be said to be complete, and the fertilized egg is called a **zygote**.

What is Accomplished by Fertilization?

The process of fertilization ties together many biological loose ends, as follows:

1. It stimulates the egg to complete the second meiotic division.
2. It restores to the zygote the normal diploid number of chromosomes (46 in humans).
3. The genetic sex of the future embryo is determined by the chromosomal complement of the spermatozoon. (If the sperm contains 22 autosomes and an X chromosome, the embryo is a genetic female, and if it contains 22 autosomes and a Y chromosome, the embryo is a male. See Chapter 16 for further details.)
4. Through the mingling of maternal and paternal chromosomes, the zygote is a genetically unique product of chromosomal reassortment, which is important for the viability of any species.
5. The process of fertilization causes metabolic activation of the egg, which is necessary for cleavage and subsequent embryonic development to occur.

Clinical Vignette

A 33-year-old woman who has had her uterus surgically removed desperately wants her own child. She is capable of producing eggs because her ovaries remain functional. She and her husband want to attempt in vitro fertilization and embryo transfer. They find a woman who, for $20,000, is willing to allow the couple's embryo to be transferred to her uterus and to serve as a surrogate mother during the pregnancy. Induction of superovulation is successful, and the physicians are able to fertilize eight eggs in vitro. Three embryos are implanted into the surrogate mother. The remaining embryos are frozen for possible future use. The embryo transfer is successful, and the surrogate mother becomes pregnant with twins. The twins are born, but the surrogate mother feels that she has bonded with them so much that she should have the right to raise them. The extremely wealthy genetic parents take the case to court, but before the case comes to trial they are both killed in an airplane accident. The surrogate mother now claims that she should get the large inheritance in the name of her twins, but the father's sister, equally aware of the financial implications, claims that she should care for the twins. The issue of what to do with the remaining five frozen embryos also comes up.

This case is fictitious, but all of its elements have occurred on an isolated basis. How would you deal with the following legal and ethical issues?

1. To whom should the twins be awarded?
2. What should be done with the remaining frozen embryos?

Summary

- Ovulation is stimulated by a surge of LH and FSH in the blood. Expulsion of the ovum from the graafian follicle involves local edema, ischemia, and collagen breakdown, with a possible contribution by fluid pressure and smooth muscle activity in rupturing the follicular wall.
- The ovulated egg is swept into the uterine tube and transported through it by ciliary action and smooth muscle contractions as it awaits fertilization by a sperm cell.
- Sperm transport in the male reproductive tract involves a slow exit from the seminiferous tubules, maturation in the epididymis, and rapid expulsion at ejaculation, where the spermatozoa are joined by secretions from the prostate and seminal vesicles to form semen.
- In the female reproductive tract, sperm transport involves entry into the cervical canal from the vagina, passage through the cervical mucus, and transport through the uterus into the uterine tubes, where capacitation occurs. The meeting of egg and sperm typically occurs in the upper third of the uterine tube.
- The fertilization process consists of several sequential events:
  1. Penetration of the corona radiata
  2. Attachment to the zona pellucida
  3. Acrosomal reaction and penetration of the zona pellucida
  4. Binding and fusion of sperm and egg
  5. Prevention of polyspermy
  6. Metabolic activation of the egg
  7. Decondensation of the sperm nucleus
  8. Completion of meiosis in the egg
  9. Development and fusion of male and female pronuclei
- Attachment of the spermatozoon to the zona pellucida is mediated by the ZP_3 protein, which also stimulates the acrosomal reaction.
- The acrosomal reaction involves fusion of the outer acrosomal membrane with the plasma membrane of the sperm cell and the fragmentation of the fused membranes, thus leading to the release of the acrosomal enzymes. One of the acrosomal enzymes, acrosin, is a serine proteinase, which digests components of the zona pellucida and assists the penetration of the swimming spermatozoa through the zona.
- After fusion of the spermatozoa to the egg membrane, a rapid electrical depolarization produces the first block to polyspermy in the egg. This is followed by a wave of Ca^{++} that causes the cortical granules to release their contents into the perivitelline space and ultimately to inactivate the sperm receptors in the zona pellucida.
- Sperm penetration stimulates a rapid intensification of respiration and metabolism of the egg.
- Within the egg, the nuclear material of the spermatozoon decondenses and forms the male pronucleus. At the same time, the egg completes the second meiotic division, and the remaining nuclear material becomes surrounded by a membrane, to form the female pronucleus.
- After DNA replication, the male and female pronuclei join, and their chromosomes become organized for a mitotic division. Fertilization is complete, and the fertilized egg is properly called a *zygote*.

■ Treatment of infertility by in vitro fertilization and embryo transfer is a multistage process involving stimulating gamete production by drugs such as clomiphene citrate, obtaining eggs by laparoscopic techniques in the woman, storing gametes by freezing, performing in vitro fertilization and culture of embryos, preserving the embryo, and transferring the embryo to the mother (**Clinical Correlation 2.1**).

■ Other techniques used for the treatment of infertility are gamete intrafallopian transfer (GIFT), which is the transfer of gametes directly into the uterine tube, and zygote intrafallopian transfer (ZIFT), with is the transfer of zygotes into the uterine tube. These techniques can be used with biological and surrogate mothers.

CLINICAL CORRELATION 2.1
Treatment of Infertility by In Vitro Fertilization and Embryo Transfer

Certain types of infertility caused by inadequate numbers or mobility of spermatozoa or by obstruction of the uterine tubes are now treatable by fertilizing an ovum in vitro and transferring the cleaving embryo into the reproductive tract of the woman. The sequential application of various techniques that were initially developed for the assisted reproduction of domestic animals, such as cows and sheep, is required. The relevant techniques are (1) stimulating gamete production, (2) obtaining male and female gametes, (3) storing gametes, (4) fertilizing eggs, (5) culturing cleaving embryos in vitro, (6) preserving embryos, and (7) introducing embryos into the uterus (**Fig. 2.8**).

Stimulation of Gamete Production

Ovulation is stimulated by altering existing hormonal relationships. For women who are **anovulatory** (do not ovulate), these techniques alone may be sufficient to allow conception.

Several methods have been used to stimulate gamete production. Earlier methods employed **clomiphene citrate**, a nonsteroidal antiestrogen that suppresses the normal negative feedback by estrogens on gonadotropin production by the pituitary (see Fig. 1.15). This method has been largely supplanted by the administration of various combinations of recombinant gonadotropin (follicle-stimulating hormone or luteinizing hormone, or both) preparations, sometimes in combination with gonadotropin-releasing hormone

agonists. These treatments result in multiple ovulation, a desired outcome for artificial fertilization because fertilizing more than one egg at a time is more efficient. Sometimes women who use these methods for the induction of ovulation produce multiple children, however, and many quintuplet to septuplet births have been recorded. Other methods of inducing ovulation are the application of **human menopausal gonadotropins** or the pulsatile administration of gonadotropin-releasing hormone. These techniques are more expensive than the administration of clomiphene.

Obtaining Gametes

For artificial insemination in vivo or artificial fertilization in vitro, spermatozoa are typically collected by masturbation. The collection of eggs requires technological assistance. Ongoing monitoring of the course of induced ovulation is accomplished by the application of imaging techniques, especially diagnostic ultrasound.

The actual recovery of oocytes involves their aspiration from ripe follicles. Although originally accomplished by **laparoscopy** (direct observation by inserting a laparoscope through a small slit in the woman's abdominal wall), visualization is now done with the assistance of ultrasound. An aspiration needle is inserted into each mature follicle, and the ovum is gently sucked into the needle and placed into culture medium in preparation for fertilization in vitro.

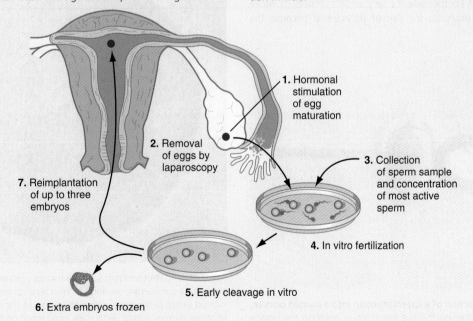

1. Hormonal stimulation of egg maturation
2. Removal of eggs by laparoscopy
3. Collection of sperm sample and concentration of most active sperm
4. In vitro fertilization
5. Early cleavage in vitro
6. Extra embryos frozen
7. Reimplantation of up to three embryos

Fig. 2.8　Schematic representation of a typical in vitro fertilization and embryo transfer procedure in humans.

CLINICAL CORRELATION 2.1
Treatment of Infertility by In Vitro Fertilization and Embryo Transfer—cont'd

Storing Gametes

Although eggs and sperm are usually placed together shortly after they are obtained, in some circumstances the gametes (especially spermatozoa) are stored for various periods before use. By bringing glycerinated preparations of spermatozoa down to the temperature of liquid nitrogen, spermatozoa can be kept for years without losing their normal fertilizing power. The freezing of eggs is possible, but much more problematic.

In Vitro Fertilization and Embryo Culture

Three ingredients for successful in vitro fertilization are as follows: (1) mature eggs; (2) normal, active spermatozoa; and (3) an appropriate culture environment. Having oocytes that are properly mature is one of the most important factors in obtaining successful in vitro fertilization. The eggs aspirated from a woman are sometimes at different stages of maturity. Immature eggs are cultured for a short time to become more fertilizable. The aspirated eggs are surrounded by the zona pellucida, the corona radiata, and a varying amount of cumulus oophorus tissue.

Fresh or frozen spermatozoa are prepared by separating them as much as possible from the seminal fluid. Seminal fluid reduces their fertilizing capacity, partly because it contains decapacitating factors. After capacitation, which in a human can be accomplished by exposing spermatozoa to certain ionic solutions, defined numbers of spermatozoa are added to the culture in concentrations of 10,000 to 500,000/mL. Rates of fertilization in vitro vary from one center to another, but 75% represents a realistic average.

In cases of infertility caused by **oligospermia** (too few spermatozoa) or excessively high percentages of abnormal sperm cells, multiple ejaculates may be obtained over an extended period. These are frozen and pooled to obtain adequate numbers of viable spermatozoa. In some cases, a few spermatozoa are microinjected into the **perivitelline space** inside the zona pellucida. Although this procedure can compensate for very small numbers of viable spermatozoa, it introduces the risk of polyspermy because the

normal gating function of the zona pellucida is bypassed. A more recent variant on in vitro fertilization is direct injection of a spermatozoon into an oocyte (**Fig. 2.9**). This technique has been used in cases of severe sperm impairment.

The initial success of in vitro fertilization is determined the next day by examination of the egg. If two pronuclei are evident (**Fig. 2.10**), fertilization is assumed to have occurred.

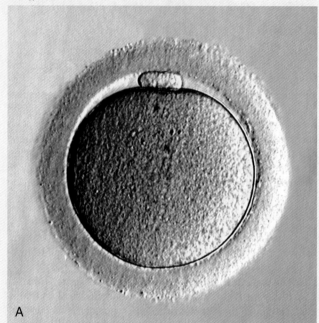

A

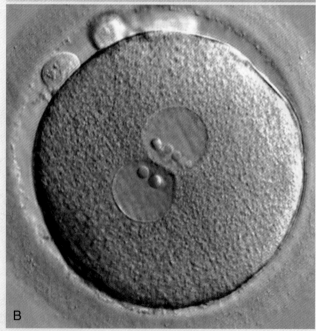

B

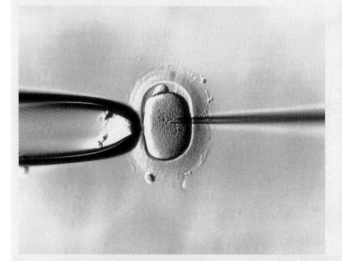

Fig. 2.9   **Microinjection of a spermatozoon into a human oocyte.** The micropipet containing the spermatozoon is entering the oocyte from the right side. (*From Veeck LL:* Atlas of the human oocyte fetal and early conceptus, *vol 2, Baltimore, 1991, Williams & Wilkins.*)

Fig. 2.10   **A,** Photomicrograph of a mature human oocyte arrested at metaphase II in culture awaiting in vitro fertilization. The polar body is located beneath the zona pellucida on top of the oocyte. **B,** A newly fertilized human oocyte, with male and female pronuclei in the center and two polar bodies on top of the oocyte. (*From Veeck LL, Zaninovic N:* An atlas of human blastocysts, *Boca Raton, Fla, 2003, Parthenon.*)

CLINICAL CORRELATION 2.1
Treatment of Infertility by In Vitro Fertilization and Embryo Transfer—cont'd

The cleavage in vitro of human embryos is more successful than that of most other mammalian species. The embryos are usually allowed to develop to the two- to eight-cell stage before they are considered ready to implant into the uterus.

Typically, all the eggs obtained from the multiple ovulations of the woman are fertilized in vitro during the same period. Practical reasons exist for doing this. One is that because of the low success rate of embryo transfer, implanting more than one embryo (commonly up to three) into the uterus at a time is advisable. Another reason is financial and also relates to the low success rate of embryo transfer. Embryos other than those used during the initial procedure are stored for future use if the first embryo transfer proves unsuccessful. Such stockpiling saves a great deal of time and thousands of dollars for the patient.

Embryo Preservation

Embryos preserved for potential future use are treated with cryoprotectants (usually glycerol or dimethyl sulfoxide) to reduce ice crystal damage. They are slowly brought to very low temperatures (usually ≤100°C) to halt all metabolic activity. The length of time frozen embryos should be kept and the procedure for handling them if the first implantation attempt is successful are questions with both technical and ethical aspects.

Embryo Transfer into the Mother

Transfer of the embryo into the mother is technically simple; yet this is the step in the entire operation that is subject to the greatest failure rate. Typically, only 30% of embryo transfer attempts result in a viable pregnancy.

Embryo transfer is commonly accomplished by introducing a catheter through the cervix into the uterine cavity and expelling the embryo or embryos from the catheter. The patient remains quiet, preferably lying down for several hours after embryo transfer.

The reasons for the low success rate of embryo transfers are poorly understood, but the number of completed pregnancies after normal fertilization in vivo is also likely to be only about one third. If normal implantation does occur, the remainder of the pregnancy is typically uneventful and is followed by a normal childbirth.

Intrafallopian Transfer

Certain types of infertility are caused by factors such as hostile cervical mucus and pathological or anatomical abnormalities of the upper ends of the uterine tubes. A simpler method for dealing with these conditions is to introduce male and female gametes directly into the lower end of a uterine tube (often at the junction of its isthmic and ampullary regions). Fertilization occurs within the tube, and the early events of embryogenesis occur naturally. The method of **gamete intrafallopian transfer (GIFT)** has resulted in slightly higher percentages of pregnancies than the standard in vitro fertilization and embryo transfer methods.

A variant on this technique is **zygote intrafallopian transfer (ZIFT)**. In this variant, a cleaving embryo that has been produced by in vitro fertilization is implanted into the uterine tube.

Surrogacy

Sometimes a woman can produce fertile eggs but cannot become pregnant. An example would be a woman whose uterus has been removed but who still possesses functioning ovaries. One option in this case is in vitro fertilization and embryo transfer, but the embryo is transferred into the uterus of another woman (**surrogate mother**). From the biological perspective, this procedure differs little from embryo transfer into the uterus of the biological mother, but it introduces a host of social, ethical, and legal issues.

Review Questions

1. Of the barriers to sperm survival and transport within the female reproductive tract, low pH is most important in the:
A. Upper uterine tube
B. Lower uterine tube
C. Uterine cavity
D. Cervix
E. Vagina

2. The principal energy source for ejaculated spermatozoa is:
A. Prostatic acid phosphatase
B. Internal glucose
C. Prostatic citric acid
D. Fructose in seminal vesicle fluid
E. Glycogen released from the vaginal epithelium

3. What is the principal hormonal stimulus for ovulation?

4. What is capacitation?

5. Where does fertilization occur?

6. Name two functions of the ZP_3 protein of the zona pellucida.

7. What is polyspermy, and how is it prevented after a spermatozoon enters the egg?

8. A woman gives birth to septuplets. What is the likely reason for the multiple births?

9. When multiple oocytes obtained by laparoscopy are fertilized in vitro, why are up to three embryos implanted into the woman's uterus, and why are the other embryos commonly frozen?

10. Why do some reproductive technology centers insert spermatozoa under the zona pellucida or even directly into the oocyte?

References

Austin CR: *Human embryos: the debate on assisted reproduction*, Oxford, 1989, Oxford University.

Barroso G and others: Developmental sperm contributions: fertilization and beyond, *Fertil Steril* 92:835-848, 2009.

Braden TD, Belfiore CJ, Niswender GD: Hormonal control of luteal function. In Findlay JK, ed: *Molecular biology of the female reproductive system*, New York, 1994, Academic Press, pp 259-287.

Chang MC: Experimental studies of mammalian fertilization, *Zool Sci* 1:349-364, 1984.

Devoto L and others: The human corpus luteum: life cycle and function in natural cycles, *Fertil Steril* 92:1067-1079, 2009.

Ducibella T, Fissore R: The roles of Ca^{2+}, downstream protein kinases, and oscillatory signaling in regulating fertilization and the activation of development, *Dev Biol* 315:257-279, 2008.

Dun MD and others: Sperm-zona pellucida interaction: molecular mechanisms and the potential for contraceptive intervention, *Handb Exp Pharmacol* 198:139-178, 2010.

Eisenbach M, Giojalas LC: Sperm guidance in mammals: an unpaved road, *Nat Rev Mol Cell Biol* 7:276-285, 2006.

Espey LL, Lipner H: Ovulation. In Knobil E, Neill JD, eds: *The physiology of reproduction*, ed 2, New York, 1994, Raven, pp 725-780.

Familiari G, Makabe S, Motta PM: The ovary and ovulation: a three-dimensional study. In Van Blerkom J, Motta PM, eds: *Ultrastructure of human gametogenesis and early embryogenesis*, Boston, 1989, Kluwer Academic, pp 85-124.

Florman HM, Ducibella T: Fertilization in mammals. In Neill JD, ed: *Physiology of reproduction*, ed 3, San Diego, 2006, Academic Press, pp 55-112.

Florman HM, Jungnickel MK, Sutton KA: Regulating the acrosome reaction, *Int J Dev Biol* 52:503-510, 2008.

Fraser LR: The "switching on" of mammalian spermatozoa: molecular events involved in promotion and regulation of capacitation, *Mol Reprod Dev* 77:197-208, 2010.

Fukuda M and others: Right-sided ovulation favours pregnancy more than left-sided ovulation, *Hum Reprod* 15:1921-1926, 2000.

Gadella BM: The assembly of a zona pellucida binding protein complex in sperm, *Reprod Domest Anim* 43(Suppl 5):12-19, 2008.

Geerling JH: Natural family planning, *Am Fam Physician* 52:1749-1756, 1995.

Holt WV, Fazeli A: The oviduct as a complex mediator of mammalian sperm function and selection, *Mol Reprod Dev* 77:934-943, 2010.

Ikawa M and others: Fertilization: a sperm's journey to and interaction with the oocyte, *J Clin Invest* 120:984-994, 2010.

Jones GS: Corpus luteum: composition and function, *Fertil Steril* 54:21-26, 1990.

Kaji K, Kudo A: The mechanism of sperm-oocyte fusion in mammals, *Reproduction* 127:423-429, 2004.

Kim E and others: Sperm penetration through cumulus mass and zona pellucida, *Int J Dev Biol* 52:677-682, 2008.

Knoll M, Talbot P: Cigarette smoke inhibits oocyte cumulus complex pick-up by the oviduct in vitro independent of ciliary beat frequency, *Reprod Toxicol* 12:57-68, 1998.

Miyazaki S: Thirty years of calcium signals at fertilization, *Semin Cell Dev Biol* 17:233-243, 2006.

Myles DG, Koppel DE, Primakoff P: Defining sperm surface domains. In Alexander NJ and others, eds: *Gamete interaction: prospects for immunocontraception*, New York, 1990, Wiley-Liss, pp 1-11.

Oh JS, Susor A, Conti M: Protein tyrosine kinase Wee1B is essential for metaphase II exit in mouse oocytes, *Science* 332:462-465, 2011.

Ozil J-P and others: Ca^{2+} oscillatory pattern in fertilized mouse eggs affects gene expression and development to term, *Dev Biol* 300:534-544, 2006.

Parrington J and others: Flipping the switch: how a sperm activates the egg at fertilization, *Dev Dyn* 236:2027-2038, 2007.

Primakoff P, Myles DG: Penetration, adhesion, and fusion in mammalian sperm-egg interaction, *Science* 296:2183-2185, 2002.

Rath D and others: Sperm interactions from insemination to fertilization, *Reprod Domest Anim* 43(Suppl 5):2-11, 2008.

Richards JAS and others: Ovulation: new dimensions and new regulators of the inflammatory-like response, *Annu Rev Physiol* 64:69-92, 2002.

Rubenstein E and others: The molecular players of sperm-egg fusion in mammals, *Semin Cell Dev Biol* 17:254-263, 2006.

Schuh M, Ellenberg J: A new model for asymmetric spindle positioning in mouse oocytes, *Curr Biol* 18:1986-1992, 2008.

Spehr M and others: Identification of a testicular odorant receptor mediating sperm chemotaxis, *Science* 299:2054-2057, 2003.

Suarez SS: Gamete and zygote transport. In Neill JD, ed: *Physiology of reproduction*, ed 3, San Diego, 2006, Academic Press, pp 113-146.

Suarez SS: Regulation of sperm storage and movement in the mammalian oviduct, *Int J Dev Biol* 52:455-462, 2008.

Talbot P and others: Oocyte pickup by the mammalian oviduct, *Mol Biol Cell* 10:5-8, 1999.

Tanghe S and others: Minireview: functions of the cumulus oophorus during oocyte maturation, ovulation, and fertilization, *Mol Reprod Dev* 61:414-424, 2002.

Tsafriri A, Reich R: Molecular aspects of mammalian ovulation, *Exp Clin Endocrinol Diabetes* 107:1-11, 1999.

Turner TT: De Graaf's thread: the human epididymis, *J Androl* 29:237-250, 2008.

Veeck LL, Zaninovic N: *An atlas of human blastocysts*, Boca Raton, Fla, 2003, Parthenon.

Wassarman PM: Zona pellucida glycoproteins, *J Biol Chem* 283:24285-24289, 2008.

Wassarman PM, Litscher ES: Mammalian fertilization: the egg's multifunctional zona pellucida, *Int J Dev Biol* 52:665-676, 2008.

Wassarman PM, Litscher ES: Towards the molecular basis of sperm and egg interaction during mammalian fertilization, *Cells Tiss Organs* 168:36-45, 2001.

Whitaker M: Calcium at fertilization and in early development, *Physiol Rev* 86:25-88, 2006.

Wood C, Trounson A, eds: *Clinical in vitro fertilization*, ed 2, London, 1989, Springer-Verlag.

Yanagimachi R: Mammalian fertilization. In Knobil E, Neill J, eds: *The physiology of reproduction*, ed 2, New York, 1994, Raven, pp 189-317.

Yeung C-H, Cooper TG: Developmental changes in signalling transduction factors in maturing sperm during epididymal transit, *Cell Mol Biol* 49:341-349, 2003.

Yu Y and others: The extracellular protein coat of the inner acrosomal membrane is involved in zona pellucida binding and penetration during fertilization: characterization of its most prominent polypeptide (IAM38), *Dev Biol* 290:32-43, 2006.

Chapter 3

Cleavage and Implantation

The act of fertilization releases the ovulated egg from a depressed metabolism and prevents its ultimate disintegration within the female reproductive tract. Immediately after fertilization, the zygote undergoes a pronounced shift in metabolism and begins several days of **cleavage**. During this time, the embryo, still encased in its zona pellucida, is transported down the uterine tube and into the uterus. Roughly 6 days later, the embryo sheds its zona pellucida and attaches to the uterine lining.

With intrauterine development and a placental connection between the embryo and mother, higher mammals, including humans, have evolved greatly differing modes of early development from those found in most invertebrates and lower vertebrates. The eggs of lower animals, which are typically laid outside the body, must contain all the materials required for the embryo to attain the stage of independent feeding. Two main strategies have evolved. One is to complete early development as rapidly as possible, a strategy that has been adopted by *Drosophila,* sea urchins, and many amphibians. This strategy involves storing a moderate amount of yolk in the oocyte and preproducing much of the molecular machinery necessary for the embryo to move rapidly through cleavage to the start of gastrulation. The oocytes of such species typically produce and store huge amounts of ribosomes, messenger RNA (mRNA), and transfer RNA (tRNA). These represent maternal gene products, and this means that early development in these species is controlled predominantly by the maternal genome. The other strategy of independent development, adopted by birds and reptiles, consists of producing a very large egg containing enough yolk that early development can proceed at a slower pace. This strategy eliminates the need for the oocyte to synthesize and store large amounts of RNAs and ribosomes before fertilization.

Mammalian embryogenesis employs some fundamentally different strategies from those used by the lower vertebrates. Because the placental connection to the mother obviates the need for the developing oocyte to store large amounts of yolk, the eggs of mammals are very small. Mammalian cleavage is a prolonged process that typically coincides with the time required to transport the early embryo from its site of fertilization in the uterine tube to the place of implantation in the uterus. A prominent innovation in early mammalian embryogenesis is the formation of the **trophoblast**, the specialized tissue that forms the trophic interface between the embryo and the mother, during the cleavage period. The placenta represents the ultimate manifestation of the trophoblastic tissues.

Cleavage

Morphology

Compared with most other species, mammalian cleavage is a leisurely process measured in days rather than hours. Development proceeds at the rate of roughly one cleavage division per day for the first 2 days (**Figs. 3.1** and **3.2**). After the 2-cell stage, mammalian cleavage is asynchronous, with 1 of the 2 cells (**blastomeres**) dividing to form a 3-cell embryo. When the embryo consists of approximately 16 cells, it is called a **morula** (derived from the Latin word meaning "mulberry").

Starting after the eight-cell stage, the embryos of placental mammals enter into a phase called **compaction**, during which the individual outer blastomeres tightly adhere through gap and tight junctions and lose their individual identity when viewed from the surface. Compaction is mediated by the concentration of calcium (Ca^{++})–activated cell adhesion molecules, such as **E-cadherin**, in a ring around the apical surface of the blastomeres. Through the activity of a sodium (Na^+), potassium (K^+)–adenosine triphosphatase (ATPase)–based Na^+ transport system, Na^+ and water (H_2O) move across the epitheliumlike outer blastomeres and accumulate in spaces among the inner blastomeres. This process, which occurs about 4 days after fertilization, is called **cavitation**, and the fluid-filled space is known as the **blastocoele** (**blastocyst cavity**). At this stage, the embryo as a whole is known as a **blastocyst** (**Fig. 3.3**).

At the blastocyst stage, the embryo, which is still surrounded by the zona pellucida, consists of two types of cells: an outer epithelial layer (the **trophoblast**) that surrounds a small inner group of cells called the **inner cell mass** (see Fig. 3.1). Each blastomere at the two-cell and the four-cell stage contributes cells to both the inner cell mass and the trophoblast. The end of the blastocyst that contains the inner cell mass is known as the **embryonic pole**, and the opposite end is called the **abembryonic pole**. The appearance of these two cell types reflects major organizational changes that have occurred within the embryo and represents the specialization of the blastomeres into two distinct cell lineages. Cells of the inner cell mass give rise to the body of the embryo itself in addition to several

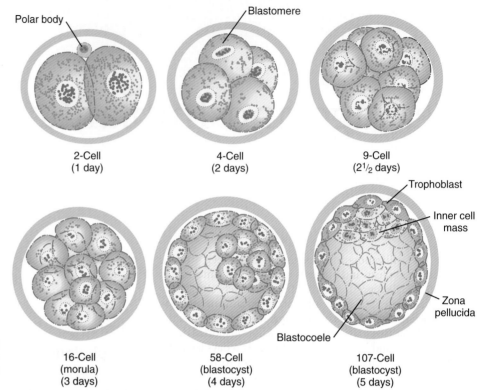

Fig. 3.1 Drawings of early cleavage stages in human embryos. The drawings of the 58-cell and 107-cell stages represent bisected embryos.

2-Cell
(1 day)

4-Cell
(2 days)

9-Cell
(2½ days)

16-Cell
(morula)
(3 days)

58-Cell
(blastocyst)
(4 days)

107-Cell
(blastocyst)
(5 days)

Polar body

Blastomere

Trophoblast

Inner cell mass

Zona pellucida

Blastocoele

extraembryonic structures, whereas cells of the trophoblast form only extraembryonic structures, including the outer layers of the placenta. There is increasing evidence that **fibroblast growth factor-4**, a growth factor secreted by cells of the inner cell mass, acts to maintain mitotic activity in the overlying trophoblast.

Molecular, Genetic, and Developmental Control of Cleavage

Along with the increase in cell numbers, mammalian cleavage is a period dominated by several critical developmental events. The earliest is the transition from maternally to zygotically produced gene products. Another is the polarization of individual blastomeres, which sets the stage for the developmental decision that results in the subdivision of the cleaving embryo into two distinct types of cells: the trophoblast and the inner cell mass (see Fig. 3.1). Most studies of the molecular biology and genetics of early mammalian development have been done on mice. Until more information on early primate embryogenesis becomes available, results obtained from experimentation on mice must be used as a guide.

Because of the lack of massive storage of maternal ribosomes and RNAs during oogenesis, development of the mammalian embryo must rely on the activation of zygotic gene products at a very early stage. Most maternal transcription products are degraded by the two-cell stage (**Fig. 3.4**). Some of these, however, stimulate the activation of the embryonic genome, which begins producing RNAs from a significant number of genes (>1500) by the time cleavage has advanced to the four-cell stage. There does not seem to be a sharp transition between the cessation of reliance on purely maternal gene products and the initiation of transcription from the

embryonic genome. Some paternal gene products (e.g., isoforms of β-glucuronidase and β$_2$-microglobulin) appear in the embryo very early, while maternal actin and histone mRNAs are still being used for the production of corresponding proteins. As an indication of the extent to which the early embryo relies on its own gene products, development past the two-cell stage does not occur in the mouse if mRNA transcription is inhibited. In contrast, similar treatment of amphibian embryos does not disrupt development until late cleavage, at which time the embryos begin to synthesize the mRNAs required to control morphogenetic movements and gastrulation.

Mature eggs and sperms are transcriptionally inactive. A major reason for this is that their DNA is highly methylated. **Methylation**, which occurs on CpG dinucleotides, normally inactivates the associated gene. Such inactivation is often called **epigenetic regulation** because it does not alter the fundamental DNA sequence. Methylation can inactivate informational genes or their regulators (e.g., enhancers or promoters). Pronounced cycles of global methylation and demethylation occur during the life span of an individual (**Fig. 3.5**). Within 4 hours after fertilization, the paternally derived genome undergoes rapid, massive demethylation. Demethylation of the maternally derived genome occurs more gradually until the early morula, at which stage all the DNA is maximally demethylated. Remethylation ensues in the inner cell mass, until by the late blastocyst stage it returns to maximal levels. Within the germ cell line, the high methylation levels characteristic of the early embryo fall after the primordial germ cells have entered the genital ridge. During later gametogenesis, remethylation occurs. This remethylation imprints (see p. 43) maternal or paternal characteristics on the gametes and for some genes has profound effects on the embryos produced

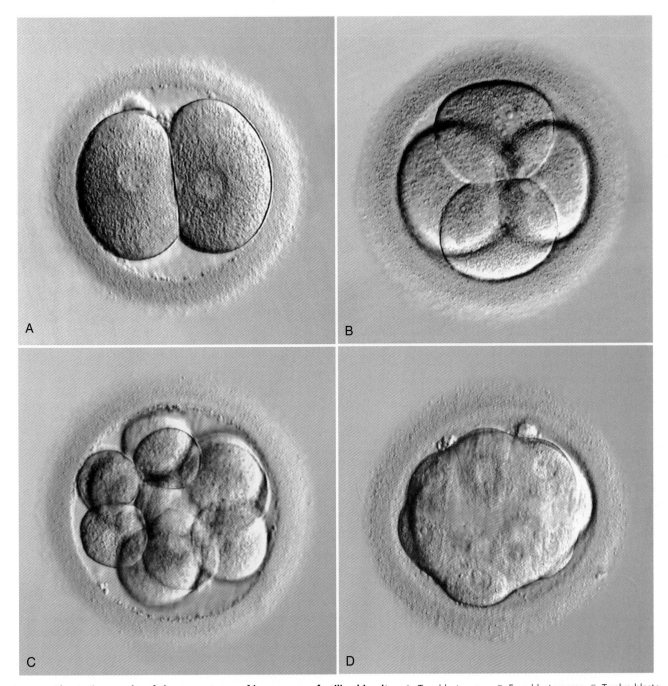

Fig. 3.2   Photomicrographs of cleavage stages of human eggs fertilized in vitro. A, Two blastomeres. B, Four blastomeres. C, Twelve blastomeres. D, Morula in late stage of compaction (5 days). Note the indistinct cell outlines. *(From Veeck LL, Zaninovic N: An atlas of human blastocysts, Boca Raton, Fla, 2003, Parthenon.)*

from these gametes. Epigenetic control is not confined to methylation patterns. Even as early as the zygote, different patterns of histone association with the chromatin account for pronounced differences in gene expression between the male and female pronuclei.

For the first couple of days after fertilization, transcriptional activity in the cleaving embryo is very low. Similarly, fertilized eggs and early mammalian embryos possess a limited capacity for the translation of mRNAs. The factor limiting translational efficiency may be the small number of ribosomes stored in the egg. As cleavage proceeds, products from maternally and

paternally derived chromosomes are active in guiding development. Haploid embryos commonly die during cleavage or just after implantation. There is increasing evidence, however, that the control of early development involves more than simply having a diploid set of chromosomes in each cell.

One of the first manifestations of embryonic gene expression is the polarization of the blastomeres of the 8- and 16-cell embryo so that they have clearly recognizable apical and basal surfaces. The polarization of blastomeres leads to one of the most important steps in early mammalian development, namely, the decision that results in the appearance of two

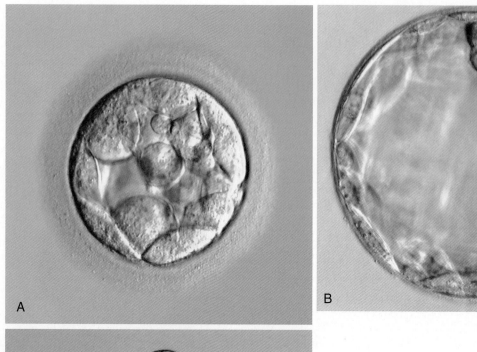

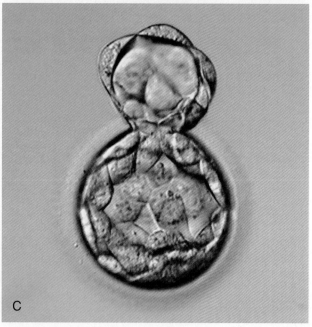

Fig. 3.3 **Human embryos resulting from in vitro fertilization.** **A,** Morula, showing the beginning of cavitation. **B,** Blastocyst, showing a well-defined inner cell mass *(arrow)* and blastocoele. At this stage, the zona pellucida is very thin. **C,** A hatching blastocyst beginning to protrude through the zona pellucida. *(From Veeck LL, Zaninovic N: An atlas of blastocysts, Boca Raton, Fla, 2003, Parthenon.)*

separate lines of cells—the trophoblast and the inner cell mass—from the early homogeneous blastomeres. In mice, up to the 8-cell stage all blastomeres are virtually identical. In the 8-cell embryo, the surfaces of the cells are covered with microvilli, and intercellular connections, mediated by **E-cadherin**, form. Shortly thereafter, differences are noted between polarized cells that have at least one surface situated on the outer border of the embryo and nonpolarized cells that are completely enclosed by other blastomeres. The polarized outer cells are destined to become trophoblast, whereas those cells located in the interior are destined to become the inner cell mass, from which the body of the embryo arises.

The relationship between the position of the blastomeres and their ultimate developmental fate was incorporated into the **inside-outside hypothesis**. The essence of this hypothesis

is that the fate of a blastomere derives from its position within the embryo, rather than from its intrinsic properties. The outer blastomeres ultimately differentiate into the trophoblast, whereas the inner blastomeres form the inner cell mass. If marked blastomeres from disaggregated embryos are placed on the surface of another early embryo, they typically contribute to the formation of the trophoblast. Conversely, if the same marked cells are introduced into the interior of the host embryo, they participate in the formation of the inner cell mass (**Fig. 3.6**).

The **cell polarity model** offers an alternative explanation for the conversion of generic blastomeres to trophoblast or inner cell mass. According to this hypothesis, if the plane of cell division of a blastomere at the eight-cell stage is parallel to the outer surface of the embryo, the outer daughter cell

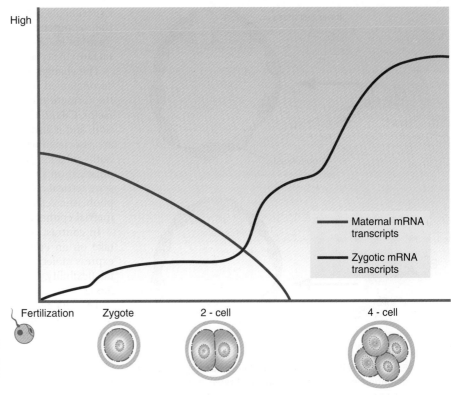

Fig. 3.4 Relative abundance of maternal versus zygotic transcription products in early cleaving embryos. *Blue line,* maternal mRNAs; *red line,* zygotic mRNAs.

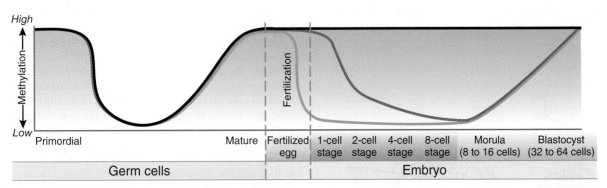

Fig. 3.5 Methylation of various classes of genes during gamete maturation and cleavage. Migrating primordial germ cells are highly methylated, but they lose their methylation on entering the primitive gonad. Methylation is then lost and later reacquired during late stages of gamete maturation. After fertilization, methylation remains high in imprinted genes *(black line),* but DNA in the male pronucleus undergoes rapid enzymatically mediated demethylation in the zygote *(blue line),* whereas demethylation in the female chromosomes occurs more slowly (over several days) *(red line).* This, in addition to changes in histone patterns, accounts for the greater levels of transcription in the paternal genome during very early development. By the blastocyst stage, high methylation levels have returned. *(Modified from Santos F, Dean W: Reproduction 127:643-651, 2004.)*

develops a polarity, with its apical surface facing the zona pellucida (**Fig. 3.7**). The inner daughter cell remains apolar and goes on to form part of the inner cell mass. Experimental evidence suggests that a key element underlying a daughter cell's becoming an outer cell is inheritance of a patch of outer cell membrane containing microvilli and the actin microfilament-stabilizing protein, **ezrin**. The proteins that produce polarity in the outer cells are postulated to direct their differentiation toward the trophoblastic lineage. Common to the inside-outside hypothesis and the cell polarity model is the recognition that a cell that does not contact the surface does not differentiate into trophoblast, but rather becomes part of the inner cell mass.

Even though by the 16-cell stage the embryo consists of clearly recognizable polar outer cells and nonpolar inner cells, cells of either type still can become transformed into cells of the other type. Thus, cells of the inner cell mass, if transplanted to the outer surface of another embryo, can become trophoblast, and at least some of the outer cells can turn into inner cell mass if transplanted into the interior. By the 32-cell stage, this capability for phenotypic transformation has become largely lost. Investigators have shown that cells of the inner cell mass of 16-cell embryos still retain the molecular machinery to turn into trophoblastic cells, because if the cells are exposed to the surface, they undergo the transformation into trophoblastic cells without new mRNA synthesis.

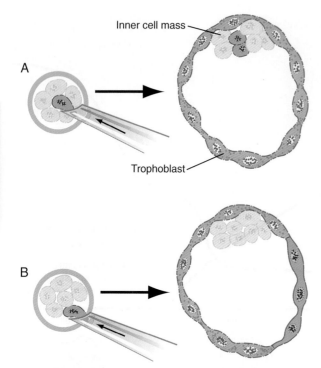

Fig. 3.6 **Experiments illustrating the inside-outside hypothesis of cell determination in early mammalian embryos.** **A,** If a marked blastomere is inserted into the interior of a morula, it and its progeny become part of the inner cell mass. **B,** If a marked blastomere is placed on the outside of a host morula, it and its descendants contribute to the trophoblast.

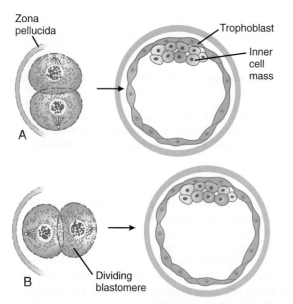

Fig. 3.7 **The cell polarity model of differentiation of blastomeres.** **A,** If the plane of cleavage of a blastomere is perpendicular to the surface of the embryo, each daughter cell becomes trophoblast. **B,** If the plane of cleavage is parallel to the surface, the daughter blastomere located at the surface becomes trophoblast, whereas the daughter cell located on the interior becomes part of the inner cell mass.

Experiments of this type show that the **developmental potential,** or **potency** (the types of cells that a precursor cell can form) of many cells is greater than their normal **developmental fate** (the types of cells that a precursor cell normally forms).

The changes in phenotype of the inner and outer cells are accompanied by important molecular differences. Critical to the formation of cells of the trophoblast is the transcription factor, **Cdx-2.** Cdx-2 is essential for trophoblastic differentiation, and it also antagonizes the expression of molecules that are associated with the inner cell mass. Increased Cdx-2 levels both enhance the formation of molecules associated with polarization and increase the proportion of cells that undergo symmetrical cell division, thus increasing the number of trophoblastic cells. *Cdx2* mutants fail to implant into the endometrial epithelium.

In contrast to cells of the trophoblast, which increasingly take on an epithelial character, cells of the inner cell mass express molecules that are associated with great developmental flexibility. Three such molecules are **oct-4**, **Nanog**, and **Sox-2**.

The *oct4* gene codes for a specific transcription factor that binds the octamer ATTTGCAT on DNA. There is a close relationship between the expression of the *oct4* gene and the highly undifferentiated state of cells. In mice, maternally derived oct-4 protein is found in developing oocytes and is active in the zygote. After the experimentally induced loss of oct-4 protein, development is arrested at the one-cell stage. This shows that maternally derived oct-4 protein is required to permit development to proceed to the two-cell stage, when transcription of the embryonic genes begins.

Oct-4 is expressed in all blastomeres up to the morula stage. As various differentiated cell types begin to emerge in the embryo, the levels of *oct4* gene expression in these cells decrease until it is no longer detectable. Such a decrease is first noted in cells that become committed to forming extraembryonic structures and finally in cells of the specific germ layers as they emerge from the primitive streak (see Chapter 5). Even after virtually all cells of the embryo have ceased to express the *oct4* gene, it is still detectable in the primordial germ cells as they migrate from the region of the allantois to the genital ridges. Because of its pattern of distribution, oct-4 protein is suspected to play a regulatory role in maintenance of the undifferentiated state and in establishing and maintaining the pluripotency of the germ cells.

Two other important genes in early development are *Nanog* and *Sox2*. Inner cells resulting from the division of cells in the eight-cell embryo begin to produce Sox-2, which binds onto DNA in partnership with oct-4 to regulate the expression of genes that control cellular differentiation. Nanog first appears in the late morula and along with Oct-4 functions to maintain the integrity of the inner cell mass. In the absence of Nanog function, cells of the inner cell mass differentiate into primitive endoderm (hypoblast), whereas lack of function of oct-4 causes inner cell mass cells to differentiate into trophoblast. Overall, but through different mechanisms, cells of both the trophoblast and inner cell mass are normally inhibited from becoming transformed into the other type.

Parental Imprinting

Experimentation, coupled with observations on some unusual developmental disturbances in mice and humans, has shown

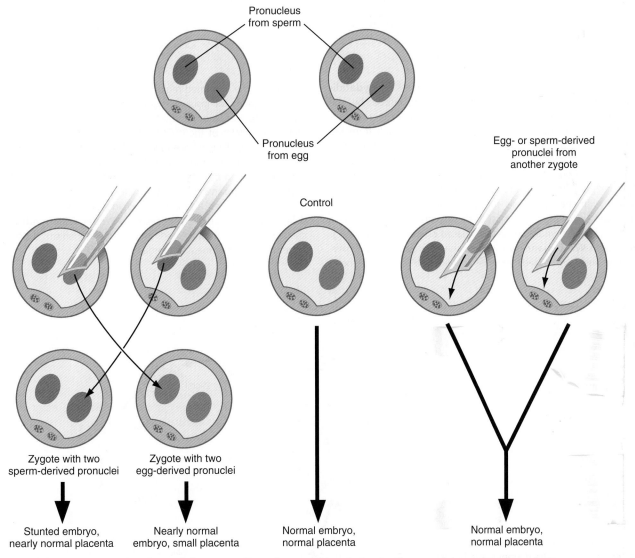

Fig. 3.8   **Experimental demonstrations of parental imprinting by the use of pronuclear transplants.**

that the expression of certain genes derived from the egg differs from the expression of the same genes derived from the spermatozoon. Called **parental imprinting**, the effects are manifest in different ways. It is possible to remove a pronucleus from a newly inseminated mouse egg and replace it with a pronucleus taken from another inseminated egg at a similar stage of development (**Fig. 3.8**). If a male or female pronucleus is removed and replaced with a corresponding male or female pronucleus, development is normal. If a male pronucleus is removed and replaced with a female pronucleus (resulting in a zygote with two female pronuclei), however, the embryo itself develops fairly normally, but the placenta and yolk sac are poorly developed. Conversely, a zygote with two male pronuclei produces a severely stunted embryo, whereas the placenta and yolk sac are nearly normal.

Parental imprinting occurs during gametogenesis. Methylation of DNA, effected through specific imprinting centers, is one of the major means of imprinting and results in the differential expression of paternal and maternal alleles of the imprinted genes. Imprinted genes are transcriptionally silenced. The imprinted genes are maintained during development and possibly into adulthood, but a given imprint is not passed onto that individual's progeny. Instead, the parental imprints on the genes are erased, and new imprints, corresponding to the sex of that individual, are established in the oocytes and sperm during gametogenesis.

Not all genes are parentally imprinted. Present estimates suggest that up to 2100 human genes are imprinted. **Clinical Correlation 3.1** discusses some conditions and syndromes associated with disturbances in parental imprinting.

X-Chromosome Inactivation

Another example of the inequality of genetic expression during early development is the pattern of X-chromosome inactivation in female embryos. It is well known from cytogenetic studies that one of the two X chromosomes in the cells of females is inactivated by extreme condensation. This is the basis for the **sex chromatin**, or **Barr body**, which can be shown in cells of females, but not in the cells of normal males.

CLINICAL CORRELATION 3.1
Conditions and Syndromes Associated with Parental Imprinting

A striking example of paternal imprinting in humans is a **hydatidiform mole** (see Fig. 7.16), which is characterized by the overdevelopment of trophoblastic tissues and the extreme underdevelopment of the embryo. This condition can result from the fertilization of an egg by two spermatozoa and the consequent failure of the maternal genome of the egg to participate in development or from the duplication of a sperm pronucleus in an "empty" egg. This form of highly abnormal development is consistent with the hypothesis that paternal imprinting favors the development of the trophoblast at the expense of the embryo.

Several other syndromes are also based on parental imprinting. **Beckwith-Wiedemann syndrome**, characterized by fetal overgrowth and an increased incidence of childhood cancers, has been mapped to the imprinted region on chromosome 11, which contains the genes for insulinlike growth factor-II (IGF-II, which promotes cell proliferation) and H19 (a growth suppressor). It occurs when both alleles of the *IGF2* gene express a paternal imprinting pattern. Another instructive example involves deletion of regions in the long arm of chromosome 15, specifically involving the gene **UBE3A**. Children of either sex who inherit the maternal deletion develop **Angelman's syndrome**, which includes severe mental retardation, seizures, and ataxia. A child who inherits a paternal deletion of the same region develops **Prader-Willi syndrome**, characterized by obesity, short stature, hypogonadism, a bowed upper lip, and mild mental retardation.

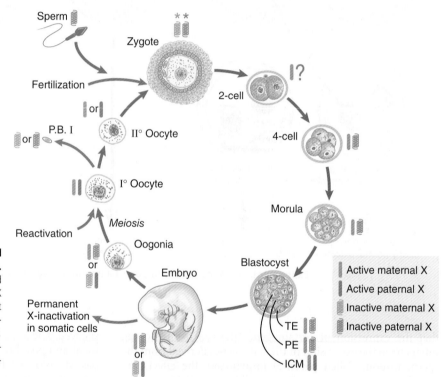

Fig. 3.9 **X-chromosomal inactivation and reactivation during the mammalian life cycle.** The *red* and *green symbols* refer to inactivated paternally *(red)* and maternally *(green)* derived X chromosomes. ICM, inner cell mass; P.B. I, first polar body; PE, primitive (extraembryonic) endoderm; TE, trophectoderm. *(Based on Gartler SM, Riggs AD: Annu Rev Genet 17:155-190, 1983; and Thorvaldsen JL, Verona RI, Bartolomei MS: Dev Biol 298:344-353, 2006.)*

The purpose of X-chromosome inactivation is dosage compensation, or preservation of the cells from an excess of X-chromosomal gene products.

X-chromosome inactivation is initiated at the **X-inactivation center**, a unique locus on the X chromosome. *XIST* (**X**-**i**nactive **s**pecific **t**ranscript), one of the genes in the X-inactivation center, produces a large RNA with no protein coding potential. XIST RNA remains in the nucleus and coats the entire inactive X chromosome, thus not allowing any further transcription from that chromosome. In the inactivated X chromosome, the *XIST* gene is unmethylated and expressed, whereas in the active X chromosome, this gene is methylated and silent.

Genetic studies show a complex ontogenetic history of X-chromosome inactivation (**Fig. 3.9**). In the female zygote, both X chromosomes are transcriptionally inactive, although not through the actions of XIST, because of the global inactivation of transcription in the early cleaving embryo. By the four-cell stage and into the morula stage, the paternally derived X chromosome becomes inactivated as the result of parental imprinting. Then, as the embryo forms the blastocyst, the paternally derived X chromosomes in the trophoblast and the hypoblast (see Fig. 5.1) remain inactivated, but within the cells of the inner cell mass both X chromosomes become active. As the cells of the inner cell mass begin to differentiate, the somatic cells undergo random permanent XIST-based X-chromosome inactivation of either the maternal or the paternal X chromosome. Within the germ cell line, activation of both X chromosomes occurs during the first meiotic division.

Developmental Properties of Cleaving Embryos

Early mammalian embryogenesis is considered to be a highly regulative process. **Regulation** is the ability of an embryo or organ primordium to produce a normal structure if parts have been removed or added.* At the cellular level, this means that the fates of cells in a regulative system are not irretrievably fixed, and the cells can still respond to environmental cues. Because the assignment of blastomeres to different cell lineages is one of the principal features of mammalian development, identifying the environmental factors that are involved is important.

Of the experimental techniques used to show regulative properties of early embryos, the simplest is to separate the blastomeres of early cleavage-stage embryos and determine whether each one can give rise to an entire embryo. This method has been used to show that single blastomeres from 2-cell and sometimes 4-cell embryos can form normal embryos, although blastomeres from later stages cannot do so. In mammalian studies, a single cell is more commonly taken from an early cleavage-stage embryo and injected into the blastocoele of a genetically different host. Such injected cells become incorporated into the host embryo, to form cellular **chimeras** or **mosaics**. When genetically different donor blastomeres are injected into host embryos, the donor cells can be identified by histochemical or cytogenetic analysis, and their fate (the tissues that they form) can be determined. **Fate mapping** experiments are important in embryology because they allow one to follow the pathways along which a particular cell can differentiate. Fate mapping experiments have shown that all blastomeres of an 8-cell mouse embryo remain **totipotent**; that is, they retain the ability to form any cell type in the body. Even at the 16-cell stage of cleavage, some blastomeres are capable of producing progeny that are found in both the inner cell mass and the trophoblastic lineage.

Another means of showing the regulative properties of early mammalian embryos is to dissociate mouse embryos into separate blastomeres and to combine the blastomeres of two or three embryos (**Fig. 3.10**). The combined blastomeres soon aggregate and reorganize to become a single large embryo, which goes on to become a normal-appearing **tetraparental** or **hexaparental mouse**. By various techniques of making chimeric embryos, it is possible to combine blastomeres to produce interspecies chimeras (e.g., a sheep-goat). It is likely that many human genetic mosaics (**chimeras**), most commonly recognized when some regions of the body are male and others are female, are the result of the fusion of two early fraternal twin embryos. Other possibilities for chimerism involve the exchange of cells through common vascular connections.

A significant question in early mammalian embryology is whether any of the three major body axes are represented in the egg or early embryo. Research on mouse embryos has resulted in dramatically different views. According to one view, the position of the second polar body, which after

*In contrast to regulative development is mosaic development, which is characterized by the inability to compensate for defects or to integrate extra cells into a unified whole. In a mosaic system, the fates of cells are rigidly determined, and removal of cells results in an embryo or a structure missing the components that the removed cells were destined to form. Most regulative systems have an increasing tendency to exhibit mosaic properties as development progresses.

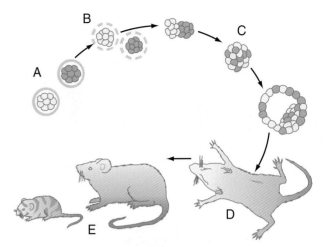

Fig. 3.10 Procedure for producing tetraparental embryos. A, Cleavage stages of two different strains of mice. B, Removal of the zona pellucida (*dashed circles*). C, Fusion of the two embryos. D, Implantation of embryos into a foster mother. E, Chimeric offspring obtained from the implanted embryos.

fertilization is typically found in line with the first cleavage plane, is a marker for the future anteroposterior axis. This would suggest that the egg before or just after fertilization possesses at least one predetermined axis, as is the case in many animals. Based on time-lapse photography, a contrary viewpoint posits that there is no predetermined axial plane within the egg, and the plane of the first cleavage division lies perpendicular to a line drawn between the final positions of the male and female pronuclei. Similarly, conflicting experimental data have not allowed researchers to determine whether there is any predetermined relationship between structures in the two- or four-cell embryo and the definitive body axes that become apparent at the time of early gastrulation. The bulk of evidence suggests that the early mammalian embryo is a highly regulative system and that the body axes do not become fixed until the end of cleavage or early gastrulation.

Experimental Manipulations of Cleaving Embryos

Much of the knowledge about the developmental properties of early mammalian embryos is the result of more recently devised techniques for experimentally manipulating them. Typically, the use of these techniques must be combined with other techniques that have been designed for in vitro fertilization, embryo culture, and embryo transfer (see Chapter 2).

Classic strategies for investigating the developmental properties of embryos are (1) removing a part and determining the way that the remainder of the embryo compensates for the loss (such experiments are called **deletion** or **ablation experiments**) and (2) adding a part and determining the way that the embryo integrates the added material into its overall body plan (such experiments are called **addition experiments**). Although some deletion experiments have been done, the strategy of addition experiments has proved to be more fruitful in elucidating mechanisms controlling mammalian embryogenesis.

Blastomere deletion and addition experiments (**Fig. 3.11**) have convincingly shown the regulative nature (i.e., the strong

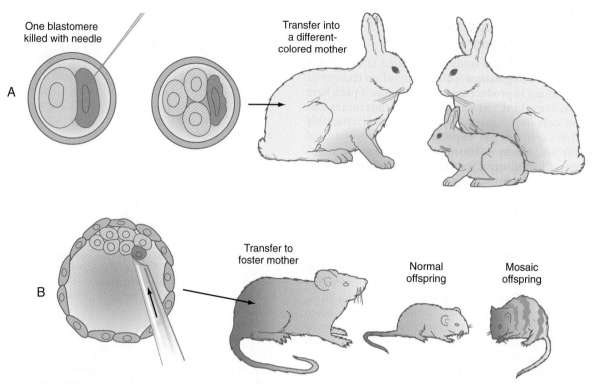

Fig. 3.11 **Blastomere addition and deletion experiments.** **A,** If one blastomere is killed with a needle, and the embryo is transferred into a different-colored mother, a normal offspring of the color of the experimentally damaged embryo is produced. **B,** If a blastomere of a different strain is introduced into a blastocyst, a mosaic offspring with color markings characteristic of the strain of the introduced blastomere is produced.

tendency for the system to be restored to wholeness) of early mammalian embryos. Such knowledge is important in understanding why the exposure of early human embryos to unfavorable environmental influences typically results in either death or a normal embryo.

One of the most powerful experimental techniques has been the injection of genetically or artificially labeled cells into the blastocyst cavity of a host embryo (see Fig. 3.11B). This technique has been used to show that the added cells become normally integrated into the body of the host embryo, thus providing additional evidence for embryonic regulation. An equally powerful use of this technique has been in the study of cell lineages in the early embryo. By identifying the progeny of the injected marked cells, investigators have been able to determine the developmental potency of the donor cells.

A technique that provides great insight into the genetic control mechanisms of mammalian development is the production of **transgenic embryos**. Transgenic embryos (commonly mice) are produced by directly injecting foreign DNA into the pronuclei of zygotes (**Fig. 3.12A**). The DNA, usually recombinant DNA for a specific gene, can be fused with a different regulatory element that can be controlled by the investigator.

Transgenic mice can be created by injecting the rat growth hormone gene coupled with a metallothionein promoter region (MT-I) into the pronuclei of mouse zygotes. The injected zygotes are transplanted into the uteri of foster mothers, which give birth to normal-looking transgenic mice. Later in life, when these transgenic mice are fed a diet rich in zinc, which stimulates the MT-I promoter region, the rat

growth hormone gene is activated and causes the liver to produce large amounts of the polypeptide growth hormone. The function of the transplanted gene is obvious; under the influence of the rat growth hormone that they are producing, the transgenic mice grow to a much larger size than their normal littermates (**Fig. 3.13**).

In addition to adding genes to embryos, several powerful techniques have been developed to inactivate specific genes or gene products. At the DNA level, it is now common to **knock out** a gene of interest as a way to determine its function in normal development. Some genes have multiple functions at various times and in various tissues throughout embryogenesis. Their function in early development may be so critical that in the absence of its function the embryo dies even as early as gastrulation. To deal with this problem, techniques have been devised to interfere with tissue-specific promotors, so that the function of a gene in a given organ (e.g., the eye) can be disrupted in the primordium of that structure alone. Other techniques operate at the RNA level. For example, noncoding RNAi (RNA interference), injected into an embryo, knocks down, rather than blocks, gene expression. At the protein level, genetically engineered nonfunctional receptor molecules injected into an embryo can displace their normal counterparts and bind a signaling molecule without the ability to transduce the signal into the interior of the cell. There are situations in which each of these techniques is particularly useful in investigating a question in development.

Some types of twinning represent a natural experiment that shows the highly regulative nature of early human embryos, as described in **Clinical Correlation 3.2.**

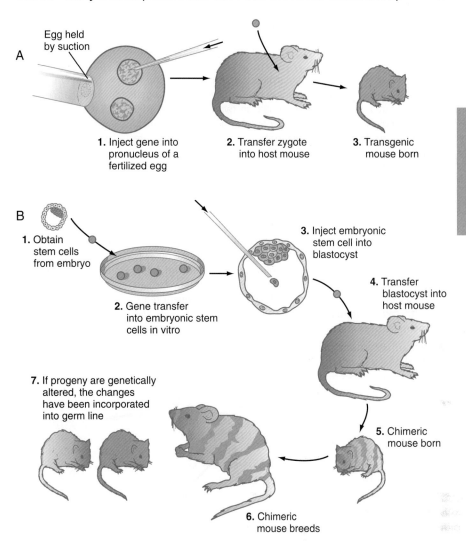

Fig. 3.12   **A,** Procedure for creating transgenic mice by pronuclear injection. **B,** Procedure for inserting genes into mice by first introducing them into embryonic stem cells and then inserting the transfected stem cells into an otherwise normal blastocyst.

Fig. 3.13   **Photograph of two 10-week-old mice.** The one on the left (normal mouse) weighs 21.2 g. The one on the right (a transgenic littermate of the normal mouse) carries a rat gene coding for growth hormone. It weighs 41.2 g. *(From Palmiter RD and others:* Nature *300:611-615, 1982.)*

Stem Cells and Cloning

A major development in biomedical research at the turn of the 21st century was the realization that certain cells (**stem cells**) in both human embryos and adults have the capacity to develop into a variety of cell and tissue types in response to specific environments. In embryos, stem cells can be derived from the inner cell mass (**embryonic stem cells [ES cells]**) or primordial germ cells (**embryonic germ cells**). In adults, stem cells have been isolated from tissues as diverse as bone marrow, skeletal muscle, brain tissue, and fat. Regardless of their origin, stem cells are maintained and propagated in an undifferentiated state in culture. Characteristically, stem cells express *oct4, Sox2,* and *Nanog* (see p. 42), which are involved in maintaining the undifferentiated state.

In response to specific combinations of exogenous agents (e.g., cocktails of growth factors) added to the culture medium, stem cells can be induced to differentiate into specific adult cell types, for example, red and white blood cells, neurons, skeletal and cardiac muscle, or cartilage. When introduced into living tissues, poorly defined local factors can direct the differentiation of adult or embryonic stem cells into specific adult cell types. These techniques have tremendous potential for the treatment of a variety of conditions, including diabetes,

CLINICAL CORRELATION 3.2
Twinning

Some types of twinning represent a natural experiment that shows the highly regulative nature of early human embryos. In the United States, about 1 pregnancy in 90 results in twins, and 1 in 8000 results in triplets. Of the total number of twins born, approximately two thirds are **fraternal**, or **dizygotic**, twins and one third are **identical**, or **monozygotic**, twins. Dizygotic twins are the product of the fertilization of 2 ovulated eggs, and the mechanism of their formation involves the endocrine control of ovulation. Monozygotic twins and some triplets are the product of a single fertilized egg. They arise by the subdivision and splitting of a single embryo. Although monozygotic twins could theoretically arise by the splitting of a 2-cell embryo, it is commonly accepted that most arise by the subdivision of the inner cell mass in a blastocyst, or possibly even splitting of the epithelial epiblast a few days later (**Fig. 3.14**). Because most monozygotic twins are normal, the early human embryo can obviously be subdivided, and each component regulates to form a normal embryo. Inferences on the origin and relationships of multiple births can be made from the arrangement of the extraembryonic membranes at the time of birth (see Chapter 7).

Apparently, among many sets of twins, one member does not survive to birth. This is a reflection of the increasing recognition that perhaps most conceptuses do not survive. According to some estimates, as many as one in eight live births is a surviving member of a twin pair. Quadruplets or higher orders of multiple births occur very rarely. In previous years, these could be combinations of multiple ovulations and splitting of single embryos. In the modern era of reproductive technology, most multiple births, sometimes up to septuplets, can be attributed to the side effects of fertility drugs taken by the mother.

The separation of portions of an embryo is sometimes incomplete, and although two embryos take shape, they are joined by a tissue bridge of varying proportions. When this occurs, the twins are called **conjoined twins** (sometimes colloquially called *Siamese twins*). The extent of bridging between the twins varies from a relatively thin connection in the chest or back to massive fusions along much of the body axis. Examples of the wide variety of types of conjoined twins are illustrated in **Figures 3.15 and 3.16**. With the increasing sophistication of surgical techniques, twins with more complex degrees of fusion can be separated. A much less

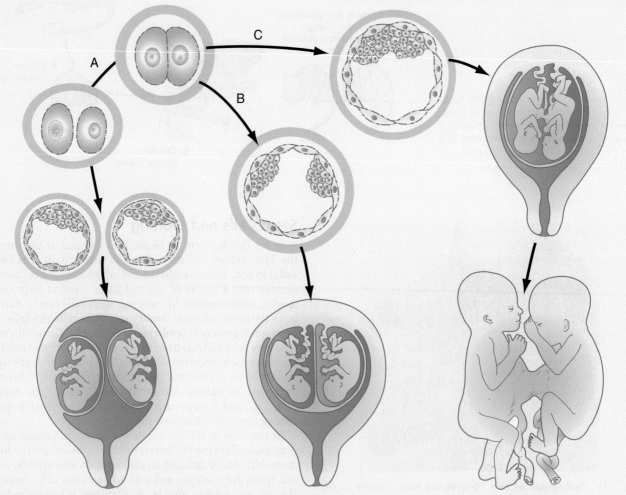

Fig. 3.14 Modes of monozygotic twinning. A, Cleavage of an early embryo, with each half developing as a completely separate embryo. **B,** Splitting of the inner cell mass of a blastocyst and the formation of two embryos enclosed in a common trophoblast. This is the most common mode of twinning. **C,** If the inner cell mass does not completely separate, or if portions of the inner cell mass secondarily rejoin, conjoined twins may result.

CLINICAL CORRELATION 3.2
Twinning—cont'd

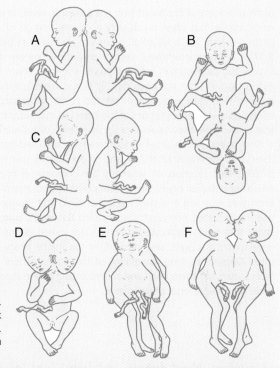

Fig. 3.15 **Types of conjoined twins.** **A,** Head-to-head fusion (cephalopagus). **B** and **C,** Rump-to-rump fusion (pygopagus). **D,** Massive fusion of head and trunk that results in a reduction in the number of appendages and a single umbilical cord. **E,** Fusion involving head and thorax (cephalothoracopagus). **F,** Chest-to-chest fusion (thoracopagus).

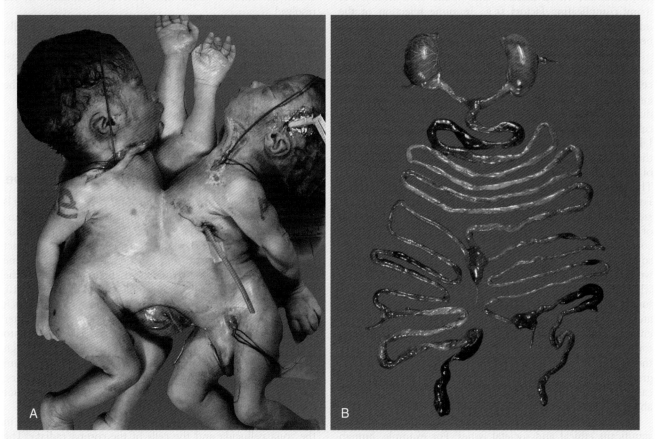

Fig. 3.16 **A,** Conjoined twins with broad truncal attachment (thoracopagus). **B,** Dissected intestinal tracts from the same twins showing partial fusion of the small intestine and mirror image symmetry of the stomachs. *(Courtesy of M. Barr, Ann Arbor, Mich.)*

Continued

common variety of conjoined twin is a **parasitic twin**, in which a much smaller but often remarkably complete portion of a body protrudes from the body of an otherwise normal host twin (**Fig. 3.17**). Common attachment sites of parasitic twins are the oral region, the mediastinum, and the pelvis. The mechanism of conjoined twinning has not been directly shown experimentally, but possible theoretical explanations are the partial secondary fusion of originally separated portions of the inner cell mass or the formation of two primitive streaks in a single embryo (see Chapter 5).

One phenomenon often encountered in conjoined twins is a reversal of symmetry of the organs of one of the pair (see Fig. 3.16B). Such reversals of symmetry are common in duplicated organs or entire embryos. More than a century ago, this phenomenon was recorded in a large variety of biological situations and was incorporated into what is now called **Bateson's rule**, which states that when duplicated structures are joined during critical developmental stages, one structure is the mirror image of the other. Despite the long recognition of this phenomenon, only in recent years has there been any understanding of the mechanism underlying the reversal of symmetry.

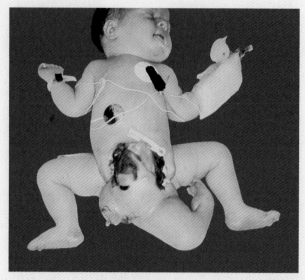

Fig. 3.17 **Parasitic twin arising from the pelvic region of the host twin.** One well-defined leg and some hair can be seen on the parasitic twin. *(Courtesy of M. Barr, Ann Arbor, Mich.)*

parkinsonism, blood diseases, and spinal cord injury, but many complicating factors (e.g., immune rejection of the implanted cells) must be dealt with before these techniques become practical and safe for human application.

An important development in stem cell technology has been the production of **induced pluripotent stem cells** (**IPS cells**) from somatic cells of adults. If genes characteristic of embryonic stem cells (e.g., *Oct4*, *Sox2*, and *Nanog*) are introduced into a differentiated adult cell (e.g., a fibroblast), the cell will then assume the properties of an embryonic stem cell. Like an embryonic stem cell, an artificially created stem cell that is exposed to an appropriate environment will be capable of differentiating into a wide variety of other adult cell types. This technique has great potential for patient-specific therapy. For example, in the treatment of a genetic disease characterized by the inability to manufacture a specific molecule, cells of a patient could be converted into IPS cells, subjected to corrective gene therapy, and then reintroduced into that person's body. Under ideal conditions, the introduced IPS cells would then begin to produce the deficient molecule. **Cloning**, which is often confused with stem cell technology, consists of fusing or introducing an adult cell or nucleus into an enucleated oocyte and allowing the hybrid cell to develop into an embryo and ultimately to mature into an adult. Although forms of cloning have been successfully accomplished since the 1960s, the creation of the sheep Dolly in 1996 had the greatest influence on the public imagination. Cloning is not easily accomplished, and there is a significant incidence of abnormal development among cloned individuals.

Cloning and stem cell technology have brought to light significant ethical and societal issues. For example, human embryonic stem cells have been introduced into mouse blastocysts in an attempt to determine the influences that control their differentiation. It will be fascinating to see how these issues, all sides of which have profound implications, are resolved.

Genetic engineering of specific genes is possible in ES cells. When such genetically manipulated cells are introduced into blastocysts, they can become incorporated into the host embryo (see Fig. 3.12B). If the progeny of a genetically engineered ES cell become incorporated into the germline, the genetic trait can be passed to succeeding generations.

Embryo Transport and Implantation

Transport Mechanisms by the Uterine Tube

The entire period of early cleavage occurs while the embryo is being transported from the place of fertilization to its implantation site in the uterus (see Fig. 2.2). It is increasingly apparent that the early embryo and the female reproductive tract influence one another during this period of transport. One such influence is **early pregnancy factor**, a molecule of the heat shock protein family and homologous to chaperonin 10, an intramitochondrial protein. Early pregnancy factor, which is detectable in maternal blood within 36 to 48 hours after fertilization, is an immunosuppressant and is postulated to provide immunological protection to the embryo. Although this factor is produced by the embryo, its presence in serum seems to result from its synthesis and secretion by the ovary. Because the assay for this protein is cumbersome, it has not found wide use in pregnancy testing.

At the beginning of cleavage, the zygote is still encased in the zona pellucida and the cells of the corona radiata. The corona radiata is lost within 2 days of the start of cleavage. The zona pellucida remains intact, however, until the embryo reaches the uterus.

The embryo remains in the ampullary portion of the uterine tube for approximately 3 days. It then traverses the isthmic portion of the tube in as little as 8 hours. Under the influence of progesterone, the uterotubal junction relaxes, thus allowing the embryo to enter the uterine cavity. A couple of days later (6 to 8 days after fertilization), the embryo implants into the midportion of the posterior wall of the uterus.

Zona Pellucida

During the entire period from ovulation until entry into the uterine cavity, the ovum and the embryo are surrounded by the zona pellucida. During this time, the composition of the zona changes, through contributions from the blastomeres and the maternal reproductive tissues. These changes facilitate the transport and differentiation of the embryo. After the embryo reaches the uterine cavity, it begins to shed the zona pellucida in preparation for implantation. This is accomplished by a process called **blastocyst hatching**. A small region of the zona pellucida, usually directly over the inner cell mass in the primate, dissolves, and the blastocyst emerges from the hole. In rodents, blastocyst hatching is accomplished through the action of cysteine protease enzymes that are released from long microvillous extensions (**trophectodermal projections**) protruding from the surfaces of the trophoblastic cells. Over a narrow time window (4 hours in rodents), the zona pellucida in this area is digested, and the embryo begins to protrude. In the uterus, the trophectodermal projections then make contact with the endometrial epithelial cells as the process of implantation begins. Enzymatic activity around the entire trophoblast soon begins to dissolve the rest of the zona pellucida. Only a few specimens of human embryos have been taken in vivo from the period just preceding implantation, but in vitro studies on human embryos suggest a similar mechanism, which probably occurs 1 to 2 days before implantation (see Fig. 3.3C). **Box 3.1** summarizes the functions of the zona pellucida.

Implantation into the Uterine Lining

Approximately 6 to 7 days after fertilization, the embryo begins to make a firm attachment to the epithelial lining of the endometrium. Soon thereafter, it sinks into the endometrial stroma, and its original site of penetration into the endometrium becomes closed over by the epithelium, similar to a healing skin wound.

Successful implantation requires a high degree of preparation and coordination by the embryo and the endometrium (**Table 3.1**). The complex hormonal preparations of the endometrium that began at the close of the previous menstrual period all are aimed at providing a suitable cellular and nutritional environment for the embryo. Even before actual contact is made between the embryo and endometrium, the uterine epithelium secretes into the uterine fluid certain cytokines and chemokines that facilitate the implantation process. At the same time, cytokine receptors appear on the surface of the trophoblast. Dissolution of the zona pellucida signals the readiness of the embryo to begin implantation.

The first stage in implantation consists of attachment of the expanded blastocyst to the endometrial epithelium. The apical surfaces of the hormonally conditioned endometrial epithelial cells express various adhesion molecules (e.g., integrins) that allow implantation to occur in the narrow window of 20 to 24 days in the ideal menstrual cycle. Correspondingly, the trophoblastic cells of the preimplantation blastocyst also express adhesion molecules on their surfaces. The blastocyst attaches to the endometrial epithelium through the mediation of bridging ligands. Some studies have stressed the importance of the cytokine **leukemia-inhibiting factor** (**LIF**) on the endometrial surface and LIF receptors on the trophoblast during implantation. In vivo and in vitro studies have shown that attachment of the blastocyst occurs at the area above the inner cell mass (**embryonic pole**), a finding suggesting that the surfaces of the trophoblast are not all the same.

Box 3.1 Summary of Functions of the Zona Pellucida

1. It promotes maturation of the oocyte and follicle.
2. The zona pellucida serves as a barrier that normally allows only sperm of the same species access to the egg.
3. It initiates the acrosomal reaction.
4. After fertilization, the modified zona pellucida prevents any additional spermatozoa from reaching the zygote.
5. During the early stages of cleavage, it acts as a porous filter through which certain substances secreted by the uterine tube can reach the embryo.
6. Because it lacks histocompatibility (human leukocyte) antigens, the zona pellucida serves as an immunological barrier between the mother and the antigenically different embryo.
7. It prevents the blastomeres of the early cleaving embryo from dissociating.
8. It facilitates the differentiation of trophoblastic cells.
9. It normally prevents premature implantation of the cleaving embryo into the wall of the uterine tube.

Table 3.1 Stages in Human Implantation

Age (Days)	Developmental Event in Embryos
5	Maturation of blastocyst
5	Loss of zona pellucida from blastocyst
6?	Attachment of blastocyst to uterine epithelium
6-7	Epithelial penetration
7½-9	Trophoblastic plate formation and invasion of uterine stroma by blastocyst
9-11	Lacuna formation along with erosion of spiral arteries in endometrium
12-13	Primary villus formation
13-15	Secondary placental villi, secondary yolk sac formation
16-18	Branching and anchoring villus formation
18-22	Tertiary villus formation

Modified from Enders AC: Implantation, embryology. In *Encyclopedia of human biology*, vol 4, New York, 1991, Academic Press.

The next stage of implantation is penetration of the uterine epithelium. In primates, the cellular trophoblast undergoes a further stage in its differentiation just before it contacts the endometrium. In the area around the inner cell mass, cells derived from the cellular trophoblast (**cytotrophoblast**) fuse to form a multinucleated **syncytiotrophoblast**. Although only a small area of syncytiotrophoblast is evident at the start of implantation, this structure (sometimes called the **syntrophoblast**) soon surrounds the entire embryo. Small projections of syncytiotrophoblast insert themselves between uterine epithelial cells. They spread along the epithelial surface of the basal lamina that underlies the endometrial epithelium to form a flattened **trophoblastic plate**. Within a day or so, syncytiotrophoblastic projections from the small trophoblastic plate begin to penetrate the basal lamina. The early syncytiotrophoblast is a highly invasive tissue, and it quickly expands and erodes its way into the endometrial stroma (**Fig. 3.18A** and **B**). Although the invasion of the syncytiotrophoblast into the endometrium is obviously enzymatically mediated, the biochemical basis in humans is not well understood. By 10 to 12 days after fertilization, the embryo is completely embedded in the endometrium. The site of initial penetration is first marked by a bare area or a noncellular plug and is later sealed by migrating uterine epithelial cells (**Fig. 3.18C** and **D**).

As early implantation continues, projections from the invading syncytiotrophoblast envelop portions of the maternal endometrial blood vessels. They erode into the vessel walls, and maternal blood begins to fill the isolated lacunae that have been forming in the trophoblast (see Fig. 3.18C and D). Trophoblastic processes enter the blood vessels and even share junctional complexes with the endothelial cells. By the time blood-filled lacunae have formed, the trophoblast changes character, and it is not as invasive as it was during the first few days of implantation. Leakage of blood from the uterus at this stage can produce "spotting," which is sometimes misinterpreted to be an abnormal menstrual period.

While the embryo burrows into the endometrium, and some cytotrophoblastic cells fuse into syncytiotrophoblast, the fibroblastlike stromal cells of the edematous endometrium swell, with the accumulation of glycogen and lipid droplets (see Fig. 7.6). These cells, called **decidual cells**, are tightly adherent and form a massive cellular matrix that first surrounds the implanting embryo and later occupies most of the endometrium. Concurrent with the **decidual reaction**, as this transformation is called, the leukocytes that have infiltrated the endometrial stroma during the late progestational phase of the endometrial cycle secrete **interleukin-2**, which prevents maternal recognition of the embryo as a foreign body during the early stages of implantation. An embryo is antigenically different from the mother and consequently should be rejected by a cellular immune reaction similar to the type that rejects an incompatible heart or kidney transplant. A primary function of the decidual reaction apparently is to provide an immunologically privileged site to protect the developing embryo from being rejected, but a real understanding of how this is accomplished has resisted years of intensive research.

Frequently, a blastocyst fails to attach to the endometrium, and implantation does not occur. Failure of implantation is a particularly vexing problem in in vitro fertilization and embryo transfer procedures, for which the success rate of implantation of transferred embryos remains at about 25% to 30% (see Clinical Correlation 2.1).

Clinical Vignette

Within a week, 2 young women in their 20s come to the emergency department of a large city hospital. Each woman complains of acute pain in the lower right abdominal quadrant. On physical examination, both patients are exquisitely sensitive to mild pressure in that area.

Further questioning of the first woman reveals that she had a menstrual period 2 weeks previously. Emergency surgery was performed and found that the woman had a ruptured appendix.

The second young woman gives a history of gonorrhea and has been treated for pelvic inflammation. Her last menstrual period was 9 weeks previously. During emergency surgery, her right uterine tube is removed.

What was the likely reason for doing so?

Embryo Failure and Spontaneous Abortion

Many fertilized eggs (>50%) do not develop to maturity and are spontaneously aborted. Most spontaneous abortions (**miscarriages**) occur during the first 3 weeks of pregnancy. Because of the small size of the embryo at that time, spontaneous abortions are often not recognized by the mother, who may equate the abortion and attendant hemorrhage with a late and unusually heavy menstrual period.

Examinations of early embryos obtained after spontaneous abortion or from uteri removed by hysterectomy during the early weeks of pregnancy have shown that many of the aborted embryos are highly abnormal. Chromosomal abnormalities represent the most common category of abnormality in abortuses (about 50% of the cases). When viewed in the light of the accompanying pathological conditions, spontaneous abortion can be viewed as a natural mechanism for reducing the incidence of severely malformed infants.

Summary

- Early human cleavage is slow, with roughly a single cleavage division occurring per day for the first 3 to 4 days. The cleaving embryo passes through the morula stage (16 cells) and enters a stage of compaction. By day 4, a fluid-filled blastocoele forms within the embryo, and the embryo becomes a blastocyst with an inner cell mass surrounded by trophoblast.

- The zygote relies on maternal mRNAs, but by the two-cell stage, the embryonic genome becomes activated. The *oct4*, *Sox2*, and *Nanog* genes are important in very early development, and their expression is associated with the undifferentiated state of cells.

- Through parental imprinting, specific homologous chromosomes derived from the mother and father exert different effects on embryonic development. In female embryos, one X chromosome per cell becomes inactivated through the action of the *XIST* gene, thus forming the sex chromatin body. The early embryo has distinct patterns of X-chromosomal inactivation.

- The early mammalian embryo is highly regulative. It can compensate for the loss or addition of cells to the inner cell mass and still form a normal embryo. The decision to form trophoblast versus inner cell mass relates to division patterns of polarized cells, starting at the eight-cell stage. According to the inside-outside hypothesis, the position of

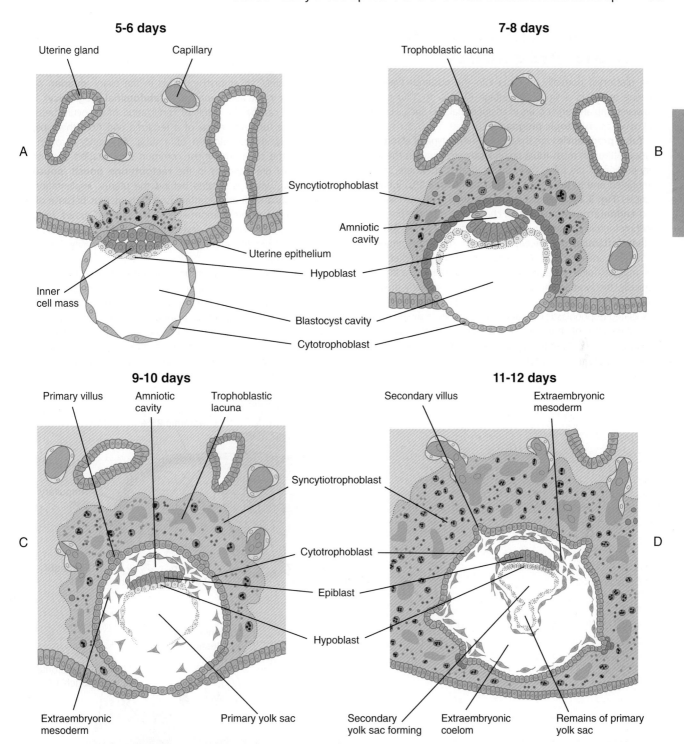

5-6 days

Uterine gland   Capillary

Syncytiotrophoblast

Uterine epithelium

Hypoblast

Inner cell mass

Blastocyst cavity

Cytotrophoblast

A

7-8 days

Trophoblastic lacuna

Amniotic cavity

B

9-10 days

Primary villus   Amniotic cavity   Trophoblastic lacuna

Syncytiotrophoblast

Cytotrophoblast

Epiblast

Hypoblast

Extraembryonic mesoderm   Primary yolk sac

C

11-12 days

Secondary villus   Extraembryonic mesoderm

Secondary yolk sac forming   Extraembryonic coelom   Remains of primary yolk sac

D

Fig. 3.18 **Major stages in implantation of a human embryo.** **A,** The syncytiotrophoblast is just beginning to invade the endometrial stroma. **B,** Most of the embryo is embedded in the endometrium; there is early formation of the trophoblastic lacunae. The amniotic cavity and yolk sac are beginning to form. **C,** Implantation is almost complete, primary villi are forming, and the extraembryonic mesoderm is appearing. **D,** Implantation is complete; secondary villi are forming.

CLINICAL CORRELATION 3.3
Ectopic Pregnancy

The blastocyst normally implants into the posterior wall of the uterine cavity. In a small percentage (0.25% to 1%) of cases, however, implantation occurs in an abnormal site. Such a condition is known as an **ectopic pregnancy**.

Tubal pregnancies are the most common type of ectopic pregnancy. Although most tubal pregnancies are found in the ampullary portion of the tube, they can be located anywhere, from the fimbriated end to the uterotubal junction (**Fig. 3.19**). Tubal pregnancies (**Fig. 3.20**) are most commonly seen in women who have had **endometriosis** (a condition characterized by the presence of endometriumlike tissue in abnormal locations), earlier surgery, or **pelvic inflammatory disease**. Scarring from inflammation or sometimes anatomical abnormalities result in blind pockets among the mucosal folds of the uterine tube; these can trap a blastocyst. Typically, the woman shows the normal signs of early pregnancy, but at about 2 to 2½ months, the implanted embryo and its associated trophoblastic derivatives have grown to the point where the stretching of the tube causes acute abdominal pain. If untreated, a tubal pregnancy typically ends with rupture of the tube and hemorrhage, often severe enough to be life-threatening to the mother.

Very rarely, an embryo implants in the ovary (**ovarian pregnancy**) or in the abdominal cavity (**abdominal pregnancy**). Such instances can be the result of fertilization of an ovum before it enters the tube, the reflux of a fertilized egg from the tube, or, very rarely, the penetration of a tubal pregnancy through the wall of the tube. The most common implantation site for an abdominal pregnancy is in the **rectouterine pouch (pouch of Douglas)**, which is located behind the uterus. Implantation on the intestinal wall or mesentery is very dangerous because of the likelihood of severe hemorrhage as the embryo grows. In some instances, an embryo has developed to full term in an abdominal location. If not delivered, such an embryo can calcify, forming a **lithopedion**.

Within the uterus, an embryo can implant close to the cervix. Although embryonic development is likely to be normal, the placenta typically forms a partial covering over the cervical canal. This condition, called **placenta previa**, can result in hemorrhage during late pregnancy and, if untreated, is likely to cause the death of the fetus, the mother, or both because of premature placental detachment with accompanying hemorrhage. Implantation directly within the cervical canal is extremely rare.

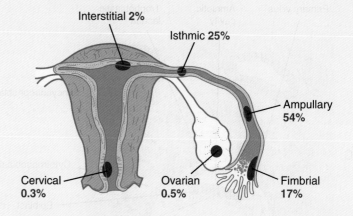

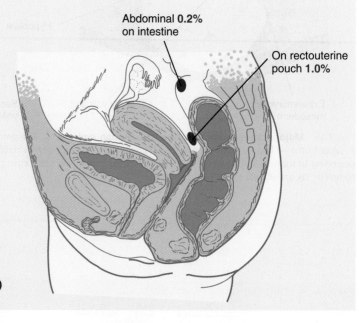

Fig. 3.19 **Sites of ectopic pregnancy (indicated by *red dots*) and the frequency of their occurrence.**

CLINICAL CORRELATION 3.3
Ectopic Pregnancy—cont'd

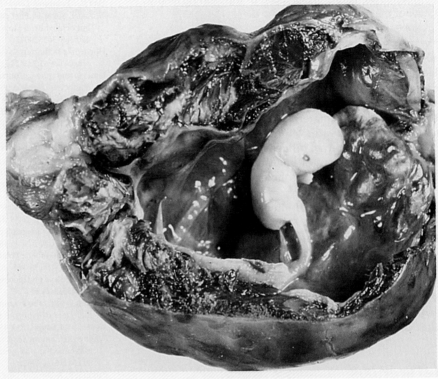

Fig. 3.20 Ruptured ectopic pregnancy in a 34-year-old woman. Because of the increasing size of the fetus and associated membranes, the uterine tube ruptured during the third month of pregnancy. *(From Rosai J: Ackerman's surgical pathology, vol 2, ed 8, St Louis, 1996, Mosby.)*

a blastomere determines its developmental fate (i.e., whether it becomes part of the inner cell mass or the trophoblast).

- Transgenic embryos are produced by injecting ribosomal DNA (rDNA) into the pronuclei of zygotes. Such embryos are used to study the effects of specific genes on development. Other techniques involve knocking out genes or interfering with the further processing of gene products.
- Monozygotic twinning, usually caused by the complete separation of the inner cell mass, is possible because of the regulative properties of the early embryo. Incomplete splitting of the inner cell mass can lead to the formation of conjoined twins.
- After fertilization, the embryo spends several days in the uterine tube before entering the uterus. During this time, it is still surrounded by the zona pellucida, which prevents premature implantation.
- Implantation of the embryo into the uterine lining involves several stages: apposition of the expanded (hatched) blastocyst to the endometrial epithelium, penetration of the uterine epithelium, invasion into the tissues underlying the epithelium, and erosion of the maternal vascular supply. Connective tissue cells of the endometrium undergo the decidual reaction in response to the presence of the implanting embryo. Implantation is accomplished through the invasive activities of the syncytiotrophoblast, which is derived from the cytotrophoblast.
- Implantation of the embryo into a site other than the upper uterine cavity results in an ectopic pregnancy (**Clinical Correlation 3.3**). Ectopic pregnancy is most often encountered in the uterine tube.

- High percentages of fertilized eggs and early embryos do not develop and are spontaneously aborted. Many of these embryos contain major chromosomal abnormalities.

Review Questions

1. What is the most common condition associated with spontaneously aborted embryos?
A. Maternal imprinting
B. Paternal imprinting
C. Ectopic pregnancy
D. Chromosomal abnormalities
E. Lack of X-chromosomal inactivation

2. What tissue from the implanting embryo directly interfaces with the endometrial connective tissue?
A. Corona radiata
B. Inner cell mass
C. Extraembryonic mesoderm
D. Epiblast
E. Syncytiotrophoblast

3. Identical twinning is made possible by what process or property of the early embryo?
A. Regulation
B. Aneuploidy
C. Paternal imprinting
D. Maternal imprinting
E. X-chromosomal inactivation

4. The zona pellucida:
A. Aids in penetration of the endometrial epithelium
B. Serves as a source of nutrients for the embryo
C. Prevents premature implantation of the cleaving embryo
D. All of the above
E. None of the above

5. What is the importance of the inner cell mass of the cleaving embryo?

6. Parental imprinting is a phenomenon showing that certain homologous maternal and paternal chromosomes have different influences on the development of the embryo. Excess paternal influences result in the abnormal development of what type of tissue at the expense of development of the embryo itself?

7. What is the function of integrins in implantation?

8. What is the cellular origin of the syncytiotrophoblast of the implanting embryo?

9. A woman who is 2 to 3 months pregnant suddenly develops severe lower abdominal pain. In the differential diagnosis, the physician must include the possibility of what condition?

References

Bateson W: *Materials for the study of variation*, London, 1894, Macmillan.

Boklage CE: Embryogenesis of chimeras, twins and anterior midline asymmetries, *Hum Reprod* 21:579-591, 2006.

Bonasio R, Tu S, Reinberg D: Molecular signals of epigenetic states, *Science* 330:612-616, 2010.

Bourc-his D, Proudhon C: Sexual dimorphism in parental imprint ontogeny and contribution to embryonic development, *Mol Cell Endocrinol* 282:87-94, 2008.

Bruce AW, Zernicka-Goetz M: Developmental control of the early mammalian embryo: competition among heterogeneous cells that biases cell fate, *Genet Dev* 20:485-491, 2010.

Carlson BM: Stem cells and cloning: what's the difference and why the fuss? *Anat Rec (New Anat)* 257:1-2, 1999.

Carlson BM, ed: *Stem cell anthology*, San Diego, 2010, Academic Press.

Cassidy SB, Schwartz S: Prader-Willi and Angelman syndromes: disorders of genomic imprinting, *Medicine* 77:140-151, 1998.

Cockburn K, Rossant J: Making the blastocyst: lessons from the mouse, *J Clin Invest* 120:995-1003, 2010.

Dard N and others: Morphogenesis of the mammalian blastocyst, *Mol Cell Endocrinol* 282:70-77, 2008.

Dey SK and others: Molecular cues to implantation, *Endocr Rev* 25:341-373, 2004.

Diedrich K and others: The role of the endometrium and embryo in human implantation, *Hum Reprod Update* 13:365-377, 2007.

Dimitriadis E and others: Local regulation of implantation at the human fetal-maternal interface, *Int J Dev Biol* 54:313-322, 2010.

Enders AC: Formation of monozygotic twins: when does it occur? *Placenta* 23:236-238, 2002.

Enders AC: Implantation, embryology. In *Encyclopedia of human biology*, ed 2, vol 4, New York, 1997, Academic Press, pp 799-807.

Enders AC: Trophoblast differentiation during the transition from trophoblastic plate to lacunar stage of implantation in the rhesus monkey and human, *Am J Anat* 186:85-98, 1989.

Erwin JA, Lee JT: New twists in X-chromosome inactivation, *Curr Opin Cell Biol* 20:349-355, 2008.

Flemming TP and others: Assembly of tight junctions during early vertebrate development, *Semin Cell Dev Biol* 11:291-299, 2000.

Gardner RL: The initial phase of embryonic patterning in mammals, *Int Rev Cytol* 203:233-290, 2001.

Gardner RL, Davies TJ: The basis and significance of pre-patterning in mammals, *Phil Trans R Soc Lond B* 358:1331-1339, 2003.

Gray D and others: First cleavage of the mouse embryo responds to change in egg shape at fertilization, *Curr Biol* 14:397-405, 2004.

Hall JG: Twinning, *Lancet* 362:735-743, 2003.

Hiiragi T, Solter D: First cleavage plane of the mouse egg is not predetermined but defined by the topology of the two opposing pronuclei, *Nature* 430:360-364, 2004.

Horstemke B, Buiting K: Genomic imprinting and imprinting defects in humans, *Adv Genet* 61:225-246, 2008.

James D and others: Contribution of human embryonic stem cells to mouse blastocysts, *Dev Biol* 295:90-102, 2006.

Johnson MH: From mouse egg to mouse embryo: polarities, axes and tissues, *Annu Rev Cell Dev Biol* 25:483-512, 2009.

Johnson MH, McConnell JML: Lineage allocation and cell polarity during mouse embryogenesis, *Semin Cell Dev Biol* 15:583-597, 2004.

Kurotaki Y and others: Blastocyst axis is specified independently of early cell lineage but aligns with ZP shape, *Science* 316:719-723, 2007.

Latham KE, Schultz RM: Preimplantation embryo development. In Fauser BCJM, ed: *Reproductive medicine*, Boca Raton, Fla, 2003, Parthenon, pp 421-438.

Li L, Zheng P, Dean J: Maternal control of early mouse development, *Development* 137:859-870, 2010.

Morton H: Early pregnancy factor: an extracellular chaperonin 10 homologue, *Immunol Cell Biol* 76:483-496, 1998.

Nafee TM and others: Epigenetic control of fetal gene expression, *Br J Obstet Gynecol* 115:158-168, 2007.

Niwa H: How is pluripotency determined and maintained? *Development* 134:635-646, 2007.

Oestrup O and others: From zygote to implantation: morphological and molecular dynamics during embryo development in the pig, *Reprod Dom Anim* 44(Suppl 3):39-49, 2009.

Palmiter RD and others: Dramatic growth of mice that develop from eggs microinjected with metallothionein-growth hormone fusion genes, *Nature* 300:611-615, 1982.

Panning B: X-chromosome inactivation: the molecular basis of silencing, *J Biol* 7:30-33, 2008.

Pederson RA, Burdsal CA: Mammalian embryogenesis. In Knobil E, Neill J, eds: *The physiology of reproduction*, ed 2, New York, 1988, Raven, pp 319-390.

Pfeifer K: Mechanisms of genomic imprinting, *Am J Hum Genet* 67:777-787, 2000.

Piotrowska-Nitsche K and others: Four-cell stage mouse blastomeres have different developmental properties, *Development* 132:479-490, 2005.

Reik W, Walter J: Imprinting mechanisms in animals, *Curr Opin Genet Dev* 8:154-164, 1998.

Rivera R: Epigenetic aspects of fertilization and preimplantation development in mammals: lessons from the mouse, *Systems Biol Reprod Med* 56:388-404, 2010.

Rivera-Perez JA: Axial specification in mice: ten years of advances and controversies, *J Cell Physiol* 213:654-660, 2007.

Rossant J: Lineage development and polar asymmetries in the pre-implantation mouse blastocyst, *Semin Cell Dev Biol* 15:573-581, 2004.

Rossant J, Tam PPL: Emerging asymmetry and embryonic patterning in early mouse development, *Dev Cell* 7:155-164, 2004.

Santos F, Dean W: Epigenetic reprogramming during early development in mammals, *Reproduction* 127:643-651, 2004.

Senner CE, Brockdorff N: Xist gene regulation at the onset of X inactivation, *Curr Opin Genet Dev* 19:122-126, 2009.

Seshagiri PB and others: Cellular and molecular regulation of mammalian blastocyst hatching, *J Reprod Immunol* 83:79-84, 2009.

Shaw JLV and others: Current knowledge of the aetiology of human tubal ectopic pregnancy, *Hum Reprod Update* 16:432-444, 2010.

Solter D: Imprinting today: end of the beginning or beginning of the end? *Cytogenet Genome Res* 113:12-16, 2006.

Solter D and others: Epigenetic mechanisms in early mammalian development, *Cold Spring Harbor Symp Quant Biol* 69:11-17, 2004.

Spencer R: Theoretical and analytical embryology of conjoined twins, part 1: embryogenesis, *Clin Anat* 13:36-53, 2000.

Starmer J, Magnuson T: A new model for random X chromosome inactivation, *Development* 136:1-10, 2009.

Surani MA: Imprinting and the initiation of gene silencing in the germ line, *Cell* 93:309-312, 1998.

Suwinska A and others: Blastomeres of the mouse embryo lose totipotency after the fifth cleavage division: expression of Cdx-2 and Oct-4 and developmental potential of inner and outer blastomeres of 16- and 32-cell embryos, *Dev Biol* 322:133-144, 2008.

Tadros W, Lipshitz HD: The maternal-to-zygotic transition: a play in two acts, *Development* 136:3033-3042, 2009.

Tarkowski AK, Ozdzendki W, Czolowska R: How many blastomeres of the 4-cell embryo contribute cells to the mouse body? *Int J Dev Biol* 45:811-816, 2001.

Tarkowski AK, Wroblewska J: Development of blastomeres of mouse eggs isolated at the 4- and 8-cell stage, *J Embryol Exp Morphol* 18:155-180, 1967.

Thorvaldsen JL, Verona RI, Bartolomei MS: X-tra! X-tra!! News from the mouse X chromosome, *Dev Biol* 298:344-353, 2006.

Uchida IA: Twinning in spontaneous abortions and developmental abnormalities, *Issues Rev Teratol* 5:155-180, 1990.

Veeck LL, Zaninovic N: *An atlas of human blastocysts*, Boca Raton, Fla, 2003, Parthenon.

Weitlauf HM: Biology of implantation. In Knobil E, Neill J, eds: *The physiology of reproduction*, New York, 1988, Raven, pp 231-262.

Yamanaka Y and others: Cell and molecular regulation of the mouse blastocyst, *Dev Dynam* 235:2301-2314, 2006.

Molecular Basis for Embryonic Development

The application of new techniques in molecular biology continues to revolutionize the understanding of the mechanisms underlying both normal and abnormal embryonic development. It is impossible to have a contemporary understanding of embryonic development without integrating fundamental molecular and morphological aspects of embryology. This chapter introduces the most important families of molecules known to direct embryonic development.

One of the most important realizations has been the conservatism of the genes that guide development. Sequencing studies have shown remarkably few changes in the nucleotide bases of many developmentally regulated genes that are represented in species ranging from worms to *Drosophila* to humans. Because of this phylogenetic conservatism, it has been possible to identify mammalian counterparts of genes that are known from genetic studies to have important developmental functions in other species (**Box 4.1**).* It is also clear that the same gene may function at different periods of development and in different organs. Such reuse greatly reduces the total number of molecules that are needed to control development. Before and after birth, specific genes may be expressed in normal and abnormal processes. One of the principal themes in contemporary cancer research is the role of mutant forms of developmentally important genes (e.g., proto-oncogenes) in converting normal cells to tumor cells.

Fundamental Molecular Processes in Development

From a functional standpoint, many of the important molecules that guide embryonic development can be grouped into relatively few categories. Some of them remain in the cells that produced them and act as **transcription factors** (**Fig. 4.2**). Transcription factors are proteins possessing domains that bind to the DNA of promoter or enhancer regions of specific genes. They also possess a domain that interacts with RNA polymerase II or other transcription factors and consequently regulates the amount of messenger RNA (mRNA) produced by the gene.

Other molecules act as intercellular **signaling molecules**. Such molecules leave the cells that produce them and exert their effects on others, which may be neighboring cells or cells located at greater distances from the cells that produce the signaling molecules. Many signaling molecules are members of large families of related proteins, called **growth factors**. To exert their effect, signaling molecules typically bind as **ligands** to **receptor molecules** that are often transmembrane proteins protruding through the plasma membrane of the cells that they affect. When these receptor molecules form complexes with signaling molecules, they set off a cascade of events in a **signal transduction pathway** that transmits the molecular signal to the nucleus of the responding cell. This signal influences the nature of the gene products produced by that cell and often the cell's future course of development.

Transcription Factors

Many families of molecules act as **transcription factors**. Some transcription factors are general ones that are found in virtually all cells of an organism. Other transcription factors are specific for certain types of cells and stages of development. Specific transcription factors are often very important in initiating patterns of gene expression that result in major developmental changes. They typically do so by acting on promoters or enhancers to activate or repress the transcription of specific genes. Based on their structure and how they interact with DNA, transcription factors can be subdivided into several main groups, the most important of which are introduced here.

Homeobox-Containing Genes and Homeodomain Proteins

One of the most important types of transcription factors is represented by the **homeodomain proteins**. These proteins contain a highly conserved **homeodomain** of 60 amino acids; a homeodomain is a type of helix-loop-helix region (**Fig. 4.3**). The 180 nucleotides in the gene that encode the homeodomain are collectively called a **homeobox**. Homeobox regions were first discovered in the homeotic genes of the *antennapedia* and *bithorax* complex in *Drosophila* (see Fig. 4.1), hence their name. This designation sometimes confuses students because, since their initial description, homeoboxes have been found in several more distantly related genes outside the

---

*By convention, names of genes are italicized, whereas the products of genes are printed in roman type. Abbreviations of human genes are all capitalized (e.g., *HOX*); those for other species are written with only the first letter capitalized (e.g., *Hox*).

Box 4.1 Early Developmental Genetics in *Drosophila*

Despite the discovery and characterization of many developmentally important genes in mammals, the basic framework for understanding the molecular basis of embryonic development still rests largely on studies of developmental genetics in *Drosophila*. Although the earliest stages of human development occur under less rigid genetic control than those of *Drosophila*, an exposure to the fundamental aspects of early *Drosophila* development nevertheless sets the stage for a deeper understanding of molecular embryogenesis in mammals.

Embryonic development of *Drosophila* is under tight genetic control. In the earliest stages, the dorsoventral and anteroposterior axes of the embryo are established by the actions of batteries of **maternal-effect genes** (**Fig. 4.1**). When these broad parameters have been established, the oval embryo undergoes a series of three sequential steps that result in the segmentation of the entire embryo along its anteroposterior axis. The first step in segmentation, under the control of what are called **gap genes**, subdivides the embryo into broad regional domains. Loss-of-function gap mutants result in loss of structure, or gaps, in the body pattern

several segments in width. In the second step, a group of **pair-rule genes** is involved in the formation of seven pairs of stripes along the craniocaudal axis of the embryo. The third level in the segmentation process is controlled by the **segment-polarity genes**, which work at the level of individual segments and are involved in their anteroposterior organization.*

The segmentation process results in a regular set of subdivisions along the anteroposterior axis of the early *Drosophila* embryo, but none of the previously mentioned developmental controls imparts specific or regional characteristics to the newly formed segments. This function is relegated to two large clusters of **homeotic genes** found in the **antennapedia** complex and the **bithorax** complex. The specific genes in these two complexes determine the morphogenetic character of the body segments, such as segments bearing antennae, wings, or legs. Mutations of homeotic genes have long been known to produce bizarre malformations in insects, such as extra sets of wings or legs instead of antennae (hence the term *antennapedia*).

Genetic hierarchy	Functions	Representative genes	Effects of mutation
Maternal-effect genes 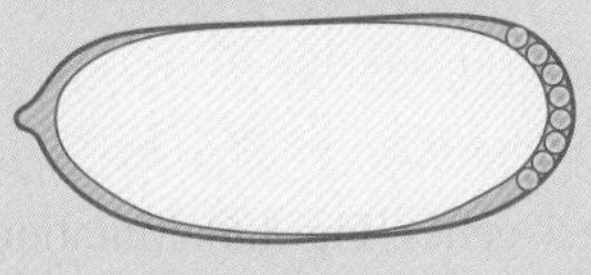	Establish gradients from anterior and posterior poles of the egg	Bicoid Swallow Oskar Caudal Torso Trunk	Major disturbances in anteroposterior organization
Segmentation genes 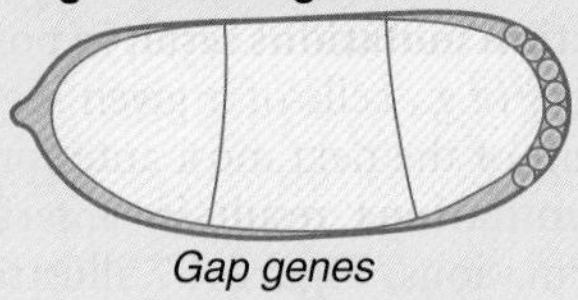 *Gap genes*	Define broad regions in the egg	Empty spiracles Hunchback Krüppel Knirps Orthodenticle Tailless	Adjacent segments missing in a major region of the body
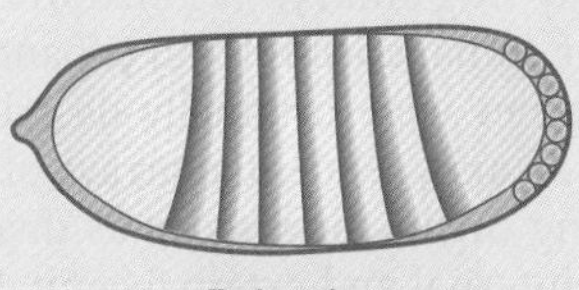 *Pair-rule genes*	Define 7 segments	Hairy Even skipped Runt Fushi tarazu Odd paired Odd skipped Paired	Part of pattern deleted in every other segment
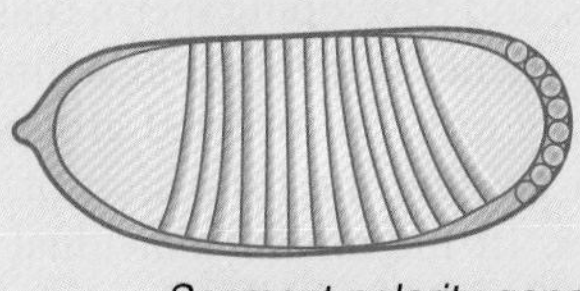 *Segment-polarity genes*	Define 14 segments	Engrailed Gooseberry Hedgehog Patched Smoothened Wingless	Segments replaced by their mirror images
Homeotic genes 	Determine regional characteristics	Antennapedia complex Bithorax complex	Inappropriate structures form for a given segmental level

Fig. 4.1 **Sequence of genetic control of early development in *Drosophila*.** Within each level of genetic control are listed representative genes.

*In *Drosophila*, each stripe (segment) is subdivided into anterior and posterior halves. The posterior half of one segment and the anterior half of the next are collectively known as a **parasegment**. The genetics and developmental aspects of insect parasegments are beyond the scope of this text, but in Chapter 6, when formation of the vertebral column is discussed, a similar set of divisions of the basic body segments in vertebrate embryos is introduced.

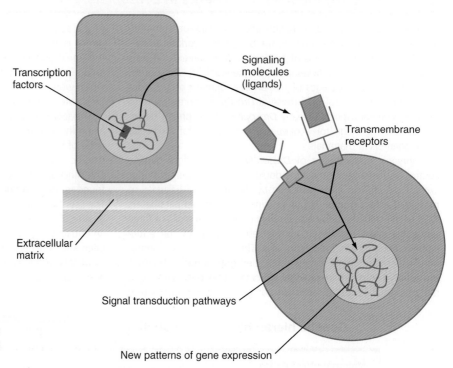

Fig. 4.2   Schematic representation of types of developmentally important molecules and their sites of action.

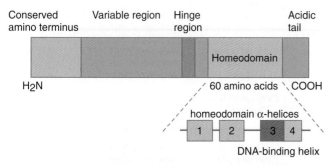

Fig. 4.3   Structure of a typical homeodomain protein.

homeotic gene cluster. Many other gene families contain not only a homeobox but also other conserved sequences (**Fig. 4.4**).

HOX GENES

The *Drosophila* **antennapedia-bithorax** complex consists of 8 homeobox-containing genes located in 2 clusters on one chromosome. Mice and humans possess at least 39 homologous homeobox genes (called ***Hox*** genes in vertebrates [***HOX*** in humans]), which are found in 4 clusters on 4 different chromosomes (**Fig. 4.5**). The *Hox* genes on the 4 mammalian chromosomes are arranged in 13 **paralogous groups**.

Vertebrate *Hox* genes play a prominent role in the craniocaudal segmentation of the body, and their spatiotemporal expression proceeds according to some remarkably regular rules. The genes are activated and expressed according to a strict sequence in the 3′ to 5′ direction, corresponding to their positions on the chromosomes. Consequently, in *Drosophila* and mammals, 3′ genes are expressed earlier and more

anteriorly than are 5′ genes (**Fig. 4.6**). Mutations of *Hox* genes result in morphological transformations of the segmental structures in which a specific gene is normally expressed. Generally, **loss-of-function mutations** result in posterior-to-anterior transformations (e.g., cells of a given segment form the structural equivalent of the next most anterior segment), and **gain-of-function mutations** result in anterior-to-posterior structural transformations. **Figure 4.7** illustrates an experiment in which injection of an antibody to a homeodomain protein into an early frog embryo resulted in the transformation of the anterior spinal cord into an expanded hindbrain.

Although *Hox* genes were originally described to operate along the main body axis, sequential arrays of expression are found in developing organs or regions as diverse as the gut, the limbs, and the internal and external genitalia. The expression of isolated *Hox* genes also occurs in locations such as hair follicles, blood cells, and developing sperm cells. The principal function of the *Hox* genes is involved in setting up structures along the main body axis, but ordered groups of *Hox* genes are later reused in guiding the formation of several specific nonaxial structures. In mammals, individual members of a paralogous group often have similar functions, so that if one *Hox* gene is inactivated, the others of that paralogous group may compensate for it. If all members of a paralogous group are inactivated, profound morphological disturbances often result (see p. 171 in Chapter 9).

The regulation of *Hox* gene expression is complex. A major regulator along parts of the anteroposterior axis of the developing central nervous system is retinoic acid, but this effect is mediated by other genes. At a different level, *Hox* expression is influenced by modifications of chromatin and the three-dimensional organization of the chromosomes. Even after the transcription has occurred, microRNAs (miRNAs) may cleave Hox mRNAs and inactivate them.

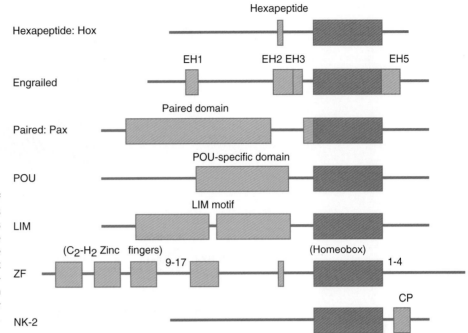

Fig. 4.4 **Schematic representation of classes of homeobox-containing genes also possessing conserved motifs outside the homeodomain.** Names of the different classes of genes are listed on the *left*. The *red boxes* represent the homeobox within each gene class. The other boxes represent conserved motifs specific to each class of genes. *(Modified from Duboule D, ed: Guidebook to the homeobox genes, Oxford, 1994, Oxford University Press.)*

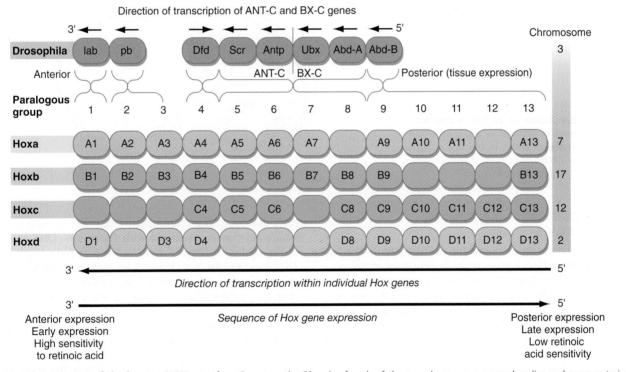

Fig. 4.5 **Organization of the human *HOX* complex.** Genes on the 3′ ends of each of the complexes are expressed earlier and more anteriorly than those on the 5′ end *(right). (Based on Scott MP: Cell 71:551-553, 1992.)*

PAX GENES

The ***Pax* gene family**, consisting of 9 known members, is an important group of genes that are involved in many aspects of mammalian development (**Fig. 4.8**). The *Pax* genes are homologous to the *Drosophila* pair-rule segmentation genes (see Fig. 4.1). All Pax proteins contain a paired domain of 128 amino acids that binds to DNA. Various members of this group also contain entire or partial homeobox domains and a conserved octapeptide sequence. *Pax* genes play a variety of important roles in the sense organs and developing nervous system, and outside the nervous system they are involved in cellular differentiative processes when epithelial-mesenchymal transitions occur.

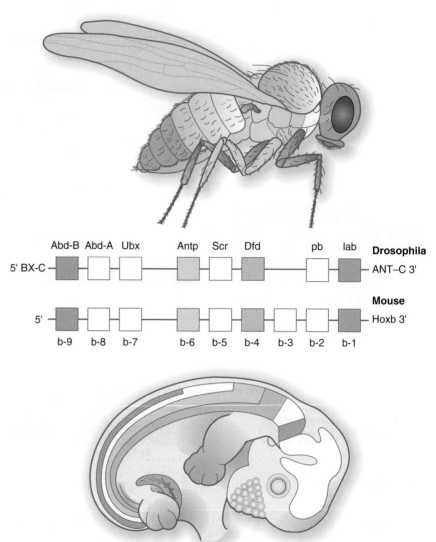

Fig. 4.6 **Organization of certain homeobox-containing genes of *Drosophila* and mouse and their segmental expression in the body.** *(Based on DeRobertis EM, Oliver G, Wright CVE: Sci Am 263:46-52, 1990. Copyright Patricia J. Wynne.)*

OTHER HOMEOBOX-CONTAINING GENE FAMILIES

The POU gene family is named for the acronym of the first genes identified: *Pit1*, a gene uniquely expressed in the pituitary; *Oct1* and *Oct2*; and *Unc86*, a gene expressed in a nematode. Genes of the POU family contain, in addition to a homeobox, a region encoding 75 amino acids, which also bind to DNA through a helix-loop-helix structure. As described in Chapter 3 (see p. 42), Oct-4 plays an important role during early cleavage.

The **Lim proteins** constitute a large family of homeodomain proteins, some of which bind to the DNA in the nucleus and others of which are localized in the cytoplasm. Lim proteins are involved at some stage in the formation of virtually all parts of the body. As discussed in Chapter 5 (see p. 83), the absence of certain Lim proteins results in the development of headless mammalian embryos.

The **Dlx gene** family, similar to the *Hox* gene family, is a group of genes that have been phylogenetically conserved. The six members of this group in mammals are related to the single *distalless* gene in *Drosophila*, and they play important roles in patterning, especially of outgrowing structures, in early embryos. The mammalian *Dlx* genes operate in pairs, which are closely associated with *Hox* genes. *Dlx5* and *Dlx6* are located 5′ to *Hoxa13*; *Dlx3* to *Dlx7* are 5′ to *Hoxb13*; and *Dlx1* and *Dlx2* are 5′ to *Hoxd13*. In addition to being involved in appendage development, *Dlx* gene products are involved in morphogenesis of the jaws and inner ear and in early development of the placenta.

The **Msx** genes (homologous to the muscle-segment homeobox [*msh*] gene in *Drosophila*) constitute a small, highly conserved family of homeobox-containing genes, with only two representatives in humans. Nevertheless, the Msx proteins play important roles in embryonic development, especially in epitheliomesenchymal interactions in the limbs and face. Msx proteins are general inhibitors of cell differentiation in prenatal development, and in postnatal life they maintain the proliferative capacity of tissues.

T-Box Gene Family

The **T-box (Tbx)** genes take their names from the **brachyury (T) locus,** which was recognized as early as 1927 to cause short tails in heterozygotic mice. In 1990, the gene was cloned and

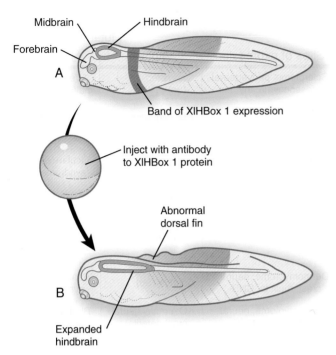

Fig. 4.7 **Effect of interference of *XlHbox 1* (~Hoxc-6) function on development in *Xenopus*.** **A,** Normal larva, showing a discrete band *(green)* of *XlHbox 1* expression. **B,** Caudal expansion of the hindbrain after antibodies to *XlHbox 1* protein are injected into the early embryo. *(Based on Wright CV and others: Cell 59:81-93, 1989.)*

found to contain a conserved region (the T-box), coding 180 to 200 amino acids, which binds to a specific nucleotide sequence in the DNA. Initially thought to be a single gene, an entire family of T-box genes with more than 100 members (18 genes in the human genome) has been described. Genes of this family play important roles in development, such as inducing the mesodermal germ layer and in coordinating outgrowth of either the arm or the leg.

Helix-Loop-Helix Transcription Factors
BASIC HELIX-LOOP-HELIX PROTEINS

The transcription factors of the **basic helix-loop-helix** type are proteins that contain a short stretch of amino acids in which two α-helices are separated by an amino acid loop. This region, with an adjacent basic region, allows the regulatory protein to bind to specific DNA sequences. The basic regions of these proteins bind to DNA, and the helix-loop-helix domain is involved in homodimerization or heterodimerization. This configuration is common in numerous transcription factors that regulate myogenesis (see Fig. 9.33).

FORKHEAD GENE FAMILY

Another large family of transcription factors (>100 members, with 30 in mice) constitutes the *forkhead (Fox)* genes. As a variant of the helix-loop-helix theme, a common element among the forkhead proteins is the forkhead DNA-binding region, which is constituted as a **winged helix** structure. The

Fox genes are expressed in many developing organs throughout the body. They tend to have microscopically distinct domains within a developing organ and can work together to direct the morphogenesis of a structure.

Zinc Finger Transcription Factors

The **zinc finger** family of transcription factors consists of proteins with regularly placed cystidine and histidine units that are bound by zinc ions to cause the polypeptide chain to pucker into fingerlike structures (**Fig. 4.9**). These "fingers" can be inserted into specific regions in the DNA helix.

SOX GENES

The *Sox genes* comprise a large family (>20 members) that have in common an **HMG (high-mobility group) domain** on the protein. This domain is unusual for a transcription factor in that, with a partner protein, it binds to 7 nucleotides on the minor instead of the major groove on the DNA helix and causes a pronounced conformational change in the DNA. Sox proteins were first recognized in 1990, when the *SRY* gene was shown to be the male-determining factor in sex differentiation (see p. 389), and the name of this group, Sox, was derived from Sry HMG **box**. One characteristic of Sox proteins is that they work in concert with other transcription factors to influence expression of their target genes (**Fig. 4.10**). As may be expected from their large number, Sox proteins are expressed by most structures at some stage in their development.

WT1

WT1 (Wilms tumor suppressor gene) is an isolated gene that in prenatal life plays a prominent role in formation of both the kidneys and gonads. It is crucial for the development of the early forms of the kidney and for the formation of the definitive adult kidney. In addition, *WT1* is necessary for formation of the gonads. Its name derives from Wilms tumor, a prominent type of kidney tumor in young children.

Signaling Molecules

Much of embryonic development proceeds on the basis of chemical signals sent from one group of cells and received and acted on by another. A significant realization is that the same signaling molecule can be used at many different times and places as the embryo takes shape. Locally controlled factors, such as the concentration or duration of exposure to a signaling molecule, are often important determinants of the fate of a group of responding cells. This situation reduces greatly the number of signaling molecules that need to be employed. Most signaling molecules are members of several, mostly large, families. The specific sequence of signaling molecule (ligand) → receptor → signal transduction pathway is often called a **signaling pathway**. This section outlines the major families of signaling molecules that guide embryonic development.

Transforming Growth Factor-β Family

The **transforming growth factor-β (TGF-β) superfamily** consists of numerous molecules that play a wide variety of

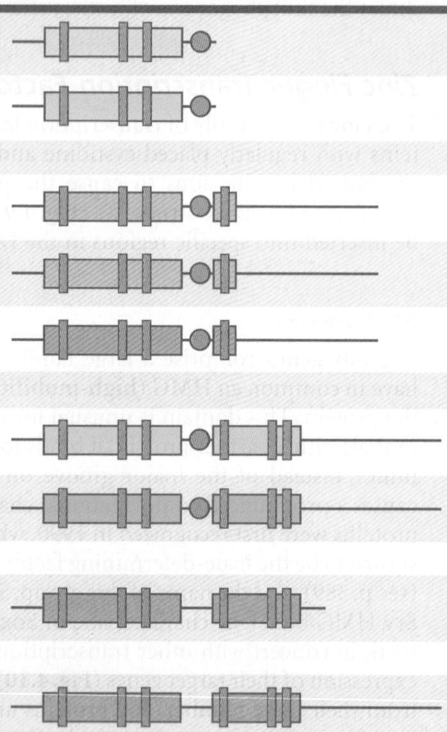

Chromosomal localization

Gene	Human	N PD Oct HD C	Sites of expression	Mutants Mouse	Human
Pax-1	20p11		Sclerotome, perivertebral mesenchyme, thymus	*Undulated* (un)	Vertebral malformations
Pax-9	14q12 -q13		Sclerotome, perivertebral mesenchyme	KO: oligodontia	Oligodontia
Pax-2	10p25		Urogenital, CNS, eye, inner ear	KO: no kidneys	Ocular-renal syndromes
Pax-5	9p13		Pro-B cells, CNS	KO: no B cells	
Pax-8	2q12 -q14		Thyroid, kidney, CNS	KO: no follicular cells in thyroid	Congenital hypothyroidism
Pax-3	2q35		Dermomyotome, neural crest, muscle, CNS	*Splotch* (Sp)	Waardenburg syndrome
Pax-7	1p36		Dermomyotome, neural crest, muscle, CNS	KO: cranial neural crest defects	Rhabdomyo-sarcoma
Pax-4	7q32		Pancreas	KO: no insulin	Type II diabetes
Pax-6	11p13		Eye, pancreas, CNS	*Small eye* (Sey)	Aniridia

Fig. 4.8   Summary diagram of the members of the *Pax* gene family, showing their location on human chromosomes, sites of expression, and known effects of mutants in human and mouse. The structures of conserved elements of these genes are schematically represented. CNS, central nervous system; KO, knockout. *(Modified from Wehr R, Gruss P:* Int J Dev Biol *40:369-377, 1996; and Epstein JC:* Trends Cardiovasc Med *6:255-260, 1996.)*

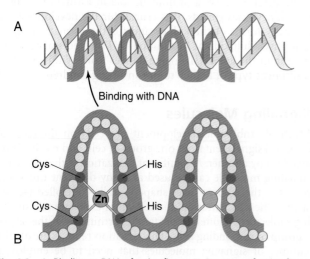

Fig. 4.9   A, Binding to DNA of a zinc finger. **B,** Structure of a zinc finger DNA-binding sequence.

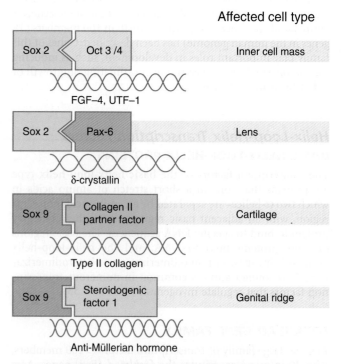

Fig. 4.10   Examples of Sox proteins forming complexes with other transcription factors as they influence the expression of specific genes (*labels below the helix representing DNA*). Tissues influenced by the Sox-based gene regulation *(right).*

roles during embryogenesis and postnatal life. The TGF family was named because its first-discovered member (TGF-β₁) was isolated from virally transformed cells. Only later was it realized that many signaling molecules with greatly different functions during embryonic and postnatal life bear structural similarity to this molecule. **Table 4.1** summarizes some of these molecules and their functions.

The formation, structure, and modifications of TGF-β₁ are representative of many types of signaling molecules and are used as an example (**Fig. 4.11**). Similar to many members of this family, TGF-β₁ is a disulfide-linked dimer, which is synthesized as a pair of inactive 390-amino acid precursors. The glycosylated precursor consists of a small N-terminal signal sequence, a much larger proregion, and a 112-amino acid C-terminal bioactive domain. The proregion is enzymatically

cleaved off the bioactive domain at a site of 4 basic amino acids adjoining the bioactive domain. After secretion from the cell, the proregion of the molecule remains associated with the bioactive region, thus causing the molecule to remain in a latent form. Only after dissociation of the proregion from the bioactive region does the bioactive dimer acquire its biological activity.

Among the most important subfamilies of the TGF-β family are the **bone morphogenetic proteins** (**BMPs**). Although BMP was originally discovered to be the active agent in the induction of bone during fracture healing, the 15 members of this group play important roles in the development of most structures in the embryo. BMPs often exert their effects by inhibiting other processes in the embryo. To make things even more complicated, certain very important

Table 4.1 Members of the Transforming Growth Factor-β Superfamily Mentioned in This Text

Member	Representative Functions	Chapters
TGF-β1 to TGF-β5	Mesodermal induction	5
	Myoblast proliferation	9
	Invasion of cardiac jelly by atrioventricular endothelial cells	17
Activin	Granulosa cell proliferation	1
	Mesodermal induction	5
Inhibin	Inhibition of gonadotropin secretion by hypophysis	1
Müllerian inhibiting substance	Regression of paramesonephric ducts	16
Decapentaplegic	Signaling in limb development	10
Vg1	Mesodermal and primitive streak induction	5
BMP-1 to BMP-15	Induction of neural plate, induction of skeletal differentiation, and other inductions	5, 9, 10
Nodal	Formation of mesoderm and primitive streak, left-right axial fixation	5
Glial cell line–derived neurotrophic factor	Induction of outgrowth of ureteric bud, neural colonization of gut	16, 12
Lefty	Determination of body asymmetry	5

BMP, bone morphogenetic protein; TGF-β, transforming growth factor-β.

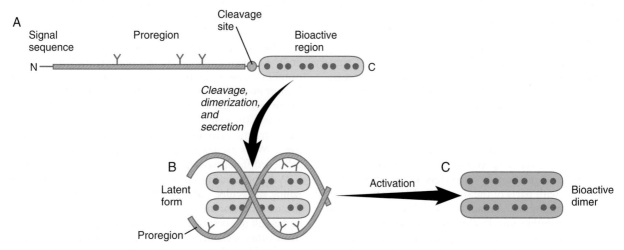

Fig. 4.11 **Steps in the activation of the growth factor, TGF-b1. A,** The newly synthesized peptide consists of a C-terminal bioactive region, to which is attached a long glycosylated proregion and an N-terminal signal sequence. **B,** The proregion is cleaved off from the bioactive region, and two secreted bioactive regions form a dimer that is maintained in a latent form by being complexed with the separated proregions. **C,** Through an activation step, the bioactive dimer is released from the proregions and can function as a signaling molecule.

Table 4.2 Major Molecular Antagonists of Growth Factors

BMPs	WNT
Noggin	Axin-1, Axin-2
Chordin	Dickkopf
Chordinlike	Cerberus
Twisted gastrulation	Wif (Wnt inhibitory
Follistatin	factor-1)
FSRP (follistatin-related protein)	Sfrp (Secreted frizzled-
DAN/cerberus	related protein)
Gremlin	Wise (Wnt modulator
Ectodin	in surface ectoderm)
Coco	

SHH	FGF (FGFR)
Cyclopamine (in plants)	Sprouty

	NODAL
	Lefty-1
	Cerberus-like

interactions in embryonic development (e.g., induction of the central nervous system; see p. 84) occur because of the inhibition of BMP by some other molecule. The net result is an effect caused by the inhibition of an inhibitor. Molecules that inhibit or antagonize the action of BMPs are listed in **Table 4.2**. These molecules bind to secreted BMP dimers and interfere with their binding to specific receptors.

Fibroblast Growth Factor Family

Fibroblast growth factor (FGF) was initially described in 1974 as a substance that stimulates the growth of fibroblasts in culture. Since then, the originally described FGF has expanded into a family of 22 members, each of which has distinctive functions. Many members of the FGF family play important roles in a variety of phases of embryonic development and in fulfilling functions, such as the stimulation of capillary growth, in the postnatal body. Some of the functions of the FGFs in embryonic development are listed in **Table 4.3**. Secreted FGFs are closely associated with the extracellular matrix and must bind to heparan sulfate to activate their receptors.

Similar to other signaling molecules, FGF activity is regulated in many ways. In contrast to the BMPs, which are regulated by several molecules that bind to them in the extracellular space, FGFs are mainly regulated farther downstream. Means of FGF regulation include the following: (1) modifications of their interaction with heparan proteoglycans in the receptor complex; (2) regulation at the membrane of the responding cell through the actions of transmembrane proteins; and (3) intracellular regulation by molecules, such as **sprouty**, which complex with parts of the signal transduction machinery of the responding cell. A main theme in the role of signaling molecules in embryonic development is variation, both in the variety of forms of signal molecules in the same family and in the means by which their activity is regulated. Most of the details of these are beyond the scope of this book, but for the beginning student it is important to recognize that they exist.

Hedgehog Family

The **hedgehog** signaling molecules burst on the vertebrate embryological scene in 1994 and are among the most important signaling molecules known (**Table 4.4**). Related to the segment-polarity molecule, hedgehog, in *Drosophila*, the three mammalian hedgehogs have been given the whimsical names of desert, Indian, and sonic hedgehog. The name hedgehog arose because mutant larvae in *Drosophila* contain thick bands of spikey outgrowths on their bodies.

Sonic hedgehog (**shh**) is a protein with a highly conserved N-terminal region and a more divergent C-terminal region. After its synthesis and release of the propeptide from the rough endoplasmic reticulum, the signal peptide is cleaved off, and glycosylation occurs on the remaining peptide (**Fig. 4.12**). Still within the cell, the shh peptide undergoes autocleavage through the catalytic activity of its C-terminal portion. During cleavage, the N-terminal segment becomes covalently bonded with cholesterol. The 19-kD N-terminal peptide is secreted from the cell, but it remains bound to the surface of the cell that produced it. All the signaling activity of shh resides in the N-terminal segment. Through the activity of another gene product (disp [dispatched] in *Drosophila*), the N-terminal segment of shh, still bound with cholesterol, is released from the cell. The C-terminal peptide plays no role in signaling.

At the surface of a target cell, shh, still complexed with cholesterol, binds to a receptor, **Patched** (**Ptc**), closely associated with another transmembrane protein, **smoothened** (**smo**). Ptc normally inhibits the signaling activity of smo, but shh inhibits the inhibitory activity of Ptc, thus allowing smo to give off an intracellular signal. Through the mediation of several other molecules, which are normally bound to microtubules, smo ultimately activates the 5-zinc finger transcription factor, **Gli**, which moves to the nucleus, binds to specific sites on the DNA of that cell, and thereby affects gene expression of the target cell.

Wnt Family

The **Wnt** family of signaling molecules is complex, with 18 members represented in the mouse. Related to the segment-polarity gene *Wingless* in *Drosophila*, Wnts play dramatically different roles in different classes of vertebrates. In amphibians, Wnts are essential for dorsalization in the very early embryo, whereas their role in preimplantation mouse development seems to be minimal. In mammals, Wnts play many important roles during the period of gastrulation. As many organ primordia begin to take shape, active Wnt pathways stimulate the cellular proliferation that is required to bring these structures to their normal proportions. Later in development, Wnts are involved in a variety of processes relating to cellular differentiation and polarity.

Wnts have been described as being "stickier" than other signaling molecules, and they often interact with components of the extracellular matrix. Their signaling pathway is complex and is still not completely understood (see Fig. 4.16). Similar to most other signaling molecules, the activity of Wnts can be regulated by other inhibitory molecules (see Table 4.2). Some inhibitory molecules, such as **Wnt-inhibitory factor-1** (**WIF-1**) and **cerberus**, directly bind to the Wnt molecule. Others, such as **dickkopf**, effect inhibition by binding to the receptor complex.

Table 4.3 Members of the Fibroblast Growth Factor Family Mentioned in This Text

FGF	Developmental System	Chapter
FGF-1	Stimulation of keratinocyte proliferation	9
	Early liver induction	15
FGF-2	Stimulation of keratinocyte proliferation	9
	Induction of hair growth	9
	Apical ectodermal ridge in limb outgrowth	10
	Stimulation of proliferation of jaw mesenchyme	14
	Early liver induction	15
	Induction of renal tubules	16
FGF-3	Inner ear formation	13
FGF-4	Maintenance of mitotic activity in trophoblast	3
	Apical ectodermal ridge in limb outgrowth	10
	Enamel knot of developing tooth	14
	Stimulation of proliferation of jaw mesenchyme	14
FGF-5	Stimulation of ectodermal placode formation	9
FGF-8	Isthmic organizer: midbrain patterning	6
	Apical ectodermal ridge in limb outgrowth	10
	From anterior neural ridge, regulation of development of optic vesicles and telencephalon	11
	Early tooth induction	14
	Stimulation of proliferation of neural crest mesenchyme of frontonasal region	14
	Stimulation of proliferation of jaw mesenchyme	14
	Induction of filiform papillae of tongue	14
	Early liver induction	15
	Outgrowth of genital tubercle	16
FGF-9	Apical ectodermal ridge in limb outgrowth	10
FGF-10	Limb induction	10
	Branching morphogenesis in developing lung	15
	Induction of prostate gland	16
	Outgrowth of genital tubercle	16
FGF-17	Apical ectodermal ridge in limb outgrowth	10

FGF, fibroblast growth factor.

Table 4.4 Sites in the Embryo Where Sonic Hedgehog Serves as a Signaling Molecule

Signaling Center	Chapters
Primitive node	5
Notochord	6, 11
Floor plate (nervous system)	11
Intestinal portals	6
Zone of polarizing activity (limb)	10
Hair and feather buds	9
Ectodermal tips of facial processes	14
Apical ectoderm of second pharyngeal arch	14
Tips of epithelial buds in outgrowing lung	15
Patterning of retina	13
Outgrowth of genital tubercle	16

Other Actions of Signaling Molecules

An important and more recent realization in molecular embryology is how often signaling molecules act by inhibiting the actions of other signaling molecules. For example, the signaling molecules **chordin**, **noggin**, and **gremlin** all inhibit the activity of BMP, which itself often acts as an inhibitor (see Table 4.2).

Evidence from several developing organ systems indicates that some signaling molecules (e.g., shh and members of the FGF family) are positive regulators of growth, whereas others (e.g., some members of the BMP family) serve as negative regulators of growth. Normal development of a variety of organs requires a balance between the activities of these positive and negative regulators. Such interactions are described later in the text for developing organ systems as diverse as limbs, hair (or feathers), teeth, and the branching of ducts in the lungs, kidneys, and prostate gland.

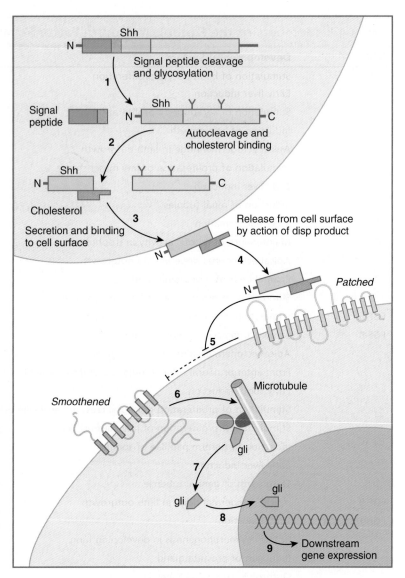

Fig. 4.12 The sonic hedgehog (shh) signaling pathway.
(*1*) The signal peptide is cleaved off the newly synthesized polypeptide, and the remainder undergoes glycosylation; (*2*) the remaining peptide undergoes autocleavage under the influence of the C-terminal portion, and cholesterol binds to the N-terminal part, which is the active part of the molecule; (*3*) the N-terminal part is secreted and bound to the cell surface; (*4*) the bound shh molecule is released from the cell surface through the action of a product of *dispersed (disp);* (*5*) the released shh inhibits the inhibitory effect of Patched on smoothened; (*6*) on release from the inhibitory influence of Patched, smoothened emits a signal that (*7*) releases the transcription factor Gli from a complex of molecules bound to microtubules; (*8*) Gli enters the nucleus and binds to the DNA, (*9*) influencing the expression of many genes.

Receptor Molecules

For intercellular signaling molecules to exert an effect on target cells, they must normally interact with receptors in these cells. Most receptors are located on the cell surface, but some, especially those for lipid-soluble molecules, such as steroids, retinoids, and thyroid hormone, are intracellular.

Cell surface receptors are typically transmembrane proteins with extracellular, transmembrane, and cytoplasmic domains (see Fig. 4.2). The extracellular domain contains a binding site for the **ligand**, which is typically a hormone, cytokine, or growth factor. When the ligand binds to a receptor, it effects a conformational change in the cytoplasmic domain of the receptor molecule. Cell surface receptors are of two main types: (1) receptors with intrinsic protein kinase activity and (2) receptors that use a second messenger system to activate cytoplasmic protein kinases. An example of the first type is the family of receptors for FGFs, in which the cytoplasmic domain possesses **tyrosine kinase** activity. Receptors for growth factors of the TGF-β superfamily are also of this type, but in them the cytoplasmic domain contains **serine/**

threonine kinase activity. In cell surface receptors of the second type, the protein kinase activity is separate from the receptor molecule itself. This type of receptor is also activated by binding with a ligand (e.g., neurotransmitter, peptide hormone, growth factor), but a series of intermediate steps is required to activate cytoplasmic protein kinases. A surface receptor, **Notch**, is introduced in greater detail in **Box 4.2** as a specific example of a receptor that plays many important roles in embryonic development.

Signal Transduction

Signal transduction is the process by which the signal provided by the **first messenger** (i.e., the growth factor or other signaling molecule) is translated into an intracellular response. Signal transduction is very complex. It begins with a response to binding of the signaling molecule to its receptor and the resulting change in the conformation of the receptor. This process sets off a chain reaction of activation or inhibition of a string of cytoplasmic molecules whose function is to carry the signal to the nucleus, where it ultimately influences gene

Box 4.2 Lateral Inhibition and the Notch Receptor

The normal development of many tissues begins with a population of developmentally equivalent cells. At some point, one of these cells begins to differentiate into a dominant mature cell type, such as a neuron, and, in doing so, it transmits to its neighboring cells a signal that prevents them from differentiating into that same cell type. As a consequence, these neighboring cells are forced to differentiate into a secondary cell type, such as a glial cell in the central nervous system (**Fig. 4.13**). This type of signaling of a dominant cell to its subservient neighbors is called **lateral inhibition**.

The common mechanism of lateral inhibition is the **Notch** signaling pathway, which is so basic that it has been preserved largely unchanged throughout the animal kingdom. Notch is a 300-kD cell surface receptor with a large extracellular domain and a smaller intracellular domain. The Notch receptor becomes activated when it combines with ligands (**Delta** or **Jagged** in vertebrates) that extend from the surface of the dominant cell. This sets off a pathway that inhibits the neighboring cell from differentiating into the dominant phenotype.

An abbreviated version of this pathway is as follows (**Fig. 4.14**): The complexing of Notch with its ligand (e.g., Delta) stimulates an intracellular protease reaction that cleaves off the intracellular domain of the Notch molecule. The liberated intracellular domain of Notch becomes translocated to the nucleus, but on its way it may become associated with regulatory proteins, such as **Deltex**. Within the nucleus, the intracellular domain of Notch combines with several helix-loop-helix transcription factors, and this complex binds to the DNA of a gene called **_enhancer of split_**. The product of this gene is another transcription factor that regulates other genes. It represses certain genes of the **_Achaete-Scute_** complex, whose function is to promote neuronal development. By this complex pathway, the subservient cells, in the nervous system for example, are denied the opportunity to differentiate into neurons and instead follow a secondary pathway, which leads to their becoming glial cells.

As complex as it seems, this description is a greatly abbreviated version of this inhibitory pathway and its controlling elements. When more is learned about all the elements involved in this pathway, it will likely look like a component of an immense network of regulatory pathways that interact in very complex ways to integrate internal and external environmental influences that determine the ultimate developmental fate of a cell.

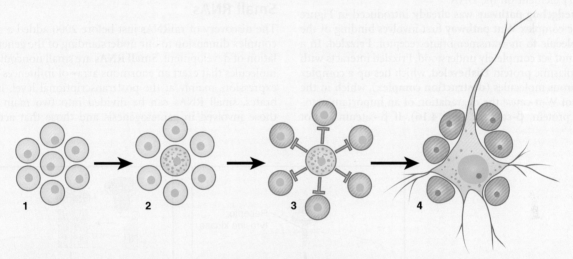

Fig. 4.13 **An example of lateral inhibition.** (*1*) A population of developmentally equivalent cells; (*2*) one cell, whether by its position or through stochastic (random) factors, begins to develop along a dominant pathway before its neighbors; (*3*) the selected cell gives off inhibitory signals (lateral inhibition) that prevent its neighbors from differentiating into the dominant cell type; (*4*) the selected cell differentiates into a mature cell type (e.g., a neuron), whereas its neighbors differentiate into secondary phenotypes (e.g., glial cells).

Fig. 4.14 **The Delta-Notch pathway.** When Delta from a dominant cell binds to Notch on the surface of the neighboring cell, proteolytic cleavage releases the intracellular domain of Notch, which complexes with Deltex and enters the nucleus. There it becomes linked to Suppressor of hairless and serves as a transcription factor, which binds to _Enhancer of split_. This sends off an inhibitory influence that represses the expression of genes, such as the _Achaete-Scute_ complex, which would otherwise promote differentiation.

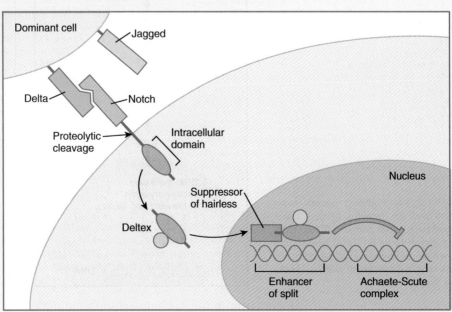

expression. It is common to speak about signal transduction pathways as though they are straight lines, but in reality signal transduction should be viewed as a massive network subject to a wide variety of modulating influences. Despite this complexity, signal transduction can be viewed as linear pathways for purposes of introduction. Several major pathways of relevance to signaling molecules treated in this text are summarized here.

Members of the FGF family connect with the **receptor tyrosine kinase (TRK) pathway** (**Fig. 4.15A**). After FGF has bound to the receptor, a G protein near the receptor becomes activated and sets off a long string of intracytoplasmic reactions, starting with RAS and ending with the entry of ERK into the nucleus, and its interaction with transcription factors. Members of the TGF-β family first bind to a type II serine/threonine kinase receptor, which complexes with a type I receptor (Fig. 4.15B). This process activates a pathway dominated by **Smad** proteins. Two different Smads (R-Smad and Co-Smad) dimerize and enter the nucleus. The Smad dimer binds with a cofactor and is then capable of binding with some regulatory element on the DNA.

The **hedgehog pathway** was already introduced in Figure 4.12. The complex **Wnt pathway** first involves binding of the Wnt molecule to its transmembrane receptor, **Frizzled**. In a manner not yet completely understood, Frizzled interacts with the cytoplasmic protein **Disheveled**, which ties up a complex of numerous molecules (**destruction complex**), which in the absence of Wnt cause the degradation of an important cytoplasmic protein, **β-catenin** (**Fig. 4.16**). If β-catenin is not

destroyed, it enters the nucleus, where it acts as a powerful adjunct to transcription factors that determine patterns of gene expression.

The more recently discovered **Hippo** pathway, highly conserved in phylogeny, is proving to be very important in regulating organ growth throughout the animal kingdom. Loss of Hippo function results in unrestrained growth of structures ranging from the cuticle of *Drosophila* to the liver of mammals. In mammals, Hippo restricts cellular proliferation and promotes the removal of excess cells through apoptosis. It is involved in maintaining the balance between stem cells and differentiated cells both prenatally and postnatally.

These and other less prominent signal transduction pathways are the intracellular effectors of the many signaling events that are necessary for the unfolding of the numerous coordinated programs that guide the orderly progression of embryonic development. Specific examples involving these signaling pathways are frequently mentioned in subsequent chapters.

Small RNAs

The discovery of miRNAs just before 2000 added a new and complex dimension to our understanding of the genetic regulation of development. **Small RNAs** are small noncoding RNA molecules that exert an enormous array of influences on gene expression, mainly at the posttranscriptional level. In vertebrates, small RNAs can be divided into two main groups: those involved in gametogenesis and those that act during

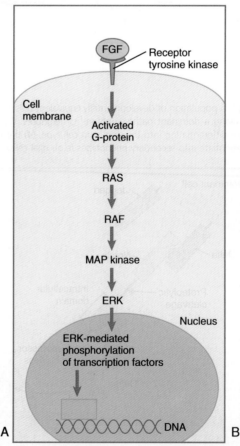
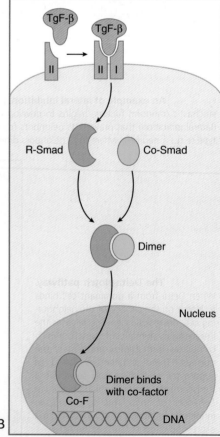

Fig. 4.15 A, Fibroblast growth factor (FGF) and the receptor tyrosine kinase signal transduction pathway. B, Transforming growth factor-β (TGF-β) binding to a type II serine/threonine kinase receptor and activating a downstream pathway involving Smad proteins.

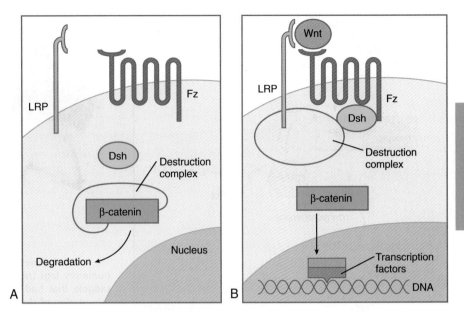

Fig. 4.16 **The Wnt signaling pathway, operating through β-catenin. A,** In the absence of a Wnt signal, β-catenin is bound in a destruction complex and is degraded. **B,** In the presence of Wnt, the receptor Frizzled (Fz) activates Disheveled (Dsh), which prevents the destruction complex from degrading β-catenin. β-catenin then enters the nucleus, where it forms complexes with transcription factors.

embryogenesis. Of those that act during gametogenesis, Piwi-interacting RNAs (**piRNAs**) are important in spermatogenesis, and endogenous small interfering RNAs (**endo-siRNAs**) play vital roles in oogenesis. **miRNAs** are expressed in somatic tissues during embryonic development.

Although small RNAs function through a bewildering array of mechanisms, one major pathway is close to being common (**Fig. 4.17**). miRNAs often begin as double-stranded molecules with a hairpin loop. Through the activity of an enzyme called **Dicer**, the miRNA precursor is cleaved, resulting in a single-stranded miRNA, which is then bound to a member of the **Argonaute (AGO) protein** family. In many cases, the AGO-siRNA complex has RNase activity and is able to disrupt a target RNA molecule enzymatically. In this way specific gene expression is modulated. By applying this principle, developmental geneticists are able to target the disruption of specific genes under investigation by interfering with the mRNAs that these genes produce.

Retinoic Acid

For years, **vitamin A** (**retinol**) and its metabolite, **retinoic acid,** have been known to play very important but equally enigmatic roles in embryonic development. In the 1960s, investigators found that either a severe deficiency or an excess of vitamin A results in a broad spectrum of severe congenital anomalies that can involve the face, eye, hindbrain, limbs, or urogenital system. It was only in the 1990s, when the binding proteins and receptors for the retinoids were characterized and the development of various knockouts was investigated, that specific clues to the function of vitamin A in embryogenesis began to emerge.

Vitamin A enters the body of the embryo as retinol and binds to a retinol-binding protein, which attaches to specific cell surface receptors (**Fig. 4.18**). Retinol is released from this complex and enters the cytoplasm, where it is bound to **cellular retinol-binding protein** (**CRBP I**). In the cytoplasm, the all-*trans* retinol is enzymatically converted first to all-*trans*

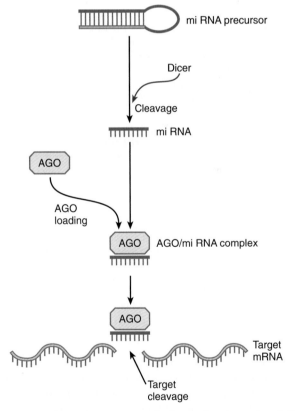

Fig. 4.17 **Schematic summary of the main elements of the microRNA (miRNA) pathway.** The double helical precursor molecule, often containing a hairpin loop, is cleaved by Dicer, resulting in a small miRNA molecule, which is then complexed with an Argonaute (AGO) protein. This complex approaches the target mRNA and through its intrinsic RNase activity, it cleaves the target mRNA molecule, thereby inactivating it.

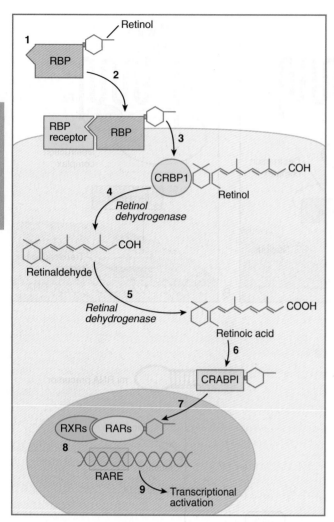

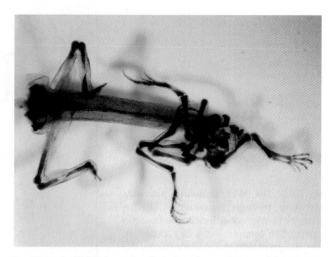

Fig. 4.19 **A skeletal preparation showing a cluster of four super-numerary legs (*right*) growing out of the regenerating tail of a tadpole that had been placed in a solution of vitamin A after amputation of the tail.** This is an example of a homeotic transformation. *(Courtesy of M. Maden, London.)*

Fig. 4.18 **The pathway of vitamin A in a cell.** (*1*) Retinol becomes bound to a retinol-binding protein (RBP) outside the cell; (*2*) this complex is bound to an RBP receptor on the cell surface; (*3*) the retinol is released into the cytoplasm and is bound to a cytoplasmic RBP (CRBP I); (*4*) through the action of retinol dehydrogenase, retinol is converted to retinaldehyde (*5*), which is converted to retinoic acid by retinal dehydrogenase; (*6*) retinoic acid is bound to a cytoplasmic receptor (CRABP I) and taken into the nucleus; (*7*) within the nucleus, retinoic acid is bound to a dimer of two nuclear retinoic acid receptors (RXR and RAR); (*8*) this complex binds to a retinoic acid response element (RARE) on the DNA and (*9*) activates transcription of target genes.

retinaldehyde and then to all-*trans* retinoic acid, the retinoid with the most potent biological activity (see Fig. 4.18). CRBP and **CRABP I (cellular retinoic acid–binding protein)** may function to control the amount of retinoids that enters the nucleus. When released from CRABP, retinoic acid enters the nucleus, where it typically binds to a heterodimer consisting of a member of the **retinoic acid receptor** (**RAR**) α, β, or γ family and a member of the **retinoid X receptor** (**RXR**) α, β, or γ family. This complex of retinoic acid and receptor heterodimer binds to a **retinoic acid response element** (**RARE**) on DNA, usually on the enhancer region of a gene, and it acts as a transcription factor, controlling the production of a gene product.

Retinoic acid is produced and used in specific local regions at various times during prenatal and postnatal life. Among its well-defined targets early in development are certain *Hox* genes (e.g., *Hoxb-1*); misexpression of these genes caused by either too little or too much retinoic acid can result in serious disturbances in the organization of the hindbrain and pharyngeal neural crest. One of the most spectacular examples of the power of retinoic acid is its ability to cause extra pairs of limbs to form alongside the regenerating tails of amphibians (**Fig. 4.19**). This is a true example of a homeotic shift in a vertebrate, similar to the formation of double-winged flies or legs instead of antennae in *Drosophila* (see p. 59).

Developmental Genes and Cancer

Many cancers are caused by mutated genes, and many of these are genes that play a role in normal embryonic development. Two main classes of genes are involved in tumor formation, and each class uses a different mechanism in stimulating tumor formation.

Proto-oncogenes, a class involving a variety of different types of molecules, induce tumor formation through dominant gain-of-function alleles that result in deregulated growth. Through several types of mechanisms, such as single-point mutations, selective amplification, or chromosomal rearrangements, proto-oncogenes can become converted to **oncogenes**, which are the actual effectors of poorly controlled cellular proliferation. Proto-oncogenes direct the normal formation of molecules including certain growth factors, growth factor receptors, membrane-bound and cytoplasmic signal proteins, and transcription factors.

The other class of genes involved in tumor formation consists of the **tumor suppressor genes**, which normally function to limit the frequency of cell divisions. Recessive loss-of-function alleles of these genes fail to suppress cell division, thus resulting in uncontrolled divisions in defined populations of cells. A good example of a tumor suppressor gene is *Patched*, already discussed as the transmembrane receptor for the signaling molecule, shh. *Patched* normally inhibits the activity of smo. Mutations of *Patched* eliminate the inhibition of smo and allow uncontrolled downstream activity from smo that stimulates the genome of the affected cell. Such a *Patched* mutation is the basis for the most common type of cancer, **basal cell carcinoma**, of the skin. shh itself is involved in tumors of the digestive tract. shh is increased in tumors of the esophagus, stomach, biliary tract, and pancreas, but the hedgehog pathway is not active in cell line tumors from the colon.

Summary

- Evidence is increasing that the basic body plan of mammalian embryos is under the control of many of the same genes that have been identified as controlling morphogenesis in *Drosophila*. In this species, the basic axes are fixed through the actions of maternal-effect genes. Batteries of segmentation genes (gap, pair-rule, and segment-polarity genes) are then activated. Two clusters of homeotic genes next confer a specific morphogenetic character to each body segment. Because of their regulative nature, mammalian embryos are not as rigidly controlled by genetic instructions as are *Drosophila* embryos.
- The homeobox, a highly conserved region of 180 base pairs, is found in multiple different genes in almost all animals. The homeobox protein is a transcription factor. Homeobox-containing genes are arranged along the chromosome in a specific order and are expressed along the craniocaudal axis of the embryo in the same order. Activation of homeobox genes may involve interactions with other morphogenetically active agents, such as retinoic acid and TGF-β.
- Many of the molecules that control development can be assigned to several broad groups. One group is the transcription factors, of which the products of homeobox-containing genes are just one of many types. A second category is signaling molecules, many of which are effectors of inductive interactions. Some of these are members of large families, such as the TGF-β and FGF families. An important class of signaling molecules is the hedgehog proteins, which mediate the activities of many important organizing centers in the early embryo. Signaling molecules interact with responding cells by binding to specific surface or cytoplasmic receptors. These receptors represent the initial elements of complex signal transduction pathways, which translate the signal to an intracellular event that results in new patterns of gene expression in the responding cells. Small RNAs play important roles in the control of gene expression, mainly at posttranscriptional levels. Retinoic acid (vitamin A) is a powerful, but poorly understood, developmental molecule. Misexpression of retinoic acid causes level shifts in axial structures through interactions with *Hox* genes.
- Many cancers are caused by mutations of genes involved in normal development. Two major classes of cancer-causing genes are proto-oncogenes, which induce tumor formation through gain-of-function mechanisms, and tumor suppressor genes, which cause cancers through loss-of-function mutations.

Review Questions

1. **What is a homeobox?**

2. **Which of the following is a transcription factor?**
A. FGF
B. Pax
C. TGF
D. Notch
E. Wnt

3. **Where in the cell is the retinoic acid receptor located?**

4. **A mutation of what receptor is the basis for basal carcinomas of the skin?**
A. Patched
B. Retinoic acid
C. Notch
D. FGF receptor
E. None of the above

5. **Zinc fingers or helix-loop-helix arrangements are characteristic of members of what class of molecules?**
A. Proto-oncogenes
B. Signaling molecules
C. Receptors
D. Transcription factors
E. None of the above

6. **Based on your knowledge of paralogous groups, which gene would be expressed most anteriorly in the embryo?**
A. *Hoxa-13*
B. *Hoxc-9*
C. *Hoxd-13*
D. *Hoxb-1*
E. *Hoxb-6*

7. **Sonic hedgehog is produced in which signaling center?**
A. Notochord
B. Intestinal portals
C. Floor plate of neural tube
D. Zone of polarizing activity in the limb bud
E. All of the above

References

Artavanis-Tsakonas S, Muskavitch MAT: Notch: the past, the present, and the future, *Curr Top Dev Biol* 92:1-29, 2010.

Attisano L, Lee-Hoeflich ST: The Smads, *Genome Biol* 2:1-8, 2001.

Bach I: The LIM domain: regulation by association, *Mech Dev* 91:5-17, 2000.

Balemans W, van Hul W: Extracellular regulation of BMP signaling in vertebrates: a cocktail of modulators, *Dev Biol* 250:231-250, 2002.

Bijlsma MF, Spek CA, Peppelenbosch MP: Hedgehog: an unusual signal transducer, *Bioessays* 26:287-304, 2004.

Böttcher RT, Niehrs C: Fibroblast growth factor signaling during early vertebrate development, *Endocr Rev* 26:63-77, 2005.

Cadigan KM, Liu YI: Wnt signaling: complexity at the surface, *J Cell Sci* 119:395-402, 2006.

Coudreuse D, Korswagen HC: The making of Wnt: new insights into Wnt maturation, sorting and secretion, *Development* 134:3-12, 2007.

Cutforth T, Harrison CJ: Ephs and ephrins close ranks, *Trends Neurosci* 25:332-334, 2002.

DeRobertis EM, Oliver G, Wright CVE: Homeobox genes and the vertebrate body plan, *Sci Am* 263:46-52, 1990.

Deschamps J, van Nes J: Developmental regulation of the *Hox* genes during axial morphogenesis in the mouse, *Development* 132:2931-2942, 2005.

Dorey K, Amaya E: FGF signalling: diverse roles during early vertebrate embryogenesis, *Development* 137:3731-3742, 2010.

Duboule D, ed: *Guidebook to the homeobox genes*, Oxford, 1994, Oxford University Press.

Epstein CJ, Erickson RP, Wynshaw-Boris A: *Inborn errors of development*, Oxford, 2004, Oxford University Press.

Fortini ME: Notch signaling: the core pathway and its posttranslational regulation, *Dev Cell* 16:633-647, 2009.

Gehring WJ: Homeobox genes, the homeobox, and the spatial organization of the embryo, *Harvey Lect* 81:153-172, 1987.

Goodman FR and others: Human *HOX* gene mutations, *Clin Genet* 59:1-11, 2001.

Gordon MD, Nusse R: Wnt signaling: multiple pathways, multiple receptors, and multiple transcription factors, *J Biol Chem* 281:22429-22433, 2006.

Graham A, Papalopulu N, Krumlauf R: The murine and *Drosophila* homeobox gene complexes have common features of organization and expression, *Cell* 57:367-378, 1989.

Halder G, Johnson RL: Hippo signaling: growth control and beyond, *Development* 138:9-22, 2011.

Hofmann C, Eichele G: Retinoids in development. In Sporn MB and others, eds: *The retinoids: biology, chemistry, and medicine*, ed 2, New York, 1994, Raven, pp 387-441.

Hunter CS, Rhodes SJ: LIM-homeodomain genes in mammalian development and human disease, *Mol Biol Rep* 32:67-77, 2005.

Ingham PW, McMahon AP: Hedgehog signaling in animal development: paradigms and principles, *Genes Dev* 15:3059-3087, 2001.

Ingham PW, Placzek M: Orchestrating ontogenesis: variations on a theme by sonic hedgehog, *Nat Rev Genet* 7:841-850, 2006.

Kamachi Y, Uchikawa M, Kondoh H: Pairing Sox off with partners in the regulation of embryonic development, *Trends Genet* 16:182-187, 2000.

Karner C, Wharton KA, Carroll TJ: Apical-basal polarity, Wnt signaling and vertebrate organogenesis, *Semin Cell Dev Biol* 17:214-222, 2006.

Kawano Y, Kypta R: Secreted antagonists of the Wnt signalling pathway, *J Cell Sci* 116:2627-2634, 2003.

Ketting RF: The many faces of RNAi, *Dev Cell* 20:148-161, 2011.

Kiefer JC: Back to basics: *Sox* genes, *Dev Dyn* 236:2356-2366, 2007.

Kmita M, Duboule D: Organizing axes in time and space: 25 years of colinear tinkering, *Science* 301:331-333, 2003.

Kraus P, Lufkin T: Mammalian *Dlx* homeobox gene control of craniofacial and inner ear morphogenesis, *J Cell Biochem* 32/33(Suppl):133-140, 1999.

Lewis MA, Steel KP: MicroRNAs in mouse development and disease, *Semin Cell Dev Biol* 21:774-780, 2010.

Louvi A, Artavanis-Tsakonas S: Notch signaling in vertebrate neural development, *Nat Rev Neurosci* 7:93-102, 2006.

Maden M: The role of retinoic acid in embryonic and post-embryonic development, *Proc Nutr Soc* 59:65-73, 2000.

Maeda RK, Karch F: The ABC of the BX-C: the bithorax complex explained, *Development* 133:1413-1422, 2006.

Mallo M, Wellik DM, Deschamps J: *Hox* genes and regional patterning of the vertebrate body plan, *Dev Biol* 344:7-15, 2010.

Marikawa Y: Wnt/β-catenin signaling and body plan formation in mouse embryos, *Semin Cell Dev Biol* 17:175-184, 2006.

Mark M, Rijli FM, Chambon P: Homeobox genes in embryogenesis and pathogenesis, *Pediatr Res* 42:421-429, 1997.

McGinnis W, Krumlauf R: Homeobox genes and axial patterning, *Cell* 68:283-302, 1992.

Miyazono K, Maeda S, Imamura T: BMP receptor signaling: transcriptional targets, regulation of signals, and signaling cross-talk, *Cytokine Growth Factor Rev* 16:251-263, 2005.

Moens CB, Selleri L: Hox cofactors in vertebrate development, *Dev Biol* 291:193-206, 2006.

Morriss-Kay GM, Ward SJ: Retinoids and mammalian development, *Int Rev Cytol* 188:73-131, 1999.

Noordermeer D and others: The dynamic architecture of *Hox* gene clusters, *Science* 334:222-225, 2011.

O'Rourke MP, Tam PPL: *Twist* functions in mouse development, *Int J Dev Biol* 46:401-413, 2002.

Østerlund T, Kogerman P: Hedgehog signalling: how to get from Smo to Ci and Gli, *Trends Cell Biol* 16:176-180, 2006.

Papaioannou VE, Silver LM: The T-box gene family, *Bioessays* 20:9-19, 1998.

Patient RK, McGhee JD: The GATA family (vertebrates and invertebrates), *Curr Opin Genet Dev* 12:416-422, 2002.

Pearson JC, Lemons D, McGinnis W: Modulating *Hox* gene functions during animal body patterning, *Nat Rev Genet* 6:893-904, 2005.

Ross SA and others: Retinoids in embryonal development, *Physiol Rev* 80:1021-1054, 2000.

Saldanha G: The hedgehog signalling pathway and cancer, *J Pathol* 193:427-432, 2001.

Scott MP: Vertebrate homeobox nomenclature, *Cell* 71:551-553, 1992.

Shen MM: Nodal signaling: developmental roles and regulation, *Development* 134:1023-1034, 2007.

Simeone A, Puelles E, Acampora D: The Otx family, *Curr Opin Genet Dev* 12:409-415, 2002.

Su N, Blelloch R: Small RNAs in early mammalian development: from gametes to gastrulation, *Development* 138:1653-1661, 2011.

Tada M, Smith JC: T-targets: clues to understanding the functions of T-box proteins, *Dev Growth Differ* 43:1-11, 2001.

Thisse B, Thisse C: Functions and regulations of fibroblast growth factor signaling during embryonic development, *Dev Biol* 287:390-402, 2005.

Tian T, Meng AM: Nodal signals pattern vertebrate embryos, *Cell Mol Life Sci* 63:672-685, 2006.

Von Bubnoff A, Cho KWY: Intracellular BMP signaling regulation in vertebrates: pathway or network? *Dev Biol* 239:1-14, 2001.

Wehr R, Gruss P: Pax and vertebrate development, *Int J Dev Biol* 40:369-377, 1996.

Wilson M, Koopman P: Matching SOX: partner proteins and co-factors of the SOX family of transcriptional regulators, *Curr Opin Genet Dev* 12:441-446, 2002.

Wu MY, Hill CS: TGF-β superfamily signaling in embryonic development and homeostasis, *Dev Cell* 16:329-343, 2009.

Yanagita M: BMP antagonists: their roles in development and involvement in pathophysiology, *Cytokine Growth Factor Rev* 16:309-317, 2005.

Chapter 5

Formation of Germ Layers and Early Derivatives

As it is implanting into the uterine wall, the embryo undergoes profound changes in its organization. Up to the time of implantation, the blastocyst consists of the inner cell mass, from which the body of the embryo proper arises, and the outer trophoblast, which represents the future tissue interface between the embryo and mother. Both components of the blastocyst serve as the precursors of other tissues that appear in subsequent stages of development. Chapter 3 discusses the way in which the cytotrophoblast gives rise to an outer syncytial layer, the syncytiotrophoblast, shortly before attaching to uterine tissue (see Fig. 3.18). Not long thereafter, the inner cell mass begins to give rise to other tissue derivatives as well. The subdivision of the inner cell mass ultimately results in an embryonic body that contains the three primary embryonic germ layers: the **ectoderm** (outer layer), **mesoderm** (middle layer), and **endoderm** (inner layer). The process by which the germ layers are formed through cell movements is called **gastrulation**.

After the germ layers have been laid down, the continued progression of embryonic development depends on a series of signals called **embryonic inductions**, which are exchanged between the germ layers or other tissue precursors. In an inductive interaction, one tissue (the **inductor**) acts on another (**responding tissue**) so that the developmental course of the latter is different from what it would have been in the absence of the inductor. The developments that can be seen with a microscope during this period are tangible reflections of profound changes in gene expression and cellular properties of implanting embryos.

Two-Germ-Layer Stage

Just before the embryo implants into the endometrium early in the second week, significant changes begin to occur in the inner cell mass and in the trophoblast. As the cells of the inner cell mass become rearranged into an epithelial configuration, sometimes referred to as the **embryonic shield**, a thin layer of cells appears ventral to the main cellular mass (see Fig. 3.18). The main upper layer of cells is known as the **epiblast**, and the lower layer is called the **hypoblast**, or **primitive endoderm** (**Fig. 5.1**).

How the hypoblast forms in human embryos is not understood, but studies on mouse embryos have shown that as early as the 64-cell stage, some cells of the inner cell mass express

the transcription factor **nanog**, whereas others express **Gata 6**. These cells are arranged in a salt and pepper pattern within the inner cell mass (**Fig. 5.2A**). The nanog-expressing cells represent the precursors of the epiblast, and those expressing Gata 6 will become the hypoblast. The basis for the differentiation of these two distinct precursor cell types is not completely understood, but according to the "time inside–time outside" hypothesis, those cells that enter the inner cell mass earliest are biased to express nanog, which perpetuates their pluripotency. Possibly because of the influence of fibroblast growth factor-4 (FGF-4), secreted by these first arrivals to the inner cell mass, later immigrants are then biased to express Gata 6. The Gata 6–expressing cells produce molecules that increase their adhesive properties, as well as their mobility, and they make their way to the lower surface of the inner cell mass to form a thin epithelium, the hypoblast. Those Gata 6 cells that fail to reach the surface of the inner cell mass undergo **apoptosis** (cell death). The nanog-expressing cells of the inner cell mass also assume an epithelial configuration as they form the epiblast. Between the epiblast and hypoblast a basal lamina forms.

A small group of hypoblast cells that becomes translocated to the future anterior end of the embryo (called **anterior visceral endoderm** by mouse embryologists) has been shown to possess remarkable signaling powers. The cells first secrete the signaling molecules, **lefty-1** and **Cerberus-1 (Cer-1)**, which inhibit the activity of the signaling molecules, **nodal** and **Wnt**, in the overlying epiblast but allow nodal and Wnt-3 expression in the posterior epiblast (see Fig. 5.8A). (Nodal signaling from the posterior epiblast stimulates the initial formation of the anterior visceral endoderm.) This represents the first clear expression of anteroposterior polarity in the embryo. It also forms two signaling domains within the early embryo. The anterior visceral endoderm soon begins to induce much of the head and forebrain and inhibits the formation of posterior structures. In the posterior part of the epiblast, nodal signaling activity stimulates the formation of the primitive streak (see next section), which is the focal point for gastrulation and germ layer formation. After the hypoblast has become a well-defined layer, and the epiblast has taken on an epithelial configuration, the former inner cell mass is transformed into a **bilaminar disk**, with the epiblast on the dorsal surface and the hypoblast on the ventral surface.

The epiblast contains the cells that make up the embryo itself, but extraembryonic tissues also arise from this layer. The

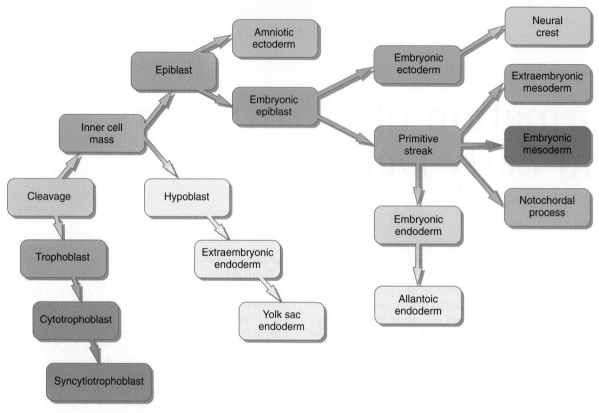

Fig. 5.1   Cell and tissue lineages in the mammalian embryo. (*Note*: The colors in the boxes are found in all illustrations involving the embryonic and extraembryonic germ layers.)

next layer to appear after the hypoblast is the **amnion**, a layer of extraembryonic ectoderm that ultimately encloses the entire embryo in a fluid-filled chamber called the **amniotic cavity** (see Chapter 7). Because of the paucity of specimens, the earliest stages in the formation of the human amnion and amniotic cavity are not completely understood. Studies on primate embryos indicate that a primordial amniotic cavity first arises by **cavitation** (formation of an internal space) within the pre-epithelial epiblast; it is covered by cells derived from the inner cell mass (see Fig. 5.2). According to some investigators, the roof of the amnion then opens, thus exposing the primordial amniotic cavity to the overlying cytotrophoblast. Soon thereafter (by about 8 days after fertilization), the original amniotic epithelium reforms a solid roof over the amniotic cavity.

While the early embryo is still sinking into the endometrium (about 9 days after fertilization), cells of the hypoblast begin to spread and line the inner surface of the cytotrophoblast with a continuous layer of extraembryonic endoderm called **parietal endoderm** (**Fig. 5.3**; see Fig. 5.2). When the endodermal spreading is completed, a vesicle called the **primary yolk sac** has taken shape (see Fig. 3.18C). At this point (about 10 days after fertilization), the embryo complex constitutes the bilaminar germ disk, which is located between the primary yolk sac on its ventral surface and the amniotic cavity on its dorsal surface (**Fig. 5.4**). Shortly after it forms, the primary yolk sac becomes constricted, forming a secondary yolk sac and leaving behind a remnant of the primary yolk sac (see Figs. 3.18D and 5.2F).

Starting at about 12 days after fertilization, another extraembryonic tissue, the **extraembryonic mesoderm**, begins to appear (see Fig. 5.2). The first extraembryonic mesodermal cells seem to arise from a transformation of parietal endodermal cells. These cells are later joined by extraembryonic mesodermal cells that have originated from the primitive streak. The extraembryonic mesoderm becomes the tissue that supports the epithelium of the amnion and yolk sac and the **chorionic villi**, which arise from the trophoblastic tissues (see Chapter 7). The support supplied by the extraembryonic mesoderm is not only mechanical, but also trophic because the mesoderm serves as the substrate through which the blood vessels supply oxygen and nutrients to the various epithelia.

Gastrulation and the Three Embryonic Germ Layers

At the end of the second week, the embryo consists of two flat layers of cells: the epiblast and the hypoblast. As the third week of pregnancy begins, the embryo enters the period of gastrulation, during which the three embryonic germ layers form from the epiblast (see Fig. 5.1). The morphology of human gastrulation follows the pattern seen in birds. Because of the large amount of yolk in birds' eggs, the avian embryo forms the primary germ layers as three overlapping flat disks that rest on the yolk, similar to a stack of pancakes. Only later do the germ layers fold to form a cylindrical body. Although the

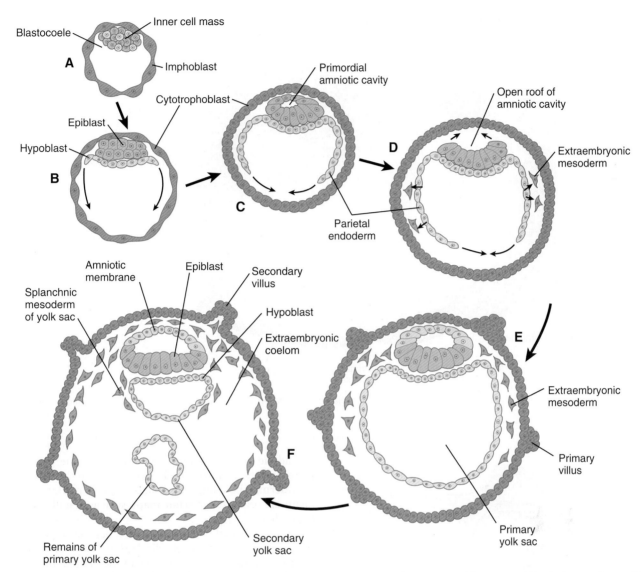

Fig. 5.2 Origins of the major extraembryonic tissues. The syncytiotrophoblast is not shown. **A,** Late blastocyst. Within the inner cell mass, blue nanog-expressing pre-epiblastic cells and yellow Gata 6–expressing prehypoblastic cells are mixed in a salt and pepper pattern. **B,** Beginning of implantation at 6 days. The hypoblast has formed and is beginning to spread beneath the cytotrophoblast as the parietal endoderm. **C,** Implanted blastocyst at 7½ days. **D,** Implanted blastocyst at 8 days. **E,** Embryo at 9 days. **F,** Late second week.

mammalian egg is essentially devoid of yolk, the morphological conservatism of early development still constrains the human embryo to follow a pattern of gastrulation similar to that seen in reptiles and birds. Because of the scarcity of material, even the morphology of gastrulation in human embryos is not known in detail. Nevertheless, extrapolation from avian and mammalian gastrulation can provide a reasonable working model of human gastrulation.

Gastrulation begins with the formation of the **primitive streak**, a linear midline condensation of cells derived from the epiblast in the posterior region of the embryo through an induction by cells at the edge of the embryonic disk in that region (see Fig. 5.4). Members of the transforming growth factor-β (TGF-β) and Wnt families of signaling molecules have been identified as likely inducing agents. Initially triangular, the primitive streak soon becomes linear and elongates, largely through a combination of proliferation and migration, as well as internal cellular rearrangements, called

convergent-extension movements. With the appearance of the primitive streak, the anteroposterior (craniocaudal) and right-left axes of the embryo can be readily identified (see Fig. 5.4).

The primitive streak is a region where cells of the epiblast converge in a well-defined spatial and temporal sequence. As cells of the epiblast reach the primitive streak, they change shape and pass through it on their way to forming new layers beneath (ventral to) the epiblast (**Fig. 5.5C**). Marking studies have shown that cells entering the primitive streak form distinct lineages as they leave. The most posterior cells both to enter and leave the streak as it is beginning to elongate form the **extraembryonic mesoderm** lining the trophoblast and yolk sac, as well as that forming the **blood islands** (see Fig. 6.19). Another wave of mesoderm, arising later and more anteriorly in the primitive streak, forms the **paraxial**, **lateral plate**, and **cardiac mesoderm**. A final wave, which enters and leaves the anteriormost end of the primitive streak, gives rise

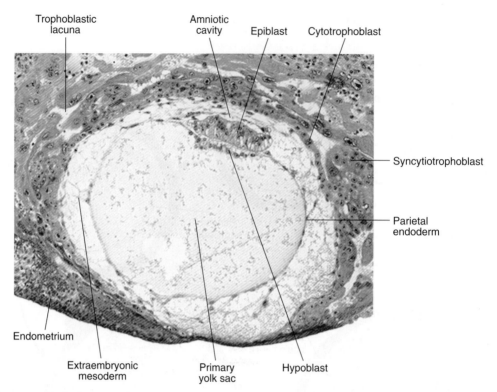

Trophoblastic lacuna

Amniotic cavity

Epiblast

Cytotrophoblast

Syncytiotrophoblast

Parietal endoderm

Endometrium

Extraembryonic mesoderm

Primary yolk sac

Hypoblast

Fig. 5.3 Digital photomicrograph of a 12-day human embryo (Carnegie No. 7700) taken just as implantation within the endometrium is completed. *(Courtesy of Dr. Ray Gasser.)*

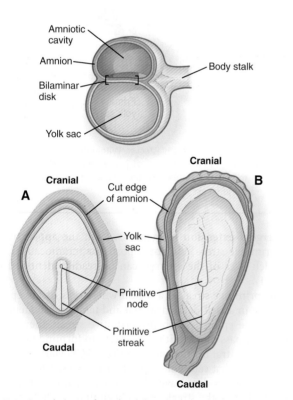

Amniotic cavity

Amnion

Bilaminar disk

Yolk sac

Body stalk

Cranial

Cranial

A

B

Cut edge of amnion

Yolk sac

Primitive node

Primitive streak

Caudal

Caudal

Fig. 5.4 Dorsal views of 16-day (**A**) and 18-day (**B**) human embryos. *Top,* Sagittal section through an embryo and its extraembryonic membranes during early gastrulation.

to midline axial structures (the **notochord**, the **prechordal plate,** and the **primitive node** itself) and also the **embryonic endoderm**. The composite results of such marking experiments are organized into **fate maps**, such as that illustrated in Figure 5.5A.

The endodermal precursor cells that pass through the anterior primitive streak largely displace the original hypoblast, but research has shown that some of the original hypoblastic cells become integrated into the newly forming embryonic endodermal layer. The displaced hypoblastic cells form extraembryonic endoderm. The movement of cells through the primitive streak results in the formation of a groove (**primitive groove**) along the midline of the primitive streak. At the anterior end of the primitive streak is a small but well-defined accumulation of cells, called **primitive node**, or **Hensen's node**.* This structure is of great developmental significance because, in addition to being the major posterior signaling center of the embryo (**Box 5.1**), it is the area through which cells migrate in a stream toward the anterior end of the embryo. These cells, called **mesendoderm**, soon segregate into a rodlike mesodermal **notochord** and the endodermal dorsal wall of the forming gut. Anterior to the notochord is a group of mesodermal cells called the **prechordal plate** (see Fig. 5.5A and B). (The important functions of the notochord and prechordal plate are discussed on p. 80.)

The specific craniocaudal characteristics of the structures arising from the newly formed paraxial mesoderm are specified by patterns of *Hox* gene expression, first in the

---

*Hensen's node is the commonly used designation for the primitive node in avian embryos, but this term is sometimes used in the mammalian embryological literature as well. This structure is the structural and functional equivalent of the dorsal lip of the blastopore in amphibians.

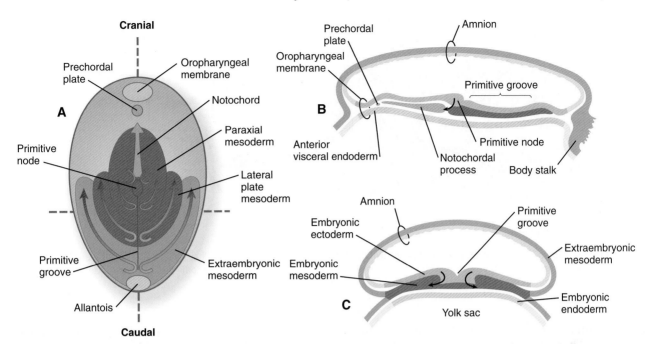

Fig. 5.5 **A,** Dorsal view through a human embryo during gastrulation. *Arrows* show the directions of cellular movements across the epiblast toward, through, and away from the primitive streak as newly formed mesoderm. The illustrated fates of the cells that have passed through the primitive streak are based on studies of mouse embryos. **B,** Sagittal section through the craniocaudal axis of the same embryo. The *curved arrow* indicates cells passing through the primitive node into the notochord. **C,** Cross section through the level of the primitive streak in **A** *(dashed lines).*

epiblast and then in the mesodermal cells themselves. The transformations of morphology and the behavior of the cells passing through the primitive streak are associated with profound changes not only in their adhesive properties and internal organization, but also in the way that they relate to their external environment. Much of the extraembryonic mesoderm forms the **body stalk**, which connects the caudal part of the embryo to the extraembryonic tissues that surround it (see Figs. 5.4 and 7.1). The body stalk later becomes the umbilical cord.

The movements of the cells passing through the primitive streak are accompanied by major changes in their structure and organization (**Fig. 5.6**). While in the epiblast, the cells have the properties of typical epithelial cells, with well-defined apical and basal surfaces, and they are associated with a basal lamina that underlies the epiblast. As they enter the primitive streak, these cells elongate, lose their basal lamina, and take on a characteristic morphology that has led to their being called **bottle cells**. When they become free of the epiblastic layer in the primitive groove, the bottle cells assume the morphology and characteristics of **mesenchymal cells**, which are able to migrate as individual cells if they are provided with the proper extracellular environment (see Fig. 5.6). Included in this transformation is the loss of specific cell adhesion molecules (CAMs), in particular **E-cadherin** (see p. 254) as the cells convert from an epithelial to a mesenchymal configuration. This transformation is correlated with the expression of the transcription factor **snail**, which is also active in the separation of mesenchymal neural crest cells from the epithelial neural tube (see p. 254). As cells in the epiblast are undergoing **epithelial-mesenchymal transition**, they begin to express the CAM **N-cadherin**, which is necessary for their spreading out from the primitive streak in the newly forming mesodermal layer.

Starting in early gastrulation, cells of the epiblast produce **hyaluronic acid**, which enters the space between the epiblast and hypoblast. Hyaluronic acid, a polymer consisting of repeating subunits of D-glucuronic acid and N-acetylglucosamine, is frequently associated with cell migration in developing systems. The molecule has a tremendous capacity to bind water (up to 1000 times its own volume), and it functions to keep mesenchymal cells from aggregating during cell migrations. Although after leaving the primitive streak the mesenchymal cells of the embryonic mesoderm find themselves in a hyaluronic acid–rich environment, hyaluronic acid alone is not enough to support their migration from the primitive streak. In all vertebrate embryos that have been investigated to date, the spread of mesodermal cells away from the primitive streak or the equivalent structure is found to depend on the presence of **fibronectin** associated with the basal lamina beneath the epiblast. The embryonic mesoderm ultimately spreads laterally as a thin sheet of mesenchymal cells between the epiblast and hypoblast layers (see Fig. 5.5C).

By the time the mesoderm has formed a discrete layer in the human embryo, the upper germ layer (remains of the former epiblast) is called the **ectoderm**, and the lower germ layer, which has displaced the original hypoblast, is called the **endoderm**. This terminology is used for the remainder of this text. As the three definitive germ layers are taking shape, **bone morphogenetic protein-4** (**BMP-4**) signals, arising from extraembryonic tissues at the caudal end of the embryo, stimulate a group of cells in the posterior region of the epiblast to become transformed into **primordial germ cells.**

Regression of the Primitive Streak

After its initial appearance at the extreme caudal end of the embryo, the primitive streak expands cranially until about 18

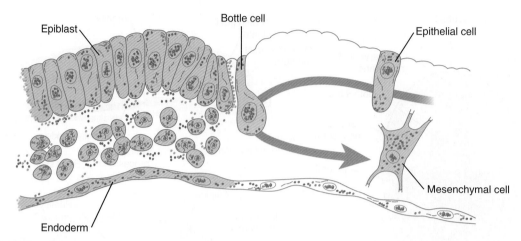

Fig. 5.6 **Cross-sectional view of an embryo during gastrulation.** Changes in the shape of a cell as it migrates along the epiblast (epithelium), through the primitive streak (bottle cell), and away from the groove as a mesenchymal cell that will become part of the mesodermal germ layer. The same cell can later assume an epithelial configuration as part of a somite.

days after fertilization (see Fig. 5.4). Thereafter, it regresses caudally (see Fig. 5.11) and strings out the notochord in its wake. Vestiges remain into the fourth week. During that time, the formation of mesoderm continues by means of cells migrating from the epiblast through the primitive groove. Regression of the primitive streak is accompanied by the establishment and patterning of the paraxial mesoderm (see p. 97), which gives rise to the somites and ultimately the segmental axial structures of the trunk and caudal regions of the body. As regression of the primitive streak comes to a close, its most caudal extent is marked by a mass of mesenchymal cells, which form the **tail bud**. This structure plays an important role in forming the most posterior portion of the neural tube (see p. 93).

The primitive streak normally disappears without a trace, but in rare instances, large tumors called teratomas appear in the sacrococcygeal region (see Fig. 1.2A). Teratomas often contain bizarre mixtures of many different types of tissue, such as cartilage, muscle, fat, hair, and glandular tissue. Because of this, sacrococcygeal teratomas are thought to arise from remains of the primitive streak (which can form all germ layers). Teratomas also are found in the gonads and the mediastinum. These tumors are thought to originate from germ cells.

Notochord and Prechordal Plate

The notochord, the structure that is the basis for giving the name *Chordata* to the phylum to which all vertebrates belong, is a cellular rod running along the longitudinal axis of the embryo just ventral to the central nervous system. Although phylogenetically and ontogenetically it serves as the original longitudinal support for the body, the notochord also plays a crucial role as a prime mover in a series of signaling episodes (inductions) that transform unspecialized embryonic cells into definitive tissues and organs. In particular, inductive signals from the notochord (1) stimulate the conversion of overlying surface ectoderm into neural tissue, (2) specify the identity of certain cells (floor plate) within the early nervous system, (3) transform certain mesodermal cells of the somites into vertebral bodies, and (4) stimulate the earliest steps in the development of the dorsal pancreas.

Cranial to the notochord is a small region where embryonic ectoderm and endoderm abut without any intervening mesoderm. Called the **oropharyngeal membrane** (see Fig. 5.5), this structure marks the site of the future oral cavity. Between the cranial tip of the notochordal process and the oropharyngeal membrane is a small aggregation of mesodermal cells closely apposed to endoderm, called the **prechordal plate** (see Fig. 5.5). In birds, the prechordal plate emits molecular signals that are instrumental in stimulating the formation of the forebrain, similar to the anterior visceral endoderm in mammals.

Both the prechordal plate and the notochord arise from the ingression of a population of epiblastic cells, which join other cells of primitive streak origin, within the primitive node. As the primitive streak regresses, the cellular precursors of first the prechordal plate and then the notochord migrate rostrally from the node, but they are left behind as a rodlike aggregation of cells (**notochordal process**; see Fig. 5.5A and B) in the wake of the regressing primitive streak. In mammals, shortly after ingression, the cells of the notochordal process temporarily spread out and fuse with the embryonic endoderm (**Fig. 5.7**). The result is the formation of a transitory **neurenteric canal** that connects the emerging amniotic cavity with the yolk sac. Later, the cells of the notochord separate from the endodermal roof of the yolk sac and form the definitive notochord, a solid rod of cells in the midline between the embryonic ectoderm and endoderm (see Fig. 5.7).

Induction of the Nervous System

Neural Induction

The inductive relationship between the notochord (**chordamesoderm**) and the overlying ectoderm in the genesis of the nervous system was recognized in the early 1900s. Although the original experiments were done on amphibians, similar experiments in higher vertebrates have shown that the essential elements of **neural** (or **primary**) **induction** are the same in all vertebrates.

Deletion and transplantation experiments in amphibians set the stage for the present understanding of neural induction. (See Chapters 6 and 11 for further details on the

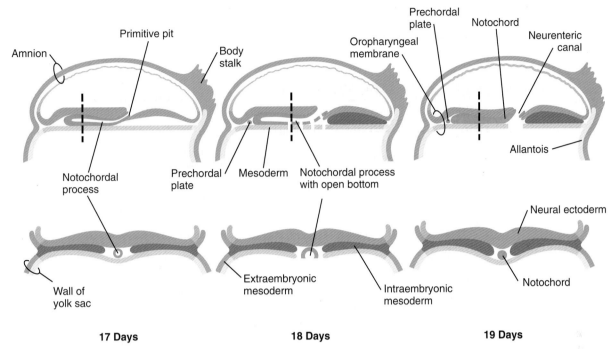

Fig. 5.7 *From left to right,* sequential stages in formation of the notochord. *Top,* Sagittal sections. *Bottom,* Cross sections at the level of the vertical line in the *upper figure.* In the *upper row,* the cranial end is on the *left.* The function of the neurenteric canal remains obscure.

Box 5.1 Molecular Aspects of Gastrulation

Many decades of research on birds and amphibians have resulted in a reasonable understanding of the cellular and molecular aspects of gastrulation in these species. More recent research suggests that despite some species differences, the basic aspects of gastrulation in mammals are fundamentally similar to those in birds.

The events of gastrulation are guided by a series of molecular inductions emanating from a succession of signaling centers, starting with the anterior visceral endoderm and progressing to the future caudal (posterior) part of the embryo. Early posterior signaling results in the formation of the primitive streak and the induction of mesoderm. When the primitive streak is established, the primitive node takes over as the center that organizes the fundamental structure of the body axis. As the notochord takes shape from cells that flow through the primitive node, it becomes an important signaling center. In humans, the role of the cells of the **prechordal plate** is not well understood. In birds, the prechordal plate acts as an anterior signaling center, similar to the anterior visceral endoderm in mice. Whether anterior signaling in humans is confined to anterior hypoblast (anterior visceral endoderm) or the prechordal plate, or both, remains to be determined.

Establishment of the Anterior Visceral Endoderm and Induction of the Primitive Streak (The Early Gastrula Organizer)

For this aspect of early development, we must rely almost entirely on studies of mouse embryos. The original symmetry of the embryo is broken by the displacement of the future anterior visceral endoderm to the anterior side of the embryonic disk. This is a function of proliferation and later migration of the cells that will constitute the anterior visceral endoderm. Migration of these cells (and the

resulting establishment of the anteroposterior axis) depends on the activation of the Wnt antagonist **Dkk 1 (Dickkopf 1)** in the future anterior part of the embryo. This confines **Wnt** activity to the future posterior part of the embryo, where it induces the expression of the signaling molecule **Nodal** (**Fig. 5.8A**). When the anterior visceral endoderm has become stabilized in the anterior part of the embryonic disk, it produces the Nodal inhibitors **lefty-1** and **Cer-1**, which confine Nodal activity to the posterior end of the embryo where, responding to extraembryonic Wnt signals, it establishes a posterior signaling center, which induces the formation of the primitive streak, the definitive endoderm, and the mesoderm. In the chick embryo, the ectopic application of two other signaling molecules, **chordin** and **Vg1**, induces the formation of an ectopic primitive streak.

Primitive Node (Organizer)

As the primitive streak elongates, migrating cells of the epiblast join the tip of the streak, and a dynamic mass of cells, called the **primitive node**, becomes evident at the tip of the primitive streak. Cells of the node express many genes, including three classic molecular markers of the organizer region in many vertebrates—**chordin, goosecoid,** and **hepatic nuclear factor-3β** (now called **Foxa-2**). Not only is the winged helix transcription factor, Foxa-2, important for the formation of the node itself, but also it is vital for the establishment of midline structures cranial to the node. Foxa-2 is required for the initiation of notochord function. In its absence, the notochord and the floor plate of the neural tube (see Chapter 11) fail to form. In contrast, endoderm, the primitive streak, and intermediate mesoderm do develop. **Goosecoid,** a homeodomain transcription factor, is prominently expressed in the

Continued

Box 5.1 Molecular Aspects of Gastrulation—cont'd

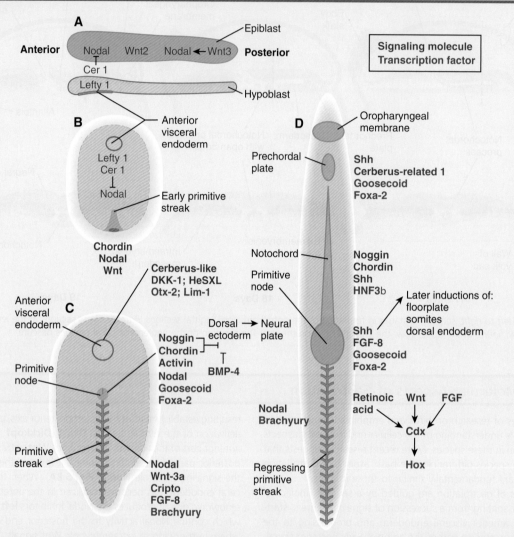

Fig. 5.8 Summary of major genes involved in various stages of early embryonic development. A, Preprimitive streak (sagittal section). B, Early formation of the primitive streak. C, Gastrulation (period of germ layer formation). D, Late gastrulation and neural induction. The molecules in *red* are signaling molecules, and the molecules in *blue* are transcription factors. Names of specific molecules *(bold)* are placed by the structures in which they are expressed.

organizer region of all vertebrates studied. Goosecoid activates *chordin, noggin,* and other genes of the organizer region. If ectopically expressed, it stimulates the formation of a secondary body axis. **Chordin** and **noggin**, signaling molecules associated with the node, are involved with neural induction, and expression of **nodal** on the left side of the embryo is a key element in the setting of left-right asymmetry.

Two genes, *T* and *nodal,* play prominent roles in the function of the primitive streak and posterior mesoderm formation. Expression of the ***T* gene** seems to be activated by products of the *Foxa-2* and *goosecoid* genes. In *T* mutants (**brachyury**), the notochord begins to form through the activity of Foxa-2, but it fails to complete development. Studies on *T* mutants have shown that activity of the *brachyury* gene is necessary for normal movements of future mesodermal cells through the primitive streak during gastrulation. In brachyury (short tail) mutant mice, mesodermal cells pile up at a poorly formed primitive streak, and the embryos show defective elongation of the body axis (including a short tail) posterior to the

forelimbs. *T* gene mutants may be responsible for certain gross caudal body defects in humans. ***Nodal****,* a member of the transforming growth factor-β (TGF-β) family of growth factor genes (see Table 4.1), is expressed throughout the posterior epiblast before gastrulation, but its activity is concentrated at the primitive node during gastrulation. Similar to the *brachyury* gene, the effects of *nodal* are strongly seen in the caudal region of the embryo. In the null mutant of *nodal,* the primitive streak fails to form, and the embryo is deficient in mesoderm. Similarly, mutants of *cripto,* an early-acting member of the epidermal growth factor family and an essential cofactor in the *nodal* signaling pathway, produce a trunkless phenotype.

As cells pass through the primitive streak, a region of *Hox* gene expression begins to form around the streak. The pattern of *Hox* gene expression in the future trunk and posterior part of the embryo is based on signaling by three molecules—retinoic acid, Wnt, and FGF—that act on the transcription factor Cdx (the mammalian equivalent of caudal in *Drosophila*) in the area of the

Box 5.1 Molecular Aspects of Gastrulation—cont'd

regressing primitive streak just behind the last-forming somites. Cdx acts on the *Hox* genes, which impose unique characteristics to the segmental structures that form along the anteroposterior axis of the embryo (Fig. 5.8D).

Prechordal Plate and Notochord

The first cells passing through the primitive node form a discrete midline mass of cells, the **prechordal plate**, which is closely associated with endoderm in the region just caudal to the oropharyngeal membrane. The next generation of cells passing through the node forms the notochord.

The **notochord** is a major axial signaling center of the trunk in the early embryo, and it is important in the formation of many axial structures. Under the influence of Foxa-2 and goosecoid, cells of the forming notochord produce **noggin** and **chordin**, molecules known to be potent neural inducers in many species. The notochord also produces **sonic hedgehog** (shh), the effector molecule for many notochordal inductions of axial structures after the neural plate is induced. Despite inducing the neural plate within the overlying ectoderm, however, the notochord does not stimulate the formation of anterior parts of the brain or head structures. This function is reserved for the anterior visceral endoderm.

The **prechordal plate**, sometimes called the **head organizer**, consists of early mesendodermal cells passing through the primitive node. These cells are structurally and functionally closely associated with cells of the underlying anterior endoderm. Along with anterior visceral endoderm (see later), the prechordal plate is the source of important signals, especially shh, that are involved in ventral patterning of the forebrain. In addition, the prechordal plate is the source of signals that are important for the survival of neural crest cells that emigrate from the early forebrain.

Anterior Visceral Endoderm (Hypoblast)

In mammals, even before mesodermal cells begin migrating through the primitive node, the anterior hypoblast (called the **anterior visceral endoderm** by mouse embryologists) expresses genes characteristic of the prechordal plate and initiates head formation. The anterior visceral endoderm itself is subdivided into an anterior part, which serves as a signaling center for early heart formation (see p. 104), and a more posterior part, which becomes part of the prechordal plate complex and induces formation of the head. According to one model, induction of the head and forebrain in mammals is a two-step process, in which an early induction by the anterior visceral endoderm confers a labile anterior character to the head and brain, and a later induction by the prechordal plate mesoderm reinforces and maintains this induction.

A major function of the anterior visceral endoderm is to emit molecular signals that inhibit the development of posterior embry-

onic structures. To produce a head, it is necessary to block the bone morphogenetic protein-4 (BMP-4) signal (by noggin) and a Wnt signal (by Dkk-1). Signaling molecules and transcription factors are produced in the head signaling centers. In mice bearing mutants of **Lim-1** (Lhx-1), a homeobox-containing transcription factor, and **cereberus-like 1**, a signaling molecule, headless mice are born (**Fig. 5.9**). The headless mice are born without neural structures anterior to rhombomere 3 (see Fig. 6.3). Otx-2, another transcription factor present in the head signaling center, is also a general marker of the induced anterior region of the central nervous system. Many other molecules are also expressed in the head signaling center. How these orchestrate the formation of the head remains to be determined.

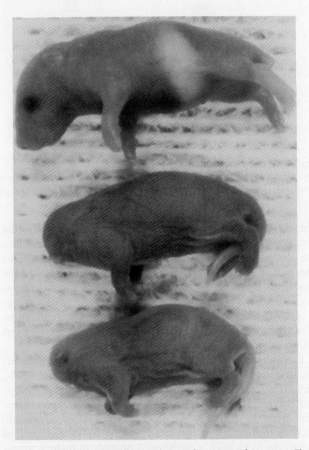

Fig. 5.9 Newborn headless mice and a normal mouse. The headless mice have a null mutant of the *Lim-1* gene. *(From Shawlot W, Behringer RR: Nature 374:425-430, 1994.)*

formation of the nervous system.) In the absence of chordamesoderm moving from the dorsal lip of the blastopore (the amphibian equivalent of the primitive node), the nervous system does not form from the dorsal ectoderm. In contrast, if the dorsal lip of the blastopore is grafted beneath the belly ectoderm of another host, a secondary nervous system and body axis form in the area of the graft (**Fig. 5.10**).

The dorsal lip has been called the **organizer** because of its ability to stimulate the formation of a secondary body axis. Subsequent research has shown that the interactions occurring in the region of the dorsal lip in amphibians are far more complex than a single induction between chordamesoderm and ectoderm. Deletion and transplantation experiments have also been conducted on embryos of birds and mammals (see

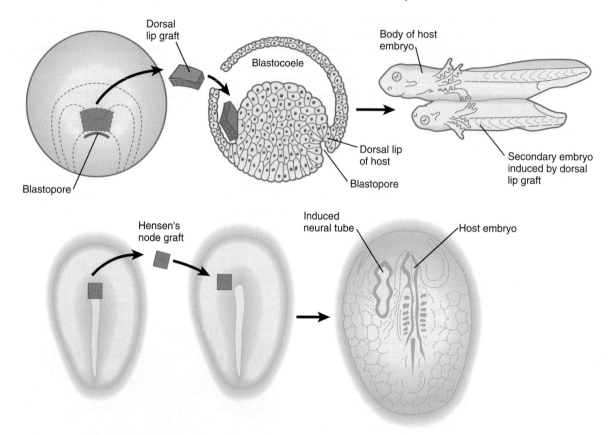

Fig. 5.10 **Early experiments showing neural induction.** *Top,* Graft of the dorsal lip of the blastopore in a salamander embryo induces a secondary embryo to form. *Bottom,* Graft of Hensen's node from one avian embryo to another induces the formation of a secondary neural tube. *(Top based on studies by Spemann H:* Embryonic development and induction, *New York, 1938, Hafner;* Bottom based on studies by Waddington C: J Exp Biol *10:38-46, 1933.)*

Fig. 5.10); clearly, the primitive node and the notochordal process in birds and mammals are homologous in function to the dorsal lip and chordamesoderm in amphibians. This means that, in higher vertebrates, the primitive node and the notochordal process act as the neural inductor, and the overlying ectoderm is the responding tissue. Over the years, embryologists have devoted an enormous amount of research to identifying the nature of the inductive signal that passes from the chordamesoderm to the ectoderm.

Early attempts to uncover the nature of the inductive stimulus were marked by great optimism. As early as the 1930s, various laboratories had proposed that molecules as diverse as proteins and steroids were the inductive stimulus. Soon thereafter came the discovery that an even wider variety of stimuli, such as inorganic ions or killed tissues, could elicit neural induction. With such a plethora of possible inductors, attention turned to the properties of the responding tissue (the dorsal ectoderm) and ways that it could react, through a final common pathway, to the inductive stimulus. The quest for the neural inductive molecules and their mode of action has been arduous and frustrating, with many blind alleys and wrong turns along the way.

Many laboratories found that isolated ectoderm could respond in vitro to inductive stimuli and become transformed into neural tissue. A useful technique for studying induction in vitro involved separating the responding tissue from the inducing tissue by a filter with pores that permitted the passage of molecules, but not cells. This technique has been used in the analysis of a variety of mammalian inductive systems.

Various experimental manipulations have shown that neural induction is not a simple all-or-nothing process. Rather, considerable regional specificity exists. (For example, certain artificial inductors stimulate the formation of more anterior neural structures, and others stimulate the formation of more posterior ones.) In amphibian embryos, anterior chordamesoderm has inducing properties different from those of posterior chordamesoderm.

More recent research has identified specific molecules that bring about neural induction. In amphibians, three signaling molecules—**noggin**, **follistatin**, and **chordin**—given off by the notochord are the inductive agents. It was first thought that these molecules directly stimulate uncommitted cells of the dorsal ectoderm to form neural tissue, but subsequent research on amphibians has shown that these inductors act by blocking the action of an inhibitor, **BMP-4**, in the dorsal ectoderm. In the absence of BMP-4 activity, dorsal ectoderm forms neural tissue as a default state.

In mammals, our current understanding of neural induction presents a more complex picture, with both the location and timing of inductive interactions playing a role in defining the initiation and organization of the central nervous system. According to a more contemporary model, during the early primitive streak stage, the precursor of the primitive node, called the **gastrula organizer**, secretes **Cer-1**, a BMP inhibitor. In the absence of BMP activity, the anterior epiblast is induced to become anterior neural tissue by default. At subsequent stages of gastrulation the anterior character of the induced neural tissue is maintained first through signals emanating

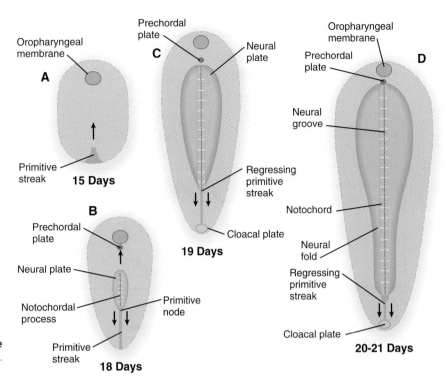

Fig. 5.11 **Relationships between the neural plate and primitive streak.** A, Day 15. B, Day 18. C, Day 19. D, Days 20 to 21.

from the anterior visceral endoderm (or its equivalent in the human) and then by signaling from anterior mesendoderm (the notochord and prechordal plate). The signals are Cer-1, a BMP inhibitor, and lefty-1, an inhibitor of nodal, which exerts a posteriorizing influence.

As gastrulation proceeds and the primitive node takes shape, the node induces epiblast to form neural tissue through a similar BMP-inhibition mechanism. This induced neural tissue is posteriorized through the action of nodal, which is concentrated in the posterior end of the embryo.

Early Formation of the Neural Plate

The first obvious morphological response of the embryo to neural induction is the transformation of the dorsal ectoderm overlying the notochordal process into an elongated patch of thickened epithelial cells called the **neural plate** (**Fig. 5.11**). The border of the neural plate is specified by exposure of those cells to a certain concentration of BMP. This is the region from which the neural crest (see p. 254) arises.

With the formation of the neural plate, the ectodermal germ layer becomes subdivided into two developmental lineages: neural and non-neural. This example illustrates several fundamental developmental concepts: **restriction**, **determination**, and **differentiation**. The zygote and blastomeres resulting from the first couple of cleavage divisions are **totipotent** (i.e., capable of forming any cell in the body).

As development progresses, certain decisions are made that narrow the developmental options of cells (**Fig. 5.12**). For example, at an early stage in cleavage, some cells become committed to the extraembryonic trophoblastic line and are no longer capable of participating in the formation of the embryo itself. At the point at which cells are committed to becoming trophoblast, a restriction event has occurred. When a group of cells has gone through its last restriction event (e.g., the transition from cytotrophoblast to syncytiotrophoblast), their

fate is fixed, and they are said to be determined.* These terms, which were coined in the early days of experimental embryology, are now understood to reflect limitations in gene expression as cell lineages follow their normal developmental course. The rare instances in which cells or tissues strongly deviate from their normal developmental course, a phenomenon called **metaplasia**, are of considerable interest to pathologists and individuals who study the control of gene expression.

Restriction and determination signify the progressive limitation of the developmental capacities in the embryo. Differentiation describes the actual morphological or functional expression of the portion of the genome that remains available to a particular cell or group of cells. Differentiation commonly connotes the course of phenotypic specialization of cells. One example of differentiation occurs in spermatogenesis, when spermatogonia, relatively ordinary-looking cells, become transformed into highly specialized spermatozoa.

Cell Adhesion Molecules

In the early 1900s, researchers determined that suspended cells of a similar type have a strong tendency to aggregate. If different types of embryonic cells are mixed together, they typically sort according to tissue type. Their patterns of sorting even give clues to their properties and behavior in the mature organism. For example, if embryonic ectodermal and mesodermal cells are mixed, they come together into an aggregate with a superficial layer of ectodermal cells surrounding a central aggregate of mesodermal cells.

Contemporary research has provided a molecular basis for many of the cell aggregation and sorting phenomena described by earlier embryologists. Of several families of **CAMs** that

---

*The term **specified** (**specification**) is becoming increasingly used as a near synonym to determination in referring to the fixation of the future fate of a cell.

Potential of Cells

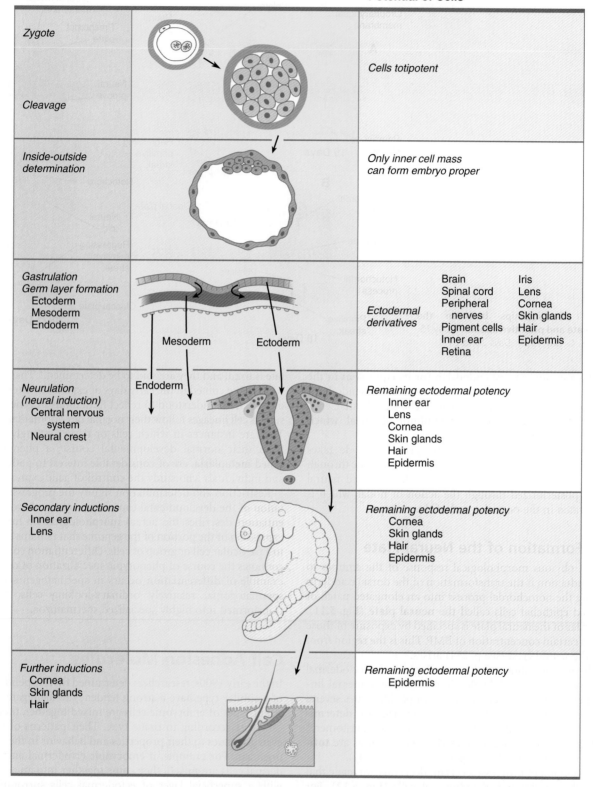

Zygote		
		Cells totipotent
Cleavage		
Inside-outside determination		*Only inner cell mass can form embryo proper*

Gastrulation
Germ layer formation
  Ectoderm
  Mesoderm
  Endoderm

Ectodermal derivatives

Brain	Iris
Spinal cord	Lens
Peripheral	Cornea
nerves	Skin glands
Pigment cells	Hair
Inner ear	Epidermis
Retina	

Mesoderm   Ectoderm

Endoderm

Neurulation
(neural induction)
  Central nervous
    system
  Neural crest

Remaining ectodermal potency
  Inner ear
  Lens
  Cornea
  Skin glands
  Hair
  Epidermis

Secondary inductions
  Inner ear
  Lens

Remaining ectodermal potency
  Cornea
  Skin glands
  Hair
  Epidermis

Further inductions
  Cornea
  Skin glands
  Hair

Remaining ectodermal potency
  Epidermis

Fig. 5.12 **Restriction during embryonic development.** The labels on the *right* illustrate the progressive restriction of the developmental potential of cells that are in the line leading to the formation of the epidermis. On the *left* are developmental events that remove groups of cells from the epidermal track.

Box 5.2   Molecular Basis for Left-Right Asymmetry

Up to the time of gastrulation, the embryo is bilaterally symmetrical, but at that time mechanisms are set in place that ultimately result in the right-sided looping of the heart, followed by asymmetric looping of the gut and the asymmetric positioning of the liver, spleen, and lobation of the lungs. In mammalian embryos, the earliest known manifestation of asymmetry involves the beating of cilia around the primitive node (**Fig. 5.13**). This beating results in a directional current leading to the expression of two signaling molecules in the transforming growth factor-β (TGF-β) family—**nodal**, a symmetry-breaking molecule in the left side of the embryo, and **lefty-1** along the left side of the primitive streak—over a very restricted developmental time (from the two-somite to the six-somite stage in the mouse). Lefty-1 may function to prevent the diffusion of left-determining molecules to the right side of the embryo. A sequence of molecular interactions downstream of nodal results in the activation of the **_Pitx2_** gene, a transcription factor, also on the left side. The Pitx2 protein leads to later asymmetric development, such as rotation of the gut and stomach, position of the spleen, and the asymmetric lobation of the lungs. Although the left-sided expression of nodal in the lateral mesoderm seems to be a point of commonality in the determination of left-right asymmetry in all vertebrates, earlier (upstream) molecular events differ among the classes of vertebrates. In the chick, important signaling molecules, such as sonic hedgehog (shh) and fibroblast growth factor-8 (FGF-8), are asymmetrically distributed around the node, whereas in the mouse, the distribution is uniform.

How the anteroposterior polarity, exemplified by the primitive streak, is translated via ciliary currents into left-right asymmetry is the subject of considerable research. A likely candidate is **planar cell polarity**, which is a mechanism directing cells to orient themselves along an axis in the plane of a flat epithelial tissue. This is accomplished by the asymmetric distribution of several planar cell polarity proteins along this axis. In the node, **Dishevelled** is concentrated in the posterior region of the cells, and a counterpart, **Prickle**, is arranged along the anterior border (**Fig. 5.14**). The basal body in each of the 200 to 300 monociliated cells of the node is associated with Dishevelled, and the cilium that protrudes from the cell does so at an angle that produces the leftward fluid current when the cilium beats. There is speculation that a Wnt gradient

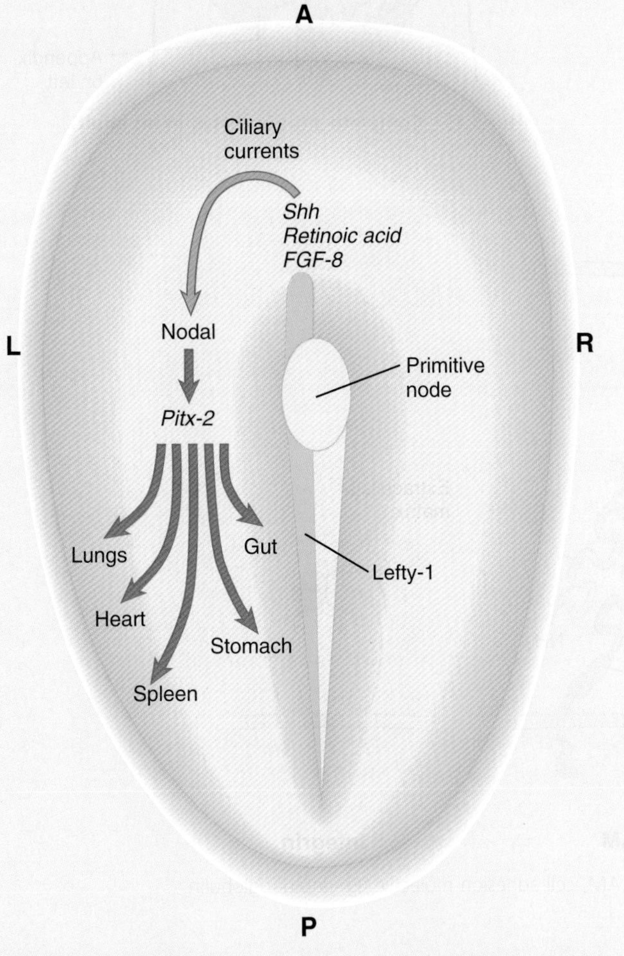

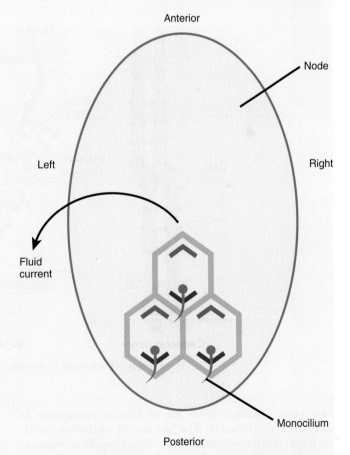

Fig. 5.13   Summary of the molecular basis for body asymmetry. Ciliary currents at the primitive node sweep the symmetry-breaking molecule nodal toward the left side of the embryo, where it stimulates an asymmetric cascade of gene expression via Pitx-2. Lefty-1, expressed along the left side of the embryo, may prevent diffusion of molecules to the right side. Only the most important molecules in a complex cascade are shown. FGF-8, fibroblast growth factor-8; Shh, sonic hedgehog.

Fig. 5.14 The relationship between the planar cell polarity proteins Dishevelled _(red)_ and Prickle _(green)_ and the location of the monocilium in cells of the primitive node. The posterior location of the monocilia is such that their beat leads to a leftward fluid current around the node.

Continued

Box 5.2   Molecular Basis for Left-Right Asymmetry—cont'd

lies behind the asymmetric distribution of Dishevelled and Prickle, but this remains to be confirmed.

In roughly 1 in 10,000 individuals, the left-right asymmetry of the body is totally reversed, a condition called **situs inversus (Fig. 5.15).** This condition is often not recognized until the individual is examined relatively late in life by an astute diagnostician. Several mutations and syndromes are associated with this condition, but one of the most instructive is **Kartagener's syndrome**, in which situs inversus is associated with respiratory symptoms (sinusitis and bronchiectasis) resulting from abnormalities of the dynein arms in cilia (immotile cilia). In a similar mouse mutant, the cilia around the primitive node do not function properly, and the lack of directionality of the resulting fluid currents around the node is suspected to result in the random localization of nodal and other asymmetry-producing molecules to the right side of the embryo. Partial situs inversus, such as an isolated right-sided heart (**dextrocardia**), can also occur. With more than 24 genes currently known to be involved in left-right asymmetry, such isolated occurrences of organ asymmetry are probably the result of mutations of genes farther downstream in the asymmetry cascade.

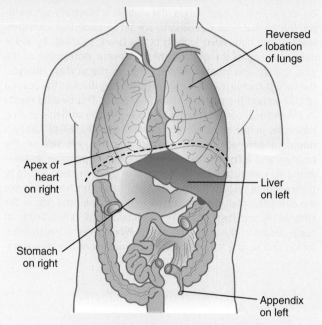

Fig. 5.15   Complete situs inversus in an adult.

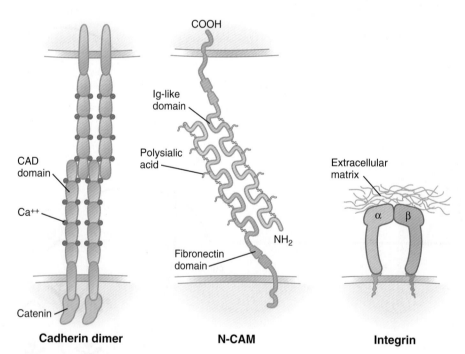

Fig. 5.16   Three major cell adhesion molecules. CAM, cell adhesion molecule; Ig, immunoglobulin.

have been described, three are of greatest importance to embryonic development. The first are the **cadherins**, which are single transmembrane glycoproteins typically arranged as homodimers that protrude from the cell surface. In the presence of calcium (Ca^{++}), cadherin dimers from adjacent cells adhere to one another and cause the cells to become firmly attached to one another (**Fig. 5.16**). One of the most ubiquitous is E-cadherin, which binds epithelial cells to one another

(see Fig. 16.6). During epithelial-mesenchymal transformations, such as that shown in Figure 5.8, the epithelial cells lose their E-cadherins as they transform into mesenchymal cells, but if these cells reform an epithelium later in development, they re-express E-cadherins.

The immunoglobulin **Ig (immunoglobulin)-CAMs** are characterized by having varying numbers of immunoglobulinlike extracellular domains. These molecules adhere to similar (**homophilic binding**) or different (**heterophilic binding**) CAMs on neighboring cells, and they do so without the mediation of calcium ions (see Fig. 5.16). One of the most prominent members of this family is **N-CAM**, which is strongly expressed within the developing nervous system. Ig-CAMs do not bind cells as tightly as cadherins, and they provide for fine-tuning of intercellular connections. N-CAM is unusual in having a high concentration of negatively charged sialic acid groups in the carbohydrate component of the molecule, and embryonic forms of N-CAM have three times as much sialic acid as the adult form of the molecule.

In the early embryo, before primary induction of the central nervous system, the ectoderm expresses N-CAM and E-cadherin (formerly known as L-CAM). After primary induction, cells within the newly formed neural tube continue to express N-CAM, but they no longer express E-cadherin. They also strongly express N-cadherin. In contrast, the ectoderm ceases to express N-CAM, but it continues to express E-cadherin (**Fig. 5.17**).

The third major family of CAMs, the **integrins**, attaches cells to components of basal laminae and the extracellular matrix (see Fig. 5.16). Integrins form heterodimers consisting of 1 of 16 α chains and 1 of 8 β chains. The matrix molecules to which they bind cells include fibronectin, laminin, and tenascin (see Fig. 12.3).

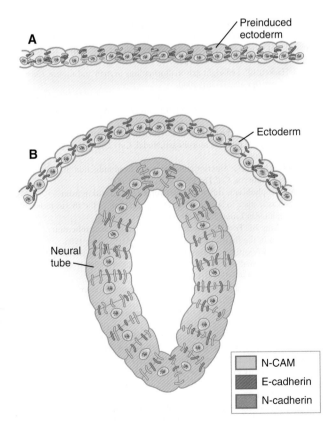

Fig. 5.17 Distribution of cell adhesion molecules in early ectoderm. Preinduced ectoderm (A) after induction of the neural tube (B). CAM, cell adhesion molecule.

A — Preinduced ectoderm

B — Ectoderm

Neural tube

N-CAM
E-cadherin
N-cadherin

Clinical Vignette

A 35-year-old married man with a history of chronic respiratory infections is found on a routine x-ray examination to have dextrocardia. Further physical examination and imaging studies reveal that he has complete situs inversus. He has also been going to another clinic for a completely different problem, which is related to the same underlying defects.

Which is the most likely clinic?
A. Urology
B. Dermatology
C. Infertility
D. Orthopedic
E. Oncology

Summary

- Just before implantation, the inner cell mass becomes reorganized as an epithelium (epiblast), and a second layer (hypoblast) begins to form beneath it. Within the epiblast, the amniotic cavity forms by cavitation; outgrowing cells of the hypoblast give rise to the endodermal lining of the yolk sac. Extraembryonic mesoderm seems to form by an early transformation of parietal endodermal cells and cells migrating through the primitive streak.

- The pregastrula embryo sets up two signaling centers. The anterior visceral endoderm induces the head and inhibits anterior extension of the primitive streak. The posterior center induces the primitive streak and the formation of mesoderm.

- During gastrulation, a primitive streak forms in the epiblast at the caudal end of the bilaminar embryo. Cells migrating through the primitive streak form the mesoderm and endoderm, and the remaining epiblast becomes ectoderm.

- The primitive node, located at the cranial end of the primitive streak, is the source of the cells that become the notochord. It also functions as the organizer or primary inductor of the future nervous system.

- As they pass through the primitive streak, future mesodermal cells in the epiblast change in morphology from epithelial epiblastic cells to bottle cells and then to mesenchymal cells. Extraembryonic mesodermal cells form the body stalk. The migration of mesenchymal cells during gastrulation is facilitated by extracellular matrix molecules such as hyaluronic acid and fibronectin.

- Late in the third week after fertilization, the primitive streak begins to regress caudally. Normally, the primitive streak disappears, but sacrococcygeal teratomas occasionally form in the area of regression.

- The essential elements of neural induction are the same in all vertebrates. In mammals, the primitive node and the notochordal process act as the primary inductors of the nervous system. Mesodermal induction occurs even earlier than neural induction. Growth factors such as Vg1 and activin are the effective agents in mesodermal induction.

- Numerous signaling centers control the organization of many important embryonic structures during early development. Each is associated with a constellation of important developmental genes. The early gastrula organizer is involved in initiation of the primitive streak. The primitive node organizes the formation of the notochord and nervous system and many aspects of cellular behavior associated with the primitive streak. The notochord is important in the induction of many axial structures, such as the nervous system and somites. Formation of the head is coordinated by the anterior visceral endoderm (hypoblast) and the prechordal plate.

■ Early blastomeres are totipotent. As development progresses, cells pass through restriction points that limit their differentiation. When the fate of a cell is fixed, the cell is said to be determined. Differentiation refers to the actual expression of the portion of the genome that remains available to a determined cell, and the term connotes the course of phenotypic specialization of a cell.

■ Left-right asymmetry in the early embryo is accomplished by the action of ciliary currents at the node carrying nodal to the left side of the embryo. This releases a cascade of molecules, with Pitx-2 prominent, that causes the asymmetric formation of structures, such as the heart, liver, lungs, and stomach.

■ Embryonic cells of the same type adhere to one another and reaggregate if separated. The molecular basis for cell aggregation and adherence is the presence of adhesion molecules on their surfaces. The three main families are the cadherins and the Ig-CAMs, which mediate cell-to-cell adhesion, and the integrins, which bind cells to the surrounding extracellular matrix.

Review Questions

1. The principal inductor in primary neural induction is the:
A. Hypoblast
B. Primitive streak
C. Extraembryonic mesoderm
D. Notochordal process
E. Embryonic ectoderm

2. Which of the following tissues arises from cells passing through the primitive streak?
A. Embryonic endoderm
B. Hypoblast
C. Cytotrophoblast
D. Primary yolk sac
E. Amnion

3. Cells of which germ layer are not present in the oropharyngeal membrane?
A. Ectoderm
B. Mesoderm
C. Endoderm
D. All are present

4. The prechordal plate plays an important role in regionalization of the:
A. Notochord
B. Forebrain
C. Embryonic mesoderm
D. Primitive node
E. Hindbrain

5. Brachyury, a deficiency in caudal tissues in the body, is caused by a mutation in what gene?
A. Lim-1
B. Noggin
C. T
D. Sonic hedgehog
E. Activin

6. Which layer of the bilaminar (two-layered) embryo gives rise to all of the embryonic tissue proper?

7. Of what importance is the primitive node in embryonic development?

8. The migration of mesodermal cells from the primitive streak is facilitated by the presence of what molecules of the extracellular matrix?

9. What molecules can bring about mesodermal induction in an early embryo?

10. At what stage in the life history of many cells are cell adhesion molecules lost?

References

Arnold SA, Robertson EJ: Making a commitment: cell lineage allocation and axis patterning in the early mouse embryo, *Nat Rev Mol Cell Biol* 10:91-103, 2009.

Aw S, Levin M: Is left-right asymmetry a form of planar cell polarity? *Development* 136:355-366, 2009.

Barrow JR and others: *Wnt3* signaling in the epiblast is required for proper orientation of the anteroposterior axis, *Dev Biol* 312:312-320, 2007.

Brennan J and others: *Nodal* signalling in the epiblast patterns the early mouse embryo, *Nature* 411:965-969, 2001.

Casey B, Hackett BP: Left-right axis malformations in man and mouse, *Curr Opin Genet Dev* 10:257-261, 2000.

Chen C and others: The Vg1-related protein Gdf3 acts in a Nodal signaling pathway in the pre-gastrulation mouse embryo, *Development* 133:319-329, 2006.

Constam DB: Running the gauntlet: an overview of the modalities of travel employed by the putative morphogen Nodal, *Curr Opin Genet Dev* 19:302-307, 2009.

De Souza FSJ, Niehrs C: Anterior endoderm and head induction in early vertebrate embryos, *Cell Tiss Res* 300:207-217, 2000.

Enders AC: Trophoblastic differentiation during the transition from trophoblastic plate to lacunar stage of implantation in the rhesus monkey and human, *Am J Anat* 186:85-98, 1989.

Enders AC, King BF: Formation and differentiation of extraembryonic mesoderm in the rhesus monkey, *Am J Anat* 181:327-340, 1988.

Ferrer-Vaquer A, Viotti M, Hadjantonakis A-K: Transitions between epithelial and mesenchymal states and the morphogenesis of the early mouse embryo, *Cell Adh Migr* 4:447-457, 2010.

Gardner RL: The initial phase of embryonic patterning in mammals, *Int Rev Cytol* 203:233-290, 2001.

Goodrich LV, Strutt D: Principles of planar polarity in animal development, *Development* 138:1877-1892, 2011.

Hammerschmidt M, Wedlich D: Regulated adhesion as a driving force of gastrulation movements, *Development* 135:3625-3641, 2008.

Hashimoto M, Hamada H: Translation of anterior-posterior polarity into left-right polarity in the mouse embryo, *Curr Opin Genet Dev* 20:433-437, 2010.

Herrmann BG, ed: The brachyury gene, *Semin Dev Biol* 6:381-435, 1995.

Idkowiak J and others: Hypoblast controls mesoderm generation and axial patterning in the gastrulating rabbit embryo, *Dev Genes Evol* 214:591-605, 2004.

Kavka AI, Green JBA: Tales of tails: brachyury and the T-box genes, *Biochim Biophys Acta* 1333:F73-F84, 1997.

Kimura-Yoshida C and others: Canonical Wnt signaling and its antagonist regulate anterior-posterior axis polarization by guiding cell migration in mouse visceral endoderm, *Dev Cell* 9:639-650, 2005.

Levin M: Left-right asymmetry in embryonic development: a comprehensive review, *Mech Dev* 122:3-25, 2005.

Levine AJ, Brivanlou AH: Proposal of a model of mammalian neural induction, *Dev Biol* 308:247-256, 2007.

Limura T, Pourquié O: Collinear activation of *Hoxb* genes during gastrulation is linked to mesoderm cell ingression, *Nature* 442:568-571, 2006.

Limura W and others: Fate and plasticity of the endoderm in the early chick embryo, *Dev Biol* 289:283-295, 2006.

Luckett WP: Origin and differentiation of the yolk sac and extraembryonic mesoderm in presomite human and rhesus monkey embryos, *Am J Anat* 152:59-98, 1978.

Luckett WP: The development of primordial and definitive amniotic cavities in early rhesus monkey and human embryos, *Am J Anat* 144:149-168, 1975.

Müller F, O'Rahilly R: The primitive streak, the caudal eminence and related structures in staged human embryos, *Cells Tiss Organs* 177:2-20, 2004.

Müller F, O'Rahilly R: The prechordal plate, the rostral end of the notochord and nearby median features in staged human embryos, *Cells Tiss Organs* 173:1-20, 2003.

Norris D: Breaking the left-right axis: do nodal parcels pass a signal to the left? *Bioessays* 27:991-994, 2005.

Nowotschin S, Hadjantonakis A-K: Cellular dynamics in the early mouse embryo: from axis formation to gastrulation, *Curr Opin Genet Dev* 20:420-427, 2010.

Ohta S and others: Cessation of gastrulation is mediated by suppression of epithelial-mesenchymal transition at the ventral ectodermal ridge, *Development* 134:4315-4324, 2007.

Patthey C, Edlund T, Gunhaga L: Wnt-regulated temporal control of BMP exposure directs the choice between neural plate border and epidermal fate, *Development* 136:73-83, 2009.

Plusa B and others: Distinct sequential cell behaviours direct primitive endoderm formation in the mouse blastocyst, *Development* 135:3081-3091, 2008.

Robb L, Tam PPL: Gastrula organizer and embryonic patterning in the mouse, *Semin Cell Dev Biol* 15:543-554, 2004.

Rozario T, DeSimone DW: The extracellular matrix in development and morphogenesis, *Dev Biol* 341:126-140, 2010.

Shawlot W, Behringer RR: Requirement for Lim1 in head organizer function, *Nature* 374:425-430, 1994.

Shiratori H, Hamada H: The left-right axis in the mouse: from origin to morphology, *Development* 133:2095-2104, 2006.

Smith JL, Schoenwolf GC: Getting organized: new insights into the organizer of higher vertebrates, *Curr Top Dev Biol* 40:79-110, 1998.

Spemann H: *Embryonic development and induction*, New York, 1938, Hafner.

Spemann H, Mangold H: Ueber Induktion von Embryonenanlagen durch Implantation ortfremder Organisatoren, *Arch Microskop Anat Entw Mech* 100:599-638, 1924.

Stern C: Neural induction: 10 years on since the "default model," *Curr Opin Cell Biol* 18:692-697, 2006.

Stern C: Neural induction: old problem, new findings, yet more questions, *Development* 132:2007-2021, 2005.

Stern C and others: Head-tail patterning of the vertebrate embryo: one, two or many unresolved problems? *Int J Dev Biol* 50:3-15, 2006.

Sulik K and others: Morphogenesis of the murine node and notochordal plate, *Dev Dyn* 201:260-278, 1994.

Tabin C: Do we know anything about how left-right asymmetry is first established in the vertebrate embryo? *J Mol Histol* 36:317-323, 2005.

Takeichi M: The cadherins: cell-cell adhesion molecules controlling animal morphogenesis, *Development* 102:639-655, 1988.

Tam PPL, Behringer RR: Mouse gastrulation: the formation of a mammalian body plan, *Mech Dev* 68:3-25, 1997.

Townes PL, Holtfreter J: Directed movements and selective adhesion of embryonic amphibian cells, *J Exp Zool* 128:53-120, 1955.

Viebahn C: The anterior margin of the mammalian gastrula: comparative and phylogenetic aspects of its role in axis formation and head induction, *Curr Top Dev Biol* 46:63-102, 1999.

Wallingford JB: Planar cell polarity signaling, cilia and polarized ciliary beating, *Curr Opin Cell Biol* 22:597-604, 2010.

Wallingford JB, Mitchell B: Strange as it may seem: the many links between Wnt signaling, planar cell polarity, and cilia, *Genes Dev* 25:201-213, 2011.

Watson CM, Tam PPL: Cell lineage determination in the mouse, *Cell Struct Funct* 26:123-129, 2001.

Wittler L, Kessel M: The acquisition of neural fate in the chick, *Mech Dev* 121:1031-1042, 2004.

Yamamoto M and others: Nodal antagonists regulate formation of the anteroposterior axis of the mouse embryo, *Nature* 428:387-392, 2004.

Yamanaka Y and others: Cell and molecular regulation of the mouse blastocyst, *Dev Dyn* 235:2301-2314, 2006.

Zernicka-Goetz M: Patterning of the embryo: the first spatial decisions in the life of a mouse, *Development* 129:815-892, 2002.

Establishment of the Basic Embryonic Body Plan

After gastrulation is complete, the embryo proper consists of a flat, three-layered disk containing the ectodermal, mesodermal, and endodermal germ layers. Its cephalocaudal axis is defined by the location of the primitive streak. Because of the pattern of cellular migration through the primitive streak and the regression of the streak toward the caudal end of the embryo, a strong **cephalocaudal gradient** of maturity is established. This gradient is marked initially by the formation of the notochord and later by the appearance of the neural plate, which results from the primary induction of the dorsal ectoderm by the notochord.

As seen in Chapter 5, despite the relatively featureless appearance of the gastrulating embryo, complex patterns of gene expression set up the basic body plan of the embryo. One of the earliest morphological manifestations of this pattern is the regular segmentation that becomes evident along the craniocaudal axis of the embryo. Such a segmental plan, which is a dominant characteristic of all early embryos, becomes less obvious as development progresses. Nonetheless, even in an adult the regular arrangement of the vertebrae, ribs, and spinal nerves persists as a reminder of humans' highly segmented phylogenetic and ontogenetic past.

Another major change crucial in understanding the fundamental organization of the body plan is the lateral folding of the early embryo from three essentially flat, stacked, pancake-like disks of cells (the primary embryonic germ layers) to a cylinder, with the ectoderm on the outside, the endoderm on the inside, and the mesoderm between them. The cellular basis for lateral folding still remains better described than understood.

This chapter concentrates on the establishment of the basic overall body plan. In addition, it charts the appearance of the primordia of the major organ systems of the body from the undifferentiated primary germ layers (see Fig. 6.27).

Development of the Ectodermal Germ Layer

Neurulation: Formation of the Neural Tube

The principal early morphological response of the embryonic ectoderm to neural induction is an increase in the height of the cells that are destined to become components of the nervous system. These transformed cells are evident as a thickened **neural plate** visible on the dorsal surface of the early embryo (**Figs. 6.1A** and **6.2A**). Unseen but also important is the restricted expression of cell adhesion molecules (Ig-CAMs), from N-CAM and E-cadherin in the preinduced ectoderm to N-CAM and N-cadherin in the neural plate.

The first of four major stages in the formation of the neural tube is transformation of the general embryonic ectoderm into a thickened neural plate. The principal activity of the second stage is further shaping of the overall contours of the neural plate so that it becomes narrower and longer. To a great extent, this is accomplished by **convergent extension**, during which the ectodermal cells forming the neural plate migrate toward the midline and also become longer along the anteroposterior axis and narrower laterally. This process, which is guided by planar cell polarity (see p. 87), results in the formation of a key-shaped neural plate (see Fig. 6.1A).

The third major stage in the process of **neurulation** is the lateral folding of the neural plate that results in the elevation of each side of the neural plate along a midline **neural groove** (see Figs. 6.1B and 6.2B). Many explanations have been proposed for lateral folding of the neural plate and ultimate closure of the neural tube. Most of these explanations have invoked a single or dominant mechanism, but it is now becoming apparent that lateral folding is the result of numerous region-specific mechanisms intrinsic and extrinsic to the neural plate.

The ventral midline of the neural plate, sometimes called the **median hinge point**, acts like an anchoring point around which the two sides become elevated at a sharp angle from the horizontal. At the median angle, bending can be largely accounted for by notochord-induced changes in the shape of the neuroepithelial cells of the neural plate. These cells become narrower at their apex and broader at their base (see Fig. 6.2B) through a combination of a basal position of the nuclei (thus causing a lateral expansion of the cell in that area) and a purse string–like contraction of a ring of actin-containing microfilaments in the apical cytoplasm. Throughout the lateral folding of the neural plate in the region of the spinal cord, much of the wall area of the neural plate initially remains flat (see Fig. 6.2B), but in the brain region, a **lateral hinge point** forms as a result of apical constriction of cells in a localized area (see Fig. 6.2C). Elevation of the **neural folds** seems to be accomplished largely by factors extrinsic to the neural epithelium, in particular, pushing forces generated by the expanding surface epithelium lateral to the neural plate.

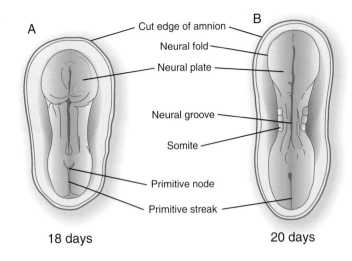

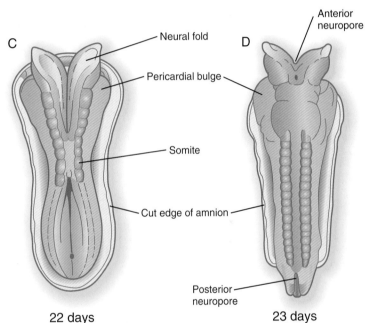

Fig. 6.1 **Early stages in the formation of the human central nervous system.** **A,** At 18 days. **B,** At 20 days. **C,** At 22 days. **D,** At 23 days.

The fourth stage in the formation of the neural tube consists of apposition of the two most lateral apical surfaces of the neural folds, their fusion (mediated by cell surface glycoconjugates), and the separation of the completed segment of the neural tube from the overlying ectodermal sheet (see Fig. 6.2C and D). At the same time, cells of the **neural crest** begin to separate from the neural tube.

Closure of the neural tube begins almost midway along the craniocaudal extent of the nervous system of a 21- to 22-day-old embryo (see Fig. 6.1C). Over the next couple of days, closure extends caudally in a zipperlike fashion, but cranially there are commonly two additional discontinuous sites of closure. The unclosed cephalic and caudal parts of the neural tube are called the **anterior** (cranial) and **posterior** (caudal) **neuropores.** The neuropores also ultimately close off so that the entire future central nervous system resembles an irregular cylinder sealed at both ends. Occasionally, one or both neuropores remain open, and serious birth defects result (see p. 248).

Caudal to the posterior neuropore, the remaining neural tube (more prominent in animals with large tails) is formed by the process of **secondary neurulation.** Secondary neurulation in mammals begins with the formation of a rodlike condensation of mesenchymal cells, the **medullary cord,** beneath the dorsal ectoderm of the **tail bud.** Within the mesenchymal rod, a central canal forms directly by **cavitation** (the formation of a space within a mass of cells). This central canal becomes continuous with the one formed during primary neurulation by the lateral folding of the neural plate and closure of the posterior neuropore. Because of the poor development of the tail bud, secondary neurulation in humans is not a prominent process.

Segmentation in the Neural Tube
Morphological Manifestations of Segmentation

Soon after the neural tube has taken shape, the region of the future brain can be distinguished from the spinal cord. The brain-forming region undergoes a series of subdivisions that constitute the basis for the fundamental gross organization of the adult brain. Segmentation by subdivision of an existing

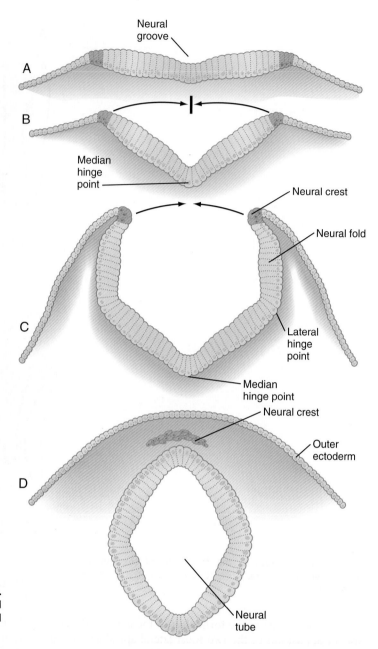

Fig. 6.2 Cross sections through the forming neural tube. A, Neural plate. B, Neural fold. C, Neural folds apposing. D, Neural tube complete. (Neural crest before and after its exit from the neural epithelium is shown in *green*.)

structure (in the case of the neural tube) contrasts with segmentation by adding terminal segments, as is the case in the formation of somites (see p. 99). An early set of subdivisions results in a three-part brain, consisting of a forebrain (**prosencephalon**), midbrain (**mesencephalon**), and hindbrain (**rhombencephalon**). Later, the prosencephalon becomes subdivided into a **telencephalon** and **diencephalon**, and the rhombencephalon is subdivided into a **metencephalon** and **myelencephalon** (see Fig. 11.2).

Superimposed on the traditional gross morphological organization of the developing brain is another, more subtle, level of segmentation, which subdivides certain regions of the brain into transiently visible series of regular segments called **neuromeres** (**Fig. 6.3**). In the hindbrain, the neuromeres, often

called **rhombomeres**, are visible from early in the fourth to late in the fifth week (Fig. 6.3B). The midbrain does not seem to be segmented, but the prosencephalon contains a less regular series of **prosomeres.**

Rhombomeres are arranged as odd and even pairs, and when established, they act like isolated compartments in insect embryos. Because of specific surface properties, cells from adjacent rhombomeres do not intermingle across boundaries between even and odd segments; however, marked cells from two even or two odd rhombomeres placed side by side do intermingle. During their brief existence, rhombomeres provide the basis for the fundamental organization of the hindbrain. In an adult, the segmental organization of the rhombomeres is manifest in the rhombomere-specific origin

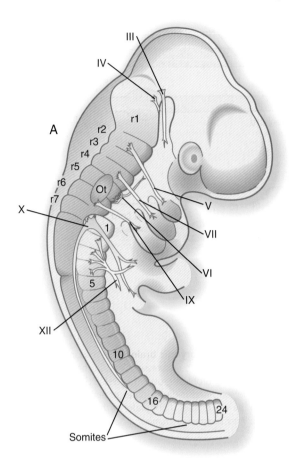

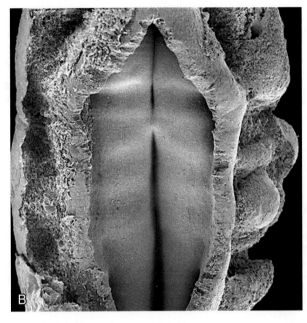

Fig. 6.3 Neuromeres in 3-day old chick brain (**A**) and a 5-week human embryo (**B**). The scanning electron micrograph in **B** looks down onto the rhombencephalon through the dissected roof. The neuromeres (rhombomeres) are the paired white horizontal stripes on either side of the midline groove. (*B, From Steding G:* The Anatomy of the human embryo, *Basel, 2009, Karger. Courtesy of Dr. J. Männer.*)

of many cranial nerves and parts of the reticular formation within the brainstem (see Fig. 11.13).

Mechanisms of Early Segmentation of the Neural Tube

While gastrulation is still taking place, the newly induced neural tube is subjected to vertical inductions from the notochord and head organizing regions (anterior visceral endoderm and prechordal plate), which are important in inducing the forebrain region. These inductions, together with a gradient of **Wnt-8** (product of a gene homologous with *Wingless*, a segment polarity gene in *Drosophila* [see Fig. 4.1]) signaling, effectively subdivide it into forebrain/midbrain and hindbrain/spinal cord segments. This subdivision is marked by the expression of two transcription factors, **Otx-2** (orthodenticle homologue 2) in the forebrain/midbrain region, and in the hindbrain, **Gbx-2** (gastrulation brain homeobox 2), whose boundaries sharply define the midbrain-hindbrain border (**Fig. 6.4A**). Fibroblast growth factors (FGFs), produced in the early primitive streak, are known to exert a posteriorizing effect on the newly forming neural plate.

The midbrain-hindbrain border becomes a powerful local signaling center, called the **isthmic organizer. Wnt-1** is synthesized in the neural ectoderm anterior, and **FGF-8** is formed posterior to the isthmic organizer (Fig. 6.4B). The transcription factors **Pax-2** and **Pax-5** and **engrailed** (**En-1** and **En-2**)

are expressed on both sides of the isthmic organizer as gradients that are crucial in organizing the development of the midbrain and the cerebellum, a hindbrain derivative.

Two additional organizing or signaling centers are established early in the formation of the forebrain region. One, the **anterior neural ridge**, is located at the anterior pole of the brain (see Fig. 6.4B). It is a site of sonic hedgehog and FGF-8 signaling activity and is important in organizing the formation of the telencephalon, parts of the diencephalon, the olfactory area, and the pituitary gland. A third signaling center, the **zona limitans** (see Fig. 6.4B), is a sonic hedgehog–secreting group of cells that organize the border between the future dorsal and ventral thalamus. Chapter 11 presents additional information on the organization and segmentation of the forebrain.

Segmentation in the Hindbrain Region

Segmentation of the hindbrain into seven rhombomeres in humans, and eight in some other animals, is the result of the expression of several categories of genes, which operate in a manner remarkably reminiscent of the way in which the early *Drosophila* embryo becomes subdivided into segments (see Fig. 4.1). Individual rhombomeres are initially specified through the ordered expression of unique combinations of transcription factors; this patterning is then translated into cellular behavior by the patterned expression of cell surface molecules.

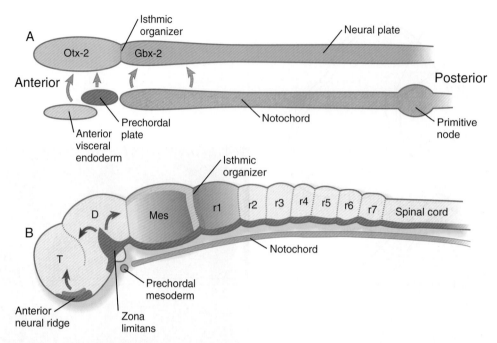

Fig. 6.4 **Schematic representation of signaling centers acting on and within the early embryonic brain. A,** In response to signals *(green arrows)* from the anterior visceral endoderm, the prechordal plate, and the notochord, the neural tube expresses Otx-2 in the future forebrain and midbrain regions and Gbx-2 in the hindbrain and spinal cord. **B,** Later in development, signals (fibroblast growth factor-8 [FGF-8] *[green]* and Wnt-1 *[yellow]*) from the isthmic organizer induce decreasing gradients of En-1 and En-2 *(blue)* on either side. Another organizer—the anterior neural ridge—secretes sonic hedgehog *(red)* and FGF-8 *(green)*, and both the zona limitans and the ventral part (floor plate) of the neural tube secrete sonic hedgehog. D, diencephalon; Mes, mesencephalon; r, rhombomere; T, telencephalon. *(B, After Lumsden A, Krumlauf R: Science 274:1109-1115, 1996.)*

After the Gbx-2–expressing area defines the rough limits of the hindbrain, several **segmentation genes** are involved in setting up the basic pattern of segmentation that leads to rhombomere formation. **Krox 20,** a zinc finger transcription factor, is expressed in and guides the formation of rhombomeres 3 and 5 (r3 and r5) (see Fig. 11.12), whereas **kreisler,** another transcription factor, and **Hoxa-1** are also involved in the formation of r5. A decreasing gradient of **retinoic acid,** produced by the anterior somites, plays an important role in the formation of the posterior rhombomeres (r4 to r7). These molecules are not involved in the specification of r1 to r3, which is regulated by Gbx-2.

The *Hox* genes are principally involved in specifying segmental identity, but before any molecular marker of morphological segmentation exists, the previously mentioned gradient of retinoic acid stimulates the expression of *Hoxa-1* and *Hoxb-1*. The influence of these two *Hox* genes and of the segmentation genes, *Krox 20* and *kreisler,* initiates the expression of the various *Hox* paralogues in a highly specific sequence along the hindbrain and spinal cord (see Fig. 11.12). As seen in Chapters 11 and 14, the pattern of *Hox* gene expression determines the morphological identity of the cranial nerves and other pharyngeal arch derivatives that arise from specific rhombomeres. At successive times during the formation of the hindbrain, different regulatory networks controlling *Hox* gene expression come into play, but details of these networks are not presented in this text. The orderly expression of *Hox* gene paralogues extends anteriorly through r2. Hox proteins are not found in r1 largely because of the antagonistic action of FGF-8, which is produced in response to signals from the isthmic organizer at the anterior end of r1. In the absence of FGF-8, Hox

proteins are expressed in r1. Another rhombencephalic protein, **sprouty 2,** acts as an antagonist of FGF-8, and this protein, in addition to the presence of Hoxa-2 in r2, confines FGF-8 mostly to r1 and contains the primordium of the cerebellum to the anterior part of r1.

Another family of genes, the **ephrins** and their receptors, determines the behavioral properties of the cells in the rhombomeres. The action of ephrins, which are expressed in even-numbered rhombomeres (2, 4, and 6), and of ephrin receptors, which are expressed in odd-numbered rhombomeres (3 and 5), seems to account for the lack of mixing behavior of cells from adjacent rhombomeres and maintains the separation of the various streams of neural crest cells that emigrate from the rhombomeres (see Fig. 12.8).

Formation and Segmentation of the Spinal Cord

Although neuromeres are not seen in the region of the neural tube that gives rise to the spinal cord, the regular arrangement of the exiting motor and sensory nerve roots is evidence of a fundamental segmental organization in this region of the body as well. In contrast to the brain, however, segmentation of the spinal cord is to a great extent imposed by signals emanating from the paraxial mesoderm, rather than from molecular signals intrinsic to the neural tube.

As the body axis is elongating, and somites are forming, the caudalmost part of the newly induced neural plate possesses the properties of a stem cell zone (**Fig. 6.5**). Under the influence of FGF-8, secreted by the adjacent presomitic paraxial mesoderm, these cells, which go on to form the spinal cord,

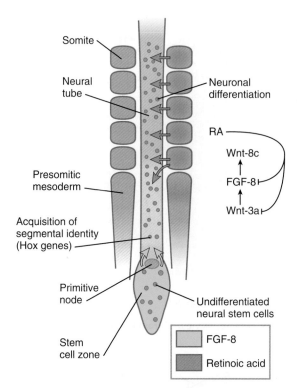

Fig. 6.5 Elongation of the spinal cord in the early embryo. Under the influence of fibroblast growth factor-8 (FGF-8) secreted by presomitic paraxial mesoderm, cells in the most posterior region continue to proliferate, whereas retinoic acid (RA), secreted by newly formed somites, stimulates the neuronal differentiation.

proliferate without undergoing differentiation. Some of the daughter cells are left behind by the posteriorly advancing stem cell zone. These cells fall under the influence of retinoic acid, produced by the newly formed somites, which are also being formed in a posterior direction (see Fig. 6.8). Retinoic acid stimulates these cells to differentiate into neurons. Elongation of the tail bud region comes to a close when the caudal extent of presomitic mesoderm is reduced, thus allowing the retinoic acid produced in the area to diffuse farther posteriorly and inhibit the action of FGF-8. As a result, proliferation of tail bud mesenchyme is greatly reduced, causing growth to cease.

The opposing actions of retinoic acid, which promotes differentiation, and FGF, which fosters proliferation at the expense of differentiation, represent a recurring theme in the development of other structures. For example, the spread of FGF-8 from the isthmic organizer (see Fig. 6.4B) antagonizes the influence of retinoic acid in r1. This permits the exuberant proliferation of the cells in this rhombomere, which is necessary for the formation of the large cerebellum from this structure. Interactions between FGF-8 and retinoic acid in the forming spinal cord and paraxial mesoderm help to set the Hox code that confers anteroposterior identity to regions of the spinal cord and the adjacent somites.

Neural Crest

As the neural tube is closing and separating from the general cutaneous ectoderm, a population of cells called the **neural**

crest leaves the dorsal part of the neural tube and begins to spread throughout the body of the embryo (see Fig. 6.2). The neural crest produces an astonishing array of structures in the embryo (see Table 12.1), and its importance is such that the neural crest is sometimes called the fourth germ layer of the body. (The neural crest is discussed further in Chapter 12.)

Sensory Placodes and Secondary Inductions in the Cranial Region

As the cranial region begins to take shape, several series of **ectodermal placodes** (thickenings) appear lateral to the neural tube and neural crest (**Fig. 6.6**). These placodes arise from a horseshoe-shaped preplacodal domain around the anterior neural plate that is established during the gastrulation and early neurulation periods, and the individual placodes result from a variety of secondary inductive processes between neural or mesenchymal tissues and the overlying ectoderm (see Table 13.1). In several cases, cells from the placodes and neural crest interact closely to form the sensory ganglia of cranial nerves (V, VII, IX, and X). Deficiencies of one of these two components can often be made up by an increased contribution by the other component. Further details of placodes and their developmental fate are given in Chapter 13.

Development of the Mesodermal Germ Layer

Basic Plan of the Mesodermal Layer

After passing through the primitive streak, the mesodermal cells spread laterally between the ectoderm and endoderm as a continuous layer of mesenchymal cells (see Fig. 5.6). Subsequently, three regions can be recognized in the mesoderm of cross-sectioned embryos (**Fig. 6.7B**). Nearest the neural tube is a thickened column of mesenchymal cells known as the **paraxial mesoderm**, or **segmental plate**. This tissue soon becomes organized into somites. Lateral to the paraxial mesoderm is a compact region of **intermediate mesoderm**, which ultimately gives rise to the urogenital system. Beyond that, the **lateral plate mesoderm** ultimately splits into two layers and forms the bulk of the tissues of the body wall, the wall of the digestive tract, and the limbs (see Fig. 6.27).

Paraxial Mesoderm

As the primitive node and the primitive streak regress toward the caudal end of the embryo, they leave behind the notochord and the induced neural plate. Lateral to the neural plate, the paraxial mesoderm appears to be a homogeneous strip of closely packed mesenchymal cells. However, if scanning electron micrographs of this mesoderm are examined with stereoscopic techniques, a series of regular pairs of segments can be discerned. These segments, called **somitomeres**, have been most studied in avian embryos, but they are also found in mammals. New pairs of somitomeres form along the primitive node as it regresses toward the caudal end of the embryo (**Fig. 6.8**). Not until almost 20 pairs of somitomeres have formed, and the primitive node has regressed quite far caudally, does the first pair of **somites** (brick-shaped masses of paraxial mesoderm) form behind the seventh pair of somitomeres.

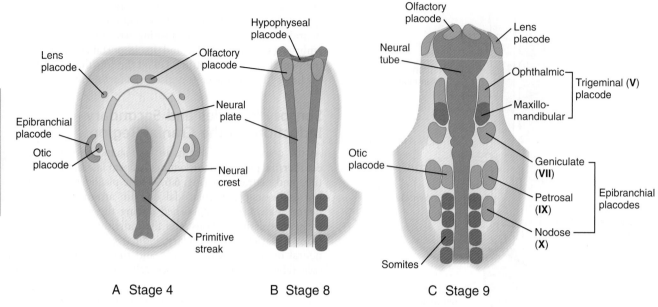

Fig. 6.6   Early stages in the formation of cranial ectodermal placodes in the chick embryo, as viewed from the dorsal aspect. The placodes are shown in *blue*.

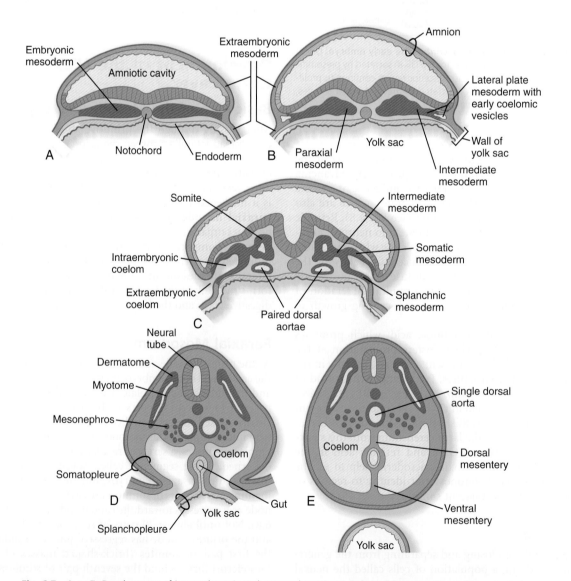

Fig. 6.7   **A** to **E,** Development of intraembryonic and extraembryonic mesoderm in cross sections of human embryos.

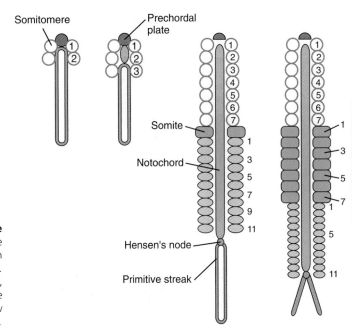

Fig. 6.8 Relationship between somitomeres and somites in the early chick embryo. Cranial somitomeres *(open circles)* take shape along Hensen's node until 7 pairs have formed. Caudal to the seventh somitomere, somites *(rectangles)* form from caudal somitomeres *(ovals)*. As the most anterior of the caudal somitomeres transform into somites, additional caudal somitomeres take shape posteriorly. For a while, the equilibrium between transformation into somites anteriorly and new formation posteriorly keeps the number of caudal somitomeres at 11.

After the first pair of somites has been established (approximately 20 days after fertilization), a regular relationship develops between the regression of the primitive streak and the formation of additional somites and somitomeres. The first 7 pairs of somitomeres in the cranial region do not undergo further separation or segmentation. Cells from these somitomeres (cranial mesoderm) will form most of the skeletal musculature of the head, and they have quite different cellular and molecular properties from those derived from somites of the trunk. The first pair of somites forms at the expense of the eighth pair of somitomeres. In the types of embryos studied to date, there is a constant relationship between the caudalmost pair of definitive somites and the number of somitomeres (usually 10 to 11) that can be shown behind them. Every few hours, the pair of somitomeres located caudal to the last-formed somites becomes transformed into a new pair of somites, and a new pair of somitomeres is laid down at the caudal end of the paraxial mesoderm near the primitive node (see Fig. 6.8). As regression of the primitive streak comes to a close, the formation of paraxial mesoderm continues through the cells contributed by the tail bud. The cervical, thoracic, and lumbar vertebrae and associated structures are derived from cells migrating through the primitive streak, whereas the cellular precursors of the sacrum and coccyx come from the tail bud.

Formation of Individual Somites

The formation of individual somites from a seemingly homogeneous strip of paraxial mesoderm is a complex process that involves a variety of levels of molecular control and changes in cellular behavior within the paraxial mesoderm. Our basic understanding of **somitogenesis** (somite formation) comes from studies on the chick. The first significant step in somitogenesis is **segmentation** of the paraxial mesoderm. In contrast to segmentation in the hindbrain (see p. 95), somite formation occurs by the sequential addition of new segments in a craniocaudal sequence.

Somitogenesis involves two mechanisms in what is often referred to as a **clock and wavefront model**. The first step (the **wavefront**) is associated with the elongation of the caudal end of the body through proliferative activity of mesenchymal cells in the most posterior nonsegmented part of the primitive streak (**Fig. 6.9A**). Cells in this area divide actively under the influence of a high local concentration of **FGF-8**. More anteriorly, where the cells are older, the concentration of FGF-8 decreases as the FGF molecules become broken down over time. Conversely, the cells closer to the last-formed somite become exposed to increasing concentrations of **retinoic acid**, which is produced in the most posterior somites and whose action opposes that of FGF. At some point in their life history, the mesenchymal cells are exposed to a balance of FGF-8 and retinoic acid concentrations that results in their crossing a developmental threshold (the wavefront, or **determination front**) that prepares them for entering the process of segmentation (somite formation). This is characterized by the expression of a transcription factor, **Mesp-2**, which prefigures a future somite. With the continued caudal elongation of the embryo and the addition of new somites, the location of the wavefront extends caudally in the growing embryo, but it remains a constant distance from the last-formed somite pair.

Next, the **segmentation clock** is initiated in those presomitic cells that have passed over the previously mentioned threshold and are expressing Mesp-2. The exact mechanism that starts the clock is still not fully defined, but many molecules in the interacting **Notch, Wnt,** and **FGF** pathways are known to be synthesized at regular periodic intervals and become localized at critical locations in the forming somite. In the chick, in which a new somite forms every 90 minutes, **lunatic fringe** becomes concentrated at the future anterior border of the somite, and **c-hairy** (a homologue of a segmentation gene in *Drosophila*) becomes concentrated along the future posterior border (Fig. 6.9B).

At the level of cellular behavior, cells at the anterior border of the forming somite express the ephrin receptor **Eph A**. Because the cells on the posterior border of the previously

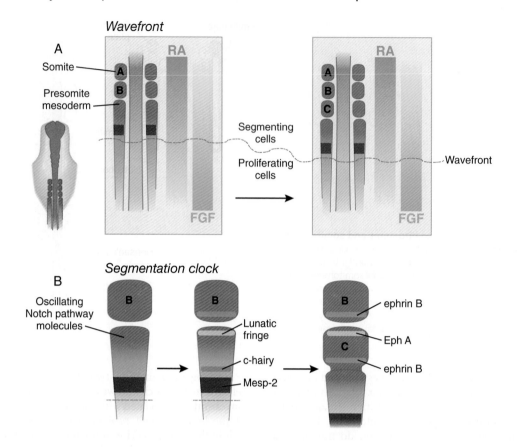

Fig. 6.9 **Aspects of the clock and wavefront model of somitogenesis.** **A,** The wavefront, consisting of opposing gradients of retinoic acid (RA) and fibroblast growth factor-8 (FGF). **B,** The segmentation clock, in which oscillating molecules in the Notch pathway stimulate the expression of lunatic fringe at the anterior and c-hairy at the posterior border of a future somite. Later interactions between Eph A and ephrin B maintain the intersomitic space.

formed somite express the ephrin ligand **ephrin B**, the cells of the two adjacent somites are prevented from mixing (as is the case with adjacent rhombomeres in the developing hindbrain), and a fissure forms between the two somites. Finally, the action of **Wnt-6** from the overlying ectoderm stimulates the expression of the transcription factor **paraxis** in the newly forming somite. This, along with the downregulation of *Snail*, results in the transformation of the mesenchymal cells of the anterior part of the somite, and later all the mesenchymal cells, into an epithelial cell type (**Fig. 6.10A**). While in the earliest stages of its formation, a somite also undergoes an internal subdivision into anterior and posterior halves. Differences in cellular properties stemming from this subdivision are of great significance in the formation of vertebrae and in guiding the migration of neural crest cells and outgrowing axons.

The continued development of a somite involves the complete transformation of the segmented blocks of mesenchymal cells into a sphere of epithelial cells through the continued action of paraxis (see Fig. 6.10A). The cells of the epithelial somite are arranged so that their apical surfaces surround a small central lumen, the **somitocoel** (which contains a few core cells), and their outer basal surfaces are surrounded by a basal lamina (containing laminin, fibronectin, and other components of the extracellular matrix).

Shortly after the formation of the epithelial somite, the cells of its ventromedial wall are subjected to an inductive stimulus in the form of the signaling molecules **sonic hedgehog** and **noggin**, originating from the notochord and the ventral wall of the neural tube. The response is the expression of *Pax1* and *Pax9* in the ventral half of the somite, which is now called the **sclerotome** (**Fig. 6.11**). This leads to a burst of mitosis, the loss of intercellular adhesion molecules (**N-cadherin**), the dissolution of the basal lamina in that region, and the transformation of the epithelial cells in that region back to a mesenchymal morphology (these cells are called **secondary mesenchyme**). These secondary mesenchymal cells migrate or are otherwise displaced medially from the remainder of the somite (see Fig. 6.10B) and begin to produce **chondroitin sulfate proteoglycans** and other molecules characteristic of cartilage matrix as they aggregate around the notochord.

Under the influence of secreted products of *Wnt* genes produced by the dorsal neural tube and the surface ectoderm, the dorsal half of the epithelial somite becomes transformed into the **dermomyotome** (see Fig. 6.10B) and expresses its own characteristic genes (*Pax3, Pax7, paraxis*). Mesenchymal cells arising from the dorsomedial and ventrolateral borders of the dermomyotome form a separate layer, the **myotome**, beneath the remaining somitic epithelium, which is now called the **dermatome** (see Fig. 6.10C). As their names imply, cells of the myotome produce muscle, and cells of the dermatome contribute to the dermis.

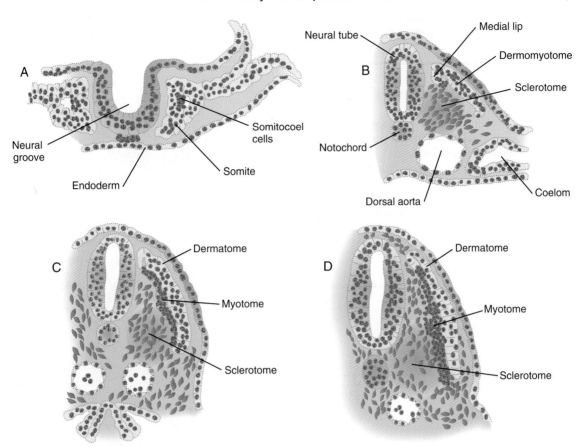

Fig. 6.10   Stages in the life history of a somite in a human embryo. A, Epithelial stage of a somite in the preneural tube stage. **B,** Epithelio-mesenchymal transformation of the ventromedial portion into the sclerotome. **C,** Appearance of a separate myotome from the original dermomyotome. **D,** Early stage of breakup of the epithelial dermatome into dermal fibroblasts.

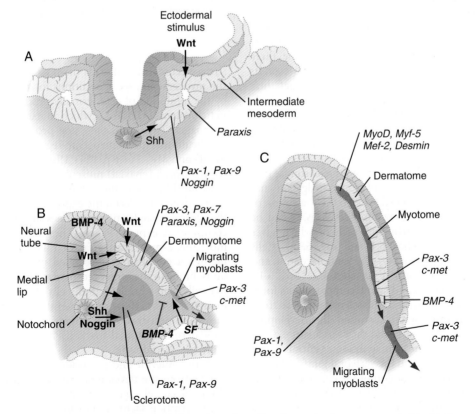

Fig. 6.11   Molecular events involved in the differentiation of somites. Signaling molecules are represented by *black arrows*. Inhibitory signals are represented by *red lines*. Genes expressed in responding tissues are indicated in italics. BMP, bone morphogenetic protein; SF, scatter factor; Shh, sonic hedgehog. *(Adapted from Brand-Saberi B and others: Int J Dev Biol 40:411-420, 1996.)*

Box 6.1 Somite Compartments and Their Derivatives

Sclerotome
  Ventral: vertebral bodies and their intervertebral disks
  Lateral: distal ribs, some tendons
  Dorsal: dorsal part of neural arch, spinous process
  Central: pedicles and ventral parts of neural arches, proximal ribs, or transverse processes of vertebrae
  Medial (meningotome): meninges and blood vessels of meninges
Arthrotome
  Intervertebral disks, vertebral joint surfaces, and proximal ribs
Dermatome
  Dermis, blade of scapula
Myotome
  Dorsomedial: intrinsic back muscles (epaxial)
  Ventrolateral: limb muscles or muscles of ventrolateral body wall (hypaxial)
Neurotome
  Endoneurial and perineurial cells
Syndetome
  Tendons of epaxial musculature

Adapted from Christ B, Huang R, Scaal M: *Dev Dyn* 236:2383, 2007.

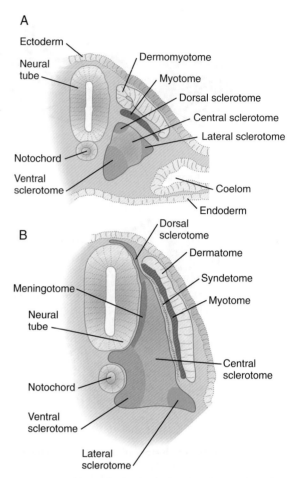

Fig. 6.12 Organization of somites at earlier (**A**) and later (**B**) stages of development. *(Based on Christ B, Huang R, Scaal M: Anat Embryol 208:333-350, 2004.)*

Organization of the Somite and the Basic Segmental Body Plan

The fates of the cells in the most recently formed somites are not fixed; if such a somite is rotated 180 degrees in a dorso-ventral direction, the cells respond to their new environment and form perfectly oriented derivatives. By the time three other new somites have formed behind a somite, however, its cells have received sufficient environmental input that their developmental course is set in place. Even within the early epithelial somite (see Fig. 6.10A), the structures that form from cells of the main epithelial sectors of the somite and from the mesenchymal somitocoel cells in the center of the somite can be mapped out.

The breakup of the ventral part of the epithelial somite into mesenchyme under the influence of sonic hedgehog and noggin coming from the notochord leads to the formation of the early sclerotome. As the sclerotome develops, it can be subdivided into many compartments, each of which gives rise to specific derivatives (**Box 6.1** and **Fig. 6.12**). Cells from several somitic compartments—the ventral, central, and dorsal—come together to form a vertebra (see Box 6.1), whereas cells from the central and lateral compartments form the ribs. Late in the development of the sclerotome, cells from its medial edge (**meningotome**) surround the developing spinal cord to form the meninges and their vasculature. Cells of the somitocoel (**arthrotome**) join with some ventral cells to form the intervertebral disks and the vertebral joint surfaces.

After the Wnt-mediated formation of the dermomyotome, cells in its dorsomedial sector find themselves exposed to a balance of sonic hedgehog signaling from the notochord and Wnt signaling from the dorsal neural tube and overlying surface ectoderm that leads them to becoming committed to the myogenic lineage. Conditions for myogenesis here are set by the inhibition of ectodermally produced bone morphogenetic protein-4 (BMP-4) (which in itself inhibits myogenesis) by noggin. These cells then stop producing Pax-3 and Pax-7 and begin to express myogenic regulatory molecules, such as MyoD and Myf-5 (see p. 184). Ultimately, these cells form the intrinsic back (epaxial) musculature.

Meanwhile, under the influence of BMP-4, produced by the lateral plate mesoderm, the expression of myogenic factors in the ventrolateral dermomyotome is suppressed, and these cells continue to express **Pax-3**. They also produce a receptor molecule, c-met. **Scatter factor** (also called **hepatic growth factor**), a growth factor secreted in the region of the limb buds, binds to the c-met receptor of the lateral dermomyotomal cells. This stimulates these cells (30 to 100 cells per somite) to migrate out of the somite and into the limb bud even before the myotome forms. While migrating, they continue to express their dermomyotomal marker, Pax-3, and the cell adhesion molecule, N-cadherin.

In the anterior and posterior borders of the somite, FGF signals from the developed myotome induce a layer of cells along the lateral edge of the sclerotome to produce **scleraxis**, a transcription factor found in tendons. These cells form a discrete layer, called the **syndetome**, and they represent the

Box 6.2 Mature Cell Types Derived from Somites

Adipocytes
Chondrocytes
Osteocytes
Endothelial cells
    Arteries
    Veins
    Capillaries
    Lymphatics
Pericytes
Fibrocytes
    Connective tissues
    Dermal
    Tendons and ligaments
Muscle cells
    Skeletal
    Smooth
Nervous system
    Arachnoid cells
    Epineurial cells
    Perineurial cells
    Endoneurial cells
    Fibrocytes of dura mater

Adapted from Christ B, Huang R, Scaal M: *Dev Dyn* 236:2383, 2007.

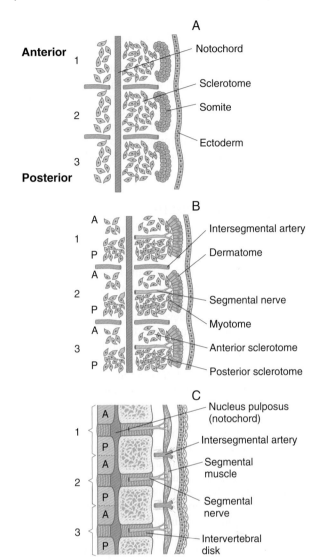

Fig. 6.13 **A**, Early movement of seemingly homogeneous sclerotome from the somite. **B**, Breakup of the sclerotomal portions of the somites into anterior (A) and posterior (P) halves, and the coalescence of the posterior portion of one somite with the anterior portion of the one caudal to it to form the body of a vertebra. **C**, With this rearrangement, the segmental muscles (derived from the myotomes) extend across intervertebral joints and are supplied by spinal nerves that grow out between the anterior and posterior halves of the somites.

precursors of the tendons that connect the epaxial muscles to their skeletal origins and insertions. Research with cell markers has shown that almost all components of the somites are able to give rise to blood vessels that nourish the various structures derived from the somitic mesoderm. **Box 6.2** lists the types of mature cells that are derived from somites.

Within a single somite, cells of the posterior sclerotome multiply at a greater rate than do those of the anterior part, and the result is higher cellular density in the posterior sclerotome (**Fig. 6.13B**). Properties of these cells and their extracellular matrix (see p. 255) do not permit the passage of either outgrowing nerve fibers or neural crest cells, which instead pass through the anterior sclerotome. Because of the outgrowing neural structures that either pass through or are derived from the anterior sclerotome, it has sometimes been called the **neurotome**.

As the cells of the sclerotome disperse around the notochord, cells of the anterior half of one somite aggregate with cells of the posterior half of the more cranial somite. Ultimately, this aggregate forms a single vertebra. Such an arrangement, which depends on interactions with the neural tube, places the bony vertebrae out of phase with the myotomally derived segmental muscles of the trunk (Fig. 6.13C). This structure allows the contracting segmental muscles to move the vertebral column laterally. The relationship between the anterior half of one somite and the adjoining posterior half of its neighboring somite is reminiscent of the parasegments of *Drosophila* (similarly arranged subdivisions of the segments into two parts), but whether they are functionally similar in terms of genetic control is undetermined.

Intermediate Mesoderm

Connecting the paraxial mesoderm and the lateral plate mesoderm in the early embryo is a small cord of cells called the **intermediate mesoderm**, which runs along the entire length of the trunk (see Fig. 6.7C). How the intermediate mesoderm forms remains unclear. It appears to arise as a response of early mesoderm to BMP, secreted by lateral ectoderm, and activin and other signals emanating from the paraxial mesoderm. The response to these signals is the expression of **Pax-2** within what becomes the intermediate mesoderm. The cranial and caudal extent of the intermediate mesoderm is defined by expression of members of the Hox-4 paralogue cranially and Hox-11 caudally. In experiments involving a cranial shift of expression of Hox-4, the cranial border of the intermediate mesoderm is correspondingly shifted toward the head. The intermediate mesoderm is the precursor of the urogenital system. The earliest signs of differentiation of the intermediate mesoderm are in the most cranial regions, where vestiges of the earliest form of the kidney, the **pronephros**, briefly appear. In the lateral region of the intermediate mesoderm, a longitudinal **pronephric duct** appears on each side of the embryo. The pronephric duct is important in organizing the development of much of the adult urogenital system, which forms largely from cells of the caudal portions of the intermediate mesoderm (see Chapter 16).

Lateral Plate Mesoderm

Shortly after gastrulation, the ectoderm overlying the lateral plate mesoderm produces BMP-4. Soon thereafter, the lateral plate mesoderm itself begins to produce BMP-4. Experimental studies have shown that this molecule has the ability to cause mesoderm, whether paraxial or lateral plate, to assume the molecular and cellular properties of lateral plate mesoderm. Whether early mesoderm develops the properties of paraxial or lateral plate mesoderm seems to depend on a balance between medializing influences emanating from the axial structures (neural tube and notochord) and lateralizing influences initially produced by lateral ectoderm.

The **lateral plate mesoderm** soon divides into two layers as the result of the formation and coalescence of coelomic (body cavity) spaces within it (see Fig. 6.7B and C). The dorsal layer, which is closely associated with the ectoderm, is called **somatic mesoderm**, and the combination of somatic mesoderm and ectoderm is called the **somatopleure** (see Fig. 6.7D). The ventral layer, called **splanchnic mesoderm**, is closely associated with the endoderm and is specified by the transcription factor **Foxf-1**. The combined endoderm and splanchnic mesoderm is called the **splanchnopleure**. The intraembryonic somatic and splanchnic mesodermal layers are continuous with the layers of extraembryonic mesoderm that line the amnion and yolk sac.

While the layers of somatic and splanchnic mesoderm are taking shape, the entire body of the embryo undergoes a lateral folding process that effectively transforms its shape from three flat germ layers to a cylinder, with a tube of endoderm (gut) in the center, an outer tubular covering of ectoderm (epidermis), and an intermediate layer of mesoderm. This transformation occurs before the appearance of the limbs.

Formation of the Coelom

As the embryo undergoes lateral folding, the small coelomic vesicles that formed within the lateral plate mesoderm coalesce into the coelomic cavity (see Fig. 6.7). Initially, the **intraembryonic coelom** is continuous with the **extraembryonic coelom**, but as folding is completed in a given segment of the embryo, the two coelomic spaces are separated. The last region of the embryo to undergo complete lateral folding is the area occupied by the yolk sac. In this area, small channels connecting the intraembryonic and extraembryonic coeloms persist until the ventral body wall is completely sealed.

In the cylindrical embryo, the somatic mesoderm constitutes the lateral and ventral body wall, and the splanchnic mesoderm forms the mesentery and the wall of the digestive tract. The somatic mesoderm of the lateral plate also forms the mesenchyme of the limb buds, which begin to appear late in the fourth week of pregnancy (see Fig. 10.1).

Extraembryonic Mesoderm and the Body Stalk

The thin layers of extraembryonic mesoderm that line the ectodermal lining of the amnion and the endodermal lining of the yolk sac are continuous with the intraembryonic somatic and splanchnic mesoderm (see Fig. 6.7A and B). The posterior end of the embryo is connected with the trophoblastic tissues (future placenta) by the mesodermal **body stalk** (see Fig. 7.1). As the embryo grows and a circulatory system becomes functional, blood vessels from the embryo grow through the body stalk to supply the placenta, and the body stalk itself becomes better defined as the **umbilical cord**. The extraembryonic mesoderm that lines the inner surface of the cytotrophoblast ultimately becomes the mesenchymal component of the placenta.

Early Stages in the Formation of the Circulatory System

As the embryo grows during the third week, it attains a size that does not permit simple diffusion to distribute oxygen and nutrients to all of its cells or efficiently remove waste products. The early development of the heart and circulatory system is an embryonic adaptation that permits the rapid growth of the embryo by providing an efficient means for the distribution of nutrients. The circulatory system faces the daunting task of having to grow and become continuously remodeled to keep pace with the embryo's overall growth while remaining fully functional in supplying the needs of the embryo's cells.

Heart and Great Vessels

The earliest development of the circulatory system consists of the migration of heart-forming cells arising in the epiblast through the primitive streak in a well-defined anteroposterior order. According to a commonly accepted model of heart development, the cells passing through the streak closest to the primitive node form the outflow tract, the cells passing through the midstreak form the ventricles, and the cells that form the atria enter the streak most posteriorly (**Fig. 6.14A**). After leaving the primitive streak, the precardiac cells, which are associated with endodermal cells as a splanchnic mesoderm, become arranged in the same anteroposterior order in a U-shaped region of **cardiogenic mesoderm**, called the **cardiac crescent** (Fig. 6.14B). Following an inductive influence (involving members of the BMP and FGF families) by the endoderm (likely the anterior visceral endoderm, which also serves as the head organizer in mammals), the cells in this field are committed to the heart-forming pathway. In response, these cells express genes for several sets of transcription factors (*Nkx2-5*, *MEF2*, and *GATA4*) that are important in early heart development. From the cardiogenic mesoderm, the heart and great vessels form from bilaterally paired tubes that fuse in the midline beneath the foregut to produce a single tube (**Fig. 6.15**; see Fig. 6.14C).

A **secondary (anterior) heart field** has been described in chick and mouse embryos. Located in the splanchnic

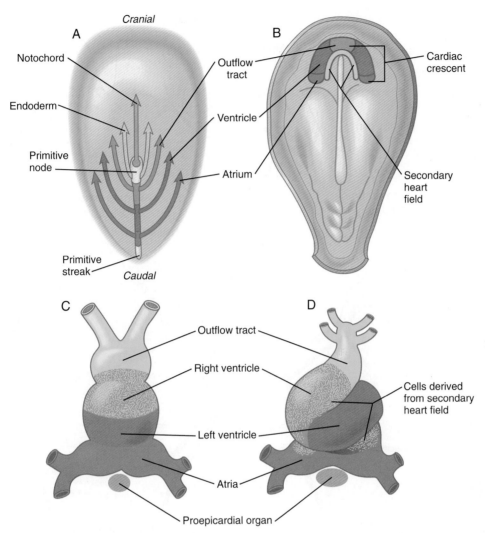

Fig. 6.14   Early stages in heart formation. A, Topographically precise movements of cardiogenic cells through the primitive streak. **B,** Horseshoe-shaped distribution of cardiogenic cells after their migration through the primitive streak. At this stage, the cardiogenic area is anterior to the rostralmost extent of the neural plate. **C,** The straight tubular heart. **D,** Ventral view of the S-shaped heart.

mesoderm on the posteromedial side of the cardiac crescent (see Fig. 6.14B), cells from the anterior part of the second heart field form most of the outflow tract and the right ventricle, and those from the posterior part of this field contribute to the formation of the atria (see Fig. 6.14D). In contrast, cells derived from the cardiac crescent form the left ventricle and most of the atria and make minor contributions to the outflow tract and right ventricle.

In human embryos, the earliest recognizable precardiac mesoderm is a crescent-shaped zone of thickened mesoderm rostral to the embryonic disk of the gastrulating embryo early in the third week (see Fig. 6.14B). As the mesoderm begins to split into the splanchnic and somatic layers, a **cardiogenic plate** is recognizable in the splanchnic mesoderm rostral to the oropharyngeal membrane (**Fig. 6.16A**). In this area, the space between the two layers of mesoderm is the forerunner of the **pericardial cavity**. The main layer of splanchnic mesoderm in the precardiac region thickens to become the **myocardial primordium**. Between this structure and the endoderm of the primitive gut, isolated mesodermal vesicles appear, which soon fuse to form the tubular **endocardial primordia**

(see Fig. 6.15A and B). The endocardial primordia ultimately fuse and become the inner lining of the heart.

As the head of the embryo takes shape by lateral and ventral folding, the bilateral cardiac primordia come together in the midline ventral to the gut and fuse to form a primitive single tubular heart. This structure consists of an inner **endocardial lining** surrounded by a loose layer of specialized extracellular matrix that has historically been called **cardiac jelly** (see Fig. 6.15C). Outside the cardiac jelly is the **myocardium**, which ultimately forms the muscular part of the heart. The outer lining of the heart, called the **epicardium**, and fibroblasts within the heart muscles are derived from the **proepicardial primordium**, which is located near the dorsal mesocardium (see Figs. 6.14C and D and 6.18). Cells migrating from the proepicardium cover the surface of the tubular heart. The entire tubular heart is located in the space known as the **pericardial coelom**. Shortly after the single tubular heart is formed, it begins to form a characteristic S-shaped loop that presages its eventual organization into the configuration of the adult heart (**Fig. 6.17**). (Additional cellular and molecular aspects of early cardiogenesis are discussed in Chapter 17.)

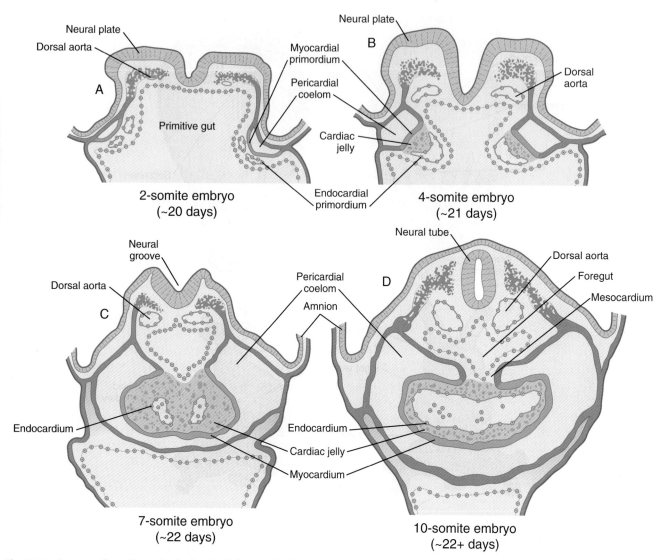

Fig. 6.15   Cross sections through the level of the developing heart from 20 to 22 days. A, Two-somite embryo. **B,** Four-somite embryo. **C,** Seven-somite embryo. **D,** Ten-somite embryo.

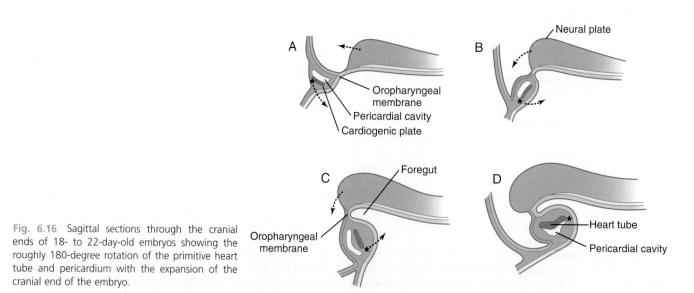

Fig. 6.16 Sagittal sections through the cranial ends of 18- to 22-day-old embryos showing the roughly 180-degree rotation of the primitive heart tube and pericardium with the expansion of the cranial end of the embryo.

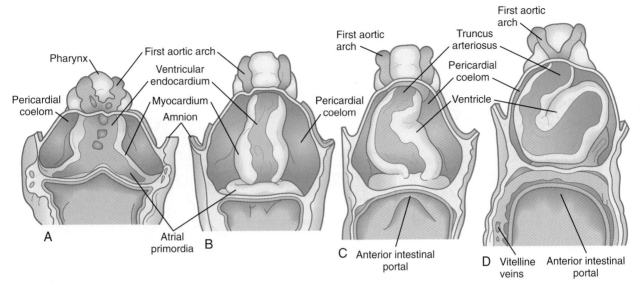

Fig. 6.17 **Formation of the S-shaped heart from fused cardiac tubes in the human embryo at about 21 to 23 days.** A, Four-somite embryo. B, Eight-somite embryo. C, Ten- to 11-somite embryo. D, Twelve-somite embryo.

The heart is formed from a variety of cell lineages. Within the cardiogenic mesoderm are cells that express N-cadherin and cells that do not (**Fig. 6.18A**). Depending on their location within the cardiogenic mesoderm, N-cadherin–positive cells go on to form either atrial or ventricular myocytes, whereas N-cadherin–negative cells form the endocardial lining and, later, cells of the endocardial cushions (see p. 428). Cells of the cardiac conduction system are derived from modified atrial and ventricular cardiac myocytes.

The early heart does not form in isolation. At its caudal end, the endocardial tubes do not fuse, but rather extend toward the posterior part of the body as the venous inflow tract of the heart (see Fig. 6.17A). Similarly, the endothelial tube leading out from the heart at its cranial end produces vascular arches that loop around the pharynx. Migrating neural crest cells form much of the walls of these vessels. By 21 or 22 days after fertilization, differentiation of cardiac muscle cells in the myocardium is sufficiently advanced to allow the heart to begin beating.

Blood and Blood Vessels

The formation of blood and blood vessels begins in the mesodermal wall of the yolk sac and in the wall of the chorion outside the embryo proper. Stimulated by an inductive interaction with the endoderm of the yolk sac and possibly also with the visceral endoderm, many small **blood islands**, consisting of stem cells called **hemangioblasts**, appear in the extraembryonic splanchnic mesoderm of the yolk sac (**Fig. 6.19**). Experimental evidence suggests that the inductive signal from the yolk sac endoderm is the signaling molecule **Indian hedgehog**. The yolk sac mesoderm responds to this signal by producing **BMP-4**, which feeds back onto itself. In a yet undefined manner, this interaction triggers the formation of blood islands within the mesoderm of the yolk sac. Within the blood islands, the central cells are blood-forming cells (**hemocytoblasts**), whereas the cells on the outside acquire the characteristics of **endothelial lining cells**, which form the inner walls of blood vessels. As the vesicular blood islands in the wall of

the yolk sac fuse, they form primitive vascular channels that extend toward the body of the embryo. Connections are made with the endothelial tubes associated with the tubular heart and major vessels, and the primitive plan of the circulatory system begins to take shape.

Development of the Endodermal Germ Layer

Starting at the time of gastrulation, the developing gut becomes specified into successively more discrete anteroposterior regions. The formation of endoderm as a germ layer depends on **nodal** signaling during gastrulation. In the high nodal environment close to the primitive node, the endodermal cells assume an anterior fate, whereas more posteriorly, the newly formed endodermal cells, which are exposed to lower levels of nodal and the presence of **FGF-4**, are specified to become more posterior structures as development proceeds. The posterior gut responds by expressing the transcription factor **Cdx-2**, which both promotes hindgut identity and suppresses the foregut differentiation program. Within the anterior domain, the gut expresses **Hex**, **Sox-2**, and **Foxa-2**. These early subdivisions of the gut set the stage for the more finely graded partitioning of the gut by the *Hox* genes (see Fig. 15.2) and the later induction of gut derivatives, such as the liver, pancreas, and lungs.

Development of the endodermal germ layer continues with the transformation of the flat intraembryonic endodermal sheet into a tubular gut as a result of the lateral folding of the embryonic body and the ventral bending of the cranial and caudal ends of the embryo into a roughly C-shaped structure (**Fig. 6.20**; see Fig. 6.7). A major morphological consequence of these folding processes is the sharp delineation of the **yolk sac** from the digestive tube.

Early in the third week, when the three embryonic germ layers are first laid down, the intraembryonic endoderm constitutes the roof of the roughly spherical yolk sac (see Fig. 6.20). Expansion of either end of the neural plate, particularly

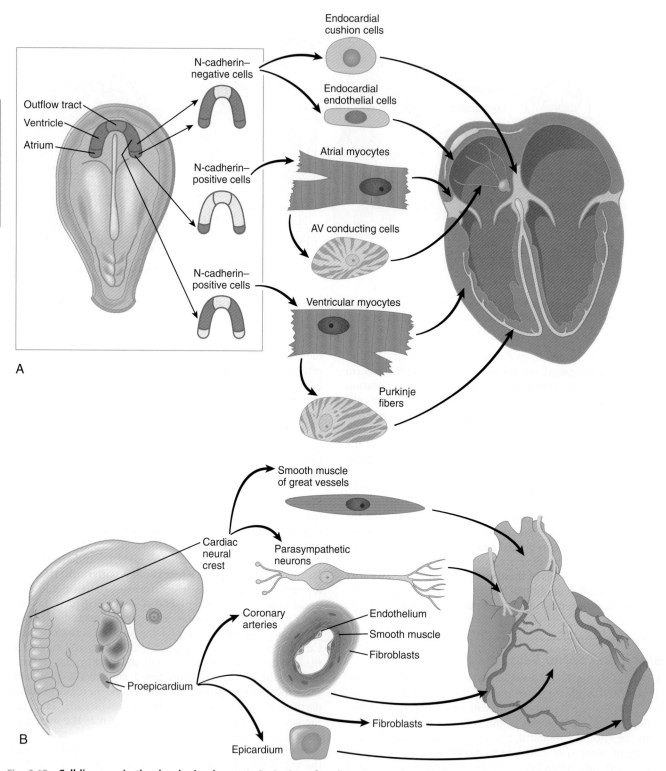

Fig. 6.18 Cell lineages in the developing heart. A, Derivatives of cardiogenic mesodermal cells. **B,** Cellular contributions of the cardiac neural crest and proepicardium to the heart. AV, atrioventricular. *(After Mikawa T: In Harvey RP, Rosenthal N, eds: Heart development, San Diego, 1999, Academic Press.)*

the tremendous growth of the future brain region, results in the formation of the **head fold** and **tail fold** along the sagittal plane of the embryo. This process, along with concomitant lateral folding, results in the formation of the beginnings of the tubular **foregut** and **hindgut**. This process also begins to delineate the yolk sac from the gut proper.

The sequence of steps in the formation of the tubular gut can be likened to a purse string constricting the ventral region of the embryo, although the actual mechanism is more related to the overall growth of the embryo than to a real constriction. The region of the imaginary purse string becomes the **yolk stalk** (also called the **omphalomesenteric** or **vitelline duct**),

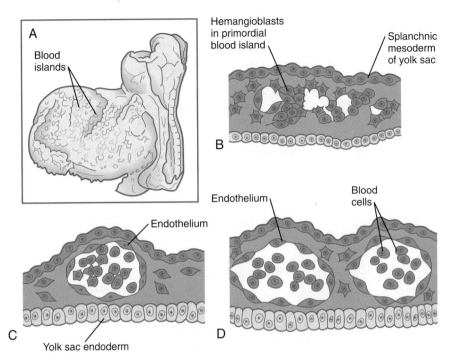

Fig. 6.19 **Development of blood islands in the yolk sac of human embryos.** **A,** Gross view of a 10-somite human embryo showing the location of blood islands on the yolk sac. **B** to **D,** Successive stages in the formation of blood islands. *(Redrawn from Corner G: Carnegie Contr Embryol 20:81-102, 1929.)*

with the embryonic gut above and the yolk sac below it (see Figs. 6.7D and 6.20D). The portion of the gut that still opens into the yolk sac is called the **midgut**, and the points of transition between the open-floored midgut and the tubular anterior and posterior regions of the gut are called the **anterior** and **posterior intestinal portals** (see Fig. 6.20B).

The endodermal edges of the anterior and posterior intestinal portals are also sites of expression of the signaling molecule **sonic hedgehog**. In the posterior intestinal portal, the appearance of sonic hedgehog in the endoderm is followed shortly by the expression of another signaling molecule, BMP-4. This is followed by the appearance of a gradient of mesodermal expression of paralogous groups 9 through 13 of the *Hox* genes (see Fig. 4.5 for an illustration of paralogous groups), with *Hoxa-d9* being expressed most cranially and *Hoxa-d13* being expressed most caudally, near the cloaca. This distribution of *Hox* gene expression associated with hindgut formation is reminiscent of that already described for the early hindbrain region (see p. 96).

In some cases, normal development of the gut and its related structures can proceed only when sonic hedgehog signaling is repressed. As discussed further on page 355, the dorsal pancreatic bud (see Fig. 6.20D) is induced by the notochord. A direct result of this induction is repression of sonic hedgehog signaling within the gut endoderm in the area of the dorsal pancreas. This repression allows the expression of the genes associated with the formation of the pancreas. At roughly the same anteroposterior level, but on the ventral side of the gut where the liver will form, hepatic endoderm expresses **albumin** in response to signals from the adjacent precardiac mesoderm.

The anterior end of the foregut remains temporarily sealed off by an ectodermal-endodermal bilayer called the **oropharyngeal membrane** (see Fig. 6.20B). This membrane separates

the future mouth (**stomodeum**), which is lined by ectoderm, from the **pharynx**, the endodermally lined anterior part of the foregut. Without an intervening layer of mesoderm, this bilayer of two epithelial sheets is inherently unstable and eventually breaks down. As discussed in Chapter 14, the endoderm of the anterior foregut acts as a powerful signaling center. Pharyngeal gut-derived molecular signals guide the formation and specific morphology of the pharyngeal arches.

The rapid bulging of the cephalic region, in conjunction with the constriction of the ventral region, has a major topographic effect on the rapidly developing cardiac region. In the early embryo, the cardiac primordia are located cephalic to the primitive gut. The forces that shape the tubular foregut cause the bilateral cardiac primordia to turn 180 degrees in the craniocaudal direction while the paired cardiac tubes are moving toward one another in the ventral midline (see Fig. 6.16).

In the region of the hindgut, the expansion of the embryo's body is not as prominent as it is in the cranial end, but nevertheless a less exaggerated ventral folding also occurs in that region. Even as the earliest signs of the tail fold are taking shape, a tubular evagination of the hindgut extends into the mesoderm of the body stalk. This evagination is called the **allantois** (see Fig. 6.20B). In most mammals and birds, the allantois represents a major structural adaptation for the exchange of gases and the removal of urinary wastes. Because of the efficiency of the placenta, however, the allantois never becomes a prominent structure in the human embryo. Nevertheless, because of the blood vessels that become associated with it, the allantois remains a vital part of the link between the embryo and the mother (see Chapter 7).

Caudal to the allantois is another ectodermal-endodermal bilayer called the **cloacal plate**, or **proctodeal membrane** (see Fig. 6.20C). This membrane, which ultimately breaks down,

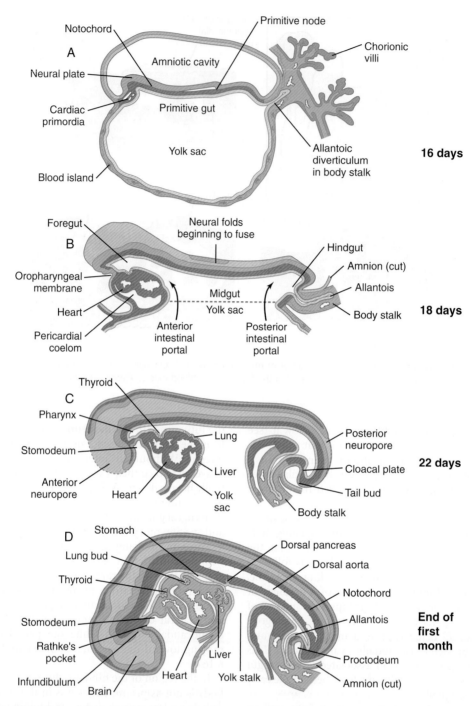

Fig. 6.20 **Sagittal sections through human embryos showing the early establishment of the digestive system.** A, At 16 days. B, At 18 days. C, At 22 days. D, At the end of the first month. *(After Patten. From Carlson BM: Patten's foundations of embryology, ed 6, New York, 1996, McGraw-Hill.)*

covers the cloaca, which in the early embryo represents a common outlet for the digestive and the urogenital systems. The shallow depression outside the proctodeal membrane is called the **proctodeum**.

As the gut becomes increasingly tubular, a series of local inductive interactions between the epithelium of the digestive tract and the surrounding mesenchyme initiates the formation

of most of the major digestive and endocrine glands (e.g., thyroid gland, salivary glands, pancreas), the respiratory system, and the liver. In the region of the stomodeum, an induction between forebrain and stomodeal ectoderm initiates the formation of the anterior pituitary gland. (Further development of these organs is discussed in Chapters 14 and 15.)

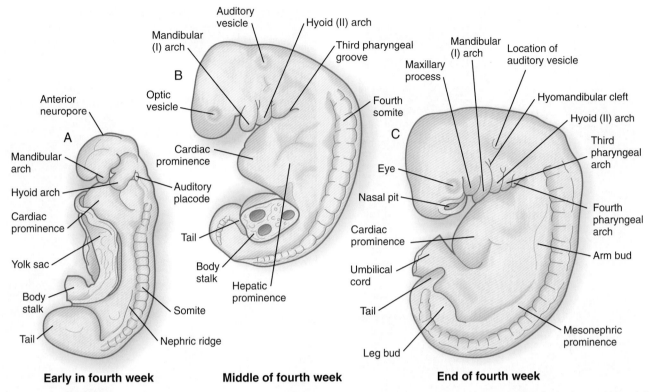

Fig. 6.21 **Gross development of human embryos during the period of early organogenesis.** **A,** Early in the fourth week. **B,** Middle of the fourth week. **C,** End of the fourth week.

Basic Structure of a 4-Week-Old Embryo

Gross Appearance

By the end of the fourth week of pregnancy, the embryo, which is still only about 4 mm long, has established the rudiments of most of the major organ systems except for the limbs (which are still absent) and the urogenital system (which has developed only the earliest traces of the embryonic kidneys). Externally, the embryo is C-shaped, with a prominent row of somites situated along either side of the neural tube (**Figs. 6.21** and **6.22**). Except for the rudiments of the eyes and ears and the oropharyngeal membrane, which is beginning to break down (**Fig. 6.23**), the head is relatively featureless. In the cervical region, **pharyngeal arches** are prominent (**Fig. 6.24**; see Fig. 6.21B and C). The body stalk still occupies a significant part of the ventral body wall, and cephalic to the body stalk, the heart and liver make prominent bulges in the contours of the ventral body wall. Posterior to the body stalk, the body tapers to a spiraled tail, which is prominent in embryos of this age.

Another prominent but little understood feature of embryos of this age is a ring of thickened ectoderm, called the **wolffian ridge**, which encircles the lateral aspect of the body (**Fig. 6.25**). Its function is not well understood, but it spans the primordia of many structures (e.g., nose, eye, inner ear, pharyngeal arches, limbs) that require tissue interactions for their early development. The wolffian ridge is marked molecularly by the expression of members of the Wnt signaling pathway. What role the thickened ectoderm plays in early organogenesis remains to be determined.

Circulatory System

At 4 weeks of age, the embryo has a functioning two-chamber heart and a blood vascular system that consists of three separate circulatory arcs (**Fig. 6.26**). The first, the **intraembryonic circulatory arc**, is organized in a manner similar to that of a fish. A ventral aortic outflow tract from the heart splits into a series of aortic arches passing around the pharynx through the pharyngeal arches and then collecting into a cephalically paired dorsal aorta that distributes blood throughout the body. A system of cardinal veins collects the blood and returns it to the heart via a common inflow tract.

The second arc, commonly called the **vitelline** or **omphalomesenteric arc**, is principally an extraembryonic circulatory loop that supplies the yolk sac (see Fig. 6.26). The third circulatory arc, also extraembryonic, consists of the vessels associated with the allantois. In humans, this third arc consists of the **umbilical vessels**, which course through the body stalk and spread in an elaborate network in the placenta and chorionic tissues. This set of vessels represents the real lifeline between the embryo and mother. Although the two extraembryonic circulatory loops do not persist as such after birth, the intraembryonic portions of these arcs are retained as vessels or ligaments in the adult body.

Derivatives of the Embryonic Germ Layers

By the end of the fourth week of development, primordia of most of the major structures and organs in the body have been laid down, many of them the result of local inductive interactions. Each of the embryonic germ layers contributes to the formation of many of these structures. **Figure 6.27**

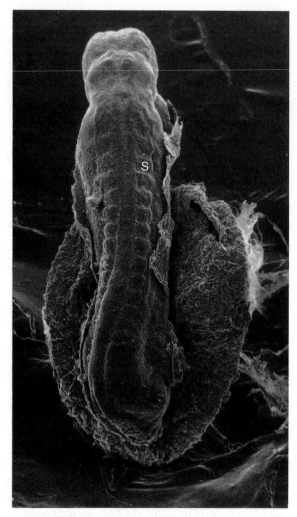

Fig. 6.22 **Scanning electron micrograph of a 3-mm human embryo approximately 26 days old.** S, Somite. *(From Jirásek JE: Atlas of human prenatal morphogenesis, Amsterdam, 1983, Martinus Nijhoff.)*

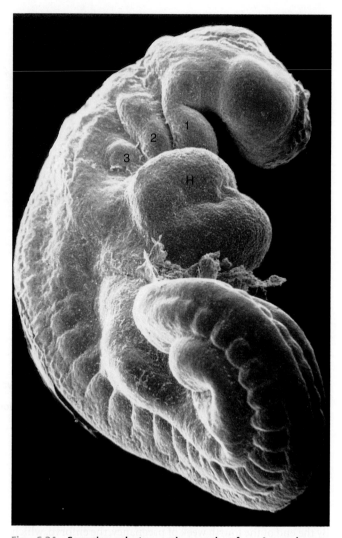

Fig. 6.24 **Scanning electron micrograph of a 4-mm human embryo 30 days old.** H, heart. Numbers 1 to 3 indicate pharyngeal arches. *(From Jirásek JE: Atlas of human prenatal morphogenesis, Amsterdam, 1983, Martinus Nijhoff.)*

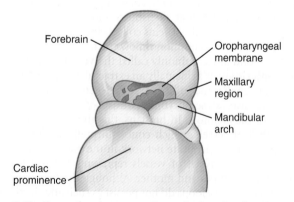

Fig. 6.23 **Face of a human embryo during the fourth week showing the breakdown of the oropharyngeal membrane.**

summarizes the germ layer origins of most of the major structures in the embryonic body. This figure is designed to be a guide that allows specific structures that are being studied to be viewed in the context of the whole body, rather than something that should be memorized at this stage. Students have found that a flow chart such as this is useful for review at the end of an embryology course.

Summary

- The response of dorsal ectodermal cells to primary induction is to thicken, thus forming a neural plate. Neurulation consists of lateral folding of the neural plate at hinge points to form a neural groove. Opposing sides of the thickened epithelium of the neural groove join to form a neural tube. The temporarily unclosed cranial and caudal ends of the neural tube are the anterior and posterior neuropores.
- Cranially, the neural tube subdivides into a primitive three-part brain consisting of the prosencephalon, mesencephalon, and rhombencephalon. The caudal part of the early brain also becomes subdivided into segments called neuromeres, of which the rhombomeres are most prominent. Specific homeobox genes are expressed in a regular order in the rhombomeres. A signaling center, the isthmic organizer, located at the midbrain and hindbrain junction acts via the production of Wnt-1 anteriorly and FGF-8 posteriorly.
- As the neural tube closes, neural crest cells emigrate from the neural epithelium and spread through the body along

well-defined paths. Secondary inductions acting on ectoderm in the cranial region result in the formation of several series of ectodermal placodes, which are the precursors of sense organs and sensory ganglia of cranial nerves.

- The embryonic mesoderm is subdivided into three craniocaudal columns: the paraxial, intermediate, and lateral plate mesoderm. Paraxial mesoderm is the precursor tissue to the paired somites and somitomeres. Segmentation of the paraxial mesoderm into somites occurs through the action of a clock mechanism that leads to the periodic expression of c-hairy and a variety of downstream molecules. As the result of a complex series of inductive interactions involving numerous signaling molecules, the epithelial somites become subdivided into sclerotomes (precursors of vertebral bodies) and dermomyotomes, which form dermatomes (dermal precursors) and myotomes (precursors of axial muscles). In further subdivisions, precursor cells of limb muscles are found in the lateral halves of the somites, and precursor cells of axial muscles are found in the medial halves. The posterior half of one sclerotome joins with the anterior half of the next caudal somite to form a single vertebral body.

- Intermediate mesoderm forms the organs of the urogenital system. The lateral plate mesoderm splits to form somatic mesoderm (associated with ectoderm) and splanchnic mesoderm (associated with endoderm). The space between becomes the coelom. The limb bud arises from lateral plate mesoderm, and extraembryonic mesoderm forms the body stalk.

- Blood cells and blood vessels form initially from blood islands located in the mesodermal wall of the yolk sac. The

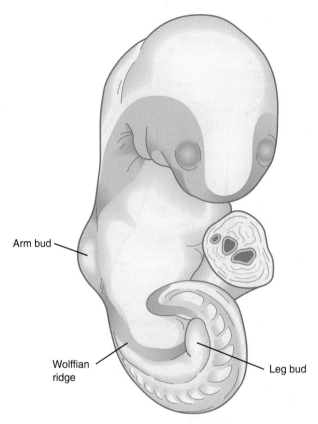

Fig. 6.25 **Ventrolateral view of a 30-somite (4.2-mm) human embryo showing the thickened ectodermal ring** *(blue)*. The portion of the ring between the upper and lower limb buds is the wolffian ridge. *(Based by O'Rahilly R, Gardner E: Anat Embryol 148:1-23, 1975.)*

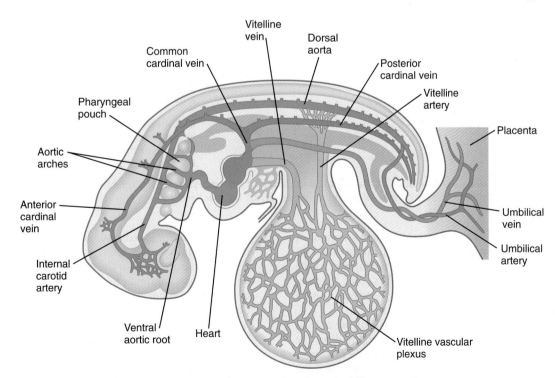

Fig. 6.26 **Basic circulatory arcs in a 4-week-old human embryo.**

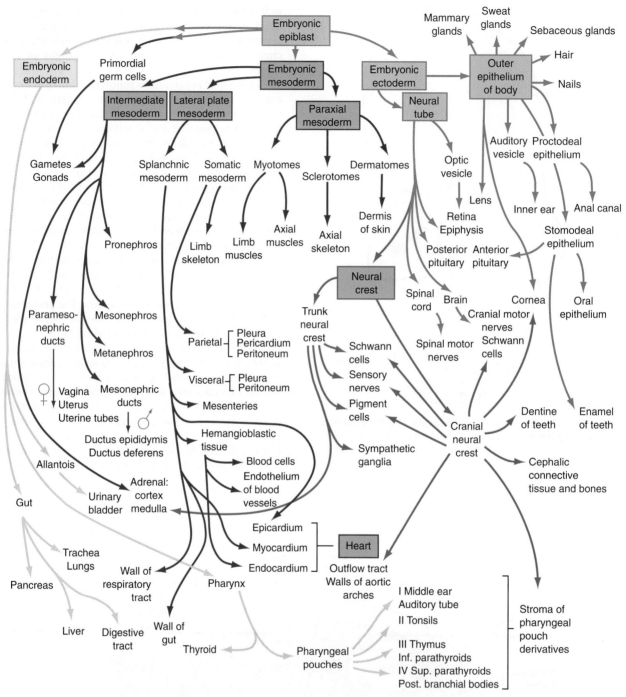

Fig. 6.27   Flow chart showing the formation of the organs and tissues of the embryo from the fundamental germ layers. The *arrows* are color-coded according to the germ layer of origin of the structure (see Fig. 4.1 for color code).

heart, originating from a horseshoe-shaped region of splanchnic mesoderm anterior to the oropharyngeal membrane, forms two tubes on either side of the foregut. As the foregut takes shape, the two cardiac tubes come together to form a single tubular heart, which begins to beat around 22 days after fertilization.

■ The embryonic endoderm initially consists of the roof of the yolk sac. As the embryo undergoes lateral folding, the endodermal gut forms cranial and caudal tubes (foregut and hindgut), but the middle region (midgut) remains open to the yolk sac ventrally. Regional specification of the gut begins with sonic hedgehog signals from the endoderm of the intestinal portals, which are translated to gradients of *Hox* gene expression in the neighboring mesoderm. As the tubular gut continues to take shape, the connection to the yolk sac becomes attenuated to form the yolk stalk. The future mouth (stomodeum) is separated from the foregut by an oropharyngeal membrane, and the hindgut is separated from the proctodeum by the cloacal plate. A ventral evagination from the hindgut is the allantois, which in many animals is an adaptation for removing urinary and respiratory wastes.

■ In a 4-week-old embryo, the circulatory system includes a functioning two-chamber heart and a blood vascular system that consists of three circulatory arcs. In addition to the intraembryonic circulation, the extraembryonic vitelline circulatory arc, which supplies the yolk sac, and the umbilical circulation, which is associated with the allantois and supplies the placenta, are present.

Review Questions

1. The sclerotome arises from cells that were located in the:
A. Notochord
B. Paraxial mesoderm
C. Intermediate mesoderm
D. Lateral plate mesoderm
E. None of the above

2. The cardiogenic plate arises from:
A. Embryonic endoderm
B. Somatic mesoderm
C. Splanchnic mesoderm
D. Intermediate mesoderm
E. Neural crest

3. An inductive stimulus from which structure stimulates the transformation of the epithelial sclerotome into secondary mesenchyme?
A. Neural crest
B. Somite
C. Ectodermal placodes
D. Embryonic endoderm
E. Notochord

4. Which of these structures in the embryo is unsegmented?
A. Somitomeres
B. Neuromeres
C. Notochord
D. Somites

5. The intermediate mesoderm is the precursor of the:
A. Urogenital system
B. Heart
C. Somites
D. Body wall
E. Vertebral bodies

6. What forces are involved in the folding of the neural plate to form the neural tube?

7. What role do neuromeres play in the formation of the central nervous system?

8. From what structures do the cells that form skeletal muscles arise?

9. Where do the first blood cells of the embryo form?

References

Abu-Issa R, Kirby ML: Heart field: from mesoderm to heart tube, *Annu Rev Cell Dev Biol* 23:45-58, 2007.

Alexander T, Nolte C, Krumlauf R: *Hox* genes and segmentation of the hindbrain and axial skeleton, *Annu Rev Cell Dev Biol* 25:431-456, 2009.

Andrew DJ, Ewald AJ: Morphogenesis of epithelial tubes: insights into tube formation, elongation, and elaboration, *Dev Biol* 341:34-55, 2010.

Baker CVH, Bronner-Fraser M: Vertebrate cranial placodes, I: embryonic induction, *Dev Biol* 232:1-61, 2001.

Bergquist H: Studies on the cerebral tube in vertebrates: the neuromeres, *Acta Zool* 33:117-187, 1952.

Bothe I and others: Extrinsic versus intrinsic cues in avian paraxial mesoderm patterning and differentiation, *Dev Dyn* 236:2397-2409, 2007.

Brent AE and others: A somitic compartment of tendon progenitors, *Cell* 113:235-248, 2003.

Buckingham M, Meilhac S, Zaffran S: Building the mammalian heart from two sources of myocardial cells, *Nat Rev Genet* 6:826-835, 2005.

Christ B, Huang R, Scaal M: Amniote somite derivatives, *Dev Dyn* 236:2382-2396, 2007.

Christ B, Huang R, Scaal M: Formation and differentiation of the avian sclerotome, *Anat Embryol* 208:333-350, 2004.

Christ B, Scaal M: Formation and differentiation of avian somite derivatives, *Adv Exp Med Biol* 638:1-41, 2008.

Colas J-F, Schoenwolf GC: Towards a cellular and molecular understanding of neurulation, *Dev Dyn* 221:117-145, 2001.

Dequéant M-L and others: A complex oscillating network of signaling genes underlies the mouse segmentation clock, *Science* 314:1595-1598, 2006.

Deschamps J, van Nes J: Developmental regulation of the *Hox* genes during axial morphogenesis in the mouse, *Development* 132:2931-2942, 2005.

Deyer LA, Kirby ML: The role of secondary heart field in cardiac development, *Dev Biol* 336:137-144, 2009.

Diez del Corral R, Storey KG: Opposing FGF and retinoid pathways: a signalling switch that controls differentiation and patterning onset in the extending vertebrate body axis, *Bioessays* 26:857-869, 2004.

Gibb S, Maroto M, Dale JK: The segmentation clock mechanism moves up a notch, *Trends Cell Biol* 20:593-600, 2010.

Glover JS, Renaud J-S, Rijli FM: Retinoic acid and hindbrain patterning, *J Neurobiol* 66:705-725, 2006.

Harvey RP, Rosenthal N, eds: *Heart development*, San Diego, 1999, Academic Press.

Heath JK: Transcriptional networks and signaling pathways that govern vertebrate intestinal development, *Curr Top Dev Biol* 90:159-192, 2010.

Holley SA: Vertebrate segmentation: snail counts the time until morphogenesis, *Curr Biol* 16:R367-R369, 2006.

Iimura T, Denans N, Porquié O: Establishment of Hox vertebral identities in the embryonic spine precursors, *Curr Top Dev Biol* 88:201-234, 2009.

Jacobson AG: Somitomeres: mesodermal segments of the head and trunk. In Hanken J, Hall BK, eds: *The skull, vol 1, Development*, Chicago, 1993, University of Chicago Press.

Kulesa PM and others: From segment to somite: segmentation to epithelialization analyzed within quantitative frameworks, *Dev Dyn* 236:1392-1402, 2007.

Larson KC, Füchtbauer E-M, Brand-Saberi B: The neural tube is required to maintain primary segmentation in the sclerotome, *Cells Tiss Organs* 182:12-21, 2006.

Lewis SL, Tam PPL: Definitive endoderm of the mouse embryo: formation, cell fates, and morphogenetic function, *Dev Dyn* 235:2315-2329, 2006.

Lumsden A, Krumlauf R: Patterning the vertebrate neuraxis, *Science* 274:1109-1115, 1996.

Martinez S: The isthmic organizer and brain regionalization, *Int J Dev Biol* 45:367-371, 2001.

Mittapalli VP and others: Arthrotome: a specific joint forming compartment in the avian somite, *Dev Dyn* 234:48-53, 2005.

Müller F, O'Rahilly R: The timing and sequence of appearance of neuromeres and their derivatives in staged human embryos, *Acta Anat* 158:83-99, 1997.

Nakatsu T, Uwabe C, Shiota K: Neural tube closure in humans initiates at multiple sites: evidence from human embryos and implications for the pathogenesis of neural tube defects, *Anat Embryol* 201:455-466, 2000.

O'Rahilly R, Müller F: The origin of the ectodermal ring in staged human embryos of the first 5 weeks, *Acta Anat* 122:145-157, 1985.

Ordahl CP, ed: Somitogenesis, part 1, *Curr Top Dev Biol* 47:1-316, 2000.

Ordahl CP, ed: Somitogenesis, part 2, *Curr Top Dev Biol* 48:1-388, 2000.

Özbudak EM, Pourquié O: The vertebrate segmentation clock: the tip of the iceberg, *Curr Opin Genet Dev* 18:317-323, 2008.

Pourquié O: Vertebrate segmentation: from cyclic gene networks to scoliosis, *Cell* 145:650-663, 2011.

Pourquié O: The vertebrate segmentation clock, *J Anat* 199:169-175, 2001.

Raya A, Izpisúa Balmonte JC: Left-right asymmetry in the vertebrate embryo: from early information to higher-level integration, *Nat Rev Genet* 7:283-293, 2006.

Rhinn M, Brand M: The midbrain-hindbrain boundary organizer, *Curr Opin Neurobiol* 11:34-42, 2001.

Rhinn M, Picker A, Brand M: Global and local mechanisms of forebrain and midbrain patterning, *Curr Opin Neurobiol* 16:5-12, 2006.

Roberts DJ and others: Sonic hedgehog is an endodermal signal inducing *Bmp-4* and *Hox* genes during induction and regionalization of the chick hindgut, *Development* 121:3163-3174, 1995.

Rossant JU, Tam PPL, eds: *Mouse development: patterning, morphogenesis, and organogenesis,* San Diego, 2002, Academic Press.

Rubenstein JLR and others: The embryonic vertebrate forebrain: the prosomeric model, *Science* 266:578-580, 1994.

Sawyer JM and others: Apical constriction: a cell shape change that can drive morphogenesis, *Dev Biol* 341:5-19, 2010.

Scaal M, Christ B: Formation and differentiation of the avian dermomyotome, *Anat Embryol* 208:411-424, 2004.

Schoenwolf GC: Histological and ultrastructural studies of secondary neurulation in mouse embryos, *Am J Anat* 169:361-376, 1984.

Skoglund P, Keller R: Integration of planar cell polarity and ECM signaling in elongation of the vertebrate body plan, *Curr Opin Cell Biol* 22:589-596, 2010.

Takahashi Y: Common mechanisms for boundary formation in somitogenesis and brain development: shaping the "chic" chick, *Int J Dev Biol* 49:221-230, 2005.

Tam PPL, Quinlan GA, Trainor PA: The patterning of progenitor tissues for the cranial region of the mouse embryo during gastrulation and early organogenesis, *Adv Dev Biol* 5:137-200, 1997.

Trainor PA, Krumlauf R: Patterning the cranial neural crest: hindbrain segmentation and *Hox* gene plasticity, *Nat Rev Neurosci* 1:116-124, 2000.

Tremblay KD: Formation of the murine endoderm: lessons from the mouse, frog, fish, and chick, *Prog Mol Biol Transl Sci* 96:1-34, 2010.

Tümpel S, Wiedemann LM, Krumlauf R: *Hox* genes and segmentation of the vertebrate hindbrain, *Curr Top Dev Biol* 88:103-137, 2009.

Vincent SD, Buckingham ME: How to make a heart: the origin and regulation of cardiac progenitor cells, *Curr Top Dev Biol* 90:1-41, 2010.

Wilson V, Olivera-Martinez I, Storey KG: Stem cells, signals and vertebrate body axis extension, *Development* 136:1591-1604, 2009.

Winslow BB, Takimoto-Kimura R, Burke AC: Global patterning of the vertebrate mesoderm, *Dev Dyn* 236:2371-2381, 2007.

Wurst W, Bally-Cuif L: Neural plate patterning: upstream and downstream of the isthmic organizer, *Nat Rev Neurosci* 2:99-108, 2001.

Young T, Deschamps J: *Hox, Cdx,* and anteroposterior patterning in the mouse embryo, *Curr Top Dev Biol* 88:235-255, 2009.

Zorn AM, Wells JM: Vertebrate endoderm development and organ formation, *Annu Rev Cell Dev Biol* 25:221-251, 2009.

Chapter 7

Placenta and Extraembryonic Membranes

One of the most characteristic features of human embryonic development is the intimate relationship between the embryo and the mother. The fertilized egg brings little with it except genetic material. To survive and grow during intrauterine life, the embryo must maintain an essentially parasitic relationship with the body of the mother for acquiring oxygen and nutrients and eliminating wastes. It must also avoid being rejected as a foreign body by the immune system of its maternal host. These exacting requirements are met by the placenta and extraembryonic membranes that surround the embryo and serve as the interface between the embryo and the mother.

The tissues that compose the fetal-maternal interface (**placenta** and **chorion**) are derivatives of the **trophoblast**, which separates from the inner cell mass and surrounds the cellular precursors of the embryo proper even as the cleaving zygote travels down the uterine tube on its way to implanting into the uterine wall (see Fig. 3.18). Other extraembryonic tissues are derived from the inner cell mass. These include the following: the **amnion** (an ectodermal derivative), which forms a protective fluid-filled capsule around the embryo; the **yolk sac** (an endodermal derivative), which in mammalian embryos no longer serves a primary nutritive function; the **allantois** (an endodermal derivative), which is associated with the removal of embryonic wastes; and much of the **extraembryonic mesoderm**, which forms the bulk of the umbilical cord, the connective tissue backing of the extraembryonic membranes, and the blood vessels that supply them.

Extraembryonic Tissues

Amnion

The origin of the amniotic cavity within the ectoderm of the inner cell mass in the implanting embryo is described in Chapter 5 (see Figs. 3.18 and 5.2). As the early embryo undergoes cephalocaudal and lateral folding, the amniotic membrane surrounds the body of the embryo like a fluid-filled balloon (**Fig. 7.1**), thus allowing the embryo to be suspended in a liquid environment for the duration of pregnancy. The amniotic fluid serves as a buffer against mechanical injury to the fetus; in addition, it accommodates growth, allows normal fetal movements, and protects the fetus from adhesions.

The thin amniotic membrane consists of a single layer of extraembryonic ectodermal cells lined by a nonvascularized layer of extraembryonic mesoderm. Keeping pace with fetal growth, the amniotic cavity steadily expands until its fluid content reaches a maximum of nearly 1 L by weeks 33 to 34 of pregnancy (**Fig. 7.2**).

In many respects, amniotic fluid can be viewed as a dilute transudate of maternal plasma, but the origins and exchange dynamics of amniotic fluid are complex and not completely understood. There are two phases in amniotic fluid production. The first phase encompasses the first 20 weeks of pregnancy, during which the composition of amniotic fluid is quite similar to that of fetal fluids. During this period, the fetal skin is unkeratinized, and there is evidence that fluid and electrolytes are able to diffuse freely through the embryonic ectoderm of the skin. In addition, the amniotic membrane itself secretes fluid, and components of maternal serum pass through the amniotic membrane.

As pregnancy advances (especially after week 20, when the fetal epidermis begins to keratinize), changes occur in the source of amniotic fluid. There is not complete agreement on the sources (and their relative contributions) of amniotic fluid in the second half of pregnancy. Nonetheless, there are increasing contributions from fetal urine, filtration from maternal blood vessels near the chorion laeve (which is closely apposed to the amniotic membrane at this stage), and possibly filtration from fetal vessels in the umbilical cord and chorionic plate.

In the third trimester of pregnancy, the amniotic fluid turns over completely every 3 hours, and at term, the fluid-exchange rate may approach 500 mL/hour. Although much of the amniotic fluid is exchanged across the amniotic membrane, fetal swallowing is an important mechanism in late pregnancy, with about 20 mL/hour of fluid swallowed by the fetus. Swallowed amniotic fluid ultimately enters the fetal bloodstream after absorption through the gut wall. The ingested water can leave the fetal circulation through the placenta. During the fetal period, excreted urine from the fetus contributes to amniotic fluid. **Clinical Correlation 7.1** discusses conditions related to the amount of amniotic fluid or substance concentrations in the fluid.

Traditionally, the amniotic membrane has been discarded along with the placenta and other extraembryonic tissues after the child has been born. In more recent years, however, important medical uses have been found for amniotic membranes. Because of the anti-inflammatory and antiangiogenic properties of amnion, sheets of amnion have been used to cover a

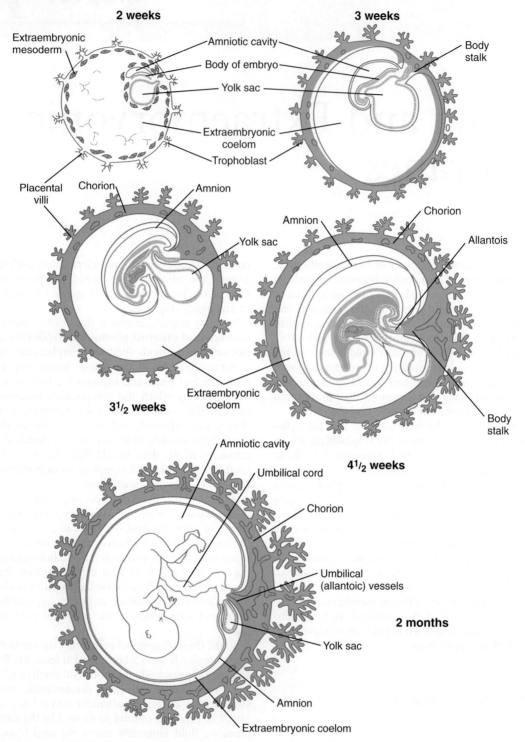

Fig. 7.1 **Human embryos showing the relationships of the chorion and other extraembryonic membranes.** *(Adapted from Carlson BM: Patten's foundations of embryology, ed 6, New York, 1996, McGraw-Hill.)*

variety of wounds or burn surfaces, especially in ophthalmic surgery. The amnion, as well as amniotic fluid and other placental tissues, has proven to be a major source of stem cells, which have the capability of differentiating into cell types from each of the three germ layers.

Yolk Sac

The yolk sac, which is lined by extraembryonic endoderm, is formed ventral to the bilayered embryo when the amnion appears dorsal to the embryonic disk (see Fig. 5.2). In contrast to birds and reptiles, the yolk sac of mammals is small and

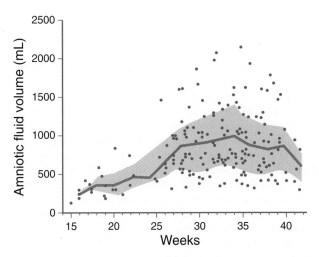

Fig. 7.2 **Volumes of amniotic fluid in women at various weeks of pregnancy.** The *lined and shaded area* represents the mean ± standard deviation. *Dots* represent outlying values. *(Data from Queenan JT and others:* Am J Obstet Gynecol *114:34-38, 1972.)*

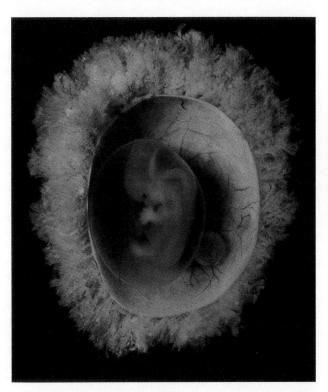

Fig. 7.3 **A 7-week-old human embryo surrounded by its amnion.** The embryo was exposed by cutting open the chorion. The small sphere to the right of the embryo is the yolk sac. *(Carnegie embryo No. 8537A, Courtesy of Chester Reather, Baltimore.)*

devoid of yolk. Although vestigial in terms of its original function as a major source of nutrition, the yolk sac remains vital to the embryo because of other functions that have become associated with it. The human yolk sac has traditionally been considered to be a vestigial structure with respect to nutrition; some evidence indicates that before the placental circulation is established, nutrients, such as folic acid and vitamins A, B$_{12}$, and E, are concentrated in the yolk sac and are absorbed by endocytosis. Because this form of **histiotrophic nutrition** occurs during the time of neurulation, it may play a role in the prevention of neural tube defects (see p. 138 in Chapter 8).

When it first appears, the yolk sac is in the form of a hemisphere bounded at the equatorial region by the dorsal wall of the primitive gut (see Fig. 7.1). As the embryo grows and undergoes lateral folding and curvature along the craniocaudal axis, the connection between the yolk sac and the forming gut becomes attenuated in the shape of a progressively narrowing stalk attached to a more spherical yolk sac proper at its distal end. In succeeding weeks, the yolk stalk becomes very long and attenuated as it is incorporated into the body of the umbilical cord. The yolk sac itself moves nearer the chorionic plate of the placenta (**Fig. 7.3**).

The endoderm of the yolk sac is lined on the outside by well-vascularized extraembryonic mesoderm. Cells found in each of these layers contribute vital components to the body of the embryo. During the third week, **primordial germ cells,** which arise in the extraembryonic mesoderm near the base of the allantois, become recognizable in the lining of the yolk sac (see Fig. 1.1). Soon these cells migrate into the wall of the gut and the dorsal mesentery as they make their way to the gonads, where they differentiate into oogonia or spermatogonia.

In the meantime, groups of extraembryonic mesodermal cells in the wall of the yolk sac become organized into **blood islands** (see Fig. 6.19), and many of the cells differentiate into primitive blood cells. **Extraembryonic hematopoiesis** continues in the yolk sac until about the sixth week, when blood-forming activity transfers to intraembryonic sites, especially the liver.

As the tubular gut forms, the attachment site of the yolk stalk becomes progressively less prominent, until by 6 weeks it has effectively lost contact with the gut. In a small percentage of adults, traces of the yolk duct persist as a fibrous cord or an outpouching of the small intestine known as **Meckel's diverticulum** (see Fig. 15.15A). The yolk sac itself may persist throughout much of pregnancy, but it is not known to have a specific function in the fetal period. The proximal portions of the blood vessels of the yolk sac (the vitelline circulatory arc) persist as vessels that supply the midgut region.

Allantois

The allantois arises as an endodermally lined ventral outpocketing of the hindgut (see Fig. 7.1). In the human embryo, it is just a vestige of the large, saclike structure that is used by the embryos of many mammals, birds, and reptiles as a major respiratory organ and repository for urinary wastes. Similar to the yolk sac, the allantois in a human retains only a secondary function, in this case respiration. In humans, this function is served by the blood vessels that differentiate from the mesodermal wall of the allantois. These vessels form the umbilical circulatory arc, consisting of the arteries and veins that supply the placenta (see Fig. 6.26). (The postnatal fate of these vessels is discussed in Chapter 18.)

The allantois proper, which consists of little more than a cord of endodermal cells, is embedded in the umbilical cord. Later in development, the proximal part of the allantois (called the **urachus**) is continuous with the forming urinary bladder (see Fig. 16.2). After birth, it becomes transformed into a dense fibrous cord (median umbilical ligament), which runs

CLINICAL CORRELATION 7.1
Conditions Related to Amniotic Fluid

The normal amount of amniotic fluid at term is typically 500 to 1000 mL. An excessive amount (>2000 mL) is **hydramnios**. This condition is frequently associated with multiple pregnancies and **esophageal atresia** or **anencephaly** (a congenital anomaly characterized by gross defects of the head and often the inability to swallow [see Fig. 8.4]). Such circumstantial evidence supports the important role of fetal swallowing in the overall balance of amniotic fluid exchange. Too little amniotic fluid (<500 mL) is **oligohydramnios**. This condition is often associated with bilateral **renal agenesis** (absence of kidneys) and points to the role of fetal urinary excretion in amniotic fluid dynamics. Oligohydramnios also can be a consequence of preterm rupture of the amniotic membrane, which occurs in about 10% of pregnancies.

There are many components, both fetal and maternal, in amniotic fluid; more than 200 proteins of maternal and fetal origin have been detected in amniotic fluid. With the analytical tools available, much can be learned about the condition of the fetus by examining the composition of amniotic fluid. **Amniocentesis** involves removing a small amount of amniotic fluid by inserting a needle through the mother's abdomen and into the amniotic cavity. Because of the small amount of amniotic fluid in early embryos, amniocentesis is usually not performed until the thirteenth or fourteenth week of pregnancy. Amniotic fluid has bacteriostatic properties, which may account for the low incidence of infections after amniocentesis is performed.

Fetal cells present in the amniotic fluid can be cultured and examined for various chromosomal and metabolic defects. More recent techniques now permit the examination of chromosomes in the cells immediately obtained, instead of having to wait up to 2 to 3 weeks for cultured amniotic cells to proliferate to the point of being suitable for genetic analysis. In addition to the detection of chromosomal defects (e.g., trisomies), it is possible to determine the sex of the fetus by direct chromosomal analysis. Many cells of the amniotic fluid have been shown to possess stem cell properties. Whether amniotic stem cells have as broad a capacity to differentiate into as wide a variety of mature cell types as embryonic stem cells remains to be established.

A high concentration of **α-fetoprotein** (a protein of the central nervous system) in amniotic fluid is a strong indicator of a neural tube defect. Fetal maturity can be assessed by determining the concentration of creatinine or the **lecithin-to-sphingomyelin ratio** (which is a reflection of the maturity of the lungs). The severity of **erythroblastosis fetalis** (Rh disease) can also be assessed by examination of amniotic fluid.

from the urinary bladder to the umbilical region (see Fig. 18.19).

Chorion and Placenta

Formation of the placental complex represents a cooperative effort between the extraembryonic tissues of the embryo and the endometrial tissues of the mother. (Early stages of implantation of the embryo and the decidual reaction of the uterine lining are described in Chapter 3.) After implantation is complete, the original trophoblast surrounding the embryo has undergone differentiation into two layers: the inner **cytotrophoblast** and the outer **syncytiotrophoblast** (see Fig. 3.18D). Lacunae in the rapidly expanding trophoblast have filled with maternal blood, and the connective tissue cells of the endometrium have undergone the decidual reaction (containing increased amounts of glycogen and lipids) in response to the trophoblastic invasion.

Formation of Chorionic Villi

In the early implanting embryo, the trophoblastic tissues have no consistent gross morphological features; consequently, this is called the period of the **previllous embryo**. Late in the second week, defined cytotrophoblastic projections called **primary villi** begin to take shape (see Fig. 5.2). Shortly thereafter, a mesenchymal core appears within an expanding villus, at which point it is properly called a **secondary villus** (**Fig. 7.4**). Surrounding the mesenchymal core of the secondary villus is a complete layer of cytotrophoblastic cells, and outside of that is the syncytiotrophoblast. By definition, the secondary villus becomes a **tertiary villus** when blood vessels penetrate its mesenchymal core and newly formed branches. This event occurs toward the end of the third week of pregnancy.

Although individual villi undergo considerable branching, most of them retain the same basic structural plan throughout pregnancy. While the placental villi are becoming established, the homeobox-containing genes, *Msx2* and *Dlx4* (*distalless-4*), are expressed near the interface between the trophoblast and underlying extraembryonic mesenchyme. These transcription factors are often seen at sites of epitheliomesenchymal interactions. The transcription factor **Gem-1**, which promotes an exit from the cell cycle, is expressed at branching points on the villi. Cytotrophoblastic cells on either side of the region of Gem-1 expression continue to proliferate. These cells form the cellular basis of new villous buds.

The terminal portion of a villus remains trophoblastic, consisting of a solid mass of cytotrophoblast called a **cytotrophoblastic cell column** (see Fig. 7.4) and a relatively thin covering of syncytiotrophoblast over that. The villus is bathed in maternal blood. A further development of the tip of the villus occurs when, under the influence of the local hypoxic environment, the cytotrophoblastic cell column expands distally and penetrates the syncytiotrophoblastic layer (**Fig. 7.5**). These cytotrophoblastic cells abut directly on maternal decidual cells and spread over them to form a complete cellular layer known as the **cytotrophoblastic shell**, which surrounds the embryo complex. The villi that give off the cytotrophoblastic extensions are known as **anchoring villi** (see Fig. 7.4) because they represent the real attachment points between the embryo complex and the maternal tissues.

It is important to understand the overall relationships of the various embryonic and maternal tissues at this stage of development (see Fig. 7.5). The embryo, attached by the **body stalk**, or **umbilical cord**, is effectively suspended in the **chorionic cavity**. The chorionic cavity is bounded by the **chorionic plate**, which consists of extraembryonic mesoderm overlaid with trophoblast. The chorionic villi extend outward

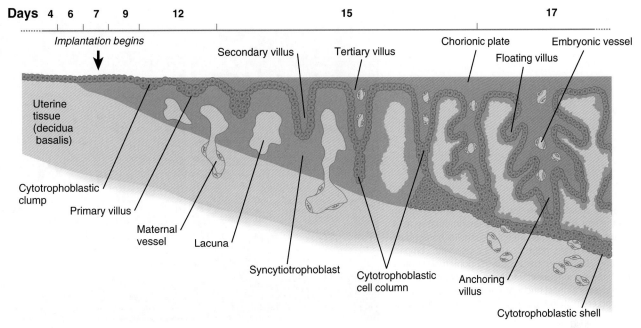

Fig. 7.4 Stages in the formation of a chorionic villus, starting with a cytotrophoblastic clump at the far left and progressing over time to an anchoring villus at *right*.

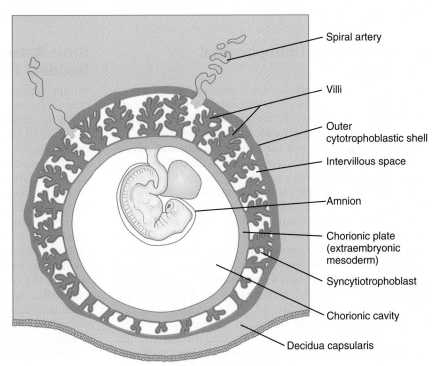

Fig. 7.5 Overall view of a 5-week-old embryo in addition to membranes showing the relationships of the chorionic plate, villi, and outer cytotrophoblastic shell.

from the chorionic plate, and their trophoblastic covering is continuous with that of the chorionic plate. The villi and the outer surface of the chorionic plate are bathed in a sea of continually exchanging maternal blood. Because of this, the human placenta is designated the **hemochorial type.***

Although chorionic villi are structurally very complicated, it is convenient to liken the basic structure of a villus complex to the root system of a plant. The anchoring villus is equivalent to the central tap root; by means of the cytotrophoblastic cell columns, it attaches the villus complex to the outer cytotrophoblastic shell. The unattached branches of the **floating villi** (see Fig. 7.12) dangle freely in the maternal blood that fills the space between the chorionic plate and the outer cytotrophoblastic shell. All surfaces of the villi, chorionic plate, and cytotrophoblastic shell that are in contact with

---

*Other mammals have various arrangements of tissue layers through which materials must pass to be exchanged between mother and fetus. For example, in an epitheliochorial placenta, which is found in pigs, the fetal component of the placenta (chorion) rests on the uterine epithelium instead of being directly bathed in maternal blood.

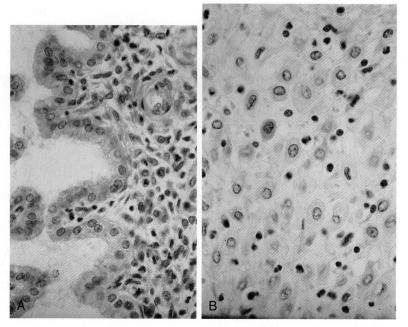

Fig. 7.6   A, Histological section through the endometrium during the late secretory stage of the endometrial cycle. A large uterine gland with an irregular epithelial border is on the *left.* On the *right,* note the stromal cells with compact nuclei and scanty cytoplasm. **B,** Endometrial stroma, showing the decidual reaction. Note the expanded cytoplasm and less compact nuclei of the decidual cells. (Hematoxylin and eosin stain.) *(Courtesy of D. MacCallum, Ann Arbor, Mich.)*

maternal blood are lined with a continuous layer of syncytiotrophoblast.

Establishing the Uteroplacental Circulation

One of the critical features of the developing embryonic-maternal interface is the establishment of a **uteroplacental circulation** that serves as the medium for bringing food and oxygen to and removing wastes from the embryo. This is accomplished by erosion of the walls of the **spiral arteries** of the uterus and their modification so that, as the embryo grows, these arteries can provide an increasing flow of blood at low pressure to bathe the syncytiotrophoblastic surface of the placenta (see Fig. 7.10). Specialized **invasive cytotrophoblastic cells**, migrating out from the anchoring villi, invade the spiral arteries (but not the veins) and cause major modifications of their walls by secreting a specialized extracellular matrix and displacing many of the normal cellular elements of the spiral arteries. As a result, the arteries become wider, but the blood escaping from their open ends leaves at a much lower pressure than normal arterial pressure.

The first maternal fluid that bathes the embryonic trophoblast is not highly cellular, and the oxygen tension is low. During this period, the fetal erythrocytes contain embryonic hemoglobin, which is adapted to bind oxygen under low tension.

Hypoxia stimulates cytotrophoblastic cells to undergo mitosis. This may be one of the environmental conditions that underlies the rapid growth of the cytotrophoblast during the early embryonic period. After 12 weeks, when the maternal blood in the placental space contains large numbers of erythrocytes and is more highly oxygenated, the fetal erythrocytes, through an isoform switch, begin to produce fetal hemoglobin, which requires higher oxygen tension to bind oxygen efficiently. The maternal blood that leaves the spiral arteries freely percolates throughout the intervillous spaces and bathes the surfaces of the villi. The maternal blood is then picked up by the open ends of the uterine veins, which also penetrate the cytotrophoblastic shell (see Fig. 7.10).

Gross Relationships of Chorionic and Decidual Tissues

Within days after implantation of the embryo, the stromal cells of the endometrium undergo a striking transformation called the **decidual*** **reaction.** After the stromal cells swell as the result of the accumulation of glycogen and lipid in their cytoplasm, they are known as **decidual cells** (**Fig. 7.6**). The decidual reaction spreads throughout stromal cells in the superficial layers of the endometrium. The maternal decidua are given topographic names based on where they are located in relation to the embryo.

The decidual tissue that overlies the embryo and its chorionic vesicle is the **decidua capsularis**, whereas the decidua that lies between the chorionic vesicle and the uterine wall is the **decidua basalis** (**Fig. 7.7**). With continued growth of the embryo, the decidua basalis becomes incorporated into the maternal component of the definitive placenta. The remaining decidua, which consists of the decidualized endometrial tissue on the sides of the uterus not occupied by the embryo, is the **decidua parietalis**.

In human embryology, the **chorion** is defined as the layer consisting of the trophoblast and the underlying extraembryonic mesoderm (see Fig. 7.1). The chorion forms a complete covering (**chorionic vesicle**) that surrounds the embryo, amnion, yolk sac, and body stalk. During the early period after implantation, primary and secondary villi project almost uniformly from the entire outer surface of the chorionic vesicle. The formation of tertiary villi is asymmetric, however, and the invasion of the cytotrophoblastic core of the primary villi by mesenchyme and embryonic blood vessels occurs

---

*The term "deciduum" refers to tissues that are shed at birth. These include the extraembryonic tissues in addition to the superficial layers of the endometrial connective tissue and epithelium.

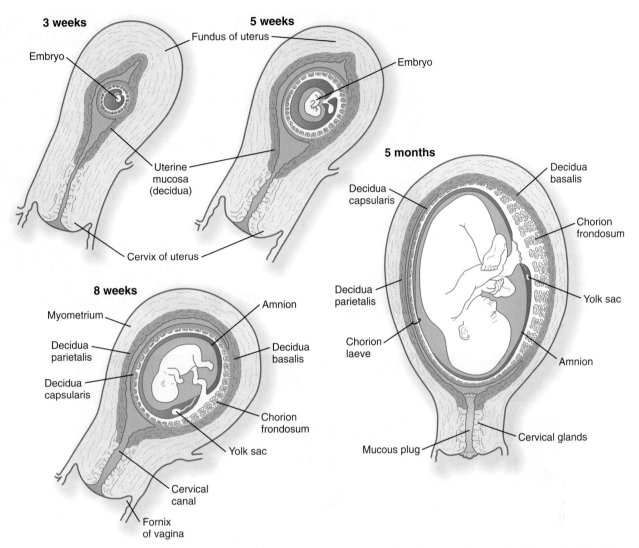

Fig. 7.7 Relationships between the embryo and maternal decidua *(pink)* from the early weeks of pregnancy through the fifth month. In the 5-month-old fetus, the placenta is represented by the white tissue to the right of the fetus. *(Adapted from Carlson BM:* Patten's foundations of embryology, *ed 6, New York, 1996, McGraw-Hill.)*

preferentially in the primary villi located nearest the decidua basalis. As these villi continue to grow and branch, the villi located on the opposite side (the abembryonic pole) of the chorionic vesicle fail to keep up and eventually atrophy as the growing embryo complex bulges into the uterine cavity (see Fig. 7.7). The region that contains the flourishing chorionic villi and that ultimately becomes the placenta is the **chorion frondosum**. The remainder of the chorion, which ultimately becomes smooth, is the **chorion laeve** (**Fig. 7.8**).

One suggested mechanism for the formation of the chorion laeve is based on oxidative stress. The normal early embryonic environment is roughly equivalent to 3% oxygen, as opposed to the normal 21% atmospheric oxygen level. The uterine spiral arteries in the region of the future chorion laeve are not as tightly sealed by cytotrophoblastic plugs as those under the area of the future placenta. This situation leads to a significant local increase in oxygen concentration, thus causing

degeneration of the syncytiotrophoblast that covers the villi and the regression of the capillary circulation within them as a result of oxidative stress.

The overall growth of the chorionic vesicle (**Fig. 7.9**), with its bulging into the uterine lumen, pushes the decidua capsularis progressively farther from the endometrial blood vessels. By the end of the first trimester, the decidua capsularis itself undergoes pronounced atrophy. Within the next month, portions of the atrophic decidua capsularis begin to disappear and leave the chorion laeve in direct contact with the decidua parietalis on the opposite side of the uterus (see Fig. 7.7). By midpregnancy, the decidua capsularis has fused with the tissues of the decidua parietalis, thereby effectively obliterating the original uterine cavity. While the chorion laeve and decidua capsularis are undergoing progressive atrophy, the placenta takes shape in its definitive form and acts as the main site of exchange between the mother and embryo.

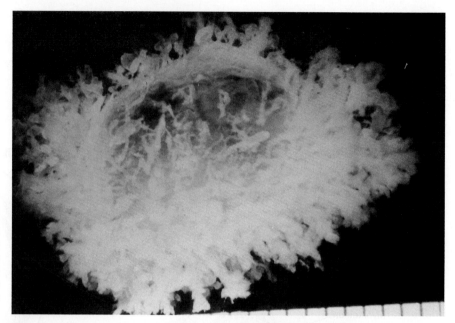

Fig. 7.8 **Early formation of the chorion laeve.** The small, bare area in this photograph of a human chorionic vesicle is a region where the chorionic villi have atrophied. This area will enlarge in succeeding weeks. *(From Gilbert-Barness E, ed: Potter's pathology of the fetus and infant, St Louis, 1997, Mosby.)*

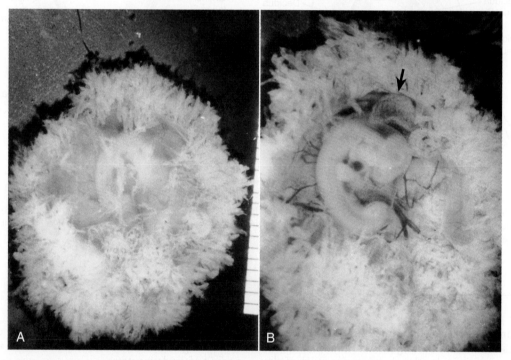

Fig. 7.9 **A,** Intact chorionic vesicle containing an embryo in the fourth week of development. The outline of the embryo can be seen through the thinned chorion laeve region. **B,** Opened chorionic vesicle, showing the disposition of the embryo inside. The yolk sac is indicated by the *arrow. (From Gilbert-Barness E, ed:* Potter's pathology of the fetus and infant, *St Louis, 1997, Mosby.)*

Formation and Structure of the Mature Placenta

As the distinction between the chorion frondosum and chorion laeve becomes more prominent, the limits of the placenta proper can be defined. The placenta consists of a fetal and a maternal component (**Fig. 7.10**). The fetal component is the part of the chorionic vesicle represented by the chorion frondosum. It consists of the wall of the chorion, called the **chorionic plate**, and the chorionic villi that arise from that region. The maternal component is represented by the decidua basalis, but covering the decidua basalis is the fetally derived outer cytotrophoblastic shell. The intervillous space between the fetal and maternal components of the placenta is occupied by freely circulating maternal blood. In keeping with its principal function as an organ-mediating exchange between the fetal and maternal circulatory systems, the overall structure of the placenta is organized to provide a very large surface area (>10 m^2) for that exchange.

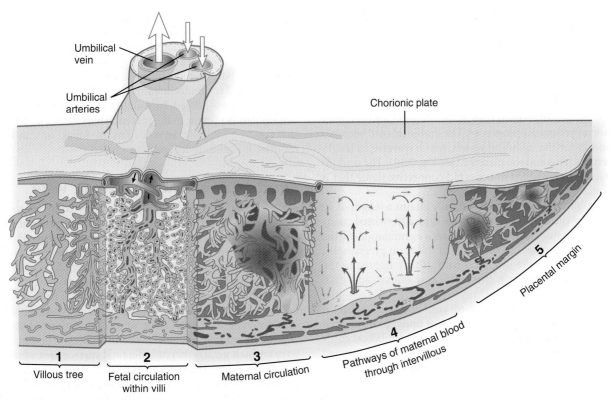

Fig. 7.10 **Structure and circulation of the mature human placenta.** Blood enters the intervillous spaces from the open ends of the uterine spiral arteries. After bathing the villi, the blood *(blue)* is drained via endometrial veins. *(From Bloom W, Fawcett DW:* Textbook of histology, *Philadelphia, 1986, Saunders.)*

Table 7.1 Developing Placenta

Age of Embryo (Weeks after Fertilization)	Placental Diameter (mm)	Placental Weight (g)	Placental Thickness (mm)	Length of Umbilical Cord (mm)	Embryo Weight (g)/ Placental Weight (g)	Villous Mass (g)	Total Villous Surface Area (cm²)	Diffusion Distance from Maternal to Fetal Circulation (μm)	Mean Trophoblastic Thickness on Villi (μm)
6	—	6	—	—	0.18	5	830	55.9	15.4
10	—	26	—	—	0.65	18	3,020	—	—
14	70	65	12	180	0.92	28	5,440	40.2	9.6
18	95	115	15	300	2.17	63	14,800	27.7	9.9
22	120	185	18	350	3.03	102	28,100	21.6	7.4
26	145	250	20	400	4.00	135	42,200	—	—
30	170	315	22	450	4.92	191	72,200	20.6	6.9
34	195	390	24	490	5.90	234	101,000	11.7	5.2
38	220	470	25	520	7.23	273	125,000	.8	4.1

Adapted from Kaufmann P, Scheffen I: In Polin R, Fox W, eds: *Fetal and neonatal physiology*, vol 1, Philadelphia, 1992, Saunders, p 48.

Structure of the Mature Placenta

The mature placenta is disklike in shape, 3 cm thick, and about 20 cm in diameter (**Table 7.1**). A typical placenta weighs about 500 g. The fetal side of the placenta is shiny because of the apposed amniotic membrane. From the fetal side, the attachment of the umbilical cord to the chorionic plate and

the large placental branches of the umbilical arteries and vein radiating from it are evident.

The maternal side of the placenta is dull and is subdivided into as many as 35 lobes. The grooves between lobes are occupied by placental septa, which arise from the decidua basalis and extend toward the basal plate. Within a placental lobe are several cotyledons, each of which consists of a main stem villus and all its branches. The intervillous space in each lobe represents a nearly isolated compartment of the maternal circulation to the placenta.

Umbilical Cord

The originally broad-based body stalk elongates and becomes narrower as pregnancy progresses. The umbilical cord becomes the conduit for the umbilical vessels, which traverse its length between the fetus and the placenta (see Fig. 7.10). The umbilical vessels are embedded in a mucoid connective tissue that is often called **Wharton's jelly**.

The umbilical cord, which commonly attains a length of 50 to 60 cm by the end of pregnancy, is typically twisted many times. The twisting can be seen by gross examination of the umbilical blood vessels. In about 1% of full-term pregnancies, true knots occur in the umbilical cord. If they tighten as the result of fetal movements, they can cause anoxia and even death of the fetus.

Occasionally, an umbilical cord contains two umbilical veins if the right umbilical vein does not undergo its normal degeneration. Approximately 0.5% of mature umbilical cords contain only one umbilical artery. This condition is associated with a 15% to 20% incidence of associated cardiovascular defects in the fetus.

Placental Circulation

Both the fetus and the mother contribute to the placental circulation (see Fig. 7.10). The fetal circulation is contained in the system of umbilical and placental vessels. Fetal blood reaches the placenta through the two umbilical arteries, which ramify throughout the chorionic plate. Smaller branches from these arteries enter the chorionic villi and break up into capillary networks in the terminal branches of the chorionic villi, where the exchange of materials with the maternal blood occurs (see Fig. 7.14). From the villous capillary beds, the blood vessels consolidate into successively larger venous branches. These retrace their way through the chorionic plate into the large single umbilical vein and to the fetus.

In contrast to the fetal circulation, which is totally contained within blood vessels, the maternal blood supply to the placenta is a free-flowing lake that is not bounded by vessel walls. As a result of the trophoblast's invasive activities, roughly 80 to 100 spiral arteries of the endometrium open directly into the intervillous spaces and bathe the villi in about 150 mL of maternal blood, which is exchanged 3 to 4 times each minute.

The maternal blood enters the intervillous space under reduced pressure because of the cytotrophoblastic plugs that partially occlude the lumens of the spiral arteries. Nevertheless, the maternal blood pressure is sufficient to force the oxygenated maternal arterial blood to the bases of the villous trees at the chorionic plate (see Fig. 7.10). The overall pressure of the maternal placental blood is about 10 mm Hg in the relaxed uterus. From the chorionic plate, the blood percolates over the terminal villi as it returns to venous outflow pathways located in the decidual (maternal) plate of the placenta. An adequate flow of maternal blood to the placenta is vital to the growth and development of the fetus, and a reduced maternal blood supply to the placenta leads to a small fetus.

In the terminal (floating) villi, the fetal capillaries are located next to the trophoblastic surface to facilitate exchange between the fetal and maternal blood (**Fig. 7-11**). The placental barrier of the mature placenta consists of the syncytiotrophoblast, its basal lamina, the basal lamina of the fetal capillary, and the capillary endothelium. Often the two basal laminae seem to be consolidated. In younger embryos, a layer of cytotrophoblast is present in the placental barrier, but by 4 months the cytotrophoblastic layer begins to break up, and by 5 months, it is essentially gone.

Structure of a Mature Chorionic Villus

Mature chorionic villi constitute a very complex mass of seemingly interwoven branches (**Fig. 7.12**). The core of a villus consists of blood vessels and mesenchyme that is similar in composition to the mesenchyme of the umbilical cord (see Fig. 7.11). Scattered among the mesenchymal cells are large **Hofbauer cells**, which function as fetal macrophages.

The villus core is covered by a continuous layer of syncytiotrophoblast, with minimum numbers of cytotrophoblastic cells beneath it. The surface of the syncytiotrophoblast is covered by immense numbers of microvilli (>1 billion/cm^2 at term), which greatly increase the total surface area of the placenta (**Fig. 7.13**). The size and density of the microvilli are not constant, but they change with the increasing age of the placenta and differing environmental conditions. Under conditions of poor maternal nutrition or oxygen transport, the microvilli increase in prominence. Poor adaptation of the microvilli to adverse conditions can lead to newborns with low birth weight.

The trophoblastic surface is not homogeneous, but rather seems to be arranged into territories. Among the many functional components of the microvillous surface are (1) numerous transport systems for substances ranging from ions to macromolecules, (2) hormone and growth factor receptors, (3) enzymes, and (4) numerous proteins with poorly understood functions. The placental surface is deficient or lacking in major histocompatibility antigens, the absence of which presumably plays a role in protecting against maternal immune rejection of the fetus and fetal membranes. In keeping with its active role in synthesis and transport, the syncytiotrophoblast is well supplied with a high density and a wide variety of subcellular organelles.

Placental Physiology

The transport of substances between the placenta and the maternal blood that bathes it is facilitated by the great surface area of the placenta, which expands from 5 m^2 at 28 weeks to almost 11 m^2 at term. Approximately 5% to 10% of the human placental surface consists of scattered areas where the barrier between fetal and maternal blood is extremely thin, measuring only a few micrometers. These areas, sometimes called **epithelial plates**, are apparently morphological adaptations designed to facilitate the diffusion of substances between the fetal and maternal circulations (**Fig. 7.14**).

The transfer of substances occurs both ways across the placenta. The bulk of the substances transferred from mother to fetus consists of oxygen and nutrients. The placenta represents the means for the final elimination of carbon dioxide and

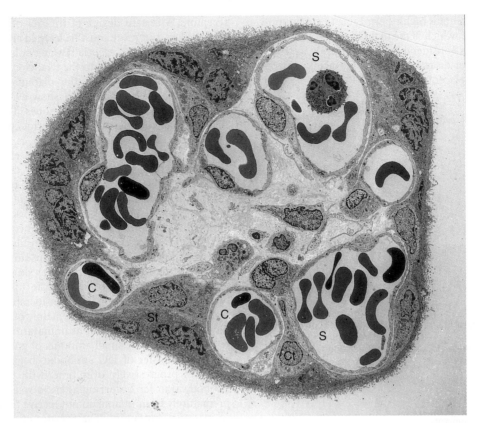

Fig. 7.11 Low-power transmission electron micrograph through a typical terminal villus of a human placenta. C, capillary; Ct, cytotrophoblast; S, sinusoid (dilated capillary); St, syncytiotrophoblast. *(From Benirschke K, Kaufmann P:* Pathology of the human placenta, *ed 2, New York, 1990, Springer.)*

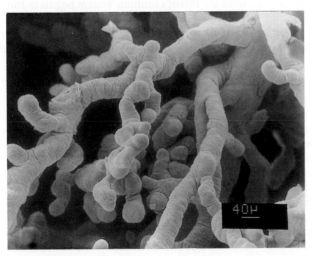

Fig. 7.12 Scanning electron micrograph of long, intermediate, knoblike terminal (floating) villi from a normal placenta near the termination of pregnancy. *(From Benirschke K, Kaufmann P:* Pathology of the human placenta, *ed 2, New York, 1990, Springer.)*

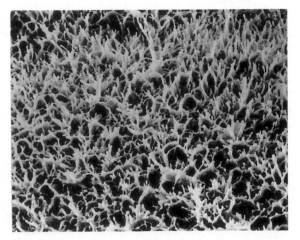

Fig. 7.13 Scanning electron micrograph of the surface of the syncytiotrophoblast of a human placenta in the twelfth week of pregnancy. The numerous microvilli increase the absorptive surface of the placenta. (×9000.) *(Courtesy of S. Bergström, Uppsala, Sweden.)*

other fetal waste materials into the maternal circulation. Under some circumstances, other substances, some of them harmful, can be transferred across the placenta. **Clinical Correlation 7.2** describes abnormal placental transfer.

Gases, principally oxygen from the mother and carbon dioxide from the fetus, readily cross the placental barrier by diffusion. The amount of exchange is limited more by blood flow than by the efficiency of diffusion. The placenta is also permeable to carbon monoxide and many inhalational anesthetics. Inhalational anesthetics can interfere with the transition of the newborn to independent function (e.g., breathing) if these agents are used during childbirth.

Like gases, water and electrolytes are readily transferred across the placenta. The rates of transfer are modified by colloid osmotic pressure in the case of water and the function of ion channels in the case of electrolytes. Fetal wastes (e.g., urea, creatinine, bilirubin) are rapidly transferred across the placenta from the fetal circulation to the maternal blood bathing the villi.

Although the placenta is highly permeable to certain nutrients, such as glucose, which is the main energy source for the fetus, the placenta is considerably less permeable to fructose and several common disaccharides. Amino acids are transported across the placenta through the action of specific

From mother to fetus

O_2
Water, electrolytes
Nutrients
    Carbohydrates
    Amino acids
    Lipids
Hormones
Antibodies
Vitamins
Iron, trace elements
Drugs
Toxic substances
    Alcohol
    Some viruses

From fetus to mother

CO_2
Water, electrolytes
Urea, uric acid
Creatinine
Bilirubin
Hormones
Red blood cell antigens

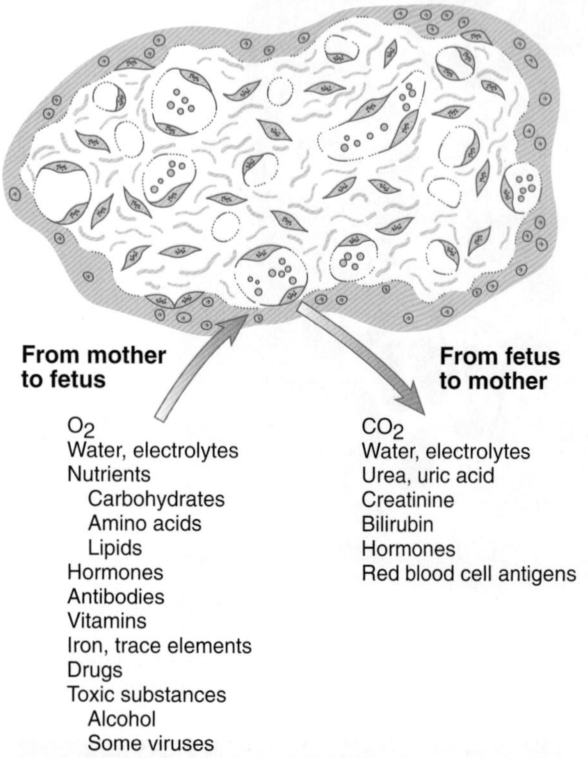

Fig. 7.14 **Exchange of substances across the placenta between the fetal and maternal circulation.**

receptors. A certain degree of transfer of maternal free fatty acids occurs, but more must be learned about the mechanism of transfer. Vitamins, especially water-soluble ones, are transferred from the maternal to the fetal circulation.

Steroid hormones cross the placental barrier from the maternal blood. Newborn boys show evidence of the effects of exposure to maternal sex hormones. The prostatic utricle, the vestigial rudiment of the uterine primordium (fused paramesonephric ducts [see p. 394]), is slightly enlarged in newborn boys. Conversely, female fetuses exposed to testosterone or certain synthetic progestins (especially during the 1950s and 1960s, before the effects were recognized) undergo masculinization of the external genitalia. Protein hormones are, in general, poorly transported across the placenta, although symptoms of maternal diabetes may be reduced during late pregnancy because of insulin produced by the fetus. Maternal thyroid hormone gains slow access to the fetus.

Some proteins are transferred very slowly through the placenta, mainly by means of pinocytosis (uptake by membrane-bound vesicles in the cells). Of considerable importance is the transfer of maternal antibodies, mainly of the immunoglobulin G (IgG) class. Because of its immature immune system, the fetus produces only small amounts of antibodies. The transplacental transfer of IgG antibodies begins at 12 weeks and increases progressively over time, with the greatest rate of antibody transfer occurring after 34 weeks. For this reason, prematurely born infants do not receive fully protective levels of maternal antibodies. The transfer of antibodies from the mother provides passive immunity of the newborn to certain common childhood diseases, such as smallpox, diphtheria, and measles, until the infant's immune system begins to function more efficiently.

Another maternal protein, transferrin, is important because, as its name implies, it carries iron to the fetus. The placental

CLINICAL CORRELATION 7.2
Abnormal Placental Transfer

The placenta is permeable to substances that can be damaging to the embryo. Numerous maternally ingested drugs readily cross the placental barrier. Certain drugs can cause major birth defects if they reach the embryo during critical periods of morphogenesis. (Several classic examples of these are described in Chapter 8.) The placenta is highly permeable to alcohol, and excessive alcohol ingestion by the mother can produce **fetal alcohol syndrome** (see p. 146). Infants born addicted to heroin or crack cocaine are common in contemporary society.

In addition to drugs, certain infectious agents can penetrate the placental barrier and infect the fetus. Some agents (e.g., rubella virus) can cause birth defects if they infect the embryo at critical periods in development. Normally, bacteria cannot penetrate the placental barrier. Common viruses that can infect the fetus are rubella virus, cytomegalovirus, poliovirus, varicella virus, variola virus, human immunodeficiency virus, and coxsackieviruses. The spirochete *Treponema pallidum,* which causes syphilis, can cause devastating fetal infections. The protozoan parasite *Toxoplasma gondii* can cross the placental barrier and cause birth defects.

Cellular Transfer and Rh Incompatibility

Small quantities of fetal blood cells often escape into the maternal circulation, either through small defects in the placental vascula-

ture or through hemorrhage at birth. If the fetal erythrocytes are positive for the Rh antigen, and the mother is Rh negative, the presence of fetal erythrocytes in the maternal circulation can stimulate the formation of anti-Rh antibody by the immune system of the mother. The fetus in the first pregnancy is usually spared the effects of the maternal antibody (often because it has not formed in sufficient quantities), but in subsequent pregnancies, Rh-positive fetuses are attacked by the maternal anti-Rh antibodies, which make their way into the fetal bloodstream. This antibody causes hemolysis of the Rh-positive fetal erythrocytes, and the fetus develops **erythroblastosis fetalis**, sometimes known as **hemolytic disease**. In severe cases, the bilirubin released from the lysed red blood cells causes water accumulation in the fetus (**hydrops fetalis**), with accompanying jaundice and brain damage in addition to anemia. When recognized, this condition is treated by exchange transfusions of Rh-negative donor blood into either the fetus or the newborn. An indication of the severity of this condition can be gained by examining the amniotic fluid.

surface contains specific receptors for this protein. The iron apparently is dissociated from its transferrin carrier at the placental surface and then is actively transported into the fetal tissues.

Placental Hormone Synthesis and Secretion

The placenta, specifically the syncytiotrophoblast, is an important endocrine organ during much of pregnancy. It produces protein and steroid hormones.

The first protein hormone produced is **human chorionic gonadotropin** (**HCG**), which is responsible for maintaining the corpus luteum and its production of progesterone and estrogens. With HCG synthesis beginning even before implantation, the presence of this hormone in maternal urine is the basis for many common tests for pregnancy. The production of HCG peaks at approximately the eighth week of gestation and then gradually declines. By the end of the first trimester, the placenta produces enough progesterone and estrogens so that pregnancy can be maintained even if the corpus luteum is surgically removed. The placenta can independently synthesize progesterone from acetate or cholesterol precursors, but it does not contain the complete enzymatic apparatus for the synthesis of estrogens. For estrogen to be synthesized, the placenta must operate in concert with the fetal adrenal gland and possibly the liver; these structures possess the enzymes that the placenta lacks.

Another placental protein hormone is **chorionic somato-mammotropin**, sometimes called **human placental lactogen**. Similar in structure to human growth hormone, it influences growth, lactation, and lipid and carbohydrate metabolism. The placenta also produces small amounts of **chorionic thyrotropin** and **chorionic corticotropin**. When secreted into the maternal bloodstream, some placental hormones stimulate changes in the metabolism and cardiovascular function of the mother. These changes ensure that appropriate types and amounts of fundamental nutrients and substrates reach the placenta for transport to the fetus.

A good example of a placental hormone that influences the mother is **human placental growth hormone**. This hormone, which differs by 13 amino acids from pituitary growth hormone, is produced by the syncytiotrophoblast. Placental growth hormone is not detectable in fetal serum, although it seems to influence growth of the placenta in a paracrine manner. This fetal hormone exerts a profound effect on the mother. During the first 15 to 20 weeks of pregnancy, maternal pituitary growth hormone is the main form present in the maternal circulation, but from 15 weeks to term, placental growth hormone gradually replaces maternal pituitary growth hormone to the extent that the maternally derived hormone becomes undetectable in the mother's serum. A major function of this hormone seems to be the regulation of maternal blood glucose levels so that the fetus is ensured of an adequate nutrient supply. Placental growth hormone secretion is stimulated by low maternal glucose levels. The increased hormone levels then stimulate gluconeogenesis in the maternal liver and other organs, thus increasing the supply of glucose available for fetal use.

In certain respects, the placenta duplicates the multilevel control system that regulates hormone production in the postnatal body. Cells of the cytotrophoblast produce a homologue of gonadotropin-releasing hormone (GnRH), as is normally done by the hypothalamus. GnRH passes into the syncytiotrophoblast, where it, along with certain opiate peptides and their receptors (which have been identified in the syncytiotrophoblast), stimulates the release of HCG from the syncytiotrophoblast. The opiate peptides and their receptors are also involved in the release of chorionic somatomammotropin from the syncytiotrophoblast. Finally, HCG seems to be involved in regulating the synthesis and release of placental steroids from the syncytiotrophoblast.

In addition to hormones, the placenta produces a wide variety of other proteins that have principally been identified immunologically. The functions of many of the placental proteins that have been discovered are still very poorly understood.

Placental Immunology

One of the major mysteries of pregnancy is why the fetus and placenta, which are immunologically distinct from the mother, are not recognized as foreign tissue and rejected by the mother's immune system. (Immune rejection of foreign tissues normally occurs by the activation of cytotoxic lymphocytes, but humoral immune responses are also possible.) Despite considerable research, the answer to this question is still unknown. Several broad explanations have been suggested to account for the unusual tolerance of the mother to the prolonged presence of the immunologically foreign embryo during pregnancy.

The first possibility is that the fetal tissues, especially those of the placenta, which constitute the direct interface between fetus and mother, do not present foreign antigens to the mother's immune system. To some extent, this hypothesis is true because neither the syncytiotrophoblast nor the nonvillous cytotrophoblast (**cytotrophoblastic shell**) expresses the two major classes of major histocompatibility antigens that trigger the immune response of the host in the rejection of typical foreign tissue grafts (e.g., a kidney transplant). These antigens are present, however, on cells of the fetus and in stromal tissues of the placenta. The expression of minor histocompatibility antigens (e.g., the HY antigen in male fetuses [see Chapter 16]) follows a similar pattern. Nevertheless, other minor antigens are expressed on trophoblastic tissues. In addition, because of breaks in the placental barrier, fetal red and white blood cells are frequently found circulating in the maternal blood. (In addition, maternal cells can colonize the fetus.) These cells should be capable of sensitizing the mother's immune system.

A second major possibility is that the mother's immune system is somehow paralyzed during pregnancy so that it does not react to the fetal antigens to which it is exposed. Yet the mother is capable of mounting an immune response to infections or foreign tissue grafts. There still remains the possibility of a selective repression of the immune response to fetal antigens, although the Rh incompatibility response shows that this is not universally the case.

A third possibility is that local decidual barriers prevent either immune recognition of the fetus by the mother or the reaching of competent immune cells from the mother to the fetus. Again, there is evidence for a functioning decidual immune barrier, but in a significant number of cases that barrier is known to be breached through trauma or disease.

A fourth possibility is that molecules formed on the fetal placental surface are able to inactivate the T cells or other immune cells locally that could reject the embryo, or that they paralyze the local cellular immune response. In mice, inactivation of a complement regulator results in immune rejection of the fetus. Whether a similar system operates during human pregnancy is unknown.

Currently, studies are being directed toward conditions such as recurrent spontaneous abortion with the hope of finding further clues to the complex immunological interrelationships between the fetus and mother. What is abundantly clear is that this is not a simple relationship. Nevertheless, the solution to this problem may yield information that may be applied to the problem of reducing the host rejection of tissue and organ transplants.

Placenta After Birth

About 30 minutes after birth, the placenta, embryonic membranes, and remainder of the umbilical cord, along with much of the maternal decidua, are expelled from the uterus as the **afterbirth**. The fetal surface of the placenta is smooth, shiny, and grayish because of the amnion that covers the fetal side of the chorionic plate. The maternal surface is a dull red and may be punctuated with blood clots. The maternal surface of the placenta must be examined carefully because if a

cotyledon is missing and is retained in the uterine wall, it could cause serious postpartum bleeding. Recognition of certain types of placental diseases can provide valuable clues to intrauterine factors that could affect the well-being of the newborn (**Clinical Correlation 7.3**).

Placenta and Membranes in Multiple Pregnancies

Several different configurations of the placenta and extraembryonic membranes are possible in multiple pregnancies. Dizygotic twins or monozygotic twins resulting from complete separation of blastomeres very early in cleavage can have completely separate placentas and membranes if the two embryos implant in distant sites on the uterine wall (**Fig. 7.17A**). In contrast, if the implantation sites are closer together, the placentas and chorions (which were initially separate at implantation) can fuse, although the vascular systems of the two embryos remain separate (Fig. 7.17B).

When monozygotic twins form by splitting of the inner cell mass in the blastocyst, it is usual to have a common placenta and a common chorion, but inside the chorion the twin embryos each develop within separate amnions

CLINICAL CORRELATION 7.3
Placental Pathological Conditions

Placental pathological conditions cover a wide spectrum, ranging from the abnormalities of implantation site to neoplasia to frank bacterial infections. Much can be learned about the past history and future prospects of a newborn by examining the placenta. This box deals only with the aspects of placental disorders that are relevant to developmental mechanisms.

Abnormal Implantation Sites

An abnormal implantation site within the uterine cavity is known as **placenta previa**. (Ectopic pregnancy is covered in Chapter 3.) When part of the placenta covers the cervical outlet of the uterine cavity, its presence is a mechanical obstacle in the birth canal. In addition, hemorrhage, which can be fatal to the fetus or the mother, is a common consequence of placenta previa as a result of the premature separation of part of the placenta from the uterus.

Gross Placental Anomalies

Many variations in shape of the placenta have been described, but few seem to be of any functional significance. One variation involves marginal rather than central attachment of the umbilical cord (**Fig. 7.15A**). If the umbilical cord attaches to the smooth membranes outside the boundaries of the placenta itself, the condition is known as a **velamentous insertion** of the umbilical cord (Fig. 7.15B).

The placenta itself can be subdivided into **accessory lobes** (Fig. 7.15C). It can also be completely divided into two parts, with smooth membrane between them (Fig. 7.15D).

Hydatidiform Mole

A **hydatidiform mole** is a noninvasive condition in which many of the chorionic villi are characterized by nodular swellings that

give them an appearance similar to bunches of grapes. Commonly, much of the villous surface of the placenta takes on this appearance; in addition, the embryo is either absent or not viable (**Fig. 7.16**). The villi show no evidence of vascularization.

Genetic analysis has determined that hydatidiform moles represent the results of paternal imprinting in which the female pronucleus of the egg does not participate in development (see Chapter 3). Instead, the chromosomal material is derived from two sperm that had penetrated the egg or by duplication of a single sperm pronucleus within the egg. The chromosomes of hydatidiform moles are paternally derived 46,XX because the number of lethal genes in 46,YY embryos is not compatible with tissue survival.

Choriocarcinoma

Choriocarcinomas are malignant tumors derived from embryonic cytotrophoblast and syncytiotrophoblast. These tumors are highly invasive into the maternal decidual tissues and blood vessels. As with hydatidiform moles, most choriocarcinomas contain only paternally derived chromosomes and are products of paternal imprinting.

Biopsy of Chorionic Villi

Biopsies of chorionic villi during the latter half of the second embryonic month are sometimes performed instead of sampling amniotic fluid. With the assistance of ultrasonography, chorionic villus biopsy specimens are obtained for analysis of possible chromosomal disorders or for diagnosis of certain metabolic disorders.

CLINICAL CORRELATION 7.3
Placental Pathological Conditions—cont'd

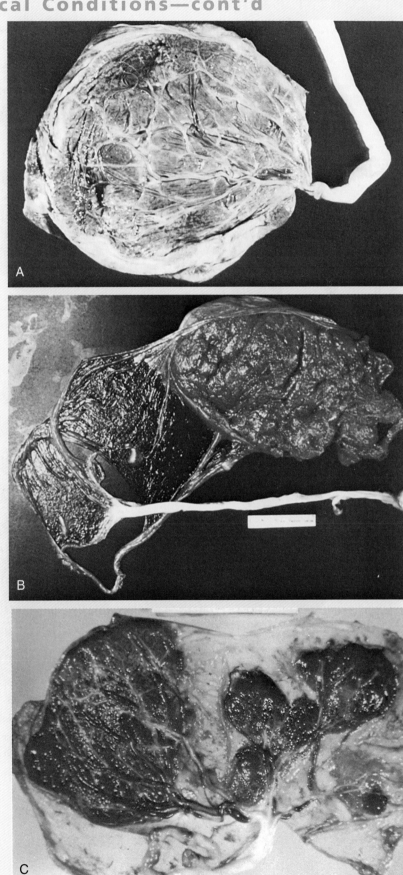

Fig. 7.15 **Variations in placental shape.** A, Marginal insertion of the umbilical cord. B, Velamentous insertion of the umbilical cord. C, Placenta with accessory (succenturiate) lobes.

Continued

CLINICAL CORRELATION 7.3
Placental Pathological Conditions—cont'd

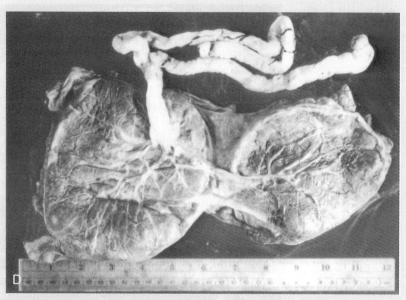

Fig. 7.15, cont'd **D,** Completely bilobed placenta. *(From Naeye RL: Disorders of the placenta, fetus, and neonate, St Louis, 1992, Mosby.)*

Fig. 7.16 **A,** Distended uterus containing a hydatidiform mole. The ovaries *(top* and *bottom)* contain bilateral theca lutein cysts. **B,** View at greater magnification that shows swollen villi. *(**A,** From Benirschke K, Kaufmann P: Pathology of the human placenta, ed 2, New York, 1990, Springer;* ***B,*** *Courtesy of K. Benirschke, San Diego.)*

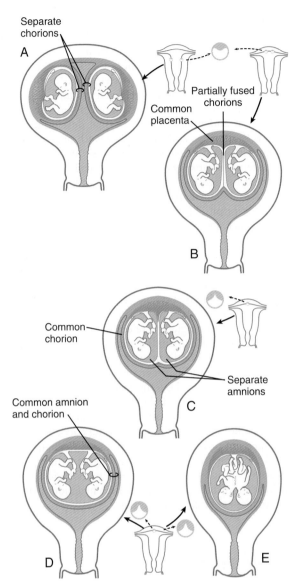

Fig. 7.17 **Extraembryonic membranes in multiple pregnancies.**
A, Completely separate membranes in dizygotic or completely separated monozygotic twins. B, Common fused placenta, separate amnions, and partially fused chorions. C, Common placenta with separate or common fused vessels and separate amnions enclosed in a common chorion. D and E, Common placenta and amniotic cavity in separate or conjoined twins.

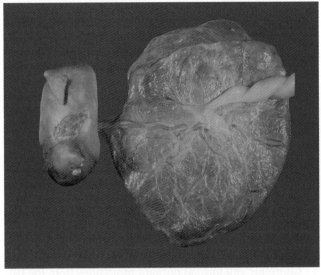

Fig. 7.18 Fused-twin placenta with an umbilical cord coming from its center and connecting to an anatomically normal fetus *(right)*. A shapeless acardiac monster is on the *left*. This condition is related to the siphoning of blood through a common circulation from the acardiac embryo to the other member of the pair. *(Photograph 7702 from the Arey-Depeña Pediatric Pathology Photographic Collection, Human Development Anatomy Center, National Museum of Health and Medicine, Armed Forces Institute of Pathology, Washington, D.C.)*

(Fig. 7.17C). In this case, there can be separate or fused vascular systems within the common placenta. When the vascular systems are fused, one twin may receive a greater proportion of the placental blood flow than the other (**twin-to-twin transfusion syndrome**). This situation may result in mild to severe stunting of growth of the embryo that receives the lesser amount of blood from the placenta. The twin from which the blood is siphoned is often highly misshapen and is commonly called an **acardiac monster** (**Fig. 7.18**).

In conjoined twins and, rarely, in monozygotic twins with minimal separation of the inner cell mass, the embryos develop within a single amnion and chorion and have a common placenta with a common blood supply (Fig. 7.17D and E). This and the previously described conditions can be determined by examination of the membranes of the

afterbirth. It was previously thought that it could be determined whether twins were monozygotic or dizygotic by simple examination of the membranes. Although in most cases the correct inference can be made, this method is not foolproof. Other methods, ranging from simple observation of gender, eye color, and fingerprint patterns to determination of blood types or even DNA fingerprinting, should be used for a definitive determination. In the current age of organ and cell transplantation, it can be vital to know whether twins are monozygotic in the event that one develops a condition that can be treated by a transplant.

Clinical Vignette

A 32-year-old woman's obstetrician notes that her weight gain during late pregnancy is excessive. At least part of her weight gain seems to be the result of a greater than normal volume of amniotic fluid. The patient lives in a remote rural area far from an imaging center. Amniocentesis is performed, and the laboratory report indicates the presence of a high level of α-fetoprotein in the amniotic fluid. The obstetrician is concerned that this pregnancy will not result in a normal single birth.

What condition does the obstetrician suspect and why?
A. Esophageal atresia
B. Renal agenesis
C. Triplets
D. Anencephaly
E. Placenta previa

Summary

■ The extraembryonic membranes consist of the chorion (the combination of trophoblast plus underlying extraembryonic mesoderm), amnion, yolk sac, and allantois.

■ The amnion, a thin ectodermal membrane lined with mesoderm, grows to enclose the embryo like a balloon. It is filled with a clear fluid, which is generated from many sources, such as the fetal skin, the amnion itself, the fetal kidneys, and possibly the fetal vessels. At term, the volume of amniotic fluid approaches 1 L. Amniotic fluid is removed by exchange across the amniotic membrane and by fetal swallowing.

- The yolk sac is a ventral, endodermally lined structure that does not serve a nutritive function in mammalian embryos. Mesodermal blood islands in the wall of the yolk sac form the first blood cells and vessels. Primordial germ cells are recognizable in the wall of the yolk sac, but they originate in extraembryonic mesoderm at the base of the allantois.

- The allantois is a small, endodermally lined diverticulum off the ventral side of the hindgut. It does not serve a direct function of respiration or storage of wastes in humans. These functions are carried out through the placenta and the umbilical vessels that arise in conjunction with the allantois.

- Chorionic villi form as outward projections from the trophoblast. Primary villi consist of projections of trophoblast alone. When a mesenchymal core forms within a villus, it is a secondary villus, and when the mesenchyme becomes vascularized, the villus is a tertiary villus. As villi mature, the cytotrophoblast in some villi grows through the syncytiotrophoblast as cytotrophoblastic cell columns and makes contact with the maternal endometrial tissue. Cytotrophoblast continues to grow around the blood-filled space surrounding the chorion to form a cytotrophoblastic shell, which is the direct interface between the fetal and maternal tissues. Villi that make direct contact with maternal tissues are anchoring villi; villi that do not make such contact are floating villi. Because chorionic villi float in a pool of maternal blood, the human placenta is designated a hemochorial placenta.

- Stimulated by the implanting embryo, endometrial stromal cells undergo the decidual reaction. Maternal tissues that are lost at childbirth are, collectively, the decidua. The decidua basalis underlies the placenta; the decidua capsularis encircles the remainder of the chorion like a capsule; portions of the uterine wall not occupied by the fetal chorion are the decidua parietalis. As the fetal chorion matures, it becomes subdivided into a chorion laeve, in which the villi regress, and the chorion frondosum, which is the region of chorion nearest the basal tissues of the endometrium. The chorion frondosum ultimately develops into the placenta.

- The mature placenta consists of the wall of the chorion (the chorionic plate) and numerous villi protruding from it. The fetal surface of the placenta is smooth and shiny because of the apposed amniotic membrane. The maternal surface is dull and lobulated, with cotyledons of numerous placental villi and their branches. The umbilical cord (formerly the body stalk) enters the middle of the placenta. Blood from the fetus reaches the placenta via the umbilical arteries. These arteries branch out into numerous small vessels that terminate into capillary loops in the ends of the placental villi. There, oxygen, nutrients, and wastes are exchanged between fetal and maternal blood, which bathes the villi. Fetal blood returns to the body of the mature fetus via a single umbilical vein. Maternal blood exiting from open-ended spiral arteries of the endometrium bathes the placental villi.

- The transfer of substances from fetal to maternal blood must occur across the endothelium of the fetal capillaries, a basal lamina, and trophoblastic tissues before reaching the maternal blood. The transfer of substances is accomplished by passive and active mechanisms. In addition to normal substances, alcohol, certain drugs, and some infectious agents can pass from the maternal blood into the fetal circulation and interfere with normal development. If a fetus is Rh positive and the mother is Rh negative, maternal anti-Rh antibodies from a previous pregnancy can pass to the fetus to cause erythroblastosis fetalis.

- The placenta produces a wide variety of hormones, many of which are normally synthesized in the hypothalamus and anterior pituitary gland. The first hormone released is HCG, which serves as the basis of many pregnancy tests. Other placental hormones are chorionic somatomammotropin (human placental lactogen), steroid hormones, human placental growth hormone, and chorionic thyrotropin and corticotropin.

- The fetal and placental tissues are immunologically different from those of the mother, but the placenta and fetus are not immunologically rejected. The reason is still unclear, but some explanations involve reduced antigenicity of the trophoblastic tissues, paralysis of the mother's immune system during pregnancy, and local immunological barriers between the fetus and mother.

- The placenta is delivered about 30 minutes after the fetus as the afterbirth. Inspection of the placenta can reveal placental pathological conditions, missing cotyledons, or the arrangement of membranes in multiple pregnancies. The last finding can help to determine whether a multiple birth is monozygotic in origin. Placental pathological findings include abnormal gross shape, benign hydatidiform moles, and malignant choriocarcinomas.

Review Questions

1. In the mature placenta, which fetal tissue directly interfaces with the maternal uterine connective tissue?
A. Cytotrophoblast
B. Syncytiotrophoblast
C. Extraembryonic mesoderm
D. Decidual cells
E. None of the above

2. Which condition is related to paternal imprinting?
A. Accessory placental lobes
B. Placenta previa
C. Oligohydramnios
D. Single umbilical artery
E. Hydatidiform mole

3. Blood vessels associated with which structure enter the fetal component of the placenta?
A. Decidua basalis
B. Allantois
C. Amnion
D. Yolk sac
E. Decidua parietalis

4. What type of cells invades the maternal spiral arteries and reduces the flow of blood from their open ends?
A. Hofbauer cells
B. Syncytiotrophoblast
C. Fetal erythrocytes
D. Cytotrophoblast
E. Amniotic epithelium

5. Which condition of the extraembryonic membranes can be found in uteri containing identical twins?
A. Common placenta and amniotic membrane
B. Common placenta and chorion, separate amnions
C. Separate placentas and extraembryonic membranes
D. Common placenta, partially fused chorions
E. All of the above

6. A 28-year-old Rh-negative woman's second son is born severely jaundiced. Which characteristic most likely describes her first child?
A. Male
B. Female
C. Rh positive
D. Rh negative
E. Hydramnios

7. Why is the human placenta designated a hemochorial type of placenta?

8. Through what layers of a placental villus must a molecule of oxygen pass to go from the maternal blood into the embryonic circulation?

9. What embryonic hormone has served as the basis for many standard pregnancy tests and why?

10. Why must a pregnant woman be very careful of what she eats and drinks?

References

Alsat E and others: Physiological role of human placental growth hormone, *Mol Cell Endocrinol* 140:121-127, 1998.

Aplin JD: Developmental biology of human villous trophoblast: current research problems, *Int J Dev Biol* 54:323-329, 2010.

Benirschke K, Kaufmann P: *Pathology of the human placenta*, ed 4, New York, 2000, Springer.

Boyd JD, Hamilton WJ: *The human placenta*, Cambridge, UK, 1970, Heffer & Sons.

Bressan FF and others: Unearthing the roles of imprinted genes in the placenta, *Placenta* 30:823-834, 2009.

Burton GJ, Jauniaux E, Charnock-Jones DS: The influence of the intrauterine environment on human placental development, *Int J Dev Biol* 54:303-311, 2010.

Chucri TM and others: A review of immune transfer by the placenta, *J Reprod Immunol* 87:14-20, 2010.

Coan PM, Burton GJ, Ferguson-Smith AC: Imprinted genes in the placenta: a review, *Placenta* 26(Suppl A):S10-S20, 2004.

Cross JC and others: Branching morphogenesis during development of placental villi, *Differentiation* 74:393-401, 2006.

Dallaire L, Potier M: Amniotic fluid. In Milunsky A, ed: *Genetic disorders and the fetus*, New York, 1986, Plenum, pp 53-97.

Demir R and others: Classification of human placental stem villi: review of structural and functional aspects, *Microsc Res Tech* 38:29-41, 1997.

Dobreva MP and others: On the origin of amniotic stem cells: of mice and men, *Int J Dev Biol* 54:761-777, 2010.

Dzierzak E, Robin C: Placenta as a source of hematopoietic stem cells, *Trends Mol Med* 16:361-367, 2010.

El Kateb A, Ville Y: Update on twin-to-twin transfusion syndrome, *Best Pract Res Clin Obstet Gynaecol* 22:63-75, 2008.

Enders AC: Trophoblast differentiation during the transition from trophoblastic plate to lacunar stage of implantation in the rhesus monkey and human, *Am J Anat* 186:85-98, 1989.

Faber JJ, Thornburg KL, eds: *Placental physiology*, New York, 1983, Raven.

Gammill HS, Nelson JL: Naturally acquired microchimerism, *Int J Dev Biol* 54:531-543, 2010.

Garnica AD, Chan W-Y: The role of the placenta in fetal nutrition and growth, *J Am Coll Nutr* 15:206-222, 1996.

Genbacev O and others: Regulation of human placental development by oxygen tension, *Science* 277:1669-1672, 1997.

Hunt JS, Pace JL, Gill RM: Immunoregulatory molecules in human placentas: potential for diverse roles in pregnancy, *Int J Dev Biol* 54:457-467, 2010.

Huppertz B, Gauster M: Mechanisms regulating human trophoblast fusion, *Adv Exp Med Biol* 713:81-95, 2011.

Johnson PM, Christmas SE, Vince GS: Immunological aspects of implantation and implantation failure, *Hum Reprod* 14(Suppl 2):26-36, 1999.

Juriscova A, Detmar J, Caniggia I: Molecular mechanisms of trophoblast survival: from implantation to birth, *Birth Defects Res C Embryo Today* 75:262-280, 2005.

Kaufmann P: Basic morphology of the fetal and maternal circuits in the human placenta, *Contrib Gynecol Obstet* 13:5-17, 1985.

Kaufmann P, Burton G: Anatomy and genesis of the placenta. In Knobil E, Neill JD, eds: *The physiology of reproduction*, ed 2, New York, 1994, Raven, pp 441-484.

Kliman HJ: Uteroplacental blood flow, *Am J Pathol* 157:1759-1768, 2000.

Knipp GT, Audus KL, Soares MJ: Nutrient transport across the placenta, *Adv Drug Deliv Rev* 38:41-58, 1999

Lavrey JP, ed: *The human placenta: clinical perspectives*, Rockville, Md, 1987, Aspen.

Loke YW, King A: *Human implantation*, Cambridge, UK, 1995, Cambridge University Press.

Maltepe E and others: The placenta: transcriptional, epigenetic, and physiological integration during development, *J Clin Invest* 120:1016-1025, 2010.

Marin JJG, Macias RIR, Serrano MA: The hepatobiliary-like excretory function of the placenta: a review, *Placenta* 24:431-438, 2003.

Mold JE and others: Fetal and adult hematopoietic stem cells give rise to distinct T cell lineages in humans, *Science* 330:1695-1699, 2010.

Mold JE and others: Maternal alloantigens promote the development of tolerogenic fetal regulatory T cells in utero, *Science* 322:1562-1565, 2008.

Morriss FJ, Boyd RDH, Mahendran D: Placental transport. In Knobil E, Neill JD, eds: *The physiology of reproduction*, ed 2, New York, 1994, Raven, pp 813-861.

Murphy VE and others: Endocrine regulation of human fetal growth: the role of the mother placenta, and fetus, *Endocr Rev* 27:141-169, 2006.

Naeye RL: *Disorders of the placenta, fetus, and neonate*, St Louis, 1992, Mosby.

Pijnenborg R, Vercruysse L, Hanssens M: The uterine spiral arteries in human pregnancy: facts and controversies, *Placenta* 27:939-957, 2006.

Quinn LM, Latham SE, Kalionis B: The homeobox genes *Msx2* and *Mox2* are candidates for regulating epithelial-mesenchymal cell interactions in the human placenta, *Placenta* 21(Suppl A 14):S50-S54, 2000.

Ramsey EM: *The placenta: human and animal*, New York, 1982, Praeger.

Red-Horse K and others: Trophoblast differentiation during embryo implantation and formation of the maternal-fetal interface, *J Clin Invest* 114:744-754, 2004.

Schneider H: The role of the placenta in nutrition of the human fetus, *Am J Obstet Gynecol* 164:967-973, 1991.

Sibley CP, Boyd RDH: Mechanisms of transfer across the human placenta. In Polin R, Fox W, eds: *Fetal and neonatal physiology*, vol 1, Philadelphia, 1992, Saunders, pp 62-74.

Tang Z and others: Placental Hofbauer cells and complications of pregnancy, *Ann N Y Acad Sci* 1221:103-108, 2011.

Zohn IE, Sarkar AA: The visceral yolk sac endoderm provides for absorption of nutrients to the embryo during neurulation, *Birth Defects Res A Clin Mol Teratol* 88:593-600, 2010.

Chapter 8

Developmental Disorders: Causes, Mechanisms, and Patterns

Congenital malformations have attracted attention since the dawn of human history. When seen in humans or animals, malformations were often interpreted as omens of good or evil. Because of the great significance attached to congenital malformations, they were frequently represented in folk art as sculptures or paintings. As far back as the classical Greek period, people speculated that maternal impressions during pregnancy (e.g., being frightened by an animal) caused development to go awry. In other cultures, women who gave birth to malformed infants were assumed to have had dealings with the devil or other evil spirits.

Early representations of some malformed infants are remarkable in their anatomical accuracy, and it is often possible to diagnose specific conditions or syndromes from the ancient art (**Fig. 8.1A**). By the Middle Ages, however, representations of malformations were much more imaginative, with hybrids of humans and other animals often represented (Fig. 8.1B).

Among the first applications of scientific thought to the problem of congenital malformations were those of the sixteenth-century French surgeon Ambrose Paré, who suggested a role for hereditary factors and mechanical influences, such as intrauterine compression, in the genesis of birth defects. Less than a century later, William Harvey, who is also credited with first describing the circulation of blood, elaborated the concept of developmental arrest and further refined thinking on mechanical causes of birth defects.

In the early nineteenth century, Etienne Geoffroy de St. Hilaire coined the term **teratology**, which literally means "the study of monsters," as a descriptor for the newly emerging study of congenital malformations. Late in the nineteenth century, scientific study of teratology was put on a firm foundation with the publication of several encyclopedic treatises that exhaustively covered anatomical aspects of recognized congenital malformations.

After the flowering of experimental embryology and genetics in the early twentieth century, laboratory researchers began to produce specific recognizable congenital anomalies by means of defined experimental genetic or laboratory manipulations on laboratory animals. This work led to the demystification of congenital anomalies and to a search for rational scientific explanations for birth defects. Nevertheless, old beliefs are tenacious, and even today patients may adhere to traditional beliefs.

The first of two major milestones in human teratology occurred in 1941, when Gregg in Australia recognized that the **rubella** virus was a cause of a recognizable **syndrome** of abnormal development, consisting of defects in the eyes, ears, and heart. About 20 years later, the effects of **thalidomide** sensitized the medical community to the potential danger of certain drugs and other environmental **teratogens** (agents that produce birth defects) to the developing embryo.

Thalidomide is a very effective sedative that was widely used in West Germany, Australia, and other countries during the late 1950s. Soon, physicians began to see infants born with extremely rare birth defects. One example is **phocomelia** (which means "seal limb"), a condition in which the hands and feet seem to arise almost directly from the shoulder and hip (**Fig. 8.2**). Another is **amelia**, in which a limb is entirely missing. Thalidomide was identified as the certain cause after some careful epidemiological detective work involving the collection of individual case reports and sorting of the drugs taken by mothers during the early period of their pregnancies. Thalidomide, which is an inhibitor of **tumor necrosis factor-α**, is still a drug of choice in the treatment of leprosy and multiple myeloma. With the intense investigations that followed the thalidomide disaster, modern teratology came of age. Despite much effort, however, the causes of most congenital malformations are still unknown.

General Principles

According to most studies, approximately 2% to 3% of all living newborns show at least one recognizable congenital malformation. This percentage is doubled when one considers anomalies diagnosed in children during the first few years after birth. With the decline in infant mortality caused by infectious diseases and nutritional problems, congenital malformations now rank high among the causes of infant mortality (currently >20%), and increasing percentages (≤30%) of infants admitted to neonatology or pediatric units come as a result of various forms of genetic diseases or congenital defects.

Congenital defects range from enzyme deficiencies caused by single nucleotide substitutions in the DNA molecule to very complex associations of gross anatomical abnormalities. Although medical embryology textbooks traditionally cover

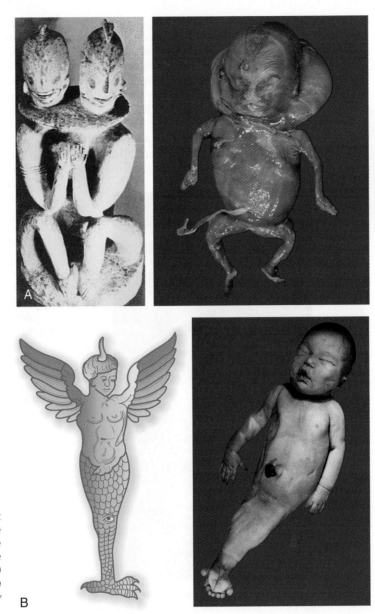

Fig. 8.1 **A,** Chalk carving from New Ireland in the South Pacific showing dicephalic, dibrachic conjoined twins *(left)*. Note also the "collar" beneath the heads, which is a representation of the malformation cystic hygroma colli *(right)*. **B,** The bird-boy of Paré (about 1520) *(left)*. Stillborn fetus with sirenomelia (fused legs) *(right)*. Compare with the lower part of the bird-boy. *(A [left,] From Brodsky I: Med J Aust 1:417-420, 1943; **A** [right] and **B** [right], Courtesy of M. Barr, Ann Arbor, Mich.)*

principally structural defects—congenital malformations—there is a continuum between purely biochemical abnormalities and defects that are manifested as abnormal structures. This continuum includes defects that constitute abnormal structure, function, metabolism, and behavior.

Birth defects present themselves in a variety of forms and associations, ranging from simple abnormalities of a single structure to often grotesque deformities that may affect an entire body region. Some of the common classes of malformations are listed in **Table 8.1.**

The genesis of congenital defects can be viewed as an interaction between the genetic endowment of the embryo and the environment in which it develops. The basic information is encoded in the genes, but as the genetic instructions unfold, the developing structures or organs are subjected to microenvironmental or macroenvironmental influences that either are compatible with or interfere with normal development. In the case of genetically based malformations or anomalies based on chromosomal aberrations, the defect is intrinsic and is commonly expressed even in a normal environment. Purely environmental causes can interfere with embryological processes in the face of a normal genotype. In other cases, environment and genetics interact. Penetrance (the degree of manifestation) of an abnormal gene or expression of one component of a genetically multifactorial cascade can sometimes be profoundly affected by environmental conditions.

Studies on mice have shown that defective function of many genes leads to some sort of developmental disturbance. Some of these defects are purely mutational, residing in the structure of the DNA itself, whereas others result from interference in transcription or translation or from regulatory elements of the gene.

Several factors are associated with various types of congenital malformations. At present, they are understood more at the level of statistical associations than as points of interference with specific developmental controls, but they are important clues to why development can go wrong. Among the factors associated with increased incidences of congenital malformations are (1) parental age, (2) season of the year, (3) country of residence, (4) race, and (5) familial tendencies.

Well-known correlations exist between parental age and the incidence of certain malformations. A classic correlation is the increased incidence of **Down syndrome** (**Fig. 8.3**; see Fig. 8.9) in children born to women older than 35 years of age. Other conditions are related to paternal age (see Fig. 8.3).

Some types of anomalies have a higher incidence among infants born at certain seasons of the year. **Anencephaly (Fig. 8.4)** occurs more frequently in January. Recognizing that the

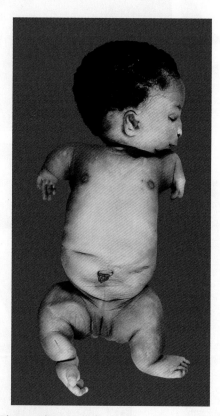

Fig. 8.2 Phocomelia in all four limbs. This fetus had not been exposed to thalidomide. *(Courtesy of M. Barr, Ann Arbor, Mich.)*

primary factors leading to anencephaly occur during the first month of embryonic life, researchers must seek the potential environmental causes that are more prevalent in April. Anencephaly has been shown to be highly correlated with maternal **folic acid deficiency**. The high incidence of this anomaly in pregnancies beginning in the early spring may relate to nutritional deficiencies of mothers during the late winter months. Folic acid supplementation in the diet of women of childbearing age significantly reduces the incidence of neural tube defects, such as anencephaly.

The relationship between the country of residence and an increased incidence of specific malformations can be related to various factors, including racial tendencies, local environmental factors, and even governmental policies. A classic example of the last is the incidence of severely malformed infants as a result of exposure to thalidomide. These cases were concentrated in West Germany and Australia because the drug was commonly sold in these locations. Because thalidomide was not approved by the Food and Drug Administration, the United States was spared from this epidemic of birth defects. Another classic example of the influence of country as a factor in the incidence of malformations is seen in neural tube defects (**Table 8.2**). The reason neural tube defects (especially anencephaly) were historically so common in Ireland has been the topic of much speculation. In light of the recognition of the importance of folic acid in the prevention of neural tube defects, it is possible that the high incidence of anencephaly in Ireland resulted from poor nutrition in pregnant women during the winter. A greater than threefold decrease in the incidence of neural tube defects in Ireland from 1980 to 1994 may be related to both better nutrition and folic acid supplementation by a certain percentage of pregnant women.

Race is a factor in many congenital malformations and a variety of diseases. In humans and mice, there are racial differences in the incidence of cleft palate. The incidence of cleft palate among whites is twice as high as it is among blacks and twice as high among Korean, Chinese, and Japanese persons as among whites.

Many malformations, particularly those with a genetic basis, are found more frequently within certain families, especially if there is any degree of consanguinity in the marriages

Table 8.1 Types of Abnormal Development

Abnormalities of Individual Structures	
Malformation	A structural defect of part of or an entire organ or larger part of a body region that is caused by an abnormal process intrinsic to its development (e.g., coloboma) (see p. 285)
Disruption	A defect in an organ or body part caused by process that interferes with an originally normal developmental process (e.g., thalidomide-induced phocomelia) (see p. 149)
Deformation	A structural abnormality caused by mechanical forces (e.g., amniotic band constriction) (see Fig. 8.16)
Dysplasia	An abnormality of a tissue due to an abnormal intrinsic developmental process (e.g., ectodermal dysplasia) (see p. 150)
Defects Involving More Than One Structure	
Sequence	A pattern of multiple malformations stemming from a disturbance of a prior developmental process or mechanical factor (e.g., Potter sequence) (see p. 384)
Syndrome	A group of malformations of different structures due to a single primary cause, but acting through multiple developmental pathways (e.g., trisomy 13 syndrome) (see Fig. 8.10)
Association	A group of anomalies seen in more than one individual that cannot yet be attributed to a definitive cause

Based on Spranger J and others: *J Pediatr* 100:160-165, 1982.

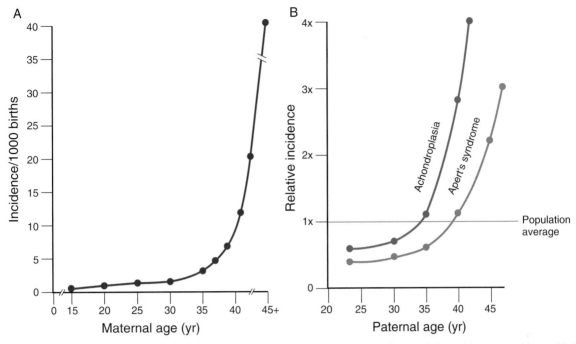

Fig. 8.3 The increased incidence of Down syndrome with increasing maternal age (A) and achondroplasia and Apert's syndrome with increasing paternal age (B). Apert's syndrome (acrocephalosyndactyly) is characterized by a towering skull and laterally fused digits.

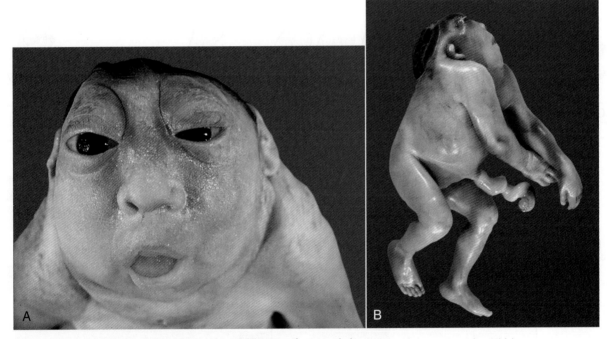

Fig. 8.4 Frontal (A) and lateral (B) views of anencephaly. *(Courtesy of M. Barr, Ann Arbor, Mich.)*

Table 8.2 Incidence of Neural Tube Defects	
Site	**Incidence***
India	0.6
Ireland	10†
United States	1
Worldwide	2.6

*Per 1000 live births.
†The present incidence in Ireland is much decreased.

over the generations. A good example is the increased occurrence of extra digits among some families within the Amish community in the United States.

Periods of Susceptibility to Abnormal Development

At certain critical periods during pregnancy, embryos are more susceptible to agents or factors causing abnormal development than at other times. The results of many investigations have allowed the following generalization: Insults to the embryo during the first 3 weeks of embryogenesis (the early period before organogenesis begins) are unlikely to result in defective development because they either kill the embryo or are compensated for by the powerful regulatory properties of the early embryo. The period of maximal susceptibility to abnormal development occurs between weeks 3 and 8, which is the period when most of the major organs and body regions are first being established.

Major structural anomalies are unlikely to occur after the eighth week of pregnancy because, by this point, most organs have become well established. Anomalies arising from the third to the ninth month of pregnancy tend to be functional (e.g., mental retardation) or involve disturbances in the growth of already formed body parts. Such a simplified view of susceptible periods does not take into account, however, the possibility that a teratogen or some other harmful influence may be applied at an early stage of development, but not be expressed as a developmental disturbance until later during embryogenesis. Certain other influences (e.g., intrauterine diseases, toxins) may result in the destruction of all or parts of structures that have already been formed.

Typically, a developing organ has a curve of susceptibility to teratogenic influences similar to that illustrated in **Figure**

8.5. Before the critical period, exposure to a known teratogen has little influence on development. During the first days of the critical period, the susceptibility, measured as incidence or severity of malformation, increases sharply and then declines over a much longer period.

Different organs have different periods of susceptibility during embryogenesis (**Fig. 8.6**). Organs that form the earliest (e.g., heart) tend to be sensitive to the effects of teratogens

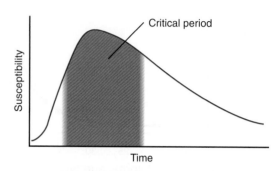

Fig. 8.5 Generalized susceptibility curve to teratogenic influences by a single organ.

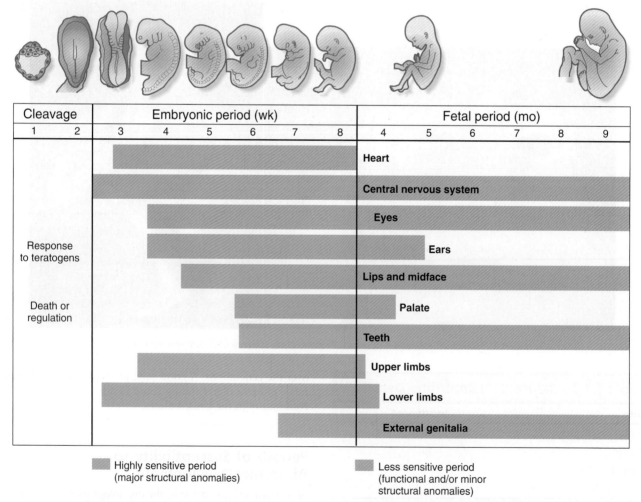

Fig. 8.6 Periods and degrees of susceptibility of embryonic organs to teratogens. *(Adapted from Moore KL, Persaud TVN: The developing human, ed 5, Philadelphia, 1993, Saunders.)*

Table 8.3 Developmental Times at Which Various Human Teratogens Exert Their Effects

Teratogens	Critical Periods (Gestational Days)	Common Malformations
Rubella virus	0-60 0-120+	Cataract or heart malformations Deafness
Thalidomide	21-40	Reduction defects of limbs
Androgenic steroids	Earlier than 90 Later than 90	Clitoral hypertrophy and labial fusion Clitoral hypertrophy only
Warfarin (Coumadin) anticoagulants	Earlier than 100 Later than 100	Nasal hypoplasia Possible mental retardation
Radioiodine therapy	Later than 65-70	Fetal thyroid deficiency
Tetracycline	Later than 120 Later than 250	Staining of dental enamel in primary teeth Staining of crowns of permanent teeth

Adapted from Persaud TVN, Chudley AE, Skalko RG, eds: *Basic concepts in teratology,* New York, 1985, Liss.

earlier than organs that form later (e.g., external genitalia). Some very complex organs, especially the brain and major sense organs, show prolonged periods of high susceptibility to disruption of normal development.

Not all teratogenic influences act in the same developmental periods (**Table 8.3**). Some influences cause anomalies if the embryo is exposed to them early in development, but they are innocuous at later periods of pregnancy. Others affect only later developmental periods. A good example of the former is thalidomide, which has a very narrow and well-defined danger zone during the embryonic period (4 to 6 weeks). In contrast, tetracycline, which stains bony structures and teeth, exerts its effects after hard skeletal structures in the fetus have formed.

Patterns of Abnormal Development

Although isolated structural or biochemical defects are not rare, it is also common to find multiple abnormalities in the same individual. This can result for many reasons. One possibility is that a single teratogen acted on the primordia of several organs during susceptible periods of development. Another is that a genetic or chromosomal defect spanned genes affecting a variety of structures, or that a single metabolic defect affected different developing structures in different ways.

Causes of Malformations

Despite considerable research since the 1960s, the cause of at least 50% of human congenital malformations remains unknown (**Fig. 8.7**). Roughly 18% of malformations can be attributed to genetic causes (chromosomal defects or mutations based on mendelian genetics), and 7% of malformations are caused by environmental factors, such as physical or chemical teratogens. Of all malformations, 25% are multifactorial, for example, caused by environmental factors acting on genetic susceptibility.

The high percentage of unknown causes is the result of having to work retrospectively to identify the origin of a malformation. Many of these causes are likely to result from some environmental factor influencing the expression of a developmentally critical gene.

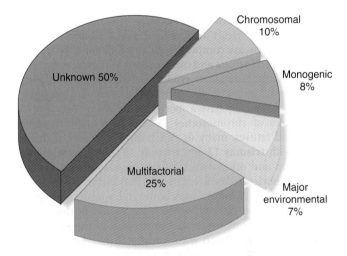

Fig. 8.7 Major causes of congenital malformations. *(Data from Persaud TVN, Chudley AE, Skalko RG, eds:* Basic concepts in teratology, *New York, 1985, Liss.)*

Genetic Factors

Genetically based malformations can be caused by abnormalities of chromosomal division or by mutations of genes. Chromosomal abnormalities are usually classified as structural or numerical errors. These arise during cell division, especially meiosis. Numerical errors of chromosomes result in **aneuploidy**, defined as a total number of chromosomes other than the normal 46.

Abnormal Chromosome Numbers

POLYPLOIDY

Polyploidy is the condition in which the chromosomal number is a higher multiple than 2 of the haploid number (23) of chromosomes. In most cases, polyploid embryos abort spontaneously early in pregnancy. High percentages of spontaneously aborted fetuses show major chromosomal abnormalities. Polyploidy, especially triploidy, is likely to be caused by either the fertilization of an egg by more than one sperm or the lack of separation of a polar body during meiosis.

MONOSOMY AND TRISOMY

Monosomy (the lack of one member of a chromosome pair) and **trisomy** (a triplet instead of the normal chromosome pair) are typically the result of nondisjunction during meiosis (see Fig. 1.7). When this happens, one gamete shows monosomy, and the other shows trisomy of the same chromosome.

In most cases, embryos with monosomy of the autosomes or sex chromosomes are not viable. Some individuals with monosomy of the sex chromosomes (45XO genotype) can survive, however (**Fig. 8.8**). Such individuals, who are said to have **Turner's syndrome**, exhibit a female phenotype, but the gonads are sterile.

Three autosomal trisomies produce infants with characteristic associations of anomalies. The best known is **trisomy 21**, also called **Down syndrome**. Individuals with Down syndrome are typically mentally retarded and have a characteristic broad face with a flat nasal bridge, wide-set eyes, and prominent epicanthic folds. The hands are also broad, and the palmar surface is marked by a characteristic transverse **simian crease** (**Fig. 8.9**). Heart defects, especially atrial and ventricular septal defects, are common, with an incidence approaching 50%. Duodenal atresia and other intestinal anomalies are also seen in patients with Down syndrome. Individuals with Down syndrome are prone to the early appearance of Alzheimer's disease and typically have a shortened life span.

Trisomies of chromosomes 13 and 18 result in severely malformed fetuses, many of which do not survive to birth. Infants with **trisomy 13** and **trisomy 18** show severe mental retardation and other defects of the central nervous system. Cleft lip and cleft palate are common. Polydactyly is often seen in trisomy 13, and infants with both syndromes exhibit other anomalies of the extremities, such as "**rocker bottom feet**," meaning a rounding under and protrusion of the heels (**Fig. 8.10**). Most infants born with trisomy 13 or trisomy 18 die within the first 1 or 2 months after birth.

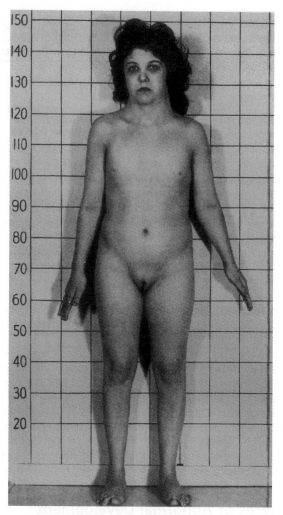

Fig. 8.8 Woman with Turner's syndrome. Note the short stature, webbed neck, and infantile sexual characteristics. *(From Connor J, Ferguson-Smith M: Essential medical genetics, ed 2, Oxford, 1987, Blackwell Scientific.)*

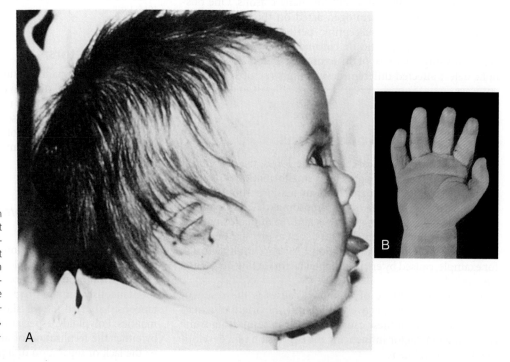

Fig. 8.9 A, Profile of a child with Down syndrome. Note the flat profile, protruding tongue, saddle-shaped bridge of nose, and low-set ears. **B,** Hand of an infant with Down syndrome shows the prominent simian crease that crosses the entire palm. *(**A,** From Garver K, Marchese S:* Genetic counseling for clinicians, *Chicago, 1986, Mosby;* ***B,*** *Courtesy of M. Barr, Ann Arbor, Mich.)*

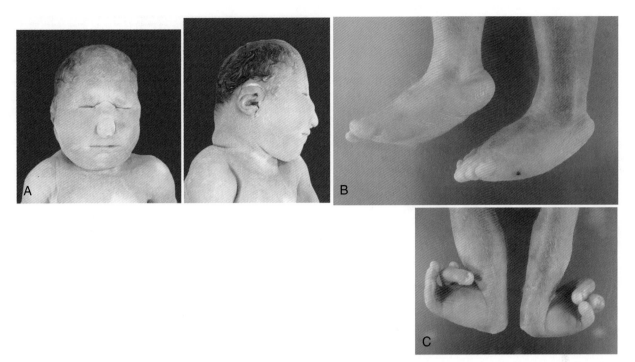

Fig. 8.10 **A,** Front and lateral views of the head of a 34-week fetus with trisomy 13. This fetus shows pronounced cebocephaly with a keel-shaped head, a flattened nose, abnormal ears, and a reduction of forebrain and upper facial structures. **B,** Rocker-bottom feet from a fetus with trisomy 18. Note the prominent heels and convex profile of the soles of the feet. **C,** Pronounced radial deviation of hands (club hands) of the same infant as in **B.** (*Courtesy of M. Barr, Ann Arbor, Mich.*)

Table 8.4 Variations in Numbers of Sex Chromosomes

Sex Chromosome Complement	Incidence	Phenotype	Clinical Factors
XO	1:3000	Immature female	Turner's syndrome: short stature, webbed neck, high and arched palate (see Fig. 8-8)
XX		Female	Normal
XY		Male	Normal
XXY	1:1000	Male	Klinefelter's syndrome: small testes, infertility, often tall with long limbs
XYY	1:1000	Male	Tall, normal appearance; reputed difficulty with impulsive behavior
XXX	1:1000	Female	Normal appearance, mental retardation (up to one third of cases), fertile (in many cases)

Abnormal numbers of the sex chromosomes are relatively common and can be detected by examination of the sex chromatin (X chromosome) or the fluorescence reactions of the Y chromosomes. **Table 8.4** summarizes some of the various types of deletions and duplications of the sex chromosomes.

Abnormal Chromosome Structure

Various abnormalities of chromosome structure can give rise to malformations in development. Some chromosomal abnormalities result from chromosome breakage induced by environmental factors such as radiation and certain chemical teratogens. This type of structural error is usually unique to a given individual and is not transmitted to succeeding generations.

Other types of structural abnormalities of chromosomes are generated during meiosis and, if present in the germ cells, can be inherited. Common types of errors in chromosome structure are **reciprocal translocations, isochromosome formation,** and **deletions** and **duplications** (**Fig. 8.11**). One well-defined congenital malformation resulting from a deletion in the short arm of chromosome 5 is the **cri du chat syndrome.** Infants with this syndrome are severely mentally retarded, have microcephaly, and make a cry that sounds like the mewing of a cat.

Genetic Mutations

Many genetic mutations are expressed as morphological abnormalities. These mutations can be of dominant or recessive genes of either the autosomes or the sex chromosomes.

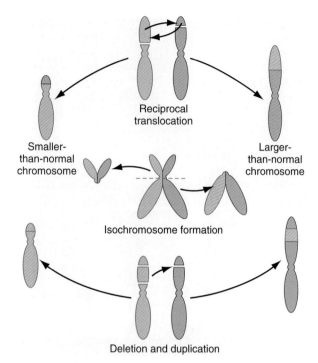

Fig. 8.11 **Different types of structural errors of chromosomes.**

For some of these conditions (e.g., hemophilia, Lesch-Nyhan syndrome, muscular dystrophy, cystic fibrosis), the molecular or biochemical lesion has been identified, but the manner in which these defects are translated into abnormal development is unclear. Many of these conditions are discussed extensively in textbooks of human genetics, and only representative examples are listed here (**Table 8.5**).

Environmental Factors

Various environmental factors are linked with birth defects. These influences range from chemical teratogens and hormones to maternal infections and nutritional factors. Although the list of suspected teratogenic factors is long, relatively few are unquestionably teratogenic in humans.

Maternal Infections

Since the recognition in 1941 that rubella was the cause of a spectrum of developmental anomalies, several other maternal diseases have been implicated as direct causes of birth defects. With infectious diseases, it is important to distinguish diseases that cause malformations by interfering with early stages in the development of organs and structures from diseases that interfere by destroying structures already formed. The same pathogenic organism can cause lesions by interference with embryonic processes or by destruction of differentiated tissues, depending on when the organism attacks the embryo.

Most infectious diseases that cause birth defects are viral, with **toxoplasmosis** (caused by the protozoan *Toxoplasma gondii*) and syphilis (caused by the spirochete *Treponema pallidum*) being notable exceptions. (A summary of the infectious diseases known to cause birth defects in humans is given in **Table 8.6**.)

Table 8.5 Genetic Mutations Leading to Abnormal Development

Condition	Characteristics
Autosomal Dominant	
Achondroplasia	Dwarfism caused mainly by shortening of limbs
Aniridia	Absence of iris (usually not complete)
Crouzon's syndrome (craniofacial dysostosis) (see Fig. 9.30)	Premature closure of certain cranial sutures leading to flat face and towering skull
Neurofibromatosis	Multiple neural crest–derived tumors on skin, abnormal pigment areas on skin
Polycystic kidney disease (adult onset, type III)	Numerous cysts in kidneys
Autosomal Recessive	
Albinism	Absence of pigmentation
Polycystic kidney disease (perinatal type I) (see Fig. 16.17)	Numerous cysts in kidneys
Congenital phocomelia syndrome (see Fig. 8.2)	Limb deformities
X-Linked Recessive	
Hemophilia	Defective blood clotting
Hydrocephalus (see Fig. 11.38)	Enlargement of cranium
Ichthyosis	Scaly skin
Testicular feminization syndrome	Female phenotype caused by inability to respond to testosterone

Table 8.6 Infectious Diseases That Can Cause Birth Defects

Infectious Agent	Disease	Congenital Defects
Viruses		
Rubella virus	German measles	Cataracts, deafness, cardiovascular defects, fetal growth retardation
Cytomegalovirus	Cytomegalic inclusion disease	Microcephaly, microphthalmia, cerebral calcification, intrauterine growth retardation
Spirochetes		
Treponema pallidum (syphilis)	Syphilis	Dental anomalies, deafness, mental retardation, skin and bone lesions, meningitis
Protozoa		
Toxoplasma gondii	Toxoplasmosis	Microcephaly, hydrocephaly, cerebral calcification, microphthalmia, mental retardation, prematurity

The time of infection is very important in relation to the types of effects on the embryo. Rubella causes a high percentage of malformations during the first trimester, whereas cytomegalovirus infections usually kill the embryo during the first trimester. The agents of syphilis and toxoplasmosis cross the placental barrier during the fetal period and, to a large extent, cause malformations by destroying existing tissues.

Chemical Teratogens

Many substances are known to be teratogenic in animals or are associated with birth defects in humans, but convincing evidence that links the substance directly to congenital malformations in humans exists for only a relatively small number (**Table 8.7**). Testing drugs for teratogenicity is difficult because what can cause a high incidence of severe defects in animal fetuses (e.g., cortisone and cleft palate in mice) may not cause malformations in other species of animals or in humans. Conversely, the classic teratogen thalidomide is highly teratogenic in humans, rabbits, and some primates, but not in commonly used laboratory rodents.

FOLIC ACID ANTAGONISTS

Previously, folic acid antagonists, which are known to be highly embryolethal, were used in clinical trials as **abortifacients** (agents causing abortion). Although three fourths of the pregnancies were terminated, almost one fourth of the embryos that survived to term were severely malformed. A classic example of an embryotoxic folic acid antagonist is **aminopterin**, which produces multiple severe anomalies such as anencephaly, growth retardation, cleft lip and palate, hydrocephaly, hypoplastic mandible, and low-set ears. These dramatic effects of folic acid antagonists underscore the importance of adequate amounts of folic acid in the diet to promote normal development.

ANDROGENIC HORMONES

The administration of androgenic hormones to pregnant women either to treat tumors or to prevent threatened abortion resulted in the birth of hundreds of female infants with various degrees of masculinization of the external genitalia. The anomalies consisted of clitoral hypertrophy and often varying amounts of fusion of the genital folds to form a scrotumlike structure (**Fig. 8.12**).

ANTICONVULSANTS

Several commonly used anticonvulsants are known or strongly suspected to be teratogenic. Phenytoin (previously known as diphenylhydantoin) produces a "fetal hydantoin syndrome" of anomalies, including growth anomalies, craniofacial defects, nail and digital hypoplasia, and mental retardation in up to one third of embryos exposed to this drug during pregnancy (**Fig. 8.13**). Trimethadione also produces a syndrome of anomalies involving low-set ears, cleft lip and palate, and skeletal and cardiac anomalies.

SEDATIVES AND TRANQUILIZERS

Thalidomide is highly teratogenic when administered even as infrequently as once during a very narrow window of pregnancy, especially between days 25 and 50, when a single dose of 100 mg can be sufficient to cause birth defects. This represents the period when the primordia of most major organ systems are being established. The most characteristic lesions produced are gross malformations of the limbs, but the

Table 8.7 Chemical Teratogens in Humans

Agent	Effects
Alcohol	Growth and mental retardation, microcephaly, various malformations of face and trunk
Androgens	Masculinization of females, accelerated genital development in males
Anticoagulants (warfarin, dicumarol)	Skeletal abnormalities; broad hands with short fingers; nasal hypoplasia; anomalies of eye, neck, central nervous system
Antithyroid drugs (e.g., propylthiouracil, iodide)	Fetal goiter, hypothyroidism
Chemotherapeutic agents (methotrexate, aminopterin)	Variety of major anomalies throughout body
Diethylstilbestrol	Cervical and uterine abnormalities
Lithium	Heart anomalies
Organic mercury	Mental retardation, cerebral atrophy, spasticity, blindness
Phenytoin (Dilantin)	Mental retardation, poor growth, microcephaly, dysmorphic face, hypoplasia of digits and nails
Isotretinoin (Accutane)	Craniofacial defects, cleft palate, ear and eye deformities, nervous system defects
Streptomycin	Hearing loss, auditory nerve damage
Tetracycline	Hypoplasia and staining of tooth enamel, staining of bones
Thalidomide	Limb defects, ear defects, cardiovascular anomalies
Trimethadione and paramethadione	Cleft lip and palate, microcephaly, eye defects, cardiac defects, mental retardation
Valproic acid	Neural tube defects

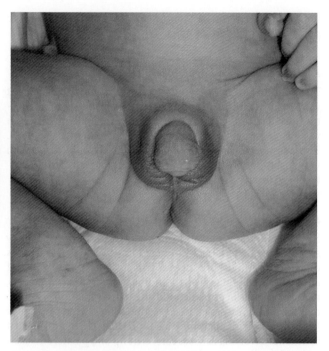

Fig. 8.12 Genetically normal female patient with congenital adrenal hyperplasia. The clitoris is enlarged, and the labia majora show scrotalization. *(From Jorde LB, Carey JC, Bamshad MJ: Medical genetics, ed 4, Philadelphia, 2010, Mosby.)*

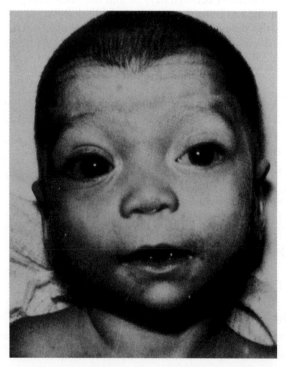

Fig. 8.13 Face of an infant with fetal hydantoin syndrome. This infant has prominent eyes, hypertelorism (increased space between the eyes), micrognathia, and microcephaly. *(From Wigglesworth JS, Singer DB: Textbook of fetal and perinatal pathology, 2 vols, Oxford, 1991, Blackwell Scientific.)*

thalidomide syndrome also includes malformations of the cardiovascular system, absence of the ears, and assorted malformations of the urinary system, gastrointestinal system, and face. Despite years of intensive research, the mechanism by which thalidomide produces malformations is still unknown. Lithium carbonate, a commonly used agent for certain psychoses, is known to cause malformations of the heart and great vessels if the drug is administered early during pregnancy.

ANTINEOPLASTIC AGENTS

Several antineoplastic agents are highly teratogenic, in large part because they are designed to kill or incapacitate rapidly dividing cells. Aminopterin is one such agent. Methotrexate and the combination of busulfan and 6-mercaptopurine cause severe anomalies of multiple organ systems. The use of these drugs during pregnancy is a difficult medical decision that must consider the lives of both the mother and the fetus.

ALCOHOL

Accumulated evidence now leaves little doubt that maternal consumption of alcohol during pregnancy can lead to a well-defined constellation of developmental abnormalities that includes poor postnatal growth rate, microcephaly, mental retardation, heart defects, and hypoplasia of facial structures (**Fig. 8.14**). This constellation of abnormalities is now popularly known as **fetal alcohol syndrome,** and estimates suggest that some form of fetal alcohol syndrome may affect as many as 1% to 5% of all live births. Ingestion of 3 oz of alcohol in a day during the first 4 weeks of pregnancy can lead to extremely severe malformations of the **holoprosencephaly** type (see p. 309).

Exposure to alcohol later in pregnancy is less likely to cause major anatomical defects in the fetus, but because of the complex course of physiological maturation in the brain throughout pregnancy, more subtle behavioral defects can result. Nevertheless, there are often striking differences from normal in the size and shape of the corpus callosum, the main connecting link between the right and left sides of the brain, and in the cerebellum, which may be hypoplastic. Much of the abnormal development of the face and forebrain can be attributed to the death of cells in the anterior neural ridge (see Fig. 6.4B), which serves as a signaling center in the early embryo. Although the intelligence quotient (IQ) of an individual with fetal alcohol syndrome may be normal, such individuals may show deficits in recognition of the consequences of actions or in planning into the future.

RETINOIC ACID (VITAMIN A)

Derivatives of retinoic acid are used in the treatment of acne, but investigators have established that retinoic acid acts as a potent teratogen when it is taken orally. Retinoic acid can produce a wide spectrum of defects, most of which are related to derivatives of the cranial neural crest (see p. 259). These involve a variety of facial structures, the outflow tract of the heart, and the thymus (**Fig. 8.15**).

Through a complex sequence of cytoplasmic binding proteins and nuclear receptors (see Fig. 4.18), retinoic acid affects *Hox* genes, especially genes expressed in the cranial and

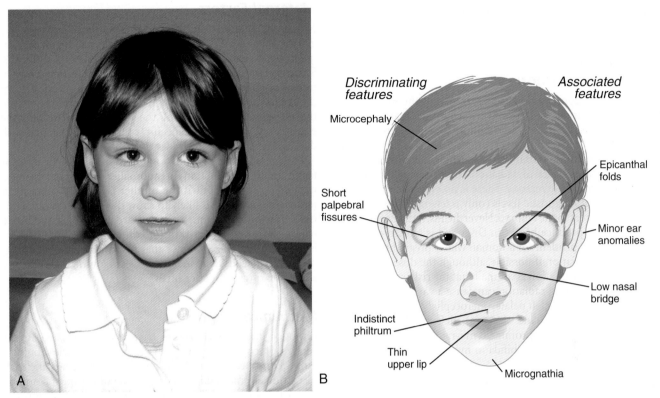

Fig. 8.14 **A,** Face of a young girl with fetal alcohol syndrome showing a long, thin upper lip; shortened, upwardly slanting palpebral fissures; and epicanthic folds. **B,** Commonly expressed facial characteristics of fetal alcohol syndrome in a young child. *(A, From Turnpenny P, Ellard S: Emery's elements of medical genetics, Philadelphia, 2012, Elsevier; B, Adapted from Streissguth AP: Alcohol Health Res World 18:74081, 1994.)*

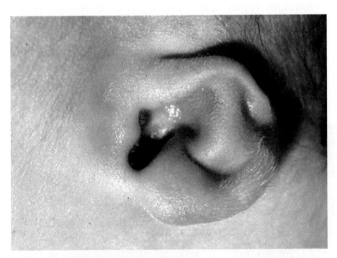

Fig. 8.15 **Etretinate embryopathy.** Among multiple facial anomalies, this infant had a highly deformed ear. *(From the Robert J. Gorlin collection, Division of Oral and Maxillofacial Pathology, University of Minnesota Dental School, courtesy of Dr. Ioannis Koutlas.)*

pharyngeal region (see Fig. 11.12), with resulting alterations of the anterior rhombomeres and the neural crest cells derived from them. As discussed later, neural crest cells emanating from rhombomeres are instrumental in patterning many structures of the face and neck and contribute to the developing heart and thymus, hence the pattern of retinoic acid–induced defects previously outlined. In view of the increasing recognition of the important role of retinoic acid or its metabolites in pattern formation during early development, extreme caution is recommended when vitamin A is used in doses greater than those needed for basic nutritional requirements.

ANTIBIOTICS

The use of two antibiotics during pregnancy is associated with birth defects. Streptomycin in high doses can cause inner ear deafness. Tetracycline given to the mother during late pregnancy crosses the placental barrier and seeks sites of active calcification in the teeth and bones of the fetus. Tetracycline deposits cause a yellowish discoloration of teeth and bones and, in high doses, can interfere with enamel formation.

OTHER DRUGS

Numerous other drugs, such as the anticoagulant warfarin, are known to be teratogenic, and other agents are strongly suspected. Firm proof of a drug's teratogenicity in humans is not easy to obtain, however. Several drugs, such as Agent Orange and some of the social drugs (e.g., lysergic acid diethylamide [LSD], marijuana), have often been claimed to cause birth defects, but the evidence to date is not entirely convincing. Several studies have shown a variety of complications in pregnancy resulting from the use of cocaine, which can readily cross the placental barrier. In addition to structural malformations in organs such as the brain, cocaine use has been linked to intrauterine growth retardation, premature labor, and spontaneous abortion, and postnatal behavioral disturbances, such as attention deficit.

Physical Factors
IONIZING RADIATION

Ionizing radiation is a potent teratogen, and the response is both dependent on the dose and related to the stage at which the embryo is irradiated. In addition to numerous animal studies, there is direct human experience based on survivors of the Japanese atomic bomb blasts and pregnant women who were given large doses of radiation (up to several thousand rads) for therapeutic reasons. There is no evidence that doses of radiation at diagnostic levels (only a few millirads) pose a significant threat to the embryo. Nevertheless, because ionizing radiation can produce breaks in DNA and is known to cause mutations, it is prudent for a woman who is pregnant to avoid exposure to radiation if possible, although the dose in a diagnostic x-ray examination is so small that the risk is minimal.

Although ionizing radiation can cause a variety of anomalies in embryos (e.g., cleft palate, microcephaly, malformations of the viscera, limbs, and skeleton), defects of the central nervous system are very prominent in irradiated embryos. The spectrum runs from spina bifida to mental retardation.

OTHER PHYSICAL FACTORS

Numerous studies on the teratogenic effects of extremes of temperature and different concentrations of atmospheric gases have been conducted on experimental animals, but the evidence relating any of these factors to human malformations is still equivocal. One exception is the effect of excess concentrations of oxygen on premature infants. When this practice was common, **retrolental fibroplasia** developed in more than 10% of premature infants weighing less than 3 lb and in about 1% of premature infants weighing 3 to 5 lb. When this connection was recognized, the practice of maintaining high concentrations of oxygen in incubators ceased, and this problem is now of only historical interest.

Maternal Factors

Numerous maternal factors have been implicated in the genesis of congenital malformations. **Maternal diabetes** is frequently associated with high birth weight and with stillbirths. Structural anomalies occur several times more frequently in infants of diabetic mothers than in infants of mothers from the general population. Although there is a correlation between the duration and severity of the mother's disease and the effects on the fetus, the specific cause of interference with development has not been identified.

In general, maternal nutrition does not seem to be a major factor in the production of anomalies (folic acid being a notable exception), but if the mother is severely deficient in iodine, the newborn is likely to show the symptoms of cretinism (growth retardation, mental retardation, short and broad hands, short fingers, dry skin, and difficulty breathing). There is now considerable evidence that **heavy smoking** by a pregnant woman leads to an increased risk of low birth weight and a low rate of growth after birth.

Mechanical Factors

Although mechanical factors have been implicated in the genesis of congenital malformations for centuries, only in more recent years has it been possible to relate specific malformations to mechanical causes. Many of the most common anomalies, such as **clubfoot**, **congenital hip dislocations**, and even certain deformations of the skull, can be attributed in large measure to abnormal intrauterine pressures imposed on the fetus. This situation can often be related to uterine malformations or a reduced amount of amniotic fluid (**oligohydramnios**).

Amniotic bands constricting digits or extremities of the fetus have been implicated as causes of intrauterine amputations (**Fig. 8.16**). These bands form as the result of tears to the extraembryonic membranes during pregnancy. Chorionic villus sampling results in a low percentage of transverse limb defects, but the mechanism underlying the defective limb development is not well understood.

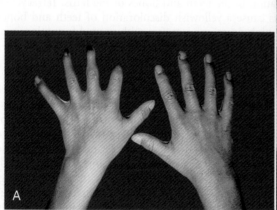
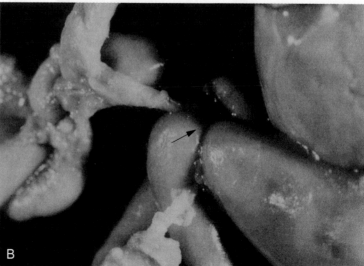

Fig. 8.16   **A,** Digital amputations of the left hand presumably caused by amniotic bands. **B,** Amniotic bands involving the umbilical cord and limbs of a fetus. The *arrow* shows a constriction ring around the thigh. *(**A**, Courtesy of M. Barr, Ann Arbor, Mich, **B**, from Wigglesworth JS, Singer DB: Textbook of fetal and perinatal pathology, 2 vols, Oxford, 1991, Blackwell Scientific.)*

Developmental Disturbances Resulting in Malformations

Duplications and Reversal of Asymmetry

The classic example of duplication is identical twinning. Under normal circumstances, both members of the twin pair are completely normal, but rarely the duplication is incomplete, and **conjoined twins** result (see Figs. 3.15 and 3.16). Twins can be conjoined at almost any site and to any degree. With modern surgical techniques, it is now possible to separate members of some conjoined pairs. A type of conjoined twinning is the condition of **parasitic twinning**, in which one member of the pair is relatively normal, but the other is represented by a much smaller body, often consisting of just the torso and limbs, attached to an area such as the mouth or lower abdomen of the host twin (see Fig. 3.17). In numerous conjoined twins, one member of the pair has reversed asymmetry in relation to the other (see Fig. 3.16).

In rare instances (approximately 1 in 10,000 births), an otherwise normal individual is found to have a partial or complete reversal of the asymmetry of the internal organs, a condition called **situs inversus** (see Fig. 5.15). Molecular research on early embryonic stages (see Fig. 5.13) has begun to provide a mechanistic explanation for this condition.

Faulty Inductive Tissue Interactions

Absent or faulty induction early in development (e.g., induction of the central nervous system) is incompatible with life, but disturbances in later inductions can cause malformations. Absence of the lens (**aphakia**) or of a kidney (**renal agenesis**) can result from an absent or abnormal inductive interaction.

Absence of Normal Cell Death

Genetically or **epigenetically** (environmental influences imposed on the genetic background) controlled cell death is an important mechanism in sculpting many regions of the body. The absence of normal interdigital cell death has been implicated in **syndactyly** (webbed digits) (see Fig. 10.23A) and abnormal persistence of the tail (see Fig. 9.25A for normal tail). The latter phenomenon has sometimes been considered an example of **atavism** (the persistence of phylogenetically primitive structures).

Failure of Tube Formation

The formation of a tube from an epithelial sheet is a fundamental developmental mechanism. A classic case of failure of tube formation is seen in the spina bifida anomalies, which are based on the incomplete fusion of the neural tube (see Fig. 11.42). (Some of the possible mechanisms involved in normal formation of the neural tube are discussed in Chapter 11.)

Disturbances in Tissue Resorption

Some structures present in the early embryo must be resorbed for subsequent development to proceed normally. Examples are the membranes that cover the future oral and anal openings. These membranes are composed of opposing sheets of ectoderm and endoderm, but if mesodermal cells become interposed between the two and this tissue becomes vascularized, breakdown typically fails to occur. **Anal atresia** is a common anomaly of this type (see Fig. 15.19).

Failure of Migration

Migration is an important developmental phenomenon that occurs at the level of cells or entire organs. The neural crest is a classic example of massive migrations at the cellular level, and disturbances in migration can cause abnormalities in any of the structures for which the neural crest is a precursor (e.g., thymus, outflow tracts of the heart, adrenal medulla). At the organ level, the kidneys undertake a prominent migration into the abdominal cavity from their origin in the pelvic region, and the testes migrate from the abdominal cavity into the scrotum. **Pelvic kidneys** (see Fig. 16.15) and undescended testes (**cryptorchidism**) are relatively common.

Developmental Arrest

Early in the history of teratology, some malformations were recognized as the persistence of structures in a state that was normal at an earlier stage of development. Many of the patterns of **cleft lip** and **cleft palate** (see Figs. 14.16 and 14.17) are examples of developmental arrest, although it is incorrect to assume that development has been totally arrested since the sixth to eighth weeks of embryogenesis. Another example of the persistence of an earlier stage in development is a **thyroglossal duct** (see Fig. 14.45), in which persisting epithelial cells mark the path of the thyroid gland as it migrates from the base of the tongue to its normal position.

Destruction of Formed Structures

Many teratogenic diseases or chemicals produce malformations by the destruction of structures already present. If the structure is in the early primordial stage, any tissues to which the primordium would normally give rise are missing or malformed. Interference with the blood supply of a structure can cause unusual patterns of malformations. In the genesis of **phocomelia** (see Fig. 8.2), damage to proximal blood vessels could destroy the primordia of the proximal limb segments, but the cells of the distal limb bud that give rise to the hands or feet could be spared if the distal microvasculature of the limb bud remained intact.

Failure to Fuse or Merge

If two structures such as the palatal shelves fail to meet at the critical time, they are likely to remain separate. Similarly, the relative displacements of mesenchyme (**merging**) that are involved in the shaping of the lower jaw may not occur on schedule or in adequate amounts. This accounts for some malformations of the lower face.

Hypoplasia and Hyperplasia

The normal formation of most organs and complex structures requires a precise amount and distribution of cellular proliferation. If cellular proliferation in a forming organ is abnormal, the structure can become too small (**hypoplastic**) (see Fig. 16.12B) or too large (**hyperplastic**). Even minor growth

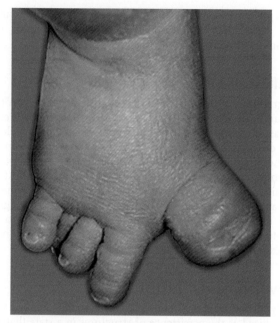

Fig. 8.17 **Gigantism (macrodactyly) of the great toe.** *(From the Robert J. Gorlin collection, Division of Oral and Maxillofacial Pathology, University of Minnesota Dental School, courtesy of Dr. Ioannis Koutlas.)*

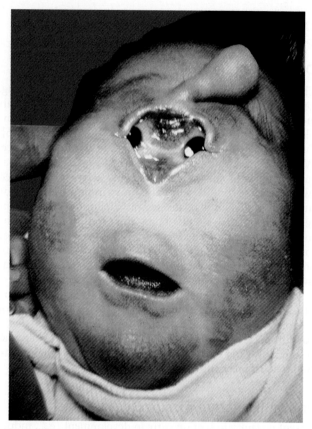

Fig. 8.18 **Cyclopia in a newborn.** Note the fleshy proboscis above the partially fused eye. *(Courtesy of M. Barr, Ann Arbor, Mich.)*

disturbances can cause severe problems in complex regions such as the face. Occasionally, **gigantism** of a structure such as a digit (**Fig. 8.17**) or whole limb occurs. The mechanism underlying this excessive growth remains obscure.

Receptor Defects

Some congenital malformations can be attributed to defects in specific receptor molecules. One of the earliest recognized is the **testicular feminization syndrome**, in which the lack of testosterone receptors results in the development of a typical female phenotype in a genetic male (see Fig. 9.13A).

Defective Fields

Proper morphogenesis of many regions of the body is under the control of poorly understood morphogenetic fields. These regions of the body are under the control of an overall developmental blueprint. Disturbances in the boundaries or overall controls of fields can sometimes give rise to massive anomalies. One example is the fusion of lower limb fields, which is probably associated with a larger defect in the field controlling the development of the caudal region of the body. This mermaid-like anomaly is called **sirenomelia** (see Fig. 8.1B), and it is an extreme example of what is called the **caudal regression syndrome**, resulting from abnormal *T* gene function (see p. 82).

Effects Secondary to Other Developmental Disturbances

Because so much of normal development involves the tight interlocking of individual processes or building on completed structures, it is not surprising that many malformations are secondary manifestations of other disturbed embryonic processes. There are numerous examples in craniofacial development. Some cases of cleft palate have been attributed to a widening of the cranial base so that the palatal shelves, which may have been normal, are unable to make midline contact.

The single or widely separated tubular probosces that appear in certain major facial anomalies, such as **cyclopia** (**Fig. 8.18**), are very difficult to explain unless it is understood that one of several primary defects, whether too much or too little tissue of the midface, prevented the two nasal primordia from joining in the midline. In the case of cyclopia, the primary defect is usually a deficiency of forebrain tissue that results from deficient sonic hedgehog signaling (see Clinical Correlation 14.1), and the facial defects are secondary to that.

Germ Layer Defects

An understanding of normal development can explain the basis for a seemingly diverse set of anomalies (**Clinical Correlation 8.1**). **Ectodermal dysplasias**, which are based on abnormalities in the ectodermal germ layer, can include malformations as diverse as thin hair, poorly formed teeth, short stature, dry and scaly skin, and hypoplastic nails (**Fig. 8.19**). Other syndromes with diverse phenotypic abnormalities are related to defects of the neural crest (see Chapter 12).

CLINICAL CORRELATION 8.1
Diagnosis and Treatment of Birth Defects

Only a few decades ago, birth defects were diagnosed only after the fact, and sometimes it was years after birth before certain defects could be discovered and treated. Although this can still happen today, technological changes have permitted earlier diagnosis and treatment of certain congenital malformations.

One of the first advances was the technology associated with karyotyping and sex chromosome analysis. Initially, these techniques were applied after birth to diagnose conditions based on abnormalities in chromosome number or structure. After the development of amniocentesis (the removal of samples of amniotic fluid during early pregnancy), chromosomal analysis could be applied to cells in the amniotic fluid. This approach was particularly useful in the diagnosis of Down syndrome, and it also permitted the prenatal diagnosis of the gender of the infant. Biochemical analysis of amniotic fluid has permitted the diagnosis of numerous inborn errors of metabolism and neural tube defects (the latter through

the detection of **S-100 protein**, which leaks through the open neural tube into the amniotic fluid).

More recently, techniques have been developed for the direct sampling of tissue from the chorionic villi. Molecular genetic analysis of the cells obtained from these samples can now be used to diagnose a wide variety of conditions. The risk-to-benefit ratio of this technique is still being debated.

With the development of imaging techniques such as ultrasound, computed tomography, and magnetic resonance imaging, visualization of fetal morphological structures became possible (see Figs. 18.11 to 18.14). These images can serve as a direct guide to surgeons who are attempting to correct certain malformations by intrauterine surgery. Because surgical wounds in fetuses typically heal without scarring, fetal corrective surgery has distinct advantages (see Chapter 18).

Clinical Vignette

A woman in her early 40s who has chronic alcoholism, who smokes heavily, and who also occasionally uses cocaine gives birth to an infant with severe anencephaly. She had previously given birth to a child who had a less severe form of spina bifida. Another child, although small in stature, seemed normal, but had a behavioral problem in school.

What is a likely basis for such a history?

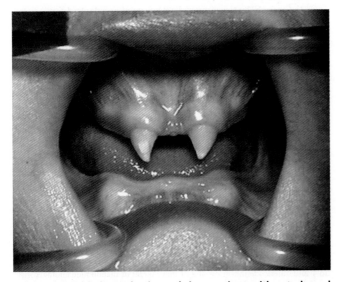

Fig. 8.19   Highly hypoplastic teeth in a patient with ectodermal dysplasia. This patient also had sparse hair. *(From the Robert J. Gorlin Collection, Division of Oral and Maxillofacial Pathology, University of Minnesota Dental School, courtesy of Dr. Ioannis Koutlas.)*

Summary

- Developmental disorders have been recognized for centuries, but a direct connection between environmental teratogens and human birth defects was not shown until 1941.

- Abnormal development is often the result of environmental influences imposed on genetic susceptibility. The factors involved in abnormal development include age, race, country, nutrition, and time of year. The study of abnormal development is teratology, and an agent that causes abnormal development is a teratogen.

- Genetic factors cause a significant number of birth defects. Abnormal chromosome numbers are associated with prenatal death and syndromes of abnormal structures. Common causes of abnormalities are monosomies and trisomies, which are often the result of nondisjunction during meiosis. Other malformations are based on abnormalities of chromosome structure. Certain malformations are based on genetic mutations.

- Environmental factors leading to defective development include maternal infections, chemical teratogens, physical factors such as ionizing radiation, maternal factors, and mechanical factors.

- A variety of disturbed developmental mechanisms may be involved in the production of a given congenital malformation, including duplications, faulty inductive tissue interactions, absence of normal cell death, failure of tube formation, disturbances in tissue resorption, failure of migration, developmental arrest, destruction of an already formed structure, failure to fuse or merge, hypoplasia or hyperplasia, receptor defects, defective fields, effects secondary to other developmental disturbances, and germ cell layer defects.

- With technological developments, it is now possible to diagnose increasing numbers of birth defects in utero. Diagnostic techniques include karyotyping and sex chromosome analysis on cells obtained from amniotic fluid, biochemical analysis of amniotic fluid, biochemical and molecular analysis of cells obtained from amniotic fluid or chorionic villus sampling, and imaging techniques, especially ultrasonography. There have been a few attempts to correct malformations by surgery in utero.

Review Questions

1. Phocomelia is most likely to be seen after maternal exposure to which teratogenic agent during the first trimester of pregnancy?
A. Alcohol
B. Aminopterin
C. Androgens
D. Ionizing radiation
E. Thalidomide

2. Which of these anomalies can be attributed to a disturbance in tissue resorption?
A. Pelvic kidney
B. Cleft lip
C. Anal atresia
D. Renal agenesis
E. Amputated digit in utero

3. Which of the following is responsible for the largest percentage of congenital malformations?
A. Maternal infections
B. Chemical teratogens
C. Genetically based conditions
D. Ionizing radiations
E. Unknown factors

4. Folic acid deficiency is now believed to be a major cause of what class of malformations?
A. Trisomies
B. Neural tube defects
C. Ambiguous genitalia
D. Polyploidy
E. Duplications

5. Cleft palate is the result of a defect in what developmental mechanism?
A. Failure to fuse
B. Failure to merge
C. Faulty inductive tissue interaction
D. Disturbance in tissue resorption
E. Absence of normal cell death

6. An increased incidence of what condition is strongly associated with increasing maternal age?
A. Trisomy 18
B. Trisomy 21
C. Trisomy 13
D. Anencephaly
E. Ambiguous external genitalia

7. A woman who was in a car accident and sustained abdominal bruising during the fourth month of pregnancy gave birth to an infant with a cleft palate. She sued the driver of the other car for expenses associated with treatment of the birth defect and claimed that the defect was caused by the accident. You are asked to be a witness for the defense. What is your case?

8. A woman who took a new sedative during the second month of pregnancy felt nauseated after ingestion of the drug and stopped taking it after a couple of weeks. She gave birth to an infant who had a septal defect of the heart and sued the manufacturer of the drug. She said that the defect was caused by the drug that made her nauseated. You are asked to be a witness for the manufacturer. What is your case?

9. What is a likely cause for a badly turned-in ankle in a newborn?

10. A 3-year-old child is much smaller than normal, has sparse hair, and has irregular teeth. What is a likely basis for this constellation of defects?

References

Bookstein FL and others: Geometric morphometrics of corpus callosum and subcortical structures in the fetal alcohol-affected brain, *Teratology* 63:4-32, 2001.

Botto LD and others: Neural-tube defects, *N Engl J Med* 341:1509-1519, 1999.

Butterworth CE, Bendich A: Folic acid and the prevention of birth defects, *Annu Rev Nutr* 16:73-97, 1996.

Buyse ML, ed: *Birth defects encyclopedia*, Dover, Mass, 1990, Centre for Birth Defects Information Services.

Catilla EE and others: Thalidomide, a current teratogen in South America, *Teratology* 54:273-277, 1996.

Cohen MM: *The child with multiple birth defects*, ed 2, New York, 1997, Oxford.

Cooper MK and others: Teratogen-mediated inhibition of target tissue response to Shh signaling, *Science* 280:1603-1607, 1998.

Copp AJ, Greene NDE: Genetics and development of neural tube defects, *J Pathol* 220:217-230, 2009.

Czeizel AE, Dudas I: Prevention of the first occurrence of neural-tube defects by periconceptional vitamin supplementation, *N Engl J Med* 327:1832-1835, 1992.

Entezami M and others: *Ultrasound diagnosis of fetal anomalies*, Stuttgart, 2004, Thieme.

Erickson JD, ed: Congenital malformations surveillance report, *Teratology* 56:1-175, 1997.

Fanaroff A, Martin RJ: *Neonatal-perinatal medicine*, ed 6, St. Louis, 1997, Mosby.

Frazer C: Of mice and children: reminiscences of a teratogeneticist, *Issues Rev Teratol* 5:1-75, 1990.

Ganapathy V, Leibach FH: Current topic: human placenta. A direct target for cocaine action, *Placenta* 15:785-795, 1994.

Goodman FR: Congenital abnormalities of body patterning: embryology revisited, *Lancet* 362:651-662, 2003.

Gorlin RJ, Cohen MM, Hennekam RCM: *Syndromes of the head and neck*, ed 4, Oxford, 2001, Oxford University Press.

Hansen DK: The embryotoxicity of phenytoin: an update on possible mechanisms, *Proc Soc Exp Biol Med* 197:361-368, 1991.

Ikonomidou C and others: Ethanol-induced apoptotic neurodegeneration and fetal alcohol syndrome, *Science* 287:1056-1060, 2000.

Ito T and others: Identification of a primary target of thalidomide teratogenicity, *Science* 327:1345-1350, 2010.

Jones KL: The effects of alcohol on fetal development, *Birth Defects Res C Embryo Today* 93:3-11, 2011.

Jones KL: *Smith's recognizable patterns of human malformation*, ed 4, Philadelphia, 1988, Saunders.

Kallen B, Mastroiacovo P, Robert E: Major congenital malformations in Down syndrome, *Am J Med Genet* 65:160-166, 1996.

Lacombe D: Transcription factors in dysmorphology, *Clin Genet* 55:137-143, 1999.

Lloyd ME and others: The effects of methotrexate on pregnancy, fertility and lactation, *Q J Med* 92:551-563, 1999.

Miller RK and others: Periconceptional vitamin A use: how much is teratogenic? *Reprod Toxicol* 12:75-88, 1998.

Mills JL and others: Vitamin A and birth defects, *Am J Obstet Gynecol* 177:31-36, 1997.

Munger RG and others: Maternal alcohol use and risk of orofacial cleft birth defects, *Teratology* 54:27-33, 1996.

Naeye RL: *Disorders of the placenta, fetus, and neonate*, St. Louis, 1992, Mosby.

Nieuwenhuis E, Hui C-C: Hedgehog signaling and congenital malformations, *Clin Genet* 67:193-208, 2004.

Nishimura H, Okamoto N: *Sequential atlas of human congenital malformations*, Baltimore, 1976, University Park Press.

Nyberg D, Mahony B, Pretorious D: *Diagnostic ultrasound of fetal anomalies*, St. Louis, 1990, Mosby.

Packham EA, Brook JD: T-box genes in human disorders, *Hum Mol Genet* 12:R37-R44, 2003.

Paré A: *On monsters and marvels*, Chicago, 1982, University of Chicago Press.

Persaud TVN, Chudley AE, Skalko RG, eds: *Basic concepts in teratology*, New York, 1985, Liss.

Polifka JE, Friedman JM: Clinical teratology: identifying teratogenic risks in humans, *Clin Genet* 56:409-420, 1999.

Pont SJ and others: Congenital malformations among liveborn infants with trisomies 18 and 13, *Am J Med Genet A* 140:1749-1756, 2006.

Rajkumar SV: Thalidomide: tragic past and promising future, *Mayo Clin Proc* 79:899-903, 2004.

Reed GB, Claireaux AE, Bain AD: *Diseases of the fetus and newborn*, St. Louis, 1989, Mosby.

Riley EP, McGee CL: Fetal alcohol spectrum disorders: an overview with emphasis on changes in brain and behavior, *Exp Biol Med* 230:357-365, 2005.

Spranger J and others: Errors of morphogenesis: concepts and terms. Recommendations of an international working group, *J Pediatr* 100:160-165, 1982.

Stephens TD, Fillmore BJ: Hypothesis: thalidomide embryopathy. Proposed mechanism of action, *Teratology* 61:189-195, 2000.

Stoler JM, Holmes LB: Recognition of facial features of fetal alcohol syndrome in the newborn, *Am J Med Genet C Semin Med Genet* 127C:21-27, 2004.

Sulik KK: Genesis of alcohol-induced craniofacial dysmorphism, *Exp Biol Med* 230:366-375, 2005.

Sulik KK, Alles AJ: Teratogenicity of the retinoids. In Saurat J-H, ed: *Retinoids: 10 years on*, Basel, 1991, Karger, pp 282-295.

Twining P, McHugo JM, Pilling DW: *Textbook of fetal anomalies*, London, 2000, Churchill Livingstone.

Volpe JJ: Effect of cocaine use on the fetus, *N Engl J Med* 327:399-407, 1992.

Warkany J: *Congenital malformations*, St. Louis, 1971, Mosby.

Wigglesworth JS, Singer DB: *Textbook of fetal and perinatal pathology*, 2 vols, Oxford, 1991, Blackwell Scientific.

Willis RA: *The borderland of embryology and pathology*, ed 2, London, 1962, Butterworth.

Wilson GN: Genomics of human dysmorphogenesis, *Am J Med Genet* 42:187-196, 1992.

Wilson JG, Fraser FC, eds: *Handbook of teratology*, vols 1-4, New York, 1977, Plenum.

Zinn AR, Ross JL: Turner syndrome and haploinsufficiency, *Curr Opin Genet Dev* 8:322-327, 1998.

Part II

Development of the Body Systems

Chapter 9

Integumentary, Skeletal, and Muscular Systems

The construction of the tissues of the body involves developmental phenomena at two levels of organization. One is the level of individual cells, in which the cells that make up a tissue undergo increasing specialization through a process called **cytodifferentiation** (see discussion of restriction, determination, and differentiation, [p. 85]). At the next level of complexity, various cell types develop in concert to form specific tissues through a process called **histogenesis**. This chapter discusses the development of three important tissues of the body: skin, bone, and muscle. The histogenesis of each of these tissues exemplifies important aspects of development.

Integumentary System

The skin, consisting of the epidermis and dermis, is one of the largest structures in the body. The epidermis represents the interface between the body and its external environment, and its structure is well adapted for local functional requirements. Simple inspection of areas such as the scalp and palms shows that the structure of the integument varies from one part of the body to another. These local variations result from inductive interactions between the ectoderm and underlying mesenchyme. Abnormalities associated with the integumentary system are presented later in **Clinical Correlation 9.1**.

Epidermis
Structural Development

The outer layer of the skin begins as a single layer of ectodermal cells (**Fig. 9.1A**). As development progresses, the ectoderm becomes multilayered, and regional differences in structure become apparent.

The first stage in epidermal layering is the formation of a thin outer layer of flattened cells known as the **periderm** at the end of the first month of gestation (Fig. 9.1B). Cells of the periderm, which is present in the epidermis of all amniote embryos, seem to be involved in the exchange of water, sodium, and possibly glucose between the amniotic fluid and the epidermis.

By the third month, the epidermis becomes a three-layered structure, with a mitotically active **basal** (or **germinative**) **layer**, an intermediate layer of cells (Fig. 9.1D) that represent the progeny of the dividing stem cells of the basal layer, and a

superficial layer of peridermal cells bearing characteristic surface blebs (**Fig. 9.2**). Peridermal cells contain large amounts of glycogen, but the function of this glycogen remains uncertain.

During the sixth month, the epidermis beneath the periderm undergoes differentiation into the definitive layers characteristic of the postnatal epidermis. Many of the peridermal cells undergo programmed cell death (**apoptosis**) and are sloughed into the amniotic fluid. The epidermis becomes a barrier between the fetus and the outside environment instead of a participant in exchanges between the two. The change in function of the fetal epidermis may have adaptive value because it occurs at about the time when urinary wastes begin to accumulate in the amniotic fluid.

Immigrant Cells in the Epidermis

Despite its homogeneous histological appearance, the epidermis is really a cellular mosaic, with contributions from cells derived not only from surface ectoderm, but also from other precursors, such as the neural crest or mesoderm. These cells play important specific roles in the function of the skin.

Early in the second month, **melanoblasts** derived from the neural crest migrate into the embryonic dermis; slightly later, they migrate into the epidermis. Although melanoblasts can be recognized early by staining with a monoclonal antibody (HMB-45, which reacts with a cytoplasmic antigen common to melanoblasts and **melanomas** [pigment cell tumors]), these cells do not begin to produce recognizable amounts of pigment until midpregnancy. This production occurs earlier in heavily pigmented individuals than in individuals with light complexions. The differentiation of melanoblasts into mature **melanocytes** involves the formation of pigment granules called **melanosomes** from **premelanosomes**.

The number of pigment cells in the skin does not differ greatly among the various races, but the melanocytes of dark-skinned individuals contain more pigment granules per cell. **Albinism** is a genetic trait characterized by the lack of pigmentation, but albinos typically contain normal numbers of melanocytes in their skin. The melanocytes of albinos are generally unable to express pigmentation because they lack the enzyme **tyrosinase**, which is involved in the conversion of the amino acid tyrosine to **melanin**.

Late in the first trimester, the epidermis is invaded by **Langerhans' cells**, which arise from precursors in the bone marrow.

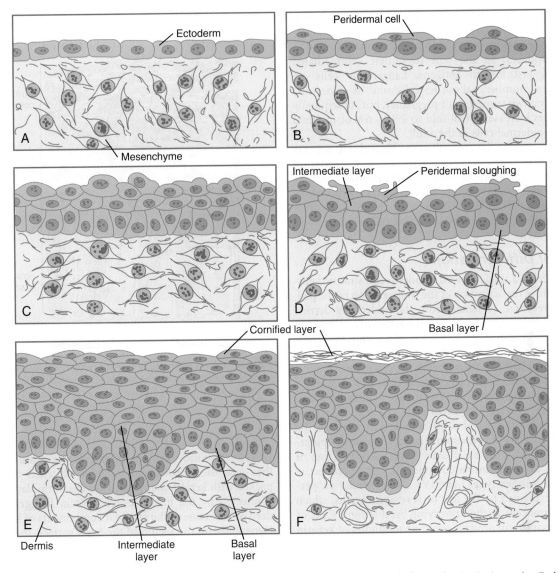

Fig. 9.1 **Stages in the histogenesis of human skin.** **A,** At 1 month. **B,** At 2 months. **C,** At 2½ months. **D,** At 4 months. **E,** At 6 months. **F,** After birth. *(Adapted from Carlson B:* Patten's foundations of embryology, *ed 6, New York, 1996, McGraw-Hill.)*

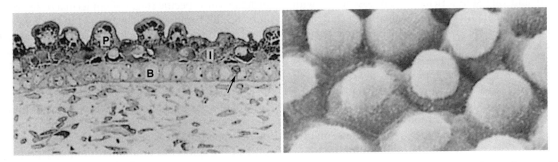

Fig. 9.2 Light micrograph *(left)* and scanning electron micrograph *(right)* of the epidermis of a 10-week human embryo. The prominent surface blebs seen in the scanning micrograph are represented by the irregular surface of the periderm (P) in the light micrograph. The *arrow* in the figure on the *left* points to a melanocyte in the basal layer (B) of the epidermis. I, intermediate layer of epidermis. *(From Sybert VP, Holbrook KA: In Reed G, Claireaux A, Bain A:* Diseases of the fetus and newborn, *St. Louis, 1989, Mosby.)*

These cells are peripheral components of the immune system and are involved in the presentation of antigens; they cooperate with T lymphocytes (white blood cells involved in cellular immune responses) in the skin to initiate cell-mediated responses against foreign antigens. Langerhans' cells are present in low numbers (about 65 cells/mm^2 of epidermis) during the first two trimesters of pregnancy, but subsequently their numbers increase several-fold to 2% to 6% of the total number of epidermal cells in the adult.

A third cell type in the epidermis, the **Merkel cell**, appears in palmar and plantar epidermis at 8 to 12 weeks of gestation and is associated with free nerve terminals. Derived from the neural crest, these cells function as slow-adapting mechanoreceptors in the skin, but cytochemical evidence suggests that they may also function as neuroendocrine cells at some stage.

Epidermal Differentiation

Progression from a single-layered ectoderm to a stratified epithelium requires the activation of the transcription factor **p63**, possibly in response to signals from the underlying dermal mesenchyme. Subsequently, through the action of a microRNA (**miR-203**), p63 must be turned off for cells within the stratified epidermis to embark on their terminal differentiation program, which involves their leaving the cell cycle.

When the multilayered epidermis becomes established, a regular cellular organization and sequence of differentiation appear within it (**Fig. 9.3**). **Stem cells*** of the basal layer (**stratum basale**) divide and contribute daughter cells to the next layer, the **stratum spinosum**. The movement of epidermal cells away from the basal layer is preceded by a loss of adhesiveness to basal lamina components (e.g., fibronectin, laminin, and collagen types I and IV). These cellular properties can be explained by the loss of several **integrins**, which attach the basal cells to the underlying basal lamina. Cells of the stratum spinosum produce prominent bundles of **keratin** filaments, which converge on the patchlike desmosomes binding the cells to each other.

Keratohyalin granules, another marker of epidermal differentiation, begin to appear in the cytoplasm of the outer, postmitotic cells of the stratum spinosum and are prominent components of the stratum granulosum. Keratohyalin granules are composed of two types of protein aggregates—one histidine rich and one sulfur rich—closely associated with bundles of keratin filaments. Because of their high content of keratin, epidermal cells are given the generic name **keratinocytes**. As the keratinocytes move into the stratum granulosum, their nuclei begin to show characteristic signs of terminal differentiation, such as a flattened appearance, dense masses of nuclear chromatin, and early signs of breaking up of the nuclear membrane. In these cells, the bundles of keratin become more prominent, and the molecular weights of keratins that are synthesized are higher than in less mature keratinocytes.

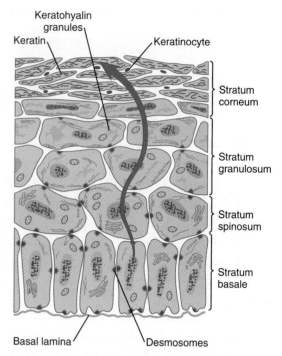

Fig. 9.3 **Layers of the fully formed human epidermis.** Cells arising in the stratum basale undergo terminal differentiation into keratinocytes as they move toward the surface. *(Adapted from Carlson B: Patten's foundations of embryology, ed 6, New York, 1996, McGraw-Hill.)*

As the cells move into the outer layer, the **stratum corneum**, they lose their nuclei and resemble flattened bags densely packed with keratin filaments. The cells of this layer are interconnected by the histidine-rich protein **filaggrin**, which is derived from one of the granular components of keratohyalin. Depending on the region of the body surface, the cells of the stratum corneum accumulate to form approximately 15 to 20 layers of dead cells. In postnatal life, whether through friction or the degradation of the desmosomes and filaggrin, these cells are eventually shed (e.g., about 1300 cells/cm^2/ hour in the human forearm) and commonly accumulate as house dust.

Biochemical work has correlated the expression of keratin proteins with specific stages of epidermal differentiation. Keratins K5, K14, and K15 are expressed in the basal layer, but as the stratum spinosum develops, the cells in that layer express K1 and K10. Loricrin and filaggrin, an intracellular binding protein, appear as the stratum granulosum and the stratum corneum develop in the early fetus.

The proliferation of basal epidermal cells is under the control of a variety of growth factors. Some of these stimulate and others inhibit mitosis. Postnatally, keratinocytes commonly spend about 4 weeks in their passage from the basal layer of the epidermis to ultimate desquamation, but in some skin diseases, such as **psoriasis**, epidermal cell proliferation is poorly controlled, and keratinocytes may be shed within 1 week after their generation.

A prominent feature of the skin, particularly the thick skin of the palms and soles, is the presence of epidermal ridges and creases. On the tips of the digits, the ridges form loops and whorls in fingerprint patterns that are unique to the individual. These patterns form the basis for the science of

---

*Many types of tissues contain a population of stem cells, which have a high capacity for proliferation. Some of the daughter cells remain as stem cells, but other daughter cells become what in the epidermis are called transit-amplifying cells. These cells, which are located in the stratum basale and to some extent in the stratum spinosum, are capable of a few more mitotic divisions before permanently withdrawing from the cell cycle. The postmitotic cells are sometimes called committed cells. In the epidermis, these are the cells that undergo keratinization.

dermatoglyphics, in which the patterns constitute the foundation for genetic analysis or criminal investigation.

The formation of epidermal ridges is closely associated with the earlier appearance of **volar pads** on the ventral surfaces of the fingers and toes (**Fig. 9.4**). Volar pads first form on the palms at about 6½ weeks, and by 7½ weeks, they have formed on the fingers. The volar pads begin to regress by about 10½ weeks, but while they are present, they set the stage for the formation of the epidermal ridges, which occurs between 11 and 17 weeks. Similar events in the foot occur approximately 1 week later than those in the hand.

The pattern of the epidermal ridges is correlated with the morphology of the volar pads when the ridges first form. If a volar pad is high and round, the epidermal ridges form a whorl; if the pad is low, an arch results. A pad of intermediate height results in a loop configuration of the digital epidermal ridges. The timing of ridge formation also seems to influence the morphology: Early formation of ridges is associated with whorls, and late formation is associated with arches.

The primary basis for dermatoglyphic patterns is still not understood.

When the epidermal ridges first form, the tips of the digits are still smooth, and the fetal epidermis is covered with peridermal cells. Beneath the smooth surface, however, epidermal and dermal ridges begin to take shape (**Fig. 9.5**). Late in the fifth month of pregnancy, the epidermal ridges become recognizable features of the surface landscape.

Dermis

The dermis arises from several sources. In the trunk, dorsal dermis arises from the dermatome of the somites, whereas ventral and lateral dermis and dermis of the limbs is derived from the lateral plate mesoderm. In the face, much of the cranial skin, and anterior neck, dermal cells are descendants of cranial neural crest ectoderm (see Fig. 12.9).

Ectodermal Wnt signaling, acting through the β-catenin pathway, specifies the dermomyotomal cells, as well as

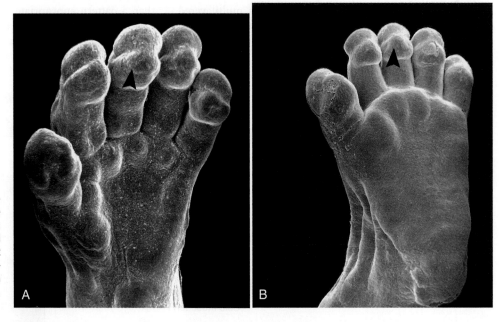

Fig. 9.4 Scanning electron micrographs of the ventral surfaces of the hand (A) and the foot (B) of a human embryo at the end of the second month. Volar pads are prominent near the tips of the digits *(arrowheads)*. *(From Jirásek J:* Atlas of human prenatal morphogenesis, *Amsterdam, 1983, Martinus Nijhoff.)*

Fig. 9.5 **Scanning electron micrographs of human digital palmar skin in a 14-week fetus.** A, Low-power view of the palmar surface of a digit. B, Epidermal surface of the dermis of the fingertip showing the primary dermal ridges. C, Basal surface of the epidermis showing the epidermal ridges. *Bars,* 100 μm. *(From Misumi Y, Akiyoshi T:* Am J Anat *119:419, 1991.)*

mesenchymal cells of the ventral somatopleure, closest to the ectoderm to become dermal cells, which express the dermal marker, **Dermo 1** (see Fig. 9.8A). The future dermis is initially represented by loosely aggregated mesenchymal cells that are highly interconnected by focal tight junctions on their cellular processes. These early dermal precursors secrete a watery intercellular matrix rich in glycogen and hyaluronic acid.

Early in the third month, the developing dermis undergoes a transition from the highly cellular embryonic form to a state characterized by the differentiation of the mesenchymal cells into fibroblasts and the formation of increasing amounts of a fibrous intercellular matrix. The principal types of fibers are types I and III collagen and elastic fibers. The dermis becomes highly vascularized, with an early capillary network transformed into layers of larger vessels. Shortly after the eighth week, sensory nerves growing into the dermis and epidermis help complete reflex arcs and thus allow the fetus to respond to pressure and stroking.

Dermal-Epidermal Interactions

The transformation of simple ectoderm into a multilayered epidermis depends on continuing inductive interactions with the underlying dermis. Dermal-epidermal interactions are also the basis for the formation of a wide variety of epidermal appendages and the appearance of regional variations in the structure of the epidermis. Early in development, the epidermis covering the palms and soles becomes significantly thicker than the epidermis elsewhere on the body. These regions also do not produce hairs, whereas hairs of some sort, whether coarse or extremely fine, form in regular patterns from the epidermis throughout most of the rest of the body.

Tissue recombination experiments on a variety of vertebrate species have shown that the underlying dermis determines the course of development of the epidermis and its derivatives, and that the ectoderm also influences the developmental course of the dermis. If the early ectodermal and mesenchymal components of the skin are enzymatically dissociated and grown separately, the ectodermal component remains simple ectoderm without differentiating into a multilayered epidermis with appropriate epidermal appendages. Similarly, isolated subectodermal mesenchyme retains its embryonic character without differentiating into dermis.

If ectoderm from one part of the body is combined with dermis from another area, the ectoderm differentiates into a regional pattern characteristic of underlying dermis, rather than a pattern appropriate for the site of origin of the ectoderm (**Fig. 9.6**). Cross-species recombination experiments have shown that, even in distantly related animals, skin

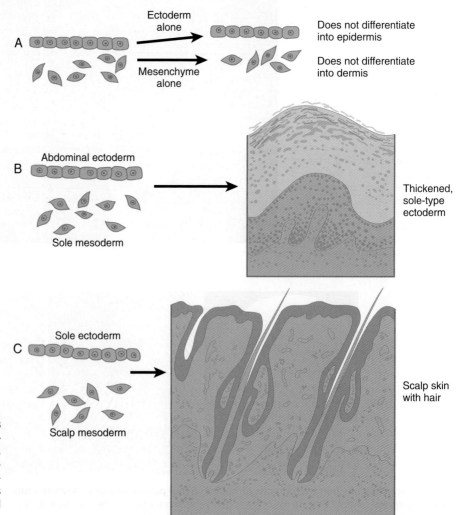

Fig. 9.6 **Recombination experiments illustrating the importance of tissue interactions in the differentiation of the skin.** When separated (**A**), ectoderm and underlying mesenchyme do not differentiate. Recombinations (**B** and **C**) show that the dermis determines the nature of the ectodermal differentiation.

ectoderm and mesenchyme can respond to each other's inductive signals.

As in many other parts of the body, inductive interactions and subsequent morphogenesis of the skin and its appendages are mediated by the production and secretion of common signaling molecules. Specific regional morphogenesis of the skin and its appendages is patterned through the actions of transcription factors, functioning in concert with the still poorly understood influences of retinoic acid, which exerts powerful effects on the skin.

Epidermal Appendages

As a result of inductive influences by the dermis, the epidermis produces a wide variety of appendages, such as hair, nails, sweat and sebaceous glands, mammary glands, and the enamel component of teeth. (The development of teeth is discussed in Chapter 14.)

Hair

Hairs are specialized epidermal derivatives that arise as the result of inductive stimuli from the dermis. There are many types of hairs, ranging from the coarse hairs of the eyelashes and eyebrows to the barely visible hairs on the abdomen and back. Regional differences in morphology and patterns of distribution are imposed on the epidermis by the underlying dermis.

Hair formation is first recognizable at about the twelfth week of pregnancy as regularly spaced epidermal placodes associated with small condensations of dermal cells called **dermal papillae** (**Fig. 9.7**). Under the continuing influence of a dermal papilla, the placode forms an epidermal downgrowth (**hair germ**), which over the next few weeks forms an early **hair peg**. In succeeding weeks, the epidermal peg overgrows the dermal papilla, and this process results in the shaping of an early **hair follicle**. At this stage, the hair follicle still does not

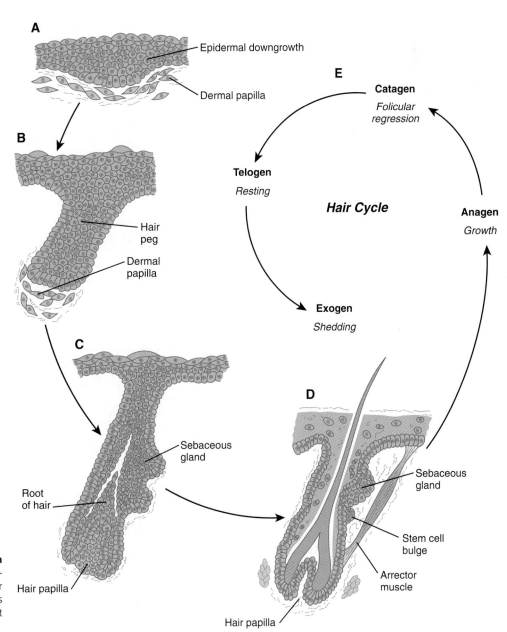

Fig. 9.7 **Differentiation of a human hair follicle.** A, Hair primordium (12 weeks). B, Early hair peg (15 to 16 weeks). C, Bulbous hair follicle (18 weeks). D, Adult hair. E, The adult hair cycle.

protrude beyond the outer surface of the epidermis, but in the portion of the follicle that penetrates deeply into the dermis, two bulges presage the formation of **sebaceous glands**, which secrete an oily skin lubricant (**sebum**), and are the attachment site for the tiny **arrector pili muscle**. The arrector pili is a mesodermally derived smooth muscle that lifts the hair to a nearly vertical position in a cold environment. In many animals, this increases the insulation properties of the hair. The developing hair follicle induces the adjacent dermal mesoderm to form the smooth muscle cells of this muscle. As the developing hair matures, a small bulge below the sebaceous gland marks an aggregation of epidermal stem cells (Fig. 9.7D).

The formation of a hair involves a series of inductive interactions mediated by signals that are only partly understood. When a dense condensation of dermal cells has formed beneath the ectoderm (**Fig. 9.8A**), the first of two dermal inductions results in the thickening of ectoderm in very regularly arranged locations to form epidermal placodes (Fig. 9.8B). Fibroblast growth factor (FGF) and Wnt (mainly Wnt-11) signaling from the dermis, along with the inactivation of local bone morphogenetic proteins (BMPs), stimulates the activation of other Wnts in the ectoderm to form an epidermal placode. The response of the ectoderm is to produce

other Wnts, acting through β-catenin intermediates, and **Edar**, the receptor for the signaling molecule **ectodysplasin**. In the areas where hairs will not develop (interfollicular areas), placode formation is inhibited by locally produced BMPs and by the inhibition of Wnts by Dickkopf. How the epidermal placodes are spaced in such a geometrically regular fashion is still not well understood.

The newly formed epidermal placodes become the inducing agent and stimulate the aggregation of mesenchymal cells beneath the placode to form the **dermal papilla** (Fig. 9.8C). Sonic hedgehog, produced by the epidermal placode, seems to be involved in this induction, but the identity of other signals is unknown. Next, the dermal papilla initiates the second dermal induction by stimulating downgrowth of the cells of the epidermal placode into the dermis (Fig. 9.8D). Epidermal downgrowth, which involves considerable epidermal cellular proliferation, is stimulated by expression of sonic hedgehog by the epidermal cells and the subsequent expression of cyclin D1, part of the cell cycling pathway. Later formation of a hair is structurally and biochemically an extremely complex process, which, among other things, involves the expression of a range of *Hox* genes in specific locations and at specific times along the length of each developing hair.

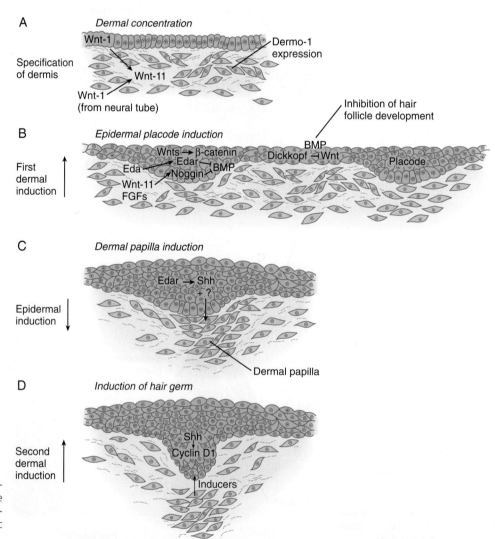

Fig. 9.8   **A to D,** Steps in the formation of a hair follicle. BMP, bone morphogenetic protein; FGF, fibroblast growth factor; Shh, sonic hedgehog.

Once formed, an individual hair follows a regular cycle of growth and shedding (see Fig. 9.7). During **anagen**, the first phase in the cycle, the hair is actively growing (around 10 cm per year). This phase can last up to 5 to 6 years. Then it enters **catagen**, a phase lasting 1 or 2 weeks, during which the hair follicle regresses to only a fraction of its original length. The hair stops growing in the resting phase (**telogen**), which lasts 5 to 6 weeks, after which time the hair is shed (**exogen**). Adjacent hairs are frequently in different phases of the hair cycle.

Erupted hairs are first seen on the eyebrows shortly after 16 weeks. Within a couple of weeks, hairs cover the scalp. The eruption of hairs follows a cephalocaudal gradient over the body. During the later stages of hair formation, the hair bulb becomes infiltrated with melanocytes, which provide color to the hair. Starting around the fifth month, the epidermal cells of the hair shaft begin to undergo keratinization, forming firm granules of **trichohyalin**, which imparts hardness to the hair.

Products of the fetal sebaceous glands accumulate on the surface of the skin as **vernix caseosa**. This substance may serve as a protective coating for the epidermis, which is continually exposed to amniotic fluid.

The first fetal hairs are very fine in texture and are close together. Known as **lanugo**, they are most prominent during the seventh and eighth months. Lanugo hairs are typically shed just before birth and are replaced by coarser definitive hairs, which arise from newly formed follicles.

The pattern of epidermal appendages such as hairs has been shown experimentally to relate to patterns generated in the dermis. Other studies have compared patterns of scalp hairs between normal embryos and embryos with cranial malformations (**Fig. 9.9**) and have shown a correlation between whorls and the direction of hair growth and the tension on the epidermis at the time of formation of the hair follicles.

Nails

Toward the end of the third month, epidermal thickenings (**primary nail field**) on the dorsal surfaces of the digits mark the beginnings of nail development. Cells from the primary nail field expand proximally to undercut the adjacent epidermis (**Fig. 9.10**). Proliferation of cells in the proximal part of the nail field results in the formation of a **proximal matrix**, which gives rise to the nail plate that grows distally to cover the **nail bed**. The nail plate itself consists of highly keratinized epidermal cells. A thin epidermal layer, the

eponychium, initially covers the entire nail plate, but it eventually degenerates, except for a thin persisting rim along the proximal end of the nail. The thickened epidermis underlying the distalmost part of the nail is called the **hyponychium**, and it marks the border between dorsal and ventral skin.

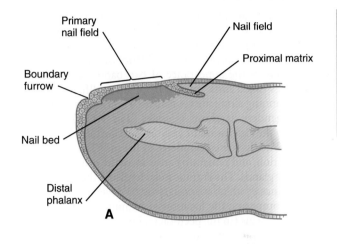

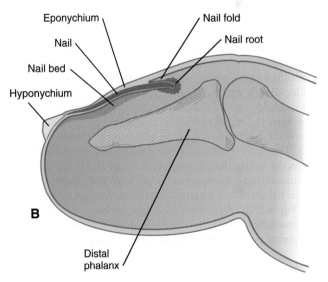

Fig. 9.10 **Fingernail development.** A, In the fourth month, the primary nail field overlies a mesenchymal nail bed and extends proximally as the proximal matrix. B, Close to term, the nail has grown close to the end of the fingertip. Much of the nail is covered by a thin eponychium, most of which will eventually degenerate.

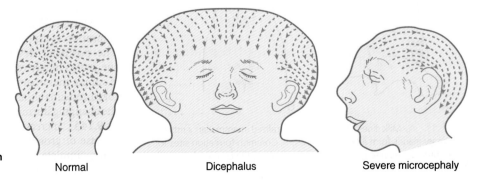

Fig. 9.9 **Patterns of whorls of hair in normal and abnormal fetuses.**

Normal   Dicephalus   Severe microcephaly

Outgrowing fingernails reach the ends of the digits by about 32 weeks, whereas in toenails, this does not occur until 36 weeks.

Mammary Glands

As with many glandular structures, the mammary glands arise as epithelial (in this case, ectodermal) downgrowths into mesenchyme in response to inductive influences by the mesenchyme. The first morphological evidence of mammary gland development is the appearance of two bands of ectodermal thickenings called **milk lines** (part of the wolffian ridge [see p. 111]) running along the ventrolateral body walls in embryos of both genders at about 6 weeks (**Fig. 9.11A**). They are marked by the expression of various **Wnts** within the ectodermal cells. The thickened ectoderm of the milk lines undergoes fragmentation, and remaining areas form the primordia of the mammary glands. The craniocaudal level and the extent along the milk lines at which mammary tissue develops vary among species. Comparing the location of mammary tissue in cows (caudal), humans (in the pectoral region), and dogs (along the length of the milk line) shows the wide variation in location and number of mammary glands. In humans, supernumerary mammary tissue or nipples can be found anywhere along the length of the original milk lines (Fig. 9.11B). Individual mammary placodes form from aggregation and proliferation of ectodermal cells of the milk line under the inductive influence of the signaling molecule **neuregulin-3**. Their dorsoventral location is marked by the expression of the transcription factor **Tbx-3**.

Mammary ductal epithelial downgrowths (**Fig. 9.12**) are associated with two types of mesoderm: fibroblastic and fatty. The early epithelial downgrowth secretes **parathyroid hormone–related hormone**, which increases the sensitivity of the underlying mesenchymal cells to **BMP-4**. BMP-4 signals within the underlying mesenchyme have two principal effects (see Fig. 9.12B). First, they stimulate further downgrowth of the mammary epithelial bud. Second, they stimulate the expression of the transcription factor **Msx-2**, which inhibits

the formation of hair follicles in the region of the nipple. Experimental evidence suggests that inductive interactions with the fatty component of the connective tissue are responsible for the characteristic shaping of the mammary duct system. As with many developing glandular structures, the inductive message seems to be mediated to a great extent by the extracellular matrix of the connective tissue.

Although the mesoderm controls the branching pattern of the ductal epithelium, the functional properties of the mammary ducts are intrinsic to the epithelial component. An experiment in which mouse mammary ectoderm was combined with salivary gland mesenchyme illustrates this point. The mammary ducts developed a branching pattern characteristic of salivary gland epithelium, but despite this, the mammary duct cells produced one of the milk proteins, **α-lactalbumin**.

In keeping with their role as secondary sexual characteristics, mammary glands are extremely responsive to the hormonal environment. This has been shown by experiments conducted on mice. In contrast to the continued downgrowth of ductal epithelium in female mice, the mammary ducts in male mice respond to the presence of testosterone by undergoing a rapid involution. Female mammary ducts react similarly if they are exposed to testosterone. Further analysis has shown that the effect of testosterone is mediated through the mammary mesenchyme, rather than acting directly on the ductal epithelium. Conversely, if male mammary ducts are allowed to develop in the absence of testosterone, they assume a female morphology. Mammary duct development in male humans does not differ from that in female humans until puberty. Even then, a rudimentary duct system remains, which is why men can develop gynecomastia or breast cancer in later life.

The role of the mesoderm and **testosterone receptors** is well illustrated in experiments involving mice with a genetic mutant, **androgen insensitivity syndrome**. This is the counterpart of a human condition called the **testicular feminization syndrome**, in which genetic male individuals lack testosterone receptors. Despite having high circulating levels of testosterone, these individuals develop female phenotypes, including typical female breast development (**Fig. 9.13A**), because without receptors, the tissue cannot respond to the testosterone.

In vitro recombination experiments on mice with androgen insensitivity have been instrumental in understanding the role of the mesoderm in mediating the effects of testosterone on mammary duct development (Fig. 9.13B). If mutant mammary ectoderm is combined with normal mesoderm in the presence of testosterone, the mammary ducts regress, but normal ectoderm combined with mutant mesoderm continues to form normal mammary ducts despite being exposed to high levels of testosterone. This shows that the genetic defect in testicular feminization is expressed in the mesoderm.

The postnatal development of female mammary gland tissue is also highly responsive to its hormonal environment. The simple mammary duct system that was laid down in the embryo remains in an infantile condition until it is exposed to the changing hormonal environment at the onset of puberty (**Fig. 9.14A**). Increasing levels of circulating estrogens, acting on a base of growth hormone and insulinlike growth factor activity, stimulate the proliferation of the mammary ducts and enlargement of the pad of fatty tissue that underlie it (Fig.

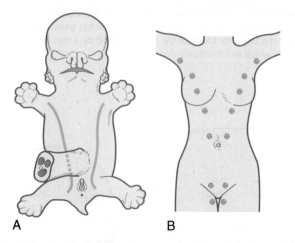

Fig. 9.11 A, Milk lines *(blue)* in a generalized mammalian embryo. Mammary glands form along these lines. **B,** Common formation sites for supernumerary nipples or mammary glands along the course of the milk lines in the human.

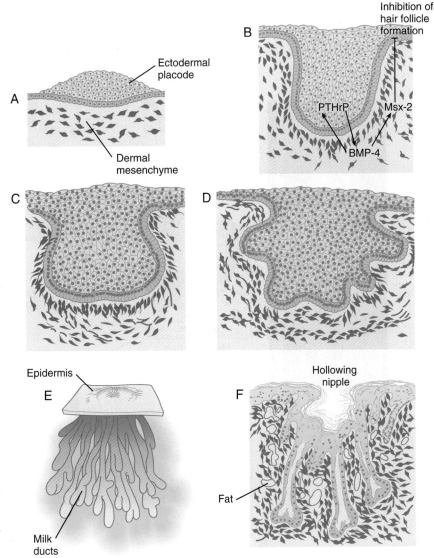

Fig. 9.12 **Stages in the embryonic development of the human mammary gland. A,** Sixth week. **B,** Seventh week. BMP-4, bone morphogenetic protein-4; PTHrP, parathyroid hormone–related protein. **C,** Tenth week. **D,** Fourth month. **E,** Sixth month. **F,** Eighth month.

9.14B). As is the case with testosterone effects, estrogen effects on the epithelium of the mammary ducts are mediated via paracrine influences from the mammary connective tissue stroma, which contains the estrogen receptors.

The next major change in the complete cycle of mammary tissue development occurs during pregnancy, although minor cyclical changes in mammary tissue are detectable in each menstrual cycle. During pregnancy, increased amounts of progesterone, along with prolactin and placental lactogen, stimulate the development of secretory alveoli at the ends of the branched ducts (Fig. 9.14C). With continuing development of the alveoli, the epithelial cells build up increased numbers of the cytoplasmic organelles, such as rough endoplasmic reticulum and the Golgi apparatus, which are involved in protein synthesis and secretion.

Lactation involves numerous reciprocal influences between the mammary glands and the brain; these are summarized in Figure 9.14D. Stimulated by prolactin secretion from the anterior pituitary, the alveolar cells synthesize milk proteins (**casein** and α-lactalbumin) and lipids. In a rapid response to the suckling stimulus, the ejection of milk is triggered by the

release of **oxytocin** by the posterior portion of the pituitary. Oxytocin causes the contraction of **myoepithelial cells,** which surround the alveoli. Suckling also causes an inhibition of the release of luteinizing hormone–releasing hormone by the hypothalamus that results in the inhibition of ovulation and a natural form of birth control.

With cessation of nursing, reduced prolactin secretion and the inhibitory effects of nonejected milk in the mammary alveoli result in the cessation of milk production. The mammary alveoli regress, and the duct system of the mammary gland returns to the nonpregnant state (Fig. 9.14E).

Clinical Correlation 9.1 summarizes several types of anomalies that affect the integumentary system.

Skeleton

Skeletal tissue is present in almost all regions of the body, and the individual skeletal elements are quite diverse in morphology and tissue architecture. Despite this diversity, however, there are some fundamental embryological commonalities.

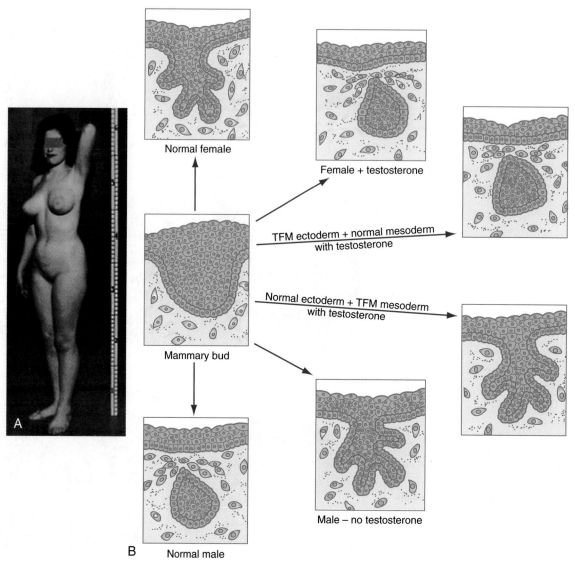

Fig. 9.13 **A,** Testicular feminization, showing the female phenotype of a genetically male individual who had primary amenorrhea. Examination of the gonads after removal revealed immature testicular tubules. **B,** Roles of genetic specificity and testosterone in the development of mouse mammary gland tissue. With normal female mammary tissue *(top center),* the addition of testosterone causes prospective duct tissue to detach and regress as in normal male development. Conversely, in the absence of testosterone, male ductal primordia *(bottom center)* assume a female configuration. In the testicular feminization mutant (TFM), if normal mammary ectoderm is cultured with TFM mammary mesoderm in the presence of testosterone, mammary duct epithelium continues to develop *(lower right).* If normal male mammary mesoderm is combined with TFM ectoderm in the presence of testosterone, the normal male pattern of separation and regression of mammary duct epithelium occurs *(upper right),* showing that the genetic defect is expressed in TFM mesoderm. *(A, From Morris JM, Mahesh VB:* Am J Obstet Gynecol 87:731, 1963; *B, based on studies by Kratochwil K:* J Embryol Exp Morphol 25:141-153, 1971.)

All skeletal tissue arises from cells with a mesenchymal morphology, but the origins of the mesenchyme vary in different regions of the body. In the trunk, the mesenchyme that gives rise to the segmented **axial skeleton** (i.e., vertebral column, ribs, sternum) originates from the sclerotomal portion of the mesodermal somites, whereas the appendicular skeleton (the bones of the limbs and their respective girdles) is derived from the mesenchyme of the lateral plate mesoderm.

The origins of the head skeleton are more complex. Some cranial bones (e.g., the bones making up the roof and much of the base of the skull) are mesodermal in origin, but the facial bones and some of the bones covering the brain arise from mesenchyme derived from the ectodermal neural crest.

The deep skeletal elements of the body typically first appear as cartilaginous models of the bones that will ultimately be formed (**Fig. 9.16**). At specific periods during embryogenesis, the cartilage is replaced by true bone through the process of **endochondral ossification**. In contrast, the superficial bones of the face and skull form by the direct ossification of mesenchymal cells without an intermediate cartilaginous stage (**intramembranous bone formation**). Microscopic details of intramembranous and endochondral bone formation are presented in standard histology texts and are not repeated here.

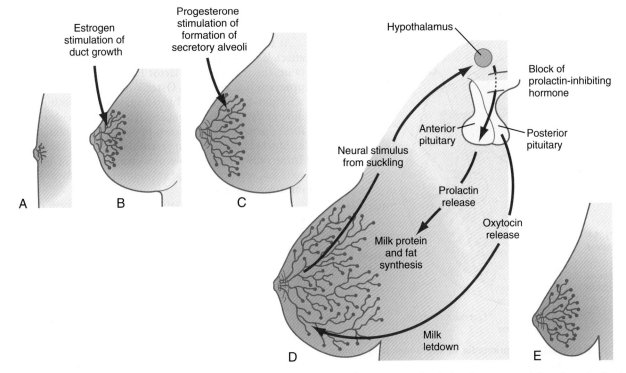

Fig. 9.14 **Development of the mammary ducts and hormonal control of mammary gland development and function. A,** Newborn. **B,** Young adult. **C,** Adult. **D,** Lactating adult. **E,** Adult after lactation.

CLINICAL CORRELATION 9.1
Abnormalities of Skin Development

Several types of anomalies affect the integumentary system. **Ectodermal dysplasia** is a germ layer defect that can affect many ectodermal derivatives depending on the type and severity of the condition. In addition to abnormalities of the epidermis itself, this syndrome can include the absence or abnormalities of hair and teeth (see Fig. 8.19) and short stature (caused by anterior pituitary gland maldevelopment).

Numerous relatively rare conditions are included among the genetically transmitted disorders of keratinization. **Ichthyosis** is characterized by scaling and cracking of a hyperkeratinized epidermis. Disorders of sweat glands are commonly associated with this condition. A more severe autosomal recessive disorder is a **harlequin fetus**, in which epidermal platelike structures form, with deep cracks between the structures, because the skin cannot expand to accommodate the increasing bulk of the fetus. Infants with this condition typically do not survive longer than a few weeks.

Many abnormalities of hair are known to have a genetic basis. These range from complete hair loss (**atrichia**), caused by mutations in the hairless gene *(HR)*, to **hypertrichosis**, a condition characterized by excessive of ectopic hair growth. Because of the complex structure and signal-calling within the hair follicle itself, mutations in a variety of genes underlie conditions involving fragility or abnormal structure of hairs themselves.

Several specific defects in the synthesis of types I and III collagen are lumped together as variants of the **Ehlers-Danlos syndrome**. Individuals with this condition typically have hypermobile joints. Affected skin can be characterized by sagging or hyperelasticity (**Fig. 9.15**).

Angiomas of the skin (birthmarks) are vascular malformations characterized by localized red or purplish spots ranging in size from tiny dots to formations many inches in diameter. Angiomas consist of abnormally prominent plexuses of blood vessels in the dermis, and they may be raised above the level of the skin or a mucous membrane (see Fig. 17-49).

Fig. 9.15 **Hyperelasticity of the skin in Ehlers-Danlos syndrome, characterized by defects in types I and III collagen.** *(From Turnpenny P, Ellard S: Emery's elements of medical genetics, ed 14, Philadelphia, 2012, Churchill Livingstone.)*

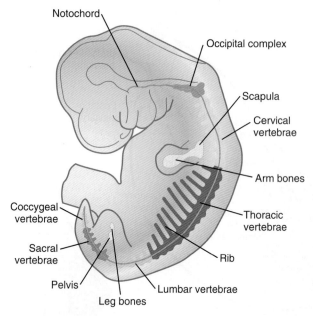

Fig. 9.16 **Precartilaginous primordia in the 9-mm long human embryo.**

A common element of many mesenchymal cell precursors of skeletal elements is their migration or relative displacement from their site of origin to the area where the bone ultimately forms. The displacement can be relatively minor, such as the aggregation of cells from the sclerotome of the receding somite around the notochord to form the body (**centrum**) of a vertebra, or it can involve extensive migrations of cranial neural crest cells to their final destinations as membrane bones of the face.

To differentiate into defined skeletal elements, the mesenchymal precursor cells must often interact with elements of their immediate environment—typically epithelia with associated basal laminae—or components of the neighboring extracellular matrix. Details of the interactions vary among regions of the body. In the limb, a continuous interaction between the **apical ectodermal ridge** (see Chapter 10) and the underlying limb bud mesoderm is involved in the specification of the limb skeleton. An inductive interaction between the sclerotome and notochord or neural tube initiates skeletogenesis of the vertebral column. In the head, preskeletal cells of the neural crest may receive information at levels ranging from the neural tube itself, to sites along their path of migration, to the region of their final destination. Inductive interactions between regions of the brain and the overlying mesenchyme stimulate formation of the membrane bones of the cranial vault.

Regardless of the nature of the initial induction, the formation of skeletal elements begins along a common path, which diverges into osteogenic or chondrogenic programs, depending on the nature of the immediate environment. Shortly after induction, the preskeletal mesenchymal cells produce the cellular adhesion molecule **N-cadherin**, which promotes their transformation from a mesenchymal to an epitheliumlike morphology and their forming cellular condensations (**Fig. 9.17**). The growth factor **transforming growth factor-β** stimulates the synthesis of **fibronectin** and finally **N-CAM**, which maintains the aggregated state of the cells in the preskeletal condensation.

At this point, specific differentiation programs come into effect. If the skeletal element is destined to form membranous bone, the transcription factor **Runx-2** sets off an **osteogenic program** (see Fig. 9.17). **Osterix (Osx)** is a downstream transcription factor from Runx-2 and is also required for the differentiation of osteoblasts. The protein encoded by the *Runx2* gene has been shown to control the differentiation of mesenchymal cells into **osteoblasts** (bone-forming cells). These cells produce molecules characteristic of bone (type I collagen, osteocalcin, and osteopontin) and form spicules of intramembranous bone.

If the cellular condensation is destined to form cartilage, it follows the **chondrogenic program**. Under the influence of **Sox-9**, the chondroblasts begin to form type II collagen and secrete a cartilaginous matrix (see Fig. 9.17). Some embryonic cartilage (e.g., in the nose, ear, and intervertebral surfaces) remains as permanent cartilage and continues to express Sox-9. The cartilage that forms the basis for endochondral bone formation undergoes specific changes that ultimately promote bone formation around it. A first step is hypertrophy, which occurs under the influence of Runx-2 and the signaling factors **Indian hedgehog** and **BMP-6**. The formation of type X collagen is characteristic of hypertrophying cartilage. Then the hypertrophic chondrocytes themselves begin to produce bone proteins, such as **osteocalcin, osteonectin,** and **osteopontin**. They also express **vascular endothelial growth factor,** which stimulates the ingrowth of blood vessels into the hypertrophic cartilage. This sets the stage for the replacement of the eroded hypertrophic cartilage by true bone as osteoblasts accompany the invading capillaries. **FGF-18,** produced by the perichondrium, inhibits the maturation of the chondrocytes around the periphery of the mass of cartilage as those at the center are undergoing hypertrophy.

Axial Skeleton
Vertebral Column and Ribs

The earliest stages in establishing the axial skeleton are introduced in Chapter 6. Formation of the axial skeleton is more complex, however, than the simple subdivision of the paraxial mesoderm into somites and the medial displacement of sclerotomal cells to form primordia of the vertebrae. Each vertebra has a complex and unique morphology specified by controls operating at several levels and during several developmental periods.

According to the traditional view of vertebral development (see Fig. 6.13), the sclerotomes split into cranial and caudal halves, and the densely packed caudal half of one sclerotome joins with the loosely packed cranial half of the next to form the centrum of a vertebra. More recent morphological research suggests that vertebral development is more complex than this model.

The vertebral column is divided into several general areas (see Fig. 9.16): (1) an **occipital region,** which is incorporated into the bony structure of the base of the skull; (2) a **cervical region,** which includes the highly specialized **atlas** and **axis** that link the vertebral column to the skull; (3) the **thoracic region,** from which the true ribs arise; (4) the **lumbar region;** (5) a **sacral region,** in which the vertebrae are fused into a single **sacrum;** and (6) a **caudal region,** which represents the tail in most mammals and the rudimentary **coccyx** in humans. A typical vertebra arises from the fusion of several

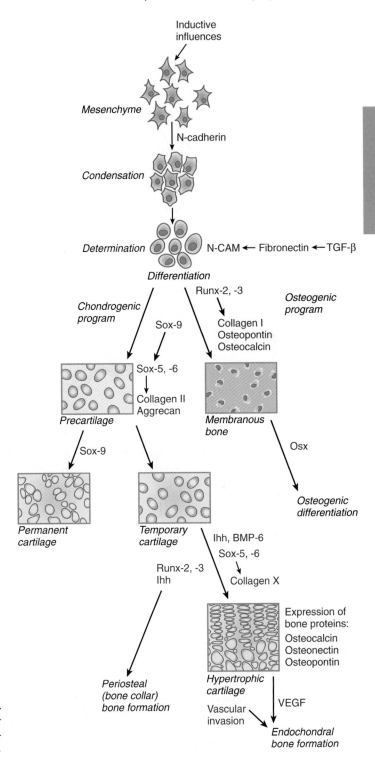

Fig. 9.17 Major steps in the differentiation of bone and cartilage. BMP-6, bone morphogenetic protein-6; CAM, cellular adhesion molecule; Ihh, Indian hedgehog; Osx, osterix; TGF-β, transforming growth factor-β; VEGF, vascular endothelial growth factor.

cartilaginous primordia. The centrum, which is derived from the ventromedial sclerotomal portions of the paired somites (see Fig. 6.10), surrounds the notochord and serves as a bony floor for the spinal cord (**Fig. 9.18**). The **neural arches**, arising from dorsal sclerotomal cells, fuse on either side with the centrum and, along with other neural arches, form a protective roof over the spinal cord. Incomplete closure of the bony roof results in a common anomaly called **spina bifida occulta** (see Fig. 11.42). The **costal process** forms the true ribs at the level of the thoracic vertebrae. At other levels along the

vertebral column, the costal processes become incorporated into the vertebrae proper.

Development of an individual vertebra begins with a sonic hedgehog–mediated induction by the notochord on the early somite to form the sclerotome. Under the continuing stimulus of sonic hedgehog, influencing the expression of Pax-1, the ventromedial portion of the somite ultimately forms the centrum of the vertebra. Formation of the dorsal part of the vertebra (the neural arch) is guided by a different set of developmental controls. An initial induction from the roof plate of

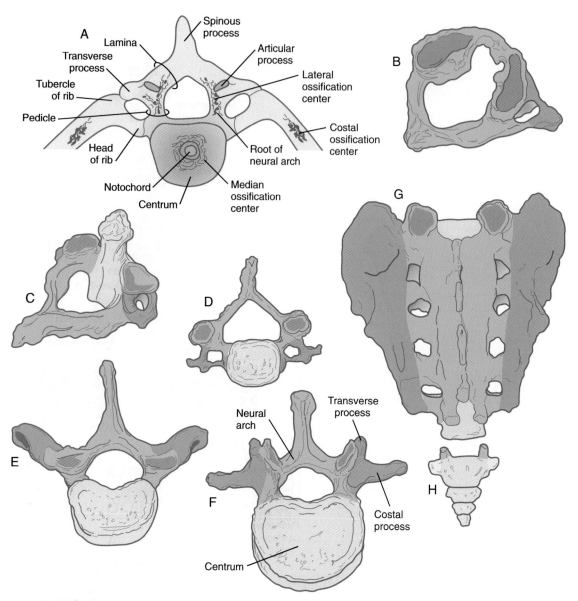

Fig. 9.18   A, Structure of a thoracic vertebra. **B** to **H,** Specific types of vertebrae, with homologous structures shown in the same color. **B,** Atlas with axis shown in its normal position beneath. **C,** Axis. **D,** Cervical vertebra. **E,** Thoracic vertebra. **F,** Lumbar vertebra. **G,** Sacrum. **H,** Coccyx.

the neural tube results in the expression of Pax-9 and the homeobox-containing genes *Msx-1* and *Msx-2*, which guide cells of the dorsal sclerotome to form the neural arch.

The fundamental regional characteristics of the vertebrae are specified by the actions of discrete combinations of homeobox-containing genes (**Fig. 9.19**). Expression of the *Hox* genes begins with the first appearance of the presomitic mesoderm and for most genes persists until chondrification begins in the primordia of the vertebrae. Formation of the normal segmental pattern along the craniocaudal axis of the vertebral column is ensured by the fact that most vertebrae are specified by a unique combination of *Hox* genes. For example, in the mouse, the atlas (C1) is characterized by the expression of *Hoxa1, Hoxa3, Hoxb1,* and *Hoxd4.* The axis (C2) is specified by these four genes plus *Hoxa4* and *Hoxb4.*

A clear association exists between major regional boundaries in the axial skeleton and the anterior expression boundaries of certain *Hox* paralogues (**Table 9.1**). **Retinoic acid**

Table 9.1 Relationship between Anterior Expression Boundaries of Key *Hox* Gene Paralogues and Major Regional Boundaries in the Axial Skeleton

Regional Boundary	*Hox* Paralogue
Occipital-cervical	*Hox3*
Cervical-thoracic	*Hox6*
Attached-floating ribs	*Hox9*
Thoracic-lumbar	*Hox10*
Lumbar-sacral	*Hox11*
Sacral-coccygeal	*Hox13*

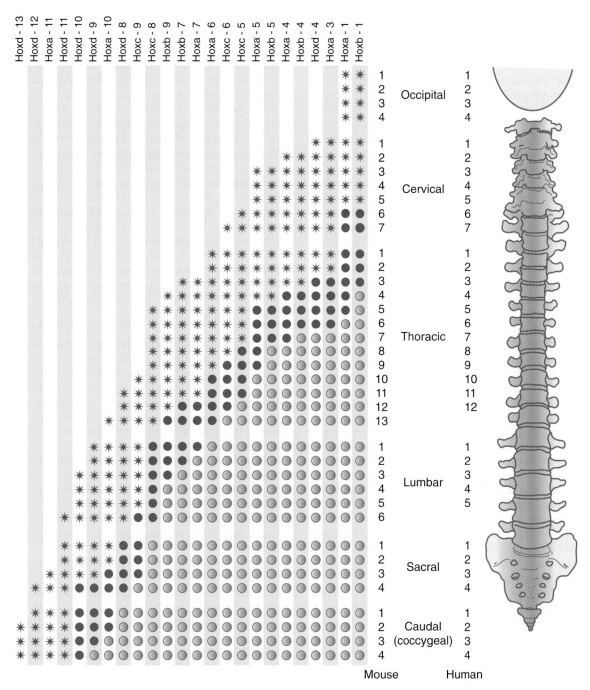

Fig. 9.19 *Hox gene expression in relation to the development of the vertebral column of the mouse.* The vertebral column of the mouse *(left)* has one more thoracic and one more lumbar vertebra than the vertebral column of the human. *Green asterisks* indicate levels at which there is definite expression of the *Hox* gene indicated at the top of the column. *Purple circles* represent the caudal border where expression fades out. *Tan circles* represent areas of no expression of the *Hox* gene. *(Based on studies by Kessel M, Balling R, Gruss P: Cell 61:301-308, 1990.)*

(vitamin A) can cause shifts in cranial or caudal levels in the overall segmental organization of the vertebrae if it is applied at specific developmental periods. For example, if administered early, retinoic acid results in a cranial shift (the last cervical vertebra is transformed into the first thoracic vertebra), and later administration causes a caudal shift (thoracic vertebrae extend into the levels of the first two lumbar vertebrae). Such shifts in level are called **homeotic transformations** and are representative of the broad family of homeotic mutants described in Chapter 4.

Both the degree of control of axial level and the degree of redundancy of this control by paralogues of the *Hox* genes are illustrated by experiments in which either some or all of the components of a specific *Hox* paralogue are knocked out.

When single *Hox* genes are knocked out, only minor morphological effects are noted. When all the members of a paralogous group are inactivated, however, profound effects appear. When all the *Hox10* paralogues are knocked out, ribs form on all the lumbar and sacral vertebrae. This finding suggests that *Hox10* represses the influence of the more anterior *Hox* genes. Similarly, *Hox11* suppresses the influence of *Hox10* and allows the sacrum to form. One of the striking features of axial development is the redundancy of the genes that pattern the vertebrae. A mutant of a specific *Hox* gene is likely to produce only a minor anatomical defect, whereas the nonfunction of an entire paralogous group produces major effects. Overall, a single paralogous group is involved in the patterning of 6 to 10 consecutive vertebrae, and the actions of at least 2 paralogous groups are involved in the formation of any individual vertebra. Control of posterior extension of the axial skeleton is balanced by the actions of the *Cdx* family of homeobox genes, which promote extension of the axial skeleton, and the *Hox13* paralogue, which exerts a braking effect on such extension. In other structures, (e.g., limbs and external genitalia) Hox-13 gene products are associated with terminal growth. Some common anomalies of the vertebrae are discussed in **Clinical Correlation 9.2**.

Among the vertebrae, the axis and atlas have an unusual morphology and a distinctive origin (**Fig. 9.22**). The centrum of the atlas is deficient, but the area of the centrum is penetrated by the protruding **odontoid process** of the axis. The odontoid process consists of three fused centra that are presumably equivalent: (1) a half-segment from the centrum of a transitional bone (the **proatlas**) not found in humans, (2) the centrum that should have belonged to the atlas, and (3) the normal centrum of the axis. This arrangement permits a greater rotation of the head about the cervical spine. When the ubiquitously expressed *Hoxa7* transgene was introduced into the germline of mice, the cranial part of the vertebral column was posteriorized. The base of the occipital bone was transformed into an occipital vertebra (the proatlas), and the atlas was combined with its centrum; the result was an axis that did not possess an odontoid process.

The ribs arise from zones of condensed mesenchymal cells lateral to the centrum. The proximal part of a rib (head, neck, and tubercle) arises from the central sclerotome (see Box 6.1). Because of the resegmentation of the somites as they form the vertebrae (see Fig. 6.13), the distal part (shaft) of the rib is derived from the lateral part of the adjacent cranial somite. By the time ossification in the vertebrae begins, the ribs separate from the vertebrae.

The formation of the proximal portions of the ribs depends on gene expression in the myotome. Products of the Hox-6 paralogous group promote the expression of two myogenic regulatory factors, **Myf-5** and **Myf-6** (see p. 184), in the myotomes of the thoracic level somites. Myf-5 and Myf-6 stimulate the formation of the growth factors, **platelet-derived growth factor** (**PDGF**) and **FGF**, which promote proximal rib growth in the sclerotome. Formation of the distal portion of the ribs requires **BMP** signals from the adjacent somatopleural mesoderm.

Accessory ribs, especially in the upper lumbar and lower cervical levels, are common, but estimates of the incidence vary widely from one series to the next. Less than 1% may be a realistic estimate. These and other common rib anomalies (**forked** or **fused ribs**) are typically asymptomatic and are usually detected on x-ray examination. They are mainly the result of misexpression of specific *Hox* genes.

The **sternum**, which along with the connective tissue surrounding the distal ribs is derived from lateral plate mesoderm, arises as a pair of cartilaginous bands that converge at the ventral midline as the ventral body wall consolidates (**Fig. 9.23**). After the primordial sternal bands come together, they reveal their true segmental nature by secondarily subdividing into craniocaudal elements. Such secondary segmentation follows an early morphological and molecular course that closely parallels the formation of synovial joints (see p. 206). These segments ultimately fuse as they ossify to form a common unpaired body of the sternum. Several common anomalies of the sternum (e.g., **split xiphoid process** or **split sternum** [see Fig. 15.39]) are readily understood from its embryological development. Anterior borders of the *Hox* code that guide the development of the sternum from lateral plate mesoderm are offset from those in the paraxial mesoderm. Malformations of the xiphoid process are seen in mice mutant for *Hoxc4* and *Hoxa5*, and mice mutant for *Hoxb2* and *Hoxb4* have split sternums.

The clavicle, which arises from neural crest and forms by an intramembranous mechanism, is one of the first bones in the body to become ossified, and ossification is well advanced by the eighth week. Studies of mice heterozygous for the *Runx2* gene have shed light not only on the nature of the clavicle, but also on a poorly understood human syndrome. Such heterozygotes exhibit hypoplasia of the clavicle, delayed ossification of membrane bones (e.g., of the skull), and open anterior and posterior fontanelles in the skull. **Cleidocranial dysplasia** in humans exhibits all these conditions, as well as supernumerary teeth. Without clavicles, affected individuals can approximate their shoulders in the anterior midline (**Fig. 9.24**). The findings in this mutant suggest that the clavicle is a purely membranous bone and may be a unique class of bone, being neither truly axial nor appendicular in the usual sense.

Another late development is the disappearance of the notochord from the bodies of the vertebrae. Between the vertebrae, the notochord expands into the condensed mesenchymal primordia of the intervertebral disks. In an adult, the notochord persists as the **nucleus pulposus**, which constitutes the soft core of the disk. The bulk of the **intervertebral disk** consists of layers of fibrocartilage that differentiate from the rostral half of the sclerotome in the somite. *Pax1* is expressed continuously during the development of intervertebral disks. In a mouse mutant, *undulated*, *Pax1* expression is deficient, and fusion of the vertebral bodies results. *Pax1* expression and the subsequent formation of the intervertebral disks seem to be important mechanisms in maintaining the individual segmental character of the vertebral column.

The caudal end of the axial skeleton is represented by a well-defined, tail-like appendage during much of the second month (**Fig. 9.25A**). During the third month, the tail normally regresses, largely through cell death and differential growth, to persist as the coccyx, but rarely a well-developed tail persists in newborns (Fig. 9.25B).

Skull

The skull is a composite structure consisting of two major subdivisions: the **neurocranium**, which surrounds the brain;

CLINICAL CORRELATION 9.2
Anomalies Involving Vertebral Segmentation

Certain conditions are characterized by abnormal segmentation of the vertebrae. A striking example is **spondylocostal dysostosis 2**, which is characterized by the presence of multiple ossified fragments (resulting from incomplete fusion of right and left sclerotome pairs) of vertebral centra in the thoracic region (**Fig. 9.20**). The genetic defect in this condition is a homozygous mutant in the *MESP2* gene. This is the gene that, early in development, marks the position of a future somite (see Fig. 6.9). Other segmentation anomalies take the form of isolated bony wedges (**hemivertebrae** [**Fig. 9.21A**]), sagittally cleft vertebrae (**butterfly vertebrae** [Fig. 9.21B]) or fused vertebrae (**block vertebrae** [Fig. 9.21C]). Hemivertebrae, a common cause of congenital **scoliosis** (lateral curvature of the vertebral column), are sometimes related to mutations of genes associated with the segmentation clock mechanism (e.g., *lunatic fringe*, *MESP2*, the Notch ligand *DLL3*) that produces the somites. Butterfly vertebrae are thought to result from a midline fusion defect that reduces the connection between the right and left sclerotomes. The **Klippel-Feil syndrome**, sometimes called **brevicollis**, is characterized by a short neck, a low hairline, and limited neck motion. The fundamental defect is fusion of one or more cervical vertebrae. Other anomalies of vertebral type, such as incorporation of the last lumbar vertebra into the sacrum, are likely consequences of mutants of individual *HOX* genes.

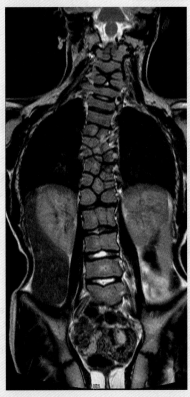

Fig. 9.20 Magnetic resonance imaging scan of the spine of an individual with spondylocostal dysostosis resulting from a homozygous mutation of *MESP-2.* In this individual, highly abnormal segmentation is seen in the thoracic vertebrae. *(From Turnpennny PD and others: Dev Dyn 236:1456-1474, 2007.)*

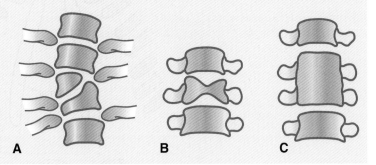

Fig. 9.21 Common segmental abnormalities of vertebrae (indicated in *purple*). A, Hemivertebra, causing lateral bending of the spine. **B,** Butterfly vertebra. **C,** Block vertebrae, caused by the fusion of two individual vertebrae.

and the **viscerocranium**, which surrounds the oral cavity, pharynx, and upper respiratory passages. Each of these subdivisions consists of two components, one in which the individual bones are first represented by cartilaginous models and are subsequently replaced by bone through endochondral ossification and another in which bone arises directly through the ossification of mesenchyme.

The phylogenetic and ontogenetic foundation of the skull is represented by the **chondrocranium**, which is the cartilaginous base of the neurocranium (**Fig. 9.26A**). The fundamental pattern of the chondrocranium has been remarkably preserved in the course of phylogeny. It is initially represented by several sets of paired cartilages. One group (parachordals, hypophyseal cartilages, and trabeculae cranii) is closely related to midline structures. Caudal to the parachordal cartilages are four **occipital sclerotomes**. Along with the parachordal cartilages, the occipital sclerotomes, which are homologous with precursors of the vertebrae, fuse to form the

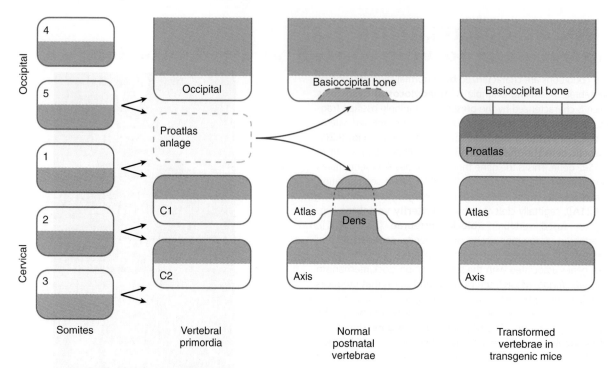

Fig. 9.22 **Formation of the atlas and axis in normal and transgenic mice.** In normal development, cells from a proatlas anlage contribute to the formation of the basioccipital bone and the dens of the axis. The normal atlas forms an anterior arch (only a transient structure in other vertebrae) instead of a centrum. The cells that would normally form the centrum at the level of the atlas instead fuse with the axis to form the dens of the axis. In mice containing the *Hoxa7* (A7) transgene, a proatlas forms, and the atlas and axis have the form of typical vertebrae *(right column). (Based on Kessel M, Balling R, Gruss P: Cell 61:301-308, 1990.)*

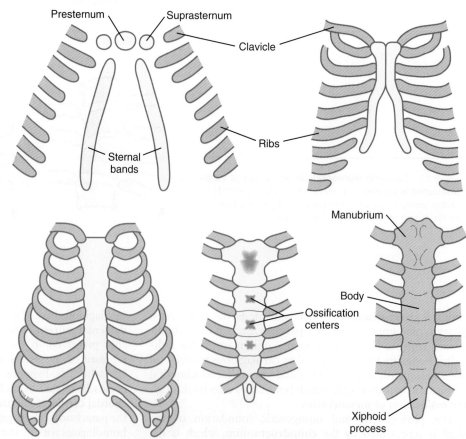

Fig. 9.23 **Successive stages in the development of the sternum and clavicle.**

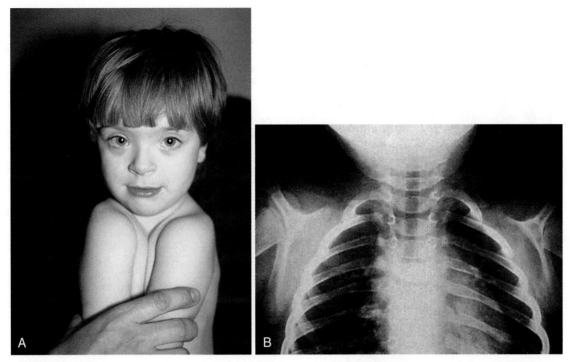

Fig. 9.24 Cleidocranial dysplasia. Absence of clavicles in individuals with mutants in Runx-2. **A,** In this boy, note the ability to approximate the shoulders without the clavicles present. **B,** In this radiograph, note the absence of clavicles. *(A, From Turnpenny P, Ellard S: Emery's elements of medical genetics, ed 14, Philadelphia, 2012, Churchill Livingstone; B, from the Robert J. Gorlin Collection, Division of Oral and Maxillofacial Pathology, University of Minnesota Dental School, courtesy of Dr. Ioannis Koutlas.)*

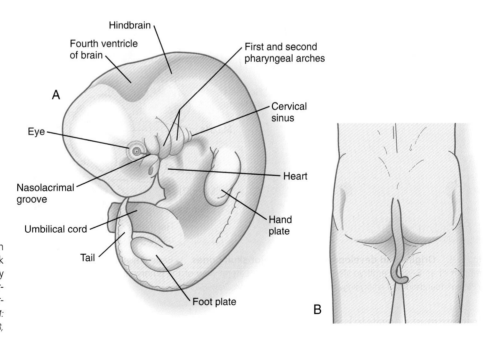

Fig. 9.25 A, Drawing of a human embryo at the end of the fifth week that shows a prominent tail. Normally the tail regresses. **B,** Drawing of a persisting tail (9 inches long) in a 12-year-old boy. *(B, Adapted from Patten BM: Human embryology, ed 3, New York, 1968, McGraw-Hill.)*

base of the occipital bone. More laterally, the chondrocranium is represented by pairs of cartilage that are associated with epithelial primordia of the sense organs (olfactory organ, eyes, and auditory organ). Molecular signals from the preoral gut endoderm are required for chondrification of the chondrocranium rostral to the tip of the notochord, which is of neural crest origin, whereas caudal to the tip of the notochord, notochordal signals promote chondrification of the mesodermally derived posterior chondrocranium.

The precursors of the chondrocranium all fuse to form a continuous cartilaginous structure that extends from the future foramen magnum to the interorbital area. This

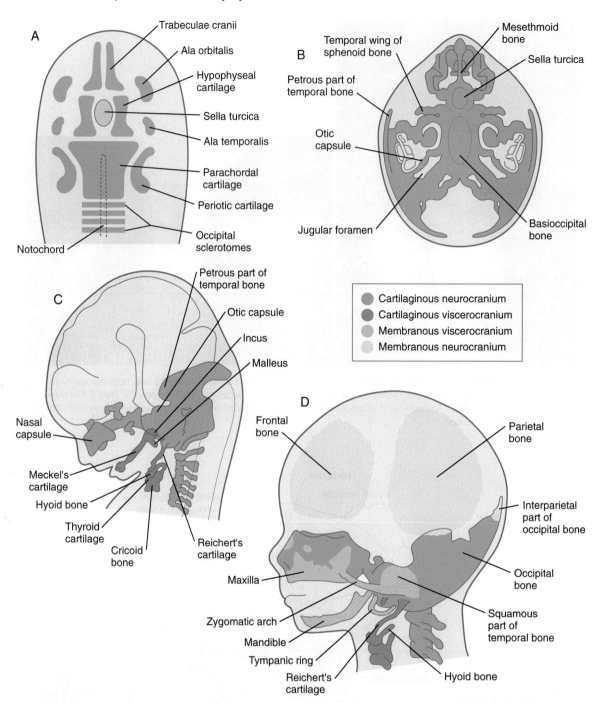

Fig. 9.26   Origins and development of the major skull bones. A, Basic skeletal elements of a 6-week embryo viewed from above. **B,** Chondrocranium of an 8-week embryo viewed from above. **C,** Lateral view of the embryo illustrated in **B. D,** Skull of a 3-month embryo. *(Adapted from Carlson B: Patten's foundations of embryology, ed 6, New York, 1996, McGraw-Hill.)*

structure must elongate to keep pace with the growing embryo. It does so by first forming numerous primary ossification centers along its length. Then the cartilage between ossification centers develops two mirror-image growth plates, similar to those at the ends of long bones, but with cartilage rather than a joint cavity between them (therefore, they are called **synchondroses**). Under the influence of Indian hedgehog, the growth plates cause the primary ossification centers to elongate, resulting in the overall elongation of the basicranium.

The individual primordial elements of the chondrocranium undergo several patterns of growth and fusion to form the structurally complex bones of the basicranium (the occipital, sphenoid, and temporal bones and much of the deep bony support of the nasal cavity) (Fig. 9.26B). In addition, some of these bones (e.g., the occipital and temporal bones) incorporate membranous components during their development, so in their final form, they are truly composite structures (see Fig. 9.26D). Other components of the neurocranium, such as the parietal and frontal bones, are purely membranous bones (**Box 9.1**).

Virtually all the bones of the neurocranium arise as the result of an inductive influence of an epithelial structure on

Box 9.1 Embryological Origins of Bones of the Cranium

Neurocranium

Chondrocranium

Occipital
Sphenoid
Ethmoid
Petrous and mastoid part of temporal

Membranous Neurocranium

Interparietal part of occipital
Parietal
Frontal
Squamous part of temporal

Viscerocranium

Cartilaginous Viscerocranium

Pharyngeal Arch I

Meckel's cartilage
    Malleus
    Incus

Pharyngeal Arch II

Reichert's cartilage
    Stapes
    Styloid process

Membranous Viscerocranium

Maxillary process (superficial)
    Squamous part of temporal
    Zygomatic
    Maxillary
    Premaxillary
    Nasal?
    Lacrimal?
Maxillary process (deep)
    Palatine
    Vomer
    Pterygoid laminae
Mandibular process
    Mandible
Tympanic ring

occipital bone) arise as flat, platelike aggregations of bony spicules (trabeculae) from mesenchyme that has been induced by specific parts of the developing brain. These bones remain separate structures during fetal development, and even at birth, they are separated by connective tissue sutures. Intersections between sutures where more than two bones meet are occupied by broader areas of connective tissue called **fontanelles**. The most prominent fontanelles are the **anterior fontanelle**, located at the intersection of the two frontal and two parietal bones, and the **posterior fontanelle**, located at the intersection of the parietal bones and the single occipital bone (**Fig. 9.27**).

During normal development, some sutures between cranial bones close; others remain open. Which of the sutures close and which remain open depend on an intricate interplay among several molecules. Ubiquitously expressed BMP in the embryonic cranium stimulates widespread bone formation, but the BMP antagonist, **noggin**, is expressed in all embryonic sutures. Under the influence of locally expressed FGF-2, noggin is downregulated in the sutures that fuse, thus allowing BMP-mediated bone formation to cement the two adjacent skull bones. Conversely, the absence of local FGF-2 allows noggin to repress BMP in the sutures that are destined not to fuse.

Similar to the neurocranium, the viscerocranium consists of two divisions: a **cartilaginous viscerocranium** and a **membranous viscerocranium**. In contrast to much of the neurocranium, the bones of the viscerocranium originate largely from neural crest–derived mesenchyme. Phylogenetically, the viscerocranium is related to the skeleton of the **branchial arches** (gill arches). Each branchial arch (more commonly called a pharyngeal arch in humans) is supported by a cartilaginous rod, which gives rise to numerous definitive skeletal elements characteristic of that arch (see Box 9.1). (Details of the organization and derivatives of the noncranial pharyngeal arch cartilages are discussed in Chapter 14 [see Fig. 14.36].)

The membranous viscerocranium consists of a series of bones associated with the upper and lower jaws and the region of the ear (see Fig. 9.26D). These arise in association with the first arch cartilage (**Meckel's cartilage**) and take over some of the functions originally subserved by Meckel's cartilage and many new functions, such as sound transmission in the middle ear. **Clinical Correlation 9.3** discusses conditions resulting from skull deformities.

Appendicular Skeleton

The **appendicular skeleton** consists of the bones of the limbs and limb girdles. There are fundamental differences in organization and developmental control between the axial and the appendicular skeleton. The axial skeleton forms a protective casing around soft internal tissues (e.g., brain, spinal cord, pharynx), and the mesenchyme forming the bones is induced by the organs that the bones surround. In contrast, the bones of the appendicular skeleton form a central supporting core of the limbs. Although interaction with an epithelium (the apical ectodermal ridge of the limb bud [see Chapter 10]) is required for the formation of skeletal elements in the limb, morphogenetic control of the limb is inherent in the mesoderm, with the epithelium playing only a stimulatory role. All components of the appendicular skeleton begin as cartilaginous models, which convert to true bone by endochondral

the neighboring mesenchyme. These interactions are typically mediated by growth factors and the extracellular matrix. Immunocytochemical studies have shown the transient appearance of **type II collagen** (the principal collagenous component of cartilage) at the sites and times during which the interactions leading to the formation of the chondrocranium occur. In addition to type II collagen, a **cartilage-specific proteoglycan** accumulates in areas of induction of chondrocranial elements. There is increasing evidence that epithelial elements in the head not only induce the skeleton, but also control its morphogenesis. This situation contrasts with morphogenetic control of the appendicular skeleton, which is determined by the mesoderm rather than the ectoderm of the limb bud.

Elements of the membranous neurocranium (the paired parietal and frontal bones and the interparietal part of the

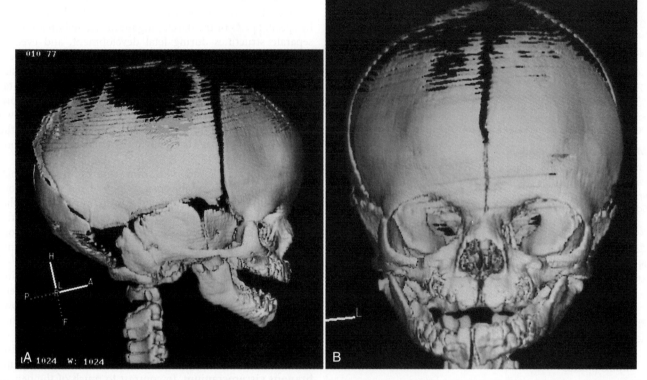

Fig. 9.27   **High-resolution computed tomography scans of the skull of a 34½ -week fetus. A,** Lateral view. **B,** Frontal view. The midline fissure in the forehead area is the metopic suture, which normally becomes obliterated after birth. The irregular black area above that is the anterior fontanelle, one of the "soft spots" in a newborn's head. *(Courtesy of R.A. Levy, H. Maher, and A.R. Burdi, Ann Arbor, Mich.)*

ossification later during embryogenesis. (Details of the formation of the appendicular skeleton are given in Chapter 10.)

Several defined genetic mutations result in prominent disturbances in the development of the appendicular skeleton. The most common form of dwarfism, **achondroplasia**, results from mutations of the **FGF receptor 3** gene (*FGFR3*). This condition is characterized by short stature secondary to limb shortening, midface hypoplasia, a disproportionately large head and pronounced lumbar lordosis (**Fig. 9.28**). A more severe consequence of the same mutation is **thanatophoric dysplasia**, in which the shortening of the extremities is even more severe. The thorax is very narrow, and death from respiratory insufficiency usually occurs in infancy. A mutation of *SOX-9* (see Fig. 9.17) causes **campomelic dysplasia**, characterized by pronounced bowing of the limbs, a variety of other skeletal anomalies, and sex reversal in XY males, resulting from a disruption of SOX-9 in sexual differentiation.

Muscular System

Three types of musculature—skeletal, cardiac, and smooth—are formed during embryonic development. Virtually all skeletal musculature is derived from the paraxial mesoderm, specifically the somites or somitomeres. Splanchnic mesoderm gives rise to the musculature of the heart (cardiac muscle) and the smooth musculature of the gut and respiratory tracts (**Table 9.2**). Other smooth muscle, such as that of the blood vessels and the arrector pili muscles, is derived from local mesenchyme.

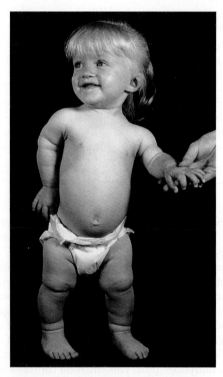

Fig. 9.28   **A young girl with achondroplasia.** *(From Turnpenny P, Ellard S: Emery's elements of medical genetics, ed 14, Philadelphia, 2012, Churchill Livingstone.)*

CLINICAL CORRELATION 9.3
Conditions Resulting from Skull Deformities

Many conditions are recognizable by gross deformities of the skull. Although many of these are true congenital malformations, others fall into the class of deformities that can be attributed to mechanical stress during intrauterine life or childbirth. Some malformations of the skull are secondary to disturbances in development of the brain. In this category are conditions such as the following: **acrania** and **anencephaly** (see Fig. 8.4), which are associated with severe malformations of the brain; **microcephaly** (see Fig. 9.9), in which the size of the cranial vault accommodates to a very small brain; and **hydrocephaly** (see Fig. 11.38), in which an extremely enlarged cranial vault represents the response of the skeleton of the head to an excessive buildup of cerebrospinal fluid and a great expansion of the brain.

One family of cranial malformations called **craniosynostosis** results from premature closure of certain sutures between major membrane bones of the neurocranium. Craniosynostosis is a feature of more than 100 human genetic syndromes and is seen in 1 in 3000 live births. Many of these syndromes are the result of gain-of-function mutants of fibroblast growth factor receptors. One type, the Boston variant, is a dominant gain-of-function mutant of the homeobox-containing gene *Msx2* expressed in the mesenchymal tissue of the early sutures and the underlying neural tissue. Exactly how this mutation is translated into premature suture closure is unknown. Premature closure of the **sagittal suture** between the two parietal bones produces a long, keel-shaped skull referred to as **scaphocephaly** (**Fig. 9.29**).

Oxycephaly, or turret skull, is the result of premature fusion of the **coronal suture**, which is located between the frontal and parietal bones. A dominant genetic condition, **Crouzon's syndrome**, has a gross appearance similar to that of oxycephaly, but the malformation of the cranial vault is typically accompanied by malformations of the face, teeth, and ears and occasional malformations in other parts of the body (**Fig. 9.30**).

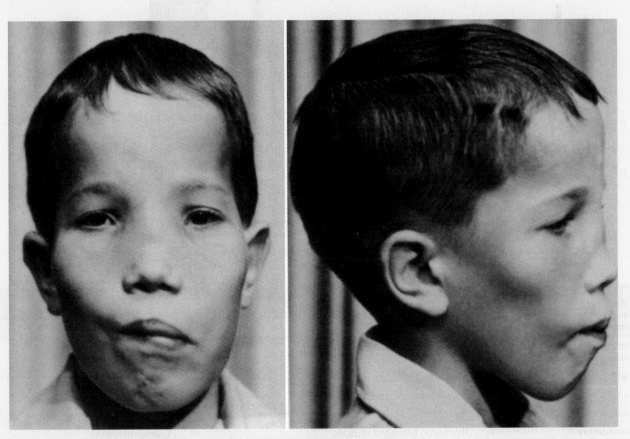

Fig. 9.29 **Frontal and lateral views of a boy with a narrow and elongated scaphocephalic skull.** Note the high forehead and flat bridge of the nose. This patient had an associated facial palsy and mixed deafness. *(From Goodman R, Gorlin R:* Atlas of the face in genetic disorders, *St. Louis, 1977, Mosby.)*

CLINICAL CORRELATION 9.3
Conditions Resulting from Skull Deformities—cont'd

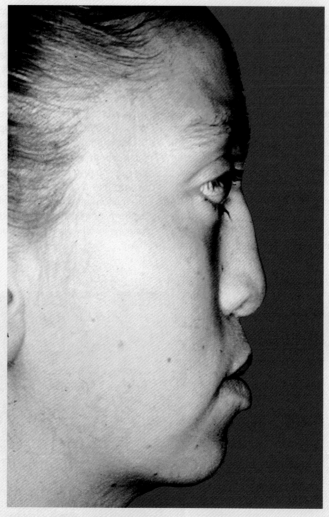

Fig. 9.30   **Lateral view of the flattened face of an individual with Crouzon's syndrome.** *(Courtesy of A.R. Burdi, Ann Arbor, Mich.)*

Table 9.2 Embryological Origins of the Major Classes of Muscle

Embryological Origin	Derived Muscle	Innervation
Somitomeres 1 through 3 and prechordal plate	Most extrinsic eye muscles	Cranial nerves III and IV
Somitomere 4	Jaw-closing muscles	Cranial nerve V (mandibular branch)
Somitomere 5	Lateral rectus muscle of eye	Cranial nerve VI
Somitomere 6	Jaw-opening and other second-arch muscles	Cranial nerve VII
Somitomere 7	Third-arch branchial muscles	Cranial nerve IX
Somites 1 and 2	Intrinsic laryngeal muscles and pharyngeal muscles	Cranial nerve X
Occipital somites (1 through 7)	Muscles of tongue, larynx, and neck	Cranial nerves XI and XII, cranial cervical nerves
Trunk somites	Trunk muscles, diaphragm, and limb muscles	Spinal nerves
Splanchnic mesoderm	Cardiac muscle	Autonomic
Splanchnic mesoderm	Smooth muscles of gut and respiratory tract	Autonomic
Local mesenchyme	Other smooth muscle: vascular, arrector pili muscles	Autonomic

From Carlson BM: Patten's foundations of embryology, ed 6, New York, 1996, McGraw-Hill.

The development of muscle can be studied at several different levels, ranging from the determination and differentiation of individual muscle cells, to the histogenesis of muscle tissue, and finally to the formation (morphogenesis) of entire muscles. Skeletal muscle is used as an example to illustrate how development occurs and is controlled at these different levels of organization.

Skeletal Muscle

There is increasing evidence that certain cells of the epiblast are determined to become myogenic cells even before the somites are completely formed, but it is convenient to begin with the emergence of muscle precursor cells in the somites.

For many decades, the origin of the skeletal musculature was in question, with the somites and lateral plate mesoderm being likely candidates. This issue was finally resolved by tracing studies involving cellular markers (**Box 9.2**), and it is now known that virtually all skeletal muscle originates in somites or somitomeres. Early steps in the determination of myogenic cells in somites are summarized in Figure 6.11.

Determination and Differentiation of Skeletal Muscle

The mature skeletal muscle fiber is a complex multinucleated cell that is specialized for contraction. Precursors of most

---

Box 9.2 Tracing Studies Involving Cellular Markers

The origins of many tissues in embryos have been identified by grafting tissues from quail embryos into homologous sites in chick embryos. The nuclei of quail cells contain a distinctive mass of dense chromatin and react with a species-specific monoclonal antibody, thus enabling researchers to distinguish quail from chick cells with great reliability (**Fig. 9.31**). If a putative precursor tissue is grafted from a quail into a chick embryo, the grafted quail tissue becomes well integrated into the chick host, and if cells migrate out of the graft, their pathway of migration into the host embryo can be clearly traced. Experiments involving this approach have been particularly useful in studies of muscle and the neural crest.

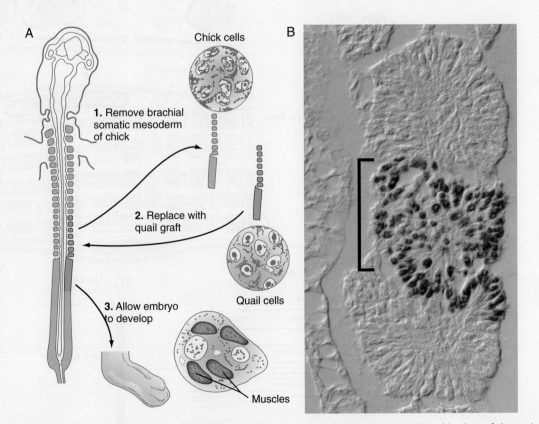

Fig. 9.31   **Principle of quail/chick grafting for the tracing of cells.** *1* and *2,* Quail tissues are transplanted in place of the equivalent tissues removed from the chick embryo. The prominent nuclear chromatin mass in quail cells provides a permanent marker that can be used to trace the fate of the cells of the graft of quail tissue. *3,* When quail somites are grafted into a chick embryo at the appropriate level, the muscles of the limb are derived from quail and not chick cells. *Inset,* Cross section of a chick/quail chimeric embryo. Transplanted quail cells *(brown stain),* which have migrated into the cardiac outflow tract, are stained with the QCPN antibody (developed by B & J Carlson) that selectively binds to quail cells. Chick cells are lightly stained. *(Inset from Sato Y, Takahashi Y: Dev Biol 282:183-191, 2005.)*

muscle lineages (**myogenic cells**) have been traced to the myotome of the somite (see Fig. 6.10). Although these cells look like the mesenchymal cells that can give rise to many other cell types in the embryo, they have undergone a restriction event committing them to the muscle-forming line. Committed myogenic cells pass through several additional mitotic divisions before completing a terminal mitotic division and becoming **postmitotic myoblasts**.

Proliferating myogenic cells are kept in the cell cycle through the action of growth factors, such as **FGF** and **transforming growth factor-β**. With the accumulation of myogenic regulatory factors (see next section), myogenic cells upregulate the synthesis of the cell cycle protein **p21**, which irreversibly removes them from the cell cycle. Under the influence of other growth factors, such as **insulinlike growth factor**, the postmitotic myoblasts begin to transcribe the mRNAs for the major contractile proteins **actin** and **myosin**, but the major event in the life cycle of a postmitotic myoblast is its fusion with other similar cells into a multinucleated **myotube** (**Fig. 9.32**). The fusion of myoblasts is a precise process involving their lining up and adhering by calcium (Ca^{++})–mediated recognition mechanisms, involving

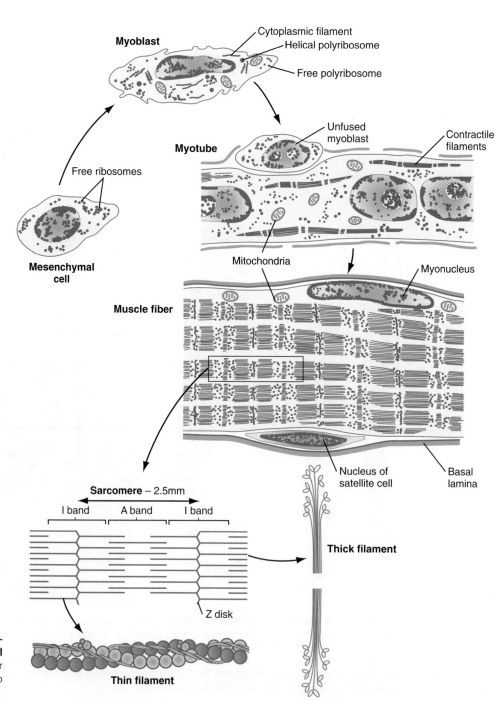

Fig. 9.32 **Stages in the morphological differentiation of a skeletal muscle fiber.** Important subcellular elements in a muscle fiber are also shown.

molecules such as M-cadherin, and the ultimate union of their plasma membranes.

Myotubes are intensively involved in mRNA and protein synthesis. In addition to forming actin and myosin, myotubes synthesize a wide variety of other proteins, including the regulatory proteins of muscle contraction—**troponin** and **tropomyosin.** These proteins assemble into myofibrils, which are precisely arranged aggregates of functional contractile units called **sarcomeres.** As the myotubes fill with myofibrils, their nuclei, which had been arranged in regular central chains, migrate to the periphery of the myotube. At this stage, the myotube is considered to have differentiated into a **muscle fiber,** the final stage in the differentiation of the skeletal muscle cell.

The development of a muscle fiber is not complete, however, with the peripheral migration of the nuclei of the myotube. The nuclei (**myonuclei**) of a multinucleated muscle fiber are no longer able to proliferate, but the muscle fiber must continue to grow in proportion to the rapid growth of the fetus and then the infant. Muscle fiber growth is accomplished by means of a population of myogenic cells, called **satellite cells,** which take up positions between the muscle fiber and the basal lamina in which each muscle fiber encases itself (see Fig. 9.32). Operating under a poorly understood control mechanism, possibly involving the **Delta/Notch** signaling system, satellite cells divide slowly during the growth of an individual. Some of the daughter cells fuse with the muscle fiber so that it contains an adequate number of nuclei to direct the continuing synthesis of contractile proteins required by the muscle fiber. After muscle fiber damage, satellite cells proliferate and fuse to form regenerating muscle fibers.

A typical muscle is not composed of homogeneous muscle fibers. Instead, usually several types of muscle fibers are distinguished by their contractile properties and morphology and by their possession of different isoforms of the contractile proteins. For the purposes of this text, muscle fibers are considered to be either fast or slow.

Muscle Transcription Factors

Myogenesis begins with a restriction event that channels a population of mesenchymal cells into a lineage of committed myogenic cells. The molecular basis for this commitment is the action of members of families of **myogenic regulatory factors,** which, acting as master genetic regulators, turn on muscle-specific genes in the premuscle mesenchymal cells.

The first-discovered family of myogenic regulatory factors is a group of four basic helix-loop-helix transcription factors, sometimes called the **MyoD family** (**Fig. 9.33**). Another regulatory factor, called **muscle enhancer factor-2** (MEF-2), works in concert with the MyoD family, but all these myogenic regulatory factors are capable of converting nonmuscle cells (e.g., fibroblasts, adipocytes, chondrocytes, retinal pigment cells) to cells expressing the full range of muscle proteins.

As with many helix-loop-helix proteins, myogenic regulatory proteins of the MyoD family form dimers and bind to a specific DNA sequence (CANNTG), called the **E box,** in the enhancer region of muscle-specific genes. The myogenic specificity of these proteins is encoded in the basic region (see Fig. 9.33).

The regulatory activities of MyoD and other members of that family are themselves regulated by other regulatory proteins, which can modify their activities (**Fig. 9.34**). Many cells contain a **transcriptional activator** designated **E12.** When a molecule of E12 forms a **heterodimer** with a molecule of MyoD, the complex binds more tightly to the muscle-enhancer region of DNA than does a pure MyoD dimer. This increases

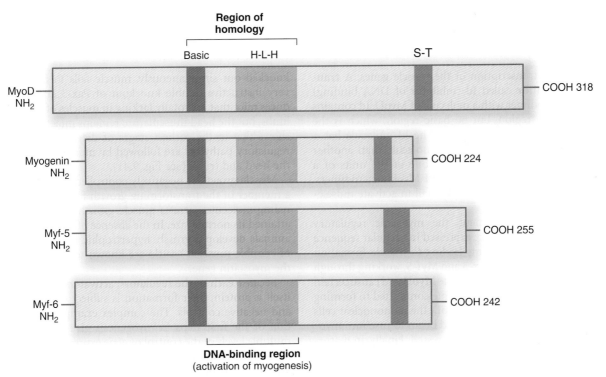

Fig. 9.33 **Structural comparison of several myogenic regulatory factors.** H-L-H, homologous helix-loop-helix regions; S-T, homologous serine/threonine-rich region.

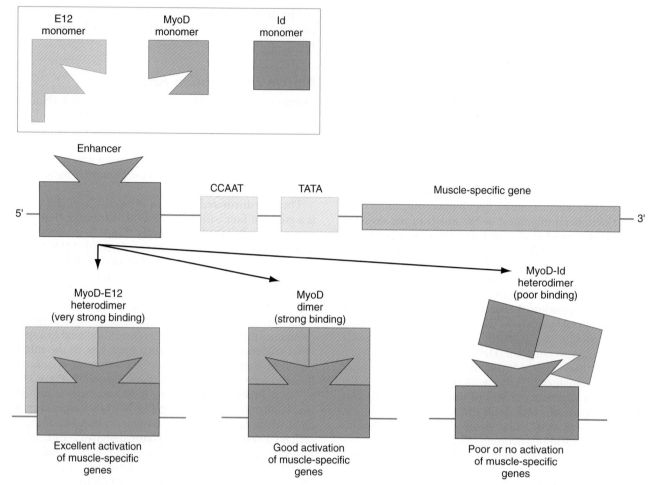

Fig. 9.34   **MyoD regulation of early myogenesis, showing interactions between MyoD and a transcriptional activator (E12) and a transcriptional inhibitor (Id).**

the efficiency of transcription of the muscle genes. A **transcriptional inhibitor** called **Id** (inhibitor of DNA binding) can form a heterodimer with a molecule of MyoD. Id contains a loop-helix-loop region, but no basic region, which is the DNA-binding part of the molecule. The Id molecule has a greater binding affinity for a MyoD molecule than another molecule of MyoD and can displace one of the units of a MyoD dimer, thus resulting in more Id-MyoD heterodimers. These bind poorly to DNA and often fail to activate muscle-specific genes.

During muscle development, the myogenic regulatory factors of the MyoD family are expressed in a regular sequence (**Fig. 9.35**). In mice, the events leading to muscle formation begin in the somite, where **Pax-3** and **Myf-5**, working through apparently separate pathways, activate **MyoD** and cause certain cells of the dermomyotome to become committed to forming muscle. With increased levels of MyoD, the mononuclear cells withdraw from the mitotic cycle and begin to fuse into myotubes. At this stage, **myogenin** is expressed. Finally, in maturing myotubes, **Myf-6** (formerly called MRF-4) is expressed.

In knockout mice, the absence of a single myogenic regulatory factor (e.g., myf-5, MyoD) alone does not prevent the formation of skeletal muscle (although there may be other minor observable defects), but when myf-5 and MyoD are knocked out simultaneously, muscle fails to form. Another very instructive double knockout of Pax-3 and myf-5 produces mice that are totally lacking in muscles of the trunk and limbs, but the head musculature remains intact. This research shows that in the earliest stages of determination, different regulatory pathways are followed by muscle-forming cells of the head and trunk (see Fig. 9.41).

Muscle growth is negatively controlled as well. **Myostatin**, a member of the transforming growth factor-β family of signaling molecules, arrests muscle growth when a muscle has attained its normal size. In the absence of myostatin function, animals develop a grossly hypertrophic musculature. Breeds of "double-muscled" cattle are known to have mutations of the myostatin gene.

Because each of the regulators, activators, and inhibitors is itself a protein, their formation is subject to similar positive and negative controls. The complex examples of the regulation of the first steps in myogenesis give some idea of the multiple levels of the control of gene expression and the stages of cytodifferentiation in mammals. Although the molecular aspects of myogenesis are better understood than are the stages underlying the differentiation of most cell types, it is likely that similar sets of interlinked regulatory mechanisms operate in their differentiation of other cells.

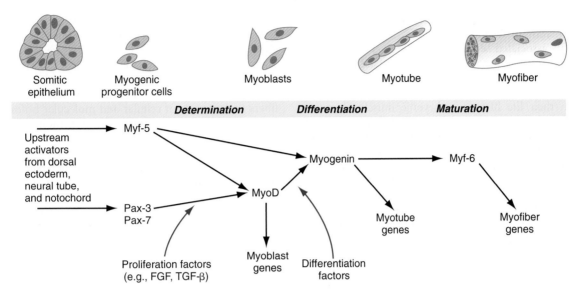

Fig. 9.35 Schematic representation of early myogenesis shows the sequence of expression of myogenic regulatory factors and other influences on the myogenic process. FGF, fibroblast growth factor; TGF-β, transforming growth factor-β.

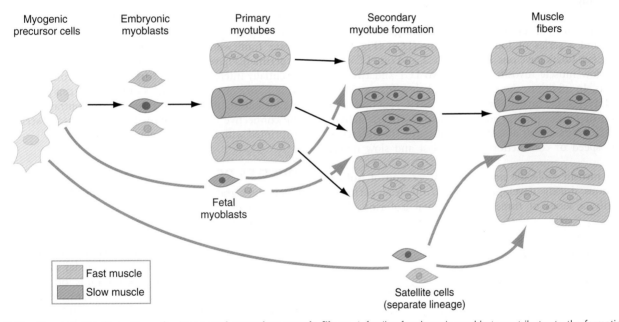

Fig. 9.36 **Stages in the formation of primary and secondary muscle fibers.** A family of embryonic myoblasts contributes to the formation of the primary myotubes, and fetal myoblasts contribute to secondary myotubes.

Histogenesis of Muscle

Muscle as a tissue consists not only of muscle fibers, but also of connective tissue, blood vessels, and nerves. Even the muscle fibers themselves are not homogeneous, but can be separated into functionally and biochemically different types.

As muscles first form, the myoblasts are intermingled with future connective tissue mesenchyme. The role of the connective tissue in the morphogenesis of a muscle is discussed in the next section. Capillary sprouts grow into the forming muscle for nourishment, and motor nerve fibers enter shortly after the first myoblasts begin forming myotubes.

At one time, it was thought that all myoblasts were essentially identical, and that their different characteristics (e.g., fast or slow) were imposed on them by their motor innervation.

Research has shown, however, that in birds and several species of mammals, distinct populations of fast and slow muscle cells exist as early as the myoblast stage, well before nerve fibers reach the developing muscles.

Not only are there fast and slow myoblasts, but also there are early and late cellular isoforms of myoblasts, which have different requirements for serum factors and nerve interactions in their differentiation. When the earliest myoblasts fuse into myotubes, they give rise to **primary myotubes**, which form the initial basis for an embryonic muscle. The differentiation of primary myotubes occurs before motor nerve axons have entered the newly forming muscle. Subsequently, smaller **secondary myotubes** arising from late myoblasts form alongside the primary myotubes (**Fig. 9.36**). By the time secondary myotubes form, early motor axons are present in the muscles, and there is evidence that the presence of nerves is required for the formation of secondary myotubes. A primary muscle fiber and its associated secondary muscle fibers are initially contained within a common basal lamina and are electrically coupled. These muscle fibers actively synthesize a wide variety of contractile proteins.

Early in their life history, embryonic muscle fibers are innervated by motoneurons. Although it has long been assumed that fast and slow motoneurons impose their own functional characteristics on the developing muscle fibers, it now seems that they may select muscle fibers of a compatible type through information contained on their cell surfaces. Initially, a motor nerve may terminate on both fast and slow muscle fibers, but ultimately, inappropriate connections are broken, so fast nerve fibers innervate only fast muscle fibers, and slow nerves innervate only slow muscle fibers.

The phenotypes of muscle fibers depend on the nature of the specific proteins that compose their contractile apparatus. There are qualitative differences in many of the contractile proteins between fast and slow muscle fibers, and within each type of muscle fiber is a succession of isoforms of major proteins during embryonic development. (The isoform transitions of **myosin** in a developing muscle fiber are used as an example.)

The myosin molecule is complex, consisting of two heavy chains and a series of four light chains (LCs) (**Fig. 9.37**). Mature fast fibers have one LC1, two LC2, and one LC3 subunits; slow muscle myosin contains two LC1 and two LC2 subunits. In addition, there are fast and slow forms (MHC_f and MHC_s) of the **myosin heavy chain** (MHC) subunits. The myosin molecules possess adenosine triphosphatase activity, and differences in this activity account partly for differences in the speed of contraction between fast and slow muscle fibers.

The myosin molecule undergoes a succession of isoform transitions during development. From the fetal period to maturity, a series of three developmental isoforms of the

MHC (embryonic [MHC_{emb}], neonatal [MHC_{neo}], and adult fast [MHC_f]) pass through a fast muscle fiber. (Developmental changes in the LC and MLC subunits are summarized in Figure 9.37.) Other contractile proteins of muscle fibers (e.g., actin, troponin) pass through similar isoform transitions. After injury to muscle in an adult, the regenerating muscle fibers undergo sets of cellular and molecular isoform transitions that closely recapitulate the transitions occurring in normal ontogenesis.

The phenotype of muscle fibers is not irreversibly fixed. Even postnatal muscle fibers possess a remarkable degree of plasticity. These fibers respond to exercise by undergoing hypertrophy or becoming more resistant to fatigue. They also adapt to inactivity or denervation by becoming atrophic. All these changes are accompanied by various changes in gene expression. Many other types of cells can also modify their phenotypes in response to changes in the environment, but the molecular changes are not always as striking as those seen in muscle fibers.

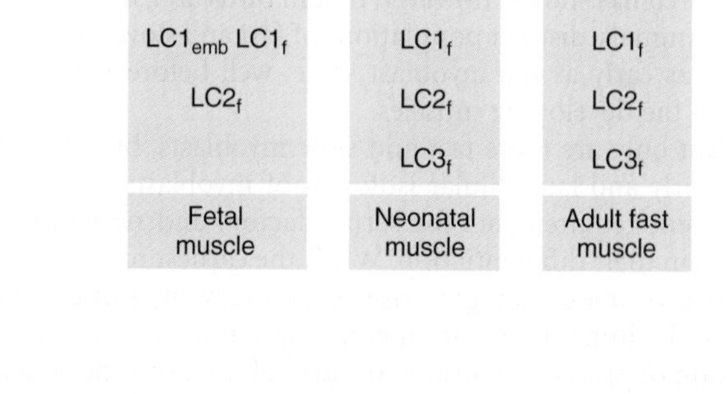

Morphogenesis of Muscle

At a higher level of organization, muscle development involves the formation of anatomically identifiable muscles. The overall form of a muscle is determined principally by its connective tissue framework rather than the myoblasts themselves. Experiments have shown that myogenic cells from somites are essentially interchangeable. Myogenic cells from somites that would normally form muscles of the trunk can participate in the formation of normal leg muscles. In contrast, the cells of the connective tissue component of the muscles seem to be imprinted with the morphogenetic blueprint.

MUSCLES OF THE TRUNK AND LIMBS

Quail/chick grafting experiments have clearly shown that the major groups of skeletal muscles in the trunk and limbs arise from myogenic precursors located in the somites. In the thorax and abdomen, the intrinsic muscles of the back (the **epaxial muscles**) are derived from cells arising in the dorsal myotomal lip, whereas ventrolateral muscles (**hypaxial muscles**) arise from epithelially organized ventral buds of the somites. Tendons of the epaxial muscles arise from the syndetome layer within the somites (see Box 6.1), whereas tendons of the limb and hypaxial musculature arise from lateral plate mesoderm. In the limb regions, myogenic cells migrate from the epithelium of the ventrolateral dermomyotome early during development. More cranial myogenic cells originating in similar regions of the occipital somites migrate into the developing tongue and diaphragm. At the lumbar levels, precursors of the abdominal muscles also move out of the epithelium of ventrolateral somitic buds.

Early specification of the future hypaxial musculature within the epithelial somite is initially regulated by dorsalizing (possibly a member of the Wnt family) and lateralizing (BMP-4) signals from the ectoderm and lateral plate mesoderm. This process activates two early transcription factors (Six [*sine oculis*] and Eya [*eyes absent*]), which leads to a more intense expression of *Pax3* and the expression of *Lbx1*, a homeobox gene that is exclusively expressed in the lateral dermomyotomal lips. *Lbx1* may prevent the premature differentiation of the hypaxial musculature. It is highly likely that

Myosin molecule

Fig. 9.37 Changes in myosin subunits during the development of a fast muscle fiber. A schematic representation of the myosin molecule is also shown.

MHC_emb	MHC_neo	MHC_f
$LC1_{emb}$ $LC1_f$	$LC1_f$	$LC1_f$
$LC2_f$	$LC2_f$	$LC2_f$
	$LC3_f$	$LC3_f$
Fetal muscle	Neonatal muscle	Adult fast muscle

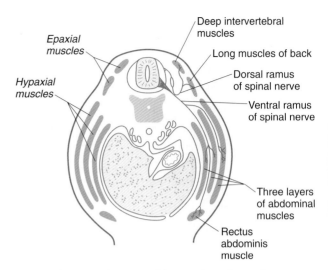

Fig. 9.39 **Groups and layers of trunk muscles.**

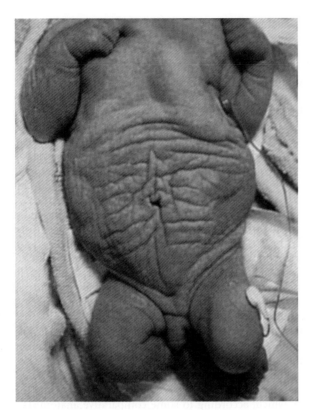

Fig. 9.38 **Infant with prune belly syndrome.** Note the wrinkled abdomen in the absence of the abdominal musculature. *(From the Robert J. Gorlin Collection, Division of Oral and Maxillofacial Pathology, University of Minnesota Dental School, courtesy of Dr. Ioannis Koutlas.)*

the **prune belly syndrome** (**Fig. 9.38**), which is characterized by the absence of the abdominal musculature, will be found to be caused by a molecular deficiency in this population of myogenic cells.

More recent experiments have shown different cellular behavior in areas of the myotomes adjacent to limb and nonlimb regions. In thoracic segments, cells of the dermatome surround the lateral edges of the myotome; this is followed by an increase in the number of myotubes formed in the myotome and the penetration of the muscle primordia into the body wall. In contrast, at the levels of the limb buds, dermatome cells die before surrounding the early myotubes that form in the myotome. These myotubes neither increase significantly in number nor move out from the myotomes to form separate muscle primordia.

After their origin from the somites, the muscle primordia of the trunk and abdomen become organized into well-defined groups and layers (**Fig. 9.39**). (Morphogenesis of the limb muscles is discussed in Chapter 10.) The results of numerous experiments have shown fundamental differences in cellular properties between the cellular precursors of limb muscles and axial muscles. These differences are summarized in **Table 9.3**.

MUSCLES OF THE HEAD AND CERVICAL REGION

The skeletal muscle of the head and neck is mesodermal in origin. Quail/chick grafting experiments have shown that the

Table 9.3 Differences between Cellular Precursors of Axial and Limb Muscles

Axial Muscles	Limb Muscles
Usual location in medial half of somite	Location in lateral half of somite
Differentiation largely in situ	Migration into limb buds before differentiation
Initial differentiation into mononucleate myocytes	Initial differentiation into multinucleate myotubes
Myogenic determination factors (Myf-5, MyoD) expressed at or before the onset of myotome formation	Expression of myogenic determination genes delayed until limb muscle masses begin to coalesce
Differentiation appearing to be strongly influenced by neural tube and notochord	Migration and differentiation little influenced by axial structures

paraxial mesoderm, specifically the somitomeres, constitutes the main source of the cranial musculature. The cells that make up some of the extraocular muscles arise from the prechordal plate of the early embryo.

Myogenesis in the head differs significantly from that in the trunk (**Fig. 9.40**). Much of the cranial musculature, especially that associated with mastication, arises from cranial unsegmented paraxial mesoderm, equivalent to the somites. Other craniofacial muscles, especially those in the lower jaw and neck, arise from lateral splanchnic mesoderm, as does cardiac muscle. In the early postgastrulation period, the lateral splanchnic mesoderm (sometimes called the cardiocraniofacial morphogenetic field), associated with the future pharynx and probably responding to the same inductive signals from the pharyngeal endoderm, gives rise to both the lower cranial

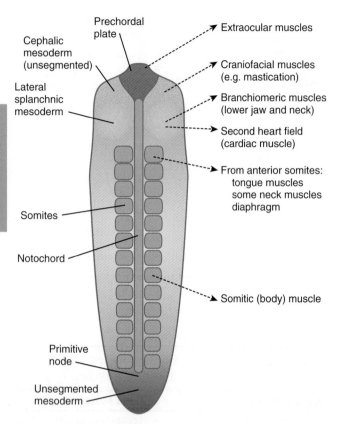

Cephalic mesoderm (unsegmented)

Prechordal plate

Lateral splanchnic mesoderm

Somites

Notochord

Primitive node

Unsegmented mesoderm

Extraocular muscles

Craniofacial muscles (e.g. mastication)

Branchiomeric muscles (lower jaw and neck)

Second heart field (cardiac muscle)

From anterior somites: tongue muscles some neck muscles diaphragm

Somitic (body) muscle

Fig. 9.40   Origins of the various groups of musculature in the body.

musculature and the secondary heart field. Early in the determination process, both these types of muscle are under the control of transcription factors (e.g., **Isl-1, Tbx-1,** and **Nkx 2.5**), which are different from those that control the early development of the trunk musculature. Different types of musculature develop under different sets of early controls before entering similar pathways of differentiation (**Fig. 9.41**).

As with muscles in the trunk, muscles in the head and neck arise by the movement of myogenic cells away from the paraxial mesoderm through mesenchyme (either neural crest–derived or mesodermal mesenchyme) on their way to their final destination. Morphogenesis of muscles in the cranial region is determined by information inherent in the neural crest–derived connective tissues that ensheathe the muscles. There is no early level specificity in the paraxial myogenic cells; this has been determined by grafting somites or somitomeres from one craniocaudal level to another. In these cases, the myogenic cells that leave the grafted structures form muscles normal for the region into which they migrate, rather than muscles appropriate for the level of origin of the grafted somites.

Certain muscles of the head, in particular muscles of the tongue, arise from the occipital somites in the manner of trunk muscles and undergo extensive migration into the enlarging head. Their more caudal level of origin is evidenced by the innervation by the **hypoglossal nerve** (cranial nerve XII), which, according to some comparative anatomists, is a series of highly modified spinal nerves. Similar to the myogenic cells of the limb, precursor cells of the tongue

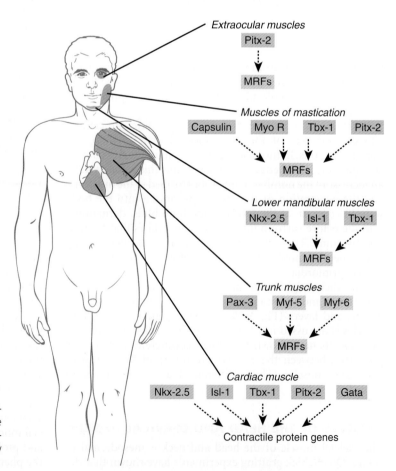

Extraocular muscles
Pitx-2
↓
MRFs

Muscles of mastication
Capsulin   Myo R   Tbx-1   Pitx-2
↓
MRFs

Lower mandibular muscles
Nkx-2.5   Isl-1   Tbx-1
↓
MRFs

Trunk muscles
Pax-3   Myf-5   Myf-6
↓
MRFs

Cardiac muscle
Nkx-2.5   Isl-1   Tbx-1   Pitx-2   Gata
↓
Contractile protein genes

Fig. 9.41   Molecular controls in the determination and differentiation of the various groups of striated muscle in the body, based on studies in the mouse. *Blue* background, transcription factors; *yellow* background, myogenic regulatory factors (MRFs).

musculature express *Pax-3* while they are migrating into the head. Despite their final location in the head, these muscles are subjected to the same types of early molecular regulation of myogenesis as trunk muscles.

Anomalies of Skeletal Muscles

Variations and anomalies of skeletal muscles are common. Some, such as the absence of portions of the pectoralis major muscle, are associated with malformations of other structures. Further discussion of anomalies of specific muscles requires a level of anatomical knowledge beyond that assumed for this text.

Muscular dystrophy is a family of genetic diseases characterized by the repeated degeneration and regeneration of various groups of muscles during postnatal life. In Duchenne's muscular dystrophy, which occurs in young boys, a membrane-associated protein called **dystrophin** is lacking from the muscle fibers. Its absence makes the muscle fibers more susceptible to damage when they are physically stressed.

Cardiac Muscle

Although a striated muscle, cardiac muscle differs from skeletal muscles in many aspects of embryonic development. Derived from the splanchnic mesoderm of the early embryo, cardiac muscle cells arise from cells present in the myocardium. Differences between the differentiation of cardiac and skeletal muscle appear early because MyoD and other common master regulators of skeletal muscle differentiation are not expressed in early cardiac muscle development. Both cardiac and skeletal muscle precursors express the MADS (MCM1, Agamous, Deficiens, and serum response factor) box-containing transcription factor MEF-2, which, in skeletal muscle, at least, dimerizes with other transcription factors to regulate the formation of some of the major contractile proteins of the myocytes. Early cardiac and skeletal muscle cells express isoforms of molecules that are characteristic of mature cells of the other type. Both cardiac and skeletal muscle cells in the embryo express high levels of cardiac α-actin; however, after birth, expression of this molecule declines in skeletal muscle, but it remains high in cardiac muscle. In cardiac hypertrophy, mature cardiac muscle cells begin to express large amounts of skeletal α-actin mRNA.

Even early cardiac myoblasts contain relatively large numbers of myofibrils in their cytoplasm, and they are capable of undergoing pronounced contractions. In the embryo, the mononucleated cardiac myocytes face a difficult problem: The cells of the developing heart must continue to contract while the heart is increasing in mass. This functional requirement necessitates that cardiac myocytes undergo mitosis even though their cytoplasm contains many bundles of contractile filaments (**Fig. 9.42**). Cells of the body often lose their ability to divide when their cytoplasm contains structures

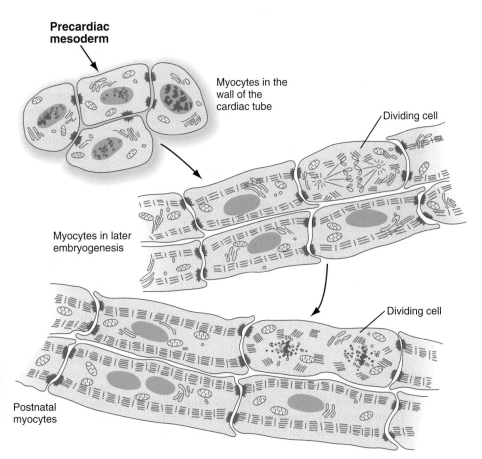

Fig. 9.42 **Stages in the histogenesis of cardiac muscle.** During mitosis, the contractile filaments undergo a partial disassembly. *(Adapted from Rumyantsev P: Cardiomyocytes in processes of reproduction, differentiation and regeneration [in Russian], Leningrad, 1982, Nauka.)*

Precardiac mesoderm

Myocytes in the wall of the cardiac tube

Dividing cell

Myocytes in later embryogenesis

Dividing cell

Postnatal myocytes

characteristic of the differentiated state. Cardiac myocytes deal with this problem by partially disassembling their contractile filaments during mitosis. In contrast to skeletal muscle, cardiac myocytes do not undergo fusion, but rather remain as individual cells, although they may become binucleated. Cardiac myocytes keep in close structural and functional contact through **intercalated disks**, which join adjacent cells to one another.

Later in development, a network of cardiac muscle cells undergoes an alternative pathway of differentiation characterized by increased size, a reduction in the concentration of myofibrils, and a greatly increased concentration of glycogen in the cytoplasm. These cells form the **conducting system**, parts of which are called **Purkinje fibers** (see p. 434). Purkinje fibers also express a different profile of contractile protein isoforms from either atrial or ventricular myocytes.

Smooth Muscle

As does cardiac muscle, much of the smooth muscle in the body arises from splanchnic mesoderm. Exceptions are the ciliary muscle and sphincter pupillae muscles of the eye, which are derived from neural crest ectoderm, and much of the vascular smooth muscle, which frequently arises from the local mesoderm. Very little is known about the morphology and mechanisms underlying the differentiation of smooth muscle cells, but the protein **myocardin** has been shown to be a master regulator of smooth muscle gene expression.

Clinical Vignette

A pediatrician notices that a new patient, a 1½-year-old boy, has a shorter than normal neck and hair farther down the neck than usual. A family history produces no evidence of other similarly affected relatives. X-ray examination reveals that the boy's neck contains only six cervical vertebrae. The pediatrician then asks whether the mother remembers taking or being exposed to certain compounds during early pregnancy.

1. Which of the following does the pediatrician suspect as possibly being related to the boy's condition?
    A. Folic acid
    B. Retinoic acid
    C. Cocaine
    D. Thalidomide
    E. Alcohol
2. A disturbance in what class of molecules is suspected to underlie this condition?
    A. *Hox* genes
    B. *Pax* genes
    C. Myogenic regulatory factors
    D. Fibroblast growth factor
    E. Hedgehog proteins

Summary

- The epidermis starts as a single layer of ectoderm, to which a single superficial layer of peridermal cells is added. As other layers are added, three cell types migrate from other sources: (1) melanoblasts (pigment cells) from the neural crest, (2) Langerhans' cells (immune cells) from precursors in the bone marrow, and (3) Merkel cells (mechanoreceptors) from the neural crest.

- In the multilayered epidermis, unspecialized cells from the stratum basale differentiate as they move through the various layers toward the surface of the epidermis. The cells produce increasing amounts of intracellular keratins and filaggrin; the latter is involved in the interconnections of the keratinocytes, the final differentiated form of the epidermal cell.

- In the trunk, the dermis arises from mesodermal cells derived from the dermatome of the somites. Dermal-epidermal interactions underlie the formation of epidermal appendages such as hairs. In mammary glands, hormonal influences are important in the development of the duct system after the ductal epithelium is induced.

- Skeletal tissue arises from the mesenchyme of either mesodermal or neural crest origin. There are two major subdivisions of the skeleton: the axial skeleton of the trunk and the appendicular skeleton of the limbs.

- The fundamental organization of the cranial components of the vertebral column is closely associated with expression of the homeobox-containing genes. Superimposed on this is the induction of many components of the axial skeleton by underlying ectodermal (usually neural) structures. Individual vertebrae are composite structures consisting of components derived from two adjoining somites.

- The skull consists of two subdivisions: the neurocranium, which surrounds the brain; and the viscerocranium, which surrounds the oral cavity. The base of the neurocranium (chondrocranium) is initially represented by several sets of paired cartilage. These later become transformed into bone. Most bones surrounding the brain are formed by intramembranous bone, which differentiates directly from mesenchyme. The viscerocranium is also derived from cartilaginous and membranous components.

- Skeletal muscle fibers undergo a sequence of differentiation from mononuclear myoblasts. First, they fuse to form multinucleated myotubes, and then they mature into skeletal muscle fibers. Mononucleated reserve cells (satellite cells) can proliferate and fuse to growing or mechanically stressed muscle fibers.

- Pax-3 and myf-5 (a member of the MyoD transcription factor family) stimulate myogenic progenitor cells of the trunk to form myoblasts. Other regulatory factors can activate (e.g., E12) or inhibit (e.g., Id) the activities of muscle regulatory factors. Early myogenic cells are kept in the cell cycle by growth factors such as FGF and transforming growth factor-β. Myoblasts are characterized by the expression of MyoD, and growth factors, such as insulinlike growth factor, promote their fusion and differentiation into myotubes, which express myogenin.

- The first multinucleated muscle fibers to form are primary myotubes. Secondary myotubes form around them. Innervation by motor nerve fibers is necessary for the full differentiation of muscle fibers. During the differentiative process, several sets of isoforms of myosin subunits and other contractile proteins appear in sequence in the muscle fibers.

- Skeletal muscles of the limbs and trunk arise from cellular precursors in the somites. The cranial musculature arises from the somitomeres. Dorsal and ventral muscles of the trunk arise from precursors located in different regions of the somites. The limb musculature also arises from cells in the ventrolateral regions of the somites. These cells migrate into the limb buds and express *Pax-3* while migrating.

■ Cardiac muscle arises from splanchnic mesoderm. Cardiomyocytes differ from skeletal muscle cells in that they can divide mitotically after they are highly differentiated and contain contractile filaments.

Review Questions

1. Satellite cells of muscle are activated under which of these conditions?
A. Normal muscle fiber growth
B. Muscle fiber regeneration
C. Muscle fiber hypertrophy
D. All of the above
E. None of the above

2. Which cellular component of the epidermis is a peripheral outpost of the immune system and functions to present antigens to other immune cells?
A. Merkel cells
B. Keratinocytes
C. Basal cells
D. Melanocytes
E. Langerhans' cells

3. Which structure is mesodermal in origin?
A. Hair shaft
B. Mammary duct
C. Sebaceous gland
D. Arrector pili muscle
E. None of the above

4. Craniosynostosis is caused by an abnormal developmental course of the:
A. Foramen magnum
B. Cranial sutures
C. Basicranium
D. Jaws
E. None of the above

5. Which myogenic regulatory factor is expressed latest in the development of a muscle fiber?
A. Myogenin
B. MyoD
C. MRF-4
D. myf-5
E. Pax-3

6. In the let-down of milk during lactation, the myoepithelial cells contract in response to:
A. Progesterone
B. Oxytocin
C. Estrogens
D. Lactalbumin
E. Casein

7. What component of the developing skin determines the nature of the hairs that form or the thickness of the epidermis in the fetus?

8. A male patient has two bilaterally symmetrical brownish spots about 8 mm in diameter located on the skin about 3 inches below each nipple. What is one explanation for them?

9. Why is cranial bone typically not found over an area where part of the brain is missing?

10. How was it determined that the limb musculature arises from the somites?

References

Skin

Awgulewitsch A: *Hox* in hair growth and development, *Naturwissenschaften* 90:193-211, 2003.

Babler WJ: Embryologic development of epidermal ridges and their configurations, *Birth Defects Orig Artic Ser* 27:95-112, 1991.

Blanpain C, Fuchs E: Epidermal stem cells of the skin, *Annu Rev Cell Dev Biol* 22:339-373, 2006.

Botchkarev VA, Sharov AA: BMP signaling in the control of skin development and hair follicle growth, *Differentiation* 72:512-526, 2004.

Cowin P, Wysolmerski J: Molecular mechanisms guiding embryonic mammary gland development, *Cold Spring Harbor Perspect Biol* 2:1-14, 2010.

Cummins H: The topographic history of the volar pads (walking pads; Tastballen) in the human embryo, *Carnegie Contr Embryol* 113:103-126, 1929.

Duernberger H, Kratochwil K: Specificity of time interaction and origin of mesenchymal cells in the androgen response of the embryonic mammary gland, *Cell* 19:465-471, 1980.

Duverger O, Morasso MI: Epidermal patterning and induction of different hair types during mouse embryonic development, *Birth Defects Res C Embryo Today* 87:263-272, 2009.

Halata Z, Grim M, Christ B: Origin of spinal cord meninges, sheaths of peripheral nerves, and cutaneous receptors, including Merkel cells, *Anat Embryol* 182:529-537, 1990.

Hens JR and others: BMP4 and PTHrP interact to stimulate ductal outgrowth during embryonic mammary development and to inhibit hair follicle induction, *Development* 134:1221-1230, 2007.

Holbrook KA: Structure and function of the developing human skin. In Goldsmith LA, ed: *Physiology, biochemistry, and molecular biology of the skin*, ed 2, New York, 1991, Oxford University Press, pp 63-110.

Imagawa W and others: Control of mammary gland development. In Knobil E, Neill JD, eds: *The physiology of reproduction*, ed 2, New York, 1994, Raven, pp 1033-1063.

Kimura S: Embryologic development of flexion creases, *Birth Defects Orig Artic Ser* 27:113-129, 1991.

Koster MI, Roop DR: Genetic pathways required for epidermal morphogenesis, *Eur J Cell Biol* 83:625-629, 2004.

Mack JA, Anand S, Maytin EV: Proliferation and cornification during development of the mammalian epidermis, *Birth Defects Res C Embryo Today* 75:314-329, 2006.

Merlob P: Congenital malformations and developmental changes of the breast: a neonatological view, *J Pediatr Endocrinol Metab* 16:471-485, 2003.

Ohtola J and others: β-catenin has sequential roles in the survival and specification of ventral dermis, *Development* 135:2321-2329, 2008.

Olivera-Martinez I and others: Dorsal dermis development depends upon a signal from the dorsal neural tube, which can be substituted by Wnt-1, *Mech Dev* 100:233-244, 2001.

Oro AE, Scott MP: Splitting hairs: dissecting roles of signaling systems in epidermal development, *Cell* 95:575-578, 1998.

Pispa J, Thesleff I: Mechanisms of ectodermal organogenesis, *Dev Biol* 262:195-205, 2003.

Robinson GW, Karpf ABC, Kratochwil K: Regulation of mammary gland development by tissue interaction, *J Mammary Gland Biol Neoplasia* 4:9-19, 1999.

Saxod R: Ontogeny of the cutaneous sensory organs, *Microsc Res Tech* 34:313-333, 1996.

Schmidt-Ullrich R, Paus R: Molecular principles of hair follicle induction and morphogenesis, *Bioessays* 27:247-261, 2005.

Sengel P: *Morphogenesis of skin*, Cambridge, 1976, Cambridge University Press.

Shimomura Y, Christiano AM: Biology and genetics of hair, *Annu Rev Genom Human Genet* 11:109-132, 2010.

Sternlicht MD and others: Hormonal and local control of mammary branching morphogenesis, *Differentiation* 74:365-381, 2006.

Strobl H and others: Epidermal Langerhans cell development and differentiation, *Immunobiology* 198:588-605, 1998.

Szeder V and others: Neural crest origin of mammalian Merkel cells, *Dev Biol* 253:258-263, 2003.

Veltmaat JM and others: Mouse embryonic mammogenesis as a model for the molecular regulation of pattern formation, *Differentiation* 71:1-17, 2003.

Watson CJ, Khaled WT: Mammary development in the embryo and adult: a journey of morphogenesis and commitment, *Development* 135:995-1003, 2008.

Yi R, Fuchs E: MicroRNA-mediated control in the skin, *Cell Death Differ* 17:229-235, 2010.

Skeleton

Abzhanov A and others: Regulation of skeletogenic differentiation in cranial dermal bone, *Development* 134:3133-3144, 2007.

Bosma JF, ed: *Symposium on development of the basicranium*, DHEW Pub. No. (NIH) 76-989, Washington, D.C., 1976, U.S. Government Printing Office.

Cohen MM: Merging the old skeletal biology with the new, I and II, *J Craniofac Genet Dev Biol* 20:84-106, 2000.

Eames BF, de la Fuente L, Helms JA: Molecular ontogeny of the skeleton, *Birth Defects Res C Embryo Today* 69:93-101, 2003.

Hall BK: *Bones and cartilage, developmental and evolutionary skeletal biology*, San Diego, 2005, Elsevier Academic Press.

Hall BK: Development of the clavicles in birds and mammals, *J Exp Zool* 289:153-161, 2001.

Hanken J, Hall BK, eds: *The skull, vol 1, Development*, Chicago, 1993, University of Chicago Press.

Hatch NE: FGF signaling in craniofacial biological control and pathological craniofacial development, *Crit Rev Eukaryot Gene Expr* 20:295-311, 2010.

Hehr U, Muenke M: Craniosynostosis syndromes: from genes to premature fusion of skull bones, *Mol Genet Metab* 68:139-151, 1999.

Helms JA, Schneider RA: Cranial skeletal biology, *Nature* 423:326-331, 2003.

Iimura T, Denans N, Pourquié O: Establishment of Hox vertebral identities in the embryonic spine precursors, *Curr Top Dev Biol* 88:201-234, 2009.

Karsenty G, Kronenberg HM, Settembre C: Genetic control of bone formation, *Annu Rev Cell Dev Biol* 25:629-648, 2009.

Kessel M: Respecification of vertebral identities by retinoic acid, *Development* 115:487-501, 1992.

Kessel M, Balling R, Gruss P: Variations of cervical vertebrae after expression of a Hox-1.1 transgene in mice, *Cell* 61:301-308, 1990.

Kjaer I, Keeling JW, Hansen BF: *The human prenatal cranium*, Copenhagen, 1999, Munksgaard.

Kornak U, Mundlos S: Genetic disorders of the skeleton: a developmental approach, *Am J Hum Genet* 73:447-474, 2003.

Lefebvre V, Bhattaram P: Vertebrate skeletogenesis, *Curr Top Dev Biol* 90:291-317, 2010.

Lefebvre V, Smits P: Transcriptional control of chondrocyte fate and differentiation, *Birth Defects Res C Embryo Today* 75:200-212, 2005.

Mallo M, Vinagre T, Carapuço M: The road to the vertebral formula, *Int J Dev Biol* 53:1469-1481, 2009.

Mallo M, Wellik DM, Deschamps J: *Hox* genes and regional patterning of the vertebrate body plan, *Dev Biol* 344:7-15, 2010.

McBratney-Owen B and others: Development and tissue origins of the mammalian cranial base, *Dev Biol* 322:121-132, 2008.

Morriss-Kay GM, Wilkie AOM: Growth of the normal skull vault and its alteration in craniosynostosis: insights from human genetics and experimental studies, *J Anat* 207:637-653, 2005.

Ridgeway EB, Weiner HL: Skull deformities, *Pediatr Clin North Am* 51:359-387, 2004.

Risbud MV, Schaer TP, Shapiro IM: Toward an understanding of the role of notochordal cells in the adult intervertebral disc: from discord to accord, *Dev Dyn* 239:2141-2148, 2010.

Schierhorn H: Ueber die Persistenz der embryonalen Schwanzknospe beim Menschen, *Anat Anz* 127:307-337, 1970.

Sensenig EC: The early development of the human vertebral column, *Carnegie Contr Embryol* 33:21-42, 1949.

Shum L and others: Morphogenesis and dysmorphogenesis of the appendicular skeleton, *Birth Defects Res C Embryo Today* 69:102-122, 2003.

Theiler K: Vertebral malformations, *Adv Anat Embryol Cell Biol* 112:1-99, 1988.

Turnpenny PD and others: Abnormal vertebral segmentation and the notch signaling pathway in man, *Dev Dyn* 236:1456-1474, 2007.

Verbout AJ: The development of the vertebral column, *Adv Anat Embryol Cell Biol* 90:1-122, 1985.

Wellik DM: *Hox* genes and vertebrate axial pattern, *Curr Top Dev Biol* 88:257-278, 2009.

Young T, Deschamps J: *Hox, Cdx*, and anteroposterior patterning in the mouse embryo, *Curr Top Dev Biol* 88:235-255, 2009.

Muscle

Biressi S, Molinaro M, Cossu G: Cellular heterogeneity during vertebrate skeletal muscle development, *Dev Biol* 308:281-293, 2007.

Buckingham M, Monterras D: Skeletal muscle stem cells, *Curr Opin Genet Dev* 18:330-336, 2008.

Currie PD, Ingham PW: The generation and interpretation of positional information within the vertebrate myotome, *Mech Dev* 73:3-21, 1998.

Dietrich S: Regulation of hypaxial muscle development, *Cell Tissue Res* 296:175-182, 1999.

Dominique J-E, Gérard C: Myostatin regulation of muscle development: molecular basis, natural mutations, physiopathological aspects, *Exp Cell Res* 312:2401-2414, 2006.

Francis-West PH, Antoni L, Anakwe K: Regulation of myogenic differentiation in the developing limb bud, *J Anat* 202:69-81, 2003.

Grifone R and others: Eya1 and Eya2 proteins are required for hypaxial somitic myogenesis in the mouse embryo, *Dev Biol* 302:6002-6016, 2007.

Gros J and others: A common somitic origin for embryonic muscle progenitors and satellite cells, *Nature* 435:954-958, 2005.

Hughes SM, Salinas PC: Control of muscle fibre and motoneuron diversification, *Curr Opin Neurobiol* 9:54-64, 1999.

Kelly RG: Core issues in craniofacial myogenesis, *Exp Cell Res* 316:3034-3041, 2010.

Murphy M, Kardon G: Origin of vertebrate limb muscle: the role of progenitor and myoblast populations, *Curr Top Dev Biol* 96:1-32, 2011.

Noden DM: The embryonic origins of avian cephalic and cervical muscles and associated connective tissues, *Am J Anat* 168:257-276, 1983.

Rumyantsev PP: *Cardiomyocytes in processes of reproduction, differentiation and regeneration* [in Russian], Leningrad, 1982, Nauka.

Sabourin LA, Rudnicki MA: The molecular regulation of myogenesis, *Clin Genet* 57:16-25, 2000.

Sambasivan R, Kuratani S, Tajbakhsh S: An eye on the head: the development and evolution of craniofacial muscles, *Development* 138:2401-2415, 2011.

Sanes JR, Donoghue MJ, Merlie JP: Positional differences among adult skeletal muscle fibers. In Kelly AM, Blau HM, eds: *Neuromuscular development and disease*, New York, 1992, Raven, pp 195-209.

Schienda J and others: Somitic origin of limb muscle satellite and side population cells, *Proc Natl Acad Sci U S A* 103:945-950, 2006.

Spiller MP and others: The myostatin gene is a downstream target gene of basic helix-loop-helix transcription factor MyoD, *Mol Cell Biol* 22:7066-7082, 2002.

Tzahor E: Heart and craniofacial muscle development: a new developmental theme of distinct myogenic fields, *Dev Biol* 327:273-279, 2009.

Wigmore PM, Dunglison GF: The generation of fiber diversity during myogenesis, *Int J Dev Biol* 42:117-125, 1998.

Limb Development

Limbs are remarkable structures that are designed almost solely for mechanical functions: motion and force. These functions are achieved through the coordinated development of various tissue components. No single tissue in the limb takes shape without reference to the other tissues with which it is associated. The limb as a whole develops according to a master blueprint that reveals itself sequentially with each successive stage in limb formation. Many of the factors that control limb development cannot be seen by examining morphology alone, but rather must be shown by experimental means or through the localization of molecules. Despite remarkable progress in understanding the molecular basis of the tissue interactions that control limb development, many fundamental questions remain poorly understood. Limb anomalies are common and highly visible. Many of these anomalies are now known to be reflections of disturbances in specific cellular or molecular interactions that are fundamental to limb development. These are discussed in Clinical Correlation 10.1 at the end of the chapter.

Initiation of Limb Development

Limb formation begins relatively late in embryonic development (at the end of the fourth week in humans) with the activation of a group of mesenchymal cells in the somatic lateral plate mesoderm (**Fig. 10.1**). The initial stimulus for limb development is incompletely understood. Experimental evidence suggests that signals from the paraxial mesoderm (probably based on the *Hox* code and ultimately dependent on retinoic acid signaling) initiates a level-specific expression of two T-box transcription factors in the lateral plate mesoderm. **Tbx5** in the area of the future forelimb and **Tbx4** (along with **Pitx-1**) in the area of the future hindlimb stimulate the expression and secretion of **fibroblast growth factor-10 (FGF-10)** by the local mesodermal cells (**Fig. 10.2A**). FGF-10 stimulates the overlying ectoderm to produce **FGF-8**. Soon a feedback loop involving FGF-10 and FGF-8 is established, and limb development begins.

The Tbx transcription factors appear to be the initial local driving forces of limb development. If Tbx5 expression in a mouse is prevented, forelimbs fail to develop (Fig. 10.2B). Similarly, in FGF-10 knockout mice, limbs (and lungs) do not form. Conversely, if a bead soaked with FGF-10 is implanted within the future flank region of a chick embryo, a supernumerary limb develops at the spot. Once the interaction between ectoderm and mesoderm has begun, the limb primordium contains sufficient developmental information to form a limb even if isolated from the rest of the body (a so-called **self-differentiating system**).

The primacy of the early limb mesoderm was shown long ago by transplantation experiments on amphibian embryos. If early limb mesoderm is removed, a limb fails to form. If the same mesoderm is transplanted to the flank of an embryo, however, a supernumerary limb grows at that site. In contrast, if the ectoderm overlying the normal limb mesoderm is removed, new ectoderm heals the defect, and a limb forms. If the original ectoderm that was removed is grafted to the flank, no limb forms. These experiments show that in *early* limb development, mesoderm is the primary bearer of the limb blueprint, and ectoderm is only secondarily co-opted into the system.

In rare instances, individuals are born without one or sometimes all limbs (**amelia**) (**Fig. 10.3**). In some cases, this situation may reflect a disturbance in the production of the transcription factors or signaling molecules that initiate limb development or the cellular receptors for these molecules.

Regulative Properties and Axial Determination

The early limb primordium is a highly regulative system, with properties similar to those described for the cleaving embryo (see p. 45). These properties can be summarized with the following experiments (**Fig. 10.4**):

1. If part of a limb primordium is removed, the remainder reorganizes to form a complete limb.
2. If a limb primordium is split into two halves, and these are prevented from fusing, each half gives rise to a complete limb (the twinning phenomenon).
3. If two equivalent halves of a limb primordium are juxtaposed, one complete limb forms.
4. If two equivalent limb disks are superimposed, they reorganize to form a single limb (see the section on tetraparental embryos [p. 45]).
5. In some species, disaggregated limb mesoderm can reorganize and form a complete limb.

The organization of the limb is commonly related to the Cartesian coordinate system. The anteroposterior* axis runs

---

*Because of different conventions in the use of axial terms, some human embryologists would take exception to the axial terminology presented here. Specifically, according to strict human terminology, anterior means "ventral," and posterior means "dorsal." However, the axial terminology used in this chapter (anterior means "cranial," and posterior means "caudal") is so uniformly used in the experimental and comparative embryological literature that a student referring to the original literature in the field of limb development would find it confusing to use human axial terminology.

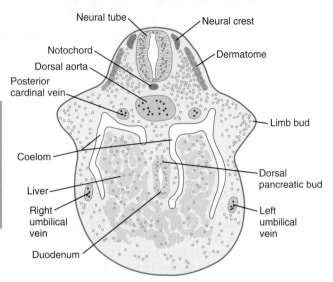

Fig. 10.1   Cross section through the trunk at an early stage of limb bud development that shows the position of the limb bud in relation to that of the somite (dermatome) and other major structures. The limb bud is an outgrowth of the body wall (lateral plate mesoderm).

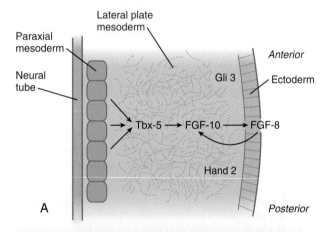

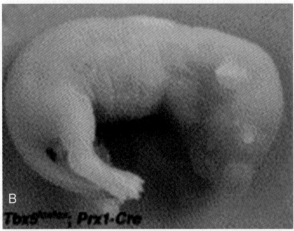

Fig. 10.2   **A,** Molecular interactions involved in the initiation of limb development. FGF, fibroblast growth factor. **B,** Absence of forelimb formation after deletion of *Tbx-5* in the limbs. (*B, From Minguillon C, Del Buono J, Logan MP: Dev Cell 8:75-84, 2005.*)

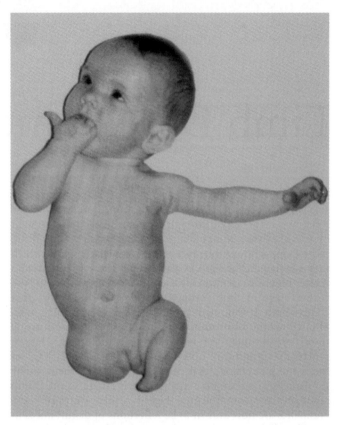

Fig. 10.3   **Amelia of the right leg in an infant.** Despite the absence of a foot, the left leg contains an upper and a lower leg segment. (*From Connor JM, Ferguson-Smith MA: Essential medical genetics, ed 3, Oxford, 1991, Blackwell Scientific.*)

from the first (anterior) to the fifth (posterior) digit. The back of the hand or foot is dorsal, and the palm or sole is ventral. The proximodistal axis extends from the base of the limb to the tips of the digits.

Experiments involving the transplantation and rotation of limb primordia in lower vertebrates have shown that these axes are fixed in a sequential order: anteroposterior to dorsoventral to proximodistal. Early fixation of the anteroposterior axis may result from the expression of the transcription factors **Gli-3** in the anterior and **Hand-2** in the posterior part of the early limb field (see Fig. 10.2A). These two molecules mutually oppose each other's actions. Before all three axes are specified, a left limb primordium can be converted into a normal right limb simply by rotating it with respect to the normal body axes. These axes are important as reference points in several aspects of limb morphogenesis. Evidence indicates a similar sequence of axial specifications in certain other primordia, such as those of the retina and inner ear.

Outgrowth of Limb Bud

Shortly after its initial establishment, the limb primordium begins to bulge from the body wall (late in the first month for the human upper extremity [**Fig. 10.5**]). At this stage, the limb bud consists of a mass of similar-looking mesodermal cells covered by a layer of ectoderm. Despite its apparently simple

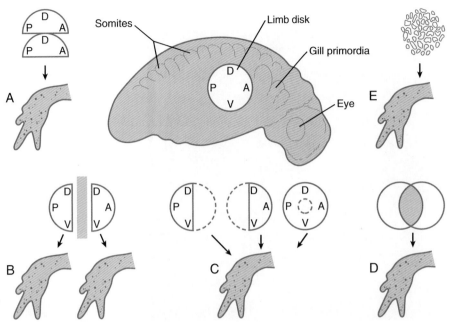

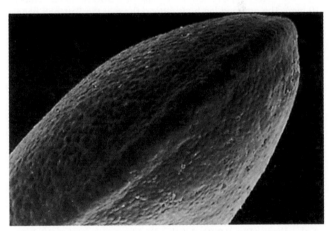

Fig. 10.4 **Experiments showing regulative properties of limb disks in amphibian embryos.** **A,** Combining two identical halves of limb disks results in a single limb. **B,** Separation of two halves of a limb disk by a barrier causes each half to form a normal limb of the same polarity. **C,** After various types of tissue removal, the remaining limb tissue regulates to form a normal limb. **D,** Combining two disks results in the formation of a single normal limb. **E,** Mechanical disruption of a limb disk is followed by reorganization of the pieces and the formation of a normal limb. A, anterior; D, dorsal; P, posterior; V, ventral. *(Data from Harrison RG: J Exp Zool 32:1-136, 1921; and Swett FH: Q Rev Biol 12:322-339, 1937.)*

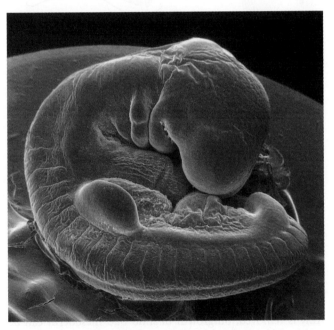

Fig. 10.5 Scanning electron micrograph of a 4-week human embryo (5 mm), with 34 pairs of somites. Toward the *lower left,* the right arm bud protrudes from the body. *(From Jirásek JE: Atlas of human prenatal morphogenesis, Amsterdam, 1983, Martinus Nijhoff.)*

Fig. 10.6 Scanning electron micrograph of the flattened limb bud of a human embryo that shows the prominent apical ectodermal ridge traversing the apical border. *(From Kelley RO, Fallon JF: Dev Biol 51:241-256, 1976.)*

structure, the limb bud contains enough intrinsic information to guide its development because if a mammalian limb bud is transplanted to another region of the body or is cultured in vitro, a recognizable limb forms.

A distinctive feature is the presence of a ridge of thickened ectoderm (**apical ectodermal ridge [AER]**) located along the anteroposterior plane of the apex of the limb bud (**Fig. 10.6**). During much of the time when the AER is present, the hand-forming and foot-forming regions of the developing limb bud are paddle shaped, with the apical ridge situated along the rim of the paddle (**Fig. 10.7**). Experiments have shown that the

AER interacts with the underlying limb bud mesoderm to promote outgrowth of the developing limb. Other aspects of limb development, such as morphogenesis (the development of form), are guided by information contained in the mesoderm.

This section outlines many of the ways in which the limb bud mesoderm and ectoderm interact to control limb development. Recognition of these developmental mechanisms is important in understanding the genesis of many limb malformations.

Apical Ectodermal Ridge

The earliest limb bud begins to form before an AER is present, but soon a thickened AER appears along the border between dorsal and ventral limb ectoderm. Molecular studies have shown that the position of the AER corresponds exactly to the

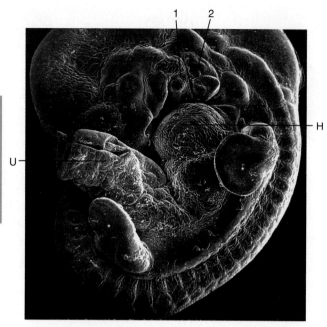

Fig. 10.7 Scanning electron micrograph of a 5-week human embryo (10 mm). The arm and leg buds *(asterisks)* are in the flattened paddle stage. H, heart; U, umbilical cord; 1, 2, pharyngeal arches 1 and 2. *(From Jirásek JE: Atlas of human prenatal morphogenesis, Amsterdam, 1983, Martinus Nijhoff.)*

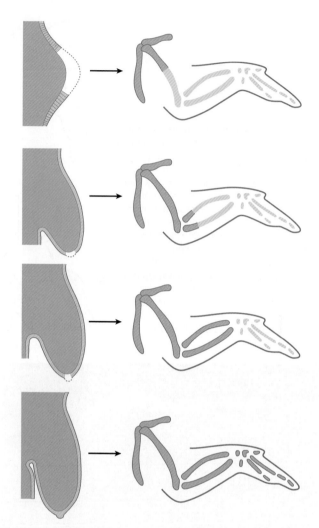

Fig. 10.8 *Top three,* Effect of removing the apical ectodermal ridge at successively later stages on the development of the avian wing bud. The more mature the wing bud, the more skeletal elements form after apical ridge removal. Missing structures are shown in *light gray. Bottom,* Normal development of an untouched wing bud. *(Based on Saunders JW: J Exp Zool 108:363-403, 1948.)*

border between **dorsal ectoderm**, which expresses the signaling molecule **radical fringe**, and **ventral ectoderm**, which expresses the transcription factor **Engrailed-1 (En-1)** (see Fig. 10.17A).

Although the AER had been recognized for years, its role in limb development was not understood until it was subjected to experimental analysis. Removal of the AER results in an arrest of limb development, thus leading to distal truncation of the limb (**Fig. 10.8**). In the *limbless* mutant in chickens, early limb development is normal; later, the AER disappears, and further limb development ceases. If mutant ectoderm is placed over normal limb bud mesoderm, limb development is truncated, whereas combining mutant mesoderm with normal ectoderm results in more normal limb development. These findings suggest that the ectoderm is defective in this mutant.

Further analysis has shown that in the *limbless* mutant, the entire ectoderm of the limb bud displays a dorsal character; that is, radical fringe and other "dorsal" molecules are expressed in dorsal and ventral ectoderm. Correspondingly, En-1 is not expressed in ventral ectoderm. In the absence of the juxtaposition of ectoderm with dorsal and ventral properties, an AER cannot be maintained.

The power of the AER is shown by experiments or mutants that result in the formation of two AERs on the limb bud. This situation leads to a supernumerary limb, as is illustrated by the mutants **eudiplopodia** in chickens and **diplopodia** in humans (**Fig. 10.9**).

The outgrowth-promoting signal produced by the AER is FGF. In the earliest stages of limb formation, the lateral ectoderm begins to produce FGF-8 as it thickens to form an AER. As the limb bud begins to grow out, the apical ridge also produces FGF-4, FGF-9, and FGF-17 in its posterior half. If the

AER is removed, outgrowth of limb bud mesoderm can be supported by the local application of FGFs. Other studies have shown that in mutants characterized by deficient or absent outgrowth of the limb, the mutant ectoderm fails to produce FGF. The effects of the FGF produced by the apical ectoderm on the underlying mesoderm are discussed later in this chapter.

Mesoderm of Early Limb Bud
Structure and Composition

The mesoderm of the early limb bud consists of homogeneous mesenchymal cells supplied by a well-developed vascular network. The mesenchymal cells are embedded in a matrix consisting of a loose meshwork of collagen fibers and ground substance, with hyaluronic acid and glycoproteins prominent constituents of the latter. There are no nerves in the early limb bud.

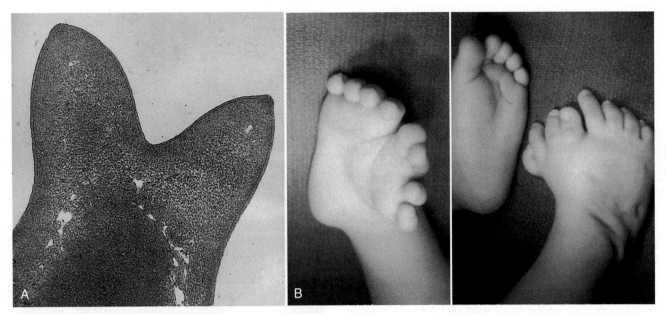

Fig. 10.9 **A,** Duplicated wing bud in a chick with eudiplopodia. Under the influence of a secondary apical ectodermal ridge, a supernumerary limb bud forms. **B,** Diplopodia in a human. Dorsal and ventral views of the right foot, in which duplication has occurred along the anteroposterior axis. (**A,** From Goetinck P: Dev Biol 10:71-79, 1964, **B,** Courtesy of D. Hootinck, Buffalo, N.Y.)

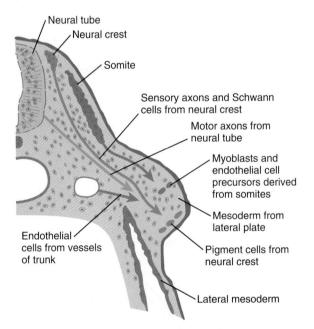

Fig. 10.10 **The different types of cells that enter the limb bud.**

It is impossible to distinguish different cell types within the early limb bud mesenchyme by their morphology alone. Nevertheless, mesenchymal cells from several sources are present (**Fig. 10.10**). Initially, the limb bud mesenchyme consists exclusively of cells derived from the lateral plate mesoderm. These cells give rise to the skeleton, connective tissue, and some blood vessels. Mesenchymal cells derived from the somites migrate into the limb bud as precursors of muscle and endothelial cells. Another population of migrating cells is that

from the neural crest; these cells ultimately form the Schwann cells of the nerves, sensory nerves, and pigment cells (**melanocytes**).

Mesodermal-Ectodermal Interactions and the Role of Mesoderm in Limb Morphogenesis

Limb development occurs as the result of continuous interactions between the mesodermal and ectodermal components of the limb bud. The apical ectoderm stimulates outgrowth of the limb bud by promoting mitosis and preventing differentiation of the distal mesodermal cells of the limb bud. Although the AER promotes outgrowth, its own existence is reciprocally controlled by the mesoderm. If an AER from an old limb bud is transplanted onto the mesoderm of a young wing bud, the limb grows normally until morphogenesis is complete. If old limb bud mesoderm is covered by young apical ectoderm, however, limb development ceases at a time appropriate for the age of the mesoderm and not that of the ectoderm.

Similar reciprocal transplantation experiments have been used to show that the overall shape of the limb is determined by the mesoderm and not the ectoderm. This is most dramatically represented by experiments done on birds because of the great differences in morphology between the extremities. If leg bud mesoderm in the chick embryo is covered with wing bud ectoderm, a normal leg covered with scales develops. In a more complex example, if chick leg bud ectoderm is placed over duck wing bud mesoderm, a duck wing covered with chicken feathers forms. Such experiments, which have sometimes involved mosaics of avian and mammalian limb bud components, show that the overall morphology of the limb is determined by the mesodermal component and not the ectoderm. In addition, the regional characteristics of ectodermal appendages (e.g., scalp hair versus body hair in the case of mammals) are also dictated by the mesoderm. Cross-species grafting experiments show, however, that the nature of the ectodermal appendages formed (e.g., hair versus feathers) is appropriate for the species from which the ectoderm was derived.

Polydactyly is a condition characterized by supernumerary digits and exists as a mutant in birds. Reciprocal transplantation experiments between mesoderm and ectoderm have shown that the defect is inherent in the mesoderm and not the ectoderm. Polydactyly in humans (**Fig. 10.11**) is typically inherited as a genetic recessive trait and is commonly found in populations such as certain Amish communities in the United States in which the total genetic pool is relatively restricted (see Clinical Correlation 10.1 for further details).

Zone of Polarizing Activity and Morphogenetic Signaling

During experiments investigating programmed cell death in the avian limb bud, researchers grafted mesodermal cells from the posterior base of the avian wing bud into the anterior margin. This manipulation resulted in the formation of a supernumerary wing, which was a mirror image of the normal wing (**Fig. 10.12**). Much subsequent experimentation has shown that this posterior region, called the **zone of polarizing activity** (ZPA), acts as a signaling center along the anteroposterior axis of the limb. The signal itself has been shown to be **sonic hedgehog** (**shh**) (see Fig. 10.16), a molecule that mediates a wide variety of tissue interactions in the embryo (see Table 4.4). As seen later in this chapter, shh not only organizes tissues along the anteroposterior axis, but also maintains the structure and function of the AER. In the absence of the ZPA or shh, the apical ridge regresses.

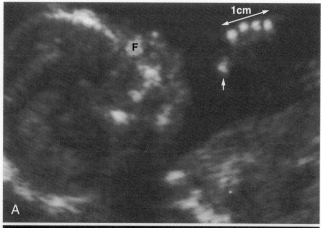

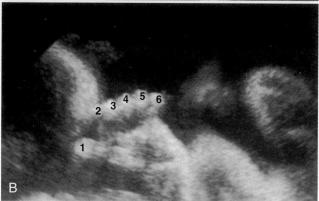

Fig. 10.11   Ultrasound images of normal (**A**) and polydactylous (six digit) (**B**) hands of human fetuses at 16 and 31 weeks. In both cases, the digits are imaged in cross section. *Arrow* indicates the thumb. F, face; 1, thumb; 2 to 6, fingers on polydactylous hand. (*A, From Bowerman R:* Atlas of normal fetal ultrasonographic anatomy, *St. Louis, 1992, Mosby;* **B,** *from Nyberg D and others:* Diagnostic ultrasound of fetal anomalies, *St. Louis, 1990, Mosby.*)

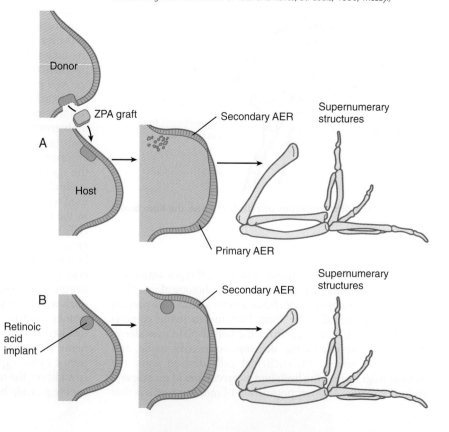

Fig. 10.12   **A,** Grafting of the zone of polarizing activity (ZPA) into the anterior border of the avian limb bud results in the formation of a secondary apical ectodermal ridge and a supernumerary limb. **B,** Implantation of a bead soaked in retinoic acid into the anterior border of the limb bud also stimulates the formation of a supernumerary limb. AER, apical epidermal ridge.

Cross-species grafting experiments have shown that mammalian (including human) limb buds also contain a functional ZPA. A transplanted ZPA acts on the AER and elicits a growth response from the mesenchymal cells just beneath the part of the ridge adjacent to the transplanted ZPA. As few as 50 cells from the ZPA can stimulate supernumerary limb formation. Other structures, such as pieces of Hensen's node, notochord, and even feather germs have been shown to stimulate the formation of supernumerary limbs if grafted into the anterior margin of the limb. Since these experiments were conducted, all the effective implanted tissues have been found to be sources of shh.

The ZPA is already established by the time the limb bud begins to grow out from the body wall. There is evidence that in the forelimb the position of the ZPA is determined by the location of the highest concentration of *Hoxb8* expression along the body axis. Experiments have shown that in response to the localized application of retinoic acid along the anterior margin of the forelimb bud, *Hoxb8* expression is induced within 30 minutes. This suggests a cascade beginning with retinoic acid signaling, leading to *Hoxb8* expression, which determines the location of the ZPA.

Shh induces the expression of the signaling molecule **gremlin**, which has two inhibitory functions. Gremlin inhibits the action of mesodermal **bone morphogenetic protein-2 (BMP-2)**, which in itself inhibits the expression of FGF-4 in the AER. Such an inhibition of a BMP inhibitor is reminiscent of the sequence of events involved in primary neural induction (see p. 84). In contrast, gremlin, which is localized in the posterior part of the limb bud, inhibits the action of Gli-3 so that Gli-3 functions only in the anterior part. Within the

anterior part of the limb bud, Gli-3 inhibits the expression of shh. In *Gli-3* mutants, shh becomes expressed ectopically in the anterior limb bud, and preaxial polydactyly results.

As the limb bud elongates, the ZPA becomes translocated more distally, and it becomes surrounded by an increasingly large zone of formerly shh-producing cells that were derived from the ZPA. Later, these cells become heavily involved in the formation of digits and in events leading to the termination of limb development.

Morphogenetic Control of Early Limb Development

Control of Proximodistal Segmentation

As a limb (e.g., an arm) grows out from a simple bud, it eventually forms three structural segments: the **stylopodium** (upper arm), the **zeugopodium** (forearm), and the **autopodium** (hand). Over the years, several hypotheses concerning the control of proximodistal segmentation have been proposed, but only more recently has a hypothesis been supported by strong experimental data. During development, more proximal segments differentiate first, followed successively by the more distal segments. The mesenchymal cells at the distal tip of the limb bud are kept in a proliferative state through the actions of **FGFs** and **Wnts**, whereas cells in the proximal part of the limb bud, under the influence of **retinoic acid** and possibly other molecules, undergo differentiation into proximal components of the limb (**Fig. 10.13**). The balance between retinoic acid and the FGFs and Wnts is

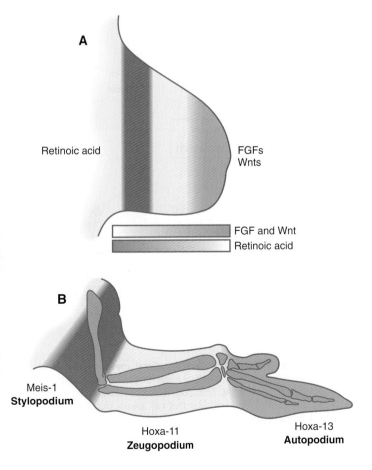

Fig. 10.13 **A,** Control of proximodistal segmentation in the chick limb by opposing gradients of proximal retinoic acid *(purple)* and distal fibroblast growth factors (FGFs) and Wnts *(green)*. **B,** Cells exposed to the highest ratio of retinoic acid to FGF *(red zone in* **A***)* differentiate into the stylopodium, characterized by Meis-1 expression; those exposed to intermediate ratios *(white zone)* differentiate into the zeugopodium, characterized by Hoxa-11 expression; and those exposed to the lowest ratio of retinoic acid to FGF differentiate into the autopodium (Hoxa-13 expression).

thought to determine the course of segmental differentiation. In the early limb bud, the proximal mesenchymal cells are exposed to a high concentration of retinoic acid because they are near the source (somites), and they differentiate into tissues of the stylopod. As the limb bud grows out, the remaining undifferentiated cells are exposed to a lesser concentration of retinoic acid because outgrowth has taken them farther from the source of retinoic acid. Thus, those remaining mesenchymal cells in later limb buds first differentiate into the zeugopodial segment and finally, in the late limb bud, into the autopodial segment. This balance between the differentiation-promoting effects of retinoic acid and the proliferation-maintaining effects of FGF is similar to that occurring in the posterior end of the early embryo (see Figs. 6-5 and 6.9A).

Cells in the distal mesenchyme are characterized by their expression of **Msx-1**, a marker of undifferentiated cells, and as they leave that region, expression of that gene ceases (**Fig. 10.14A**). Something about the distal mesenchymal environment stimulates Msx-1 expression because if mesenchymal cells that have left that region (and consequently cease production of Msx-1) are transplanted back into the distal region, they express that molecule again (Fig. 10.14B).

Molecular Signals in Limb Development

As discussed earlier, initial development of the limbs involves the establishment of a limb field by the actions of a *Hox* gene combinatorial code that acts through yet unidentified axial signals to stimulate the expression of Tbx5 in the area of the future forelimb and Tbx4 in the hindlimb. Even in later development, Tbx5 is expressed exclusively in the forelimb, and Tbx4 is expressed exclusively in the hindlimb (**Fig. 10.15**). Because of these exclusive expression regions, it was originally assumed that these two genes determined the identity of the forelimb and hindlimb. More recent research has shown this not to be the case, however, and the search for the factors that determine limb identity continues. **Pitx-1**, which is also expressed in the hindlimb, may play a more important role than Tbx-4 as a determinant of hindlimb identity. The main functions of Tbx-4 and Tbx-5 appear to be the initiation of development in a limb-specific manner.

When the limb bud takes shape, its further development depends to a great extent on the actions of three signaling centers, one for each of the cardinal axes of the limb. As already discussed, outgrowth along the proximodistal axis is largely under the control of the apical ectodermal ridge and the FGFs that it produces. FGF-8 is produced along the entire length of the AER, and FGF-4 is produced only along the posterior half. FGF-4, in particular, is an integral part of a feedback loop linking the growth center in the AER to that of the ZPA.

The second major signaling center, this time along the anteroposterior axis, is the ZPA, and the signaling molecule is **shh** (**Fig. 10.16**). Although shh is a diffusible molecule, it functions through its effects on BMP-2 and the inhibitor of BMP-2, **gremlin** (**Fig. 10.17**). Gremlin has two major functions. First, it antagonizes Gli-3, confining Gli-3 activity to the anterior part of the limb bud, where it represses the expression of posterior limb genes. As mentioned earlier, gremlin also inhibits the inhibitory action of BMP-4 on the AER, thus promoting the activity of FGF-4. FGF-4 is necessary for maintaining the activity of the ZPA.

Organization of the dorsoventral axis of the limb begins when the dorsal ectoderm produces the signaling molecule, **Wnt-7a**, which stimulates the underlying limb bud

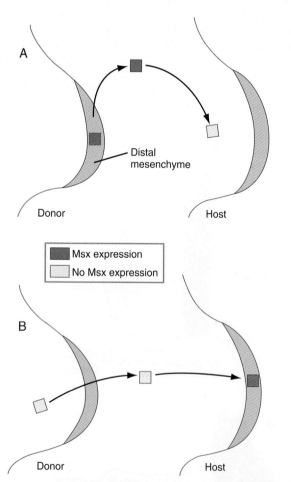

Fig. 10.14 **Msx-1 expression and the distal mesenchyme. A,** If Msx-1–expressing tissue from the distal mesenchyme is transplanted into more proximal regions of the limb bud, it soon stops expressing that molecule. **B,** If proximal mesenchyme that has already stopped expressing Msx-1 is transplanted back into the distal mesenchymal region, it expresses the molecule again.

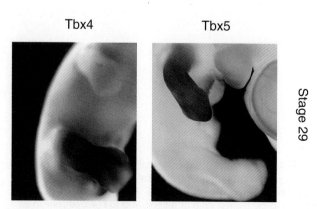

Tbx4   Tbx5

Stage 29

Fig. 10.15 Whole mount in situ hybridization preparations of stage 29 chick embryos showing the localized expression of the mRNAs to Tbx4 in the hindlimb and Tbx5 in the forelimb. *(Courtesy of H.-G. Simon, Northwestern University Medical School, Chicago.)*

mesenchyme to express the transcription factor, **Lmx-1b**, a molecule that imparts a dorsal character to the mesoderm underlying the dorsal ectoderm. Ventral ectoderm produces **En-1**, which represses the formation of Wnt-7a and consequently the formation of Lmx-1b in what will become ventral limb mesoderm, possibly by a default pathway (see Fig. 10.17A). The AER marks the border between dorsal and ventral limb bud ectoderm, and this border is characterized by an expression boundary between another signaling factor, **radical fringe**, secreted by dorsal ectoderm, and the En-1 formed in the ventral ectoderm. Further ventral spread of radical fringe expression is held in check by En-1.

All three axial signaling centers (**Table 10.1**) interact in the early limb bud. Wnt-7a from the dorsal ectoderm has a stimulating effect on the ZPA (see Fig. 10.17B), whereas shh from the ZPA is required for the production of FGFs from the AER that provide additional positive feedback to the ZPA.

Simultaneous with the establishment of the ZPA, an orderly sequence of the homeobox-containing genes *Hoxd9* to *Hoxd13* (**Fig. 10.18**) and certain of the *Hoxa* genes occurs in the early limb bud. This sequence represents a second wave of *Hox* gene expression after that involved in the initiation of

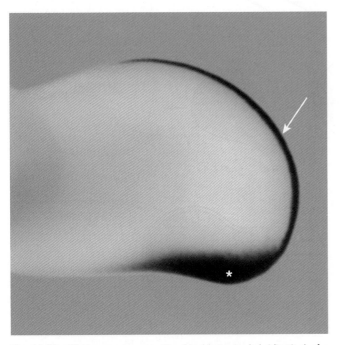

Fig. 10.16 Whole mount preparation (double in situ hybridization) of a late chick limb bud showing the expression of fibroblast growth factor-8 in the apical ectodermal ridge *(arrow)* and sonic hedgehog *(asterisk)* in the zone of polarizing activity, which has moved distally as the limb bud has grown outward. *(Courtesy of E. McGlinn and C. Tabin, Boston.)*

Table 10.1 Axial Control in the Developing Limb

Axis	Signaling Center	Molecular Signal
Proximodistal	Apical ectodermal ridge	FGF-2, FGF-4, FGF-8
Anteroposterior	Zone of polarizing activity	Sonic hedgehog
Dorsoventral	Dorsal ectoderm	Wnt-7a (dorsal)
	Ventral ectoderm	En-1 (ventral)

FGF, fibroblast growth factor.

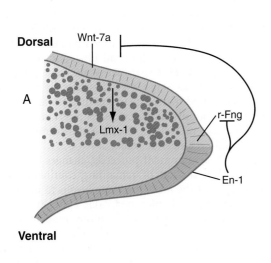

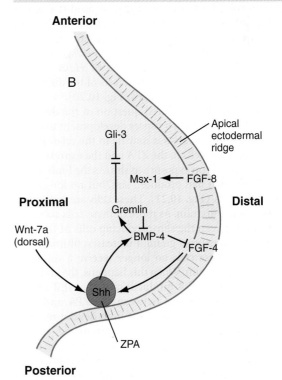

Fig. 10.17 **Schematic representations of molecular control of limb development.** A, Molecular control of the dorsoventral axis. En-1 inhibits Wnt-7a and r-Fng. B, Molecular control along the anteroposterior and proximodistal axes. BMP-2, bone morphogenetic protein-2; En-1, Engrailed-1; FGF, fibroblast growth factor; r-Fng, radical fringe; Shh, sonic hedgehog; ZPA, zone of polarizing activity.

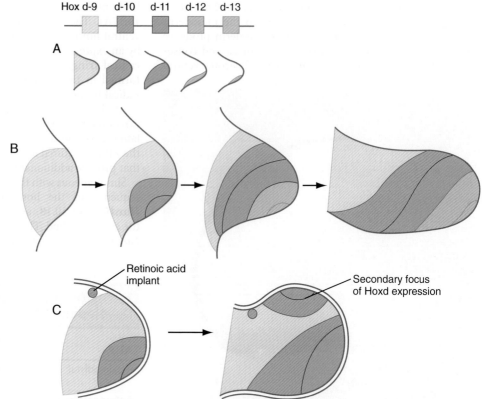

Fig. 10.18 ***Hoxd* gene expression in the chick limb bud. A,** Map of this gene family and distribution of individual gene products. **B,** Development of the aggregate pattern of *Hoxd* gene expression over time in the normal limb bud. **C,** The development of a secondary focus of *Hoxd* gene expression in the area of supernumerary limb formation caused by an implant of retinoic acid. *(Based on Tabin CJ: Development 116:289-296, 1992.)*

limb development. Shh stimulates the expression of the *Hox* genes in the limb, and Gli-3 is involved in confining *Hox* gene expression to the more posterior parts of the limb bud. The *Hox* genes are involved in patterning the proximodistal axis of the limb (**Fig. 10.19**). Studies on mice and the analysis of certain human mutants have shown that certain defects in limb regions correspond to absent expression of specific *Hox* gene paralogues. For example, mutations of *Hoxa 13* and *Hoxd13* cause characteristic reduction defects of the digits that result from shortness of the phalanges (**Fig. 10.20**).

An interesting, but little explored question in the development of many structures is what causes development to cease. In the case of the limb, the answer may lie in the relationship between shh-producing cells in the ZPA and the expression of gremlin, which depends on exposure to shh. As the limb develops, a zone of cells that had produced shh, but no longer do, forms around the ZPA (**Fig. 10.21**). These cells are not themselves able to produce gremlin. As more of these cells accumulate, the distance between the shh-producing cells of the ZPA and the cells that can express gremlin increases, ultimately to the point at which these cells no longer receive a sufficient stimulus to produce gremlin. When this happens, the gremlin-based maintenance of FGF-4 production of the AER ceases, and the entire feedback system between the ZPA and AER winds down; this results in cessation of limb development. If the intervening wedge of formerly shh-producing cells is removed from the distal tip of the mature limb bud, the more anteriorly located mesodermal cells are again exposed to above-threshold concentrations of shh, and they can again produce gremlin. This reconstitutes the ZPA-AER axis through a regulative mechanism, and limb development continues past

the point at which it usually ceases. The result is the formation of digits with more than the normal number of phalangeal segments.

Cell Death and Development of Digits

Although it may seem paradoxical, genetically programmed **cell death** (**apoptosis**) is important in the development of many structures in the body. In the forelimb, it is prominently manifested in the anterior limb margin, in the future axillary region, between the radius and ulna, and in the interdigital spaces (**Fig. 10.22**). Experiments on avian embryos showed that, to a certain stage, mesodermal cells scheduled to die could be spared by transplanting them to areas in which cell death did not normally occur. After a certain time, however, a "death clock" was set (an example of determination), and the cells could no longer be rescued.

As limb development proceeds, changes become apparent in the AER. Instead of remaining continuous around the entire apex of the limb, the ridge begins to break up, leaving intact segments of thickened ridge epithelium covering the emerging digital rays (cartilaginous models of the digital bones). Between the digits, the ridge regresses (see Fig. 10.22A). As the digital primordia continue to grow outward, cell death sculpts the interdigital spaces (see Fig. 10.22C). **BMP-2,** BMP-4, and BMP-7 and the transcription factors **Msx-1** and **Msx-2** are strongly expressed in the interdigital spaces. The exact mechanism of interdigital cell death is still unclear, but the BMPs, especially BMP-4 acting under the mediation of Msx-2, are the prime movers in initiating interdigital cell death. The FGFs produced by the AER seem to play a dual role

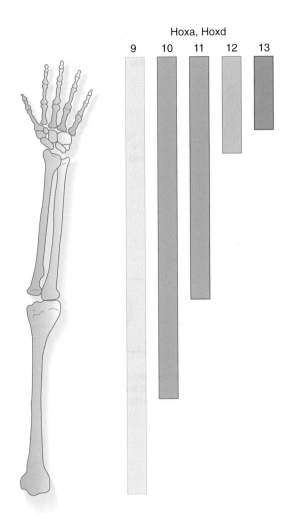

Hoxa, Hoxd

9   10   11   12   13

Fig. 10.19 **Levels of *Hox* gene expression in relation to skeletal components of the limb.** Molecular data from the mouse are superimposed on the human limb skeleton.

in interdigital cell death. Although FGF-2 antagonizes the death-inducing effects of the BMPs, FGFs promote the production of Msx-2, which cooperates with the BMPs in inducing interdigital cell death.

If interdigital cell death does not occur, a soft tissue web connects the digits on either side. This is the basis for the development of webbed feet on ducks and the abnormal formation of **syndactyly** (**Fig. 10.23A**) in humans. BMP is not found in the interdigital mesoderm in developing duck feet, although it is found in other regions of cell death in the duck limb.

There is more, however, to the development of digits than simply sculpting the interdigital spaces by cell death. Well before cell death becomes evident, other events specify the nature of each digit. A future digit is first recognizable as a longitudinal condensation of mesenchyme, which soon begins to lay down a precartilaginous matrix. The early digital ray then undergoes segmentation (see Fig. 10.26) to form specific phalangeal segments. Each digit develops its own character, as determined by the number of phalangeal segments or its specific size and shape. The underlying basis for the development of digital form is just becoming understood.

How individual digits form has long remained a mystery, but new research findings are beginning to clarify some aspects of the process. It is now evident that the identity of individual digits is not fixed until relatively late in limb formation. The driving force for the specification of most digits is shh. The exception to this rule is the first digit (thumb), which forms even in *shh*[-/-] mutants. The identity of the remainder of the digits is determined by the concentration and the duration of exposure of their cells to shh. Digit 2 is formed from cells that have been exposed to shh, but have not themselves produced this signaling factor. Digits 3 to 5 arise from cells that have produced shh. Digit 3 is actually a hybrid. The anterior half consists of cells that have been exposed to, but have not produced shh, whereas the posterior half of that digit is composed of shh-producing cells that have been exposed to shh for the

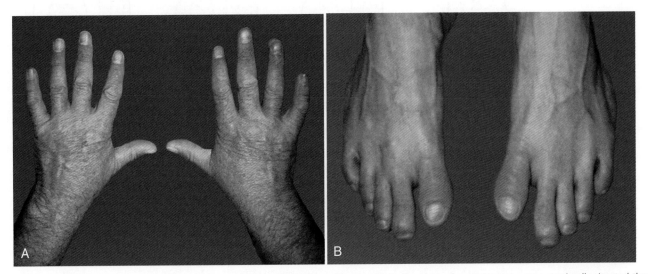

A   B

Fig. 10.20   **Hands and feet of a person with a mutation of the *Hoxa13* gene.** Both thumbs and great toes are more proximally situated than normal. In addition, some phalanges are shortened, and the nails are hypoplastic. *(Courtesy of J.W. Innis, Ann Arbor, Mich.)*

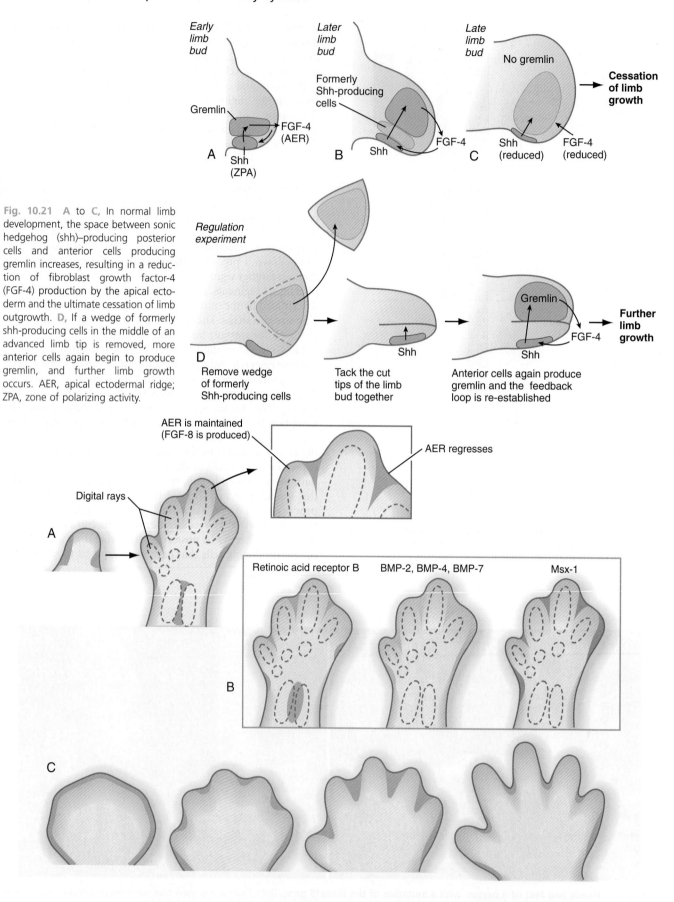

Fig. 10.21 A to C, In normal limb development, the space between sonic hedgehog (shh)–producing posterior cells and anterior cells producing gremlin increases, resulting in a reduction of fibroblast growth factor-4 (FGF-4) production by the apical ectoderm and the ultimate cessation of limb outgrowth. D, If a wedge of formerly shh-producing cells in the middle of an advanced limb tip is removed, more anterior cells again begin to produce gremlin, and further limb growth occurs. AER, apical ectodermal ridge; ZPA, zone of polarizing activity.

Fig. 10.22 Cell death in the development of the hand and digits. A, Cell death in the chick limb bud. B, Gene expression in zones of cell death of the chick embryo. C, Cell death in the developing human hand. AER, apical ectodermal ridge; BMP, bone morphogenetic protein; FGF-8, fibroblast growth factor-8.

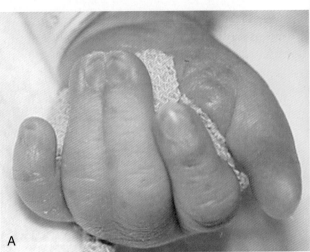 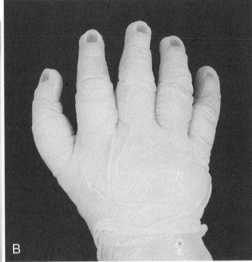

Fig. 10.23 **A,** Syndactyly in a human. **B,** Triphalangeal thumb in a human fetus. *(**A,** From the Robert J. Gorlin Collection, Division of Oral and Maxillofacial Pathology, University of Minnesota Dental School, courtesy of Dr. Ioannis Koutlas; **B,** courtesy of M. Barr, Ann Arbor, Mich.)*

shortest amount of time. A longer period of exposure and a greater concentration of shh is required to form digit 4, and digit 5 requires the longest exposure time and concentration of shh. Growth of individual digital primordia is maintained by the production of FGF-8 by the remnants of the AER overlying the tips of the digital primordia, while BMP-mediated cell death is occurring in the interdigital mesenchyme.

All human digits contain three phalangeal segments except for the first digits (thumb and great toe), which consist of only two segments. Rarely, an individual is born with a triphalangeal thumb (see Fig. 10.23B). Why digits have different numbers of phalangeal segments is still not understood.

Development of Limb Tissues

The morphogenetic events previously described occur largely during the early stages of limb development, when the limb bud consists of a homogeneous-appearing mass of mesodermal cells covered by ectoderm. The differentiation and histogenesis of the specific tissue components of the limb are later developmental events that build on the morphogenetic blueprint already established.

Skeleton

The skeleton is the first major tissue of the limb to show signs of overt differentiation. Its gross morphology, whether normal or abnormal, closely reflects the major pattern-forming events that shape the limb as a whole. Formation of the skeleton can be first seen as a condensation of mesenchymal cells in the central core of the proximal part of the limb bud. Even before undergoing condensation, these cells are determined to form cartilage, and if they are transplanted to other sites or into culture, they differentiate only into cartilage. Other mesenchymal cells that would normally form connective tissue retain the capacity to differentiate into cartilage if they are transplanted into the central region of the limb bud.

The ectoderm of the limb bud exerts an inhibitory effect on cartilage differentiation, so cartilage does not form in the region just beneath the ectoderm. On the dorsal side of the limb bud, mesenchymal cells are prevented from differentiating into cartilage by Wnt-7a, produced by the ectoderm.

The condensed cells that make up the precartilaginous aggregates express BMP-2 and BMP-4. As skeletal development continues, their expression becomes progressively restricted to the cells that become the perichondrium or periosteum surrounding the bones. BMP-3 transcripts are first seen in cartilage, rather than precartilage, but this growth factor also ultimately becomes located in the perichondrium. The translocation of expression of these BMP molecules to the perichondrium reflects their continuing role in the earliest phases of differentiation of skeletal tissues.

In contrast, BMP-6 is expressed only in areas of maturing (hypertrophying) cartilage within the limb bones. **Indian hedgehog,** a molecule related to shh, is also expressed in the same regions of hypertrophying cartilage (which is also marked by the presence of type X collagen), and this signaling molecule may induce the expression of BMP-6.

Differentiation of the cartilaginous skeleton occurs in a proximodistal sequence, and in mammals the postaxial structures of the distal limb segments differentiate before the preaxial structures. For example, the sequence of formation of the digits is from the fifth to the first (**Fig. 10.24**). The postaxial skeleton of the arm is considered to be the humerus, ulna, digits 2 through 5, and their corresponding carpal and metacarpal elements. The preaxial portion of the limb bud becomes progressively reduced during limb outgrowth and contributes only to the radius and possibly the first digital ray. Certain limb defects, sometimes called **hemimelias,** are characterized by deficiencies of preaxial or postaxial limb components (**Fig. 10.25**).

The development of the limb girdles remains incompletely investigated, but experimental work on the chick has shown that the blade of the scapula is derived from cells of the dermomyotome, whereas the remainder of the scapula arises

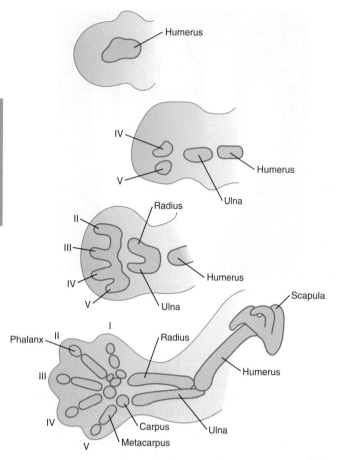

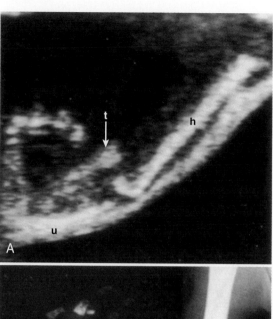

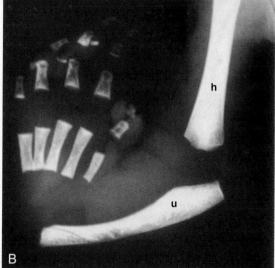

Fig. 10.24 **Formation of the skeleton in the mammalian forelimb.**

Fig. 10.25 **Radial hemimelia (absent radius) in a 27-week fetus.** **A,** Ultrasound image showing a thumb (t, *arrow*) but no radius. **B,** Postnatal x-ray image confirming the absent radius. h, humerus; u, ulna. *(From Nyberg D and others:* Diagnostic ultrasound of fetal anomalies, *St. Louis, 1990, Mosby.)*

from lateral plate mesoderm. The three bones of the pelvis all arise from lateral plate mesoderm, with no known contribution from the somites. Each of the bones of the pelvis, as well as the two developmentally different components of the scapula, is characterized by a different molecular signature. How the bones of the appendages are patterned to connect with their respective girdles is still poorly understood, but studies of mutants suggest that the transcription factors **Pbx-1** and **Pbx-2** play an important upstream role.

A characteristic feature in differentiation of the limb skeleton is the formation of joints. Joint formation occurs by the transverse splitting of precartilaginous rods, rather than by the apposition of two separate skeletal elements. Joint formation is first apparent when transverse strips of highly condensed cells cross a precartilaginous rod (**Fig. 10.26**). Formation of the zone of cell density is induced by **Wnt-14**, which stimulates the formation of **growth differentiation factor-5**, a member of the BMP family, in the region of the future joint. BMP activity, which is strongly involved in cartilage formation, must be excluded from the region of the developing joint. **Noggin**, an antagonist of BMP, plays an important role in joint formation because in its absence, BMP is expressed throughout the region where the joint should form, and the digital rays develop into solid rods of cartilage without joints. The roles of noggin and BMP in joint formation are very similar to those seen in the formation of the sutures between the cranial bones (see p. 177).

Condensation is followed by cell death in the interphalangeal joint regions and **hyaluronin** secretion and matrix changes in the region of the future joint. Then the skeletal elements on either side of the joint form articular cartilage, and a fluid-filled gap is created between them. Additional condensations of mesenchymal cells form the joint capsule, ligaments, and tendons. During later development, muscular activity is required to maintain the integrity of the joint, but early joint development is completely independent of muscular activity. A well-known mutant family, called **brachypodism**, involves a shortening of the limb and the lack of development of certain joints, specifically the interphalangeal joints. There are five major groups of brachydactylies, each of which contains several subtypes.

Musculature

The musculature of the limb is derived from myogenic cells that migrate into the very early limb bud from the ventral part

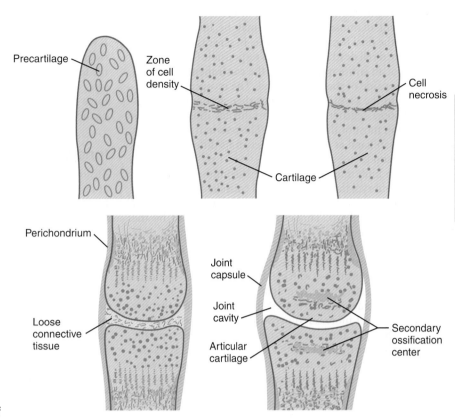

Fig. 10.26 **Sequence of the formation of the joints in the limbs.**

of the dermomyotome of the somite. Each somite in the limb region contributes 30 to 100 migratory precursor cells to the future limb musculature. These cells are stimulated to leave the somite and migrate toward the limb through the stimulus of **scatter factor** (**hepatic growth factor**), which is produced by the proximal cells of the limb-forming area. Before migrating, the premuscle cells in the somite express **c-met**, which is a specific receptor for scatter factor. The premyogenic cells, which are morphologically indistinguishable from the other mesenchymal cells, express Pax-3 and spread throughout the limb bud. In the splotch mutant, which is characterized by the absence of Pax-3 expression, muscle cells do not populate the limb bud. Migrating premuscle cells also express the cell adhesion molecule **N-cadherin**, which is important in correctly distributing them throughout the limb bud mesenchyme. The migrating premuscle cells keep pace with the elongation of the limb bud, although cells expressing characteristic muscle molecules (e.g., MyoD) are not seen in the distal mesenchyme. Some experimental evidence suggests that premyogenic cells are not present in the distal mesenchyme. The reason may be the high concentrations of BMP in the distal mesenchyme, which block the proliferation of myoblasts and can even cause the death of these cells. Actual differentiation of the premuscle cells into muscle within the limb requires signals from the ectoderm, principally Wnt-6. If the ectoderm of the limb bud is removed, cartilage and connective tissue, but not muscle, differentiate.

Shortly after the condensations of the skeletal elements take shape, the myogenic cells themselves begin to coalesce into two common muscle masses: one the precursor of the flexor muscles and the other giving rise to the extensor muscles. The

transcription factor **Tcf-4** is expressed throughout the connective tissue associated with the muscle masses. This is the connective tissue that determines the morphology of the individual muscles as they take shape.

The next stage in muscle formation is the splitting of the **common muscle masses** into anatomically recognizable precursors of the definitive muscles of the limb. Little is known about the mechanisms that guide the splitting of the common muscle masses, although more recent evidence suggests that the pattern of blood vessels defines the future sites of cleavage through the secretion of **platelet-derived growth factor** and its action on the formation of connective tissue sheaths around the forming muscles. The fusion of myoblasts into early myotubes begins to occur during these early stages of muscle development.

Considerable evidence suggests that myogenic precursor cells do not possess intrinsic information guiding their morphogenesis. Rather, the myogenic cells follow the lead of connective tissue cells, which are the bearers and effectors of the morphogenetic information required to form anatomically correct muscles. In experiments in which the somites normally associated with a limb bud are removed and replaced by somites from elsewhere along the body axis, myogenic cells are morphogenetically neutral. Muscle morphogenesis is typically normal even though the muscle fiber precursors are derived from abnormal sources.

A later function of the T-box transcription factors **Tbx-5** and **Tbx-4**, which play earlier important roles in initiating the development of the forelimbs and hindlimbs, respectively (see p. 193), is the regulation of muscle patterning. Mutations of these genes result in abnormal limb muscle patterning.

Depending on the specific muscle, the migration, fusion, or displacement of muscle primordia may be involved in the genesis of the final form of the muscle. In one case, genetically programmed cell death, apoptosis, is responsible for the disappearance of an entire muscle layer (the **contrahentes muscle**) in the flexor side of the human hand. The myogenic cells differentiate to the myotube stage; they then accumulate glycogen and soon degenerate. The contrahentes muscle layer is preserved in most of the great apes. The reason it degenerates in the human hand at such a late stage in its differentiation is not understood.

Although limb muscles assume their definitive form in the very early embryo, they must undergo considerable growth in length and cross-sectional area to keep up with the overall growth of the embryo. This growth is accomplished by the division of the satellite cells (see p. 183) and the fusion of their progeny with the muscle fibers. The added satellite cell nuclei increase the potential of the muscle fiber to produce structural and contractile proteins, which increase the cross-sectional area of each muscle fiber. Accompanying this addition to the nuclear complement of the muscle fibers is their lengthening by the addition of more sarcomeres, usually at the ends of the muscle fibers. The formation of new muscle fibers typically ceases at or shortly after birth. Although the muscles are capable of contracting in the early fetal period, their physiological properties continue to mature until after birth.

Tendons

To function properly, muscles must attach to bones through the formation of **tendons**. A tendon is a band of dense fibrous connective tissue that is attached to the muscle through the **myotendinous junction** and to the bone through the **enthesis**, a complex structure with four zones forming a gradient from type I collagen to fibrocartilage and cartilage and, finally, an actual osseous union with the bone.

Early experiments showed that when the somites adjoining the limb-forming regions were removed, the limbs developed without muscles, but rudimentary tendons did appear, although they later degenerated. These experiments showed that muscle fibers arise from somitic mesoderm, whereas tendons originate from lateral plate mesoderm. Further research has shown that all tendons are not equal. The tendons in limbs, axial structures, and the head require different conditions for their formation.

Overall, three phases are involved in tendon formation: (1) induction by FGFs, (2) early organization through the action of **transforming growth factor-β**, and (3) consolidation and differentiation, which require the expression of **scleraxis** (**Scx**). Tendons of the proximal limb arise from limb mesoderm located just under the lateral ectoderm, where they are induced by FGFs emanating from that ectoderm. Muscle is not needed for their early formation, but interactions with muscle are required for the later differentiation of tendons. The long tendons leading to the digits are more independent from muscle influences during the earlier stages of their formation than are the proximal tendons. Tendons in the head arise from cranial neural crest mesenchyme but, like limb tendons, are independent of muscles in the early stages of their formation. Conversely, the tendons of the axial muscles arise from the **syndetome** compartment of the somites and require an inductive influence from the myotome in order to form.

As the differentiating tendon approaches the developing bone, Scx in the tendon cells stimulates the production of **BMP-4**, which, in turn, stimulates the bony outgrowth or bone ridge on which the tendon attaches. The molecular basis for the formation of the myotendinous junction remains obscure. Once the muscle begins to function and exerts mechanical force through its contractions, final differentiation of the body of the tendon and of the enthesis occurs.

Innervation

Motor axons originating in the spinal cord enter the limb bud at an early stage of development (during the fifth week) and begin to grow into the dorsal and ventral muscle masses before these masses have split up into primordia of individual muscles (**Fig. 10.27**). Tracing studies have shown a high degree of order in the projection of motoneurons into the limb. Neurons located in medial positions in the spinal cord send axons to the ventral muscle mass, whereas neurons located more laterally in the spinal cord supply the dorsal muscle mass. Similarly, a correlation exists between the craniocaudal position of neurons in the cord with the anteroposterior pattern of innervation of limb muscles within the common muscle masses. For example, the rostralmost neurons innervate the most anterior muscle primordia.

Local cues at the base of the limb bud guide the entering pathways of nerve fibers into the limb bud. If a segment of the spinal cord opposite the area of limb bud outgrowth is reversed in the craniocaudal direction, the motoneurons change the direction of their outgrowth and enter the limb bud in their normal positions (**Fig. 10.28**). If larger segments of spinal cord are reversed, and the neurons are at considerable distances from the level of the limb bud, their axons do not find their way to their normal locations in the limb bud. The muscles themselves apparently do not provide specific target cues to the ingrowing axons because if muscle primordia are prevented from forming, the main patterns of innervation in the limb are still normal.

Sensory axons enter the limb bud after the motor axons. Similarly, neural crest cell precursors of Schwann cells lag slightly behind the outgrowth of motor axons into the limb bud. Cells of the neural crest surround motor and sensory nerve fibers to form the coverings of the nerves in the limbs. By the time digits have formed in developing limbs, the basic elements of the gross pattern of innervation in the adult limb have been established.

Vasculature

The earliest vasculature of the limb bud is derived from endothelial cells arising from several segmental branches of the aorta and the cardinal veins and, to some extent, from **angioblasts** (endothelial cell precursors) arising from the somites or endogenous to the limb bud mesoderm. Initially, the limb vasculature consists of a fine capillary network, but soon, some channels are preferentially enlarged, resulting in a large central artery that supplies blood to the limb bud (**Fig. 10.29**). From the central artery, the blood is distributed to the periphery via a mesh of capillaries and then collects into a **marginal sinus**, which is located beneath the AER. Blood in the marginal sinus drains into peripheral venous channels, which carry it away from the limb bud.

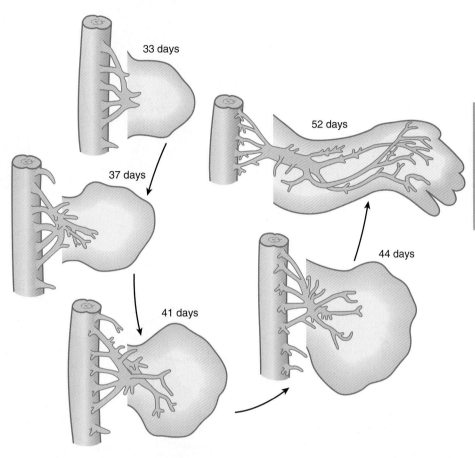

Fig. 10.27 **Development of the nerve pattern of the human upper extremity.** *(Based on Shinohara H and others: Acta Anat 138:265-269, 1990.)*

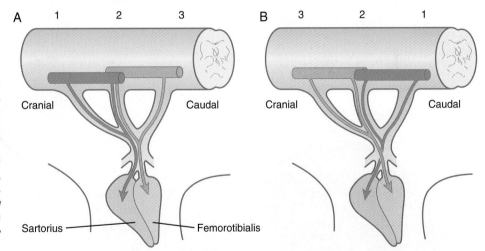

Fig. 10.28 **Pathways taken by axons from specific pools of motoneurons in the spinal cord to limb muscles in the hindlimb of the chick embryo.** **A,** Normal limb. **B,** After reversal of three segments of the embryonic spinal cord, axons originating from the spinal cord pass through abnormal pathways to innervate the muscles that they were originally destined to innervate. *(Adapted from Brown M and others: Essentials of neural development, Cambridge, 1990, Cambridge University Press.)*

Even in the earliest limb bud there is a peripheral avascular zone of mesoderm within about 100 μm of the ectoderm of the limb bud (**Fig. 10.30A**). The avascular region persists until the digits have begun to form. Angioblasts are present in the avascular zone, but they are isolated from the functional capillaries. Experimental studies have shown that the proximity of ectoderm is inhibitory to vasculogenesis in the limb bud mesoderm. If the ectoderm is removed, vascular channels form to the surface of the limb bud mesoderm, and if a piece of ectoderm is placed into the deep limb mesoderm, an avascular zone forms around it (Fig. 10.30B). Degradation products of hyaluronic acid, which is secreted by the ectoderm, seem to be the inhibitory agents.

Just before the skeleton begins to form, avascular zones appear in the areas where the cartilaginous models of the bones will take shape. Neither the stimuli for the disappearance of the blood vessels nor the fate of the endothelial cells that were present in these regions is understood at this time.

The pattern of the main vascular channels changes constantly as the limb develops, probably from the expansion of preferred channels within the capillary network that perfuses the distal part of the developing limb. With the establishment of the digital rays, the apical portion of the marginal sinus breaks up, but the proximal channels of the marginal sinus persist into adulthood as the **basilic** and **cephalic veins** of the arm (see Fig. 10.29C).

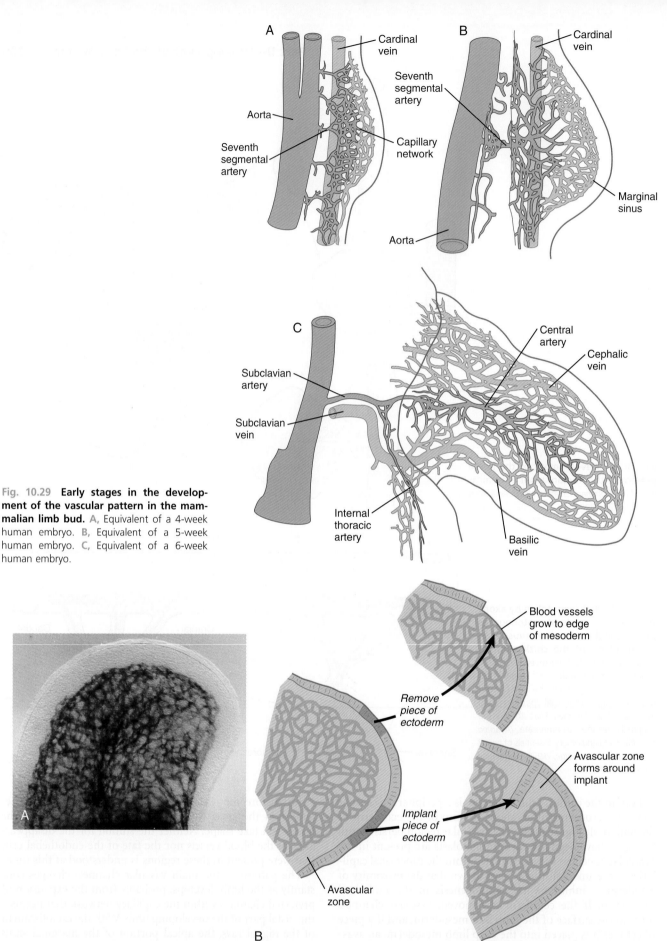

Fig. 10.29 **Early stages in the development of the vascular pattern in the mammalian limb bud.** A, Equivalent of a 4-week human embryo. B, Equivalent of a 5-week human embryo. C, Equivalent of a 6-week human embryo.

Fig. 10.30 A, Photomicrograph of a quail wing bud with ink-injected blood vessels. B, Experiments illustrating the inhibitory effect of limb ectoderm on vascularization of the subjacent mesoderm. *Left,* Normal limb bud with an avascular zone beneath the ectoderm. *Upper right,* After removal of a piece of ectoderm, capillaries grow to the edge of the mesoderm in the region of removal. *Lower right,* A zone of avascularity appears around a piece of implanted ectoderm. (*A, Courtesy of R. Feinberg. Based on Feinberg RN, Noden DM: Anat Rec 231:136-144, 1991.*)

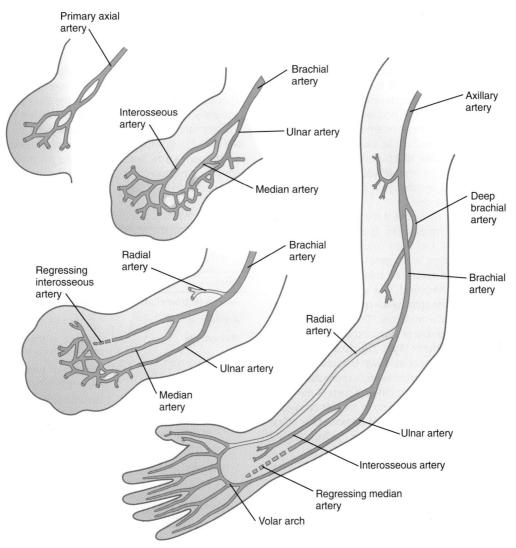

Fig. 10.31 **Formation of arteries in the human arm.**

Similar major changes occur in the arterial channels that course through the developing limb (**Fig. 10.31**). Preferred channels connected to the primary axial artery ultimately take ascendancy, especially in the forearm, thus leaving the original primary axial artery a relatively minor vessel (the **interosseous artery**) in the forearm.

Clinical Correlation 10.1 discusses limb anomalies.

Clinical Vignette

After a normal pregnancy and labor, a 32-year-old woman gives birth to a 7-lb boy who has a duplication of his right foot along the anteroposterior axis. To the left of a single big toe (hallux) are four additional toes arranged in a mirror-symmetric fashion so that the pattern of digits in that foot is 543212345, with 1 being the common hallux. Two older siblings are normal.

How can you explain this malformation on the basis of developmental mechanisms learned in this chapter?

Summary

- Limbs arise from the lateral mesoderm and the overlying ectoderm under the influence of an inductor working through the Tbx transcription factors. The early limb bud is a highly regulative system that can compensate for a variety of surgical disturbances and still form a normal limb. The axes of the limb are fixed in an anteroposterior, dorsoventral, and proximodistal sequence.

- The early limb bud mesoderm stimulates the overlying ectoderm to form an AER that stimulates the outgrowth of the limb through proliferation of the underlying mesodermal cells. FGF-4 and FGF-8 are secreted by the apical ridge and can induce outgrowth in the absence of the ridge. The overall morphogenesis of the limb is determined by properties of the mesoderm, whereas the ectoderm acts in a more permissive sense.

- Cell death is an important mechanism in normal limb development. Regions of programmed cell death include

CLINICAL CORRELATION 10.1
Limb Anomalies

Because they are so obvious, limb anomalies have attracted attention for centuries and have been subject to many systems of classification. Most earlier classification schemes were based on morphology alone, usually from the perspective of surgeons or rehabilitation specialists. Only in more recent decades has it been possible to assign genetic or mechanistic causes to some of the known limb malformations. **Table 10.2** presents a summary of the common types of morphological limb defects. Many of the most common limb malformations are the result of still poorly understood developmental disturbances.

The number of genetic conditions known to underlie limb defects is vast (**Table 10.3**). For example, as of 2010, 310 clinical entities involving polydactyly had been described. Of these, 80 were associated with mutations in 99 genes. In most cases, the means by which the gene mutations are translated into defective development are not well understood. One example of the complexity of limb anomalies is the **split hand–split foot malformation**, sometimes known as ectrodactyly. This malformation is characterized by a reduced number of digits and a wide separation between the anterior and posterior digits (**Fig. 10.32**). Mutations of at least 15 genes have been associated with the split hand–split foot malformation, and this malformation is a component of at least 25 separate syndromes that affect different parts of the body. A common developmental pathway leading to the split hand–split foot malformation is disruption of the middle portion of the apical ectodermal ridge or its functions through various mechanisms late in limb formation.

Some of the most frequently encountered limb anomalies have causes that do not involve classic development mechanisms. Several have mechanical causes. **Intrauterine amputations** by amniotic bands, presumably caused by tears in the amnion, can result in the loss of parts of digits or even hands or feet (see Fig.

8.16). Other deformities, such as **clubfoot** (**talipes equinovarus**) and some causes of congenital dislocations, have been attributed to persistent mechanical pressures of the uterine wall on the fetus, particularly in cases of **oligohydramnios** (see Chapter 7).

A very rare limb deformity is **macromelia** (or **macrodactyly**, see Fig. 8-17), in which a limb or a digit is considerably enlarged over the normal size. Such abnormalities are sometimes associated with neurofibromatosis, and the neural crest may be involved in this defect.

Table 10.2 Common Structural Types of Limb Malformations

Term	Description
Amelia (ectromelia)	Absence of an entire limb
Acheiria, apodia	Absence of hands, feet
Phocomelia	Absence or shortening of proximal limb segments
Hemimelia	Absence of preaxial or postaxial parts of limb
Meromelia	General term for absence of part of a limb
Ectrodactyly	Absence of any number of digits
Polydactyly	Excessive number of digits
Syndactyly	Presence of interdigital webbing
Brachydactyly	Shortened digits
Split hand or foot	Absence of central components of hand or foot

Table 10.3 Some Genetic Conditions Causing Primary Patterning Defects of the Limb

Type of Molecule	Gene	Syndrome	Limb Defect
Transcription factor	GLI3	Greig's	Polydactyly, syndactyly, cephalopolysyndactyly
	GLI3	Pallister-Hall	Posterior polydactyly
	TBX3	Ulnar-mammary	Upper limb deficiencies and posterior duplications
	TBX5	Holt-Oram	Upper limb deficiencies and anterior duplications
	HOXA13	Hand-foot-genital	Brachydactyly
	HOXD13	Synpolydactyly	Syndactyly, insertional polydactyly, brachydactyly
	PAX3	Waardenburg I and III	Syndactyly
	SOX9	Campomelic dysplasia	Bowing of long bones
Signaling protein	CDMP1	Hunter-Thompson	Brachydactyly
	CDMP1	Grebe's	Severe brachydactyly
	SHH	Preaxial polydactyly	Extra digits
	SHH, FBLN1	Polysyndactyly	Interdigital webbing, extra digits
Receptor protein	FGFR1 or FGFR2	Pfeiffer's	Brachydactyly, syndactyly
	FGFR2	Apert's	Syndactyly
	FGFR2	Jackson-Weiss	Syndactyly, brachydactyly
	DACTYLIN, p63	Split hand–split foot, ectrodactyly	Distal syndactyly, fusion
	SALL1	Triphalangeal thumb	Additional segment in thumb

Adapted from Bamshad M and others: *Pediatr Res* 45:291-299, 1999.

CLINICAL CORRELATION 10.1
Limb Anomalies—cont'd

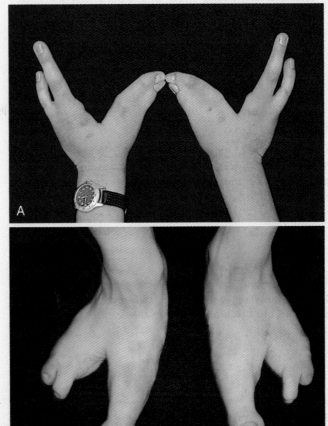

Fig. 10.32　The hands and feet of an individual with split hand–split foot malformation. In this case, the condition was caused by a mutation in *P63*, a tumor suppressor gene. In addition to the limb malformations, this patient was also afflicted with ectodermal dysplasia, characterized by enamel defects of the teeth and thin, frayed hair shafts. *(Courtesy of Piranit N. Kantaputra, Chiang Mai, Thailand.)*

the axillary region and the interdigital spaces. In the absence of interdigital cell death, syndactyly results.

■ A ZPA located in the posterior mesoderm acts as a biological signaler and plays an important role in the anteroposterior organization of the limb by releasing shh. Wnt-7a, which is released by dorsal ectoderm, is the organizer of dorsoventrality in the limb bud. According to the concept of positional information, cells in the developing limb are exposed to positional cues (e.g., the signal from the ZPA) that allow them to determine their relative position within the limb bud. The cells process this information and differentiate accordingly. Proximodistal control of morphogenesis may reside in the progress zone, a narrow band of mesoderm beneath the AER.

■ Digits 2 through 5 form on the basis of their increasing exposure to shh (time and concentration), whereas the formation of the first digit is independent of shh.

■ Retinoic acid exerts a profound effect on limb morphogenesis and can cause the formation of a supernumerary limb if applied to the anterior border of the limb bud, but its exact role in limb development remains obscure. Expression of a variety of homeobox-containing genes follows well-defined patterns in the normally developing limb. Some patterns of gene expression are profoundly altered in limbs treated with retinoic acid.

■ The skeleton of the limb arises from lateral plate mesoderm. The ectoderm of the limb bud inhibits cartilage formation in the mesoderm cells immediately beneath it. This could explain the reason the skeleton of the vertebrate limb forms in a central position.

■ Limb muscles arise from cells derived from somitic mesoderm. These cells express Pax-3 during their migration into the limb bud. Myogenic cells first form dorsal and ventral common muscle masses, which later split into primordia of individual muscles. Morphogenetic control of muscles resides in the associated connective tissue, rather than in the muscle cells themselves. Later stages in muscle development may involve cell death, the fusion of muscle primordia, and the displacement of muscle primordia to other areas.

■ Tendons arise from lateral plate mesoderm and initially form independently of the muscles. Later differentiation of tendons requires interactions with contracting muscles.

■ Nerves grow into the developing limb bud and become associated with the common muscle masses as they split into individual muscles. Local cues are important in guiding growing axons into the developing limb.

- The vasculature of the limb bud arises from cells budding off the aorta and cardinal veins and from endogenous mesodermal cells. The early vascular pattern consists of a central artery, which drains into a peripheral marginal sinus and then into peripheral venous channels. Blood vessels do not form beneath the ectoderm or in the central cartilage-forming regions.
- Limb anomalies can form as the result of genetic mutations, drug effects, disturbed tissue interactions, and purely mechanical effects.

Review Questions

1. Which of the following molecules plays an important role in the determination of the dorsoventral axis of the developing limb?
A. Msx-1
B. Wnt-7a
C. Hoxd-13
D. Pax-1
E. FGF-8

2. What molecule is associated with myogenic cells migrating into the limb bud from the somites?
A. shh
B. BMP-7
C. FGF-4
D. Pax-3
E. En-1

3. What is the principal function of the AER?
A. Stimulating outgrowth of the limb bud
B. Setting up the anteroposterior axis of the limb bud
C. Determining the specific characteristics of the ectodermal appendages of the limb
D. Determining the pattern of neural ingrowth into the limb
E. Attracting the subcutaneous plexus of capillaries in the limb bud

4. In the developing limb the sonic hedgehog (shh) gene product is produced in the:
A. Progress zone
B. Region of interdigital cell death
C. ZPA
D. AER
E. Common muscle mass

5. The connective tissue of the limb arises from the:
A. Paraxial mesoderm
B. Neural crest
C. Intermediate mesoderm
D. Somitic mesoderm
E. Lateral mesoderm

6. The formation of clubfoot (talipes equinovarus) is associated with:
A. A misplaced ZPA
B. Defective cellular migration from somites
C. Thalidomide
D. Oligohydramnios
E. A neural crest defect

7. An infant whose mother underwent chorionic villus sampling during pregnancy was born with the tips of two digits missing. What is a possible cause?

8. A woman who underwent amniocentesis during pregnancy gave birth to a child with a duplicated thumb. What is a possible cause?

9. If the somites close to a limb-forming region are experimentally removed, the limbs form without muscles. Why?

10. A child is born with webbed fingers (syndactyly). What is the reason for this anomaly?

References

Ahn S, Joyner AL: Dynamic changes in the response of cells to positive hedgehog signaling during mouse limb patterning, *Cell* 118:505-516, 2004.

Basel D, Kilpatrick MW, Tsipouras P: The expanding panorama of split hand foot malformation, *Am J Med Genet A* 140A:1359-1365, 2006.

Bastida MF, Ros MA: How do we get a perfect complement of digits? *Curr Opin Genet Dev* 18:374-380, 2008.

Biesecker LG: Polydactyly: how many disorders and how many genes? 2010 update, *Dev Dyn* 240:931-942, 2011.

Brunet LJ and others: Noggin, cartilage morphogenesis, and joint formation in the mammalian skeleton, *Science* 280:1455-1457, 1998.

Butterfield NC, McGlinn E, Wicking C: The molecular regulation of vertebrate limb patterning, *Curr Top Dev Biol* 90:319-341, 2010.

Capellini TD, Zappavigna V, Selleri L: Pbx homeodomain proteins: TALEnted regulators of limb patterning and outgrowth, *Dev Dyn* 240:1063-1086, 2011.

Čihák R: Ontogenesis of the skeleton and intrinsic muscles of the human hand and foot, *Adv Anat Embryol Cell Biol* 46:1-194, 1972.

Cooper KL and others: Initiation of proximal-distal patterning in the vertebrate limb by signals and growth, *Science* 332:1083-1086, 2011.

DeLaurier A, Schweitzer R, Logan M: *Pitx 1* determines the morphology of muscle, tendon, and bones of the hindlimb, *Dev Biol* 299:22-34, 2006.

Duboc V, Logan MPO: Regulation of limb bud initiation and limb-type morphology, *Dev Dyn* 240:1017-1027, 2011.

Duijf PHG, van Bokhoven H, Brunner HG: Pathogenesis of split-hand/split-foot malformation, *Hum Mol Genet* 12:R51-R60, 2003.

Dylevsky I: Connective tissue of the hand and foot, *Acta Univ Carol Med Monogr* 127:1-195, 1988.

Dylevsky I: Growth of the human embryonic hand, *Acta Univ Carol Med Monogr* 114:1-139, 1986.

Feinberg RN, Noden DM: Experimental analysis of blood vessel development in the avian wing bud, *Anat Rec* 231:136-144, 1991.

Fernández-Terán MA, Hinchliffe JR, Ros MA: Birth and death of cells in limb development: a mapping study, *Dev Dyn* 235:2521-2537, 2006.

Geetha-Loganathan P and others: Ectodermal Wnt-6 promotes Myf-5 dependent avian limb myogenesis, *Dev Biol* 288:221-233, 2005.

Goodman FR: Limb malformations and the human *Hox* genes, *Am J Med Genet* 112:256-265, 2002.

Harfe BD: Keeping up with the zone of polarizing activity: new roles for an old signaling center, *Dev Dyn* 240:915-919, 2011.

Harrison RG: On relations of symmetry in transplanted limbs, *J Exp Zool* 32:1-136, 1921.

Hasson P: "Soft" tissue patterning: muscles and tendons of the limb take form, *Dev Dyn* 240:1100-1107, 2011.

Kantraputra PN, Matangkasombut O, Sripathomsawat W: Split hand-split foot-ectodermal dysplasia and amelogenesis imperfecta with *TP63* mutation, *Am J Med Genet A* 158A:188-192, 2011.

King M and others: T-genes and limb bud development, *Am J Med Genet A* 140A:1407-1413, 2006.

Laufer E and others: Expression of radical fringe in limb-bud ectoderm regulates apical ectodermal ridge formation, *Nature* 386:366-373, 1997.

Minguillon C, Del Buono J, Logan MP: *Tbx5* and *Tbx4* are not sufficient to determine limb-specific morphologies but have common roles in initiating limb outgrowth, *Dev Cell* 8:75-84, 2005.

Mrázková O: Blood vessel ontogeny in upper extremity of man as related to developing muscles, *Acta Univ Carol Med Monogr* 115:1-114, 1986.

Murchison ND and others: Regulation of tendon differentiation by scleraxis distinguishes force-transmitting tendons from muscle-anchoring tendons, *Development* 134:2697-2708, 2007.

Naiche LA, Papaioannou VE: *Tbx4* is not required for hindlimb identity or post-bud hindlimb outgrowth, *Development* 134:93-103, 2007.

Nissim A and others: Regulation of *Gremlin* expression in the posterior limb bud, *Dev Biol* 299:12-21, 2006.

Provot S, Schipani E: Molecular mechanisms of endochondral bone development, *Biochem Biophys Res Commun* 328:658-665, 2005.

Robert B, Lallemand Y: Anteroposterior patterning in the limb and digit specification: contribution of mouse genetics, *Dev Dyn* 235:2337-2352, 2006.

Rodriguez-Guzman M and others: Tendon-muscle crosstalk controls muscle bellies morphogenesis, which is mediated by cell death and retinoic acid signaling, *Dev Biol* 302:267-280, 2007.

Rodriguez-Niedenfuehr M and others: Development of the arterial pattern in the upper limb of staged human embryos: normal development and anatomical variations, *J Anat* 199:407-417, 2001.

Roselló-Díez A, Ros MA, Torres M: Diffusible signals, not autonomous mechanisms, determine the main proximodistal limb subdivision, *Science* 332:1086-1088, 2011.

Rubin L, Saunders JW: Ectodermal-mesodermal interactions in the growth of limb buds in the chick embryo: constancy and temporary limits of the ectodermal induction, *Dev Biol* 28:94-112, 1972.

Saito D and others: Level-specific role of paraxial mesoderm in regulation of *Tbx5/Tbx4* expression and limb initiation, *Dev Biol* 292:79-89, 2006.

Saunders JW: The proximodistal sequence of origin on the parts of the chick wing and the role of the ectoderm, *J Exp Zool* 108:363-403, 1948.

Saunders JW, Gasseling MT: Ectodermal-mesenchymal interactions in the origin of limb symmetry. In Fleischmajer R, Billingham RE, eds: *Epithelial-mesenchymal interactions*, Baltimore, 1968, Williams & Wilkins, pp 78-97.

Scherz PJ and others: The limb bud Shh-FGF feedback loop is terminated by expansion of former ZPA cells, *Science* 305:396-399, 2004.

Schweitzer R, Zelzer E, Volk T: Connecting muscles to tendons: tendons and musculoskeletal development in flies and vertebrates, *Development* 137:2807-2817, 2010.

Shinohara H and others: Development of innervation patterns in the upper limb of staged human embryos, *Acta Anat* 138:265-269, 1990.

Stricker S, Mundlos S: Mechanisms of digit formation: human malformation syndromes tell the story, *Dev Dyn* 240:990-1004, 2011.

Swett FH: Determination of limb-axes, *Q Rev Biol* 12:322-339, 1937.

Tabin C, Wolpert L: Rethinking the proximodistal axis of the vertebrate limb in the molecular era, *Genes Dev* 21:1433-1442, 2007.

Talamillo A and others: The developing limb and the control of the number of digits, *Clin Genet* 67:143-153, 2005.

Tarchini B, Duboule D: Control of *Hoxd* genes' collinearity during early limb development, *Dev Cell* 10:93-103, 2006.

Theil T and others: *Gli* genes and limb development, *Cell Tissue Res* 296:75-83, 1999.

Tickle C: Making digit patterns in the vertebrate limb, *Nat Rev Mol Cell Biol* 7:45-53, 2006.

Toser S and others: Involvement of vessels and PDGFB in muscle splitting during chick limb development, *Development* 134:2579-2591, 2007.

Towers M, Tickle C: Growing models of vertebrate limb development, *Development* 136:179-190, 2009.

Wang G, Scott SA: Independent development of sensory and motor innervation patterns in embryonic chick hindlimbs, *Dev Biol* 208:324-336, 1999.

Wilkie AOM and others: FGFs, their receptors, and human limb malformations: clinical and molecular correlations, *Am J Med Genet* 112:266-278, 2002.

Zákány J, Duboule D: *Hox* genes in digit development and evolution, *Cell Tissue Res* 296:19-25, 1999.

Zákány J, Kmita M, Duboule D: A dual role for *Hox* genes in limb anterior-posterior asymmetry, *Science* 304:1669-1672, 2004.

Zeller R, López-Rios J, Zuniga A: Vertebrate limb bud development: moving towards integrative analysis of organogenesis, *Nat Rev Genet* 10:845-858, 2009.

Zeller R: The temporal dynamics of vertebrate limb development, teratogenesis and evolution, *Curr Opin Genet Dev* 20:384-390, 2010.

Zwilling E: Limb morphogenesis, *Adv Morphog* 1:301-330, 1961.

Chapter **11**

Nervous System

Many fundamental developmental processes are involved in the formation of the nervous system. Some of these dominate certain stages of embryogenesis; others occur only at limited times and in restricted locations. The major processes are as follows:

1. **Induction,** including primary induction of the nervous system by the underlying notochord and secondary inductions driven by neural tissues themselves
2. **Proliferation,** first as a response of the neuroectodermal cells to primary induction and later to build up critical numbers of cells for virtually all aspects of morphogenesis of the nervous system
3. **Pattern formation,** in which cells respond to genetic or environmental cues in forming the fundamental subdivisions of the nervous system
4. **Determination** of the identity of specific types of neuronal or glial cells
5. **Intercellular communication** and the adhesion of like cells
6. **Cell migration,** of which a variety of distinct patterns is found in the nervous system
7. **Cellular differentiation** of neurons and glial cells
8. Formation of specific connections or **synapses** between cells
9. **Stabilization** or **elimination** of specific interneuronal connections, sometimes associated with massive episodes of cell death of unconnected neurons
10. Progressive **development of integrated patterns** of neuronal function, which results in coordinated reflex movements

Establishment of the Nervous System

As described in Chapter 5, primary induction of the nervous system results in the formation of a thickened ectodermal **neural plate** overlying the notochord. Much of the dorsal ectoderm in gastrulating embryos produces the signaling protein **bone morphogenetic protein-4 (BMP-4),** which inhibits the dorsal ectoderm from forming neural tissue. Instead of sending positive signals to the overlying ectoderm, the neural inducers, **noggin** and **chordin,** block the inhibitory influence of BMP-4 and allow the dorsal ectoderm to form neural tissue (the neural plate [see Fig. 5.8]).

Shortly after neural induction, further signals from the notochord and head organizing regions (prechordal plate and anterior visceral endoderm) result in the expression of the transcription factor **Otx-2** in the forebrain-midbrain region and **Gbx2** in the hindbrain region. The expression boundary between these two transcription factors forms the **isthmic organizer.** The signaling molecules **fibroblast growth factor-8 (FGF-8)** and **Wnt-1** diffuse from this boundary and are instrumental in setting up the pattern for forming the midbrain and hindbrain. Then, under the influence of specific combinations of *Hox* genes and other transcription factors, the hindbrain undergoes a highly regular segmentation into **rhombomeres,** which presage the overall organization of the entire facial and cervical region (see Fig. 11.12).

The neural tube, which is the morphological manifestation of the earliest stages in establishing the nervous system, is a prominent structure. In a human, it dominates the cephalic end of the embryo (see Fig. 6.1). This chapter describes how the early neural tube develops into the major morphological and functional components of the mature nervous system.

Early Shaping of the Nervous System

Closure of the neural tube first occurs in the region where the earliest somites appear; closure spreads cranially and caudally (see Fig. 6.1). The unfused regions of the neural tube are known as the **cranial** and **caudal neuropores.** Even before the closure of the neuropores (24 days' gestation for the cranial neuropore and 26 days' gestation for the caudal neuropore), some fundamental subdivisions in the early nervous system have become manifest. The future spinal cord and brain are recognizable, and within the brain the forebrain (**prosencephalon**), midbrain (**mesencephalon**), and hindbrain (**rhombencephalon**) can be distinguished (**Fig. 11.1A**).

A prominent force in shaping the early nervous system is the overall bending of the cephalic end of the embryo into a "C" shape. Associated with this bending is the appearance at the end of the third week of a prominent **cephalic flexure** of the brain at the level of the mesencephalon (see Fig. 11.1A). Soon the brain almost doubles back on itself at the cephalic flexure. At the beginning of the fifth week, a **cervical flexure** appears at the boundary between the hindbrain and the spinal cord.

By the fifth week, the original three-part brain has become subdivided further into five parts (**Fig. 11.2**; see Fig. 11.1B). The prosencephalon gives rise to the **telencephalon** (endbrain), with prominent lateral outpocketings that ultimately

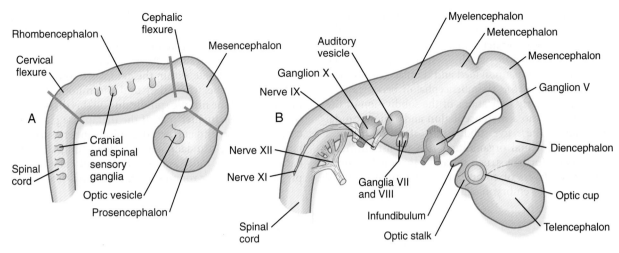

Fig. 11.1   Basic anatomy of the three-part (A) and five-part (B) human brain.

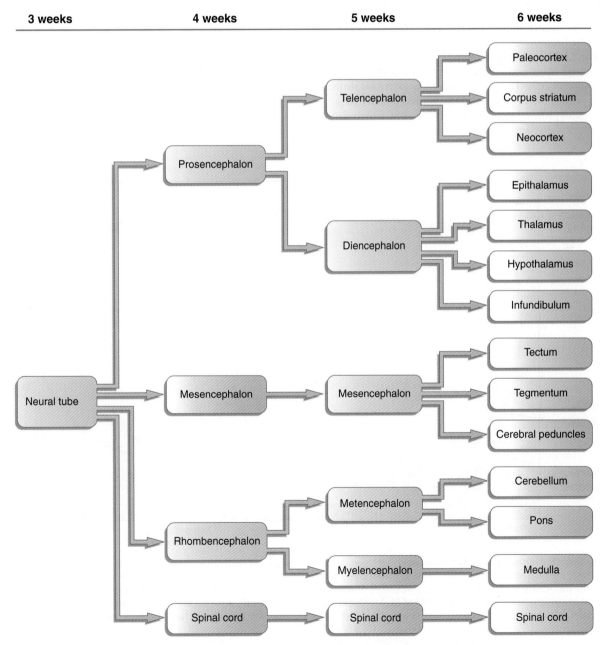

Fig. 11.2   **Increasing levels of complexity of the developing human brain.**

form the cerebral hemispheres, and a more caudal **diencephalon**. The diencephalon is readily recognizable because of the prominent lateral **optic vesicles** that extend from its lateral walls. The mesencephalon, which is sharply bent by the cephalic flexure, remains undivided and tubular in its overall structure. The roof of the rhombencephalon becomes very thin, and there are early indications of the subdivision of the rhombencephalon into a **metencephalon** and a more caudal **myelencephalon**. These five subdivisions of the early brain represent a fundamental organization that persists through adulthood. Many further structural and functional components give added layers of complexity to the brain over the next several weeks of embryonic life.

Histogenesis Within the Central Nervous System

Proliferation Within the Neural Tube

Shortly after induction, the thickening neural plate and early neural tube become organized into a pseudostratified epithelium (**Fig. 11.3**). In this type of epithelium, the nuclei appear to be located in several separate layers within the elongated neuroepithelial cells. The nuclei undertake extensive shifts of position within the cytoplasm of the neuroepithelial cells.

The neuroepithelial cells are characterized by a high degree of mitotic activity, and the position of the nuclei within the

neural tube and their stage in the mitotic cycle are closely correlated (**Fig. 11.4**). DNA synthesis occurs in nuclei located near the **external limiting membrane** (the basal lamina surrounding the neural tube). As these nuclei prepare to go into mitosis, they migrate within the cytoplasm toward the lumen of the neural tube, where they undergo mitotic division. The orientation of the mitotic spindle during this division predicts the fate of the daughter cells. If the metaphase plate (plane of cleavage) is perpendicular to the apical (inner) surface of the neural tube, the two daughter cells slowly migrate in tandem back toward the outer side of the neural tube, where they prepare for another round of DNA synthesis (see Fig. 11.4).

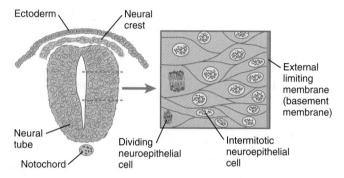

Fig. 11.3 *Left,* Cross section through the early neural tube. *Right,* Higher magnification of a segment of the wall of the neural tube.

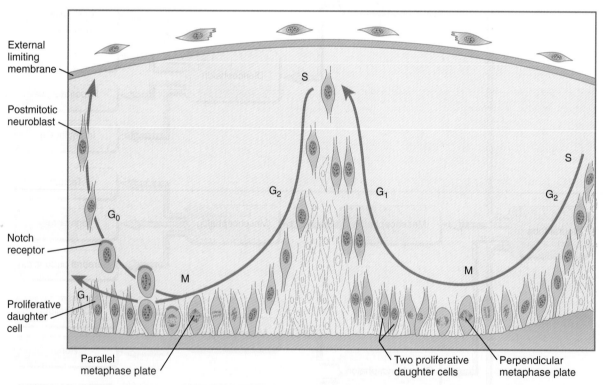

Fig. 11.4 **Mitotic events in the early neural tube.** In the pseudostratified epithelial cells that constitute the early neural tube, nuclei that synthesize DNA (S phase) are located near the external limiting membrane. These nuclei move toward the inner margin of the neural tube, where mitosis (M) occurs. If the metaphase plate is perpendicular to the inner margin, the two daughter cells remain in a proliferative state and migrate back toward the external limiting membrane for another round of DNA synthesis. If in the next mitosis the metaphase plate is oriented parallel to the inner margin, one daughter cell remains in the proliferative state. The other daughter cell, which expresses Notch, leaves the mitotic cycle to become a neuroblast.

In contrast, if the plane of cleavage is parallel to the inner surface of the neural tube, the daughter cells undergo dramatically different fates. The daughter cell that is closer to the inner surface migrates away very slowly and remains a proliferative progenitor cell that is capable of mitosis. The daughter cell that is closer to the basal surface (external limiting membrane) inherits a high concentration of the Notch receptor on its surface and quickly moves away from the apical surface as a **postmitotic neuroblast** (see Fig. 11.4). The neuroblasts, cellular precursors of neurons, begin to produce cell processes that ultimately become axons and dendrites.

Cell Lineages in Histogenesis of the Central Nervous System

The origins of most cells found in the mature central nervous system can be traced to **multipotential stem cells** within the early neuroepithelium (**Fig. 11.5**). These cells undergo numerous mitotic divisions before maturing into **bipotential progenitor cells**, which give rise to either neuronal or glial progenitor cells. Activation of the proneural genes *neurogenin 1* and *neurogenin 2* promotes the differentiation of neurons from the bipolar progenitor cells. Glial cells differentiate under the influence of other stimuli. This developmental bifurcation is accompanied by a change in gene expression. Multipotential

stem cells express an intermediate filament protein called **nestin**. Nestin is downregulated as descendants of bipolar progenitor cells separate into neuronal progenitor cells, which express **neurofilament protein**, and glial progenitor cells, which express **glial fibrillary acidic protein**.

The **neuronal progenitor cells** give rise to a series of neuroblasts. The earliest **bipolar neuroblasts** possess two slender cytoplasmic processes that contact the external limiting membrane and the central luminal border of the neural tube. By retracting the inner process, a bipolar neuroblast loses contact with the inner luminal border in the process of becoming a **unipolar neuroblast**. The unipolar neuroblasts accumulate large masses of rough endoplasmic reticulum (**Nissl substance**) in their cytoplasm and begin to send out several cytoplasmic processes. At this point, they are known as **multipolar neuroblasts**. Their principal developmental activities are to send out axonal and dendritic processes and to make connections with other neurons or end organs.

The other major lineage stemming from the bipotential progenitor cells is the glial line. **Glial progenitor cells** continue to undergo mitosis, and their progeny split into several lines. One, the **O-2A progenitor cell** (see Fig. 11.5), is a precursor to two lines of glial cells that ultimately form the **oligodendrocytes** and **type 2 astrocytes**. Another glial lineage gives rise to **type 1 astrocytes**. Human oligodendrocytes arise from progenitor cells located in the ventral ventricular zone

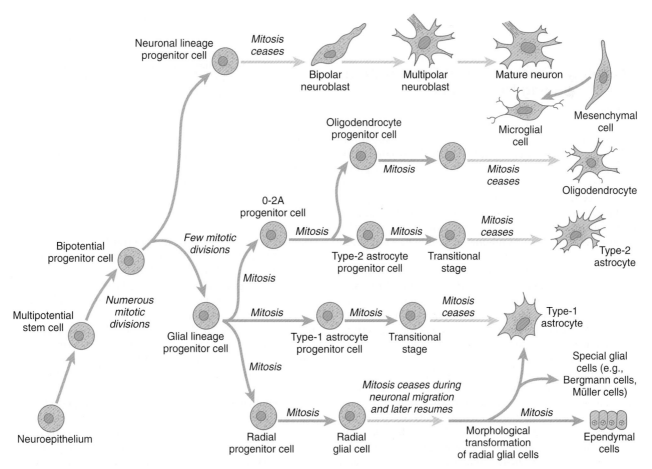

Fig. 11.5 **Cell lineages in the developing central nervous system.** *(Based on Cameron R, Rakic P: Glia 4:124-127, 1991.)*

(see MN in Fig. 11.10) alongside the floor plate. From there, they spread throughout the brain and spinal cord and ultimately form the myelin coverings around neuronal processes in the white matter. The formation of oligodendrocyte precursors depends on an inductive signal from the notochord (sonic hedgehog [shh]). If the notochord is transplanted alongside dorsal neural tube, oligodendrocyte precursors differentiate there, thus showing that cells with the potential to form oligodendrocytes reside in that area, but normally do not develop because of the lack of an adequate inductive signal.

The third glial lineage has a more complex history. Radial progenitor cells give rise to **radial glial cells**, which act as guidewires in the brain for the migration of young neurons (see Fig. 11.23). When the neurons are migrating along the radial glial cells during midpregnancy, they inhibit the proliferation of the radial glial cells. After neuronal cell migration, the radial glial cells, now free from the inhibitory influence of the neurons, reenter the mitotic cycle. Their progeny can transform into several cell types. Some can seemingly cross lineage lines and differentiate into type 1 astrocytes (see Fig. 11.5). Other progeny differentiate into various specialized glial cell types, **ependymal cells**, and even adult neural stem cells. According to some authors, the remaining neuroepithelial cells represent another source of ependymal cells.

Not all cells of the central nervous system originate in the neuroepithelium. **Microglial cells**, which serve a phagocytic function after damage to the brain, are immigrant cells derived from primitive myeloid precursors (macrophages). Microglia are not found in the developing brain until the brain is penetrated by blood vessels.

Fundamental Cross-Sectional Organization of the Developing Neural Tube

The developing spinal cord is a useful prototype for studying the overall structural and functional features of the central nervous system because it preserves its fundamental organization through much of development. With the beginning of cellular differentiation in the neural tube, the neuroepithelium thickens and appears layered. The layer of cells closest to the lumen (**central canal**) of the neural tube remains epithelial and is called the **ventricular zone** (the **ependymal zone** in older literature). This zone, which still contains mitotic cells, ultimately becomes the **ependyma**, a columnar epithelium that lines the ventricular system and central canal of the central nervous system (**Fig. 11.6**). Farther from the ventricular zone is the **intermediate** (formerly called **mantle**) **zone**, which contains the cell bodies of the differentiating postmitotic neuroblasts. As the neuroblasts continue to produce axonal and dendritic processes, the processes form a peripheral **marginal zone** that contains neuronal processes, but not neuronal cell bodies.

As the spinal cord matures, the intermediate zone becomes the gray matter, in which the cell bodies of the neurons are located. The marginal zone is called the **white matter** because of the color imparted by the numerous tracts of myelinated nerve fibers in that layer (see Fig. 11.6). During development, the proliferating progenitor cell populations in the ventricular zone become largely exhausted, but it is now known that a subpopulation persists into adulthood as neural stem cells.

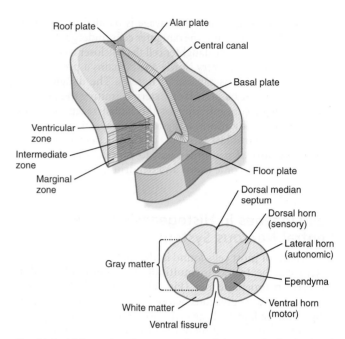

Fig. 11.6 Major regions in cross sections of the neural tube *(top)* and spinal cord *(bottom)*.

The remaining cells differentiate into the epithelium of the ependymal layer.

When the basic layers in the spinal cord are established, several important topographical features can be recognized in cross sections of the spinal cord. A **sulcus limitans** within the central canal divides the spinal cord into a dorsal **alar plate** and a ventral **basal plate** on each side of the central canal. The right and left alar plates are connected dorsally over the central canal by a thin **roof plate**, and the two basal plates are connected ventrally by a **floor plate**.

The basal plate represents the motor component of the spinal cord. Axons arising from neurons located in the **ventral horn** of the gray matter exit the spinal cord as **ventral motor roots** of the spinal nerves (see Fig. 11.15). The gray matter of the alar plate, called the **dorsal horn**, is associated with sensory functions. Sensory axons from the spinal ganglia (neural crest derivatives) enter the spinal cord as dorsal roots and synapse with neurons in the dorsal horn. A small projection of gray matter between the dorsal and ventral horns at spinal levels T1 to L2 contains cell bodies of autonomic neurons. This projection is called the **lateral horn** or sometimes the **intermediolateral gray column** (see Fig. 11.6).

The floor plate is far more than an anatomical connection between the right and left basal plates. Cells of the future floor plate are the first to differentiate in the neural plate after primary induction of the nervous system. Experimental work has shown a specific inductive influence of the notochord on the neuroepithelial cells that overly it. If an extra notochord is grafted along the lateral surface of the neural tube, the neuroepithelial cells closest to it acquire the properties of floor plate cells (**Fig. 11.7**). Conversely, if a segment of normal notochord is removed, the neuroepithelial cells overlying it do not acquire the properties of floor plate cells. Through its action on the floor plate, the notochord also exerts a profound effect on the organization of the dorsal and ventral roots that enter

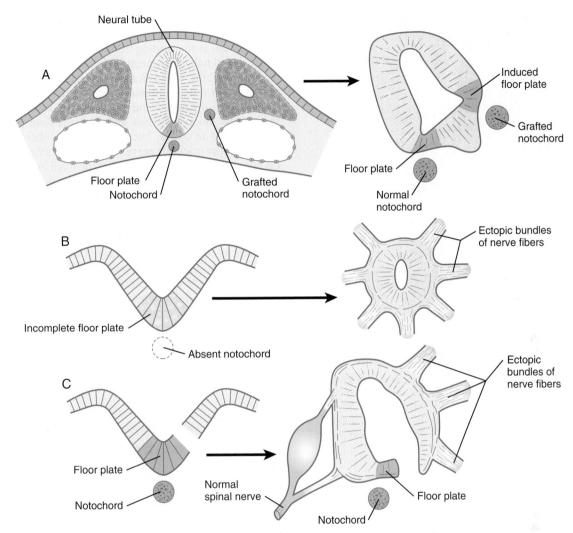

Fig. 11.7 **Experiments illustrating the influence of the notochord on development of the floor plate and exit sites of nerves from the spinal cord.** A, Grafting an extra notochord near the neural tube induces a secondary floor plate. B, In the absence of a notochord, a very incomplete floor plate forms, and nerve fibers exit from multiple sites around the spinal cord. C, Slitting the neural plate on one side of the floor plate removes the wall of the neural tube from the influence of the notochord, allowing the disorganized exit of nerve fibers from that part of the spinal cord. *(Adapted from Hirano S, Fuse S, Sohal GS: Science 251:310-313, 1991.)*

and leave the spinal cord. If the notochord is absent, the neural tube closes, but recognizable dorsal and ventral roots are absent. Numerous ectopic nerve fibers appear in their place (**Fig. 11.8**). If the future floor plate is split, the side of the neural tube on which the notochord is located develops normal dorsal and ventral roots, whereas the side lacking these structures gives off ectopic nerves (see Fig. 11.7C).

A molecular basis for cross-sectional pattern formation within the early neural plate and neural tube has been identified (**Fig. 11.9**). The homeobox-containing transcription factors, **Pax-3**, **Pax-7**, **Msx-1**, and **Msx-2**, are expressed throughout the early neural plate. Before the neural plate has folded over to become the neural tube, the notochord, which is adherent to the midline neural plate at this stage, releases **shh**. Local hedgehog signaling stimulates the neural plate cells directly above the notochord to transform into the **floor plate**. One of the first stages of this transformation is the repression of Pax-3 and Pax-7 expression, which allows the neuroectodermal cells near the midline of the neural plate to adopt a ventral fate (i.e., floor plate or basal plate). Cells of the floor plate itself then become sites of production of shh.

Development of the overall cross-sectional organization of the neural tube involves not only a ventralizing influence from the notochord, but also an opposing dorsalizing influence from the epidermal ectoderm adjacent to the developing neural tube. In the lateral regions of the neural plate (future dorsal region of the neural tube), BMP-4 and BMP-7, expressed by non-neural ectoderm at the ectodermal–lateral neural plate junction, exert a dorsalizing inductive effect on the neuroectodermal cells that results in the formation of the **roof plate**, which takes shape soon after the last neural crest cells have emigrated from the neural tube. BMP within the roof plate acts as a patterning signal and induces the further dorsalizing molecules Pax-3, Pax-7, Msx-1, and Msx-2 (see Fig. 11.9). Dorsal Wnt signaling promotes the proliferation of neural progenitor cells and also works with BMPs as an overall dorsalizing influence in the dorsoventral patterning of neurons. After closure of the neural tube, signals from the roof plate

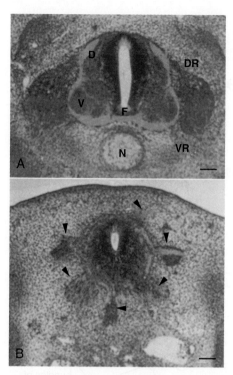

Fig. 11.8   A, Photomicrograph of a normal embryonic quail spinal cord. **B,** In an experiment in which the notochord was absent, the spinal cord is disorganized and has multiple exit sites of nerve fibers (see Fig. 11.7B). *Arrowheads* indicate ectopic spinal nerves. D, dorsal; DR, dorsal root; F, floor plate; N, notochord; V, ventral; VR, ventral root. *(From Hirano S, Fuse S, Sohal GS:* Science *251:310-313, 1991.)*

induce a series of six early and two late generated dorsal interneuronal types in a manner reminiscent of the better defined specification of ventral interneurons (see later).

While the broader regions of the cross section of the spinal cord are being set in place, a tightly controlled molecular grid forms the basis for specification of the major types of neurons found in the ventral part of the spinal cord. Within the basal plate is an array of five types of neurons—**motoneurons** and four types of **interneurons**—that are arranged in a well-defined dorsoventral pattern. These classes of neurons are specified by specific combinations of homeodomain transcription factors, whose pattern of expression is set by a gradient of shh emanating from the floor plate as modulated by the activating and repressive properties of the Gl-1 to Gl-3 proteins (**Fig. 11.10**). Some of these transcription factors (class I) are repressed at various dorsoventral levels by the shh gradient, whereas others (class II) are induced by shh (see Fig. 11.10). The net result is that a different combination of the transcription factors at each dorsoventral level specifies each of the five types of neurons, which are characterized by their own unique molecular signature. One in particular, **islet-1,** is characteristic of motoneurons. Shortly after the production of motoneurons has ceased, a shift in regulatory factors stimulates the production of **glial progenitor cells** from the ventral neuroepithelium. This leads to the formation of the oligodendrocytes that become closely associated with the neurons.

In addition to inducing motoneurons, the floor plate plays other roles in the developing nervous system. Many groups of neuronal processes cross from one side of the central nervous system to the other through the floor plate as **commissural axons.** These axons, leading from neuronal cell bodies located in the dorsal half of the neural tube, are attracted to the floor plate by specific molecules produced in that region (e.g., **netrin 1**). In mutant animals lacking netrin 1, commissural axons are disorganized and do not cross to the other side through the floor plate. Not only does the floor plate attract certain types of axons, but it also repels others. A specific example is the trochlear nerve (cranial nerve IV), whose axons do not cross to the other side from their cell bodies of origin.

Crossing the midline and proper positioning of axons after crossing are largely controlled by the **Slit-Robo** system. **Slit** proteins are axonal repellents, which are located in the ventral midline of the floor plate. **Robo** proteins (1, 2, and 3) are receptors for the Slit ligands. Robo-1 and Robo-2 combine with Slit to repel axonal extension at the midline. For commissural axons that cross the midline, Robo-3 is expressed as they approach the midline. Robo-3 interferes with the repulsive activity of the Robo-1, Robo-2/slit combination and thereby allows the commissural axons to cross the midline. Once these axons have crossed the floor plate, the expression of Robo-3 is downregulated, and the remaining Robo-1, Robo-2/Slit combinations repel the axons away from the midline that they just crossed.

Craniocaudal Pattern Formation and Segmentation

Through neural induction, the early central nervous system becomes organized into broad regions that develop cranial, middle, and caudal characteristics. This is soon followed by the appearance of the morphological subdivisions outlined in Figure 11.2. At an even finer level, segments called **rhombomeres** appear in the region of the hindbrain (see Fig. 6.3), and a less distinct series of subdivisions called **prosomeres** appears in the forebrain.

Patterning in the Hindbrain and Spinal Cord Regions

The rhombomeres (**Fig. 11.11**), introduced in Chapter 6, are the morphological reflection of a highly segmentally ordered pattern of expression of a variety of developmentally prominent transcription factors (**Fig. 11.12**). The establishment of the isthmic organizer and the pathways that set up this pattern are discussed in Chapter 6 (see pp. 95-96).

The correspondence between the rhombomeres of the developing hindbrain and other structures of the cranial and pharyngeal arch region is remarkable (see Chapter 14). The cranial nerves, which have a highly ordered pattern by which they supply structures derived from the pharyngeal arches and other structures in the head, have an equally highly ordered origin with respect to the rhombomeres (**Fig. 11.13**). Cranial nerve V innervates structures derived from the first pharyngeal arch. Cranial nerves VII and IX innervate the second and third arch structures. In embryos of birds, the species studied most extensively, cell bodies of the motor components of cranial nerves V, VII, and IX are initially found exclusively in rhombomeres 2, 4, and 6.

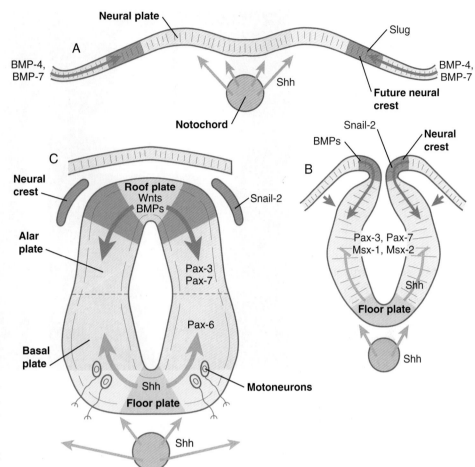

Fig. 11.9 Dorsal and ventral signaling in the early central nervous system. **A,** Signals from sonic hedgehog (Shh) *(orange arrows)* in the notochord induce the floor plate. **B,** In the dorsal part of the future neural tube, bone morphogenetic protein-4 (BMP-4) and BMP-7 *(green arrows)* from the ectoderm adjacent to the neural tube induce snail-2 in the future neural crest and maintain Pax-3 and Pax-7 expression dorsally. Ventrally, Shh, now produced by the floor plate, induces motoneurons. **C,** Shh, produced by the floor plate, suppresses the expression of dorsal *Pax* genes (*Pax-3* and *Pax-7*) in the ventral half of the neural tube. Wnts and BMPs counter this effect by exerting a dorsalizing influence.

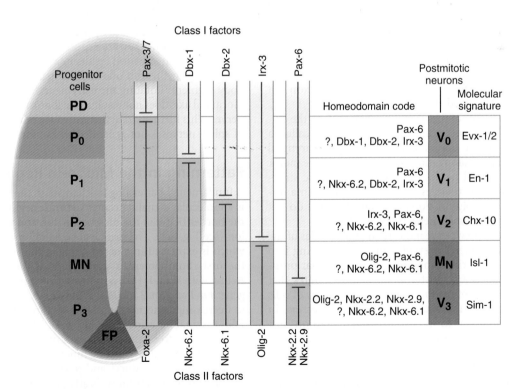

Fig. 11.10 The molecular basis for the specification of neurons throughout the cross section of the ventral neural tube. At the *left,* various classes of progenitor neurons are given labels beginning in P. At the *right* of the neural tube, a decreasing concentration gradient of sonic hedgehog is indicated by the fading red-brown background. The class I factors *(upper bars)* are repressed by sonic hedgehog, whereas the class II factors are induced by sonic hedgehog. To the *right* of the bars is the set of homeodomain codes that specify the various levels of precursor neurons, and at the *far right* is the molecular signature of the neurons. FP, floor plate; MN, motoneuron precursors.

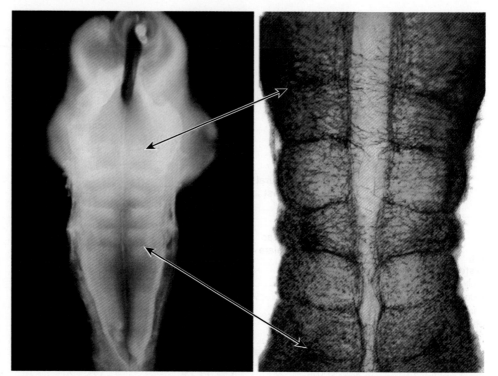

Fig. 11.11 The hindbrain of a live 3-day chick embryo *(left)*, showing segmentation (rhombomeres). On the *right* is a whole mount of a similar area stained for neurofilament protein that shows darkly stained immature neurons running along the rhombomere borders. *(From Lumsden A: Mech Dev 121:1081-1088, 2004.)*

Dye injection studies have shown that the progeny of a single neuroblast remain within the rhombomere containing the injected cell, a finding suggesting that rhombomeres have the properties of cellular compartments. Axons contributing to a cranial nerve extend laterally within the rhombomere and converge on a common exit site in the craniocaudal midpoint of the rhombomere. At a slightly later stage in development, motoneurons originating in the next more posterior rhombomere (3, 5, 7) extend axons laterally. Before the axons reach the margin of the rhombomere, however, they cross into rhombomeres 2, 4, or 6 and converge on the motor axon exit site in the even-numbered rhombomere.

The cell bodies (within the central nervous system, the collection of cell bodies of a single cranial nerve is called a **nucleus**) of the cranial nerves that innervate the pharyngeal arches arise in register along the craniocaudal axis. The motor nuclei of other cranial nerves that innervate somatic structures (e.g., extraocular muscles or the tongue) arise in a different craniocaudal column along the hindbrain and do not occupy contiguous rhombomeres (see Fig. 11.13).

Direct and indirect evidence indicates that properties of the walls of the rhombomeres prevent axons from straying into inappropriate neighboring rhombomeres. One cellular property, which is also characteristic of regions of somites that restrict the movement of neural crest cells, is the ability of cells of the wall of the rhombomere to bind specific lectins. In apparent contradiction to the compartmentalization just described, processes growing from sensory neuroblasts and from nerves of a tract called the **medial longitudinal fasciculus** are free to cross rhombomeric boundaries. Blood vessels first enter the hindbrain in the region of the floor plate soon after the emergence of the motor axons and spread within the inter-rhombomeric junctions. The way the vascular branches recognize the boundaries of the rhombomeres is unknown.

In contrast to the hindbrain, the pattern of nerves emanating from the spinal cord does not appear to be determined by craniocaudal compartmentalization within the spinal cord. Rather, the segmented character of the spinal nerves is dictated by the somitic mesoderm along the neural tube. Outgrowing motoneurons from the spinal cord and migrating neural crest cells can easily penetrate the anterior mesoderm of the somite, but they are repulsed by the posterior half of the somite. This situation results in a regular pattern of spinal nerve outgrowth, with one bilateral pair of spinal nerves per body segment. Rotating the early neural tube around its craniocaudal axis does not result in an abnormal pattern of spinal nerves. This further strengthens the viewpoint that the pattern of spinal nerves is not generated within the neural tube itself.

Patterning in the Midbrain Region

A fundamental patterning mechanism in the midbrain region is a molecular signaling center (**isthmic organizer**) located at the border between the mesencephalon and the metencephalon (see Fig. 6.4). The principal signaling molecule is FGF-8, which is expressed in a narrow ring at the anterior border of the first rhombomere, a subdivision of the metencephalon. Acting with Wnt-1, FGF-8 induces the expression of engrailed genes **En-1** and **En-2** and **Pax-2** and **Pax-5**, which are expressed in decreasing concentrations at increasing distances from the FGF-8 signaling center (see Fig. 6.4). The main function of Wnt-1 seems to be the stimulation of local cell proliferation, whereas the overall organizing function belongs to FGF-8. The isthmic organizer induces and polarizes the dorsal midbrain region and the cerebellum. Grafts of the isthmic organizer or beads releasing FGF-8 alone into more cranial regions of the forebrain of the avian embryo induce a second tectum (dorsal mesencephalon or colliculi in mammals). Similarly, isthmic

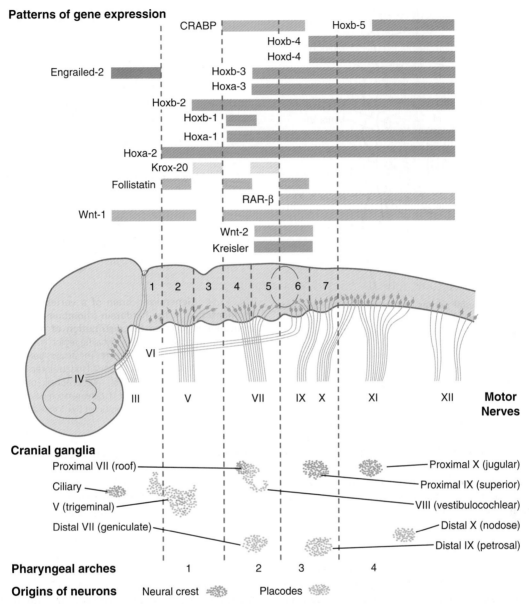

Fig. 11.12 Patterns of *Hox* and other gene expression in relation to anatomical landmarks in the early mammalian embryo. The *bars* refer to craniocaudal levels of expression of a given gene product. Cranial sensory nerves derived from the neural crest and placodal precursors are laid out in proper register. CRABP, cytoplasmic retinoic acid–binding protein; RAR, retinoic acid receptor. *(Adapted from Noden DM: J Craniofac Genet Dev Biol 11:192-213, 1991.)*

grafts into regions of the hindbrain can induce supernumerary cerebellar structures.

One byproduct of these molecular studies is the realization that in some species the boundary between the future midbrain and hindbrain does not correspond to the anatomical constriction between the mesencephalon and rhombencephalon. Instead, it is located cranial to that constriction in a plane marked by the posterior limit of expression of the homeobox-containing gene, *Otx-2*.

Similar to the spinal cord, the midbrain is also highly patterned along its cross-sectional (dorsoventral) axis, and ventrally secreted shh is the molecular basis for much of this patterning, which is represented by five arc-shaped territories of gene expression. If a point source of shh is introduced into a more dorsal region of the midbrain, an additional ectopic

set of five corresponding territorial arcs is produced. In addition to promoting neuronal development in the basal plate of the midbrain, shh restricts ventral expression of molecules, such as **Pax-7**, which are characteristic of the alar plate.

Cranially, the midbrain is separated from the forebrain (diencephalon) through a different set of molecular interactions. The alar plate of the diencephalon is characterized by the expression of **Pax-6**, a molecule that, among other things, acts as the master gene underlying eye formation (see Chapter 13). The mesencephalon is a domain of **En-1** expression. Through the action of intermediate negative regulators, Pax-6 inhibits En-1 expression, whereas En-1 directly inhibits Pax-6 expression. The craniocaudal level at which each of these molecules inhibits the other becomes a sharp diencephalic-mesencephalic border.

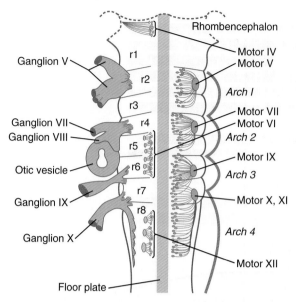

Fig. 11.13 **Origin of cranial nerves in relation to rhombomeres (r) in the developing chick brain.** *(Based on Lumsden A, Keynes R: Nature 337:424-428, 1989.)*

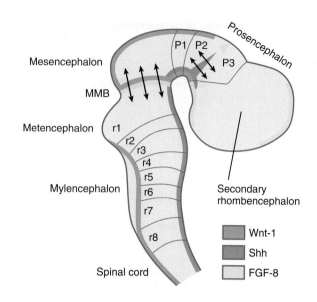

Fig. 11.14 **Generalized brain of a vertebrate embryo showing segmentation in the hindbrain (rhombomeres [r]) and forebrain (prosomeres [P]) and the distribution of major signaling molecules.** The midbrain/hindbrain signaling region is indicated by the *arrows* just rostral to the first rhombomere. The *arrows* between the second and third prosomeres represent the zona limitans interthalamica, a signaling region in the forebrain. MMB, mesencephalic-metencephalic border. *(Adapted from Bally-Cuif L, Wassef M: Determination events in the nervous system of the vertebrate embryo, Curr Opin Genet Dev 5:450-458, 1995.)*

Patterning in the Forebrain Region

Although much less apparent than in the hindbrain, the neuromeric organization of the early brain extends into the forebrain region as a set of three **prosomeres**, which extend from the midbrain-forebrain boundary through the thalamus (**Fig. 11.14**). Prosomeres 1 to 3 (p1 to p3), the most posterior prosomeres, become incorporated into the diencephalon, with p2 and p3 forming the dorsal and ventral **thalamus**, which serves as a major relay station for transmitting neural signals between the cerebral cortex and the body. An earlier representation suggested that an additional set of prosomeres (p4 to p6) formed the organizational basis for the rostral diencephalon (hypothalamus) and telencephalon. A more recent interpretation, however, places these structures in what has been called the **secondary rhombencephalon**, a developmental field that encompasses the entire prechordal portion of the neural tube. Within this domain, the basal plate develops into the major regions of the hypothalamus, the structure that integrates autonomic nervous functions and controls endocrine release from the pituitary. The alar plate in this domain is the precursor of the **cerebral cortex**, the **basal ganglia** (collectively the telencephalon), and the **optic vesicles** (diencephalic structures that take the lead in formation of the eyes). As development progresses, the secondary rhombencephalon becomes sharply folded beneath p2 to p3, and in humans the enormous outgrowth of the alar plates of the secondary rhombencephalon envelops the other prosomeres as the **telencephalic vesicles** (future cerebral cortex).

Discrete patterns of gene expression also mirror the basic regional organization of the forebrain. Early in development, **FGF-8**, secreted by the anterior neural ridge (see Fig. 6.4B), induces the expression of **Foxg-1**, formerly called BF-1 (brain factor-1), which regulates development of the telencephalon and optic vesicles. Within the forebrain, a thin rim of expression of the transcription factor **Nkx 2.2** marks the border between alar and basal plates. Along the dorsoventral axis, the alar plates are characterized by the expression of *Emx1* and *Emx2* and *Pax6*, important regulators of regional identity within the cerebral cortex. In an intermediate region of the prosencephalic alar plate, *Emx* expression drops out, leaving *Pax-6* to function as the master gene controlling eye development (see Chapter 13). In the more ventral area that eventually forms the basal ganglia, *Dlx* (distalless) is a dominant expressed gene.

Similar to the spinal cord, the ventral forebrain is induced and organized by shh, secreted by midline axial structures. In the absence of shh signaling in this area, the tissues of the ventral forebrain are greatly reduced, leading sometimes to midline fusion of the optic vesicles and a general reduction of the growth of the midface region. This situation results in a type of anomaly called **holoprosencephaly** (see p. 309), which in extreme cases is accompanied by **cyclopia**.

At the border between the future dorsal (p2) and ventral (p3) thalamus is a narrow band of shh expression called the **zona limitans interthalamica**. Signals emanating from the zona limitans interthalamica specify aspects of cellular identity and behavior in the diencephalic regions on either side (see Fig. 6.4B). The mechanisms leading to the formation of the zona limitans interthalamica are poorly understood, but this structure arises directly above the anterior tip of the notochord. This finding suggests that the formation of this structure, similar to that of the isthmus, is a reflection of an earlier molecular boundary zone.

Peripheral Nervous System

Structural Organization of a Peripheral Nerve

The formation of a peripheral nerve begins with the outgrowth of axons from motor neuroblasts located in the basal plate (the future ventral horn of the gray matter) of the spinal cord (**Fig. 11.15**). Near the dorsal part of the spinal cord, thin processes also begin to grow from neural crest–derived neuroblasts that have aggregated to form the spinal ganglia. **Dendrites**, which conduct impulses toward the nerve cell body, grow from the sensory neurons toward the periphery. **Axons**, which conduct impulses away from the cell body, penetrate the dorsolateral aspect of the spinal cord and terminate in the dorsal horn (the gray matter of the alar plate). Within the gray matter, short interneurons connect the terminations of the sensory axons to the motoneurons. These three connected neurons (motor, sensory, and interneuron) constitute a simple **reflex arc** through which a sensory stimulus can be translated into a simple motor response. Autonomic nerve fibers are also associated with typical spinal nerves.

A long-standing question concerning the organization of the nervous system involves the interface between the central nervous system and the peripheral nervous system, in particular how their respective cellular components are kept separate. Such separation at the points of exit of the motor axons from the ventral neural tube and the entry into the dorsal neural tube of spinal axons from the dorsal roots is accomplished by localized **boundary caps** of neural crest cells (**Fig. 11.16**). The boundary caps act as selective filters, which allow the free passage of outgrowing and ingrowing axons between the neural tube and the periphery, but which serve as a barrier to keep cells in their appropriate compartment. In the absence of boundary caps, many cell bodies of motoneurons translocate away from the lateral motor column (their normal location) into the space outside the neural tube.

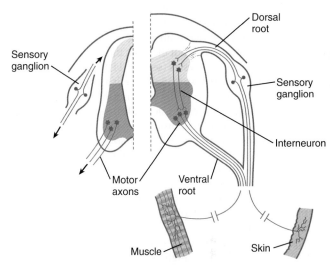

Fig. 11.15 **Development of a peripheral nerve.** *Left,* Early embryo. *Right,* Fetus.

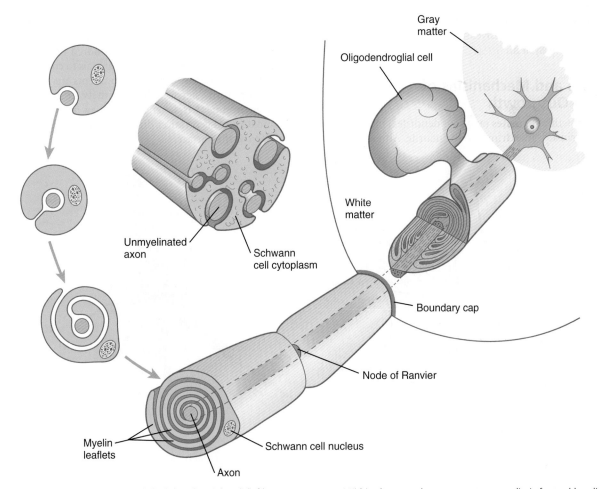

Fig. 11.16 Myelination in the central *(right)* and peripheral *(left)* nervous systems. Within the central nervous system, myelin is formed by oligodendroglial cells. In the peripheral nervous system, the Schwann cells wrap around individual axons. The *inset* shows a segment through a region of unmyelinated nerve fibers embedded in the cytoplasm of a single Schwann cell.

Within a peripheral nerve, the neuronal processes can be myelinated or unmyelinated. At the cellular level, **myelin** is a multilayered spiral sheath consisting largely of phospholipid material that is formed by individual **Schwann cells** (neural crest derivatives) that wrap themselves many times around a nerve process like a jelly roll (see Fig. 11.16). This wrapping serves as a form of insulation that largely determines the character of the electrical impulse (action potential) traveling along the neuronal process. **Unmyelinated nerve fibers** are also embedded in the cytoplasm of Schwann cells, but they lack the characteristic spiral profiles of myelinated processes (see Fig. 11.16).

The Schwann cells that surround myelinated and unmyelinated axons are different not only in their morphology, but also in their patterns of gene expression. Through the actions of a family of growth factor–like proteins (**neuregulins**), the axon associated with a Schwann cell precursor promotes the differentiation of the Schwann cell and helps determine whether it produces myelin or forms a nonmyelinating Schwann cell. Schwann cell precursors that are not associated with axons do not receive neuregulin support and undergo programmed cell death, a mechanism that preserves an appropriate ratio of Schwann cells to axons.

Within the central nervous system, the color of the white matter is the result of its high content of myelinated nerve fibers, whereas the gray matter contains unmyelinated fibers. Schwann cells are not present in the central nervous system; instead, myelination is accomplished by oligodendrocytes. Although one Schwann cell in a myelinated peripheral nerve fiber can wrap itself around only one axon or dendrite, a single oligodendroglial cell can myelinate several nerve fibers in the central nervous system.

Patterns and Mechanisms of Neurite Outgrowth

The outgrowth of **neurites** (axons or dendrites) involves many factors intrinsic and extrinsic to the neurite. Although similar in many respects, the outgrowth of axons and dendrites differs in fundamental ways.

An actively elongating neurite is capped by a **growth cone** (**Fig. 11.17**). Growth cones are characterized by an expanded region of cytoplasm with numerous spikelike projections called **filopodia**. In vitro and in vivo studies of living nerves show that the morphology of an active growth cone is in a constant state of flux, with filopodia regularly extending and retracting as if testing the local environment. Growth cones contain numerous cytoplasmic organelles, but much of the form and function of the filopodia depends on the large quantities of **actin** microfilaments that fill these processes. In a growth cone, an equilibrium exists between the extension of actin microfilaments by terminal addition and resorption at the proximal end. Under conditions favorable for growth, the balance tips toward extension, whereas an unfavorable environment results in the resorption of the microfilaments and collapse of the growth cone.

Whether growth cones progress forward, remain static, or change directions depends in large measure on their interactions with the local environment. If the environment is favorable, a filopodium remains extended and adheres to the substrate around it, whereas other filopodia on the same growth cone retract. Depending on the location of the adhering filopodia, the growth cone may lead the neurite to which it is attached straight ahead or may change its direction of outgrowth. This outgrowth seems to be guided by four broad types of environmental influences: **chemoattraction, contact attraction, chemorepulsion**, and **contact repulsion**. It now seems that outgrowing nerve processes find themselves in different environments every couple hundred micrometers, and that some environments give them signals to continue extending forward, whereas other environments may act as "stop" signals or "turn" signals. The sensitivity of growth cones to their environment is so great that they may be able to discriminate a concentration difference of as little as one molecule across the surface.

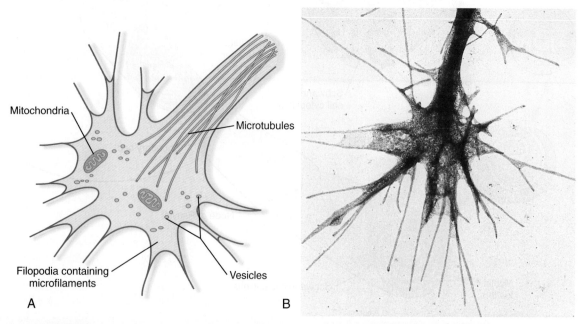

Mitochondria

Microtubules

Filopodia containing microfilaments

Vesicles

A

B

Fig. 11.17   A, Growth cone at the end of an elongating axon. **B,** High-voltage electron micrograph of a growth cone in culture. *(A, From Landis S: Annu Rev Physiol 45:567-580, 1983; B, courtesy of K. Tosney, Ann Arbor, Mich.)*

Growth cones can respond to concentration gradients of diffusible substances (e.g., nerve growth factor) or to weak local electrical fields. A major family of chemoattractant molecules is called the **netrins**. The repulsive counterparts to the netrins are members of a family of secreted proteins called **semaphorins**. The various growth cone attractants produce their effects by stimulating signal transduction machinery that results in elevated levels of cyclic nucleotide second messengers (cyclic adenosine monophosphate or cyclic guanosine monophosphate), whereas repulsants exert their effects through decreased levels of these second messengers.

Many molecules are involved in the guidance of neuronal outgrowth (**Table 11.1**). Some of these molecules can exert either attractive or repulsive actions, depending on the character of the neuron, the time in development, or some combination of intrinsic and extrinsic characteristics. To a greater or lesser extent, the nature of the reaction of the growth cone is determined locally, because in vitro studies have shown that such reactions can occur even when the neuronal process is cut off from the cell body.

Growth cones can also respond to fixed physical or chemical cues from the microenvironment immediately surrounding them. The caudal half of the somite repels the ingrowth of motor axons and of neural crest cells into that area. Repulsion is manifested by the collapse of the growth cone and the retraction of the filopodia. In contrast, extracellular matrix glycoproteins, such as **fibronectin** and especially **laminin**, strongly promote the adhesion and outgrowth of neurites. Integral membrane proteins (**integrins**) on the neurites bind specifically to arginine-glycine-asparagine sequences on the glycoproteins and promote adhesion to the substrate containing these molecules.

Other molecules, such as **N-cadherin, E-cadherin,** and **L1,** are involved in intercellular adhesion at various stages of cell migration or neurite elongation. N-cadherin, which uses calcium (Ca^{++}) as an ionic agent to bind two like molecules together, is heavily involved in the intercellular binding of cells in neuroepithelia. It also plays a role in the adhesion of parallel outgrowing neurites. In a peripheral nerve, one **pioneering axon** typically precedes the others in growing toward its target. Other axons follow, forming **fascicles** (bundles) of axons. Fasciculation is facilitated by intercellular adhesion proteins such as L1, which help bind parallel nerve fibers. If antibodies to the L1 protein are administered to an area of neurite outgrowth, fasciculation is disrupted. Neural cell adhesion molecule (N-CAM) is present on the surfaces of most embryonic nerve processes and muscle fibers and is involved in the initiation of neuromuscular contacts. Antibodies to N-CAM interfere with the development of neuromuscular junctions in embryos. Outgrowing neurites interact with many other molecules, and the full extent of these interactions is just becoming apparent.

Although the growth cone can be thought of as the director of neurite outgrowth, other factors are important for the elongation of axons. Essential to the growth and maintenance of axons and dendrites is **axonal transport**. In this intracellular process, materials produced in the cell body of the neuron are carried to the ends of these neurites, which can be several feet long in humans.

The cytoskeletal backbone of an axon is an ordered array of microtubules and neurofilaments. **Microtubules** are long, tubular polymers composed of **tubulin** subunits. As an axon extends from its cell body, tubulin subunits are transported down the axon and polymerize onto the distal end of the microtubule. The assembly of **neurofilaments** is organized in a similar polarized manner. The site of these cytoskeletal additions is close to the base of the growth cone, meaning that the axon elongates by being added to distally rather than being pushed out by an addition to its proximal end near the neuronal cell body. A characteristic accompaniment of axonal growth is the production of large amounts of **growth-associated proteins** (**GAPs**). Particularly prominent among these is **GAP-43**, which serves as a substrate for protein kinase C and is concentrated in the growth cone.

Outgrowing axons and dendrites differ in several important ways. In contrast to axons, dendrites contain microtubules with polarity running in both directions (**Fig. 11.18**). Another prominent difference is the absence of GAP-43 protein in growing dendrites. Among the first signs of polarity of a developing neuron are the concentration of GAP-43 in the outgrowing axon and its disappearance from the dendritic processes.

Neurite-Target Relations During Development of a Peripheral Nerve

Developing neurites continue to elongate until they have contacted an appropriate end organ. In the case of motoneurons, that end organ is a developing muscle fiber. Dendrites of sensory neurons relate to many types of targets. The end of the neurite first must recognize its appropriate target, and then it must make a functional connection with it.

In the case of motoneurons, evidence that very specific cues guide individual nerves and axons to their muscle targets is increasing. Tracing and transplantation studies have shown that outgrowing motor nerves to limbs supply the limb muscles in a well-defined order, and that after minor positional dislocations, they seek out the correct muscles (see Fig. 10.28). More recent evidence suggests that even at the level of neurons, "fast" axons are attracted to the precursors of fast muscle fibers, and "slow" axons are attracted to the precursors of slow muscle fibers. There are many similar examples of target specificity in dendrites in the peripheral nervous system and of dendrites and axons in the central nervous system. Even

Table 11.1 Receptor/Ligand Pairs Known to Influence Axonal Outgrowth* During Embryonic Development

Ligand	Receptor
Slit	Robo-1, Robo-4
Ephrin	Eph
Netrin	UNC-5, DCC
Semaphorin	Plexin, Neuropilin
VEGF	VEGFR, Neuropilin
Draxin	Netrin receptors

DCC, deleted in colorectal cancer; UNC-5, uncoordinated locomotion-5; VEGF, vascular endothelial growth factor; VEGFR, vascular endothelial growth factor receptor.

*Also vascular sprouting (see Chapter 17).

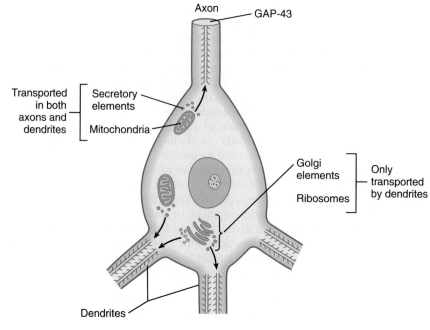

Fig. 11.18 **Polarity in a developing neuron.** In the axon, microtubules have only one polarity, but in dendrites, microtubules with opposite polarities are present.

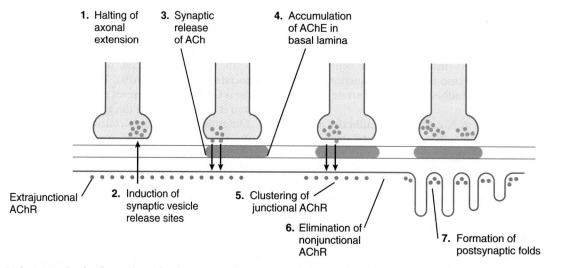

Fig. 11.19 **Major steps in the formation of a neuromuscular junction.** ACh, acetylcholine; AChE, acetylcholinesterase; AChR, acetylcholine receptors.

within a single developing muscle fiber, an aggregation of muscle-specific tyrosine kinase receptor molecules determines the exact site at which the outgrowing axonal terminal connects with the muscle fiber.

When a motor axon and a muscle fiber meet, a complex series of changes in the nerve and muscle fibers mark the formation of a functional **synapse**, in this case called a **neuromuscular junction** (**Fig. 11.19**). The early changes consist of (1) the cessation of outgrowth of the axon, (2) the preparation of the nerve terminal for the ultimate release of appropriate neurotransmitter molecules, and (3) modifications of the muscle fibers at the site of nerve contact so that the neural stimulus can be received and translated into a contractile stimulus. Both neural and muscular components of the neuromuscular junction are involved in stabilizing the

morphological and functional properties of this highly specialized synapse.

As the growth cone of a motoneuron reaches a muscle fiber, synaptic vesicles begin to accumulate in the growth cone, and an interaction with a nerve-specific form of laminin begins to stabilize the neuromuscular contact. Further stabilization is provided by a protein, **agrin**, which becomes concentrated at the neuromuscular junction. The synaptic vesicles store and ultimately release the neurotransmitter substance **acetylcholine** from the nerve terminal (see Fig. 11.19). Before the developing muscle fiber is contacted by the motoneuron, **acetylcholine receptors** (nonjunctional type) are scattered throughout the length of the muscle fiber. After initial nerve contact, myonuclei in the vicinity of the neuromuscular junction produce junction-specific acetylcholine receptors that reside on nerve-induced postjunctional folds of the muscle fiber membrane, and the scattered nonjunctional receptors disappear. Between the nerve terminal and the postsynaptic apparatus of the muscle fiber lies a basal lamina containing molecules that stabilize the acetylcholine receptors at the neuromuscular junction and **acetylcholinesterase**, an enzyme produced by the muscle fiber.

Factors Controlling Numbers and Kinds of Connections Between Neurites and End Organs in the Peripheral Nervous System

At many stages in the formation of a peripheral nerve, interactions between the outgrowing neurites and the target structure influence the numbers and quality of either the nerve fibers or the targets. The existence of such mechanisms was shown in the early 1900s by transplanting limb buds onto flank regions. The motor nerves and sensory ganglia that supplied the grafted limbs were substantially larger than the contralateral spinal nerves, which innervated only structures of the body wall. Examination of the spinal cord at the level of the transplant revealed larger ventral horns of gray matter containing more motoneurons than normal for levels of the spinal cord that supply only flank regions.

Additional experiments of this type cast light on normal anatomical relationships, which show larger volumes of gray matter and larger nerves at levels from which the normal limbs are innervated. Deletion experiments, in which a limb bud is removed before neural outgrowth, or the congenital absence of limbs resulted in deficient numbers of peripheral neurons and reduced volumes of gray matter in the affected regions.

Neuronal **cell death** (**apoptosis**) plays an important role in normal neural development. When a muscle is first innervated, far more than the normal adult number of neurons supply it. At a crucial time in development, massive numbers of neurons die. This seemingly paradoxical phenomenon appears to occur for several reasons, including the following:

1. Some axons fail to reach their normal target, and cell death is a way of eliminating them.
2. Cell death could be a way of reducing the size of the neuronal pool to something appropriate to the size of the target.
3. Similarly, cell death could compensate for a presynaptic input that is too small to accommodate the neurons in question.
4. Neuronal cell death may also be a means of eliminating connection errors between the neurons and their specific end organs.

All these reasons for neuronal cell death may be part of a general biological strategy that reduces superfluous initial connections to ensure that enough correct connections have been made. The other developmental strategy, which seems to be much less used, is to control the outgrowth and connection of neurites with their appropriate end organs so tightly that there is little room for error from the beginning. Because of the overall nature of mammalian development, such tight developmental controls would rob the embryo of the overall flexibility it needs to compensate for genetically or environmentally induced variations in other aspects of development.

The mechanisms by which innervated target structures prevent the death of the neurons that supply them are only beginning to be understood. A popular hypothesis is that the target cells release chemical **trophic factors** that neurites take up, usually by binding to specific receptors. The trophic factor sustains the growth of the neurite. The classic example of a trophic factor is **nerve growth factor**, which supports the outgrowth and prevents the death of sensory neurons. Several other well-characterized molecules are also recognized to be trophic factors.

Autonomic Nervous System

The autonomic nervous system is the component of the peripheral nervous system that subserves many of the involuntary functions of the body, such as glandular activity and motility within the digestive system, heart rate, vascular tone, and sweat gland activity. It has two major divisions—the sympathetic and parasympathetic nervous system. Components of the **sympathetic nervous system** arise from the thoracolumbar levels (T1 to L2) of the spinal cord, whereas the **parasympathetic nervous system** has a widely separated dual origin from the cranial and sacral regions. Both components of the autonomic nervous system consist of two tiers of neurons: **preganglionic** and **postganglionic**. Postganglionic neurons are derivatives of the neural crest (see Chapter 12).

Sympathetic Nervous System

Preganglionic neurons of the sympathetic nervous system arise from the **intermediate horn** (visceroefferent column) of the gray matter in the spinal cord. At levels from T1 to L2, their myelinated axons grow from the cord through the ventral roots, thus paralleling the motor axons that supply the skeletal musculature (**Fig. 11.20**). Shortly after the dorsal and ventral roots of the spinal nerve join, the preganglionic sympathetic axons, which are derived from the neuroepithelium of the neural tube, leave the spinal nerve via a **white communicating ramus**. They soon enter one of a series of **sympathetic ganglia** to synapse with neural crest–derived postganglionic neurons.

The sympathetic ganglia, the bulk of which are organized as two chains running ventrolateral to the vertebral bodies, are laid down by neural crest cells that migrate from the closing neural tube along a special pathway (see Fig. 12.4). When the **migrating sympathetic neuroblasts** have reached the site at which the **sympathetic chain ganglia** form, they spread cranially and caudally until the extent of the chains approximates that seen in an adult. Some of the sympathetic neuroblasts migrate farther ventrally than the level of the chain ganglia to form a variety of other **collateral ganglia** (e.g., **celiac** and **mesenteric ganglia**), which occupy variable positions within the body cavity. The **adrenal medulla** can be broadly viewed as a highly modified sympathetic ganglion.

The outgrowing preganglionic sympathetic neurons either terminate within the chain ganglia or pass through on their way to more distant sympathetic ganglia to form synapses with the cell bodies of the second-order postganglionic sympathetic neuroblasts (see Fig. 11.20). Axons of some postganglionic neuroblasts, which are unmyelinated, leave the chain ganglia as a parallel group and reenter the nearest spinal nerve through the **gray communicating ramus**. When in the spinal nerve, these axons continue to grow until they reach appropriate peripheral targets, such as sweat glands, arrector pili

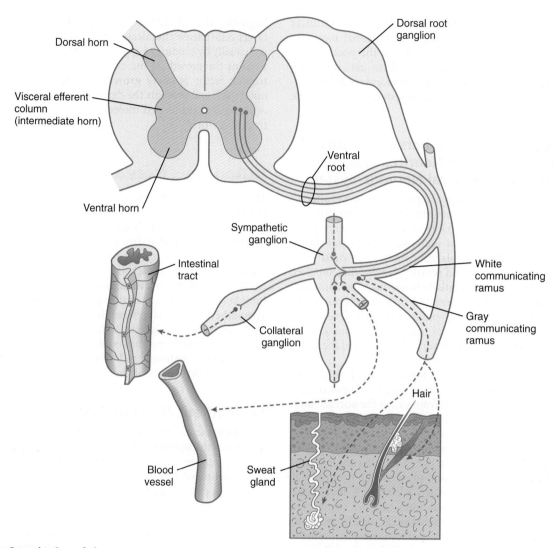

Fig. 11.20 Organization of the autonomic nervous system at the level of the thoracic spinal cord. First-order sympathetic neurons are indicated by *solid blue lines*; second-order sympathetic neurons are indicated by *dashed green lines*.

muscles, and walls of blood vessels. Axons of other postganglionic sympathetic neurons leave their respective ganglia as tangled **plexuses** of nerve fibers and grow toward other visceral targets.

Parasympathetic Nervous System

Although also organized on a preganglionic and postganglionic basis, the parasympathetic nervous system has a distribution quite different from that of the sympathetic system. Preganglionic parasympathetic neurons, similarly to those of the sympathetic nervous system, originate in the visceroefferent column of the central nervous system. The levels of origin of these neuroblasts, however, are in the midbrain and hindbrain (specifically associated with cranial nerves III, VII, IX, and X) and in the second to fourth sacral segments of the developing spinal cord. Axons from these preganglionic neuroblasts grow long distances before they meet the neural crest–derived postganglionic neurons. These are typically embedded in scattered small ganglia or plexuses in the walls of the organs that they innervate.

The neural crest precursors of the postganglionic neurons often undertake extensive migrations (e.g., from the hindbrain to final locations in the walls of the intestines). The migratory properties of the neural crest precursors of parasympathetic neurons are impressive, but this population of cells also undergoes a tremendous expansion until the final number of enteric neurons approximates the number of neurons in the spinal cord. Evidence is increasing that factors in the gut wall stimulate the mitosis of the neural crest cells migrating there. A striking demonstration of the stimulatory powers of the gut is the ability of pieces of gut wall transplanted along the neural tube to cause great expansion of the region of the neural tube closest to the graft (**Fig. 11.21**).

Differentiation of Autonomic Neurons

At least two major steps are involved in the differentiation of autonomic neurons. The first is the determination of certain migrating neural crest cells to differentiate into autonomic neurons instead of the other possible neural crest derivatives. The differentiation from generic preneuronal cells

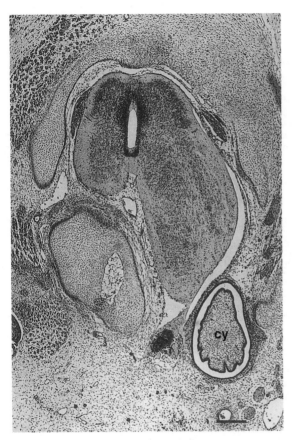

Fig. 11.21   Influence of the gut on growth of the neural tube. A graft of quail duodenum was placed between the neural tube and somites of a chick embryo host. The spinal cord on the side near the graft of gut has greatly enlarged, causing secondary distortion of the musculoskeletal structures near it. *cy*, cyst of donor endoderm. *(From Rothman TP and others:* Dev Biol *124:331-346, 1987.)*

is accomplished through the actions of locally produced BMPs, which commit the exposed cells to becoming future autonomic neurons.

At early stages, the neural crest cells have the option of becoming components of either the sympathetic or the parasympathetic system. This was shown by level-shift transplantations of neural crest cells in birds. When the cephalic neural crest, which would normally form parasympathetic neurons, was transplanted to the level of somites 18 to 24, the transplanted cells migrated and settled into the adrenal medulla as **chromaffin cells**, which are part of the sympathetic nervous system. Conversely, trunk neural crest cells transplanted into the region of the head often migrated into the lining of the gut and differentiated into postganglionic parasympathetic neurons.

A second major step in the differentiation of autonomic neurons involves the choice of the neurotransmitter that the neuron uses. Typically, parasympathetic postganglionic neurons are **cholinergic** (i.e., they use acetylcholine as a transmitter), whereas sympathetic neurons are **adrenergic** (noradrenergic) and use norepinephrine as a transmitter. Cascades of transcription factors are now known to be involved in the further differentiation of generic autonomic neuronal precursors into either sympathetic or parasympathetic neurons. For

example, **Hand-2**, which is also heavily involved in early formation of the heart, is required for the differentiation of an adrenergic neuron.

As they arrive at their final destinations, autonomic neurons are noradrenergic. They then enter a phase during which they select the neurotransmitter substance that will characterize their mature state. Considerable experimental evidence suggests that the choice of transmitter proceeds independently of other concurrent events, such as axonal elongation and the innervation of specific target organs.

At late stages in their development, autonomic neurons still retain flexibility in their choice of neurotransmitter. Sympathetic neurons in newborn rats are normally adrenergic, and if grown in standard in vitro culture conditions, these neurons produce large amounts of norepinephrine and negligible amounts of acetylcholine. If the same neurons are cultured in a medium that has been conditioned by the presence of cardiac muscle cells, they undergo a functional conversion and instead produce large amounts of acetylcholine (**Fig. 11.22**).

An example of a natural transition of the neurotransmitter phenotype from noradrenergic to cholinergic occurs in the sympathetic innervation of sweat glands in the rat. Neurotransmitter transitions depend on target-derived cues. One such cue is **cholinergic differentiation factor**, a glycosylated basic 45-kD protein. This molecule, which is present in cardiomyocyte-conditioned medium, is one of many chemical environmental factors that can exert a strong influence on late phases of differentiation of autonomic neurons.

Congenital Aganglionic Megacolon (Hirschsprung's Disease)

If a newborn exhibits symptoms of complete constipation in the absence of any demonstrable physical obstruction, the cause is usually an absence of parasympathetic ganglia from the lower (sigmoid) colon and rectum. This condition, commonly called **aganglionic megacolon** or **Hirschsprung's disease** (see p. 350), is normally attributed to the absence of colonization of the wall of the lower colon by neural crest–derived parasympathetic neuronal precursors, probably of cranial origin. In rare cases, greater parts of the colon lack ganglia.

Later Structural Changes in the Central Nervous System*

Histogenesis Within the Central Nervous System

A major difference between the brain and the spinal cord is the organization of the gray and white matter. In the spinal cord, the gray matter is centrally located, with white matter surrounding it (see Fig. 11.6). In many parts of the brain, this

---

*The later changes in the central nervous system are so extensive that an exhaustive treatment of even one aspect, such as morphology, is well beyond the scope of this book. This section instead stresses fundamental aspects of the organization of the central nervous system and summarizes the major changes in the organization of the brain and spinal cord.

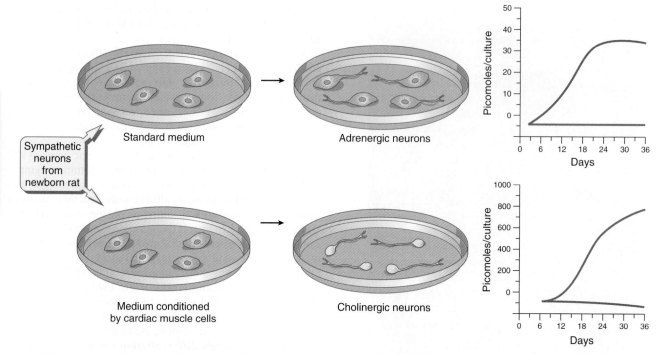

Fig. 11.22 **Experiment illustrating the effect of the environment on the choice of transmitter by differentiating sympathetic neurons.** In standard medium, the neurons become adrenergic; in medium conditioned by cardiac muscles, they become cholinergic. Levels of norepinephrine are in *red*; levels of acetylcholine are in *blue*. *(Based on Patterson PH and others: Sci Am 239:50-59, 1978.)*

arrangement is reversed, with a large core of white matter and layers of gray matter situated superficial to this core.

A fundamental process in histogenesis within the brain is cell migration. From their sites of origin close to the ventricles in the brain, neuroblasts migrate toward the periphery by following set patterns. These patterns often result in a multilayering of the gray substance of the brain tissue. Key players in the migratory phenomenon are radial glial cells, which extend long processes from cell bodies located close to the ventricular lumen toward the periphery of the developing cortex (**Fig. 11.23**). Young postmitotic neurons, which are usually simple bipolar cells, wrap themselves around the radial glial cells and use them as guides on their migrations from their sites of origin to the periphery.

In areas of brain cortex characterized by six layers of gray matter, the large neurons populating the innermost layer migrate first. The remaining layers of gray matter are formed by smaller neurons migrating through the first layer and other previously formed layers to set up a new layer of gray matter at the periphery. With this pattern of histogenesis, the outermost layer of neurons is the one formed last, and the innermost is the layer formed first. In a mouse mutant called **weaver**, specific behavioral defects are related to abnormal function of the cerebellum. The morphological basis for this mutant consists of an abnormality of the radial glial cells in the cerebellum and a consequent abnormal migration of the cells that normally form the granular layer of the cerebellar cortex. Another mutant, called **reeler**, is characterized by abnormal behavior and the absence of normal cortical layering. More recently, an extracellular protein, called **reelin**, has been shown to be defective in the reeler mutant. Reelin may serve as a stop signal for radial neuronal migration or as an insertional signal for migrating neurons.

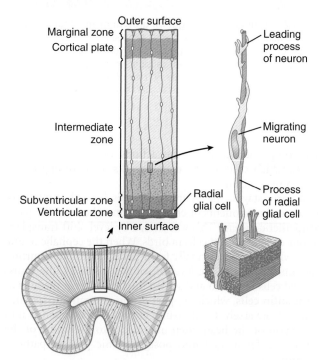

Fig. 11.23 **Radial glial cells and their association with peripherally migrating neurons during the development of the brain.** *(Based on Rakic P: Birth Defects Orig Article Series 11:95-129, 1975.)*

Not all neuronal migration within the central nervous system follows the inside-out pattern described earlier. Neuronal migration parallel to the surface occurs during early development of the cerebellum, and, in contrast to the cerebral cortex, within the three layers of gray matter in the hippocampus and superior colliculi, the neurons in the outer layer are the oldest, and the neurons of the innermost layer are the youngest. Molecular studies are beginning to define the basis for the cellular organization of different areas or even sides of the cortex.

Increasing evidence indicates that the seemingly featureless cerebral cortex is a matrix of discrete **columnar radial units** that consist of radial glial cells and the neuroblasts that migrate along them. There may be as many as 200 million radial units in the human cerebral cortex. The radial units begin as proliferative units, with most cortical neurons generated between days 40 and 125. As with many aspects of neural differentiation, the number of radial units seems to be sensitive to their own neural input. In cases of **congenital anophthalmia** (the absence of eyes), neural input from visual pathways to the area of the occipital cortex associated with vision is reduced. This condition results in gross and microscopic abnormalities of the visual cortex, principally related to a reduced number of radial units in that region.

Spinal Cord

In the spinal cord, inputs from many peripheral sensory nerves are distributed as local reflex arcs or are channeled to the brain through tracts of axons. In addition, motor messages originating in the brain are distributed to appropriate peripheral locations via motor tracts and ventral (motor) roots of individual spinal nerves.

The tracts in the spinal cord and other parts of the central nervous system do not become completely myelinated until late in development—sometimes not until years after birth. An unmyelinated tract is developmentally active, with axonal growth and branching still taking place. The appearance of myelin coverings not only facilitates the conduction of neural impulses, but also stabilizes the tract anatomically and functionally. One consequence of myelinization is that after injury to the postnatal spinal cord, axonal regeneration is poor, largely because of the inhibitory action of myelin breakdown products on axonal outgrowth. Axons in the same tracts can often regenerate if injury occurs before myelinization has occurred.

The early spinal cord is divided into alar and basal plate regions, which are precursors for the sensory and motor regions of the spinal cord (see Fig. 11.6). The mature spinal cord has a similar organization, but these regions are subdivided further into somatic and visceral components. Within the brain, still another layer of input and output is added with "special" components. These components are summarized in **Box 11.1**.

A gross change of the spinal cord that is of clinical significance is the relative shortening of the spinal cord in relation to the vertebral column (**Fig. 11.24**). In the first trimester, the spinal cord extends the entire length of the body, and the spinal nerves pass through the intervertebral spaces directly opposite their site of origin. In later months, growth of the posterior part of the body outstrips that of the vertebral column and the spinal cord, but growth of the spinal cord lags significantly behind that of the vertebral column. This disparity is barely apparent in the cranial and thoracic regions, but at birth the spinal cord terminates at the level of L3. By adulthood, the spinal cord terminates at L2.

The consequence of this growth disparity is the considerable elongation of the lumbar and sacral dorsal and ventral

---

Box 11.1 Functional Regions in the Spinal Cord and Brain

Alar Plate (Afferent or Sensory)

General somatic afferent: sensory input from skin, joints, and muscles

Special visceral afferent: sensory input from taste buds and pharynx

General visceral afferent: sensory input from viscera and heart

Basal Plate (Efferent: Motor or Autonomic)

General visceral efferent: autonomic (two-neuron) links from the intermediate horn to the viscera

Special visceral efferent: motor nerves to striated muscles of the pharyngeal arches

General somatic efferent: motor nerves to the striated muscles other than those of the pharyngeal arches

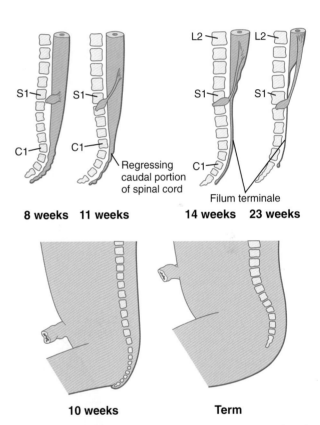

8 weeks   11 weeks   **14 weeks   23 weeks**

10 weeks   **Term**

Fig. 11.24 *Top,* Changes in the level of the end of the spinal cord in relation to bony landmarks in the vertebral column during fetal development. *Bottom,* Development of the curvature of the spine.

---

spinal nerve roots to accommodate the increased distance between their point of origin and the appropriate intervertebral space. This elongation gives these nerve roots the collective appearance of a horse's tail (hence their name—**cauda equina**). A thin, filamentlike **filum terminale** extending from the end of the spinal cord proper to the base of the vertebral column marks the original excursion of the spinal cord. This arrangement is convenient for the clinician because the space below the termination of the spinal cord is a safe place from which to withdraw cerebrospinal fluid for analysis.

Myelencephalon

The **myelencephalon**, the most caudal subdivision of the rhombencephalon (see Figs. 11.1 and 11.2), develops into the **medulla oblongata** of the adult brain (**Fig. 11.25**). It is in many respects a transitional structure between the brain and spinal cord, and the parallels between its functional organization and that of the spinal cord are readily apparent (**Fig. 11.26**). Much of the medulla serves as a conduit for tracts that link the brain with input and output nodes in the spinal cord, but it also contains centers for the regulation of vital functions such as the heartbeat and respiration.

The fundamental arrangement of alar and basal plates with an intervening sulcus limitans is retained almost unchanged in the myelencephalon. The major topographical change from the spinal cord is a pronounced expansion of the roof plate to form the characteristic thin roof overlying the expanded central canal, which in the myelencephalon is called the **fourth ventricle** (see Fig. 11.37). (Details of the ventricles and the coverings of the brain and spinal cord are presented later in this chapter.)

Special visceral **afferent** (leading toward the brain) and **efferent** (leading from the brain) columns of nuclei (aggregations of neuronal cell bodies in the brain) appear in the

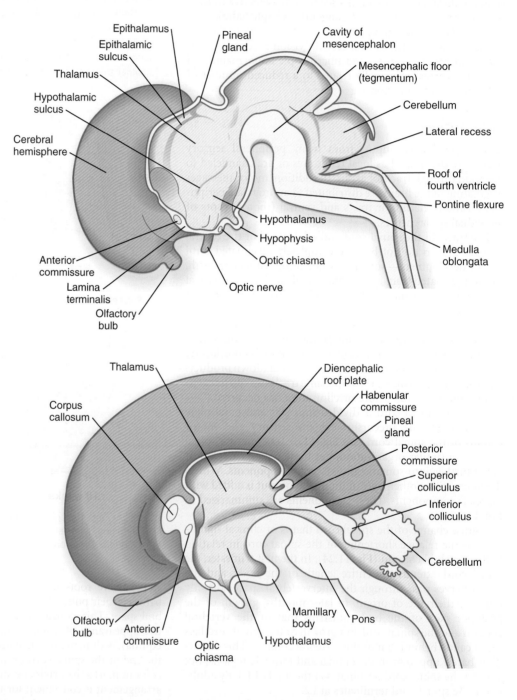

Fig. 11.25 Anatomy of the brain in 9-week *(top)* and 16-week *(bottom)* human embryos.

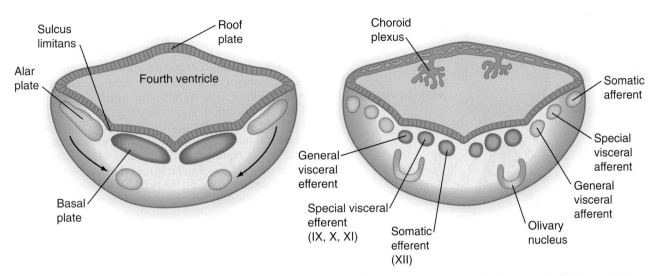

Fig. 11.26 Cross sections through the developing myelencephalon at early *(left)* and later *(right)* stages of embryonic development. Motor tracts (from the basal plate) are shown in *green*; sensory tracts (from the alar plate) are *orange*. *(Adapted from Sadler T: Langman's medical embryology, ed 6, Baltimore, 1990, Williams & Wilkins.)*

myelencephalon to accommodate structures derived from the pharyngeal arches. Even as functional neuronal connections between cranial sensory nerves and nuclei within the myelencephalon are being set up, *Hox* gene expression related to the individual rhombomeres seems to be involved in the differentiation of specific nuclei and types of connecting neurons.

Metencephalon

The **metencephalon**, the more cranial subdivision of the hindbrain, consists of two main parts: the **pons**, which is directly continuous with the medulla, and the **cerebellum**, a phylogenetically newer and ontogenetically later-appearing component of the brain (see Fig. 11.25). The formation of these structures depends on the inductive activity of FGF-8 emanating from the isthmic organizer (see Fig. 6.4).

As its name implies, the pons, derived from the basal plate, serves as a bridge that carries tracts of nerve fibers between higher brain centers and the spinal cord. Its fundamental organization remains similar to that of the myelencephalon, with three sets of afferent and efferent nuclei (**Fig. 11.27**). In addition to these nuclei, other special pontine nuclei, which originated from alar plate–derived neuroblasts, are present in the ventral white matter. The caudal part of the pons also has an expanded roof plate similar to that of the myelencephalon.

The cerebellum, an alar plate derivative, is structurally and functionally complex, but phylogenetically it arose as a specialization of the vestibular system and was involved with balance. Other functions, such as the orchestration of general coordination and involvement in auditory and visual reflexes, were later superimposed.

The future site of the cerebellum is first represented by the rhombic lips of the 5- to 6-week embryo. The rhombic lips represent the diamond (rhombus)–shaped border between the thin roof plate and the main body of the rhombencephalon (**Fig. 11.28**). The rhombic lips, situated from rhombomeres 1 to 8 (r1 to r8), are the product of an inductive interaction between the roof plate (through BMP signaling)

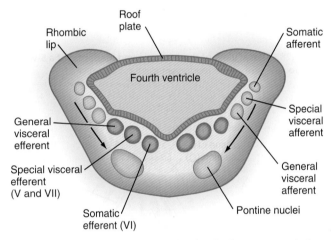

Fig. 11.27 **Cross section through the developing metencephalon.** Motor tracts are shown in *green*; sensory tracts are *orange*. *(Adapted from Sadler T: Langman's medical embryology, ed 6, Baltimore, 1990, Williams & Wilkins.)*

and the neural tube within the original rhombencephalon. The cerebellum proper arises from the anterior (cerebellar) rhombic lips (r1), whereas the posterior rhombic lips (r2 to r8) give rise to the migratory precursors of a variety of ventrally situated nuclei (e.g., olivary and pontine nuclei; see Figs. 11.26 and 11.27) that are located in the metencephalon and rhombencephalon. Soon after the induction of the rhombic lips, precursors of **granule cells** migrate anteriorly along the dorsal region of r1 from the cerebellar rhombic lips to form a transient germinal epithelium called the **external granular layer** (**Fig. 11.29**). After terminal mitotic divisions, the postmitotic external granule cells undergo a ubiquitin-mediated second radial migration toward the interior of the future cerebellum. En route, these cells pass through a layer of precursors of the larger **Purkinje cells**, which are migrating radially in the opposite direction. Once past the Purkinje cells, the

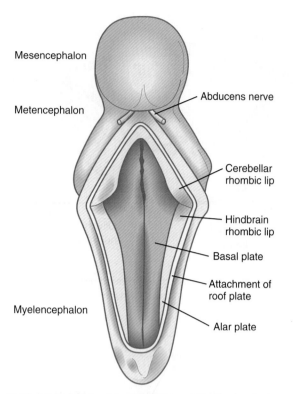

Fig. 11.28 **Dorsal view of the midbrain and hindbrain of a 5-week embryo.** The roof of the fourth ventricle has been opened.

migrating granule cells settle in the **inner granular layer,** simply called the **granular layer** in the mature cerebellum (**Fig. 11.30E**; see Fig. 11.29). A strong interaction exists between the Purkinje cells and the granule cells, which become the most numerous cell type within the entire central nervous system. Purkinje cells secrete shh, which is a key mitogen inducing proliferation of granule precursor cells.

Until the end of the third month, the expansion of the cerebellar rhombic lips is mainly anterior and inward, but thereafter the rapid growth in volume of the cerebellum is directed outward (see Fig. 11.30). As the volume of the developing cerebellum expands, the two lateral rhombic lips join in the midline, thus giving the early cerebellar primordium a dumbbell appearance. The cerebellum then enters a period of rapid development and external expansion. As the complex process of cerebellar histogenesis proceeds, many fibers emanating from the vast number of neurons generated in the cerebellar cortex leave the cerebellum through a pair of massive **superior cerebellar peduncles,** which grow into the mesencephalon.

Mesencephalon

The **mesencephalon,** or **midbrain,** is structurally a relatively simple part of the brain in which the fundamental relationships between the basal and alar plates are preserved (**Fig. 11.31**). As in the spinal cord, dorsoventral organization within the midbrain is heavily based on shh signaling from the floor,

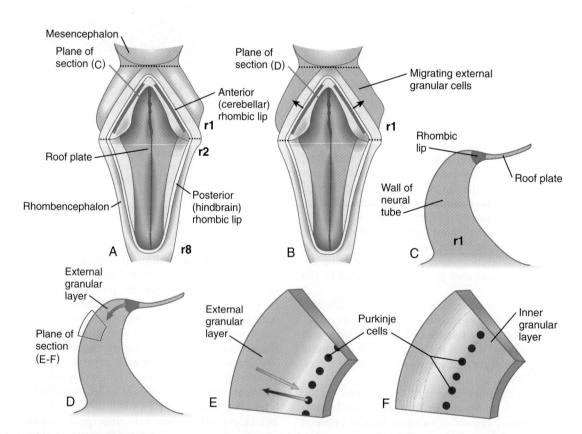

Fig. 11.29 **Origin and migrations of cerebellar precursor cells.** **A,** Dorsal view of the hindbrain. **B,** Dorsal view of the early migration of the external granular cells. **C,** Cross-sectional view of the premigratory stage, indicated by the *red line* in **A. D,** Cross-sectional view of the early migratory stage, indicated by the *red line* in **B. E** and **F,** Cross-sectional views of later stages of inward migration of the external granular cells and outward migration of the Purkinje cells.

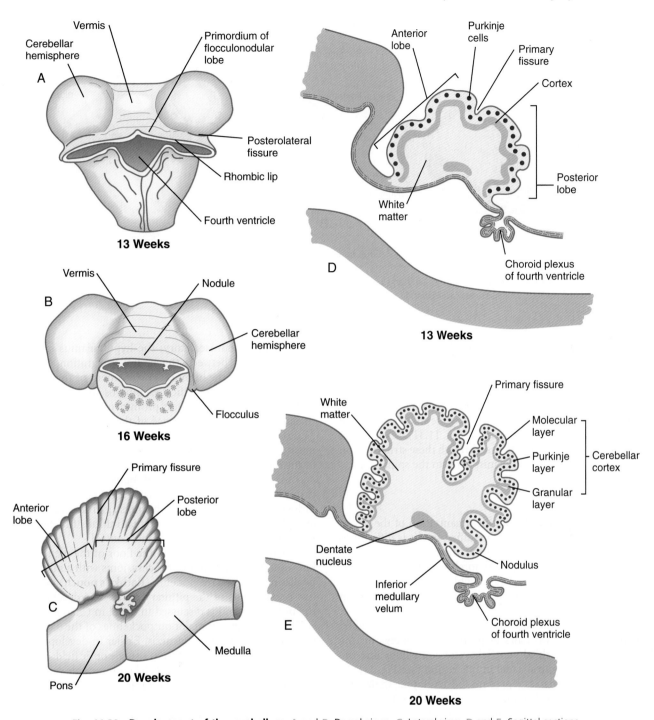

Fig. 11.30 **Development of the cerebellum.** A and B, Dorsal views. C, Lateral view. D and E, Sagittal sections.

which in addition to promoting neuronal development within the basal plate, suppresses the expression of molecules characteristic of the alar plate. A late function of Otx-2 in the region of the border between the alar and basal plates confines shh activity to the basal part of the midbrain.

The basal plates form the neuron-rich area, called the **tegmentum**, that is the location of the somatic efferent nuclei of cranial nerves III and IV, which supply most of the extrinsic muscles of the eye. A small visceroefferent nucleus, the **Edinger-Westphal nucleus**, is responsible for innervation of the pupillary sphincter of the eye.

The alar plates form the sensory part to the midbrain (**tectum**), which subserves the functions of vision and hearing. In response to the localized expression of En-1 and Pax-7, neuroblasts migrating toward the roof form two prominent pairs of bulges, collectively called the **corpora quadrigemina**. The caudal pair, called the **inferior colliculi**, is simple in structure and functionally part of the auditory system. The **superior colliculi** take on a more complex layered architecture through the migration patterns of the neuroblasts that give rise to it. The superior colliculi are an integral part of the visual system, and they serve as an important synaptic relay

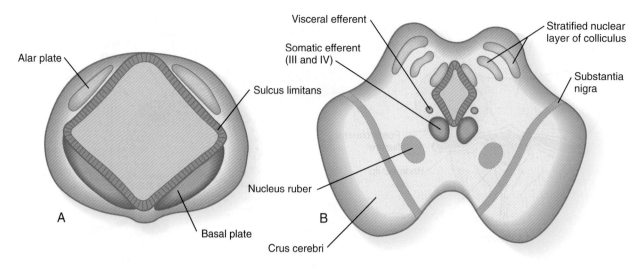

Fig. 11.31 **A** and **B,** Cross sections through the early and later developing mesencephalon. Motor tracts are shown in *green;* sensory tracts are *orange. (Adapted from Sadler T: Langman's medical embryology, ed 6, Baltimore, 1990, Williams & Wilkins.)*

station between the optic nerve and the visual areas of the cerebral (occipital) cortex. Connections between the superior and inferior colliculi help coordinate visual and auditory reflexes.

The third major region of the mesencephalon is represented by prominent ventrolateral bulges of white matter called the **cerebral peduncles** (crus cerebri; see Fig. 11.31). Many of the major descending fiber tracts pass through these structures on their way from the cerebral hemispheres to the spinal cord.

Diencephalon

Cranial to the mesencephalon, the organization of the developing brain becomes so highly modified that it is difficult to relate later morphology to the fundamental alar plate–basal plate plan. It is believed that the forebrain structures (diencephalon and telencephalon) are highly modified derivatives of the alar plates and roof plate without major representation by basal plates.

Development of the early diencephalon is characterized by the appearance of two pairs of prominent swellings on the lateral walls of the **third ventricle**. These swellings line the greatly expanded central canal in this region (see Fig. 11.25). The largest pair of masses represents the developing **thalamus**, in which neural tracts from higher brain centers synapse with the tracts of other regions of the brain and brainstem. Among the many thalamic nuclei are those that receive input from the auditory and visual systems and transmit them to the appropriate regions of the cerebral cortex. In later development, the thalamic swellings may thicken to the point where they meet and fuse in the midline across the third ventricle. This connection is called the **massa intermedia**.

Ventral to the thalamus, the swellings of the incipient **hypothalamus** are separated from the thalamus by the **hypothalamic sulcus**. As mentioned earlier, the hypothalamus receives input from many areas of the central nervous system. It also acts as a master regulatory center that controls many basic homeostatic functions, such as sleep, temperature control, hunger, fluid and electrolyte balance, emotions, and rhythms of glandular secretion (e.g., of the pituitary). Many of its

functions are neurosecretory; the hypothalamus serves as a major interface between the neural integration of sensory information and the humoral environment of the body.

In early embryos (specifically, embryos around 7 to 8 weeks' gestational age), a pair of less prominent bulges dorsal to the thalamus marks the emergence of the **epithalamus** (see Fig. 11.25), a relatively poorly developed set of nuclei relating to masticatory and swallowing functions. The most caudal part of the diencephalic roof plate forms a small diverticulum that becomes the **epiphysis** (**pineal body**), a phylogenetically primitive gland that often serves as a light receptor. Under the influence of light-dark cycles, the pineal gland secretes (mainly at night) **melatonin**, a hormone that inhibits function of the pituitary-gonadal axis of hormonal control.

A ventral downgrowth from the floor of the diencephalon, known as the **infundibular process**, joins with a midline outpocketing from the stomodeum (**Rathke's pouch**) to form the two components of the pituitary gland. The development of the pituitary gland is discussed in detail in Chapter 14.

The **optic cups** are major outpocketings of the diencephalic wall during early embryogenesis. Earlier in development, the ventral diencephalon constitutes a single optic field, characterized by the expression of Pax-6. Then the single optic field is separated into left and right optic primordia by anterior movements of ventral diencephalic cells, which depend on the expression of the gene *cyclops*. Further development of the optic cups and the optic nerves (cranial nerve II) is discussed in Chapter 13.

Telencephalon

Telencephalic development is the product of interactions among three patterning centers in the forebrain. The **rostral patterning center**, derived from the earlier anterior neural ridge (see Fig. 6.4B), secretes FGF-8, which directly affects the two other patterning centers—the **dorsal patterning center** (sometimes called the cortical hem), which produces **BMPs** and **Wnts**, and the **ventral patterning center**, which produces shh. Acting through molecules, such as **Emx-2**, FGF-8 plays a significant role in the overall growth of the telencephalon.

FGF-8 mutants are characterized by reduced telencephalic size and a shift toward sensory versus frontal functions. Wnts, produced by the dorsal patterning center, promote the formation of caudal telencephalic structures, such as the hippocampus, and BMPs pattern the dorsal midline and induce choroid plexus formation. Working through the downstream molecule, **Nkx-2.1**, FGF-8 from the rostral patterning center may provide an early step in ventralizing the telencephalon through its effect on shh. After these early patterning events, telencephalic development is marked by tremendous growth.

Development of the **telencephalon** is dominated by the massive expansion of the bilateral **telencephalic vesicles**, which ultimately become the cerebral hemispheres (see Fig. 11.25). The walls of the telencephalic vesicles surround the expanded lateral ventricles, which are outpocketings from the midline third ventricle located in the diencephalon (see Fig. 11.37). Although the cerebral hemispheres first appear as lateral structures, the dynamics of their growth cause them to approach the midline over the roofs of the diencephalon and mesencephalon (**Fig. 11.32**). The two cerebral hemispheres never meet in the dorsal midline because they are separated by a thin septum of connective tissue (part of the dura mater) known as the **falx cerebri**. Below this septum, the two cerebral hemispheres are connected by the ependymal roof of the third ventricle.

Although the cerebral hemispheres expand greatly during the early months of pregnancy, their external surfaces remain smooth until the fourteenth week. With continued growth, the cerebral hemispheres undergo folding at several levels of organization. The most massive folding involves the large **temporal lobes**, which protrude laterally and rostrally from the caudal part of the cerebral hemispheres. From the fourth to the ninth month of pregnancy, the expanding temporal lobes and the frontal and parietal lobes completely cover areas of the cortex known as the **insula** (island) (**Fig. 11.33**). While these major changes in organization are occurring, other precursors of major surface landmarks of the definitive cerebral cortex are being sculpted. Several major sulci and fissures begin to appear as early as the sixth month. By the eighth month, the **sulci** (grooves) and **gyri** (convolutions) that characterize the mature brain take shape.

Internally, the base of each telencephalic vesicle thickens to form the comma-shaped **corpus striatum** (**Fig. 11.34**). Located dorsal to the thalamus, the corpus striatum becomes more C-shaped as development progresses. With histodifferentiation of the cerebral cortex, many fiber tracts converge on the area of the corpus striatum, which becomes subdivided into two major nuclei—the **lentiform nucleus** and the **caudate nucleus**. These structures, which are components of the complex aggregation of nuclei known as the **basal ganglia**, are involved in the unconscious control of muscle tone and complex body movements.

Although the grossly identifiable changes in the developing telencephalic vesicles are very prominent, many internal cellular events determine the functionality of the telencephalon. Specific details are beyond the scope of this text, but for many parts of the brain the general sequence of events begins with early regionalization of the telencephalon. This is followed by the generation and directed migration of neuronal precursors and the formation of the various layers of the cerebral cortex or the formation of aggregates of neurons in internal structures of the telencephalon or diencephalon, such as the thalamus or the hippocampus. When the neuronal cell bodies are properly positioned, axonal or dendritic processes growing from them undergo tightly guided outgrowth to specific targets, such as the pyramidal cells of the cerebral cortex. The pyramidal cells send out long processes that may leave the telencephalon as massive fiber bundles, such as the pyramids, which are the gross manifestations of the corticospinal tracts that are part of the circuitry controlling coordinated movement.

Aside from the telencephalic vesicles, the other major component of the early telencephalon is the **lamina terminalis**, which forms its median rostral wall (**Fig. 11.35**; see Fig.

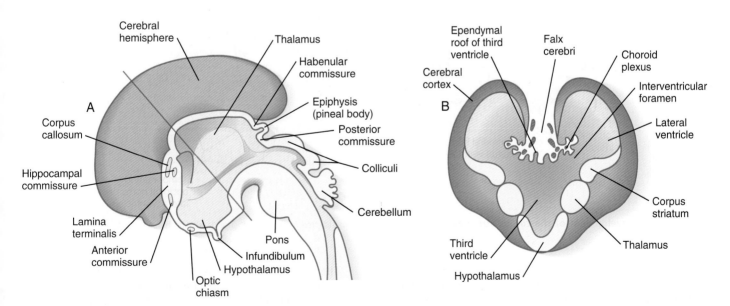

Fig. 11.32 **Early formation of the cerebral hemispheres in a 10-week embryo.** A, Sagittal section through the brain. B, Cross section through the level indicated by the *red line* in **A**. *(Adapted from Moore K, Persaud T: The developing human, ed 5, Philadelphia, 1993, WB Saunders.)*

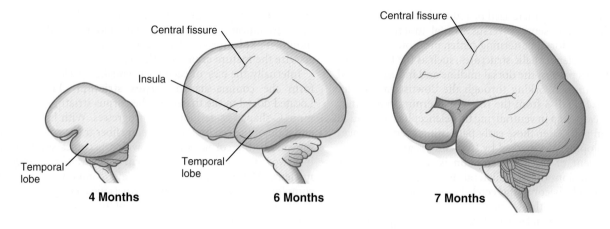

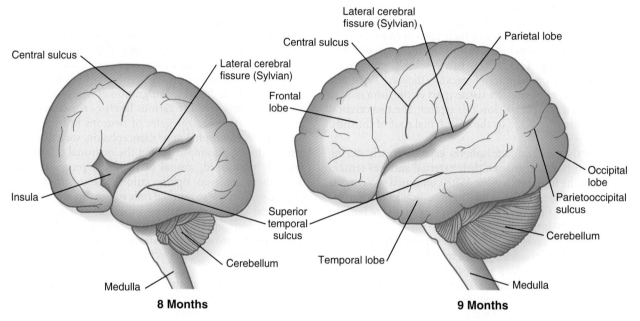

Fig. 11.33 **Lateral views of the developing brain.**

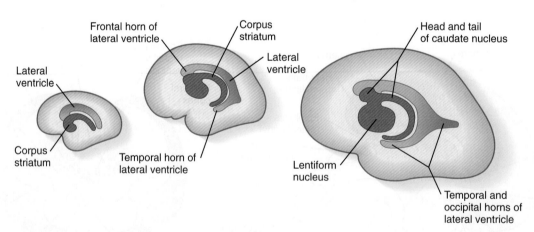

Fig. 11.34 **Development of the corpus striatum and lateral ventricles.** *(Adapted from Moore K: The developing human, ed 4, Philadelphia, 1988, Saunders.)*

11.37A). Initially, the two cerebral hemispheres develop separately, but toward the end of the first trimester of pregnancy, bundles of nerve fibers begin to cross from one cerebral hemisphere to the other. Many of these connections occur through the lamina terminalis.

The first set of connections to appear in the lamina terminalis becomes the **anterior commissure** (see Fig. 11.25B), which connects olfactory areas from the two sides of the brain. The second connection is the **hippocampal commissure** (**fornix**). The third commissure to take shape in the lamina

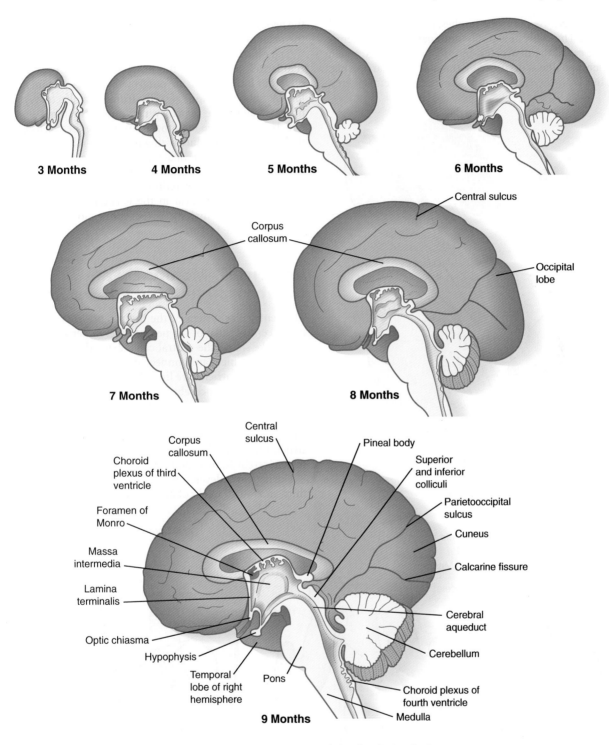

Fig. 11.35   **Medial views of the developing brain.**

terminalis is the **corpus callosum**, the most important connection between the right and left halves of the brain. It initially forms (see Fig. 11.32A) at 74 days as a small bundle in the lamina terminalis, but it expands greatly to form a broad band connecting a large part of the base of the cerebral hemispheres (see Fig. 11.35). Formation of the corpus callosum is complete by 115 days. In mutations of the homeobox gene, *EMX2*, the corpus callosum fails to form, thus leading to an anomaly sometimes called **schizencephaly** (split brain). Other

commissures not related to the lamina terminalis are the **posterior** and **habenular commissures** (see Fig. 11.32), which are located close to the base of the pineal gland, and the **optic chiasma**, the region in the diencephalon where parts of the optic nerve fibers cross to the other side of the brain.

Neuroanatomists subdivide the telencephalon into several functional components that are based on the phylogenetic development of this region. The oldest and most primitive component is called the **rhinencephalon** (also the **archicortex** and **paleocortex**). As the name implies, it is heavily involved in olfaction. The morphologically dominant cerebral hemispheres are called the **neocortex**. In early development, much of the telencephalon is occupied by rhinencephalic areas (**Fig. 11.36**), but with the expansion of the cerebral hemispheres, the neocortex takes over as the component occupying most of the mass of the brain.

The **olfactory nerves** (cranial nerve I), arising from paired ectodermal placodes in the head, send fibers back into the **olfactory bulbs**, which are outgrowths from the rhinencephalon. A subpopulation of cells from the olfactory placode migrates along the olfactory nerve into the brain and ultimately settles in the hypothalamus, where these cells become the cells that secrete luteinizing hormone–releasing hormone. Interactions between olfactory placode ectoderm and the neural crest–derived frontonasal mesenchyme, mediated to a considerable extent by retinoic acid produced by the local mesenchyme, are critical in the generation of the olfactory neurons and their making correct connections with the olfactory bulb in the forebrain.

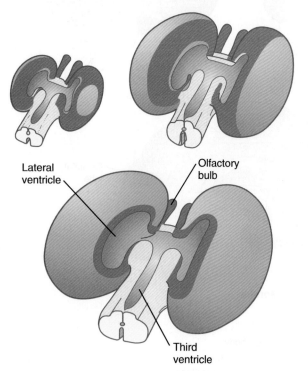

Fig. 11.36 Decrease in the prominence of the rhinencephalic areas *(green)* of the brain as the cerebrum expands.

Ventricles, Meninges, and Cerebrospinal Fluid Formation

The ventricular system of the brain represents an expansion of the central canal of the neural tube. As certain parts of the brain take shape, the central canal expands into well-defined **ventricles**, which are connected by thinner channels (**Fig. 11.37**). The ventricles are lined by ependymal epithelium and are filled with clear **cerebrospinal fluid**. Cerebrospinal fluid is formed in specialized areas called **choroid plexuses**, which are located in specific regions in the roof of the third, fourth, and lateral ventricles. Choroid plexuses are highly vascularized structures that project into the ventricles (see Fig. 11.32B) and secrete cerebrospinal fluid into the ventricular system.

During early development of the brain (equivalent to the third and fourth weeks of human development), cerebrospinal

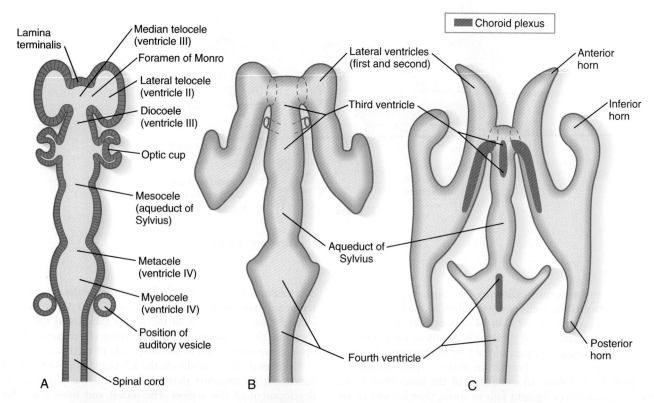

Fig. 11.37 **Development of the ventricular system of the brain.** **A,** Section from an early embryo. **B,** Ventricular system during expansion of the cerebral hemispheres. **C,** Postnatal morphology of the ventricular system.

fluid plays an important role in overall growth and development of the brain. As the amount of cerebrospinal fluid increases through an osmotic mechanism, its pressure increases on the inner surfaces of the brain. This change, along with the possible effect of growth factors in the fluid, results in increased mitotic activity within the neuroepithelium and a considerable increase in the mass of the brain. If the cerebrospinal fluid is shunted away from the ventricular cavities, overall growth of the brain is considerably reduced.

In the fetus, cerebrospinal fluid has a well-characterized circulatory path. As it forms, it flows from the lateral ventricles into the third ventricle and, ultimately, the fourth ventricle. Much of it then escapes through three small holes in the roof of the fourth ventricle and enters the **subarachnoid space** between two layers of meninges. A significant portion of the fluid leaves the skull and bathes the spinal cord as a protective layer.

If an imbalance exists between the production and resorption of cerebrospinal fluid, or if its circulation is blocked, the fluid may accumulate within the ventricular system of the brain and, through increased mechanical pressure, result in massive enlargement of the ventricular system. This condition causes thinning of the walls of the brain and a pronounced increase in the diameter of the skull, a condition known as **hydrocephalus** (**Fig. 11.38**). The blockage of fluid can result from congenital **stenosis** (narrowing) of the narrow parts of the ventricular system, or it can be the result of certain fetal viral infections.

A specific malformation leading to hydrocephalus is the **Arnold-Chiari malformation**, in which parts of the cerebellum herniate into the foramen magnum and mechanically prevent the escape of cerebrospinal fluid from the skull. This condition can be associated with some form of closure defect of the spinal cord or vertebral column. The underlying cause of the several anatomical forms of Arnold-Chiari malformation remains unknown.

In the early fetal period, two layers of mesenchyme appear around the brain and spinal cord. The thick outer layer, which is of mesodermal origin, forms the tough **dura mater** and the membrane bones of the calvarium. A thin inner layer of neural crest origin later subdivides into a thin **pia mater**, which is closely apposed to the neural tissue, and a middle **arachnoid layer**. Spaces that form within the pia-arachnoid layer fill with cerebrospinal fluid.

Cranial Nerves

Although based on the same fundamental plan as the spinal nerves, the cranial nerves (**Fig. 11.39**) have lost their regular segmental arrangement and have become highly specialized (**Table 11.2**). One major difference is the tendency of many cranial nerves to be either sensory (dorsal root based) or motor (ventral root based), rather than mixed, as is the case with the spinal nerves.

The cranial nerves can be subdivided into several categories on the basis of their function and embryological origin. Cranial nerves I and II (olfactory and optic) are often regarded as extensions of brain tracts rather than true nerves. Cranial nerves III, IV, VI, and XII are pure motor nerves that seem to have evolved from primitive ventral roots. Nerves V, VII, IX, and X are mixed nerves with motor and sensory components, and each nerve supplies derivatives of a different pharyngeal arch (**Fig. 11.40**; see Table 11.2 and Fig. 14.34).

The sensory components of the nerves supplying the pharyngeal arches (V, VII, IX, and X) and the auditory nerve (VIII) have a multiple origin from the neural crest and the ectodermal placodes, which are located along the developing brain (see Fig. 13.1). These nerves have complex, often multiple sensory ganglia. Neurons in some parts of the ganglia are of neural crest origin, and neurons of other parts of ganglia arise from placodal ectoderm. (Ectodermal placodes are discussed on p. 269.)

Development of Neural Function

During the first 5 weeks of embryonic development, there is no gross behavioral evidence of neural function. Primitive

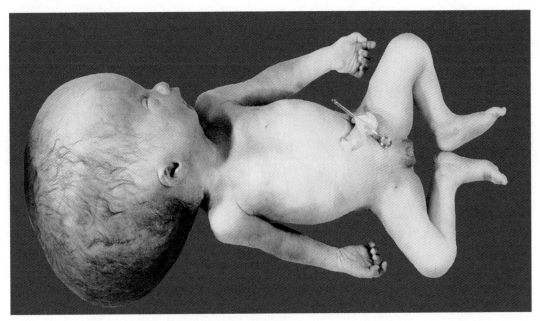

Fig. 11.38   **Fetus with pronounced hydrocephalus.** *(Courtesy of M. Barr, Ann Arbor, Mich.)*

Table 11.2 Cranial Nerves

Cranial Nerve	Associated Component of Central Nervous System	Functional Components	Distribution
Olfactory (I)	Telencephalon/olfactory placode	Special sensory (olfaction)	Olfactory area of nose
Optic (II)	Diencephalon (evagination)	Special sensory (vision)	Retina of eye
Oculomotor (III)	Mesencephalon	Motor, autonomic (minor)	Intraocular and four extraocular muscles
Trochlear (IV)	Mesencephalon (isthmus)	Motor	Superior oblique muscle of eye
Trigeminal (V)	Metencephalon (r2, r3, [pharyngeal arch 1])	Sensory, motor (some)	Derivatives of branchial arch I
Abducens (VI)	Metencephalon (r5, r6)	Motor	Lateral rectus muscle of eye
Facial (VII)	Metencephalic/myelencephalic junction (r4 [pharyngeal arch 2])	Motor Sensory (some) Autonomic (minor)	Derivatives of branchial arch II
Auditory (VIII)	Metencephalic/myelencephalic junction (r4-6, otic placode)	Special sensory (hearing, balance)	Inner ear
Glossopharyngeal (IX)	Myelencephalon (r6, r7 [pharyngeal arch 3])	Sensory, motor (some)	Derivatives of pharyngeal arch III
Vagus (X)	Myelencephalon (r7, r8 [pharyngeal arch 4])	Sensory, motor, autonomic (major)	Derivatives of pharyngeal arch IV
Accessory (XI)	Myelencephalon (r7, r8 [pharyngeal arch 4]) Spinal cord	Motor Autonomic (minor)	Gut, heart, visceral organs Some neck muscles
Hypoglossal (XII)	Myelencephalon (r8 [arch 4])	Motor	Tongue muscles

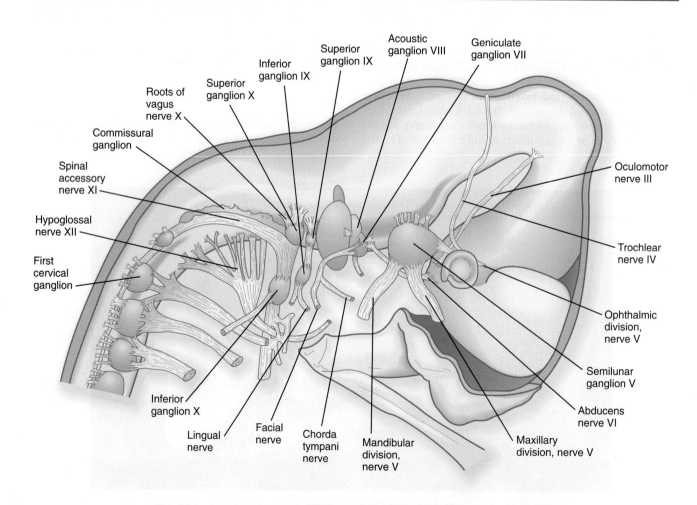

Fig. 11.39 **Reconstruction of the brain and cranial nerves of a 12-mm pig embryo.**

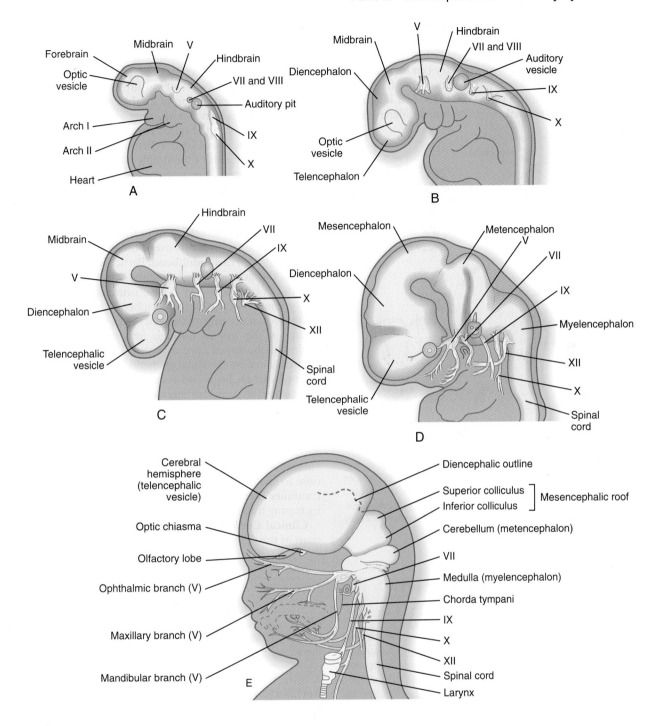

Fig. 11.40 **Development of the cranial nerves in human embryos.** A, At 3½ weeks. B, At 4 weeks. C, At 5½ weeks. D, At 7 weeks. E, At 11 weeks.

reflex activity can first be elicited at the sixth week, when touching the perioral skin with a fine bristle is followed by contralateral flexion of the neck. Over the next 6 to 8 weeks, the region of skin sensitive to tactile stimulation spreads from the face to the palms of the hands and the upper chest; by 12 weeks, the entire surface of the body except for the back and top of the head is sensitive. As these sensitive areas expand, the nature of the reflexes elicited matures from generalized

movements to specific responses of more localized body parts. There is a general craniocaudal sequence of appearance of reflex movements.

Spontaneous uncoordinated movements typically begin when the embryo is more than 7 weeks old. Later coordinated movements (see Fig. 18.7) are the result of the establishment of motor tracts and reflex arcs within the central nervous system. Behavioral development during the last trimester, which has been revealed by studying premature infants, is often subtle and reflects the structural and functional maturation of neuronal circuits.

The development of functional circuitry can be illustrated by the spinal cord. Several stages of structural and functional maturation can be identified (**Fig. 11.41**). The first is a prereflex stage, which is characterized by the initial differentiation

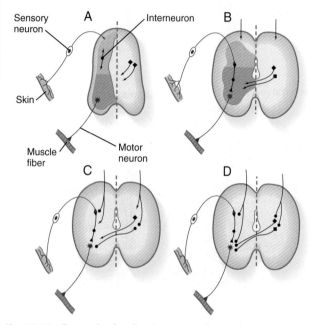

Fig. 11.41 **Stages in the development of neural circuitry. A,** Presynaptic stage. **B,** Closure of the primary reflex circuit. **C,** Connections with longitudinal and lateral inputs. **D,** Completion of circuits and myelination. *(Based on Bodian D: In Quartan GC, Melnechuk T, Adelman G, eds: The neurosciences: second study program, New York, 1970, Rockefeller University Press, pp 129-140.)*

(including axonal and dendritic growth) of the neurons according to a well-defined sequence, starting with motoneurons, followed by sensory neurons, and finally including the interneurons that connect the two (see Figs. 11.15 and 11.41A). The second stage consists of closure of the primary circuit, which allows the expression of local segmental reflexes. While the local circuit is being set up, other axons are growing down descending tracts in the spinal cord or are crossing from the other side of the spinal cord. When these axons make contact with the components of the simple reflex that was established in the second stage, the anatomical basis for intersegmental and cross-cord reflexes is set up. Later in the fetal period, these more complex circuits are completed, and the tracts are myelinated by oligodendrocytes.

The functional maturation of individual tracts, as indicated by their myelination, occurs over a broad time span and is not completed until early adulthood. Particularly in early postnatal life, the maturation of functional tracts in the nervous system can be followed by clinical neurological examination.

Myelination begins in the peripheral nervous system, with motor roots becoming myelinated before sensory roots (which occurs in the second through fifth months). Myelination begins in the spinal cord at about 11 weeks and proceeds according to a craniocaudal gradient. During the third trimester, myelination begins to occur in the brain, but there, in contrast to the peripheral nervous system, myelination is first seen in sensory tracts (e.g., in the visual system). Myelination in complex association pathways in the cerebral cortex occurs after birth. In the **corticospinal tracts,** the main direct connection between the cerebral cortex and the motor nerves emanating from the spinal cord, myelination extends caudally only to the level of the medulla by 40 weeks. Myelination continues after birth, and its course can be appreciated by the increasing mobility of infants during their first year of life.

Clinical Correlation 11.1 presents congenital malformations of the nervous system.

CLINICAL CORRELATION 11.1
Congenital Malformations of the Nervous System

In an organ system as prominent and complex as the nervous system, it is not surprising that the brain and spinal cord are subject to a wide variety of congenital malformations. These malformations range from severe structural anomalies resulting from incomplete closure of the neural tube to functional deficits caused by unknown factors acting late in pregnancy.

Defects in Closure of the Neural Tube

Failure of the neural tube to close occurs most commonly in the anterior and posterior neuropore, but failure to close in other locations is also possible. In this condition, the spinal cord or brain in the affected area is splayed open, with the wall of the central canal or ventricular system constituting the outer surface. Many closure defects can be diagnosed by the detection of elevated levels of α-fetoprotein in the amniotic fluid or by ultrasound scanning. A closure defect of the spinal cord is called **rachischisis,** and a closure defect of the brain is called **cranioschisis.** A patient with cranioschisis dies. Rachischisis (**Figs. 11.42** and **11.43A**) is associated with a wide variety of severe problems, including chronic infection, motor and sensory deficits, and disturbances in bladder function. These

defects commonly accompany anencephaly (see Fig. 8.4), in which there is a massive deficiency of cranial structures.

Other Closure Defects

A defect in the formation of the bony covering overlying either the spinal cord or the brain can result in a graded series of structural anomalies. In the spine, the simplest defect is called **spina bifida occulta** (Fig. 11.43B), which occurs in at least 5% of the population. The spinal cord and meninges remain in place, but the bony covering (neural arch) of one or more vertebrae is incomplete. Sometimes the defect goes unnoticed for many years. The neural arches are induced by the roof plate of the neural tube, with the mediation of Msx-2. Spina bifida occulta probably results from a local defect in induction. The site of the defect in the neural arches is often marked by a tuft of hair. This localized hair formation may result from exposure of the developing skin to other inductive influences from the neural tube or its coverings. Normally, the neural arches act as a barrier to such influences.

The next most severe category of defect is a **meningocele,** in which the dura mater may be missing in the area of the defect,

Fig. 11.42 **Fetus with a severe case of rachischisis.** The brain is not covered by cranial bones, and the light-colored spinal cord is totally exposed. *(Courtesy of M. Barr, Ann Arbor, Mich.)*

and the arachnoid layer bulges prominently beneath the skin (Fig. 11.43C). The spinal cord remains in place, however, and neurological symptoms are often minor. The most severe condition is a **myelomeningocele**, in which the spinal cord bulges or is entirely displaced into the protruding subarachnoid space (**Fig. 11.44**; see Fig. 11.43D). Because of problems associated with displaced spinal roots, neurological problems are commonly associated with this condition.

A similar spectrum of anomalies is associated with cranial defects (**Figs. 11.45** and **11.46**). A **meningocele** is typically associated with a small defect in the skull, whereas brain tissue alone (**meningoencephalocele**) or brain tissue containing part of the ventricular system (**meningohydroencephalocele**) may protrude through a larger opening in the skull. Depending on the nature of the protruding tissue, these malformations may be associated with neurological deficits. The mechanical circumstances may also lead to secondary hydrocephalus in some cases.

Microcephaly is a relatively rare condition characterized by the underdevelopment of the brain and the cranium (see Fig. 9.9). Primary microcephaly (in contrast to secondary microcephaly, which arises after birth) is most likely caused by a reduction in the number of neurons formed in the fetal brain. It could also arise from premature closure of the cranial sutures.

Many of the functional defects of the nervous system are poorly characterized, and their etiology is not understood. Studies of mice with genetically based defects of movement or behavior caused by abnormalities of cell migration or histogenesis in certain regions of the brain suggest the probability of a parallel spectrum of human defects. A good example is **lissencephaly**, a condition characterized by a smooth brain surface instead of the gyri and sulci that characterize the normal brain. Underlying this gross defect is abnormal layering of cortical neurons in a manner reminiscent of

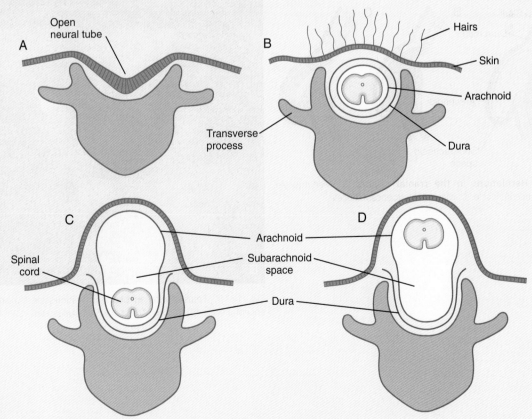

Fig. 11.43 **Varieties of closure defects of the spinal cord and vertebral column.** **A,** Rachischisis. **B,** Spina bifida occulta, with hair growth over the defect. **C,** Meningocele. **D,** Myelomeningocele.

CLINICAL CORRELATION 11.1
Congenital Malformations of the Nervous System—cont'd

the pathological features seen in reeler mice (see p. 234). At present, mutations of at least five genes affecting various aspects of neuronal migration toward the cortex are known in humans. **Mental retardation** is common and can be attributed to many genetic and environmental causes. The timing of the insult to the brain may be late in the fetal period.

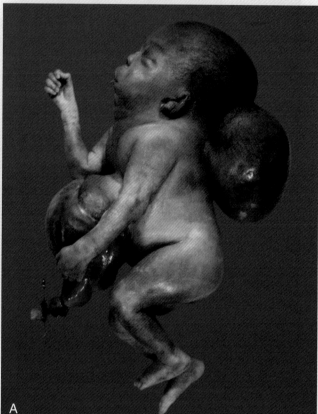

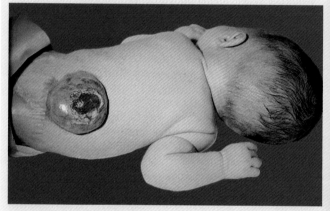

Fig. 11.44 **Infant with a myelomeningocele and secondary hydrocephalus.** *(Courtesy of M. Barr, Ann Arbor, Mich.)*

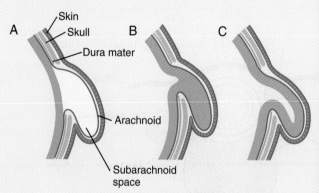

Fig. 11.45 **Herniations in the cranial region.** **A,** Meningocele. **B,** Meningoencephalocele. **C,** Meningohydroencephalocele.

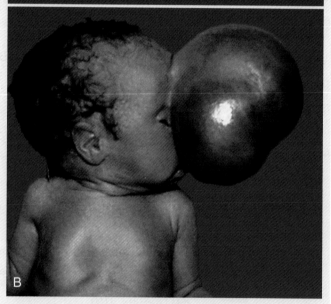

Fig. 11.46 Fetuses with an occipital meningocele (**A**) and a frontal encephalocele (**B**). *(Courtesy of M. Barr, Ann Arbor, Mich.)*

Clinical Vignette

An infant is born with rachischisis of the lower spine. During succeeding weeks, his head also begins to increase in size. Radiological imaging reveals that the infant's ventricular system is greatly dilated, and the walls of the brain itself are thinned.

Assuming that the infant lives, what are some of the clinical problems that he faces in later life?

Summary

- While the neural tube is closing, its open ends are the cranial and caudal neuropores. The newly formed brain consists of three parts: the prosencephalon, the mesencephalon, and the rhombencephalon. The prosencephalon later subdivides into the telencephalon and the diencephalon, and the rhombencephalon forms the metencephalon and myelencephalon.

- Within the neural tube, neuroepithelial cells undergo active mitotic proliferation. Their daughter cells form neuronal or glial progenitor cells. Among the glial cells, radial glial cells act as guidewires for the migration of neurons from their sites of origin to definite layers in the brain. Microglial cells arise from mesoderm.

- The neural tube divides into ventricular, intermediate, and marginal zones. Neuroblasts in the intermediate zone (future gray matter) send out processes that collect principally in the marginal zone (future white matter). The neural tube is also divided into a dorsal alar plate and a ventral basal plate. The basal plate represents the motor component of the spinal cord, and the alar plate is largely sensory.

- Through an induction by the notochord, mediated by shh, a floor plate develops in the neural tube. Further influences of shh, produced by the notochord and the floor plate, result in the induction of motoneurons in the basal plate.

- Much of the early brain is a highly segmented structure. This structure is reflected in the rhombomeres and molecularly in the patterns of expression of homeobox-containing genes. Neurons and their processes developing within the rhombomeres follow specific rules of behavior with respect to rhombomere boundaries. Nerve processes growing from the spinal cord react to external cues provided by the environment of the somites. Neurons and neural crest cells can readily penetrate the anterior but not the posterior mesoderm of the somite.

- The midbrain and metencephalic structures are specified by a signaling center, the isthmic organizer, at the midbrain-hindbrain border. FGF-8 is one of the major signaling molecules. The forebrain is divided into three segments called prosomeres. FGF-8, secreted by the anterior neural ridge, induces the expression of Foxg-1, which regulates development of the telencephalon and optic vesicles. Shh, secreted by midline axial structures, organizes the ventral forebrain.

- A peripheral nerve forms by the outgrowth of motor axons from the ventral horn of the spinal cord. The outgrowing axons are capped by a growth cone. This growing tip continually samples its immediate environment for cues that guide the amount and direction of axonal growth. The motor component of a peripheral nerve is joined by the

- sensory part, which is based on neural crest–derived cell bodies in dorsal root ganglia along the spinal cord. Axons and dendrites from the sensory cell bodies penetrate the spinal cord and grow peripherally with the motor axons. Connections between the nerve and end organs are often mediated through trophic factors. Neurons that do not establish connections with peripheral end organs often die.

- The autonomic nervous system consists of two components: the sympathetic nervous system and the parasympathetic nervous system. Both components contain preganglionic neurons, which arise from the central nervous system, and postganglionic components, which are of neural crest origin. Typically, sympathetic neurons are adrenergic, and parasympathetic neurons are cholinergic. The normal choice of transmitter can be overridden, however, by environmental factors so that a sympathetic neuron can secrete acetylcholine.

- The spinal cord functions as a pathway for organized tracts of nerve processes and an integration center for local reflexes. During the fetal period, growth in the length of the spinal cord lags behind that of the vertebral column, thus pulling the nerve roots and leaving the end of the spinal cord as a cauda equina.

- Within the brain, the myelencephalon retains an organizational similarity to the spinal cord with respect to the tracts passing through, but centers that control respiration and heart rate also form at the site. The metencephalon contains two parts: the pons (which functions principally as a conduit) and the cerebellum (which integrates and coordinates many motor movements and sensory reflexes). In the cerebellum, which forms from the rhombic lips, the gray matter forms on the outside. The ventral part of the mesencephalon is the region through which the major tracts of nerve processes that connect centers in the cerebral cortex with specific sites in the spinal cord pass. The dorsal part of the mesencephalon develops the superior and inferior colliculi, which are involved with the integration of visual and auditory signals.

- The diencephalon and the telencephalon represent modified alar plate regions. Many important nuclei and integrating centers develop in the diencephalon, among them the thalamus, hypothalamus, neural hypophysis, and pineal body. The eyes also arise as outgrowths from the diencephalon. In humans, the telencephalon ultimately overgrows other parts of the brain. Similar to the cerebellum, the telencephalon is organized with the gray matter in layers outside the white matter. Neuroblasts migrate through the white matter to these layers by using radial glial cells as their guides.

- Within the central nervous system, the central canal expands to form a series of four ventricles in the brain. Specialized vascular plexuses form cerebrospinal fluid, which circulates throughout the central nervous system. Around the brain and spinal cord, two layers of mesenchyme form the meninges.

- The cranial nerves are organized on the same fundamental plan as the spinal nerves, but they have lost their regular segmental pattern and have become highly specialized. Some are purely motor, others are purely sensory, and others are mixed.

- Many congenital malformations of the nervous system are based on incomplete closure of the neural tube or associated skeletal structures. In the spinal cord, the spectrum of defects ranges from a widely open neural tube (rachischisis) to relatively minor defects in the neural arch over the cord (spina bifida occulta). A similar spectrum of defects is seen in the brain.

■ Neural function appears in concert with the structural maturation of various components of the nervous system. The first reflex activity is seen in the sixth week. During successive weeks, the reflex movements become more complex, and spontaneous movements appear. Final functional maturation coincides with myelination of the tracts and is not completed until many years after birth.

Review Questions

1. What molecule produced by the notochord is instrumental in inducing the floor plate of the neural tube?
A. Hoxa-5
B. Retinoic acid
C. Pax-3
D. Msx-1
E. Shh

2. The cell bodies of the motoneurons of a spinal nerve arise from the:
A. Basal plate
B. Marginal zone
C. Floor plate
D. Roof plate
E. Alar plate

3. An infant with a tuft of hair over the lumbar region of the vertebral column undergoes surgery for a congenital anomaly in that region. During surgery, it was found that the dura and arachnoid layers over the spinal cord were complete, but that the neural arches of several vertebrae were missing. What condition did the infant have?
A. Meningocele
B. Meningomyelocele
C. Encephalocele
D. Spina bifida occulta
E. Rachischisis

4. Growth cones adhere strongly to a substrate containing:
A. Acetylcholine
B. Laminin
C. Epinephrine
D. Norepinephrine
E. Sonic hedgehog

5. Complete failure of the neural tube to close in the region of the spinal cord is:
A. Spina bifida occulta
B. Meningocele
C. Cranioschisis
D. Rachischisis
E. Myelomeningocele

6. Rhombomeres are segmental divisions of the:
A. Forebrain
B. Midbrain
C. Hindbrain
D. Spinal cord
E. None of the above

7. Pregnant women typically first become aware of fetal movements during what month of pregnancy?
A. Second
B. Third
C. Fourth
D. Sixth
E. Eighth

8. Rathke's pouch arises from the:
A. Diencephalon
B. Stomodeal ectoderm
C. Mesencephalon
D. Pharyngeal endoderm
E. Infundibulum

9. In the early days after birth, an infant does not pass fecal material and develops abdominal swelling. An anal opening is present. What is the probable condition?

10. What is the likely appearance of the spinal cord and brachial nerves in an infant who was born with the congenital absence of one arm (amelia)?

References

Altman J, Bayer SA: *Development of the human spinal cord*, New York, 2001, Oxford University Press.

Balmer CW, LaMantia A-S: Noses and neurons: induction, morphogenesis, and neuronal differentiation in the peripheral olfactory pathway, *Dev Dyn* 234:464-481, 2005.

Briscoe J and others: A homeodomain protein code specifies progenitor cell identity and neuronal fate in the ventral neural tube, *Cell* 101:435-445, 2000.

Butler SJ, Tear G: Getting axons onto the right path: the role of transcription factors in axon guidance, *Development* 134:439-448, 2007.

Cameron RS, Rakic P: Glial cell lineage in the cerebral cortex: a review and synthesis, *Glia* 4:124-137, 1991.

Carpenter EM: *Hox* genes and spinal cord development, *Dev Neurosci* 24:24-34, 2002.

Chandrasekhar A: Turning heads: development of vertebrate branchiomotor neurons, *Dev Dyn* 229:143-161, 2004.

Chilton JK: Molecular mechanisms of axon guidance, *Dev Biol* 292:13-24, 2006.

Chizikov VV, Millen KJ: Roof plate-dependent patterning of the vertebrate dorsal central nervous system, *Dev Biol* 277:287-295, 2005.

Ciani L, Salinas PC: Wnts in the vertebrate nervous system: from patterning to neuronal conductivity, *Nat Rev Neurosci* 6:351-362, 2005.

Copp AJ, Greene NDE: Genetics and development of neural tube defects, *J Pathol* 220:217-230, 2010.

Cordes SP: Molecular genetics of cranial nerve development in the mouse, *Nat Rev Neurosci* 2:611-623, 2001.

Cowan WM, Jessell TM, Zipursky SL: *Molecular and cellular approaches to neural development*, New York, 1997, Oxford University Press.

Dessaud E, McMahon AP, Briscoe J: Pattern formation in the vertebrate neural tube: a sonic hedgehog morphogen-regulated transcriptional network, *Development* 135:2489-2503, 2008.

Detrait ER and others: Human neural tube defects: developmental biology, epidemiology, and genetics, *Neurotoxicol Teratol* 27:515-524, 2005.

Dickson BJ: Molecular mechanisms of axon guidance, *Science* 298:1959-1964, 2002.

Eickholt BJ and others: Rhombomere interactions control segmental differentiation of hindbrain neurons, *Mol Cell Neurosci* 18:141-148, 2001.

Famulski JK and others: Siah regulation of Pard3A controls neuronal cell adhesion during germinal zone exit, *Science* 330:1834-1838, 2010.

Farrar NR, Spencer GE: Pursuing a "turning point" in growth cone research, *Dev Biol* 318:102-111, 2008.

Freeman MR: Specification and morphogenesis of astrocytes, *Science* 330:774-778, 2010.

Fucillo M, Joyner AL, Fishell G: Morphogen to mitogen: the multiple roles of hedgehog signaling in vertebrate neural development, *Nat Rev Neurosci* 7:772-783, 2006.

Gato A, Desmond ME: Why the embryo still matters: CSF and the neuroepithelium as interdependent regulators of embryonic brain growth, morphogenesis and histogenesis, *Dev Biol* 327:263-272, 2009.

Ginhoux F and others: Fate mapping analysis reveals that adult microglia derive from primitive macrophages, *Science* 330:841-8445, 2010.

Goldberg JL: How does an axon grow? *Genes Dev* 17:941-958, 2003.

Guillemot F: Spatial and temporal specification of neural fates by transcription factor codes, *Development* 134:3771-3780, 2007.

Hammond R and others: Slit-mediated repulsion is a key regulator of motor axon pathfinding in the hindbrain, *Development* 132:4483-4495, 2005.

Hatten ME: New directions in neuronal migration, *Science* 297:1660-1663, 2002.

Hidalgo-Sánchez M and others: Specification of the meso-isthmo-cerebellar region: the *Otx2/Gbx2* boundary, *Brain Res Rev* 49:134-149, 2005.

Howard MJ: Mechanisms and perspectives on differentiation of autonomic neurons, *Dev Biol* 277:271-286, 2005.

Ille F and others: Wnt/BMP signal integration regulates the balance between proliferation and differentiation of neuroepithelial cells in the dorsal spinal cord, *Dev Biol* 304:394-408, 2007.

Ingham PW, Placzek M: Orchestrating ontogenesis: variations on a theme by sonic hedgehog, *Nat Rev Genet* 7:841-850, 2006.

Islam SM and others: Draxin, a repulsive guidance protein for spinal cord and forebrain commissures, *Science* 3223:388-393, 2009.

Kaprielian Z, Imondi R, Runko E: Axon guidance at the midline of the developing CNS, *Anat Rec* 261:176-197, 2000.

Kato M, Dobyns WB: Lissencephaly and the molecular basis of neuronal migration, *Hum Mol Genet* 12:R89-R96, 2003.

Kaufman BA: Neural tube defects, *Pediatr Clin North Am* 51:389-419, 2004.

Krispin S, Nitzann E, Kalcheim C: The dorsal neural tube: a dynamic setting for cell fate decisions, *Dev Neurobiol* 70:796-812, 2010.

Lambert de Rouvroit C, Goffinet AM: Neuronal migration, *Mech Dev* 105:47-56, 2001.

Liu A, Niswander LA: Bone morphogenetic protein signaling and vertebrate nervous system development, *Nat Rev Neurosci* 6:945-954, 2005.

Long H and others: Conserved roles for Slit and Robo proteins in midline commissural axon guidance, *Neuron* 42:213-223, 2004.

Lumsden A, Krumlauf R: Patterning the vertebrate neuraxis, *Science* 274:1109-1115, 1996.

Marquardt T, Pfaff SL: Cracking the transcriptional code for cell specification in the neural tube, *Cell* 106:1-4, 2001.

McInnes RR, Michaud J: Mechanisms regulating the development of the corpus callosum and its agenesis in mouse and human, *Clin Genet* 66:276-289, 2004.

Melani M, Weinstein BM: Common factors regulating patterning of the nervous and vascular systems, *Annu Rev Cell Dev Biol* 26:639-665, 2010.

Mitchell LE and others: Spina bifida, *Lancet* 364:1885-1895, 2004.

Müller F, O'Rahilly R: The initial appearance of the cranial nerves and related neuronal migration in staged human embryos, *Cells Tissues Organs* 193:215-238, 2011.

Müller F, O'Rahilly RO: The timing and sequence of appearance of neuromeres and their derivatives in staged human embryos, *Acta Anat* 158:83-99, 1997.

Nakamura H and others: Isthmus organizer for midbrain and hindbrain development, *Brain Res Rev* 49:120-126, 2005.

Narita Y, Rijli FM: *Hox* genes in neural patterning and circuit formation in the mouse hindbrain, *Curr Top Dev Biol* 88:139-167, 2009.

Norman MG: Malformations of the brain, *J Neuropathol Exp Neurol* 55:133-143, 1996.

Oppenheim RW: Cell death during development of the nervous system, *Annu Rev Neurosci* 14:453-501, 1991.

O'Rahilly R, Müller F: *The embryonic human brain*, New York, 1994, Wiley-Liss.

Oury F and others: *Hoxa2*- and rhombomere-dependent development of the mouse facial somatosensory map, *Science* 313:1408-1413, 2006.

Placzek M, Dodd J, Jessell TM: Discussion point: the case for floor plate induction by the notochord, *Curr Opin Neurobiol* 10:15-22, 2000.

Puelles L: Forebrain development: prosomere model, *Encyclop Neurosci* 4:315-319, 2009.

Puelles L, Rubenstein JLR: Forebrain gene expression domains and the evolving prosomeric model, *Trends Neurosci* 26:469-476, 2003.

Puelles E and others: Otx dose-dependent integrated control of antero-posterior and dorso-ventral patterning of midbrain, *Nat Neurosci* 6:453-460, 2003.

Ragsdale CW, Grove EA: Patterning the mammalian cerebral cortex, *Curr Opin Neurobiol* 11:50-58, 2001.

Ramos C, Robert B: *msh/Msx* gene family in neural development, *Trends Genet* 21:624-632, 2005.

Rash BG, Grove EA: Area and layer patterning in the developing cerebral cortex, *Curr Opin Neurobiol* 16:25-34, 2006.

Rhinn M, Picker A, Brand M: Global and local mechanisms of forebrain and midbrain patterning, *Curr Opin Neurobiol* 16:5-12, 2006.

Rowitch DH: Glial specification in the vertebrate neural tube, *Nat Rev Neurosci* 5:409-419, 2004.

Sarnat HB: Molecular genetic classification of central nervous system malformations, *J Child Neurol* 15:675-687, 2000.

Sato T, Joyner AL, Nakamura H: How does Fgf signaling from the isthmic organizer induce midbrain and cerebellum development? *Dev Growth Differ* 46:487-494, 2004.

Scholpp S and others: Otx1, Otx2 and Irx1b establish and position the ZLI in the diencephalon, *Development* 134:3167-3176, 2007.

Sousa VH, Fishell G: Sonic hedgehog functions through dynamic changes in temporal competence in the developing forebrain, *Curr Opin Genet Dev* 20:391-399, 2010.

Storm EE and others: Dose-dependent functions of FGF8 in regulating telencephalic patterning centers, *Development* 133:1831-1844, 2006.

ten Donkelaar HJ: Major events in the development of the forebrain, *Eur J Morphol* 38:301-308, 2000.

ten Donkelaar HJ and others: Development and malformations of the human pyramidal tract, *J Neurol* 251:1429-1442, 2004.

Thompson J, Lovicu F, Ziman M: The role of Pax7 in determining the cytoarchitecture of the superior colliculus, *Dev Growth Differ* 46:213-218, 2004.

Ulloa F, Marti E: Wnt won the war: antagonistic role of Wnt over shh controls dorso-ventral patterning of the vertebrate neural tube, *Dev Dyn* 239:69-76, 2010.

Vermeren M and others: Integrity of developing spinal motor columns is regulated by neural crest derivatives at motor exit points, *Neuron* 37:403-415, 2003.

Wilson L, Maden M: The mechanisms of dorsoventral patterning in the vertebrate neural tube, *Dev Biol* 282:1-13, 2005.

Wingate RJT: The rhombic lip and early cerebellar development, *Curr Opin Neurobiol* 11:82-88, 2001.

Woods CG: Human microcephaly, *Curr Opin Neurobiol* 14:112-117, 2004.

Ypsilanti AR, Zagar Y, Chédotal A: Moving away from the midline: new developments for Slit and Robo, *Development* 137:1939-1952, 2010.

Zhuang B-Q, Sockanathan S: Dorsal-ventral patterning: a view from the top, *Curr Opin Neurobiol* 16:20-24, 2006.

Chapter 12

Neural Crest

The neural crest, whose existence has been recognized for more than a century, forms an exceptionally wide range of cell types and structures, including several types of nerves and glia, connective tissue, bones, and pigment cells. Its importance and prominence are such that the neural crest has often been called the fourth germ layer of the body. Not until adequate methods of marking neural crest cells became available—first with isotopic labels and subsequently with stable biological markers, monoclonal antibodies, intracellular dyes, and genetic markers—did the neural crest become one of the most widely studied components of the vertebrate embryo. Most studies on the neural crest have been conducted on the avian embryo because of its accessibility and the availability of specific markers (see Fig. 9.31). More recently, emphasis has shifted to studies on the mouse, especially for dissecting molecular controls, but it appears that most of the information on the biology of the neural crest derived from birds can be applied to mammalian embryos. Some important syndromes and malformations are based on abnormalities of the neural crest. Some of these syndromes are presented in Clinical Correlation 12.1, at the end of the chapter.

Developmental History of the Neural Crest

The neural crest originates from cells located along the lateral margins of the neural plate. Tracing the history of the neural crest in any region involves consideration of the following: (1) its origin, induction, and specification; (2) epithelial-to-mesenchymal transformation and emigration from the neural tube; (3) migration; and (4) differentiation. Each of these phases in the development of the generic neural crest is covered before neural crest development in specific regions of the body.

Origin, Induction, and Specification

According to the most recent data, the earliest stages of neural crest induction may occur as early as gastrulation, but according to the classical model, the neural crest arises as the result of inductive actions by the adjacent non-neural ectoderm and possibly nearby mesoderm on the neural plate (**Fig. 12.1**). The ectodermal inductive signals are **bone morphogenetic proteins** (**BMPs**) and **Wnts**. **Fibroblast growth factor-8** (**FGF-8**) from mesoderm plays a role in neural crest induction in

amphibians, and it seems to be involved in mammals as well. The role of BMPs is complex and relates to a concentration gradient along the ectodermal layer as neurulation proceeds. The highest concentrations of BMP are seen in the lateral ectoderm, and cells exposed to these concentrations remain ectodermal. Cells within the neural plate are exposed to the lowest concentrations of BMP because of the local inhibitory actions of noggin and chordin (see Fig. 5.8D), and, by default, they remain neural. Cells at the border of the neural plate are exposed to intermediate levels of BMP, and, in this environment, they are induced to form neural crest precursor cells.

In response to these inductive signals, cells at the border of the neural plate activate genes coding for several transcription factors, including **Msx-1** and **Msx-2**, **Dlx-5**, **Pax-3/Pax-7**, and **Gbx-2**. These and other gene products turn on a network of genes that transform the epithelial neural crest precursor cells into mobile mesenchymal cells that break free from the neuroepithelium of the neural tube.

Epitheliomesenchymal Transformation and Emigration from the Neural Tube

Within the neural tube, neural crest precursor cells are epithelial and are tightly adherent to other neuroepithelial cells through a variety of intercellular connections. Prominent among them are the **cadherins**. Among the new transcription factors upregulated in induced neural crest precursor cells are **snail-1** and **snail-2** (formerly called **slug**) and **Foxd-3**, which are instrumental in allowing the neural crest cells to break free from the neural epithelium and then migrate away as mesenchymal cells.* Under the influence of snail-1 and snail-2, the profile of cadherins produced by the neural crest precursors changes from **type I cadherins** (e.g., **N-cadherin** and **E-cadherin**), which are strongly adhesive, to **type II cadherins**, which are less adhesive.

Neural crest cells break free from the neural tube in the trunk at the level of the last-formed somite or the neural plate in the head by changing their shape and properties from those of typical neuroepithelial cells to those of mesenchymal cells. Important to this process is the loss of cell-to-cell adhesiveness. This loss is effected by the loss of cell adhesion molecules (CAMs) characteristic of the neural tube (e.g., N-CAM,

---

*Snail-2 is also expressed during gastrulation by cells of the epiblast after they have entered the walls of the primitive streak and are about to leave as mesenchymal cells of the mesoderm germ layer.

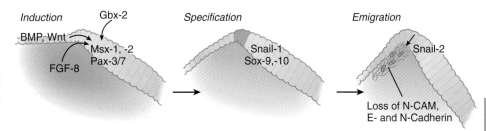

Fig. 12.1 Induction and emigration of neural crest cells from the neural tube. BMP, bone morphogenetic protein; FGF-8, fibroblast growth factor-8; N-CAM, neural cell adhesion molecule.

E-cadherin, and N-cadherin). These molecules remain downregulated during migration, but after neural crest cells have completed their migrations and have differentiated into certain structures (e.g., spinal ganglia), CAMs are often expressed again.

In the head, where closure of the neural plate has not yet occurred, neural crest cells must penetrate the basal lamina underlying the neural plate. This is accomplished by the production of enzymes that degrade components of the basal lamina and by sending out processes that penetrate the basal lamina. In the trunk, neural crest cells do not leave the neuroepithelium until after the neural tube has formed. They do not, however, have to contend with penetrating a basal lamina because the dorsal part of the neural tube does not form a basal lamina until after emigration of the crest cells.

Neural Crest Cell Migration

After leaving the neuroepithelium, the neural crest cells first encounter a relatively cell-free environment rich in extracellular matrix molecules (**Fig. 12.2**). In this environment, the cells undergo extensive migrations along several well-defined pathways. These migrations are determined by intrinsic properties of the neural crest cells and features of the external environment encountered by the migrating cells.

Neural crest migration is influenced by a variety of molecules residing in the extracellular matrix. Although the presence of a basal lamina can inhibit their emigration from the neural tube, neural crest cells often prefer to migrate along basal laminae, such as those of the surface ectoderm or neural tube, after they have left the neural tube. Components of the extracellular matrix permissive for migration include molecules found in basal laminae, such as fibronectin, laminin, and type IV collagen (**Fig. 12.3**). Attachment to and migration over these substrate molecules are mediated by the family of attachment proteins called **integrins**. Other molecules, such as chondroitin sulfate proteoglycans, are not good substrates for neural crest cells and inhibit their migration.

Neural crest cells emigrate from the neural tube or neural folds in streams, with each cell in contact with neighbors through filopodial contacts. During their migratory phase, neural crest cells are exquisitely sensitive to guidance molecules, most of which are inhibitory. Among the most important of these guidance molecules are the ligand/receptor pairs **Robo/Slit**, **Neuropilin/Semaphorin** and **Ephrin/Eph** (see Table 11.1). Much less is known about attractive influences on neural crest cell migration. During migration, neural crest cells extend protrusions that both test the environment and are part of the propulsive mechanism. If an inhibitory influence is encountered, the protrusions collapse through signals

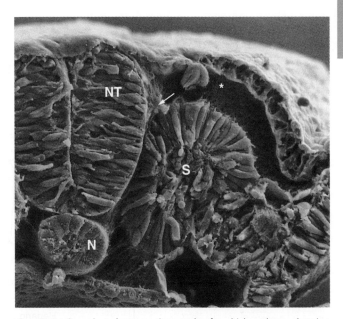

Fig. 12.2 Scanning electron micrograph of a chick embryo, showing the early migration of neural crest cells *(arrow)* out of the neural tube (NT). The subectodermal pathway of neural crest migration *(asterisk)* is relatively cell free, but it contains a fine mesh of extracellular matrix molecules. N, notochord; S, somite. *(Courtesy of K. Tosney, Ann Arbor, Mich.)*

derived from a **planar cell polarity** pathway (see p. 87). This mechanism acts as a brake when the cells encounter an inhibitory environment, but it is also involved in their forward propulsion. In a migrating stream of neural crest cells, contact with the cells behind also results in the pulling of protrusions at the trailing edge of the cells, thus resulting in a net forward motion of the leading cells. Specific examples of the environmental control of neural crest cell migrations are given later in this chapter. Much remains to be learned about what causes neural crest cells to stop migrating, but often they stop migrating in areas where repulsive signals are low.

Differentiation of Neural Crest Cells

Neural crest cells ultimately differentiate into an astonishing array of adult structures (**Table 12.1**). What controls their differentiation is one of the principal questions of neural crest biology. Two opposing hypotheses have been proposed. According to one, all neural crest cells are equal in developmental potential, and their ultimate differentiation is entirely determined by the environment through which they migrate and into which they finally settle. The other hypothesis

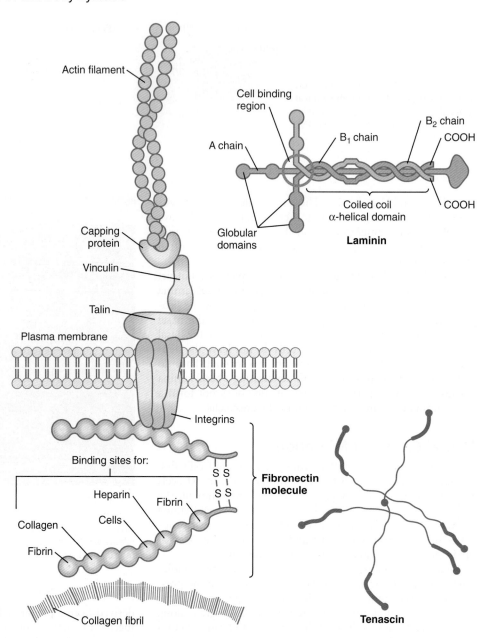

Fig. 12.3 **Structure of some common extracellular matrix molecules.**

suggests that premigratory crest cells are already programmed for different developmental fates, and that certain stem cells are favored, whereas others are inhibited from further development during migration. More recent research indicates that the real answer can be found somewhere between these two positions. Increasing evidence suggests that among migrating neural crest cells is a mix of cells whose fate has been predetermined within the neural tube and cells whose ultimate phenotype depends on environmental influences.

A correlation exists between the time of migration of neural crest cells from the neural tube and their developmental potential. Many cells that first begin to migrate have the potential to differentiate into several different types of cells. Crest cells that begin to migrate later are capable of forming only derivatives characteristic of more dorsal locations (e.g., spinal ganglia), but not sympathetic neurons or adrenal medullary cells. Crest cells that leave the neural tube last can form only pigment cells.

Several experiments have shown that the fates of some neural crest cells are not irreversibly fixed along a single pathway. One type of experiment involves the transplantation of neural crest cells from one part of the body to another. For example, many neural crest cells from the trunk differentiate into sympathetic neurons that produce norepinephrine as the transmitter. In the cranial region, however, neural crest cells give rise to parasympathetic neurons, which produce acetylcholine. If thoracic neural crest cells are transplanted into the head, some cells differentiate into cholinergic parasympathetic neurons instead of the adrenergic sympathetic neurons normally produced. Conversely, cranial neural crest cells grafted into the thoracic region respond to their new environment by forming adrenergic sympathetic neurons.

A more striking example is the conversion of cells of the periocular neural crest mesenchyme, which in birds would normally form cartilage, into neurons if they are associated with embryonic hindgut tissue in vitro. Many of the regional

Table 12.1 Major Derivatives of the Neural Crest

	Trunk Crest	**Cranial and Circumpharyngeal Crests**
NERVOUS SYSTEM		
Sensory nervous system	Spinal ganglia	Ganglia of trigeminal nerve (V), facial nerve (VII), glossopharyngeal nerve (superior ganglion) (IX), vagus nerve (jugular ganglion) (X)
	Satellite cells of sensory ganglia	Satellite cells of sensory ganglia
	Schwann cells of all peripheral nerves, enteric glial cells	Schwann cells of peripheral nerves
	Merkel cells	
Autonomic nervous system	Sympathetic chain ganglia, collateral ganglia: celiac and mesenteric	Parasympathetic ganglia: ciliary, ethmoidal, sphenopalatine, submandibular, visceral
	Parasympathetic ganglia: pelvic and visceral plexuses	
Meninges	None	Leptomeninges of prosencephalon and part of mesencephalon
Pigment cells	Melanocytes	Melanocytes
Endocrine and paraendocrine cells	Adrenal medulla, neurosecretory cells of heart and lungs	Carotid body (type I cells), parafollicular cells (thyroid)
MESECTODERMAL CELLS		
Skeleton	None	Cranial vault (squamosal and part of frontal), nasal and orbital, otic capsule (part), palate and maxillary, mandible, sphenoid (small contribution), trabeculae (part), visceral cartilages, external ear cartilage (part)
Connective tissue	None	Dermis and fat of skin; cornea of eye (fibroblasts of stroma and corneal endothelium); dental papilla (odontoblasts); connective tissue stroma of glands: thyroid, parathyroid, thymus, salivary, lacrimal; outflow tract (truncoconal region) of heart; cardiac semilunar valves; walls of aorta and aortic arch–derived arteries; adipocytes
Muscle	None	Ciliary muscles, dermal smooth muscles, vascular smooth muscle, minor skeletal muscle elements (?)

influences on the differentiation of local populations of neural crest cells are now recognized to be interactions between the migrating neural crest cells and specific tissues that they encounter during migration. Examples of tissue interactions that promote the differentiation of specific neural crest derivatives are given in **Table 12.2**.

The plasticity of differentiation of neural crest cells can be shown by cloning single neural crest cells in culture. In the same medium, and under apparently the same environmental conditions, the progeny of the single cloned cells frequently differentiate into neuronal and non-neuronal (e.g., pigment cell) phenotypes. Similarly, if individual neural crest cells are injected in vivo with a dye, greater than 50% of the injected cells will give rise to progeny with two to four different phenotypes containing the dye. By exposing cloned neural crest precursor cells to specific environmental conditions in vitro, one can begin to understand the mechanisms that determine phenotype in vivo. In one experiment, rat neural crest cells grown under standard in vitro conditions differentiated into neurons, but when they were exposed to glial growth factor, they differentiated into Schwann cells because the glial growth factor suppressed their tendency to differentiate into neurons. Similarly, the growth factors BMP-2 and BMP-4 cause cultured neural crest cells to differentiate into autonomic neurons, whereas exposure of these cells to transforming growth factor-β causes them to differentiate into smooth muscle.

Table 12.2 Environmental Factors Promoting Differentiation of Neural Crest Cells

Neural Crest Derivative	**Interacting Structure**
Bones of cranial vault	Brain
Bones of base of skull	Notochord, brain
Pharyngeal arch cartilages	Pharyngeal endoderm
Meckel's cartilage	Cranial ectoderm
Maxillary bone	Maxillary ectoderm
Mandible	Mandibular ectoderm
Palate	Palatal ectoderm
Otic capsule	Otic vesicle
Dentin of teeth	Oral ectoderm
Glandular stroma: thyroid, parathyroid, thymus, salivary	Local epithelium
Adrenal medullary chromaffin cells	Glucocorticoids secreted by adrenal cortex
Enteric neurons	Gut wall
Sympathetic neurons	Spinal cord, notochord, somites
Sensory neurons	Peripheral target tissue
Pigment cells	Extracellular matrix along pathway of migration

Not all types of transformations among possible neural crest derivatives can occur. Crest cells from the trunk transplanted into the head cannot form cartilage or skeletal elements, although this is normal for cells of the cranial neural crest. Most experiments suggest that early neural crest cells segregate into intermediate lineages that preserve the option of differentiating into several, but not all, types of individual phenotypes. In the chick embryo, some neural crest cells are antigenically different from others even before they have left the neural tube.

Many neural crest cells are bipotential, depending on signals from their local environment for cues to their final differentiation. Cultured heart cells secrete a protein that converts postmitotic sympathetic neurons from an adrenergic (norepinephrine transmitter) phenotype to a cholinergic (acetylcholine-secreting) phenotype (see Fig. 11.22). During normal development, the sympathetic neurons that innervate sweat glands are catecholaminergic until their axons actually contact the sweat glands. At that point, they become cholinergic.

Major Divisions of the Neural Crest

The neural crest arises from a wide range of craniocaudal levels, from the prosencephalon to the future sacral region. For many years, it was traditional to subdivide the neural crest into trunk and cranial components. In more recent years, however, it has become increasingly apparent that the neural crest in the posterior rhombencephalic region, often called the circumpharyngeal crest, represents another major subdivision seeding cells into the pharyngeal region, the outflow tract of the heart and great vessels, and much of the gut-associated crest derivatives.

Trunk Neural Crest

The neural crest of the trunk extends from the level of the sixth somite to the most caudal somites. Three pathways of migration are commonly described (**Fig. 12.4**). These pathways occur in different sequences and are subject to different controls. The first neural crest cells to leave the neural tube migrate around and between the somites, which are still in an epithelial configuration. Their migratory path follows the intersomitic blood vessels, and the cells rapidly reach the region of the dorsal aorta (see Fig. 12.4, pathway 1). It may be that at this early stage no other pathway is available to these migrating cells. These cells constitute the sympathoadrenal lineage.

Slightly later in development, the somites have become dissociated into sclerotomal and dermomyotomal compartments. At this stage, the neural crest cells preferentially enter

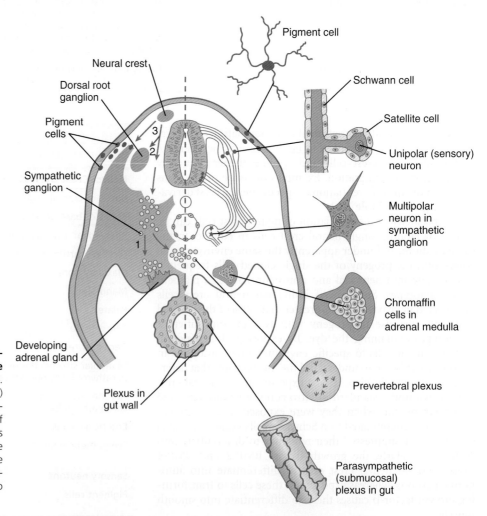

Fig. 12.4 Major neural crest migratory pathways and derivatives in the trunk. *Left,* Pathways in the early embryo. The first emigrating cells (pathway 1) follow the ventral (sympathoadrenal) pathway *(red arrows).* The second wave of emigrating cells (pathway 2) follows the ventrolateral pathway indicated by the *purple arrow.* The last cells to leave the neural tube (pathway 3) follow the dorsolateral pathway *(green arrow)* as they go on to differentiate into pigment cells.

the anterior compartment of the sclerotome. They are kept from entering the posterior compartment mainly through the repulsive action of **semaphorinA3F** (**SEMA3F**) in the posterior sclerotome, by acting through its receptor **Neuropilin-2** (**Nrp-2**) on the neural crest cells. Other molecular repulsion mechanisms are also involved, but in mammals, this mechanism is the most influential. Passage through the anterior sclerotome is facilitated by extracellular matrix molecules, in particular **thrombospondin**. These cells constitute the ventrolateral pathway, and they ultimately form the dorsal root ganglia (see Fig. 12.4, pathway 2). These cells form the ganglia in concert with the outgrowth of the motor axons from the spinal cord, which follow similar environmental cues.

The last pathway (see Fig. 12.4, pathway 3) is the dorsolateral pathway, and the cells that follow it appear to be determined even before emigration from the neural tube to become pigment cells. Other neural crest cells are not able to use this pathway. In mammals, cells that follow this pathway depend on the **Steel factor**, produced by the dermomyotome, to be able to use this pathway. The cells that take this pathway migrate just beneath the ectoderm and ultimately enter the ectoderm as pigment cells (melanocytes).

Sympathoadrenal Lineage

The sympathoadrenal lineage is derived from a committed sympathoadrenal progenitor cell that has already passed numerous restriction points so that it no longer can form sensory neurons, glia, or melanocytes. This progenitor cell gives rise to four types of cellular progenies: (1) adrenal chromaffin cells; (2) small, intensely fluorescent cells found in the sympathetic ganglia; (3) adrenergic sympathetic neurons; and (4) a small population of cholinergic sympathetic neurons.

Development of the autonomic nervous system (sympathetic and parasympathetic components) depends on the interplay of two DNA-binding proteins, **Phox-2** (a homeodomain protein) and **Mash-1** (a helix-loop-helix transcription factor). Exposure to BMPs emanating from the wall of the dorsal aorta, around which these cells aggregate, further restricts this cellular lineage into a bipotential progenitor cell that can give rise to either adrenal chromaffin cells or sympathetic neurons. The bipotential progenitor cell already possesses some neuronal traits, but final differentiation depends on the environment surrounding these cells. Differentiation into sympathetic ganglia requires signals from the ventral neural tube, the notochord, and the somites. **Norepinephrine**, produced by the notochord, and BMPs from the dorsal aorta are among the signals that promote the differentiation of sympathetic neurons. In contrast, precursor cells in the developing adrenal medulla encounter glucocorticoids secreted by adrenal cortical cells. It has long been believed that under this hormonal influence, these cells lose their neuronal properties and differentiate into chromaffin cells.

The entire length of the gut is populated by neural crest–derived parasympathetic neurons and associated cells, the enteric glia. These arise from neural crest cells in the cervical (vagal) and sacral levels and, under the influence of **glial-derived neurotrophic factor**, undertake extensive migrations along the developing gut. Sacral neural crest cells colonize the hindgut, but even there they form only a few enteric neurons. The rest are derived from the vagal crest. The autonomic innervation of the gut is covered in greater detail in the discussion of the vagal crest (see p. 264).

Sensory Lineage

Considerable uncertainty surrounds the events leading cells following the ventrolateral migratory pathway to form sensory (dorsal root) ganglia and the several cell types (neurons, Schwann cells, satellite cells) found within the ganglia. As the cells move through the somite in chains, many are interconnected by long filopodia, and even though their craniocaudal spacing seems largely determined by the segmentation of the somites, cells of adjacent ganglia precursors communicate through the filopodia and sometimes even move from one ganglion precursor to another. Exposure to the **Wnt/catenin** pathway pushes some precursor cells to form sensory neurons, whereas **glial growth factor** (neuregulin) promotes the differentiation of Schwann cells. When the primordia of the ganglia are established, the neurons send out processes linking them both to the dorsal horn of the spinal cord and to peripheral end organs.

Melanocyte Lineage

The melanocyte lineage is unusual in that it produces only one cell type, and the melanocyte precursor cells are determined either before or shortly after their emigration from the neural tube. In response to Wnt and endothelin signaling, melanocyte specification occurs relatively late in the cycle of neural crest emigration. Characteristic of these melanocyte precursors is the expression of the transcription factor **Mitf** (microphthalmia-associated transcription factor). Late-emigrating neural crest cells are stimulated to migrate along the dorsolateral pathway through eph/Ephrin signaling, and because these cells downregulate the Robo receptors for Slit, which is expressed in the dermomyotome, their passage along this pathway is not impeded. Interactions between the **Steel factor**, produced by cells of the dermomyotome, and its receptor, **c-kit**, present in the pigment cell precursors, are critical elements in the dispersal of premelanocytes in the mammalian embryo. Cells of the melanocyte lineage migrate under the ectoderm throughout the body and ultimately colonize the epidermis as pigment cells.

Compared with the cranial neural crest, the trunk neural crest has a limited range of differentiation options. The derivatives of the trunk neural crest are summarized in Table 12.1.

Cranial Neural Crest

The cranial neural crest is a major component of the cephalic end of the embryo. Comparative anatomical and developmental research suggests that the cranial neural crest may represent the major morphological substrate for the evolution of the vertebrate head. Largely because of the availability of precise cellular marking methods, the understanding of the cranial neural crest has increased dramatically. Most studies on the cranial neural crest have been conducted on avian embryos; however, the properties and role of the neural crest in mammalian cranial development are quite similar to those in birds.

In the mammalian head, neural crest cells leave the future brain well before closure of the neural folds (**Fig. 12.5**). In the area of the forebrain, no neural crest arises rostral to the

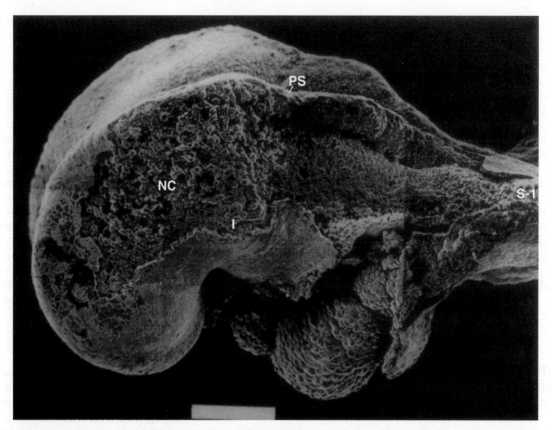

Fig. 12.5 **Neural crest migration in the head of a seven-somite rat embryo.** In this scanning electron micrograph, the ectoderm was removed from a large part of the side of the head, thus exposing migrating neural crest (NC) cells cranial *(to the left)* to the preotic sulcus (PS). Many of the cells are migrating toward the first pharyngeal arch (I). The area between the preotic sulcus and the first somite (S-1) is devoid of neural crest cells because in this region they have not begun to migrate from the closing neural folds. The *white bar* at the *bottom* represents 100 μm. *(From Tan SS, Morriss-Kay G: Cell Tissue Res 204:403-416, 1985.)*

anterior diencephalon (anterior neural ridge [see Fig. 6.4B]), but from the region marked by prosomeres 1 to 3, a continuous sheet of neural crest cells migrates over much of the head (**Fig. 12.6**). Neural crest is inhibited from forming in the anterior neural ridge by the signaling molecule **Dickkopf 1**, a Wnt inhibitor that is secreted by the nearby prechordal mesoderm. Specific streams of neural crest cells emanating from the hindbrain populate the first three pharyngeal arches. Although the streams of migrating cranial neural crest appear at first glance to be not very discrete, there is an overall very specific spatiotemporal order in their pathways to their final destinations in the head and neck.

A major functional subdivision of cranial neural crest occurs at the boundary between rhombomeres 3 (r3) and 4 (r4). Neural crest cells emerging from the diencephalon posteriorly through r3 do not express any *Hox* genes, whereas the cells emerging from the hindbrain region from r4 and posteriorly express a well-ordered sequence of *Hox* genes (see Fig. 12.8).

There is remarkable specificity in the relationship among the origins of the neural crest in the hindbrain, its ultimate destination within the pharyngeal arches, and the expression of certain gene products (**Figs. 12.7** and **12.8**). Neural crest cells associated with r1 and r2 migrate into and form the bulk of the first pharyngeal arch; those of r4, into the second arch;

Fig. 12.6 **Major cranial neural crest migration routes in the mammal.** *(Based on Morriss-Kay G, Tuckett F: J Craniofac Genet Dev Biol 11:181-191, 1991.)*

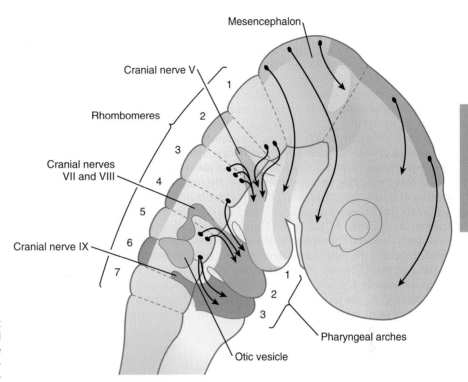

Fig. 12.7 Migration pathways of neural crest cells from the mesencephalon into the head and from rhombomeres 2, 4, and 6 into the first three pharyngeal arches. Small contributions from rhombomeres 1, 3, and 5 are indicated by *arrows*.

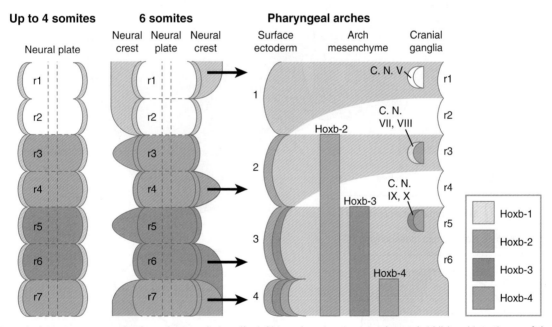

Fig. 12.8 Spread of *Hox* gene expression from the neural plate *(far left)* into the migrating neural crest *(middle)* and into tissues of the pharyngeal arches *(right)*. Arrows in the middle diagram indicate directions of neural crest migration. C. N., cranial nerve; r, rhombomere. *(Adapted from Hunt P and others:* Development *1[Suppl]:187-196, 1991.)*

and those of r6 and r7, into the third arch, as three separate streams of cells.

For many years, it was thought that neural crest cells did not migrate from r3 or r5 even though neural crest cells form in these areas. Some of the neural crest cells associated with r3 and r5 undergo apoptosis because of the presence of the apoptosis-inducing molecule **BMP-4**, but research has shown that **semaphorins** in the mesenchyme lateral to r3 and r5 exert a repulsive effect on neural crest cells that try to enter these areas. A few neural crest cells from r3 diverge into small streams that enter the first and second pharyngeal arches, and cells from r5 behave similarly, by merging with the streams of neural crest cells emanating from r4 and r6.

A close correlation exists between the pattern of migration of the rhombomeric neural crest cells and the expression of products of the Hoxb gene complex. Hoxb-2, Hoxb-3, and

Hoxb-4 products are expressed in a regular sequence in the neural tube and the neural crest–derived mesenchyme of the second, third, and fourth pharyngeal arches. Hoxb is not expressed in r1 and r2 or in the first pharyngeal arch mesenchyme. Only after the pharyngeal arches become populated with neural crest cells does the ectoderm overlying the arches express a similar pattern of Hoxb gene products (see Fig. 12.8). These *Hoxb* genes may play a role in positionally specifying the neural crest cells with which they are associated. Interactions between the neural crest cells and the surface ectoderm of the pharyngeal arches may specify the ectoderm of the arches.

The *Hox* genes play an important role in determining the identity of the pharyngeal arches. The first arch develops independently from *Hox* influence, but *Hoxa2* is critical in determining the identity of the second arch by repressing the elements that would turn it into a first arch. In the absence of *Hoxa2* function, the second arch develops into a mirror image of the first arch. Overall, members of the *Hox3* paralogous group are heavily involved in patterning the third arch and *Hox4* paralogues, the fourth, although research has produced evidence of some overlap of functions.

Emigrating cranial neural crest cells consist of a mix of cells whose fate has already been fixed and those whose fate is largely determined by their environment. As they move away from the brain, cranial crest cells migrate as sheets rostrally or streams (in the pharyngeal area) in the **dorsolateral pathway** directly beneath the ectoderm. This is in strong contrast to migratory patterns in trunk neural crest, where the first two waves of migration head directly ventrally or ventrolaterally (see Fig. 12.4, pathways 1 and 2). As they approach the pharyngeal arches, especially the second arch, the lead cells in the streams of neural crest are attracted by **vascular endothelial growth factor** (**VEGF**), a chemoattractant produced by the distal ectoderm. The trailing cells in the stream are interconnected by long filopodia and follow the lead cells as they disperse into the pharyngeal arches themselves.

Cranial neural crest cells differentiate into a wide variety of cell and tissue types (see Table 12.1), including connective tissue and skeletal tissues. These tissues constitute much of the soft and hard tissues of the face (**Fig. 12.9**). (Specific details of morphogenesis of the head are presented in Chapter 14.)

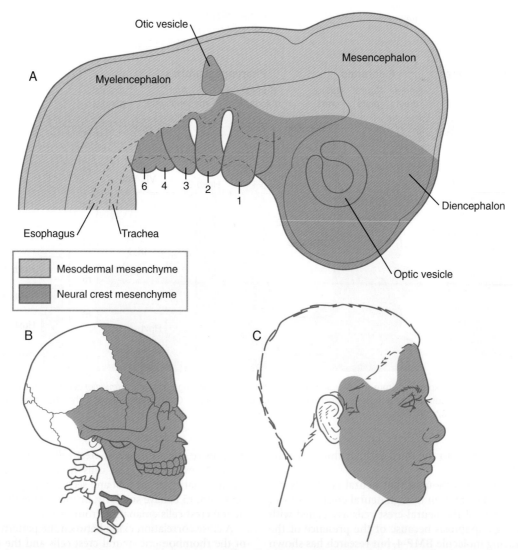

Fig. 12.9 Neural crest distribution in the human face and neck. A, In the early embryo. B and C, In the adult skeleton and dermis.

Circumpharyngeal Neural Crest

The **circumpharyngeal neural crest** arises in the posterior rhombencephalic region at the levels of somites 1 to 7 (**Fig. 12.10**). This region of neural crest represents a transition between cranial and trunk neural crest. Cells arising at the levels of the first four somites behave more like cranial crest, whereas those emigrating at the levels of somites 5 to 7 follow pathways more characteristic of trunk crest. A prominent landmark in this area is the **circumpharyngeal ridge**, an arc-shaped aggregation of cells that passes behind the sixth pharyngeal arch (**Fig. 12.11**). Ventral to the pharynx, this ridge sweeps cranially and provides the pathway through which the hypoglossal nerve (XII) and its associated skeletal muscle precursors pass. Most neural crest cells from the somite 1 to 3 level pass into either the outflow tract of the heart or into the fourth and sixth pharyngeal arches (see Fig. 12.10). These cells are considered to constitute the **cardiac crest**. Other cells from this level, as well as those arising from the level of somites 4 to 7, are called the **vagal crest**. These cells migrate into the gut as precursors of the parasympathetic innervation of the digestive tract. They also form sensory neurons and glia, as well as making some contribution to sympathetic ganglia. Like the cranial crest cells, most cells of the cardiac crest migrate along the dorsolateral pathway between the somites and the ectoderm (see Fig. 12.10), whereas those of the vagal crest, like those of the trunk, initially migrate along the ventral pathways between the neural tube and the dermomyotome.

Cardiac Crest

The cardiac crest, arising at the level of somites 1 to 3, surrounds the endothelial precursors of the third, fourth, and sixth aortic arches, and it contributes massively to the truncoconal ridges that separate the outflow tract of the heart into aortic and pulmonary segments (see Chapter 17). Under the strong influence of semaphorins, cardiac crest cells migrate toward the heart and contribute to the leaflets of the semilunar valves at the base of the outflow tract, and in birds, at least, they may penetrate the interventricular septum. The cardiac neural crest may interact with pharyngeal endoderm to modify the signals leading to the normal differentiation of myocardial cells.

Although much of the cardiac crest contributes to the outflow tract of the heart and the great vessels, portions of the cardiac neural crest population become associated with the newly forming thymus, parathyroid, and thyroid glands. Two streams of cardiac neural crest cells leave the neural tube. The earlier stream contributes principally to the cardiac outflow tract and aortic arch arteries, whereas cells of the later stream become incorporated into pharyngeal glands. On their way to the heart and pharyngeal structures, cardiac crest cells migrate along the dorsolateral pathway and reach their destinations via the circumpharyngeal ridge.

Some neural crest cells migrate ventral to the pharynx in bilateral streams accompanying the somite-derived myoblasts that are migrating cranially to form the intrinsic muscles of the tongue and the hypopharyngeal muscles. This is the only known case in which somite-derived muscles are invested with neural crest–derived connective tissue. The cardiac neural crest also supplies the Schwann cells that are present in the hypoglossal and other cranial nerves.

A disturbance in this region of neural crest can result in cardiac septation defects (**aorticopulmonary septum**) and glandular and craniofacial malformations. **DiGeorge's syndrome**, which is associated with a deletion on chromosome 22, is characterized by hypoplasia and reduced function of the thymus, thyroid, and parathyroid glands and cardiovascular defects, such as persistent truncus arteriosus and abnormalities of the aortic arches. *Hoxa3* mutant mice show a similar spectrum of pharyngeal defects. The common denominator for this constellation of pathological features is a defect of the cardiac crest supplying the third and fourth pharyngeal arches and cardiac outflow tract. Similar defects have been described in human embryos exposed to excessive amounts of retinoic acid early in embryogenesis.

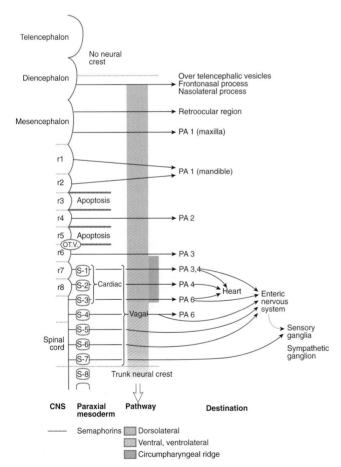

Fig. 12.10 **Schematic representation of migrations of the cranial and circumpharyngeal neural crest cells.** The *arrows* indicate migratory pathways, starting with their origin in the central nervous system. The circumpharyngeal crest arises from the level of somites 1 to 7. Note the shift in migratory pathway from one characteristic of cranial crest (*blue*) to one like that of trunk crest (*pale orange*). OT.V., otic vesicle; PA, pharyngeal arch; r, rhombomere; S, somite.

Vagal Crest

Within the gut, neural crest cells form the enteric nervous system, which in many respects acts like an independent component of the nervous system. The number of enteric neurons

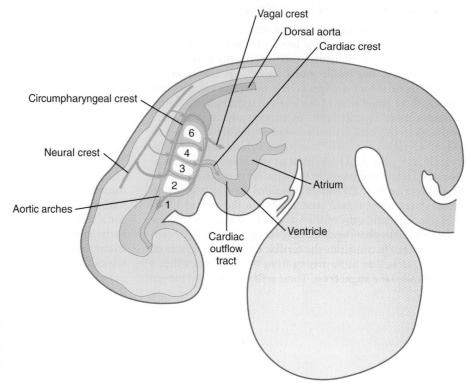

Fig. 12.11 Migration of circumpharyngeal neural crest cells (*green lines* and *arrows*) through the pharyngeal region and into the aortic arches and the outflow tract of the heart.

nearly matches the number of neurons in the spinal cord, and most of these neurons are not directly connected to either the brain or the spinal cord. This independence explains how the bowel can maintain reflex activity in the absence of direct input from the central nervous system.

The cells that form the neurons of the enteric nervous system come from the part of the circumpharyngeal crest known as the **vagal crest**. These cells exit from the levels of somites 1 to 7, follow a ventral pathway through the dorsal part of the circumpharyngeal ridge, and then exit this pathway caudal to the sixth pharyngeal arch. Most of these cells become closely associated with the embryonic gut, but some are involved in formation of sensory dorsal root ganglia and associated glia. At the level of somite 7, some cells even contribute to local sympathetic ganglia. Neural crest cells are not committed to form gut-associated nervous tissue before they leave the spinal cord. If vagal crest is replaced by neural crest of the trunk, which does not normally give rise to gut-associated derivatives, the gut is colonized by the transplanted trunk-level neural crest cells.

Despite the strong influence of the environment of the gut on the differentiation of neural crest cells exposed to its influences, neural crest cells retain a surprising degree of developmental flexibility. If crest-derived cells already in the gut of avian embryos are retransplanted into the trunk region of younger embryos, they seem to lose the memory of their former association with the gut. They enter the pathways (e.g., adrenal or peripheral nerve) common to trunk cells (except that they cannot enter the pigment cell pathway) and differentiate accordingly.

Under the influence of **glial-derived neurotrophic factor** (**GNDF**), vagal crest cells enter the anterior region of the foregut and begin to populate the gut. One potential reason that neural crest cells in the trunk are unable to enter the gut is that the cells in the mesentery near the gut express **Slit-2**, the molecule that also prevents neurons from crossing the midline of the central nervous system. Trunk neural crest cells express the Slit receptor **Robo**, thus causing them to avoid cells that express Slit. Vagal crest cells do not express Robo and are permitted access to the gut wall.

Within the gut wall, vagal crest cells undertake a major invasion that sweeps through the length of the gut and ultimately stops near the posterior end of the hindgut by the end of the seventh week of pregnancy. The vagal crest–derived enteric neuronal precursors, which later become parasympathetic neurons, advance down the gut at a rate of approximately 40 to 45 μm/hour. These cells advance as interconnected strands, and they undergo proliferation at the level of the wavefront. Advancement of the wavefront is apparently more the result of proliferation and the spilling over of cellular neural crest progeny into the unpopulated regions of the gut than of actual directed migration of individual cells. When the cellular wavefront arrives at the cecum, the cells pause for several hours because of the presence of local signaling factors. They then proceed into the future colon. Within the colon, the vagal crest cells ultimately meet with a smaller number of cells emigrating from the sacral neural crest, at which point invasive activity ceases, and further organization of the enteric ganglia continues. When they first colonize the gut, the neural crest cells express no neuronal markers, but under the influence of **Hand-2**, a wave of differentiation passes down the gut, and the cells synthesize neurofilament proteins and initially express catecholaminergic traits. These cells form myenteric plexuses.

Some common neurocristopathies are presented in **Clinical Correlation 12.1.**

CLINICAL CORRELATION 12.1
Neurocristopathies

Because of the complex developmental history of the neural crest, various congenital malformations are associated with its defective development. These malformations have commonly been subdivided into two main categories—defects of migration or morphogenesis and tumors of neural crest tissues (**Box 12.1**). Some of these defects involve only a single component of the neural crest; others affect multiple components and are recognized as syndromes.

Box 12.1 Major Neurocristopathies

Defects of Migration or Morphogenesis

Trunk Neural Crest
Hirschsprung's disease (aganglionic megacolon)

Cranial Neural Crest
Aorticopulmonary septation defects of heart
Anterior chamber defects of eye
Cleft lip, cleft palate, or both
Frontonasal dysplasia
DiGeorge's syndrome (hypoparathyroidism, thyroid deficiency, thymic dysplasia leading to immunodeficiency, defects in cardiac outflow tract and aortic arches)
Certain dental anomalies

Trunk and Cranial Neural Crest
CHARGE association (**c**oloboma, **h**eart disease, **a**tresia of nasal choanae, **r**etardation of development, **g**enital hypoplasia in males, anomalies of the **e**ar)
Waardenburg's syndrome

Tumors and Proliferation Defects

Pheochromocytoma: tumor of chromaffin tissue of adrenal medulla
Neuroblastoma: tumor of adrenal medulla, autonomic ganglia, or both
Medullary carcinoma of thyroid: tumor of parafollicular (calcitonin-secreting) cells of thyroid
Carcinoid tumors: tumors of enterochromaffin cells of digestive tract
Neurofibromatosis (von Recklinghausen's disease): peripheral nerve tumors
Neurocutaneous melanosis (multiple congenital pigmented nevi, melanocytic tumors of the central nervous system)

Differentiation Defect Involving Neural Crest Cells

Albinism

Several syndromes or associations of defects are understandable only if the wide distribution of derivatives of the neural crest is recognized. For example, one association called **CHARGE** consists of **c**oloboma (see Chapter 13), **h**eart disease, **a**tresia of nasal choanae, **r**etardation of development, **g**enital hypoplasia in males, and anomalies of the **e**ar.

Types I and III of **Waardenburg's syndrome**, which is caused by Pax-3 mutations, involve various combinations of pigmentation defects (commonly a white stripe in the hair and other pigment anomalies in the skin), deafness, cleft palate, and **ocular hypertelorism** (increased space between the eyes). One variant (type I) of Waardenburg's syndrome is also characterized by hypoplasia of the limb muscles; this is not surprising considering the important association between Pax-3 and myogenic cells migrating into the limb buds from the somites. Pax-3 is similarly expressed in migrating cardiac neural crest cells, but it is downregulated when the cells settle in the walls of the cardiac outflow tract or aortic arches. Cardiovascular defects in these areas are also seen in Pax-3 mutants.

DiGeorge syndrome, which is associated with a deletion on chromosome 22 that encompasses up to 15 genes, is characterized by hypoplasia and reduced function of the thymus, thyroid, and parathyroid glands and cardiovascular defects, including persistent truncus arteriosus and abnormalities of the aortic arches. *Hoxa3* mutant mice show a similar spectrum of pharyngeal defects. The common denominator for this constellation of pathological features is a defect of the neural crest supplying the third and fourth pharyngeal arches and the cardiac outflow tract. Similar defects have been described in human embryos exposed to excessive amounts of retinoic acid early in embryogenesis.

Neurofibromatosis (von Recklinghausen's disease) is a common genetic disease manifested by multiple tumors of neural crest origin. Common features are **café au lait spots** (light brown pigmented lesions) on the skin, multiple (often hundreds) **neurofibromas** (peripheral nerve tumors), occasional gigantism of a limb or digit, and various other conditions (Fig. 12.12). Neurofibromatosis occurs in approximately 1 of 3000 live births, and the gene is very large and subject to a high mutation rate. Some rare syndromes involving features such as disturbed pigmentation, cutaneous microvascular lesions, asymmetric enlargements of structures, and nerve lesions (including neurofibromatosis) are sometimes lumped together in the category of neural crest–based neurocutaneous syndromes.

Because of the massive contribution of the neural crest to the face and other parts of the head and neck, various malformations in the craniofacial region involve neural crest derivatives. Many different facial abnormalities lumped together under the term **frontonasal dysplasia** (see Chapter 14) heavily involve neural crest–derived tissues.

Continued

CLINICAL CORRELATION 12.1
Neurocristopathies—cont'd

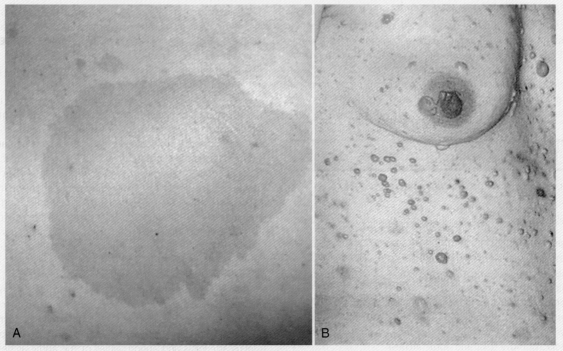

Fig. 12.12   **A,** Large café au lait spot on the skin of a patient with neurofibromatosis. **B,** Multiple neurofibromas on the skin. *(From the Robert J. Gorlin Collection, Division of Oral and Maxillofacial Pathology, University of Minnesota Dental School; courtesy of Dr. Ioannis Koutlas.)*

Clinical Vignette

A newborn is diagnosed as having an incomplete separation between the aorta and the pulmonary artery (a mild form of persistent truncus arteriosus). Later, after corrective heart surgery, she has more colds and sore throats than her siblings. After testing, the physician tells the parents that there is evidence of immunodeficiency. The physician also tells the parents that it would be a good idea to check her levels of parathyroid hormone.

What is the basis for this suggestion?

Summary

- The neural crest arises from neuroepithelial cells along the lateral border of the neural plate. Having left the neural tube, neural crest cells migrate to peripheral locations throughout the body. Some substrates, such as those containing chondroitin sulfate molecules, are not favorable for neural crest cell migration.
- Neural crest cells differentiate into many types of adult cells, such as sensory and autonomic neurons, Schwann cells, pigment cells, and adrenal medullary cells. Cells from the cranial and circumpharyngeal neural crest also differentiate into bone, cartilage, dentin, dermal fibroblasts, selected smooth muscle, the connective tissue stroma of pharyngeal glands, and several regions of the heart and great vessels.
- The control of differentiation of neural crest cells is diverse; some cells are determined before they begin to migrate, and others respond to environmental cues along their

paths of migration. Trunk neural crest cells cannot differentiate into skeletal elements.
- Neural crest cells in the trunk follow three main paths of migration: (1) a ventral path for cells of the sympathoadrenal lineage, (2) a ventrolateral pathway leading through the anterior halves of the somites for sensory ganglion-forming cells, and (3) a dorsolateral pathway for pigment cells.
- Cells of the cranial neural crest form many tissues of the facial region. In the pharyngeal region, the pathways of crest cell migration are closely correlated with regions of expression of products of the *Hoxb* gene complex. Cells of the cranial crest may be patterned with level-specific instructions, whereas cells of the trunk crest are not.
- Several genetic diseases and syndromes are associated with disturbances of the neural crest. Neurofibromatosis is often characterized by multiple tumors and pigment disturbances. Disturbances of the cardiac neural crest can result in septation defects in the heart and outflow tract.

Review Questions

1. Which of these cell and tissue types arises from cranial, but not trunk, neural crest cells?

A. Sensory ganglia
B. Adrenal medulla
C. Melanocytes
D. Schwann cells
E. None of the above

2. Which molecule is a poor substrate for migrating neural crest cells?
A. Laminin
B. Chondroitin sulfate
C. Fibronectin
D. Type IV collagen
E. Hyaluronic acid

3. Neural crest cells arise from the:
A. Somite
B. Dorsal non-neural ectoderm
C. Neural tube
D. Splanchnic mesoderm
E. Yolk sac endoderm

4. A 6-month-old infant exhibits multiple congenital defects, including a cleft palate, deafness, ocular hypertelorism, and a white forelock but otherwise dark hair on his head. The probable diagnosis is:
A. CHARGE association
B. von Recklinghausen's disease
C. Hirschsprung's disease
D. Waardenburg's syndrome
E. None of the above

5. What molecule is involved in the migration of neural crest cells from the neural tube?
A. Slug
B. BMP-2
C. Mash 1
D. Norepinephrine
E. Glial growth factor

6. Which is not a derivative of the neural crest?
A. Sensory neurons
B. Motoneurons
C. Schwann cells
D. Adrenal medulla
E. Dental papilla

7. What maintains the competence of neural crest cells to differentiate into autonomic neurons?
A. Sonic hedgehog
B. Acetylcholine
C. Mash 1
D. Glial growth factor
E. Transforming growth factor-β

8. If trunk neural crest cells are transplanted into the cranial region, they can form all of the following types of cells except:
A. Pigment cells
B. Schwann cells
C. Sensory neurons
D. Cartilage
E. Autonomic neurons

9. How does the segmental distribution of the spinal ganglia occur?

10. What are three major differences between cranial and trunk neural crests?

References

Aman A, Piotrowski T: Cell migration during morphogenesis, *Dev Biol* 341:20-33, 2010.

Anderson DJ: Genes, lineages and the neural crest: a speculative review, *Phil Trans R Soc Lond B* 355:953-964, 2000.

Aybar MJ, Mayor R: Early induction of neural crest cells: lessons learned from frog, fish and chick, *Curr Opin Genet Dev* 12:452-458, 2002.

Betancur P, Bronner-Fraser M, Sauka-Spengler T: Assembling neural crest regulatory circuits into a gene regulatory network, *Annu Rev Cell Dev Biol* 26:581-603, 2010.

Billon N and others: The generation of adipocytes by the neural crest, *Development* 134:2283-2292, 2007.

Boot MJ and others: Spatiotemporally separated cardiac neural crest subpopulations that target the outflow tract septum and pharyngeal arch arteries, *Anat Rec A Discov Mol Cell Evol Biol* 275:1009-1018, 2003.

Druckenbrod NR, Epstein ML: The pattern of neural crest advance in the cecum and colon, *Dev Biol* 287:125-133, 2005.

Clay MR, Halloran MC: Regulation of cell adhesions and motility during initiation of neural crest migration, *Curr Opin Neurobiol* 21:17-22, 2011.

Creuzet SE: Neural crest contribution to forebrain development, *Semin Cell Dev Biol* 20:751-759, 2009.

Gammill LS and others: Guidance of trunk neural crest migration requires neuropilin2/semaphorin 3F signaling, *Development* 133:99-106, 2005.

Gross JB, Hanken J: Review of fate-mapping studies of osteogenic cranial neural crest in vertebrates, *Dev Biol* 317:389-400, 2008.

Hall BK, Hörstadius S: *The neural crest*, London, 1988, Oxford University Press.

Harris ML, Erickson CA: Lineage specification in neural crest cell pathfinding, *Dev Dyn* 236:1-19, 2007.

Huber K: The sympathoadrenal cell lineage: specification, diversification, and new perspectives, *Dev Biol* 298:335-343, 2006.

Kasemeier-Kulesa JC, Kulesa PM, Lefcort F: Imaging neural crest cell dynamics during formation of dorsal root ganglia and sympathetic ganglia, *Development* 132:235-245, 2005.

Kelly Kuan C-Y and others: Somite polarity and segmental patterning of the peripheral nervous system, *Mech Dev* 121:1055-1068, 2004.

Knecht AK, Bronner-Fraser M: Induction of the neural crest: a multigene process, *Nat Rev Genet* 3:453-461, 2002.

Kulesa PM, Gammill LS: Neural crest migration: patterns, phases and signals, *Dev Biol* 344:56-568, 2010.

Kulesa PM, Ellies DL, Trainor PA: Comparative analysis of neural crest cell death, migration, and function during vertebrate embryogenesis, *Dev Dyn* 229:14-29, 2004.

Kulesa PM and others: Cranial neural crest migration: new rules for an old road, *Dev Biol* 344:543-554, 2010.

Kuo BR, Erickson CA: Regional differences in neural crest morphogenesis, *Cell Adh Migr* 4:567-585, 2010.

Kuratani S: Spatial distribution of postotic crest cells defines the head/trunk interface of the vertebrate body: embryological interpretation of peripheral nerve morphology and evolution of the vertebrate head, *Anat Embryol* 195:1-13, 1997.

Kuratani SC, Kirby ML: Initial migration and distribution of the cardiac neural crest in the avian embryo: an introduction to the concept of the circumpharyngeal crest, *Am J Anat* 191:215-227, 1991.

Le Douarin NM, Kalcheim C: *The neural crest*, ed 2, Cambridge, 1999, Cambridge University Press.

Meulemans D, Bronner-Fraser M: Gene-regulatory interactions in neural crest evolution and development, *Dev Cell* 7:291-299, 2004.

Minoux M, Rijli FM: Molecular mechanisms of cranial neural crest migration and patterning in craniofacial development, *Development* 137:2605-2621, 2010.

Noden DM: Origins and patterning of craniofacial mesenchymal tissues, *J Craniofac Genet Dev Biol* 2(Suppl):15-31, 1986.

O'Rahilly R, Müller F: The development of the neural crest in the human, *J Anat* 211:335-351, 2007.

Raible DW: Development of the neural crest: achieving specificity in regulatory pathways, *Curr Opin Cell Biol* 18:698-703, 2006.

Ruhrberg C, Schwarz Q: In the beginning: generating neural crest cell diversity, *Cell Adh Migr* 4:622-630, 2010.

Sarnat HB: Embryology of the neural crest: its inductive role in the neurocutaneous syndromes, *J Child Neurol* 20:637-643, 2005.

Sauka-Spengler T, Bronner-Fraser M: Development and evolution of the migratory neural crest: a gene regulatory perspective, *Curr Opin Genet Dev* 16:360-366, 2006.

Simpson MJ and others: Cell proliferation drives neural crest cell invasion of the intestine, *Dev Biol* 302:553-568, 2007.

Stoller JZ, Epstein JA: Cardiac neural crest, *Semin Cell Dev Biol* 18:704-715, 2005.

Tosney KW: Long-distance cue from emerging dermis stimulates neural crest melanoblast migration, *Dev Dyn* 229:99-108, 2004.

Trainor PA: Specification of neural crest cell formation and migration in mouse embryos, *Semin Cell Dev Biol* 16:683-693, 2005.

Trainor PA, Ariza-McNaughton, Krumlauf R: Role of the isthmus and FGFs in resolving the paradox of neural crest plasticity and prepatterning, *Science* 295:1288-1291, 2002.

Trainor PA, Krumlauf R: *Hox* genes, neural crest cells and branchial arch patterning, *Curr Opin Cell Biol* 13:698-705, 2001.

Trainor PA, Krumlauf R: Plasticity in mouse neural crest cells reveals a new patterning role for cranial mesoderm, *Nat Cell Biol* 2:96-102, 2000.

Tucker AS, Lumsden A: Neural crest cells provide species-specific patterning information in the developing branchial skeleton, *Evol Dev* 6:32-40, 2004.

Sense Organs

The major sense organs arise in large measure from the thickened ectodermal placodes that appear lateral to the neural plate in the early embryo (see Fig. 6.6). The following descriptions begin with the most cranial placodes and continue to the most caudal. The midline **hypophyseal placode**, located in the anterior neural ridge (see Fig. 6.6B), becomes the primordium of Rathke's pouch (the precursor of the adenohypophysis). This structure arises adjacent to the neural tissue that ultimately forms the neurohypophysis. Arising also from the anterior neural region, bilateral **olfactory placodes** (see Fig. 6.6) are the precursors of the olfactory epithelium. They give rise to olfactory neurons and their supporting cells and to glial cells and neuroendocrine cells that migrate from the placode into the brain. Closely associated with the olfactory placodes is the preneural tissue that forms the functionally associated olfactory bulbs of the brain. The bilateral **lens placodes**, associated with the optic vesicles (future retina) extending outward from the diencephalic part of the brain, are the lens precursors.

Next in line are the paired **trigeminal placodes** (cranial nerve V), each of which arises from two placodal precursors—ophthalmic and maxillomandibular (see Fig. 6.6C). The **otic placodes** (precursors of the inner ear) in the human are the remaining representatives of the **dorsolateral series** of placodes, all of which produce vibration-detecting organs. In fishes and some amphibians, the other members of the dorsolateral series give rise to the lateral line organs, which serve as vibration and electroreceptors in aquatic vertebrates.

The caudalmost group constitutes the **epibranchial placodes**, which are located just dorsal to the region where the first through third pharyngeal pouches abut the cervical ectoderm (see Fig. 6.6A). Their specification depends on signals (fibroblast growth factor [FGF] and bone morphogenetic protein [BMP]) emanating from the pharyngeal pouch endoderm. These placodes produce sensory neurons that supply visceral structures. The first epibranchial placode produces neurons (geniculate ganglion of cranial nerve VII) (**Fig. 13.1B**) that innervate taste buds. Similarly, neurons arising from the second epibranchial placode (inferior [petrosal] ganglion of cranial nerve IX) innervate taste buds, as well as the heart and other visceral organs. The third epibranchial placode contributes to the inferior (nodose) ganglion of the vagus nerve (cranial nerve X), and its neurons innervate the heart, stomach, and other viscera. Of cranial nerves V, VII, IX, and X, the proximal sensory ganglia are derived largely from neural crest cells, and the distal ganglia are principally placodal in origin (see Fig. 13.1B). The placode-derived neurons (those in the distal ganglia) begin to establish peripheral and central connections before axons emerge in the neural crest–derived neuronal precursors in the proximal ganglia.

A major function of both the epibranchial and trigeminal placodes is to produce neurons. The conversion of epithelial cells to neuroblasts in the placodes is accomplished in much the same manner as that which occurs within the neural tube (see Fig. 11.4). Through the process called **interkinetic nuclear migration**, actual cell division occurs at the apical (in this case, outer) end of the tall epithelial placodal cells. Then the nuclei migrate toward the basal (inner) surface of the cells as they become committed to a neuronal fate as **sensory neuroblasts**. They then pass through breaches in the basal lamina and migrate internally. There they join other neuroblasts from the same origin to form precursors of the appropriate sensory ganglia.

All the placodes arise from a single **preplacodal region**, which encircles the cranial neural plate. The preplacodal region is induced by cranial mesoderm, with the neural tube playing a supporting role. The inductive process involves activating the **FGF pathway**, along with inhibiting **Wnt** and **BMP** by their natural antagonists. Levels of BMP must be lower to induce placodes rather than neural crest, and there is a gradient of BMP from highest in the neural plate, to medium for neural crest, and lowest for placodal induction. Characteristic of the induced preplacodal tissue is the expression of the transcription factors, **Six** and **Eya**, which promote a generic placodal fate to the cells within the preplacodal region. This is followed by specific secondary inductive signals from different sources that specify the formation of individual placodes (**Table 13.1**).

This chapter concentrates on the development of the eyes and ears, the most complex and important sense organs in humans. The organs of smell and taste are discussed in Chapter 14 because their development is intimately associated with the development of the face and pharynx. The sensory components of the cranial nerves are discussed in Chapter 11.

Eye

The eye is a very complex structure that originates from constituents derived from several sources, including the wall of the diencephalon, the overlying surface ectoderm, and migrating cranial neural crest mesenchyme. Two basic themes characterize early ocular development. One is an ongoing series of inductive signals that result in the initial establishment of the major components of the eye. The other is the coordinated differentiation of many of these components.

For normal vision to occur, many complex structures within the eye must properly relate to neighboring structures. The cornea and lens must both become transparent and properly aligned to provide an appropriate pathway for light to reach

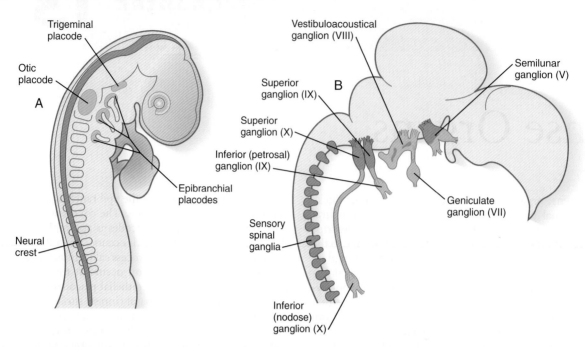

Fig. 13.1 Ectodermal placodes and neural crest in the formation of sensory ganglia of cranial and spinal nerves in the chick embryo. **A,** At 2 days. **B,** At 8 days. Neural crest is shown in *green*; placodes are shown in *blue*. *(Adapted from LeDouarin N and others: Trends Neurosci 9:175-180, 1986.)*

Table 13.1 Origins and Inducing Tissues for Cranial Placodes

Placode	Origin	Inducing Tissue	Inducers
Adenohypophysis	Anterior neural ridge Non-neural ectoderm	Anterior neural ridge Mesoderm	Shh Nodal
Lens	Anterior neural ridge Non-neural ectoderm	Neural plate Mesoderm	FGFR BMP
Olfactory	Anterior neural ridge Non-neural ectoderm	Anterior neural ridge Neural crest	FGF
Otic	Neural folds Non-neural ectoderm	Hindbrain Mesoderm	FGF FGFR Retinoic acid
Trigeminal	Non-neural ectoderm	Dorsal neural tube	PDGF Wnt
Epibranchial	Non-neural ectoderm	Hindbrain Mesoderm	FGF FGFR

Modified from McCabe KL, Bronner-Fraser M: *Dev Biol* 332:192, 2009.

BMP, bone morphogenetic protein; FGF, fibroblast growth factor; FGFR, fibroblast growth factor receptor; PDGF, platelet-derived growth factor; Shh, sonic hedgehog.

the retina. The retina must be configured to receive concrete visual images and to transmit patterned visual signals to the proper parts of the brain through neural processes extending from the retina into the optic nerve.

Early Events in the Establishment of the Eye

A single continuous eye field begins to take shape around the area of the prechordal plate during late gastrulation. Cells in the eye field express **RAX** (retina and anterior neural fold homeobox). Mutations in *RAX* are the basis for

anophthalmia, a rare condition characterized by the absence of any ocular structures in humans. Other prominent markers are **Pax-6** and **Lhx-2**, which are heavily involved in patterning the eye fields (**Fig. 13.2**). With the secretion of **sonic hedgehog** (**shh**) by the prechordal plate and the ventral midline of the diencephalon, Pax-6 expression in the midline is repressed, and the single eye field splits into two separate eye fields, located on either side of the diencephalon. Rax and another important transcription factor, **Six-3**, protect the ability of the forebrain to secrete shh by suppressing Wnt activity. If Wnt is not suppressed, the anterior region of the developing brain becomes posteriorized and is unable to secrete shh. The

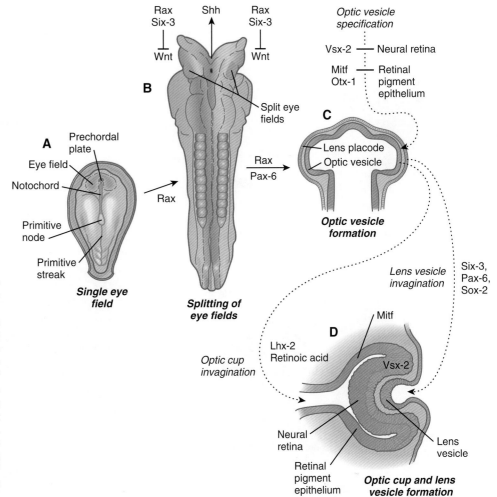

Fig. 13.2 Major steps in eye development. A, A single eye field forms during late gastrulation. B, Through the action of sonic hedgehog (Shh), the single eye field splits into bilateral fields. C, The formation of optic vesicles is heavily dependent on the expression of Pax-6 and Rax. The optic vesicles are patterned into future neural retina (through Vsx-2) and retinal pigment epithelium (through Mitf and Otx-2). D, Inductive interactions between the optic vesicle and the overlying lens placode result in invagination of both the optic cup and the lens vesicle.

absence of Six-3 activity results in the loss of shh secretion and prevents the splitting of the eye fields and leads to the condition of **holoprosencephaly** and the formation of only one eye (see p. 309 and Fig. 8.18).

Development of the eye is first evident at about 22 days' gestation, when the lateral walls of the diencephalon begin to bulge out as **optic grooves** (**Fig. 13.3**). Within a few days, the optic grooves enlarge to form **optic vesicles**, which terminate very close to the overlying lens placode in the surface ectoderm. As the optic vesicle expands, the pattern for the future **neural retina** and **retinal pigment epithelium** is laid down. Initially, the transcription factor **Mitf** is expressed throughout the optic vesicle, but the subsequent expression of **Vsx-2** in the distal optic vesicle (future neural retina) confines Mitf to more proximal regions, which will become the retinal pigment epithelium. Apposition of the outer wall of the optic vesicle to the surface ectoderm is essential for the transmission of an important inductive message that stimulates the cells of the lens placode to thicken and begin forming the lens (see Fig. 13.2; **Fig. 13.4**).

The interaction between the optic vesicle and the overlying ectoderm was one of the first recognized inductive processes. It was initially characterized by deletion and transplantation experiments conducted on amphibian embryos. When the optic vesicles were removed early, the surface ectoderm differentiated into ordinary ectodermal cells instead of lens fibers. Conversely, when optic vesicles were combined with certain types of ectoderm other than eye, the ectoderm was stimulated to form lens fibers. Subsequent research on amphibian embryos has shown that series of preparatory inductions from neural plate and underlying mesoderm condition the ectoderm for its final induction into lens by the optic vesicle. In mammals, an important mechanism underlying the severe **microphthalmia** (tiny eyes) or **anophthalmia** (absence of eyes) seen in the *small eye* and *fidget* mutants is a disturbance in the apposition of optic vesicles and surface ectoderm that interferes with lens induction.

The paired box gene *Pax6* plays a prominent role throughout early eye development and at certain later stages of development of the retina and lens. Pax-6 is initially expressed in the lens and the nasal placodes and much of the diencephalon. In *Drosophila*, *Pax6* has been called a master gene for eye development; that is, it can turn on the cascade of genes that guide development of the eye. The power of Pax-6 is shown by the formation of ectopic eyes on antennae and legs in *Drosophila* when the gene is improperly expressed. In the absence of Pax-6 expression (*eyeless* mutant), eyes do not form. In the *small eye* mutation, the mammalian equivalent of *eyeless*, the early optic vesicle forms, but, as previously noted, eye development does not progress because the surface ectoderm is unable to respond to the inductive signal emitted by the optic vesicle.

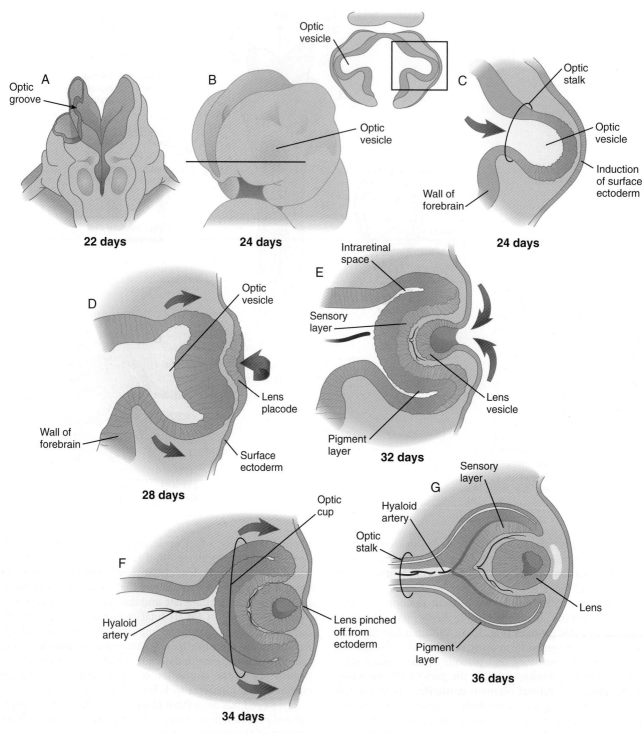

Fig. 13.3   A to **G,** Early development of the human eye.

The identification in humans of two genes (*Eya* [eyes absent] and *Six* [sine oculis]) that are activated by Pax-6 in *Drosophila* strongly suggests that despite major differences in the structure and development of the vertebrate and insect eye, the basic genetic apparatus has been conserved throughout phylogeny. In mice, **Eya-1** and **Eya-2** are expressed in the lens placodes and are required for placodal induction and early differentiation, but in the absence of Pax-6 function, they are not expressed, and eye development fails to proceed.

As the process of lens induction occurs, the surface ectoderm stimulates the outer face of the optic vesicle to flatten and ultimately to become concave. This results in the transformation of the optic vesicle to the **optic cup** (see Fig. 13.3F). The progression from optic vesicle to optic cup requires the expression of Lhx-2 and the action of retinoic acid. In their absence, eye development is arrested at the optic vesicle stage (see Fig. 13.2). Meanwhile, the induced lens ectoderm thickens and invaginates to form a **lens vesicle**, which detaches from

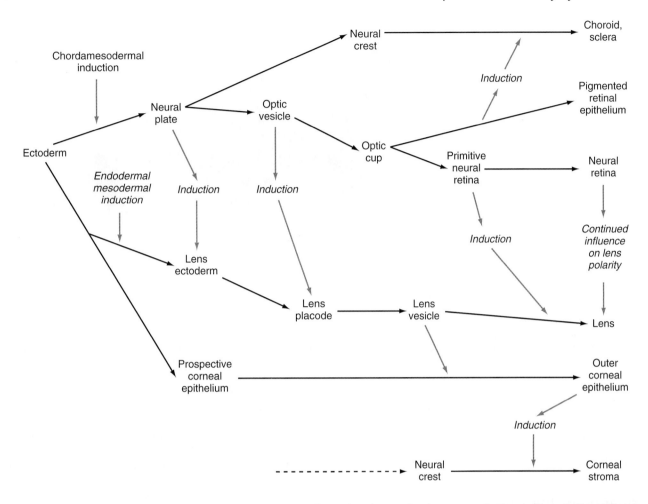

Fig. 13.4 **Flow chart of major inductive events and tissue transformations in eye development.** Inductive events are indicated by *colored arrows.*

the surface epithelium from which it originated (**Fig. 13.5**; see Fig. 13.3). Then the lens vesicle takes over and becomes the primary agent of a new inductive reaction by acting on the overlying surface ectoderm and causing it to begin corneal development (see Fig. 13.4).

Formation of the optic cup is an asymmetric process that occurs at the ventral margin of the optic vesicle, rather than at its center. This results in the formation of a gap called the choroid fissure, which is continuous with a groove in the **optic stalk** (**Fig. 13.6**). During much of early ocular development, the choroid fissure and optic groove form a channel through which the **hyaloid artery** passes into the posterior chamber of the eye. Differential expression of *Pax* genes determines which cells become optic cup (future retina), and which cells become optic stalk (future optic nerve). Through exposure to high concentrations of shh, the expression of Pax-6 is inhibited, and **Pax-2** is induced in the optic stalk, whereas a lower concentration of shh more distally permits the expression of Pax-6 in the optic vesicle, thus paving the way toward formation of the retina.

The optic stalk initially represents a narrow neck that connects the optic cup to the diencephalon, but as development progresses, it is invaded by neuronal processes emanating from the ganglion cells of the retina. Pax-2–expressing cells in the optic stalk provide guidance cues to outgrowing retinal axons that pass through the optic nerve and optic chiasm and

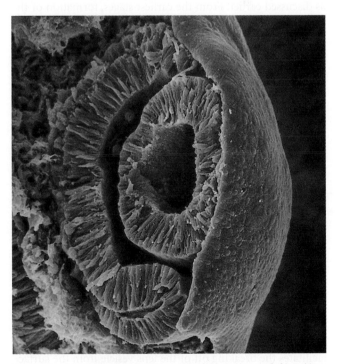

Fig. 13.5 Scanning electron micrograph of the optic cup *(left)* and lens vesicle *(center)* in the chick embryo. *(Courtesy of K. Tosney, Ann Arbor, Mich.)*

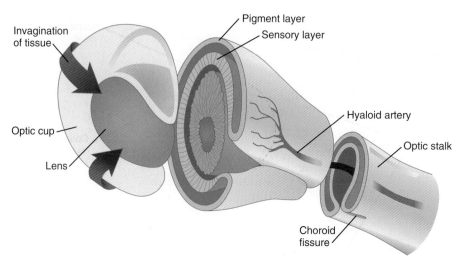

enter the contralateral optic tract. After the neuronal processes have made their way to the appropriate regions of the brain, the optic stalk is properly known as the **optic nerve**.

Later in development, the choroid fissure closes, and no trace of it is seen in the normal iris. Nonclosure of the choroid fissure results in the anomaly of coloboma (see Fig. 13.19B). In certain forms of coloboma, especially forms that are associated with anomalies of the kidneys, mutations of Pax-2 genes are seen. In Pax-2–mutant mice, retinal axons do not cross the midline through the optic chiasm, but rather remain in the ipsilateral optic tract.

Formation of the Lens

The lens is derived from cells in the generic preplacodal region, as discussed earlier. From the earliest stages, formation of the lens depends on genetic instructions provided by Pax-6. Pax-6 expression is required for the surface ectoderm to respond to inductive signals (FGF and BMP) from the underlying optic vesicle by activating and combining with another transcription factor, Sox-2. This leads to the thickening of the surface ectoderm to form the **lens placode** (see Fig. 13.3D). At the same time, migrating neural crest cells, which do not penetrate the space between optic vesicle and prospective lens, give off signals that inhibit cells in other areas of the preplacodal region from forming lens. Inhibition of lens-forming capacity is signaled by the downregulation of Pax-6 by these cells. Pax-6 expression continues as the lens placode invaginates to form the **lens vesicle**, which ultimately breaks off from the surface ectoderm. At this point, Pax-6 plays a new role in regulating the activity of the genes governing the formation of the lens crystallin proteins.

When it is breaking off from the surface ectoderm, the lens vesicle is roughly spherical and has a large central cavity (see Fig. 13.3E). At the end of the sixth week, the cells at the inner pole of the lens vesicle begin to elongate in an early step toward their transformation into the long, transparent cells called **lens fibers** (**Fig. 13.7A**). The influence of the transcriptional regulator **Foxe-3**, which operates downstream of Pax-6, facilitates the breaking off of the lens vesicle from the surface ectoderm and the transformation of posterior cells into lens fibers.

Differentiation of the lens is a very precise and well-orchestrated process involving several levels of organization. At the cellular level, relatively nonspecialized lens epithelial cells, under the influence of Sox-2, Pax-6, and other proteins paired with an oncogene called **Maf**, undergo a profound transformation into transparent, elongated cells that contain large quantities of specialized **crystallin proteins**. At the tissue level, the entire lens is responsive to signals from the retina and other structures of the eye so that its shape and overall organization are best adapted for the transmission of undistorted light rays from the corneal entrance to the light-receiving cells of the retina.

At the cellular level, cytodifferentiation of the lens consists of the transformation of mitotically active lens epithelial cells into elongated postmitotic lens fiber cells. Up to 90% of the soluble protein in these postmitotic cells consists of crystallin proteins. The mammalian lens contains three major crystallin proteins: α, β, and γ.

The formation of crystallin-containing lens fibers begins with the elongation of epithelial cells from the inner pole of the lens vesicle (see Fig. 13.3). These cells become the fibers of the **lens nucleus** (**Fig. 13.8**). The remaining lens fibers arise from the transformation of the cuboidal cells of the anterior lens epithelium. During embryonic life, mitotic activity is spread throughout the outer lens epithelial cells. Around the time of birth, mitotic activity ceases in the central region of this epithelium, thus leaving a germinative ring of mitotically active cells surrounding the central region. Daughter cells from the germinative region move into the equatorial region of cellular elongation, where they cease to divide and take on the cytological characteristics of RNA-producing cells and begin to form crystallin mRNAs. These cells soon elongate tremendously, fill up with crystallins, and transform into secondary lens fibers that form concentric layers around the primary fibers of the lens nucleus. The midline region where secondary lens fibers from opposite points on the equator join is recognized as the anterior and posterior **lens suture** (see Fig. 13.7D). With this arrangement, the lens fibers toward the periphery are successively younger. As long as the lens grows, new secondary fibers move in from the equator onto the outer cortex of the lens.

The crystallin proteins show a very characteristic pattern and sequence of appearance, with the α-crystallins appearing

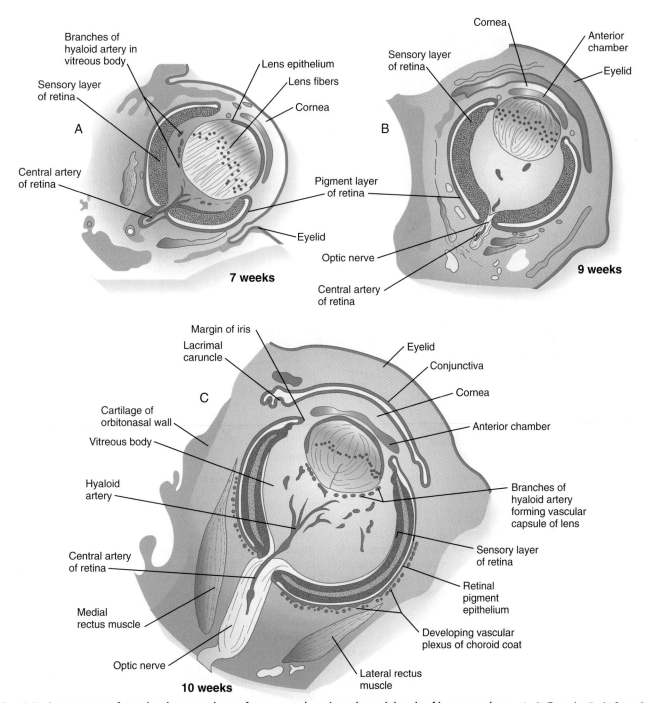

Fig. 13.7 **Later stages of eye development drawn from coronal sections through heads of human embryos.** A, At 7 weeks. B, At 9 weeks. C, At 10 weeks.

Continued

first in the morphologically undifferentiated epithelial cells. Synthesis of β-crystallins is seen when the lens fibers begin to elongate, whereas the expression of γ-crystallins is restricted to terminally differentiated lens fiber cells. Each of the crystallin protein families contains several members. They show different patterns of activation (some members of a family being coordinately activated) and different patterns of accumulation. These patterns facilitate the optical clearing of the lens to allow the efficient transmission of light.

Throughout much of its life, the lens is under the influence of the retina. After induction of the lens, secretions of the retina, of which FGF is a major component, accumulate in the vitreous humor behind the lens and stimulate the formation of lens fibers. A striking example of the continued influence of the retina on lens morphology is seen after a developing lens is rotated so that its outer pole faces the retina. Very rapidly, under the influence of retinal secretions, the low epithelial cells of the former outer pole begin to elongate and

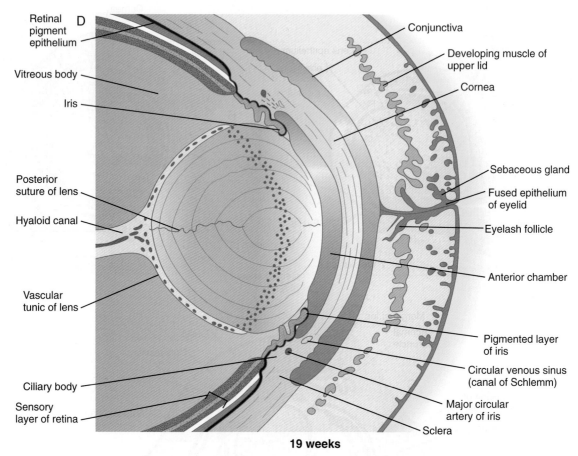

Retinal pigment epithelium D

Vitreous body

Iris

Posterior suture of lens

Hyaloid canal

Vascular tunic of lens

Ciliary body

Sensory layer of retina

Conjunctiva

Developing muscle of upper lid

Cornea

Sebaceous gland

Fused epithelium of eyelid

Eyelash follicle

Anterior chamber

Pigmented layer of iris

Circular venous sinus (canal of Schlemm)

Major circular artery of iris

Sclera

19 weeks

Fig. 13.7, cont'd   D, At 19 weeks. *(Adapted from Carlson B: Patten's foundations of embryology, ed 6, New York, 1996, McGraw-Hill.)*

form an additional set of lens fibers (**Fig. 13.9**). A new lens epithelium forms on the corneal side of the rotated lens. Such structural adaptations are striking evidence of a mechanism that ensures correct alignment between the lens and the rest of the visual system throughout development.

Formation of the Cornea

Formation of the cornea is the result of the last of the series of major inductive events in eye formation (see Fig. 13.4), with the lens vesicle acting on the overlying surface ectoderm. This induction results in the transformation of a typical surface ectoderm, consisting of a basal layer of cuboidal cells and a superficial periderm, to a transparent, multilayered structure with a complex extracellular matrix and cellular contributions from several sources. In keeping with its multifaceted role at almost all stages of eye development, Pax-6 expression in the surface ectoderm is a requirement for corneal induction.

The inductive influence of the lens stimulates a change in the basal ectodermal cells. These cells increase in height, largely as a result of the elaboration of secretory organelles (e.g., Golgi apparatus) on the basal ends of the cells. As these changes are completed, the cells begin to secrete epithelially derived collagen types I, II, and IX to form the **primary stroma** of the cornea (**Fig. 13.10**).

Using the primary stroma as a basis for migration, neural crest cells around the lip of the optic cup migrate centrally between the primary stroma and the lens capsule. Although mesenchymal in morphology during their migration, these cells become transformed into a cuboidal epithelium called the **corneal endothelium** when their migration is completed. At this point, the early cornea consists of (1) an outer epithelium, (2) a still acellular primary stroma, and (3) an inner endothelium.

The migration of neural crest cells between the lens and overlying ectoderm is subject to tight developmental control. A positive stimulus for migration is the production of **transforming growth factor-β** (**TGF-β**) by the lens. Modulating this influence is the presence of **semaphorin 3A** on the lens. The periocular neural crest cells express **neuropilin-1**, which, when combined with semaphorin, inhibits migration. At a critical time in development, a subpopulation of these neural crest cells ceases to express neuropilin. These cells are then able to migrate between the lens and corneal epithelium. A similar mechanism allows the penetration of sensory nerve fibers into the cornea.

After the corneal endothelium has formed a continuous layer, its cells synthesize large amounts of **hyaluronic acid** and secrete it into the primary stroma. Because of its pronounced water-binding capacities, hyaluronic acid causes the primary stroma to swell greatly. This provides a proper substrate for the second wave of cellular migration into the developing cornea (**Fig. 13.11**). These cells, also of neural crest origin, are fibroblastic. They migrate and proliferate in the

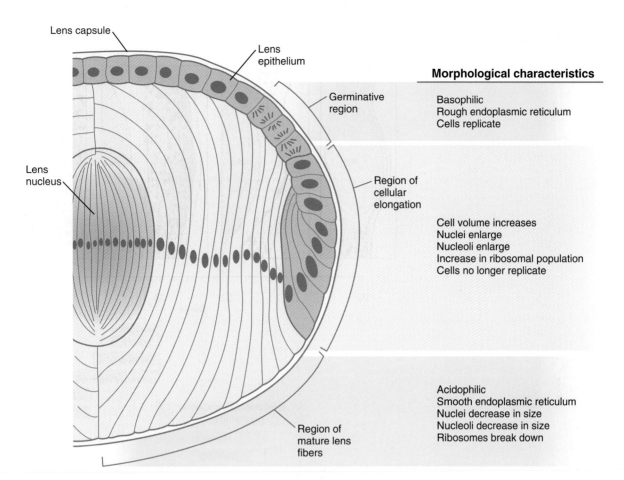

Morphological characteristics

Germinative region
- Basophilic
- Rough endoplasmic reticulum
- Cells replicate

Region of cellular elongation
- Cell volume increases
- Nuclei enlarge
- Nucleoli enlarge
- Increase in ribosomal population
- Cells no longer replicate

Region of mature lens fibers
- Acidophilic
- Smooth endoplasmic reticulum
- Nuclei decrease in size
- Nucleoli decrease in size
- Ribosomes break down

Fig. 13.8 Organization of the vertebrate lens. As the lens grows, epithelial cells from the germinative region stop dividing, elongate, and differentiate into lens fiber cells that produce lens crystallin proteins. *(From Papaconstantinou J: Science 156:338-346, 1967.)*

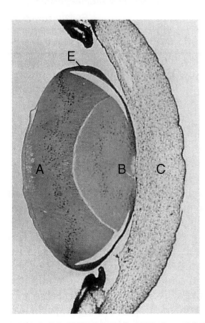

Fig. 13.9 Section through the lens of an 11-day-old chick embryo. At 5 days, the lens was surgically reversed so that the anterior epithelial cells (E) faced the vitreous body and retina. The formerly low epithelial cells elongated to form new lens fibers (A). Because of the reversal of the polarity of the equatorial zone of the lens, new epithelial cells were added over the original mass of lens fibers (B) onto the corneal face of the reversed lens. C indicates cornea. *(From Coulombre JL, Coulombre AJ: Science 142:1489-1490, 1963.)*

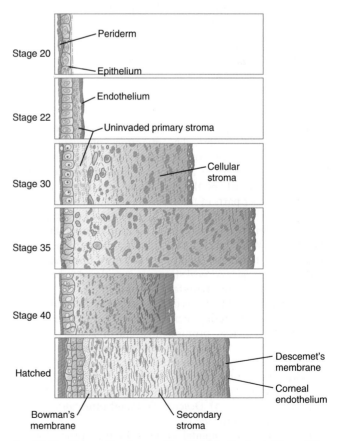

Fig. 13.10 Stages (Hamburger-Hamilton) in the formation of the cornea in the chick embryo. *(Based on studies by Hay ED, Revel JP: Fine structure of the developing avian cornea, Monogr Dev Biol 1:1-144, 1969.)*

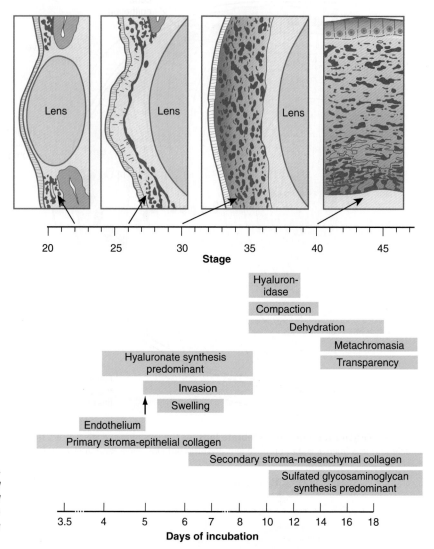

Fig. 13.11 **Major events in corneal morphogenesis in the chick embryo.** *(Based on Toole BP, Trelstad RL: Hyaluronate production and removal during corneal development in the chick, Dev Biol 26:28-35, 1971; and on studies by Hay ED, Revel JP: Fine structure of the developing avian cornea, Monogr Dev Biol 1:1-144, 1969.)*

hyaluronate-rich spaces between layers of collagen in the primary corneal stroma. The migratory phase of cellular seeding of the primary corneal stroma ceases when these cells begin to produce large amounts of **hyaluronidase**, which breaks down much of the hyaluronic acid in the primary stroma. In other parts of the embryo (e.g., limb bud), there is also a close correlation between high amounts of hyaluronic acid and cellular migration and a cessation of migration with its removal. With the removal of hyaluronic acid, the cornea decreases in thickness. When the migratory fibroblasts have settled, the primary corneal stroma is considered to have been transformed into the **secondary stroma**.

The fibroblasts of the secondary stroma contribute to its organization by secreting coarse collagen fibers to the stromal matrix. Nevertheless, prominent layers of acellular matrix continue to be secreted by epithelial and endothelial cells of the cornea. These secretions provide the remaining layers that constitute the mature cornea. Listed from outside in, they are (1) the outer epithelium, (2) **Bowman's membrane,** (3) the secondary stroma, (4) **Descemet's membrane,** and (5) the corneal endothelium (see Fig. 13.10).

The final developmental changes in the cornea involve the formation of a transparent pathway free from optical distortion, through which light can enter the eye. A major change is a great increase in transparency, from about 40% to 100% transmission of light. This increase is accomplished by removing much of the water from the secondary stroma. The initial removal of water occurs with the degradation of much of the water-binding hyaluronic acid. The second phase of dehydration is mediated by **thyroxine**, which is secreted into the blood by the maturing thyroid gland. Thyroxine acts on the corneal endothelium by causing it to pump sodium from the secondary stroma into the anterior chamber of the eye. Water molecules follow the sodium ions, thus effectively completing the dehydration of the corneal stroma. The role of the thyroid gland in this process was shown in two ways. When relatively mature thyroid glands were transplanted onto the extraembryonic membranes of young chick embryos, thereby allowing thyroid hormone to gain access to the embryonic circulation via the blood vessels supplying the membrane (**chorioallantoic membrane**), premature dehydration of the cornea occurred. Conversely, the application of thyroid inhibitors retarded the clearing of the cornea.

The other late event in the cornea is a pronounced change in its radius of curvature in relation to that of the eyeball as a whole. This morphogenetic change, which involves several mechanical events, including intraocular fluid pressure, allows the cornea to work with the lens in bringing light rays into focus on the retina. If irregularities develop in the curvature of the cornea during its final morphogenesis, the individual develops **astigmatism**, which causes distortions in the visual image.

Retina and Other Derivatives of the Optic Cup

While the lens and cornea are taking shape, profound changes are also occurring in the optic cup (see Fig. 13.3). The inner layer of the optic cup thickens, and the epithelial cells begin a long process of differentiation into neurons and light receptor cells of the **neural retina**. The outer layer of the optic cup remains thin and ultimately becomes transformed into the **retinal pigment epithelium** (see Fig. 13.7). Cells of the retinal pigment epithelium do not differentiate into neurons during normal embryogenesis, but in postnatal life some of these cells maintain stem cell properties and can differentiate into multiple cell types. At the same time, the outer lips of the optic cup undergo a quite different transformation into the iris and ciliary body, which are involved in controlling the amount of light that enters the eye (iris) and the curvature of the lens (ciliary body).

Formation of the optic cup and the distinction between neural retina and retinal pigment epithelium depend on molecular events occurring early in eye formation. Stimulated by FGFs from the overlying surface ectoderm, a cooperative interaction between Pax-2 and Pax-6 subdivides the optic vesicle into distal (closest to the surface ectoderm) and proximal (closest to the optic stalk) regions. Under the influence of Pax-6, the distal part of the optic vesicle invaginates to form the inner wall of the optic cup and begins to express the paired-like homeodomain transcription factor **Vsx-2**, which characterizes future neural retina. The helix-loop-helix transcription factor **Mitf** (microphthalmia-associated transcription factor) is initially expressed throughout the entire optic vesicle, but through the inductive effect of BMP from the surrounding neural crest mesenchyme and the action of Pax-2 and Pax-6, its expression becomes restricted to the proximal part of the optic vesicle, which is the future retinal pigment epithelium.

The neural retina is a multilayered structure; its embryonic development can be appreciated only after its adult organization is understood (**Fig. 13.12**). When seen in cross section under a microscope, the neural retina consists of alternating light-staining and dense-staining strips that correspond to layers rich in nuclei or cell processes. The direct sensory pathway in the neural retina is a chain of three neurons that traverse the thickness of the retina. The first element of the chain is the light receptor cell, either a **rod** or a **cone**. A light ray that enters the eye passes through the entire thickness of the neural retina until it impinges on the outer segment of a rod or cone cell (photoreceptor) in the extreme outer layer of the retina. The nucleus of the stimulated rod and cone cell is located in the **outer nuclear layer**. The photoreceptor cell sends a process toward the **outer plexiform layer**, where it synapses with a process from a bipolar cell located in the **inner nuclear layer**. The other process from the bipolar neuron leads into the **internal plexiform layer** and synapses with the third neuron in the chain, the **ganglion cell**. The bodies of the ganglion cells, which are located in the **ganglion cell layer**,

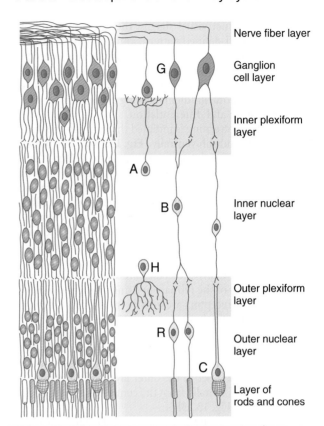

Fig. 13.12 **Tissue and cellular organization of the neural retina of a human fetus.** A, amacrine cell; B, bipolar cell; C, cone; G, ganglion cell; H, horizontal cell; R, rod.

send out long processes that course through the innermost **nerve fiber layer** toward their exit site from the eye, the optic nerve, through which they reach the brain.

If all light signals were processed only through the simple three-link series of neurons in the retina, visual acuity would be much less than it actually is. Many levels of integration have occurred by the time a visual pattern is stored in the visual cortex of the brain. The first is in the neural retina. At synaptic sites in the inner and the outer plexiform layers of the retina, other cells, such as **horizontal** and **amacrine cells** (see Fig. 13.12), are involved in the horizontal redistribution of the simple visual signal. This redistribution facilitates the integration of components of a visual pattern. Another prominent cell type in the retina is the **Müller glial cell**, which sends processes to almost all layers of the retina and seems to play a role similar to that of astrocytes in the central nervous system.

Neural Retina

From the original columnar epithelium of the inner sensory layer of the optic cup (see Fig. 13.7), the primordium of the neural retina takes on the form of a mitotically active, thickened pseudostratified columnar epithelium organized in a manner similar to that of the early neural tube. During the early stages of development of the retina, its polarity becomes fixed according to the same axial sequence as that seen in the limbs (see Chapter 10). The nasotemporal (anteroposterior) axis is fixed first; this is followed by fixation of the dorsoventral axis. Finally, radial polarity is established.

To process distinct visual signals, the retina must develop according to a well-defined pattern, and that pattern must be carried on to the brain so that distinct visual images are produced. Despite its cuplike shape, patterning in the retina is often described in a two-dimensional manner, with dorsoventral and nasotemporal gradients serving as the basis for retinal pattern formation and differentiation.

Dorsoventral patterning is initiated through the presence of **BMP-4** dorsally and shh ventrally (**Fig. 13.13A**). The presence of shh ventrally stimulates the production of **Otx-2** in the outer layer of the optic vesicle and differentiation of the retinal pigment epithelium. Within the inner layer of the optic vesicle, shh and a protein, **ventroptin**, both antagonists of BMP-4, stimulate the expression of the transcription factors **Vax-2** and **Pax-2** in the ventral retina. In the dorsal part of the future retina, BMP-4 signals the expression of **Tbx-5**, the transcription factor involved in the formation of the forelimb (see p. 193). Although many molecules are unequally distributed along the two axes of the retina, opposing gradients of specific ephrins and their receptors are heavily involved in the characterization of the retinal axes (see Fig. 13.13A).

As the number of cells in the early retina increases, the differentiation of cell types begins. There are two major gradients of differentiation in the retina. The first proceeds roughly vertically from the inner to the outer layers of the retina. The second moves horizontally from the center toward the periphery of the retina (Fig. 13.13B).

Evidence obtained on fish retinas suggests that the horizontal gradient of neurogenesis may be driven by an initial point

source of shh in a manner similar to a better-understood process that occurs in *Drosophila*. Differentiation in the horizontal gradient begins with the appearance of ganglion cells and the early definition of the ganglion cell layer (**Fig. 13.14**). As the ganglion cells differentiate, the surrounding cells are prevented from premature differentiation by the activity of the **Notch** gene. A major function of *Notch* is to maintain populations of cells in the nondifferentiated state until the appropriate local cues for their further differentiation appear. With the later differentiation of the horizontal and amacrine cells, the inner and outer nuclear layers take shape. As the cells within the nuclear layers send out processes, the inner and outer plexiform layers become better defined. The bipolar neurons and cone cells differentiate last, thus completing the first gradient.

The horizontal gradient of differentiation of the neural retina is based on the outward spread of the first vertical gradient from the center to the periphery of the retina. The retina cannot grow from within. During the phase of growth in the human eye (or throughout life in the case of continuously growing animals such as fish), developmentally immature retinal precursor cells along the edge of the retina undergo mitosis as an ever-expanding concentric ring on the periphery of the retina. Just inside the ring of mitosis, cellular differentiation occurs in a manner corresponding roughly to that of the vertical gradient.

Cell lineage experiments involving the use of retroviral or other tracers (e.g., horseradish peroxidase) introduced into neuronal precursors in the early retina have revealed two

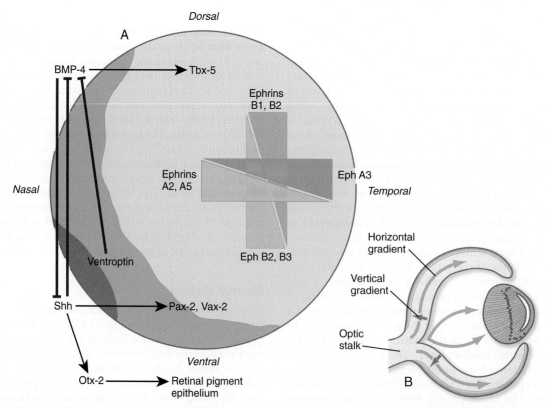

Fig. 13.13 A, Molecular basis for the specification of pattern of the neural retina. The original molecular drivers are sonic hedgehog (Shh) ventrally and bone morphogenetic protein-4 (BMP-4) dorsally. The molecular interactions indicated by the *arrows* are supplemented by dorsoventral and nasotemporal gradients of ephrins and their receptors. **B,** Horizontal and vertical gradients in the differentiation of the layers of the neural retina.

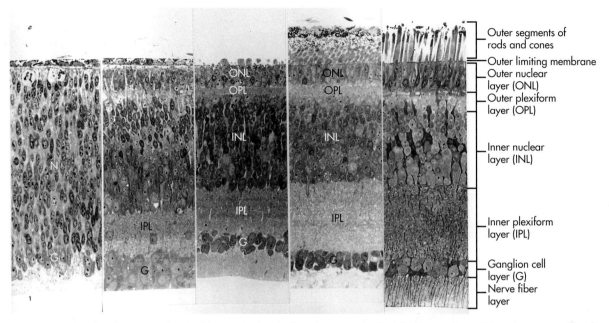

Outer segments of
rods and cones

Outer limiting membrane

Outer nuclear
layer (ONL)

Outer plexiform
layer (OPL)

Inner nuclear
layer (INL)

Inner plexiform
layer (IPL)

Ganglion cell
layer (G)

Nerve fiber
layer

Fig. 13.14 **Progressive development of retinal layers in the chick embryo.** At the *far left,* the ganglion cell layer begins to take shape from the broad neuroblast layer (N). With time, further layers take shape until all layers of the retina are represented *(far right). (From Sheffield JB, Fischman DA: Z Zellforsch Mikrosk Anat 104:405-418, 1970.)*

significant cellular features of retinal differentiation. First, progeny of a single labeled cell are distributed in a remarkably straight radial pattern following the vertical axis of retinal differentiation. There seems to be little lateral mixing among columns of retinal cells (**Fig. 13.15**). The second cellular feature of retinal differentiation is that a single labeled precursor cell can give rise to more than one type of differentiated retinal cell.

A later stage in retinal differentiation is the growth of axons from the ganglion cells along the innermost layer of the retina toward and into the optic stalk. Cells near the exit point express **netrin-1**, which acts as an attractor for the outgrowing nerve processes. When the axons reach the optic stalk, they grow into it, following cues provided by Pax-2–expressing cells. As the axons reach the area of the optic chiasm, they undergo a segregation so that the axons that arise from the temporal half of the retina remain on the same side, and the axons from the nasal half of the retina decussate through the optic chiasm over to the other side. The choice of pathways involves precise local signaling, with netrin-1 again serving as a growth cone attractor and shh playing a new role as a repulsive signal to axonal outgrowth. During this phase of retinal axonal outgrowth, the precise retinal map is maintained in the organization of the optic nerve, and it is ultimately passed on to the visual centers of the brain.

Iris and Ciliary Body

Mediated by a still poorly defined influence of the lens, differentiation of the **iris** and **ciliary body** occurs at the lip of the optic cup, where the developing neural and pigmented retinal layers meet. Rather than being sensory in function, these structures are involved in modulating the amount and character of light that ultimately impinges on the retina. In addition, the ciliary body is the source of the aqueous humor

that fills the anterior chamber of the eye. The iris partially encircles the outer part of the lens, and through contraction or relaxation, it controls the amount of light passing through the lens. The iris contains an inner unpigmented epithelial layer and an outer pigmented layer, which are continuous with the neural and pigmented layers of the retina (**Fig. 13.16**). The **stroma of the iris**, which is superficial to the outer pigmented layer of the iris, is of neural crest origin and secondarily migrates into the iris. Within the stroma of the iris lie the primordia of the **sphincter pupillae** and **dilator pupillae** muscles. These muscles are unusual because they are of neurectodermal origin; they arise from the anterior epithelial layer of the iris through the transdifferentiation of pigmented cells into smooth muscle.

Eye color results from levels and distribution of pigmentation in the iris. The bluish color of the iris in most newborns is caused by the intrinsic pigmentation of the outer pigmented layer of the iris. Pigment cells also appear in the iridial stroma in front of the pigmente epithelium. The greater the density of pigment cells is in this area, the browner the eye color will be. Definitive pigmentation of the eye gradually develops over the first 6 to 10 months of life.

Between the iris and the neural retina lies the ciliary body, a muscle-containing structure that is connected to the lens by radial sets of fibers called the **suspensory ligament of the lens.** By contractions of the ciliary musculature acting through the suspensory ligament, the ciliary body modulates the shape of the lens in focusing light rays on the retina. Factors leading to the formation of the ciliary body are uncertain, but it seems to be induced by the lens (acting through FGFs) and surrounding neural crest mesenchyme (acting through BMPs).

The ciliary body of the eye secretes aqueous humor into the posterior chamber of the eye. The fluid passes in front of the lens through the pupil into the anterior chamber, where it maintains outward pressure on the cornea. It is then resorbed

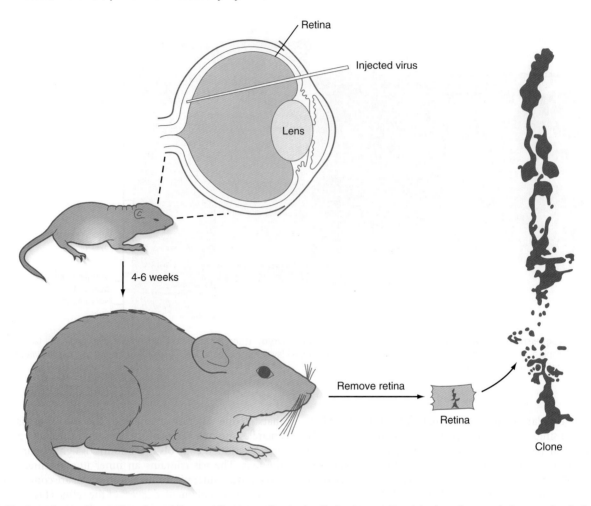

Fig. 13.15 **Experiment illustrating the origins and lineages of retinal cells in the rat.** *Top,* Injection of a retroviral vector that includes the gene for β-galactosidase into the space between the neural and pigmented retinal layers. About 4 to 6 weeks later, the retinas were removed, fixed, and histochemically reacted for β-galactosidase activity. The drawing at the *right* illustrates a vertical clone of cells derived from a virally infected precursor cell. Several cell types (rods, a bipolar cell, and a Müller glial cell) constituted this clone. *(Adapted from Turner DL, Cepko CL:* Nature *328:131-136, 1987.)*

through a trabecular meshwork and into the **canal of Schlemm** in the angle of the eye (see Fig. 13.16). This outflow area arises from the organization of neural crest cells into a trabecular meshwork of lamellae covered by flat endotheliumlike cells, which abut onto an enlarged venous sinus, designated the canal of Schlemm. Development of this network in humans is not completed until around the time of birth.

Vitreous Body and Hyaloid Artery System

During early development of the retina, a loose mesenchyme invades the cavity of the optic cup and forms a loose fibrillar mesh along with a gelatinous substance that fills the space between the neural retina and lens. This material is called the **vitreous body.**

During much of embryonic development, the vitreous body is supplied by the hyaloid artery and its branches (**Fig. 13.17**). The hyaloid artery enters the eyeball through the choroid fissure of the optic stalk (see Fig. 13.6), passes through the retina and vitreous body, and terminates in branches to the posterior wall of the lens. As development of

the retinal vasculature progresses, the portions of the hyaloid artery (and its branches supplying the lens) in the vitreous body regress through apoptosis of their endothelial cells and leave a **hyaloid canal.** The more proximal part of the hyaloid arterial system persists as the **central artery of the retina** and its branches.

Choroid Coat and Sclera

Outside the optic cup lies a layer of mesenchymal cells, largely of neural crest origin. Reacting to an inductive influence from the pigment epithelium of the retina, these cells differentiate into structures that provide vascular and mechanical support for the eye. The innermost cells of this layer differentiate into a highly vascular tunic called the **choroid coat** (see Fig. 13.7C), and the outermost cells form a white, densely collagenous covering known as the **sclera.** The opaque sclera, which serves as a tough outer coating of the eye, is continuous with the cornea. The extraocular muscles, which are derived from cranial mesoderm and provide gross movements to the eyeball, attach to the sclera.

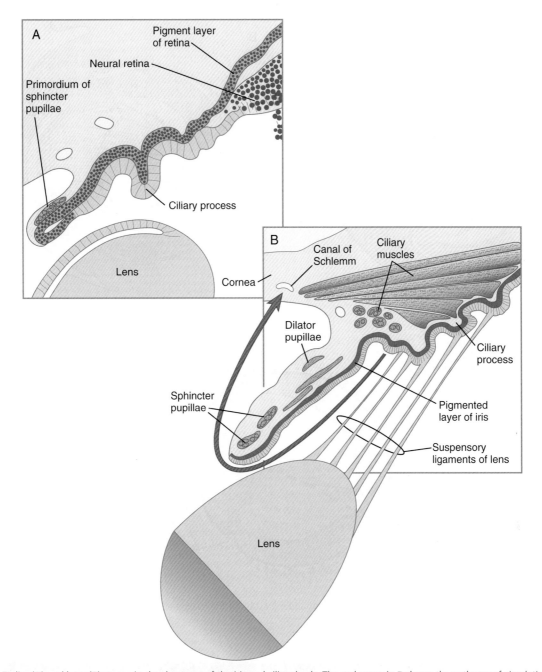

Fig. 13.16 Earlier (**A**) and later (**B**) stages in development of the iris and ciliary body. The *red arrow* in **B** shows the pathway of circulation of aqueous humor from its origin behind the iris to its removal through the canal of Schlemm.

Eyelids and Lacrimal Glands

The eyelids first become apparent during the seventh week as folds of skin that grow over the cornea (**Fig. 13.18A**; see Fig. 13.7). When their formation has begun, the eyelids rapidly grow over the eye until they meet and fuse with one another by the end of the ninth week (Fig. 13.18B; see Fig. 18.4). Many growth factors are involved in the migration of the epidermal cells across the cornea. Mutants of these result in lack of eyelid fusion in the embryo and impaired epidermal wound healing postnatally. The temporary fusion involves only the epithelial layers of the eyelids, thus resulting in a persistent epithelial lamina between them. Before the eyelids reopen, eyelashes and the small glands that lie along the margins of the lids begin to

differentiate from the common epithelial lamina. Although signs of loosening of the epithelial union of the lids can be seen in the sixth month, BMP-mediated reopening of the eyelids normally does not occur until well into the seventh month of pregnancy.

The space between the front of the eyeball and the eyelids is known as the **conjunctival sac**. Multiple epithelial buds grow from the lateral surface ectoderm at about the time when the eyelids fuse. These buds differentiate into the lacrimal glands, which produce a watery secretion that bathes the outer surface of the cornea when mature. This secretion ultimately passes into the nasal chamber by way of the **nasolacrimal duct** (see Chapter 14). The lacrimal glands are not fully mature at birth,

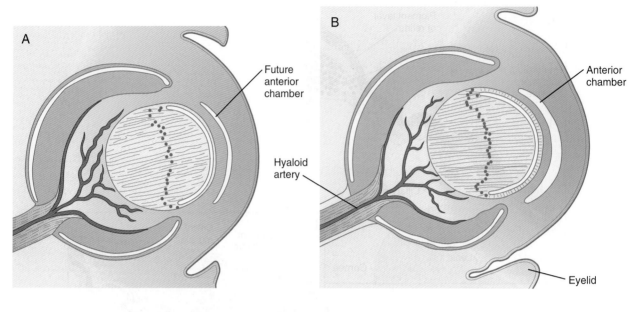

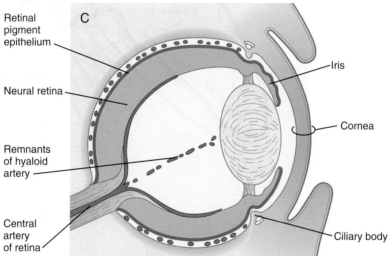

Fig. 13.17 **A** to **C,** Stages in development and regression of the hyaloid artery in the embryonic eye.

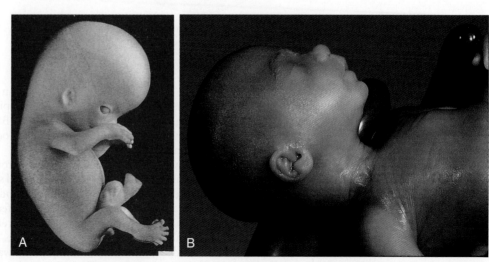

Fig. 13.18 **A,** Head of a human embryo approximately 47 days old. Upper and lower eyelids have begun to form. The external ear is still low set and incompletely formed. **B,** A 5½-month-old (crown-rump length 200 mm) human embryo. The upper and lower eyelids are fused, and the external ear is better formed. Note the receding chin. *(A, From Streeter G: Carnegie contributions to embryology, No. 230, 165-196, 1951; B, EH 1196 from the Patten Embryological Collection at the University of Michigan; courtesy of A. Burdi, Ann Arbor, Mich.)*

and newborns typically do not produce tears when crying. The glands begin to function in lacrimation at about 6 weeks.

Despite the many types of malformations of the eye and visual system, the incidence of most individual types of defects is uncommon. Mutations of almost any of the genes mentioned in this section result in a recognizable malformation of the eye. Among these, anophthalmia, microphthalmia, congenital cataracts, and colobomas are common phenotypes. Examples of a few of the many ocular malformations are given in **Clinical Correlation 13.1**.

Ear

The ear is a complex structure consisting of three major subdivisions: the external, middle, and internal ear. The **external ear** consists of the **pinna** (auricle), the **external auditory meatus** (external ear canal), and the outer layers of the **tympanic membrane** (eardrum), and it functions principally as a sound-collecting apparatus. The middle ear acts as a transmitting device. This function is served by the chain of three middle ear ossicles that connect the inner side of the tympanic membrane to the oval window of the inner ear. Other components of the middle ear are the middle ear cavity (**tympanic cavity**), the **auditory tube** (eustachian tube), the middle ear musculature, and the inner layer of the tympanic membrane. The inner ear contains the primary sensory apparatus, which is involved with hearing and balance. These functions are served by the **cochlea** and **vestibular apparatus**.

From an embryological standpoint, the ear has a dual origin. The inner ear arises from a thickened ectodermal placode at the level of the rhombencephalon. The structures of the middle and external ear are derivatives of the first and second pharyngeal arches and the intervening first pharyngeal cleft and pharyngeal pouch.

CLINICAL CORRELATION 13.1
Congenital Malformations of the Eye

Anophthalmos and Microphthalmos

Anophthalmos, the absence of an eye resulting from a mutation in *RAX*, is very rare and can normally be attributed to lack of formation of the optic vesicle. Because this structure acts as the inductive trigger for much of subsequent eye development, many local inductive interactions involved in the formation of eye structures fail to occur. **Microphthalmos**, which can range from an eyeball that is slightly smaller than normal to one that is almost vestigial, can be associated with a large number of genetic defects (e.g., aniridia) or various other causes, including intrauterine infections (**Fig. 13.19A**). Microphthalmos is one of the common components of the rubella syndrome.

Coloboma of the Iris

Nonclosure of the choroid fissure of the iris during the sixth or seventh week results in its persistence as a defect called **coloboma iridis** (Fig. 13.19B). The location of colobomas of the iris (typically at the 5 o'clock position in the right eye and the 7 o'clock position in the left eye) marks the position of the embryonic choroid fissure. Because of the enlarged pupillary space resulting from the colobomatous fissure, individuals with this condition are sometimes sensitive to bright light, given their inability to contract the pupil properly.

Congenital Cataract

Cataract is a condition characterized by opacity of the lens of the eye. Not so much a structural malformation as a dysplasia, congenital cataracts first came into prominence as one of the triad of defects resulting from exposure of the embryo to the rubella virus.

Cyclopia

Cyclopia (bold) or synophthalmia (bold) represent different degrees of or lack of splitting of the single eye field into two separate bilateral eye fields. Generally due to deficient midline shh signaling from the brain, cyclopia is a malformation secondary to formation of the midline of the brain and face. Cyclopia (see Fig. 8.18) is often accompanied by the presence of a fleshy proboscis dorsal to the eye.

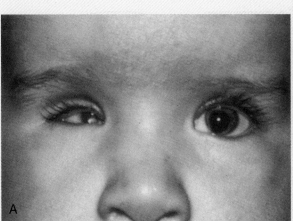

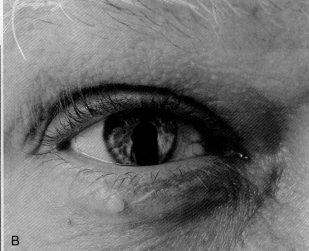

Fig. 13.19　**A,** Microphthalmos of the right eye. **B,** Congenital coloboma of the iris. The fissure is in the region of closure of the choroid fissure. (*A, From Smith B:* Ophthalmic plastic and reconstructive surgery, *vol 2, St Louis, 1987, Mosby.*)

Development of the Inner Ear

Development of the ear begins with preliminary inductions of the surface ectoderm, first by the notochord (chordamesoderm) and then by the paraxial mesoderm (**Fig. 13.20**). These inductions prepare the ectoderm for a third induction, in which **FGF-3** signals from the rhombencephalon induce the adjacent surface ectoderm to express **Pax-2**. Then **Wnt** signals greater than a certain threshold stimulate Pax-2–positive cells to form both the otic and epibranchial placodes (**Fig. 13.21**). The cells exposed to a subthreshold concentration of Wnt are destined to become epidermal cells. Late in the fourth week, possibly under the influence of **FGF-3** secreted by the adjacent rhombencephalon, the otic placode invaginates and then separates from the surface ectoderm to form the **otic vesicle**, or **otocyst** (see Fig. 13.21C). Even at its earliest stages, the development of the main components of the inner ear is under separate genetic control: Pax-2 and Sox-3 for the auditory portion (cochlea) and Nkx-5 for the vestibular portion (semicircular canals).

When the otic vesicle has formed, the next steps in its development are molecular patterning and subdivision of the otic

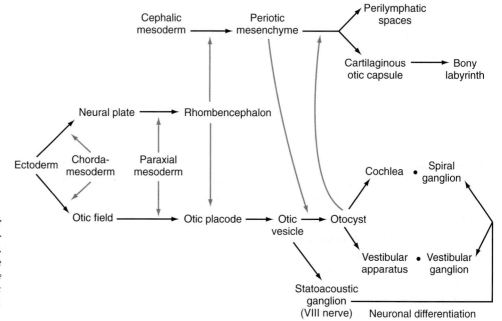

Fig. 13.20 Flow chart of major inductive events and tissue transformations in the developing ear. *Colored arrows* refer to inductive events. *(Based on McPhee JR, van de Water TR. In Jahn AF, Santos-Sacchi J, eds: Physiology of the ear, New York, 1988, Raven, pp 221-242.)*

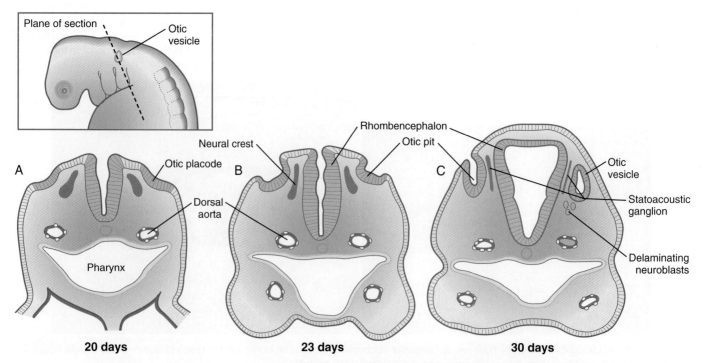

Fig. 13.21 A to **C**, Formation of the otic vesicles from thickened otic placodes.

vesicle into sensory (future hair cells) and neuronal components of the inner ear. Similar to the developing limb and eye, three cartesian axes (anteroposterior, dorsoventral, and mediolateral) become specified early in the development of the otocyst. Ventrality is specified by **shh** coming from ventral sources, and dorsality is specified by **Wnts** coming from the dorsal neural tube.

The otic vesicle soon begins to elongate, thus forming a dorsal vestibular region and a ventral cochlear region (**Fig. 13.22A**). The paired box gene, *Pax2*, is heavily involved in early development of the otic vesicle. Expression of Pax-2 in the ventral otic vesicle is important for the continued development of the endolymphatic duct and the cochlear apparatus, whereas dorsal expression of Dlx-5 and Gbx-2 is important

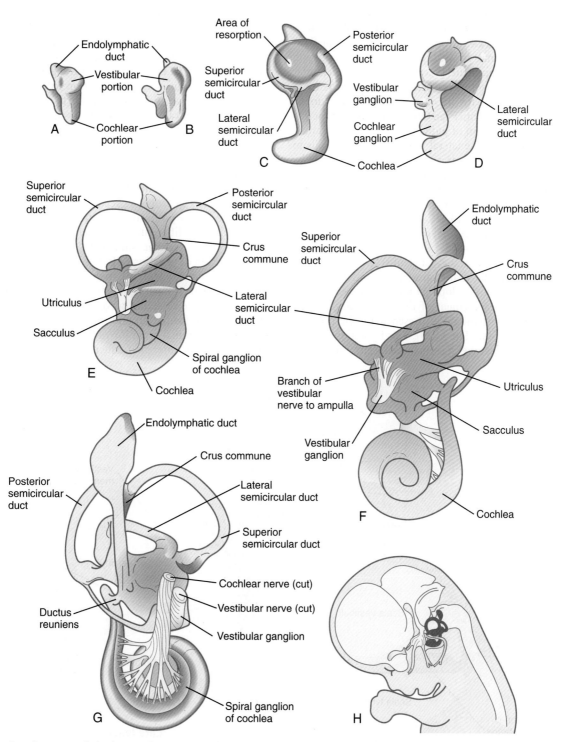

Fig. 13.22   Development of the human inner ear. A, At 28 days. **B,** At 33 days. **C,** At 38 days. **D,** At 41 days. **E,** At 50 days. **F,** At 56 days, lateral view. **G,** At 56 days, medial view. **H,** Central reference drawing at 56 days. *(From Carlson B:* Patten's foundations of embryology, *ed 6, New York, 1996, McGraw-Hill.)*

for development of the vestibular system. Quite early, the **endolymphatic duct** arises as a short, fingerlike projection from the dorsomedial surface of the otocyst (Fig. 13.22B). FGF-3, secreted by rhombomeres 5 and 6, is necessary for normal development of the endolymphatic duct. At about 5 weeks, the appearance of two ridges in the vestibular portion of the otocyst foreshadows the formation of two of the **semicircular ducts** (Fig. 13.22C).

As the ridges expand laterally, their opposing epithelial walls approximate each other, to form a **fusion plate**. Programmed cell death in the central area of epithelial fusion or epithelial cell migration away from this area converts the flangelike structures to canals by setting up a zone of resorption (see Fig. 13.22C). The epithelial precursors of the semicircular canals express the homeobox transcription factor gene *Nkx-5-1*, which is important for the development of the dorsal vestibular portion of the inner ear. Other transcription factors play a role in the formation of individual semicircular canals. In the absence of *Otx1*, the lateral semicircular canal fails to form, and the homeobox transcription factor Dlx-5 must be expressed for the anterior and posterior canals to develop. The cochlear part of the otocyst begins to elongate in a spiral; it attains one complete revolution at 8 weeks and two revolutions by 10 weeks (see Fig. 13.22C through F). The last half turn of the cochlear spiral (a total of two and one-half turns) is not completed until 25 weeks.

The inner ear (membranous labyrinth) is encased in a capsule of skeletal tissue that begins as a condensation of mesenchyme around the developing otocyst at 6 weeks' gestation. The process of encasement of the otocyst begins with an induction of the surrounding mesenchyme by the epithelium of the otocyst (see Fig. 13.20). This induction, involving BMP-4, stimulates the mesenchymal cells, mainly of mesodermal origin, to form a cartilaginous matrix (starting at about 8 weeks). The capsular cartilage serves as a template for the later formation of the true bony labyrinth. The conversion from the cartilaginous to the bony labyrinth occurs between 16 and 23 weeks' gestation.

The sensory neurons that compose the eighth cranial nerve (specifically the **statoacoustic ganglion**) arise from cells that migrate out from a portion of the medial wall of the otocyst (see Fig. 13.21). How the axons originating from the statoacoustic ganglion make their way to precise locations within the inner ear is not well understood, but guidance by neural crest cells and channeling by Slit-Robo interactions are known to play prominent roles in directing growth cones to their proper targets. The cochlear part (**spiral ganglion**) of cranial nerve VIII fans out in close association with the sensory cells (collectively known as the **organ of Corti**) that develop within the cochlea. Neural crest cells invade the developing statoacoustic ganglion and ultimately form the satellite and supporting cells within it.

The sensory cells of the organ of Corti are also derived from the epithelium of the otocyst. They undergo a very complex pattern of differentiation (**Fig. 13.23**). The generation of sensory neuroblast precursors in the inner ear seems to use the **Notch pathway** to control the proportion of epithelial cells that differentiate into neuroblasts versus supporting cells in a manner similar to that described for the differentiation of ganglion cells in the early retina (see p. 280). As in other sensory systems, highly regulated developmental controls ensure precise matching between sensory cells designed to receive sound waves at different frequencies or gravitational information and the neurons that transmit the signals to the brain. Formation of the graded array of hair cells that respond to different frequencies of sound is to a great

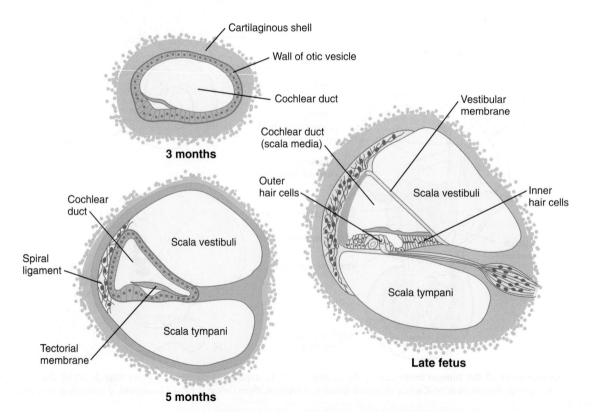

Fig. 13.23 **Cross sections through the developing organ of Corti.**

extent based on execution of the **planar cell polarity** pathway (see p. 87).

Development of the Middle Ear

Formation of the middle ear is intimately associated with developmental changes in the first and second pharyngeal arches (see Chapter 14). The **middle ear cavity** and the **auditory tube** arise from an expansion of the first pharyngeal pouch called the **tubotympanic sulcus (Fig. 13.24)**. Such an origin ensures that the entire middle ear cavity and auditory tube are lined with an endodermally derived epithelium.

By the end of the second month of pregnancy, the blind end of the tubotympanic sulcus (pharyngeal pouch 1) approaches the innermost portion of the first pharyngeal cleft. Nonetheless, these two structures are still separated by a mass of mesenchyme. Later, the endodermal epithelium of the tubotympanic sulcus becomes more closely apposed to the ectoderm lining the first pharyngeal cleft, but they are always separated by a thin layer of mesoderm. This complex, containing tissue from all three germ layers, becomes the **tympanic membrane** (eardrum). During fetal life, a prominent ring-shaped, neural crest–derived bone, called the **tympanic ring**, supports the tympanic membrane. Experiments have

shown that the tympanic ring is actively involved in morphogenesis of the tympanic membrane. Later, the tympanic ring becomes absorbed into the temporal bone.

Just dorsal to the end of the tubotympanic sulcus, a conspicuous condensation of mesenchyme of neural crest origin appears at 6 weeks and gradually takes on the form of the middle ear ossicles. These ossicles, which lie in a bed of very loose embryonic connective tissue, extend from the inner layer of the tympanic membrane to the oval window of the inner ear. Although the middle ear cavity is surrounded by the developing temporal bone, the future middle ear cavity remains filled with loose mesenchyme until late in pregnancy. During the eighth and ninth months, programmed cell death and other resorptive processes gradually clear the middle ear cavity and leave the auditory ossicles suspended within it. Even at the time of birth, remains of the middle ear connective tissue may dampen the free movement of the auditory ossicles. Free movement of the auditory ossicles is acquired within 2 months after birth. Coincident with the removal of the connective tissue of the middle ear cavity is the expansion of the endodermal epithelium of the tubotympanic sulcus, which ultimately lines the entire middle ear cavity.

The middle ear ossicles themselves have a dual origin. According to comparative anatomical evidence, the **malleus**

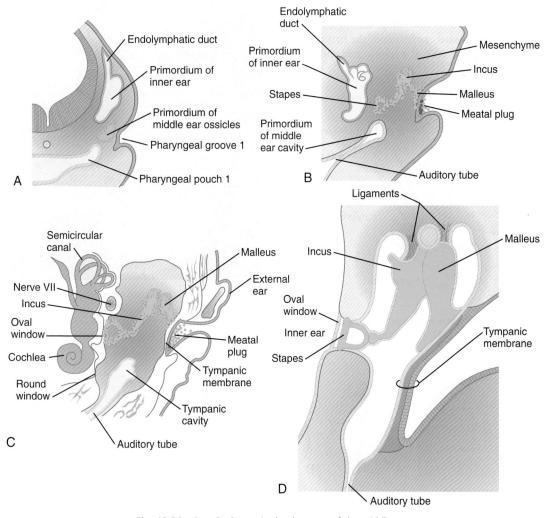

Fig. 13.24   **A** to **D,** Stages in development of the middle ear.

and **incus** arise from neural crest–derived mesenchyme of the first pharyngeal arch, whereas the **stapes** originates from second-arch mesenchyme (**Fig. 13.25**).

Two middle ear muscles help modulate the transmission of auditory stimuli through the middle ear. The **tensor tympani muscle**, which is attached to the malleus, is derived from first-arch mesoderm and is correspondingly innervated by the trigeminal nerve (cranial nerve V). The **stapedius muscle** is associated with the stapes, is of second-arch origin, and is innervated by the facial nerve (cranial nerve VII), which supplies derivatives of that arch.

Development of the External Ear

The external ear (pinna) is derived from mesenchymal tissue of the first and second pharyngeal arches that flank the first (hyomandibular) pharyngeal cleft. During the second month, three nodular masses of mesenchyme (**auricular hillocks**) take shape along each side of the first pharyngeal cleft (**Figs. 13.26** and **13.27**). The auricular hillocks enlarge asymmetrically and ultimately coalesce to form a recognizable external ear. During its formation, the pinna shifts from the base of the neck to its normal adult location on the side of the head. Because of its intimate association with the pharyngeal arches and its complex origin, the external ear is a sensitive indicator of abnormal development in the pharyngeal region. Other anomalies of the first and second arches are often attended by misshapen or abnormally located external ears.

The external auditory meatus takes shape during the end of the second month by an inward expansion of the first pharyngeal cleft. Early in the third month, the ectodermal epithelium

of the forming meatus proliferates and forms a solid mass of epithelial cells called the **meatal plug** (see Fig. 13.24C). Late during the fetal period (at 28 weeks), a channel within the meatal plug extends the existing external auditory meatus to the level of the tympanic membrane.

The external ear and external auditory meatus are very sensitive to drugs. Exposure to agents such as streptomycin, thalidomide, and salicylates during the first trimester can cause agenesis or atresia of both these structures. Congenital malformations of the ear are discussed in **Clinical Correlation 13.2**.

Clinical Vignette

A pediatrician is asked to examine a young boy who has arrived from a country with poor access to medical care. The boy has low-set, misshapen ears, an underslung lower jaw, and a severe hearing deficit. The boy's teeth are also poorly aligned.
1. What is the common denominator for this set of conditions?
2. The pediatrician orders an imaging study of the boy's kidneys and urinary tract. Why does she do that?

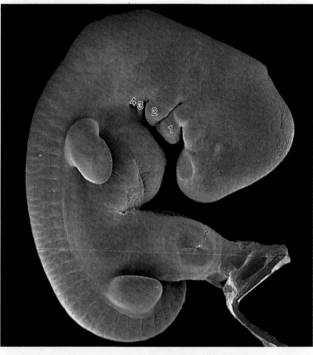

Fig. 13.27   Scanning electron micrograph of a 6-week human embryo showing development of the external ear at a stage roughly equivalent to that illustrated in Figure 13.26A. Pharyngeal arch 2 is beginning to overgrow arches 3 and 4 to form the cervical sinus. *(From Steding G: The anatomy of the human embryo, Basel, 2009, Karger. Courtesy of Dr. J. Männer.)*

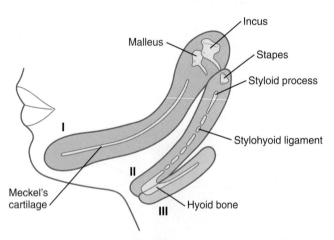

Fig. 13.25   According to the traditional theory of the formation of the middle ear ossicles, the malleus and incus are derived from arch I, and the stapes is derived from arch II.

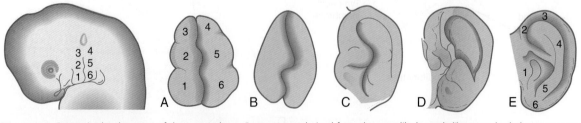

Fig. 13.26   A to E, Stages in development of the external ear. Components derived from the mandibular arch (I) are *unshaded;* components derived from the hyoid arch (II) are *shaded.*

CLINICAL CORRELATION 13.2
Congenital Malformations of the Ear

The ear is subject to a wide variety of genetically based defects, ranging from those that affect the function of specific hair cells in the inner ear to those that produce gross malformations of the middle and external ear. Defects of the middle and external ear are often associated with genetic conditions that affect broader areas of cranial tissues or regions. Many malformations involving the first or second pharyngeal arches are also accompanied by malformations of the ear and reduced acuity of hearing.

Congenital Deafness

Many disturbances of development of the ear can lead to hearing impairments, which affect 1 in every 1000 newborns. Conditions such as **rubella** can lead to maldevelopment of the organ of Corti that causes inner ear deafness. Abnormalities of the middle ear ossicles or ligaments, which can be associated with anomalies of the first and second arches, can interfere with transmission, resulting in middle ear deafness. Agenesis or major atresias of the external ear can cause deafness by interfering with the primary collection of sound waves.

The surge of molecular studies on the development of the inner ear has identified in mice and humans a variety of mutants of the inner ear that could lead to deficits in hearing, equilibrium, or both. Mutants, such as those of Pax-3 causing variants of Waardenburg's syndrome (see p. 265), can affect development at levels ranging from gross morphogenesis of the ear to specific cellular defects in the cochleosaccular complex.

Auricular Anomalies

Because of the multiple origins of its components, there is great variety in the normal form of the pinna. Variations include obvious malformations such as **auricular appendages**, or **sinuses** (**Fig. 13.28**). Many malformations of the external ears are not functionally important, but rather are associated with other developmental anomalies, such as malformations of the kidneys and pharyngeal arches. Exposure to excess retinoic acid or derivatives commonly results in anomalies of the external ear (see Fig. 8.15).

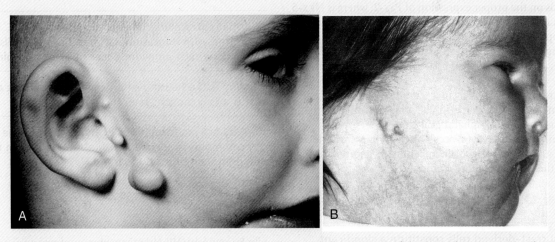

Fig. 13.28 A, Auricular anomalies and tags associated with the mandibular arch (I) component of the external ear. B, Anotia. The external ear is represented only by a couple of small tags. *(Courtesy of M. Barr, Ann Arbor, Mich.)*

Summary

- The eye begins as an outpocketing (optic groove) of the lateral wall of the diencephalon. The optic grooves enlarge to form the optic vesicles, which induce the overlying ectoderm to form the lens primordium. The optic stalk, which connects the optic cup to the diencephalon, forms a groove containing the hyaloid artery, which supplies the developing eye. The paired box gene *Pax-6* acts as a master control gene for eye development.
- Under the influence of Pax-6, the lens forms from an ectodermal thickening that invaginates to form a lens vesicle. The cells of the inner wall of the lens vesicle elongate and synthesize lens-specific crystallin proteins. In the growing lens, the inner epithelium forms a spherical mass of banana-shaped lens fibers (epithelial cells). The anterior lens epithelium consists of cuboidal epithelial cells. The

overall polarity of the lens is under the influence of the retina.
- The cornea is formed through induction of the surface ectoderm by the lens. After induction, the basal ectodermal cells secrete an extracellular matrix that serves as a substrate for the migration of neural crest cells forming the corneal endothelial layer. The corneal endothelial cells secrete large amounts of hyaluronic acid into the early cornea. This permits the migration of a second wave of neural crest cells into the cornea. These fibroblastlike cells secrete collagen fibers into the coarse corneal stroma matrix. Under the influence of thyroxine, water is removed from the corneal matrix, and it becomes transparent.
- The neural retina differentiates from the inner layer of the optic cup. The outer layer forms the pigment layer of the retina. The neural retina is a complex multilayered structure with three layers of neurons connected by cellular

processes. Cellular differentiation in the neural retina follows vertical and horizontal gradients. Under the influence of Pax-2, cell processes grow from retinal neurons through the optic stalk to make connections with optic centers within the brain.

■ The iris and ciliary body form from the outer edge of the optic cup. Sphincter and dilator pupillae muscles form within the iris. Eye color is related to the levels and distribution of pigmentation within the iris. Outside the optic cup, mesenchyme differentiates into a vascular choroid coat and a tough collagenous sclera. Eyelids begin as folds of skin that grow over the cornea and then fuse, closing off the eyes. The eyelids reopen late in the seventh month.

■ The developing eyes are sensitive to certain teratogens and intrauterine infections. Exposure to these can cause microphthalmia or congenital cataracts. Nonclosure of the choroid fissure results in coloboma.

■ The inner ear arises by an induction of the surface ectoderm by the developing hindbrain. Steps in its formation include ectodermal thickening (placode), invagination to form an otic vesicle, and, later, growth and morphogenesis into auditory (cochlea) and vestibular (semicircular canals) portions. Normal development of the cochlea depends on the proper expression of Pax-2, whereas Nkx-5 is required for the formation of the semicircular canals.

■ Development of the middle ear is associated with the first pharyngeal cleft and the arches on either side. Middle ear ossicles and associated muscles take shape within the middle ear cavity.

■ The external ear arises from six modular masses of mesenchyme that take shape in the pharyngeal arch tissue surrounding the first pharyngeal cleft.

■ Congenital deafness can occur after certain intrauterine disturbances, such as rubella infection. Structural anomalies of the external ear are common.

Review Questions

1. Neural crest–derived cells constitute a significant component of which tissue of the eye?
A. Neural retina
B. Lens
C. Optic nerve
D. Cornea
E. None of the above

2. The otic placode arises through an inductive message given off by the:
A. Telencephalon
B. Rhombencephalon
C. Infundibulum
D. Diencephalon
E. Mesencephalon

3. What molecule plays a role in guidance of advancing retinal axons through the optic nerve?
A. Pax-2
B. FGF-3
C. BMP-4
D. Pax-6
E. BMP-7

4. Surface ectoderm is induced to become corneal epithelium by an inductive event originating in the:
A. Optic cup
B. Chordamesoderm
C. Optic vesicle
D. Lens vesicle
E. Neural retina

5. The second pharyngeal arch contributes to the:
A. Cochlea and earlobe
B. Auditory tube and incus
C. Stapes and earlobe
D. Auditory tube and stapes
E. Otic vesicle and stapes

6. During a routine physical examination, an infant was found to have a small segment missing from the lower part of one iris. What is the diagnosis, what is the basis for the condition, and why may the infant be sensitive to bright light?

7. Why does a person sometimes get a runny nose while crying?

8. What extracellular matrix molecule is often associated with migrations of mesenchymal cells, and where does such an event occur in the developing eye?

9. Why is the hearing of a newborn often not as acute as it is a few months later?

10. Why are malformations or hypoplasia of the lower jaw commonly associated with abnormalities in the shape or position of the ears?

References

Adler R, Canto-Solere MV: Molecular mechanisms of optic vesicle development: complexities, ambiguities and controversies, *Dev Biol* 305:1-13, 2007.

Alsina B, Giraldez F, Pujades C: Patterning and cell fate in ear development, *Int J Dev Biol* 53:1503-1513, 2009.

Baker CVH, Bronner-Fraser M: Vertebrate cranial placodes, I: embryonic induction, *Dev Biol* 232:1-61, 2001.

Barald KF, Kelley MW: From placode to polarization: new tunes in inner ear development, *Development* 131:4119-4130, 2004.

Barishak YR: *Embryology of the eye and its adnexae*, Basel, 1992, Karger.

Bok J, Chang W, Wu DK: Patterning and morphogenesis of the vertebrate inner ear, *Int J Dev Biol* 51:521-533, 2007.

Brown ST, Wang J, Groves AK: *Dlx* gene expression during chick inner ear development, *J Comp Neurol* 483:48-65, 2005.

Brugmann SA, Moody SA: Induction and specification of the vertebrate ectodermal placodes: precursors of the cranial sensory organs, *Biol Cell* 97:303-319, 2005.

Byun TH and others: Timetable for upper eyelid development in staged human embryos and fetuses, *Anat Rec* 294:789-796, 2011.

Chalupa LM, ed: Development and organization of the retina: cellular, molecular and functional perspectives, *Semin Cell Dev Biol* 9:239-318, 1998.

Chapman SC: Can you hear me now? Understanding vertebrate middle ear development, *Front Biosci* 16:1675-1692, 2011.

Choo D: The role of the hindbrain in patterning of the otocyst, *Dev Biol* 308:257-265, 2007.

Corwin J, ed: Developmental biology of the ear, *Semin Cell Dev Biol* 8:215-284, 1997.

Cvekl A, Tamm ER: Anterior eye development and ocular mesenchyme: new insights from mouse models and human diseases, *Bioessays* 26:374-386, 2004.

Cvekl A, Wang W-L: Retinoic acid signaling in mammalian eye development, *Exp Eye Res* 89:280-291, 2009.

da Silva MRD and others: FGF-mediated induction of ciliary body tissue in the chick eye, *Dev Biol* 3-4:272-285, 2007.

Donner AL, Lachke SA, Maas RL: Lens induction in vertebrates: variations on a conserved theme of signaling events, *Semin Cell Dev Biol* 17:676-685, 2006.

Esteve P, Bovolenta P: Secreted inducers in vertebrate eye development: more functions for old morphogens, *Curr Opin Neurobiol* 16:13-19, 2006.

Fekete DM, Campero AM: Axon guidance in the inner ear, *Int J Dev Biol* 51:549-556, 2007.

Fritsch B, Barald KF, Lomax MI: Early embryology of the vertebrate ear. In Rubel EW, Popper AN, Fay RR, eds: *Development of the auditory system*, New York, 1997, Springer, pp 80-145.

Fuhrman S: Eye morphogenesis and patterning of the optic vesicle, *Curr Top Dev Biol* 93:61-84, 2010.

Graw J: Eye development, *Curr Top Dev Biol* 90:343-386, 2010.

Graw J: The genetic and molecular basis of congenital eye defects, *Nat Rev Genet* 4:876-888, 2003.

Hay ED: Development of the vertebrate cornea, *Int Rev Cytol* 63:263-322, 1980.

Kelley MW: Cellular commitment and differentiation in the organ of Corti, *Int J Dev Biol* 51:571-583, 2007.

Kelly MC, Chen P: Development of form and function in the mammalian cochlea, *Curr Opin Neurobiol* 19:395-401, 2009.

Kiernan AE, Steel KP, Fekete DM: Development of the mouse inner ear. In Rossant J, Tam PPL, eds: *Mouse development: patterning, morphogenesis, and organogenesis*, San Diego, 2002, Academic Press, pp 539-566.

Kim JH and others: Early fetal development of the human cochlea, *Anat Rec* 294:996-1002, 2011.

Kondoh H: Development of the eye. In Rossant J, Tam PPL, eds: *Mouse development: patterning, morphogenesis, and organogenesis*, San Diego, 2002, Academic Press, pp 519-539.

Ladher RK, O'Neill P, Begbie J: From shared lineage to distinct functions: the development of the inner ear and epibranchial placodes, *Development* 137:1777-1785, 2010.

Lang RA: Pathways regulating lens induction in the mouse, *Int J Dev Biol* 48:783-791, 2004.

Levin AV: Congenital eye anomalies, *Pediatr Clin North Am* 50:55-76, 2003.

Lewis WH: Experimental studies on the development of the eye in amphibia, I: on the origin of the lens, *Am J Anat* 3:505-536, 1904.

Lovicu FJ, Robinson ML: *Development of the ocular lens*, Cambridge, 2004, Cambridge University Press.

Lwigale PY, Bronner-Fraser M: Semaphorin3A/neuropilin-1 signaling acts as a molecular switch regulating neural crest migration during cornea development, *Dev Biol* 336:257-265, 2009.

Martinez-Morales JR, Wittbrodt J: Shaping the vertebrate eye, *Curr Opin Genet Dev* 19:511-517, 2009.

McCabe KL, Bronner-Fraser M: Molecular and tissue interactions governing induction of cranial ectodermal placodes, *Dev Biol* 332:189-195, 2009.

Medina-Martinez O, Jamrich M: Foxe view of lens development and disease, *Development* 134:1455-1463, 2007.

Mui SH and others: The homeodomain protein Vax2 patterns the dorsoventral and nasotemporal axes of the eye, *Development* 129:797-804, 2002.

Nishimura Y, Kumoi T: The embryologic development of the human external auditory meatus, *Acta Otolaryngol* 112:496-503, 1992.

Ohyama T and others: Wnt signals mediate a fate decision between otic placode and epidermis, *Development* 133:865-875, 2006.

O'Rahilly R: The timing and sequence of events in the development of the human eye and ear during the embryonic period proper, *Anat Embryol* 168:87-99, 1983.

Pichaud F, Desplan C: Pax genes and eye organogenesis, *Curr Opin Genet Dev* 12:430-434, 2002.

Puligilla C, Kelley MW: Building the world's best hearing aid: regulation of cell fate in the cochlea, *Curr Opin Genet Dev* 19:368-373, 2009.

Represa J, Frenz DA, van de Water TR: Genetic patterning of embryonic inner ear development, *Acta Otolaryngol* 120:5-10, 2000.

Reza HM, Yasuda K: Lens differentiation and crystallin regulation: a chick model, *Int J Dev Biol* 48:805-817, 2004.

Riley BB, Phillips BT: Ringing in the new ear: resolution of cell interactions in otic development, *Dev Biol* 261:289-312, 2003.

Rodríguez-Vázquez JF, Mérida-Velasco JR, Verdugo-López S: Development of the stapedius muscle and unilateral agenesis of the tendon of the stapedius muscle in a human fetus, *Anat Rec* 293:25-31, 2010.

Romand R, Dollé P, Hashino E: Retinoid signaling in inner ear development, *J Neurobiol* 66:687-704, 2006.

Sakuta H and others: Ventroptin: a BMP-4 antagonist expressed in a double-gradient pattern in the retina, *Science* 293:111-115, 2001.

Sanchez-Calderón H and others: A network of growth and transcription factors controls neuronal differentiation and survival in the developing ear, *Int J Dev Biol* 51:557-570, 2007.

Schlosser G: Induction and specification of cranial placodes, *Dev Biol* 294:303-351, 2006.

Schneider-Maunoury S, Pujades C: Hindbrain signals in otic regionalization: walk on the wild side, *Int J Dev Biol* 51:495-506, 2007.

Smith AN, Radice G, Lang RA: Which FGF ligands are involved in lens induction? *Dev Biol* 337:195-198, 2010.

Spemann H: *Embryonic development and induction*, New Haven, Conn, 1938, Yale University Press.

Streeter GC: Development of the auricle in the human embryo, *Contr Embryol Carnegie Inst* 14:111-138, 1922.

Streit A: The preplacodal region: an ectodermal domain with multipotential progenitors that contribute to sense organs and cranial sensory ganglia, *Int J Dev Biol* 51:447-461, 2007.

Streit A: Early development of the cranial sensory nervous system: from a common field to individual placodes, *Dev Biol* 276:1-15, 2004.

Trousse F and others: Control of retinal ganglion cell axon growth: a new role for sonic hedgehog, *Development* 128:3927-3936, 2001.

Vendrell V and others: Induction of inner ear fate by FGF3, *Development* 127:2011-2019, 2000.

Yang X-J: Roles of cell-extrinsic growth factors in vertebrate eye pattern formation and retinogenesis, *Semin Cell Dev Biol* 15:91-103, 2004.

Head and Neck

Among the earliest vertebrates, the cranial region consisted of two principal components: (1) a **chondrocranium,** associated with the brain and the major sense organs (nose, eye, ear); and (2) a **viscerocranium**, a series of branchial (pharyngeal) arches associated with the oral region and the pharynx (**Fig. 14.1A**). As vertebrates became more complex, the contributions of the neural crest to the head became much more prominent, and the face and many dermal (intramembranously formed) bones of the skull (**dermocranium**) were added. With the early evolution of the face, the most anterior of the branchial arches underwent a transformation to form the upper and lower jaws and two of the middle ear bones, the malleus and the incus. Along with an increase in complexity of the face (Fig. 14.1B) came a corresponding increase in complexity of the forebrain (telencephalon and diencephalon). From structural and molecular aspects, the rostral (anteriormost) part of the head shows distinctly different characteristics from the pharyngeal region, as follows:

1. The pharyngeal region and hindbrain are highly segmented (see Fig. 14.3), whereas segmentation is less evident in the forebrain and rostral part of the head.
2. Structural segmentation in the pharyngeal region is associated with complex segmental patterns of gene expression (see Fig. 11.12).
3. Formation of the forebrain and associated structures of the rostral part of the head depends on the actions of specific genes (e.g., *Lim1* [see Fig. 5.9], *Emx1, Emx2, Otx1,* and *Otx2*) and inductive signaling by the prechordal mesoderm or anterior visceral endoderm.
4. Much of the connective tissue and skeleton of the rostral (phylogenetically newer) part of the head is derived from the neural crest. The anterior end of the notochord, terminating at the hypophysis, constitutes the boundary between the mesodermally derived chondrocranium and the more rostral neural crest–derived chondrocranium. Neural crest cells are also prominent contributors to the ventral part of the pharyngeal region.

In previous chapters, the development of certain components of the head (e.g., the nervous system, neural crest, bones of the skull) is detailed. The first part of this chapter provides an integrated view of early craniofacial development to show how the major components are interrelated. The remainder of the chapter concentrates on the development of the face, pharynx, and pharyngeal arch system. Clinical Correlations 14.1 and

14.2, which appear later in the chapter, present malformations associated with the head and neck.

Early Development of the Head and Neck

Development of the head and neck begins early in embryonic life and continues until the cessation of postnatal growth in the late teens. Cephalization begins with the rapid expansion of the rostral end of the neural plate. Very early, the future brain is the dominant component of the craniofacial region. Beneath the brain, the face, which does not take shape until later in embryogenesis, is represented by the **stomodeum (Fig. 14.2)**. In the early embryo, the stomodeum is sealed off from the primitive gut by the **oropharyngeal membrane**, which breaks down by the end of the first embryonic month (see Fig. 6.23). Surrounding the stomodeum are several tissue prominences that constitute the building blocks of the face (see Fig. 14.6). In keeping with its origin from the anterior neural ridge and its later serving as the tissue of origin of Rathke's pouch, the ectoderm of the oropharyngeal membrane is first characterized by its expression of the homeodomain transcription factor **Pitx-2**. In the rostral midline is the frontonasal prominence, which is populated by mesenchymal cells derived from forebrain and some midbrain neural crest. On either side of the frontonasal prominence, ectodermal nasal placodes, which arose from the anterior neural ridge (see p. 95), develop into horseshoe-shaped structures, each consisting of a nasomedial process, also derived from forebrain neural crest, and a nasolateral process, derived from midbrain neural crest. Farther caudally, the stomodeum is bounded by maxillary and mandibular processes, which are also filled with neural crest–derived mesenchyme.

The future cervical region is dominated by the pharyngeal apparatus, consisting of a series of pharyngeal pouches, arches, and clefts. Many components of the face, ears, and glands of the head and neck arise from the pharyngeal region. Also prominent are the paired ectodermal placodes (see Fig. 13.1), which form much of the sensory tissue of the cranial region.

Tissue Components and Segmentation of Early Craniofacial Region

The early craniofacial region consists of a massive neural tube beneath which lie the notochord and a ventrally situated

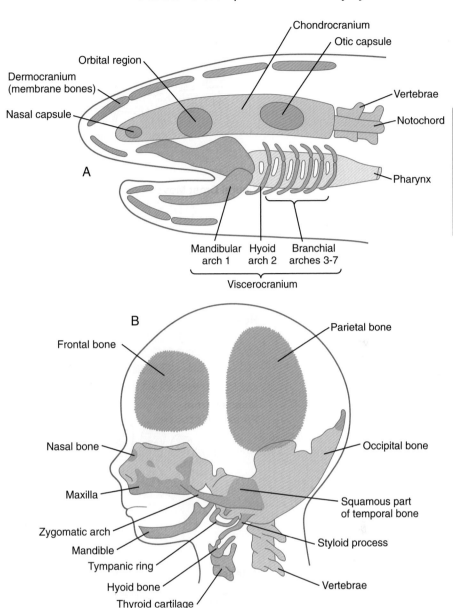

Fig. 14.1 **Organization of the major components of the vertebrate skull.** A, Skull of a primitive aquatic vertebrate, showing the chondrocranium *(green)*, viscerocranium *(orange)*, and dermocranium *(brown)*. B, Human fetal head, showing the distribution of the same components of the cranial skeleton.

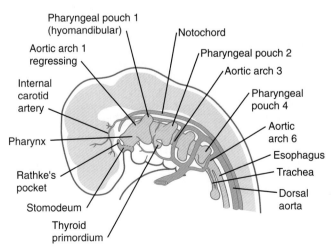

Fig. 14.2 **Basic organization of the pharyngeal region of the human embryo at the end of the first month.**

pharynx (see Fig. 14.2). The pharynx is surrounded by a series of pharyngeal arches. Many of the tissue components of the head and neck are organized segmentally. **Figure 14.3** illustrates the segmentation of the tissue components of the head. As discussed in earlier chapters, distinct patterns of expression of certain homeobox-containing genes are associated with morphological segmentation in some cranial tissues, particularly the central nervous system (see Fig. 11.12). The chain of events between segmental patterns of gene expression and the appearance of morphological segmentation in parts of the cranial region remains incompletely understood.

Fundamental Organization of the Pharyngeal Region

Because many components of the face are derived from the pharyngeal region, an understanding of the basic organization

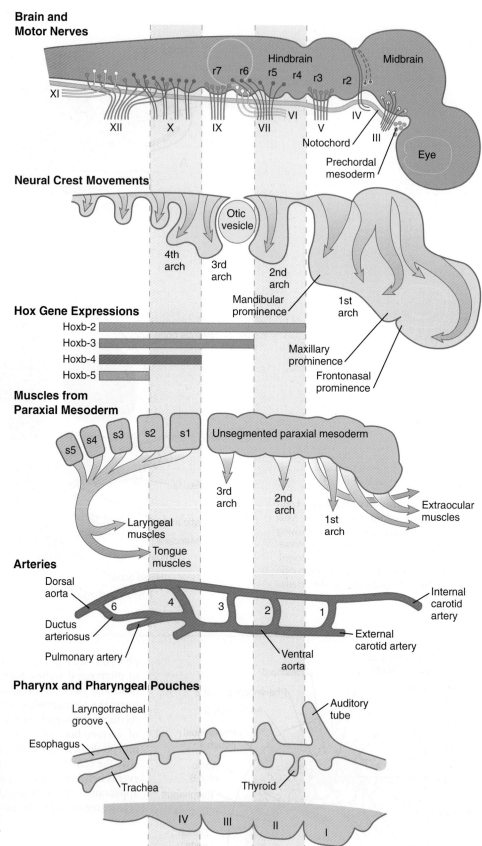

Fig. 14.3 Lateral view of the organization of the head and pharynx of a 30-day-old human embryo, with individual tissue components separated but in register through the *dashed lines*. *(Based on Noden DM: Brain Behav Evol 38:190-225, 1991; and Noden DM, Trainor PA: J Anat 207:575-601, 2005.)*

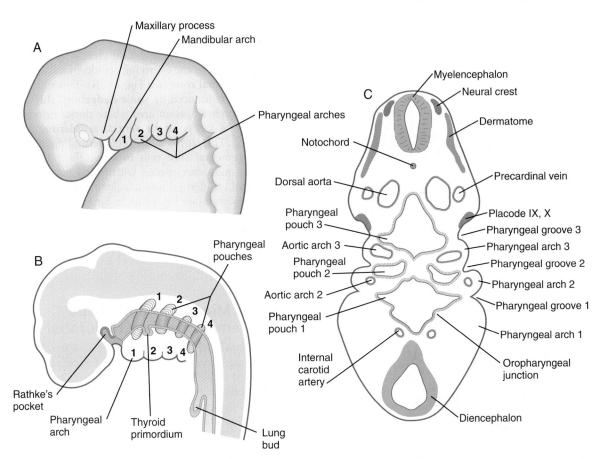

Fig. 14.4 **A** and **B,** Superficial and sagittal views of the head and pharynx of a human embryo during the fifth week. **C,** Cross section through the pharyngeal region of a human embryo of the same age. Because of the strong C curvature of the head and neck of the embryo, a single section passes through the levels of the forebrain *(bottom)* and the hindbrain *(top).*

of this region is important. In a 1-month-old embryo, the pharyngeal part of the foregut contains four lateral pairs of endodermally lined outpocketings called **pharyngeal pouches** and an unpaired ventral midline diverticulum, the **thyroid primordium** (**Fig. 14.4**). If the contours of the ectodermal covering over the pharyngeal region are followed, bilateral pairs of inpocketings called **pharyngeal grooves**, which almost make contact with the lateralmost extent of the pharyngeal pouches, are seen (see Fig. 14.4C).

Alternating with the pharyngeal grooves and pouches are paired masses of mesenchyme called **pharyngeal (branchial) arches**. Central to each pharyngeal arch is a prominent artery called an **aortic arch**, which extends from the ventral to the dorsal aorta (see Chapter 17 and Fig. 14.2). The mesenchyme of the pharyngeal arches is of dual origin. The mesenchyme of the incipient musculature originates from mesoderm, specifically the somitomeres. Much of the remaining pharyngeal arch mesenchyme, especially that of the ventral part, is derived from the neural crest, whereas mesoderm makes various contributions to the dorsal pharyngeal arch mesenchyme.

Establishing the Pattern of the Craniofacial Region

Establishment of the fundamental structural pattern of the craniofacial region is a complex process that involves

interactions among numerous embryonic tissues. Major players are as follows: the neural tube, which acts as a signaling center and gives rise to the cranial neural crest; the paraxial mesoderm; the endoderm of the pharynx; and the cranial ectoderm.

Very early in development, the cranial neural tube becomes segmented on the basis of molecular instructions, based largely on *Hox* gene expression (see Fig. 11.12), and this coding spills over into the neural crest cells that leave the neural tube (see Fig. 12.8). More recently, investigators have recognized that the pharyngeal endoderm also exerts a profound patterning influence on facial development. Patterning of the pharyngeal endoderm itself is heavily based on its exposure to retinoic acid. Formation of the first pharyngeal pouch does not require retinoic acid, but pharyngeal pouches 3 and 4 have an absolute requirement for retinoic acid, whereas pouch 2 needs some, but not so much, exposure to retinoic acid.

Formation of the pharyngeal arches depends on signals from the pharyngeal pouches. Even though neural crest cells are major contributors to the underlying tissues of the pharyngeal arches, experiments have shown that neural crest is not required for the formation or patterning of the pharyngeal arches. In almost all aspects of lower facial morphogenesis, the development of neural crest derivatives depends on signals from cranial ectoderm, but the ectoderm is prepatterned by signals (importantly **fibroblast growth factor-8 [FGF-8]**) emanating from the pharyngeal endoderm (**Fig. 14.5A**).

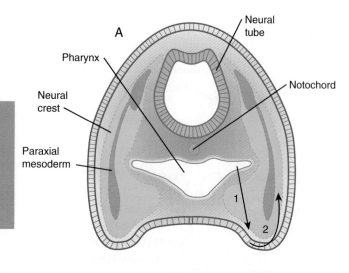

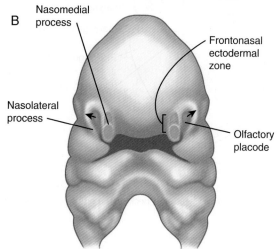

Fig. 14.5 **Signaling centers in the early craniofacial region. A,** In the pharynx, pouch endoderm signals to the ectoderm, which signals to underlying neural crest cells. **B,** The frontonasal ectodermal zone, which produces signals important for the development of the midface. Dorsal cells *(green)* express fibroblast growth factor-8 (FGF-8), and ventral cells *(orange)* express sonic hedgehog (shh). The olfactory placode *(violet)* induces the nasolateral processes through FGF signaling.

Development of individual pharyngeal arches depends on various sets of molecular instructions. The first arch, which forms the upper and the lower jaws, is not included in the overall Hox code that underlies development of the remainder of the arches and determines their anteroposterior identity (see Fig. 12.8). Within individual pharyngeal arches, a code based on the homeobox-containing *Dlx* genes heavily influences dorsoventral patterning (see p. 301). Other molecular influences also strongly affect patterning of aspects of pharyngeal arch development. A major force in the patterning of pharyngeal arch 1 is **endothelin-1 (Edn-1)**, which is secreted by the ectoderm of the arches and combines with its receptor (**Ednr**) on migrating neural crest cells. Although expressed on all of the pharyngeal arches, Edn-1 exerts its most prominent effect on the development of the first arch through its effects on *Dlx* expression.

A prominent feature of the early developing face is the unpaired **frontonasal prominence**, which constitutes the

rostralmost part of the face (see Fig. 14.6). Originating over the bulging forebrain, the frontonasal process is filled with cranial neural crest. These neural crest cells are targets of a signaling center in the overlying ectoderm, called the **frontonasal ectodermal zone** (see Fig. 14.5B). This signaling center, which itself is induced by **sonic hedgehog (shh)** emanating from the forebrain, is an area where dorsal ectodermal cells expressing FGF-8 confront ventral ectodermal cells, which express shh. This confluence of ectodermal signals acts on the underlying neural crest cells to shape the tip of the snout. Mammals and other species with broad faces have bilateral frontonasal ectodermal zones, located at the tips of the nasomedial processes (see Fig. 14.6). In birds, which have a narrow midface tapering into a beak, the two frontonasal ectodermal zones fuse into a single signaling center. In avian embryos, transplantation of the facial ectodermal zone into an ectopic region results in the formation of a second beak.

Cellular Migrations and Tissue Displacements in the Craniofacial Region

Early craniofacial development is characterized by several massive migrations and displacements of cells and tissues. The neural crest is the first tissue to exhibit such migratory behavior, with cells migrating from the nervous system even before closure of the cranial neural tube (see Chapter 12). Initially, segmental groups of neural crest cells are segregated, especially in the pharyngeal region (see Fig. 14.3). These populations of cells become confluent, however, during their migrations through the pharyngeal arches. Much of the detailed anatomy of the facial skeleton and musculature is based on the timing, location, and interactions of individual streams of neural crest and mesodermal cells. Recognition of this level of detail (which is beyond the scope of this text) is important in understanding the basis underlying many of the numerous varieties of facial clefts that are seen in pediatric surgical clinics.

The early cranial mesoderm consists mainly of the **paraxial** and **prechordal mesoderm** (see Fig. 14.3). Although the paraxial mesoderm rostral to the occipital somites has been traditionally considered to be subdivided into somitomeres (see Fig. 6-8), some embryologists now classify it as being unsegmented mesoderm (see Fig. 14.3). Mesenchymal cells originating in the paraxial mesoderm form the connective tissue and skeletal elements of the caudal part of the cranium and the dorsal part of the neck. Within the pharyngeal arches, cells from the paraxial mesoderm initially form a mesodermal core, which is surrounded by cranial neural crest cells (see Fig. 14.5A). Myogenic cells from paraxial mesoderm undergo extensive migrations to form the bulk of the muscles of the cranial region. Similar to their counterparts in the trunk and limbs, these myogenic cells become integrated with local connective tissue to form muscles. Another similarity with the trunk musculature is that morphogenetic control seems to reside within the connective tissue elements of the muscles, rather than in the myogenic cells themselves. In the face and ventral pharynx, this connective tissue is of neural crest origin.

The prechordal mesoderm, which emits important forebrain inductive signals in the early embryo, is a transient mass of cells located in the midline, rostral to the tip of the notochord. Although the fate of these cells is controversial, some investigators believe that myoblasts contributing to the extra-ocular muscles take origin from these cells. On their way to the eye, cells of the prechordal mesoderm may pass through the rostralmost paraxial mesoderm.

The **lateral plate mesoderm** is not well defined in the cranial region. Transplantation experiments have shown that it gives rise to endothelial and smooth muscle cells and, at least in birds, to some portions of the laryngeal cartilages.

Another set of tissue displacements important in the cranial region is the joining of cells derived from the ectodermal placodes with cells of the neural crest to form parts of sense organs and ganglia of certain cranial nerves (see Fig. 13.1).

Development of the Facial Region

Formation of the Face and Jaws

Development of the face and jaws is a complex three-dimensional process involving the patterning, outgrowth, fusion, and molding of various tissue masses. The forebrain acts as a mechanical substrate and a signaling center for early facial development, and the stomodeum serves as a morphological point of reference. The lower face (maxillary region and lower jaw) is phylogenetically derived from a greatly expanded first pharyngeal arch. Much of the mesenchyme of the face is neural crest, originating from the forebrain to the first two rhombomeres. Each of the tissue components of the early face is the product of a unique set of morphogenetic determinants and growth signals, and increasing evidence indicates that specific sets of molecular signals control their development along the proximodistal and the rostrocaudal axes.

At a higher level, the building blocks of the face relate to one another in highly specific ways, and clues to their origins and relationships can be derived by examination of their blood supply. Disturbances at this level regularly result in the production of craniofacial anomalies, and an understanding of the fundamental elements of facial morphogenesis is crucial to rational surgical approaches to these malformations.

Structures of the face and jaws originate from several primordia that surround the stomodeal depression of a 4- to 5-week-old human embryo (**Fig. 14.6**). These primordia consist of the following: an unpaired **frontonasal prominence**; paired **nasomedial** and **nasolateral processes**, which are components of the horseshoe-shaped olfactory (nasal) primordia; and paired **maxillary processes** and **mandibular prominences**, both components of the first pharyngeal arches. The upper jaw contains a mixed population of neural crest cells derived from the forebrain and midbrain, whereas the lower jaw contains mesenchymal cells derived from midbrain and hindbrain (rhombomeres 1 and 2) neural crest. The specific morphology of facial skeletal elements is determined by signals passed from the pharyngeal endoderm to the facial ectoderm and then to the neural crest precursors of the facial bones. Narrow zones of pharyngeal endoderm control the morphogenesis of specific portions of the skeleton of the lower face. FGF-8 signaling from the facial ectoderm plays a key role in patterning of the facial skeleton.

Another factor that strongly influences facial form is the responsiveness of the various facial processes to **Wnts**. In many developing structures, Wnt signaling stimulates cellular proliferation that increases the mass of that structure. In facial development, species with an elongated midface (e.g., the beak of birds) have a midline zone of Wnt responsiveness within the frontonasal process. Other species (e.g., humans) that have flat but broad faces have Wnt-responsive regions in the maxillary and mandibular processes, thus supporting lateral growth in the face.

The frontonasal process is a prominent structure in the earliest phases of facial development, and its formation is the result of an exquisitely sensitive signaling system that begins with the synthesis of **retinoic acid** in a localized region of ectoderm opposite the forebrain and continues with the action of shh produced by the ventral forebrain. The action of shh, through the mediation of the most rostral wave of neural crest cells, underlies the establishment of the frontonasal ectodermal zone, located at the tips of the nasomedial processes (see p. 298). The signaling molecules (FGF-8 and shh) emanating from this zone stimulate cell proliferation in the neural crest mesenchyme of the frontonasal process. In the absence of such signaling, cell death in the region increases, and cell proliferation decreases, resulting in various midfacial defects (see Clinical Correlation 14.1). Retinoic acid is unusual in that both deficiencies and excess amounts can cause very similar defects. From 4 to 5 weeks, the frontonasal process is a dominant structure of the early face (see Fig. 14.6), but with subsequent growth of the maxillary process and the nasomedial and nasolateral processes, it recedes from the oral region. The nasolateral process develops as a result of FGF signaling emanating from the nasal pit.

The maxillary and mandibular processes have traditionally been considered to be derivatives of the first pharyngeal arch. More recent research has suggested that although some of the cells that form the maxillary process arise from the first arch, many mesenchymal cells of the maxillary process are not first-arch derivatives, but instead come from other areas of cranial neural crest. How these cells are integrated into a unified structure and what controls their specific morphogenesis remain to be determined.

As with the limb buds, outgrowth of the frontonasal prominence and the maxillary and mandibular processes depends on mesenchymal-ectodermal interactions. In contrast to the limb, however, the signaling system (FGF and shh) is concentrated in the apical ectoderm of these processes, where it may act as a morphogenetic organizer and a stimulus for mesenchymal outgrowth of the facial primordia. The homeobox-containing gene Msx1 is expressed in the rapidly proliferating mesenchyme at the tips of the facial primordia. The parallel with expression of Msx1 in the subectodermal region of the limb (see p. 200) suggests that similar mechanisms operate in outgrowing limb and facial primordia. *Hox* genes are not expressed in the first arch, and the presence of **Otx-2** distally, coupled with the absence of Hox proximally, provides the molecular basis for the development of the first arch.

The subdivision of the first arch into maxillary and mandibular regions is controlled to a large extent by **endothelin-1**. Expressed at the distal (ventral) tip of the arch, endothelin-1 effectively represses the local expression of genes, such as *Dlx-1/2*, that are heavily involved in formation of the proximal maxilla. Distally, endothelin-1 promotes the expression of distal genes, such as *Dlx-5/6* and their downstream targets (*Hand-2* and *Goosecoid*), which pattern the mandible. At an intermediate dorsoventral level within the first arch, endothelin-1 stimulates the expression of *Barx-1*, which is a

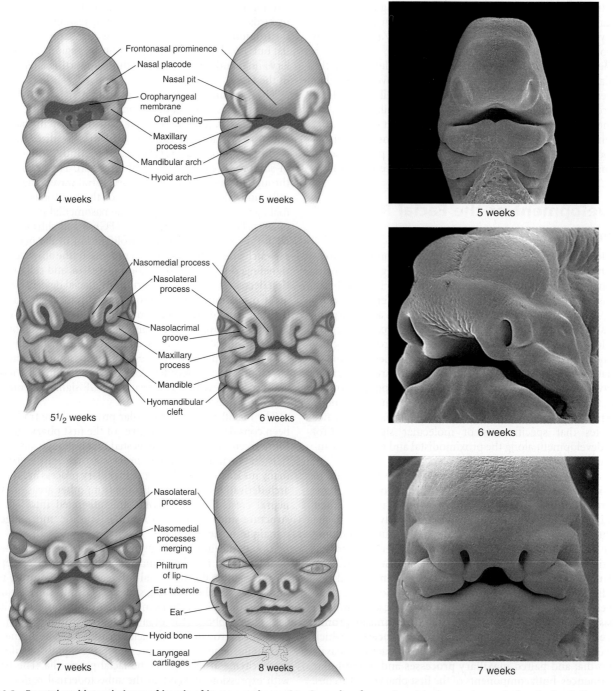

Fig. 14.6 **Frontal and lateral views of heads of human embryos 4 to 8 weeks of age.** *(Scanning electron micrographs from Steding G:* The anatomy of the human embryo, *Basel, 2009, Karger; courtesy of Dr. J. Männer.)*

prime determinant of the formation of the mandibular joint. When *endothelin-1* is mutated or inactivated, the mandible becomes transformed into a structure resembling the maxilla. If endothelin-1 is overexpressed in the proximal part of the first arch, the future maxilla becomes transformed into a mandible. This effect is transmitted through the activation of Dlx-5/6 (see later). In the proximal (dorsal) part of the first arch, the influence of endothelin-1 is reduced, and active patterning genes lay the foundation for the formation of both the maxilla and the middle ear bones (malleus, incus, and tympanic ring).

Despite the relatively featureless appearance of the early mandibular process (first pharyngeal arch), the mediolateral (oral-aboral) and proximodistal axes are tightly specified. This recognition has considerable clinical significance because increasing numbers of genetic mutants are recognized to affect only certain regions of the arch, such as the absence of distal (adult midline) versus proximal structures. The medial (oral) region of the mandibular process, which seems to be the driver of mandibular growth, responds to local epithelial signals (FGF-8) by stimulating proliferation of the underlying mesenchyme through the mediation of Msx-1, similar to the

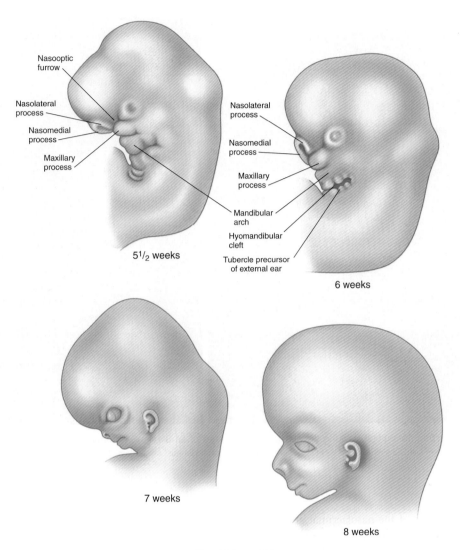

Nasooptic
furrow

Nasolateral
process

Nasomedial
process

Maxillary
process

5½ weeks

Nasolateral
process

Nasomedial
process

Maxillary
process

Mandibular
arch

Hyomandibular
cleft

Tubercle precursor
of external ear

6 weeks

7 weeks

8 weeks

Fig. 14.6, cont'd

subectodermal region of the limb bud. Growth of the jaws is influenced by various growth factors, especially the bone morphogenetic proteins (BMPs), which, at different stages, are produced in either the ectoderm or the mesenchyme and can have strikingly different effects. Experiments on avian embryos have shown that increasing the expression of **BMP-4** in first-arch mesenchyme results in the formation of a much more massive beak than that of normal embryos.

Proximodistal organization of the arch is reflected by nested expression patterns of the transcription factor **Dlx** (the mammalian equivalent of *distalless* in *Drosophila*) along the arch. The paralogues Dlx-3 and Dlx-7 (Dlx-4) are expressed most distally, expression of Dlx-5 and Dlx-6 extends more proximally, and expression of Dlx-1 and Dlx-2 extends most proximally. In mice, only Dlx-1 and Dlx-2 are expressed in the maxillary process. Although mutants of individual *Dlx* genes produce minor abnormalities, mice in which *Dlx-5* and *Dlx-6* have been knocked out develop with a homeotic transformation of the distal lower jaws into upper jaws. It seems that *Dlx-5* and *Dlx-6,* functioning downstream of endothelin-1, are selector genes that control the rostrocaudal identity of the distal segments of the first pharyngeal arch.

Through differential growth between 4 and 8 weeks (see Fig. 14.6), the nasomedial and maxillary processes become more prominent and ultimately fuse to form the upper lip and jaw (**Fig. 14.7**). As this is occurring, the frontonasal prominence, which was a prominent tissue bordering the stomodeal area in the 4- and 5-week-old embryo, is displaced as the two nasomedial processes merge. The merged nasomedial processes form the **intermaxillary segment**, which is a precursor for (1) the **philtrum** of the lip, (2) the **premaxillary component** of the upper jaw, and (3) the **primary palate**.

Between the maxillary process and the nasal primordium (nasolateral process) is a **nasolacrimal groove** (naso-optic furrow) that extends to the developing eye (see Fig. 14.6). The ectoderm of the floor of the nasolacrimal groove thickens to form a solid epithelial cord, which detaches from the groove. The epithelial cord undergoes canalization and forms the **nasolacrimal duct** and, near the eye, the **lacrimal sac**. The nasolacrimal duct extends from the medial corner of the eye to the nasal cavity (inferior meatus) and, in postnatal life, acts as a drain for lacrimal fluid. This connection explains why people can have a runny nose when crying. Meanwhile, the expanding nasomedial process fuses with the maxillary

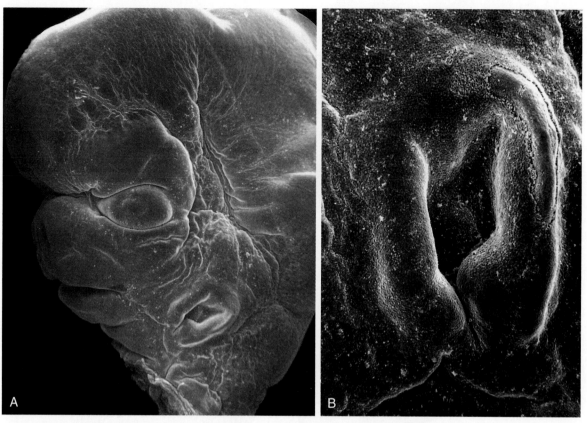

Fig. 14.7   **A,** Scanning electron micrograph showing the general facial features of an 8-week-old human embryo. **B,** Higher magnification of the ear, which is located in the neck in **A**. *(From Jirásek JE:* Atlas of human prenatal morphogenesis, *Amsterdam, 1983, Martinus Nijhoff.)*

process, and over the region of the nasolacrimal groove, the nasolateral process merges with the superficial region of the maxillary process. The region of fusion of the nasomedial and maxillary processes is marked by an epithelial seam, called the **nasal fin**. Mesenchyme soon penetrates the nasal fin, and the result is a continuous union between the nasomedial and maxillary processes.

The lower jaw is formed in a simpler manner. The bilateral mandibular prominences enlarge, and their medial components merge in the midline, to form the point of the lower jaw. The midline dimple that is seen in the lower jaw of some individuals is a reflection of variation in the degree of merging of the mandibular prominences. A prominent cartilaginous rod called **Meckel's cartilage** differentiates within the lower jaw (see Fig. 14.36D). Derived from neural crest cells of the first pharyngeal arch, Meckel's cartilage forms the basis around which membrane bone (which forms the definitive skeleton of the lower jaw) is laid down. Experimental evidence indicates that the rodlike shape of Meckel's cartilage is related to the inhibition of further chondrogenesis by the surrounding ectoderm. If the ectoderm is removed around Meckel's cartilage, large masses of cartilage form instead of a rod. These properties are similar to the inhibitory interactions between ectoderm and chondrogenesis in the limb bud. A later-acting influence in growth of the lower jaw is the planar cell polarity pathway, which influences outgrowth of Meckel's cartilage. If this pathway is disrupted, the mandible will not develop to its normal length.

Shortly after the basic facial structures take shape, they are invaded by mesodermal cells associated with the first and second pharyngeal arches. These cells form the muscles of mastication (first-arch derivatives, which are innervated by cranial nerve V) and the muscles of facial expression (second-arch derivatives, which are innervated by cranial nerve VII). At the level of individual muscles, highly coordinated spatiotemporal relationships between mesodermal and neural crest cells are very important in the determination of muscle attachments and the overall shape of the muscles.

Although the basic structure of the face is established between 4 and 8 weeks, changes in the proportionality of the various regions continue until well after birth. In particular, the midface remains underdeveloped during embryogenesis and early postnatal life.

Temporomandibular Joint and Its Relationship with the Jaw Joint of Lower Vertebrates

Of considerable clinical importance and evolutionary interest is the **temporomandibular joint**, which represents the hinge between the mandibular condyle and the squamous part of the temporal bone. The temporomandibular joint, which phylogenetically appeared with the evolution of mammals, is a complex synovial joint surrounded by a capsule and containing an **articular disk** between the two bones. Based on the early expression of Barx-1, this joint is formed late during development, first appearing as mesenchymal condensations associated with the temporal bone and mandibular condyle during the seventh week of development. The articular disk and capsule begin to take shape a week later, and the actual joint cavity forms between weeks 9 and 11.

In lower vertebrates, the jaw opens and closes on a hinge between cartilaginous portions of the mandibular process— the **articular bone** in the lower jaw and the **quadrate bone** in the upper jaw, both derivatives of Meckel's cartilage. During phylogenesis, the distal membranous bone (the **dentary bone**) associated with Meckel's cartilage increased in prominence as the jaw musculature became more massive. The dentary bone of contemporary mammals and humans constitutes most of the lower jaw, and Meckel's cartilage is seen only as a prominent cartilaginous rod within the forming jaw complex during the late embryonic stage of development.

In mammals, over many millions of years, the original jaw-opening joint became less prominent and was incorporated into the middle ear as the malleus (derivative of articular bone of the lower jaw) and the incus (derivative of the ancestral quadrate bone in the skull). The incus connects with the stapes (a derivative of the second pharyngeal arch). The **tympanic ring**, a neural crest–derived bone that surrounds and supports the tympanic membrane, is derived from the **angular bone**, one of the first-arch membrane bones that overlies the proximal part of Meckel's cartilage.

Formation of the Palate

The early embryo possesses a common oronasal cavity, but in humans the **palate** forms between 6 and 10 weeks to separate the oral from the nasal cavity. The palate is derived from three primordia: an unpaired **median palatine process** and a pair of **lateral palatine processes** (**Figs. 14.8 and 14.9**).

The median palatine process is an ingrowth from the newly merged nasomedial processes. As it grows, the median palatine process forms a triangular bony structure called the **primary palate**. In postnatal life, the skeletal component of the primary palate is referred to as the **premaxillary component of the maxilla**. The four upper incisor teeth arise from this structure (**Fig. 14.10**).

Formation of the palate involves (1) growth of the palatal shelves, (2) their elevation, (3) their fusion, and (4) removal of the epithelial seam at the site of fusion. The lateral palatine processes, which are the precursors of the **secondary palate**, first appear as outgrowths of the maxillary processes during the sixth week. At first, they grow downward on either side of the tongue (**Fig. 14.11**). Similar to other facial primordia, outgrowth of the palatal shelves involves ectodermal-mesenchymal interactions and specific growth factors. **FGF-10** produced in the mesenchyme of the forming palatal shelf is bound to an FGF receptor in the ectoderm (**Fig. 14.12**). This process stimulates the release of shh from the ectoderm. Shh causes the release of **BMP-2** in the mesenchyme. BMP-2 and **Msx-1**,

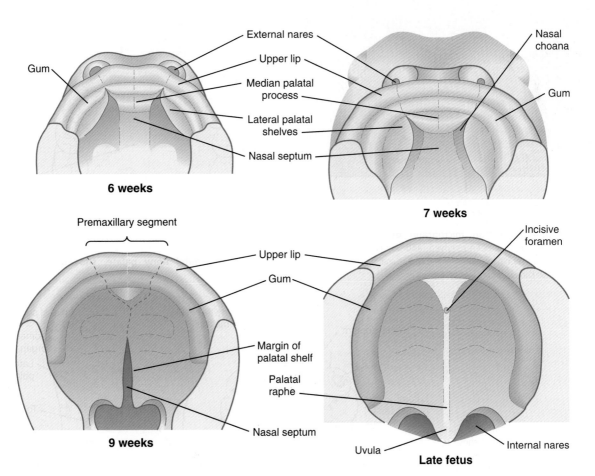

Fig. 14.8   **Development of the palate as seen from below.**

which interacts with BMP-4, stimulate proliferation of the mesenchymal cells of the palatal shelf and its resultant growth. During week 7, the lateral palatine processes (**palatal shelves**) dramatically dislodge from their positions alongside the tongue and become oriented perpendicularly to the maxillary processes. The apices of these processes meet in the midline and begin to fuse.

Despite many years of investigation, the mechanism underlying the elevation of the palatine shelves remains obscure. Swelling of the extracellular matrix of the palatal shelves seems to impart a resiliency that allows them to approximate one another shortly after they become dislodged from along the tongue. Research suggests that the rapid closure of the palatal shelves is accomplished by the flowing of the internal tissues, rather than by a reaction that approximates the closure of swinging doors.

Another structure involved in formation of the palate is the **nasal septum** (see Figs. 14.8 and 14.11). This midline structure, which is a downgrowth from the frontonasal prominence, reaches the level of the palatal shelves when the palatal shelves fuse to form the definitive secondary palate. Rostrally, the nasal septum is continuous with the primary palate.

At the gross level, the palatal shelves fuse in the midline, but rostrally they also join the primary palate. The midline point of the fusion of the primary palate with the two palatal shelves is marked by the **incisive foramen** (see Fig. 14.10).

Because of its clinical importance, fusion of the palatal shelves has been investigated intensively. When the palatal shelves first make midline contact, each is covered throughout by a homogeneous epithelium. During the process of fusion, however, the midline epithelial seam disappears. The epithelium on the nasal surface of the palate differentiates into a ciliated columnar type, whereas the epithelium takes on a stratified squamous form on the oral surface of the palate. Significant developmental questions include the following:

1. What causes the disappearance of the midline epithelial seam?
2. What signals result in the diverse pathways of differentiation of the epithelium on either side of the palate?

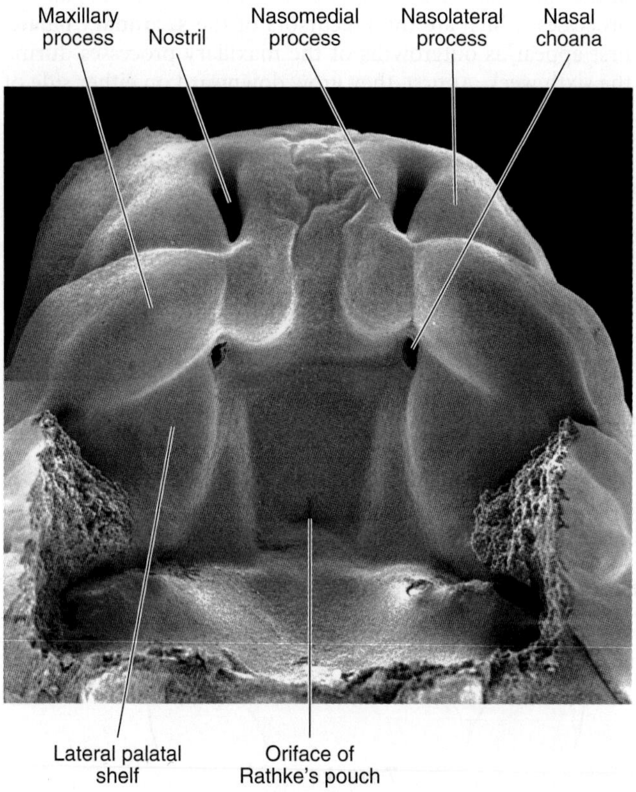

Fig. 14.9 **Scanning electron micrograph of 7-week human embryo.** The lower jaw has been removed, and one is looking toward the roof of the oronasal cavity. *(From Steding G: The anatomy of the human embryo, Basel, 2009, Karger; courtesy of Dr. J. Männer.)*

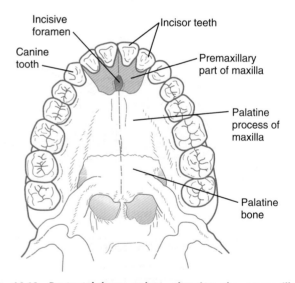

Fig. 14.10 **Postnatal bony palate, showing the premaxillary segment.**

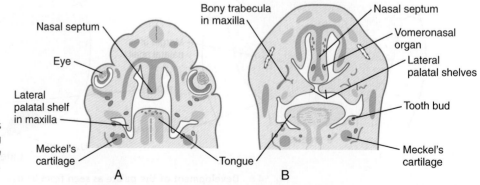

Fig. 14.11 **A** and **B,** Frontal sections through the human head, showing fusion of the palatal shelves. *(From Patten B: Human embryology, ed 3, New York, 1968, McGraw-Hill.)*

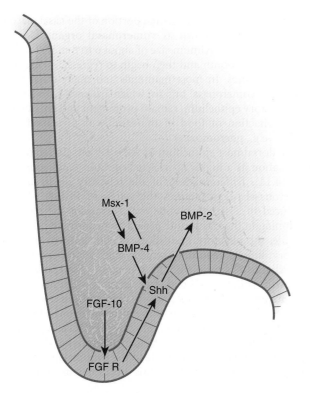

Fig. 14.12 Important signaling interactions in the developing palatal shelves. BMP, bone morphogenetic protein; FGF, fibroblast growth factor; FGF R, fibroblast growth factor receptor; Shh, sonic hedgehog.

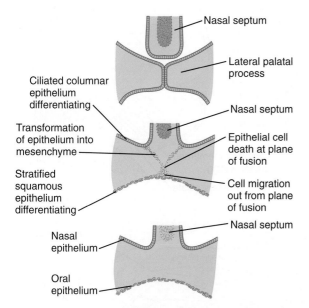

Fig. 14.13 Developmental processes associated with fusion of the palatal shelves and the nasal septum.

The disappearance of the midline epithelial seam after the approximation of the palatal shelves involves several fundamental developmental processes (**Fig. 14.13**). Some of the epithelial cells at the fusion seam undergo apoptosis and disappear. Others may migrate out from the plane of fusion and become inserted into the epithelial lining of the oral cavity.

Still other epithelial cells undergo a morphological transformation into mesenchymal cells. **Transforming growth factor-β3 (TGF-β3)** is expressed by the ectodermal cells of the distal rim of the palatal shelves just before fusion and loses prominence shortly thereafter. It plays an important role in stimulating apoptosis of the epithelial cells at the fusion seam. In TGF-β3–mutant mice, the lateral palatal shelves approximate in the midline, but the epithelial seam fails to disappear, and the mice develop isolated cleft palate.

Experiments involving the in vitro culture of a single palatal shelf of several species have shown clearly that all aspects of epithelial differentiation (cell death in the midline and different pathways of differentiation on the oral and nasal surfaces) can occur in the absence of contact with the opposite palatal shelf. These different pathways of differentiation are not intrinsic to the regional epithelia, but they are mediated by the underlying neural crest–derived mesenchyme. The mechanism of this regional specification of the epithelium remains poorly understood. According to one model, the underlying mesenchyme produces growth factors that influence the production and regional distribution of extracellular matrix molecules (e.g., type IX collagen). The way these events are received and interpreted by the epithelial cells is unknown.

Formation of the Nose and Olfactory Apparatus

The human olfactory apparatus first becomes visible at the end of the first month as a pair of thickened ectodermal **nasal placodes** located on the frontal aspect of the head (**Fig. 14.14A**). Similar to the formation of the lens placodes, the formation of the nasal placodes requires the expression of **Pax-6** and the action of retinoids produced in the forebrain. In the absence of Pax-6 expression, neither the nasal placodes nor the lens placodes form. The nasal placodes originate from the anterolateral edge of the neural plate before its closure.

Soon after their formation, the nasal placodes form a surface depression (the **nasal pits**) surrounded by horseshoe-shaped elevations of mesenchymal tissue with the open ends facing the future mouth (see Fig. 14.6). The two limbs of the mesenchymal elevations are the **nasomedial** and **nasolateral processes.** Formation of the thickened nasal processes depends on the retinoid-stimulated production of **FGF-8,** which stimulates proliferation of the mesenchymal cells within the nasal processes. The source of these retinoids is the epithelium of the nasal pit itself. Meanwhile, production of retinoids by the forebrain diminishes. As a consequence, the frontonasal prominence, which depends on forebrain retinoids for supporting proliferation of its mesenchymal cells, is reduced. As the nasal primordia merge toward the midline during weeks 6 and 7, the nasomedial processes form the tip and crest of the nose along with part of the nasal septum, and the nasolateral processes form the wings (**alae**) of the nose. The receding frontonasal process contributes to part of the bridge of the nose.

Meanwhile, the nasal pits continue to deepen toward the oral cavity and form substantial cavities themselves (see Fig. 14.14). By 6½ weeks, only a thin **oronasal membrane** separates the oral cavity from the nasal cavity. The oronasal membrane soon breaks down, thereby making the nasal cavities continuous with the oral cavity through openings behind the primary palate called **nasal choanae** (see Fig. 14.9). Shortly after breakdown of the oronasal membrane, however, the outer part of the nasal cavity becomes blocked with a plug of epithelial cells, which persists until the end of the fourth month. With the fusion of the lateral palatal shelves, the nasal cavity is considerably lengthened and ultimately communicates with the upper pharynx.

Similar to the other major sensory organs of the head, the epithelium of each nasal pit induces the surrounding neural crest mesenchyme to form a cartilaginous capsule around it. In a three-dimensionally complex manner, the medial parts of the nasal capsules combine with more centrally derived deep neural crest mesenchyme to form the midline nasal septum and ethmoid bones. The lateral region of the nasal capsule forms the nasal bones. During the third month, shelflike structures called **nasal conchae** form from the ethmoid bones on the lateral wall of the nasal cavity. These structures increase the surface area available for conditioning the air within the nasal cavity. Late in fetal life and for several years after birth, the paranasal sinuses form as outgrowths from the walls of the nasal cavities. The size and shape of these structures have a significant impact on the form of the face during its postnatal growth period.

At 6 to 7 weeks, a pair of epithelial ingrowths can be seen in each side of the nasal septum near the palate. Developing as invaginations from the medial portion of the nasal placode, these diverticula, known as **vomeronasal organs** (see Fig. 14.11B), reach a maximum size of about 6 to 8 mm at around the sixth fetal month and then begin to regress, leaving small cystic structures. In most mammals and many other vertebrates, the vomeronasal organs, which are lined with a modified olfactory epithelium, remain prominent and are involved in the olfaction of food in the mouth or sexual olfactory stimuli (e.g., pheromones).

The dorsalmost epithelium of the nasal pits undergoes differentiation as a highly specialized olfactory epithelium (see Fig. 14.14). Differentiation of the olfactory organ and the vomeronasal organ requires the action of **FGF-8**, which is produced in a signaling zone that surrounds the nasal pit. Beginning in the embryonic period and continuing throughout life, the olfactory epithelium is able to form primitive sensory bipolar neurons, which send axonal projections toward the olfactory bulb of the brain. Preceding axonal ingrowth, some cells break free from this epithelium and migrate toward the brain. Some of these cells may synthesize a substrate for the ingrowth of the olfactory axons. Other cells migrating from the olfactory placode (specifically, the vomeronasal primordium) synthesize **luteinizing hormone–releasing hormone** and migrate to the hypothalamus, the site of synthesis and release of this hormone in adults. The

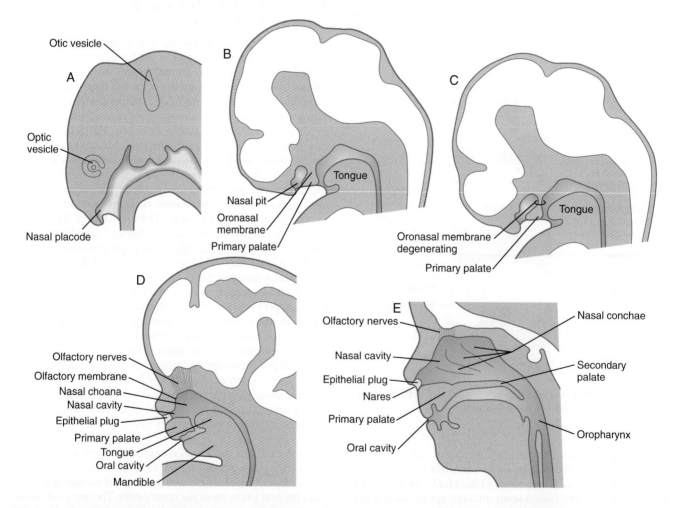

Fig. 14.14 Sagittal sections through embryonic heads with special emphasis on development of the nasal chambers. A, At 5 weeks. **B,** At 6 weeks. **C,** At 6½ weeks. **D,** At 7 weeks. **E,** At 12 weeks.

embryonic origin of these cells in the olfactory placode helps to explain the basis for **Kallmann's syndrome**, which is characterized by anosmia and hypogonadotropic hypogonadism. Cells of the olfactory placode also form supporting (sustentacular) cells and glandular cells in the olfactory region of the nose. Physiological evidence shows that the olfactory epithelium is capable of some function in late fetal life, but full olfactory function is not attained until after birth.

Formation of the Salivary Glands

Starting in the sixth week, the **salivary glands** originate as solid, ridgelike thickenings of the oral epithelium (**Fig. 14.15**). Extensive epithelial shifts in the oral cavity make it difficult to determine the germ layer origins of the salivary gland epithelium. The parotid glands are probably derived from ectoderm, whereas the submandibular and sublingual glands are thought to be derived from endoderm.

As with other glandular structures associated with the digestive tract, the development of salivary glands depends on a continuing series of epitheliomesenchymal interactions. Branching morphogenesis of the salivary glands depends heavily on shh signaling, acting on FGFs. In contrast to most other glandular structures, however, in which epithelially produced shh acts on the underlying mesenchyme, in the salivary glands the entire sequence of shh signaling and of FGF response occurs within the epithelium. The basal lamina that surrounds the early epithelial lobular ingrowths differs in composition, depending on the growth potential of the region. Around the stalk and in clefts, the basal lamina contains types I and IV collagen and a basement membrane-1 proteoglycan.

These components are not found in the regions of the lobules that undergo further growth. Under the influence of the surrounding mesenchyme, the basal lamina in growing regions loses the collagens and proteoglycans that are associated with stable structures (e.g., stalks, clefts).

In addition to alterations in the basal lamina, branching is associated with the local contraction of ordered microfilaments within the apices of epithelial cells at the branch points. Continued growth at the tips of lobules of the glands is supported by high levels of mitotic activity of the epithelium and the deposition of newly synthesized glycosaminoglycans in the area. During organogenesis, the parasympathetic innervation, acting through acetylcholine secretion, maintains the population of epithelial progenitor cells. In its absence, the amount of budding of epithelial lobules is dramatically reduced. The structural and functional differentiation of the epithelium of the salivary gland continues throughout fetal life.

Clinical Correlation 14.1 presents malformations of the face and oral regions.

Formation of the Teeth

A tooth is a highly specialized extracellular matrix consisting of two principal components—enamel and dentin—each secreted by a different embryonic epithelium. Tooth development is a highly orchestrated process involving intimate interactions between the epithelia that produce the dentin and enamel. Extending a common theme of development into the macroscopic dimension, teeth undergo an **isoform transition**, with the postnatal replacement of the deciduous teeth by their permanent adult counterparts.

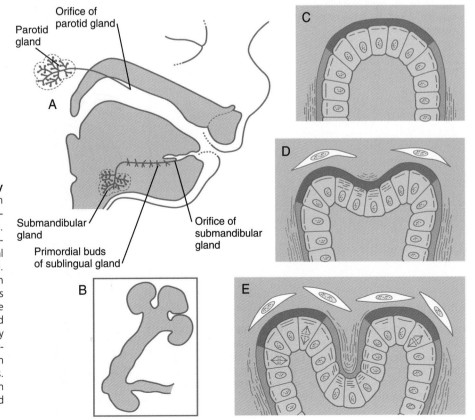

Fig. 14.15 Development of the salivary glands. A, Salivary gland development in an 11-week-old human embryo. **B,** Development of salivary gland epithelium in vitro. **C,** Accumulation of newly synthesized glycosaminoglycans (*dark green*) in the basal lamina at the end of a primary lobule. **D,** Early cleft formation is associated with the contraction of bundles of microfilaments in the apices of the epithelial cells lining the cleft. Collagen fibers (*wavy lines*) are lined up lateral to the lobule and in the newly forming cleft. **E,** As the cleft deepens, glycosaminoglycan synthesis is reduced in the cleft, and collagen deposition continues. (**C** through **E** show the relationship between disposition of the extracellular matrix and lobulation of the glandular primordium.)

CLINICAL CORRELATION 14.1
Malformations of the Face and Oral Region

Cleft Lip and Palate

Cleft lip and cleft palate are common malformations, with an incidence of approximately 1 in 1000 births (cleft lip) and 1 in 2500 births (cleft palate). Numerous combinations and degrees of severity exist, ranging from a unilateral cleft lip to a bilateral cleft lip associated with a fully cleft palate.

Structurally, **cleft lip** results from the lack of fusion of the maxillary and nasomedial processes. In the most complete form of the defect, the entire premaxillary segment is separated from both maxillae, with resulting bilateral clefts that run through the lip and the upper jaw between the lateral incisors and the canine teeth (**Fig. 14.16**). The point of convergence of the two clefts is the incisive foramen (**Fig. 14.17B**). The premaxillary segment commonly protrudes past the normal facial contours when viewed from the side. The mechanism frequently underlying cleft lip is

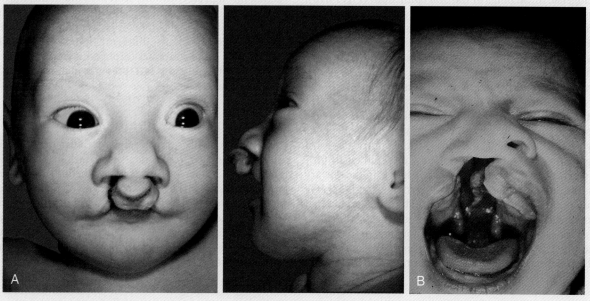

Fig. 14.16   A, Front and lateral views of an infant with bilateral cleft lip and palate. On the lateral view, note how the premaxillary segment is tipped outward. B, Unilateral cleft lip and complete cleft palate. Note the duplicated uvula at the back of the oral cavity. *(Courtesy of A. Burdi, Ann Arbor, Mich.)*

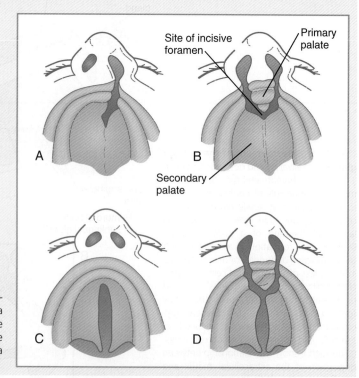

Fig. 14.17   **Common varieties of cleft lip and palate.** A, Unilateral cleft passing through the lip and between the premaxilla (primary palate) and secondary palate. B, Bilateral cleft lip and palate similar to that seen in the patient in Figure 14.19A. C, Midline palatal cleft. D, Bilateral cleft lip and palate continuous with a midline cleft of the secondary palate.

CLINICAL CORRELATION 14.1
Malformations of the Face and Oral Region—cont'd

hypoplasia of the maxillary process that prevents contact between the maxillary and nasomedial processes from being established.

Cleft palate results from incomplete or absent fusion of the palatal shelves (see Figs. 14.16B and 14.17). The extent of palatal clefting ranges from involvement of the entire length of the palate to something as minor as a bifid uvula. As with cleft lip, cleft palate is usually multifactorial. Some chromosomal syndromes (e.g., trisomy 13) are characterized by a high incidence of clefts. In other cases, cleft lip and cleft palate can be linked to the action of a chemical teratogen (e.g., anticonvulsant medications). Experiments on mice have shown that the incidence of cleft palate after exposure to a dose of cortisone is strongly related to the genetic background of the mouse. In humans, mutations of *MSX1* are strongly associated with nonsyndromic cleft palate. The higher female incidence of cleft palate may be related to the occurrence of fusion of the palatal shelves in female embryos about 1 week later than in male embryos, thus prolonging the susceptible period.

The genetic and molecular basis of normal palate closure is complex. Even closure of the anterior and posterior parts of the palate operates under different combinations of molecular interactions. Consequently, more than 300 genetic syndromes include cleft palate as part of the spectrum of disease.

Oblique Facial Cleft

Oblique facial cleft is a rare defect that results when the nasolateral process fails to fuse with the maxillary process, usually resulting from hypoplasia of one of the tissue masses (**Fig. 14.18A**). This cleft frequently manifests as an epithelially lined fissure running from the upper lip to the medial corner of the eye.

Macrostomia (Lateral Facial Cleft)

An even rarer condition called **macrostomia** (Fig. 14.18B) results from hypoplasia or poor merging of the maxillary and mandibular processes. As the name implies, this condition manifests as a very large mouth on one or both sides. In severe cases, the cleft can reach almost to the ears.

Median Cleft Lip

Another rare anomaly, **median cleft lip**, results from incomplete merging of the two nasomedial processes (Fig. 14.18C).

Holoprosencephaly

Holoprosencephaly includes a broad spectrum of defects, all based on defective formation of the forebrain (prosencephalon) and structures whose normal formation depends on influences from the forebrain. This condition has been estimated to be present in up to 1 in 250 of all embryos and 1 in 10,000 live births. The defect arises in early pregnancy when the forebrain is taking shape, and the brain defects usually involve archencephalic structures (e.g., the olfactory system). Because of the influence of the brain on surrounding structures, especially the cranial base, primary defects of the forebrain often manifest externally as facial malformations, typically a reduction in tissue of the frontonasal process.

In extreme cases, holoprosencephaly can take the form of cyclopia (see Fig. 8.18), in which the near absence of upper facial and midfacial tissue results in a convergence and fusion of the optic primordia. Reduction defects of the nose can also be components of this condition. The nose can be either absent or represented by a tubular **proboscis** (or two such structures), sometimes even located above the eye (see Fig. 8.18). Midline defects of the upper lip can also be attributed to holoprosencephaly (see Fig. 14.18C).

The root cause of holoprosencephaly occurs very early in embryonic development, with disturbances in the ability of the prechordal plate and anterior endoderm to secrete **sonic hedgehog (shh)** and other factors required for induction and early development of the ventral forebrain. In their absence, the single optic field either does not split or splits incompletely, and ventral forebrain structures do not develop. This is also reflected in a reduced rostral neural crest, which provides the cellular basis for the formation of most midface structures. Even earlier in development, disturbances in bone morphogenetic protein (BMP) levels, often

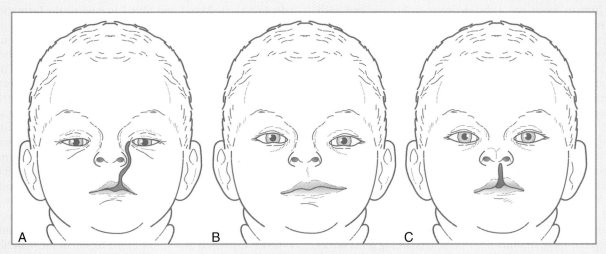

Fig. 14.18 Varieties of facial clefts. A, Oblique facial cleft combined with a cleft lip. **B,** Macrostomia. **C,** Medial cleft lip with a partial nasal cleft.

Continued

CLINICAL CORRELATION 14.1
Malformations of the Face and Oral Region—cont'd

caused by imbalances in BMP inhibitors, can influence early formation of the forebrain and lead to holoprosencephaly.

Many cases of holoprosencephaly (e.g., **Meckel's syndrome**, which includes midline cleft lip, olfactory bulb absence or hypoplasia, and nasal abnormalities) can be attributed to genetic causes. Meckel's syndrome is an autosomal recessive condition. Several types of hereditary holoprosencephaly result from mutations of the *SHH* gene, which normally induces the formation of several midline structures in the forebrain. Exposure to an excess of retinoic acid, which causes misregulation of genes in the shh pathway, also causes holoprosencephaly in laboratory animals and possibly in humans. Most cases of holoprosencephaly seem to be multifactorial, although maternal consumption of alcohol during the

first month of pregnancy is suspected to be a leading cause of this condition. One percent to 2% of infants born to diabetic mothers may develop some degree of holoprosencephaly. Trisomies of chromosomes 13 and 18 are commonly associated with holoprosencephaly.

Frontonasal Dysplasia

Frontonasal dysplasia encompasses various nasal malformations that result from excessive tissue in the frontonasal process. The spectrum of anomalies usually includes a broad nasal bridge and **hypertelorism** (an excessive distance between the eyes). In very severe cases, the two external nares are separated, often by several centimeters, and a median cleft lip can occur (**Fig. 14.19**).

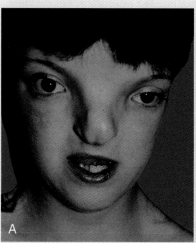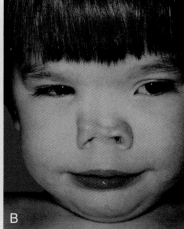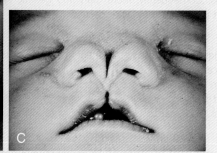

Fig. 14.19   **A to C, Varying degrees of frontonasal dysplasia.** *(Courtesy of A. Burdi, Ann Arbor, Mich.)*

Patterning of the Dentition

Each human tooth has a distinctive morphology, and each type of tooth forms in a characteristic location. For many years, virtually nothing was known about the patterning of the dentition, but analysis of certain types of genetically modified mice has provided some concrete clues to the molecular basis of dental patterning. Both the dental field overall and the patterning of the dentition take shape very early in craniofacial development, before any overt indication of tooth formation. The expression of **Pitx-2** (a *bicoid*-related transcription factor; see Fig. 4.1) outlines first the entire ectodermal dental field, and later the epithelium of the individual tooth germs. The homeobox-containing genes *Dlx-1* and *Dlx-2* are expressed in the maxillary arch and in the proximal part of the mandibular arch. When both these genes are knocked out in mice, the upper jaw develops without molar teeth, although molars develop in the lower jaws. The incisor teeth in both jaws develop normally. Another homeobox-containing gene, *Barx-1*, is induced by FGF-8 in proximal ectoderm of the mandibular process, and in the formation of molar teeth in knockouts it may compensate for the absence of Dlx-1 and Dlx-2 in the lower jaw. FGF-8 acts proximally to restrict Barx-1 and Dlx-2

to guide the molar-producing domain, and BMP-4 acts distally to activate Msx-1 and Msx-2 in guiding the formation of incisor teeth.

Not only the location, but also the type of tooth is under tight developmental control. The difference in morphology between an incisor tooth, possessing a single cusp, and a molar tooth, which contains several cusps, is related to the number of enamel knots (see later) in the developing tooth. A striking example of the molecular control of dental patterning is the conversion of incisor teeth to molars in mice. In the distal part of the mandibular arch, ectodermal BMP-4 signals normally repress the expression of *Barx-1*, but when BMP-4 signaling is inhibited by the implantation of noggin beads, Barx-1 expression is induced in the dental mesenchyme, and the developing incisor teeth are transformed into molars. The transcription factor **Islet-1** is expressed only in the oral surface ectoderm in the area where incisors will form. In contrast, *Pitx-1* is expressed only in the molar region of the mandibular mesenchyme where it acts upstream of *Barx-1*.

Mammals have only a single row of teeth in each jaw. This is controlled by two overlapping gradients of opposite polarity along the lingual-buccal axis (**Fig. 14.20**). On the buccal side

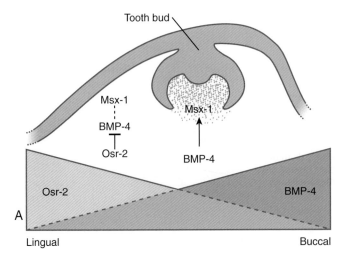

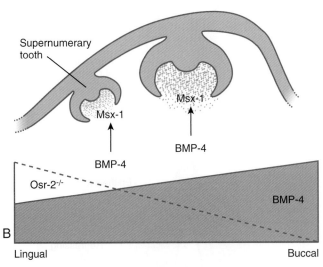

Fig. 14.20 **Experiment demonstrating control of the formation of tooth rows in the oral cavity.** Normally, a higher concentration of Osr-2 in the lingual region suppresses the tooth-inducing activity of the bone morphogenetic protein-4 (BMP-4)–Msx-1 pathway (A). In mutants (B), the absence of Osr-2 activity permits the formation of an extra row of teeth on the lingual side of the normal row. *(Based on Cobourne MT and Sharpe PT: Semin Cell Dev Biol 21:314-324, 2010.)*

Stages of Tooth Development

Tooth development begins with the migration of neural crest cells into the regions of the upper and lower jaws. The overlying oral ectoderm thickens into C-shaped bands (**dental laminae**) in the upper and lower jaws. The appearance of the dental laminae during the sixth week is the first manifestation of a series of ectodermal-mesenchymal interactions that continues until tooth formation is virtually completed.

Although each tooth has a specific time sequence and morphology of development, certain general developmental stages are common to all teeth (**Fig. 14.21**). As the dental lamina grows into the neural crest mesenchyme, epithelial primordia of the individual teeth begin to take shape as **tooth buds**. In keeping with their interactive mode of development, the tooth buds are associated with condensations of mesenchymal cells. The tooth bud soon expands, passing through a mushroom-shaped **cap stage** before entering the **bell stage** (see Fig. 14.21D).

By the bell stage, the tooth primordium already has a complex structure, even though it has not formed any components of the definitive tooth. The epithelial component, called the **enamel organ**, is still connected to the oral epithelium by an irregular stalk of dental lamina, which soon begins to degenerate. The enamel organ consists of an **outer sheath** of epithelium, a mesenchymelike **stellate reticulum**, and an inner epithelial **ameloblast layer**. Ameloblasts are the cells that begin to secrete the enamel of the tooth. The initial formation of ameloblasts depends on the actions of the transcription factor **Tbx-1**. Within the concave surface of the enamel organ is a condensation of neural crest mesenchyme called the **dental papilla**. Cells of the dental papilla opposite the ameloblast layer become transformed into columnar epithelial cells called **odontoblasts** (**Fig. 14.22**). These cells secrete the dentin of the tooth. Attached to the dental lamina close to the enamel organ is a small bud of the permanent tooth (see Fig. 14.21E and F). Although delayed, it goes through the same developmental stages as the deciduous tooth.

Late in the bell stage, the odontoblasts and ameloblasts begin to secrete precursors of dentin and enamel, beginning first at the future apex of the tooth. Over several months, the definitive form of the tooth takes shape (see Fig. 14.21). Meanwhile, a condensation of mesenchymal cells forms around the developing tooth. Cells of this structure, called the **dental sac**, produce specialized extracellular matrix components (**cementum** and the **periodontal ligament**) that provide the tooth with a firm attachment to the jaw. While these events are occurring, the tooth elongates and begins to erupt through the gums (**gingiva**).

Tissue Interactions in Tooth Development

Teeth are formed by a series of inductive interactions. Tissue recombination experiments have shown that the thickened ectoderm of the dental lamina initiates tooth formation. Early in development, dental ectoderm can induce nondental cranial neural crest mesenchyme to participate in the formation of a tooth, but predental neural crest mesenchyme cannot induce nondental ectoderm to form a tooth. Research suggests that the transcription factor **Lef-1** may stimulate the predental surface ectoderm to secrete **FGF-8**, which induces the underlying mesenchyme to express **Pax-9** (**Fig. 14.23**). In the absence of Pax-9 expression, tooth development does not proceed past the bud stage. **BMP-2** and **BMP-4**, also produced by the

of the jaw, a high concentration of **BMP-4** stimulates the expression of **Msx-1** in the dental mesenchyme in the normal process of tooth development, thus accounting for the development of the normal teeth. On the lingual side of the jaw, a high concentration of the transcription factor **Osr-2** (the mammalian equivalent of the *Drosophila* pair-rule gene *odd skipped*; see Fig. 4.1) inhibits the BMP-Msx axis and consequently tooth formation in that area. When Osr-2 is inactivated, BMP-4–Msx-1 activity on the lingual side of the jaw is not inhibited, and supernumerary teeth form on the lingual side of the normal row of teeth (see Fig. 14.20B). Enhancing or inhibiting the function of many of the other genes involved in tooth development can also lead to the formation of supernumerary teeth, but these genes do not cause expansion of the entire dental field as does *Osr-2*.

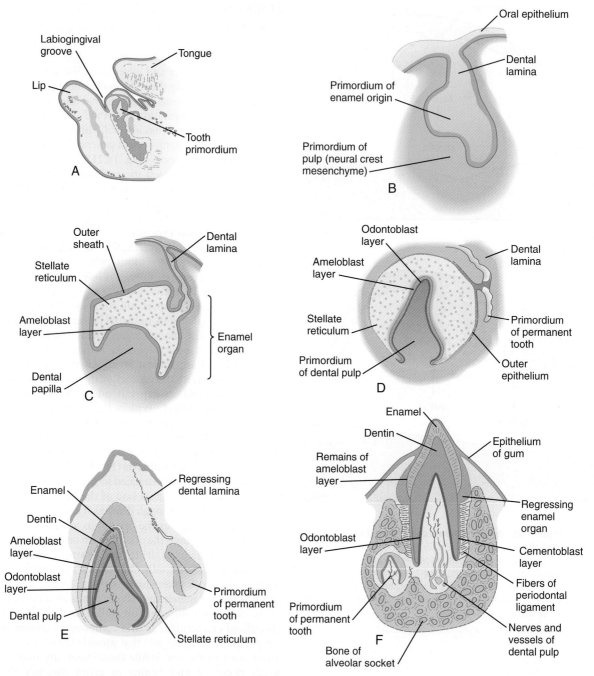

Fig. 14.21 **Development of a deciduous tooth.** **A,** Parasagittal section through the lower jaw of a 14-week-old human embryo showing the relative location of the tooth primordium. **B,** Tooth primordium in the bud stage in a 9-week-old embryo. **C,** Tooth primordium at the cap stage in an 11-week-old embryo, showing the enamel organ. **D,** Central incisor primordium at the bell stage in a 14-week-old embryo before deposition of enamel or dentin. **E,** Unerupted incisor tooth in a term fetus. **F,** Partially erupted incisor tooth showing the primordium of a permanent tooth near one of its roots. *(After Patten B:* Human embryology, *ed 3, New York, 1968, McGraw-Hill.)*

surface ectoderm, inhibit the action of FGF-8. Investigators have suggested that such inhibition is the basis for the non–tooth-forming spaces between the developing teeth. More recent research has shown that lateral inhibition through the **Delta/Notch** system is also involved in dental spacing. Another characteristic transcription factor induced in the mesenchyme beneath the thickened dental lamina is **Msx-1**.

Slightly later in dental development, BMP-4, instead of functioning as an inhibitor, acts along with FGF-8 and shh to stimulate the mesenchyme of the tooth bud to express a variety of characteristic molecules, including the following: the transcription factors **Msx-1**, **Msx-2**, and **EGR-1**; the extracellular matrix molecules **tenascin** and **syndecan**; and **BMP-4**. If the overlying ectoderm is separated from the neural crest mesenchyme, the predental mesenchyme will not develop into a dental papilla, and a tooth will not form. The role of BMP-4 as an inducer was shown by adding a small bead soaked in BMP-4 to a small mass of cultured predental neural crest mesenchyme (see Fig. 14.23B). Under the influence of the BMP-4 released from the bead, the mesenchyme began to

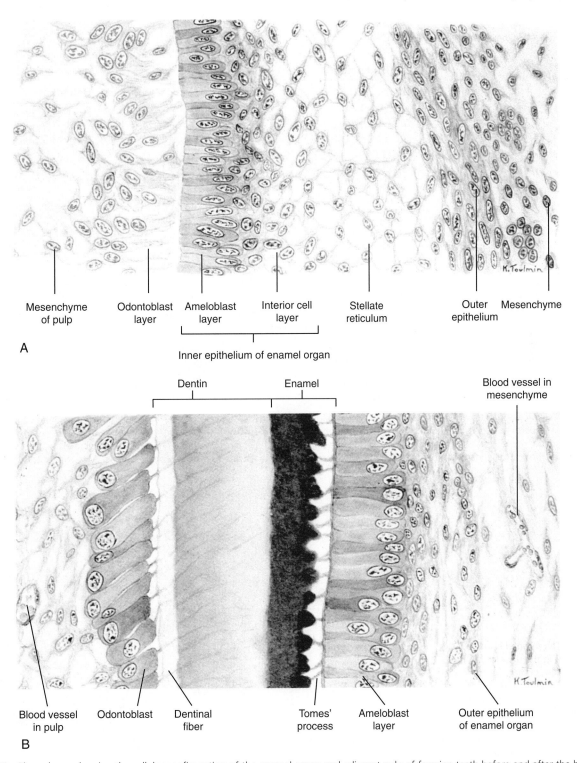

Mesenchyme
of pulp

Odontoblast
layer

Ameloblast
layer

Interior cell
layer

Stellate
reticulum

Outer
epithelium

Mesenchyme

Inner epithelium of enamel organ

A

Dentin

Enamel

Blood vessel in
mesenchyme

Blood vessel
in pulp

Odontoblast

Dentinal
fiber

Tomes'
process

Ameloblast
layer

Outer epithelium
of enamel organ

B

Fig. 14.22 Pig embryos showing the cellular configuration of the enamel organ and adjacent pulp of forming teeth before and after the beginning of deposition of enamel and dentin. **A,** Stage equivalent to a 4-month-old human embryo. **B,** Stage equivalent to a 5-month-old human embryo. *(After Patten B:* Human embryology, *ed 3, New York, 1968, McGraw-Hill.)*

express Msx-1, Msx-2, Egr-1, and BMP-4. The mesenchyme did not produce tenascin or syndecan, however, thus showing that other signals in addition to BMP-4 are required to achieve the full inductive response.

With the initial induction of the dental mesenchyme, that tissue itself becomes the next prime mover in tooth development. Inductive signals emanating from the dental mesenchyme next act on the ectoderm of the dental ledge, now in the late bud to early cap stage. Recombination experiments have shown that the dental mesenchyme determines the specific form of the tooth. When molar mesenchyme is combined in vitro with incisor epithelium, a molar tooth takes shape, whereas combining incisor mesenchyme with molar epithelium results in the formation of an incisor.

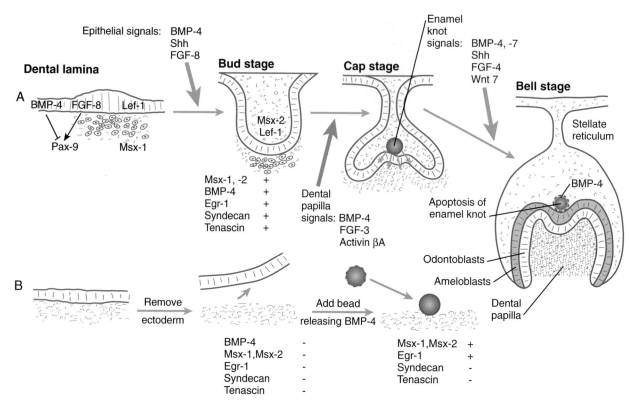

Fig. 14.23 **A,** Inductive interactions during tooth development. Molecules associated with the *green arrow* represent components of the signal from dental lamina ectoderm to underlying neural crest mesenchyme; molecules associated with the *violet arrow* are signals from the dental papilla to the overlying ectoderm; molecules associated with the *pink arrow* are signals from the enamel knot to dental papilla. **B,** In vitro experiment showing that a bead releasing bone morphogenetic protein-4 (BMP-4) can induce dental mesenchyme to express specific markers (Msx-1, Msx-2, and Egr-1). FGF, fibroblast growth factor; Shh, sonic hedgehog.

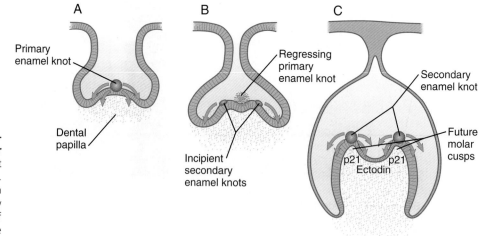

Fig. 14.24 **The enamel knot as a signaling center in the developing molar tooth.** **A,** The primary enamel knot induces proliferation on either side. **B,** Two secondary enamel knots form on each side of the regressing primary enamel knot. **C,** Under the influence of the secondary enamel knots, the future cusps of the molar begin to form.

Under the inductive influence of the dental mesenchyme, now called the **dental papilla**, a small group of ectodermal cells at the tip of the dental papilla ceases dividing. This mass of cells, called the **enamel knot** (**Fig. 14.24**), serves as a signaling center that regulates the shape of the developing tooth. Through the production of several signaling molecules, including **shh**, **FGF-4**, and **BMP-2**, **BMP-4**, and **BMP-7**, the enamel knot stimulates the proliferation of cells in the dental cap down and away from itself. By serving as a fixed point in this process,

the enamel knot determines the site of the tip of a cusp in the developing tooth. In the case of molar teeth, which have multiple cusps, secondary enamel knots form—one for each cusp. The location and spacing of secondary enamel knots are determined by two molecules, both induced by BMPs. **p21** is strongly expressed at sites where secondary enamel knots form, and **ectodin** is expressed in the intervening spaces. In the absence of ectodin, the secondary enamel knots and the resulting cusps of the molar teeth become massive because of the lack of its restraining influence. Ultimately, the cells of the enamel knot undergo apoptosis, possibly under the influence of BMP-4, which stimulates cell death in several other developmental systems. The appearance of apoptosis terminates the inductive signaling from this structure.

Formation of Dentin and Enamel

Late in their differentiation, odontoblasts withdraw from the cell cycle, elongate, and start secreting **predentin** from their apical surfaces, which face the enamel organ. The production of predentin signals a shift in the patterns of synthesis from type III collagen and fibronectin to type I collagen and other molecules (e.g., **dentin phosphoprotein, dentin osteocalcin**) that characterize the dentin matrix. The first dentin is laid against the inner surface of the enamel organ at the apex of the tooth (see Fig. 14.22B). With the secretion of additional dentin, the accumulated material pushes the odontoblastic epithelium from the odontoblast-ameloblast interface.

Terminal differentiation of the ameloblasts occurs after the odontoblasts begin to secrete predentin. In response to inductive signals from the odontoblasts, the ameloblasts withdraw from the cell cycle and begin a new pattern of synthesis, thus producing two classes of proteins: **amelogenins** and **enamelins**. About 5% of enamel consists of organic matrix, and amelogenins account for about 90% of this, with enamelins constituting most of the remainder. Enamelins are secreted before the amelogenins, and they may serve as nuclei for the formation of crystals of **hydroxyapatite**, the dominant inorganic component of enamel.

The amelogenin genes are located on the X and Y chromosomes in humans. Enamel genes have been highly conserved during vertebrate phylogeny. Investigators have suggested that, in early vertebrates, enamel once served as part of an electroreceptor apparatus.

Tooth Eruption and Replacement

Each tooth has a specific time of eruption and replacement (**Table 14.1**). With the growth of the root, the enamel-covered crown pushes through the oral epithelium. The sequence of eruption begins with the central incisor teeth, usually a few months after birth, and continues generally stepwise until the last of the deciduous molars forms at the end of the second year. A total of 20 deciduous teeth are formed.

Meanwhile, the primordium of the permanent tooth is embedded in a cavity extending into the bone on the lingual side of the alveolar socket in which the tooth is embedded (see Fig. 14.21F). As the permanent tooth develops, its increasing size causes resorption of the root of the deciduous tooth. When a sufficient amount of the root is destroyed, the deciduous tooth falls out, leaving room for the permanent tooth to take its place. The sequence of eruption of the permanent teeth is the same as that of the deciduous teeth, but an additional 12 permanent teeth (for a total of 32) are formed without deciduous counterparts.

The formation and eruption of teeth are important factors in midfacial growth, much of which occurs after birth. Tooth development and the corresponding growth of the jaw to accommodate the teeth, along with the development of the paranasal sinuses, account for much of the tissue mass of the midface.

Table 14.1 Usual Times of Eruption and Shedding of Deciduous and Permanent Teeth

Teeth	Eruption	Shedding
Deciduous		
Central incisors	6-8 mo	6-7 yr
Lateral incisors	7-10 mo	7-8 yr
Canines	14-18 mo	10-12 yr
First molars	12-16 mo	9-11 yr
Second molars	20-24 mo	10-12 yr
Permanent		
Central incisors	7-8 yr	
Lateral incisors	8-9 yr	
Canines	11-13 yr	
First premolars	10-11 yr	
Second premolars	11-12 yr	
First molars	6-7 yr	
Second molars	12-13 yr	
Third molars	15-25 yr	

Clinical Correlation 14.2 presents malformations of the teeth and dentition.

Development of the Pharynx and Its Derivatives

Considering the complexity of the structural arrangements of the embryonic pharynx, it is not surprising that many different structures originate in the pharyngeal region. This complexity provides many opportunities for abnormal development, as discussed in Clinical Correlation 14.3 at the end of this section. This section details aspects of later development that lead to the formation of specific structures. Adult derivatives of the regions of the pharynx and pharyngeal arches are summarized in **Figure 14.34**.

External Development of the Pharyngeal Region

Externally, the pharyngeal (branchial) region is characterized by four pharyngeal arches and grooves interposed between the arches (**Fig. 14.35**; see Fig. 14.4). These structures give rise to a diverse array of derivatives.

Pharyngeal Arches

As mentioned earlier, the endoderm of the foregut is the main driver in organizing development of the pharynx. Under the influence of various concentrations of retinoic acid (see p. 297), combinations of *Hox* genes determine the craniocaudal identity of the segments of the pharynx. The first pharyngeal arch develops independently of *Hox* genes, whereas expression of *Hoxa-2* and *Hoxa-3* is required for formation of the second and third pharyngeal arches. Similar to the pharyngeal arches,

Text continued on p. 320

CLINICAL CORRELATION 14.2
Dental Anomalies

By Piranit N. Kantapura, Chiang Mai University, Thailand.

Abnormal Tooth Number

Hypodontia

Hypodontia or congenital absence of teeth can be isolated (non-syndromic) or associated with certain genetic syndromes. Isolated hypodontia can be caused by mutations in the *MSX1*, *EDA*, *AXIN2*, *PAX9*, and *WNT10A* genes. Mutations in the *EDA* gene, which encodes **ectodysplasin** (see p. 162), cause alterations in its signaling pathway, which subsequently results in X-linked **hypohidrotic ectodermal dysplasia** (XLHED) or isolated hypodontia. Characteristic features of patients with XLHED consist of missing or malformed teeth, missing and sparse hair, and absent or dysfunctional exocrine glands. The normally soluble ligand ectodysplasin binds to the membrane-bound receptor **Edar**, and through its signaling pathway it activates transcription of its target genes. EDA signaling is known to suppress bone morphogenetic protein-4 (BMP-4) activity and upregulate sonic hedgehog (shh), both of which have crucial roles in tooth formation. Therefore, it is understandable that mutations of *EDA* cause missing or malformed teeth.

This disorder is X-linked; thus, most patients are male (**Fig. 14.25A**). Sporadic cases have been reported, but frequently the mutations are inherited from mothers who are heterozygous. The heterozygous mothers are usually normal or have mild manifestations, including microdontia and hypodontia (Fig. 14.25B).

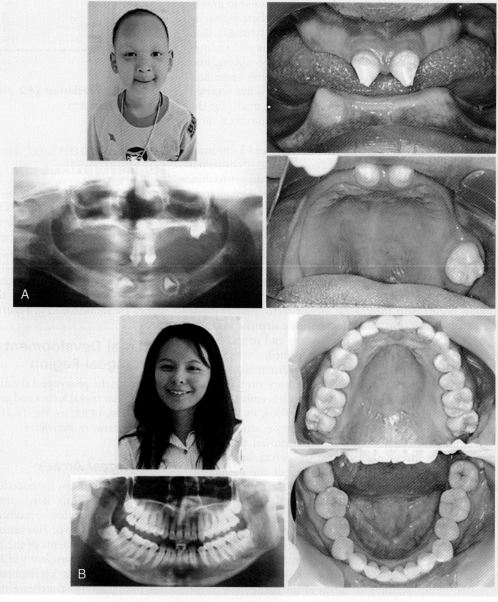

Fig. 14.25 Hypodontia in a Thai boy. A, X-linked hypohidrotic ectodermal dysplasia caused by an *EDA* mutation. Note the severe hypodontia, sparse hair, and hyperpigmented skin around his eyes. **B,** His mother, who is heterozygous for the *EDA* mutation is phenotypically normal. *(Courtesy of Dr. P. Kantaputra, Chiang Mai University, Thailand.)*

CLINICAL CORRELATION 14.2
Dental Anomalies—cont'd

Hyperdontia (Supernumerary Teeth)

Hyperdontia, an excess number of teeth, can be isolated or associated with several genetic syndromes, including **cleidocranial dysplasia** and **trichorhinophalangeal syndrome** (TRPS 1), which can be caused by mutations in *TRPS1*. TRPS 1 is a strong transcriptional repressor protein, and its mutation has been hypothesized to cause gain of function and to be responsible for supernumerary teeth (**Fig. 14.26**) and mandibular prognathism.

TRPS 1 binds to the promoter of *RUNX2*. All lines of evidence suggest that TRPS 1 and RUNX-2 share the same pathway and that TRPS 1 acts as a repressor of RUNX-2. In mice, the coexpression of *Runx2* and *Trps 1* has been demonstrated in developing bone (see Fig. 9.17) and in dental mesenchyme during early tooth development.

Abnormal Tooth Size and Shape

Smallest Teeth

Microdontia (small teeth) can be isolated or associated with genetic syndromes. The most commonly affected teeth are the maxillary permanent lateral incisors. They may be of normal shape or peg-shaped. One of the most common causes of microdontia appears to be mutations of the **MSX1** gene. Such mutations can also lead to hypodontia and orofacial clefting. The smallest known teeth have been reported in patients with **microcephalic osteodysplastic primordial dwarfism** type II (MOPD II; **Fig. 14.27**). Patients with this syndrome have severe prenatal and postnatal growth retardation, with a relatively proportionate head size at birth, but extreme microcephaly in adulthood. The striking dental anomalies consist of extremely small teeth, opalescent and abnormally shaped teeth, and rootless molars. Teeth are spontaneously exfoliated because the alveolar bone is severely hypoplastic. MOPD

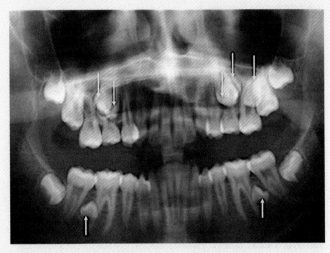

Fig. 14.26 Multiple supernumerary teeth *(arrows)* in a patient with trichorhinophalangeal syndrome resulting from a mutation in *TRPS1*. *(Courtesy of Dr. P. Kantaputra, Chiang Mai University, Thailand.)*

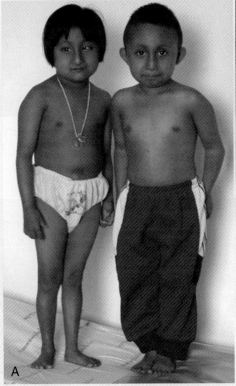

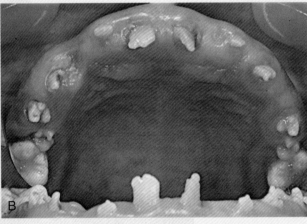

Fig. 14.27 **A** and **B,** Extremely small teeth in a Thai boy with microcephalic osteodysplastic primordial dwarfism type II, who had homozygous mutations in the *PCNT* gene. At this early age, microcephaly is not yet evident. *(Courtesy of Dr. P. Kantaputra, Chiang Mai University, Thailand.)*

Continued

CLINICAL CORRELATION 14.2
Dental Anomalies—cont'd

II, a very rare autosomal recessive disorder, is caused by mutations in *PCNT. PCNT* encodes the centrosomal protein **pericentrin**, a protein crucial for cell division. This gene is expressed in both the epithelium and the mesenchyme during early tooth development. The clinical phenotype, including the small size of the teeth, is the consequence of loss of microtubule integrity, which leads to defective centrosome function.

Largest Teeth

Macrodontia, the condition of having larger than normal teeth, is an extremely rare condition. Macrodontia of the maxillary permanent central incisors is typical in patients with **KBG syndrome**, which is characterized by intellectual disability, skeletal malformation, and macrodontia. It is caused by mutations in *ANKRD11*, whose protein plays an important role in neural plasticity. Generalized macrodontia is extremely rare. This syndrome has been reported in patients with Ekman-Westborg-Julin syndrome or multiple macrodontic multituberculism. Patients with this syndrome have the largest teeth ever reported (**Fig. 14.28**). The molecular origin of this syndrome is unknown.

Molarized Incisors

True transformation of incisors to molars is an extremely rare condition (**Fig. 14.29**). Misexpression of *Barx-1* in mice by the inhibition of BMP signaling can result in the formation of molar teeth instead of incisors (see p. 310). This finding provides strong support for a role of homeobox genes in controlling tooth type.

Abnormal Tooth Structure

Abnormal Dentin: Dentinogenesis Imperfecta

Abnormalities of dentin include **dentinogenesis imperfecta** and **dentin dysplasia**. Teeth with dentinogenesis imperfecta appear blue-gray or amber brown and are opalescent. Dentinogenesis imperfecta can be nonsyndromic or associated with **osteogenesis imperfecta** (**Fig. 14.30**). Nonsyndromic dentinogenesis imperfecta is caused by mutations in the *DSPP* (dentin sialophosphoprotein) gene, which encodes dentin sialoprotein, a noncollagenous protein of dentin.

Dentinogenesis imperfecta, which is associated with osteogenesis imperfecta, is caused by mutations in the type I collagen genes

COL1A1 or *COL1A2*. Osteogenesis imperfecta is a heterogeneous group of heritable connective tissue disorders, caused by abnormal type I collagen synthesis. Characteristic features include increased bone fragility, bone deformities (see Fig. 14.30A and B), joint hyperextensibility, blue sclerae, hearing loss, and dentinogenesis imperfecta.

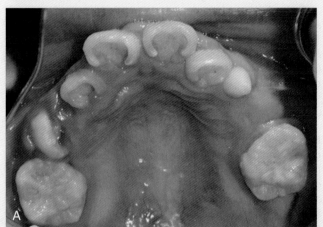

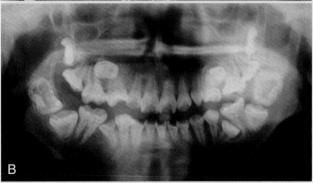

Fig. 14.28 **A** and **B,** Extremely large teeth in a Thai patient with Ekman-Westborg-Julin syndrome. Note the shovel incisors in **A.** *(Courtesy of Dr. P. Kantaputra, Chiang Mai University, Thailand.)*

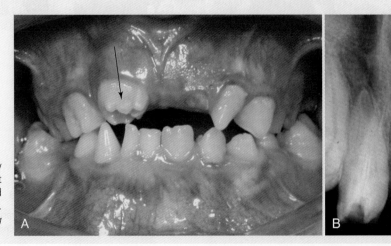

Fig. 14.29 **A,** Molarized maxillary right permanent incisor *(arrow)* at age 7. **B,** Radiograph of molarized maxillary central incisors at age 10. *(Courtesy of Dr. P. Kantaputra, Chiang Mai University, Thailand.)*

CLINICAL CORRELATION 14.2
Dental Anomalies—cont'd

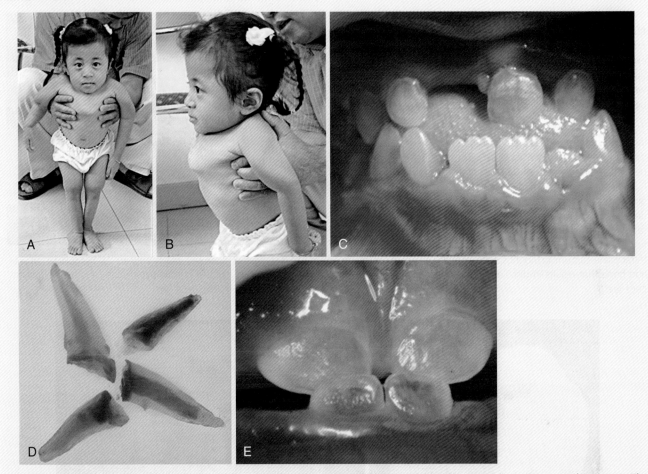

Fig. 14.30 **A Thai patient with osteogenesis imperfecta.** A and B, Note the bone deformities, especially in the pectoral region. C, The patient's teeth are affected with dentinogenesis imperfecta. D, Teeth with dentinogenesis imperfecta from a patient affected with osteogenesis imperfecta. E, Translucent teeth affected by isolated dentinogenesis imperfecta. *(Courtesy of Dr. P. Kantaputra, Chiang Mai University, Thailand.)*

Abnormal Enamel: Amelogenesis Imperfecta

Amelogenesis imperfecta is a group of clinically and genetically heterogeneous disorders that affect the development of enamel and result in abnormalities of the amount, composition, and/or structure of enamel. These disorders are caused by mutations in a variety of genes that are important for enamel formation. The enamel may be hypoplastic, hypomature, or hypocalcified (**Fig. 14.31**). Mutations in several genes, including *ENAM, AMEL, DLX3,* and *P63,* are known to cause isolated or syndromic amelogenesis imperfecta.

Dental Fluorosis

Excessive fluoride consumption during tooth formation can cause **enamel fluorosis**, which ranges from white spots or lines in the enamel to enamel hypoplasia (**Fig. 14.32**). The white opaque appearance of fluorosed enamel is caused by a hypomineralized enamel subsurface. The changes in enamel are related to cell-matrix interactions when the teeth are forming. At an early stage of enamel maturation, the amount of amelogenin in the enamel matrix of fluorosed enamel is very high, thus resulting in a delay in the removal of amelogenin by proteinases as the enamel matures. This subsequently causes abnormal mineralization of the enamel.

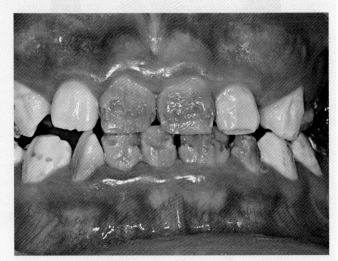

Fig. 14.31 Amelogenesis imperfecta, hypoplastic, hypomaturation, and hypocalcification type in a Thai patient who had a mutation in the *TP63* gene and was affected with the split hand–split foot-ectodermal dysplasia-amelogenesis imperfecta syndrome. (See Fig. 10.32 for split hand in a patient with the same mutation.) *(Courtesy of Dr. P. Kantaputra, Chiang Mai University, Thailand.)*

Continued

CLINICAL CORRELATION 14.2
Dental Anomalies—cont'd

Investigators have demonstrated a genetic component to dental fluorosis susceptibility. This means that some people are more prone to have enamel fluorosis than others. Some people with excessive fluoride consumption may have skeletal fluorosis, but fluorosed bone disappears later in life because bones remodel, whereas enamel does not. Therefore, enamel fluorosis is permanent.

Tetracycline-Stained Teeth

Ingestion of tetracyclines during tooth development causes abnormal tooth formation (see Table 8.7). The discoloration of teeth is permanent and ranges from yellow or gray to brown (**Fig. 14.33B**). It appears fluorescent under ultraviolet light (Fig. 14.33A). The discoloration is permanent, and the degree of discoloration depends on the dose and type of drug. Tetracyclines are incorporated into tissues that are calcifying at the time of their administration. They are able to chelate with calcium ions and form a tetracycline-calcium orthophosphate complex in teeth, cartilage, and bone that results in discoloration of the primary and permanent teeth.

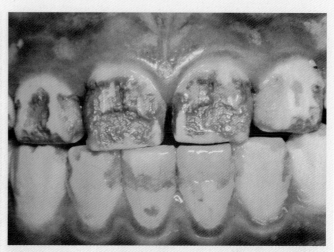

Fig. 14.32 Enamel fluorosis, which appears as hypoplastic enamel, caused by excessive fluoride consumption during tooth development. *(Courtesy of Dr. P. Kantaputra, Chiang Mai University, Thailand.)*

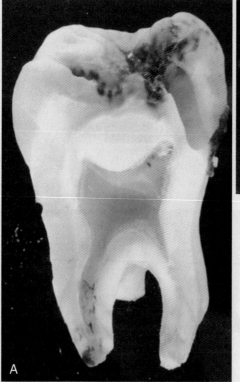

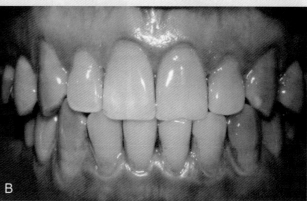

Fig. 14.33 **A**, Tetracycline deposition in dental pulp and root dentin. Under ultraviolet light, the tetracycline fluoroses yellow. Tetracycline marks in the root dentin correspond with root development at the time of drug administration. **B**, Tetracycline-stained teeth. (**A**, *Courtesy of Dr. W. Wiwatcunoopakarn;* **B**, *courtesy of Dr. P. Kantaputra, Chiang Mai University, Thailand.)*

the pharyngeal pouches, as well as having individual identities, are characterized by highly regionalized molecular patterns in the dorsoventral and the craniocaudal axes. Signals from the pharyngeal endoderm pattern the pharyngeal arches even before cranial neural crest cells enter the arches (see Fig. 14.5). **Tbx-1** expression in the early pharyngeal endoderm influences FGF-8 signaling, and in its absence pharyngeal pouches do not form normally, and a sequence of malformations reminiscent of DiGeorge's syndrome occurs.

In addition to being packed with mesenchyme (mainly of neural crest origin except for the premuscle mesoderm, which migrates from the somitomeres), each pharyngeal arch is

Pharyngeal arch structures

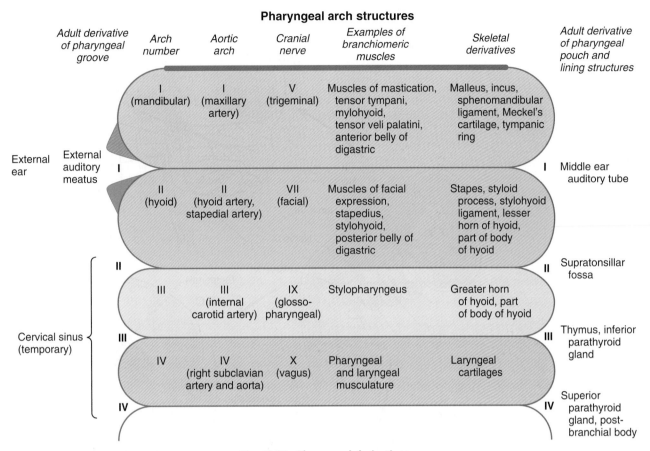

Adult derivative of pharyngeal groove		Arch number	Aortic arch	Cranial nerve	Examples of branchiomeric muscles	Skeletal derivatives	Adult derivative of pharyngeal pouch and lining structures
External ear	External auditory meatus	I (mandibular)	I (maxillary artery)	V (trigeminal)	Muscles of mastication, tensor tympani, mylohyoid, tensor veli palatini, anterior belly of digastric	Malleus, incus, sphenomandibular ligament, Meckel's cartilage, tympanic ring	
		II (hyoid)	II (hyoid artery, stapedial artery)	VII (facial)	Muscles of facial expression, stapedius, stylohyoid, posterior belly of digastric	Stapes, styloid process, stylohyoid ligament, lesser horn of hyoid, part of body of hyoid	I Middle ear auditory tube
Cervical sinus (temporary)		III	III (internal carotid artery)	IX (glosso-pharyngeal)	Stylopharyngeus	Greater horn of hyoid, part of body of hyoid	II Supratonsillar fossa
		IV	IV (right subclavian artery and aorta)	X (vagus)	Pharyngeal and laryngeal musculature	Laryngeal cartilages	III Thymus, inferior parathyroid gland
							IV Superior parathyroid gland, post-branchial body

Fig. 14.34   **Pharyngeal derivatives.**

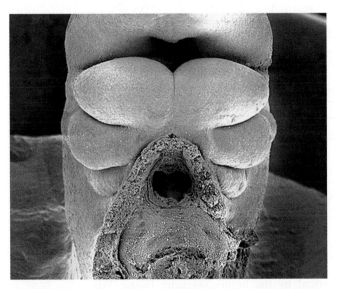

Fig. 14.35   **Scanning electron micrograph of a 5-week human embryo.** The stomodeum is the *dark area* at the *top* of the figure. *Beneath that,* three pairs of pharyngeal arches are prominent. *(From Steding G: The anatomy of the human embryo, Basel, 2009, Karger; courtesy of Dr. J. Männer.)*

associated with a major artery (aortic arch) and a cranial nerve (see Fig. 14.34). Each also contains a central rod of precartilaginous mesenchyme, which is transformed into characteristic adult skeletal derivatives. Understanding the relationship between the pharyngeal arches and their innervation

and vascular supply is very important, because tissues often maintain their relationship with their original nerve as they migrate out or become displaced from their site of origin in the pharyngeal arch system.

The **first pharyngeal arch** (mandibular) contributes mainly to structures of the face (both mandibular and maxillary portions) and ear (see Fig. 14.34). Its central cartilaginous rod, Meckel's cartilage, is a prominent component of the embryonic lower jaw until it is surrounded by locally formed intramembranous bone, which forms the definitive jaw. During later development, the distal part of Meckel's cartilage undergoes resorption because of extensive apoptosis of the chondrocytes. More dorsally, Meckel's cartilage forms the **sphenomandibular ligament**, the **anterior ligament of the malleus**, and the **malleus** (**Fig. 14.36**). In addition, the **incus** arises from a primordium of the **quadrate cartilage**. The first-arch musculature is associated with the masticatory apparatus, the pharynx, and the middle ear. A common feature of these muscles is their innervation by the trigeminal nerve (cranial nerve V).

The molecular basis for development of the first pharyngeal arch is quite different from that of the other arches, starting with the origin of the neural crest that populates it. The neural crest cells that populate the first arch are derived from rhombomeres 1 and 2 and midbrain, which are anterior to the expression domain of the *Hox* genes. Cellular precursors of the first-arch mesenchyme are instead associated with the expression of Otx-2. The role of signaling molecules and transcription factors, such as Dlx and Msx, is discussed earlier (see pp. 299-301).

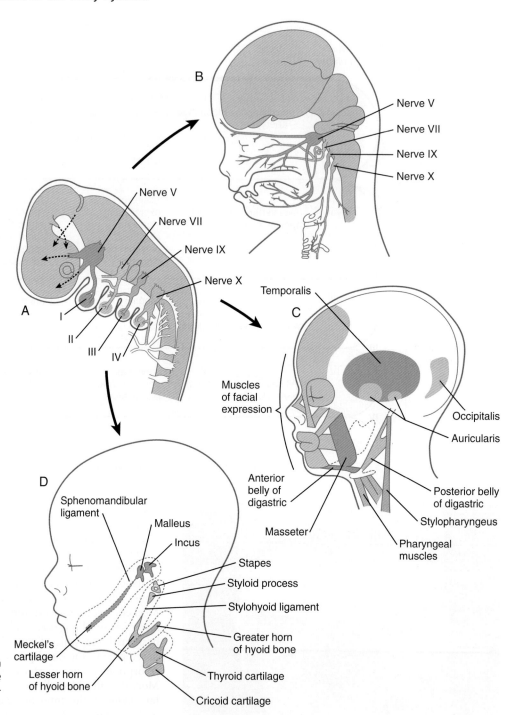

Fig. 14.36 Pharyngeal arch system (A) and adult derivatives of the neural (B), muscular (C), and skeletal (D) components of the arches.

The **second pharyngeal arch** (hyoid) also forms a string of skeletal structures from the body of the hyoid bone to the **stapes** of the middle ear. These structures are derived from Reichert's cartilage, which, instead of being a single rod, is actually two cartilaginous rodlike elements with a strand of mesenchymal tissue connecting them. Although much of the second-arch mesoderm migrates to the face to form the **muscles of facial expression**, additional muscles become associated with other second-arch skeletal derivatives; a good example is the **stapedius muscle**, which is associated with the stapes bone. These second-arch muscles are innervated by the facial nerve (cranial nerve VII).

Patterning of the second pharyngeal arch is strongly influenced by the homeobox gene *Hoxa2*. When this gene is knocked out in mice, skeletal derivatives of the second arch fail to form. The second arch in such mutant animals contains mirror-image duplicates of many of the proximal bones of the first-arch skeleton. This may be a result of the response of the mutant second-arch mesenchyme to an ectodermal signal from the first pharyngeal cleft that plays a role in patterning the first arch. The formation of mirror-image first-arch structures is analogous to the formation of mirror-image supernumerary limbs after transplantation of the zone of polarizing activity in the limb bud (see p. 198). The finding that only proximal structures are affected reflects the different genetic control of proximal and distal segments of the arches.

The **third** and **fourth pharyngeal arches** are otherwise unnamed. The third arch gives rise to structures related to the hyoid bone and upper pharynx. The third-arch skeleton becomes the greater horn of the hyoid bone. The one muscular derivative (stylopharyngeus) of the third arch is innervated by the glossopharyngeal nerve (cranial nerve IX). The fourth arch gives rise to certain muscles and cartilages of the larynx and lower pharynx. The muscles are innervated by the vagus nerve (cranial nerve X), which also grows into the thoracic and abdominal cavities.

Pharyngeal Grooves

The first pharyngeal groove is the only one that persists as a recognizable adult structure: the **external auditory meatus.**

Grooves II and III become covered by the enlarged external portion of the second arch (a phylogenetic homologue of the operculum [gill cover] of fish). The enlargement of the second arch is caused by the presence of a signaling center in the ectoderm at its tip; such a signaling center is not present in arches 3 or 4. As in the facial primordia, this signaling center produces shh, FGF-8, and BMP-7, which stimulate growth of the underlying mesenchyme. During the period of their overshadowing by the hyoid (second) arch, grooves II and III are collectively known as the **cervical sinus** (**Fig. 14.37;** see Fig. 13.27). As development progresses, the posterior ectoderm of the second arch fuses with ectoderm of a swelling (cardiac swelling) just posterior to the fourth arch, thus causing the cervical sinus to disappear and the external contours of the neck to become smooth.

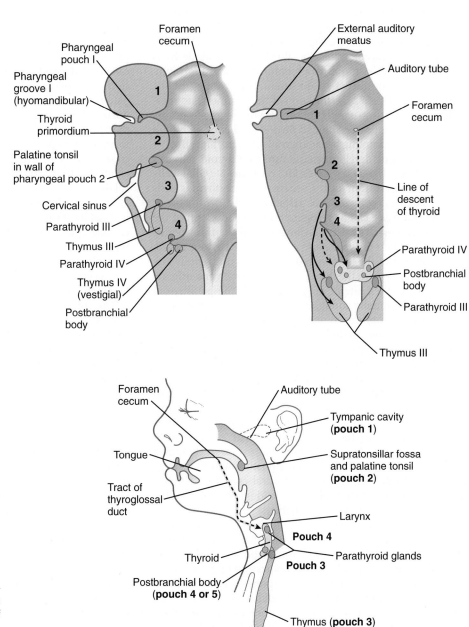

Fig. 14.37 Embryonic origins and pathways of primordia of glands derived from the pharyngeal pouches and the floor of the pharynx.

Internal Development of the Stomodeal and Pharyngeal Regions

Pharynx and Pharyngeal Pouches

The embryonic pharynx is directly converted to the smooth-walled pharynx of the adult. Of greater developmental interest is the fate of the pharyngeal pouches and their lining epithelium.

As with their corresponding first pharyngeal clefts, the **first pharyngeal pouches** become intimately involved in the formation of the ear. The end of each pouch expands to become the **tympanic cavity** of the middle ear, and the remainder becomes the **auditory (eustachian) tube**, which connects the middle ear with the pharynx (see Fig. 13.24).

The **second pharyngeal pouches** become shallower and less conspicuous as development progresses. Late in the fetal period, patches of lymphoid tissue form aggregations in the walls to create the **palatine (faucial) tonsils**. The pouches themselves are represented only as the **supratonsillar fossae**.

The **third pharyngeal pouch** is a more complex structure, consisting of a solid, dorsal epithelial mass and a hollow, elongated ventral portion (see Fig. 14.37). Through yet incompletely known mechanisms, the third pouch endoderm is specified very early into a parathyroid/thymus primordium. In a manner reminiscent of dorsoventral patterning of the neural tube (see p. 222), delineation of the common endodermal primordium into parathyroid and thymic segments occurs in response to opposing **shh** and **BMP-4** gradients, with high levels of shh promoting a parathyroid fate and high BMP-4 leading to thymic development. Future parathyroid cells can be recognized by their expression of **Gcm-2 (Glial cells missing)** transcription factor, whereas future thymus cells express **Foxn-1**. By 5 weeks of gestation, cells identifiable as **parathyroid** tissue can be recognized in the endoderm of the solid dorsal mass. The ventral elongation of the third pouch differentiates into the epithelial portion of the **thymus gland**. The primordia of the thymus and the parathyroid glands lose their connection with the third pharyngeal pouch and migrate caudally from their site of origin. Although the parathyroid III primordia initially comigrate with the thymic primordia, they ultimately continue to migrate toward the midline. There they join with the thyroid gland and pass the parathyroid primordia of the fourth pouch to form the **inferior parathyroid glands**. The third pharyngeal pouch disappears.

The **fourth pharyngeal pouch** is organized much as the third, with a solid, bulbous, dorsal parathyroid IV primordium. It also contains a small ventral epithelial outpocketing, which contributes a minor component to the thymus in some species. In humans, the thymic component of the fourth pouch is vestigial. At the ventralmost part of each fourth pouch is another structure called the **postbranchial (ultimobranchial) body** (see Fig. 14.37). Neural crest cells migrate into the postbranchial bodies and ultimately become the secretory component of these structures.

As with their counterparts from the third pouch, the parathyroid IV primordia lose their connection with the fourth pouch and migrate toward the thyroid gland as **superior parathyroid glands**. The postbranchial bodies also migrate toward the thyroid, where they become incorporated as **parafollicular** or **C cells**. The parafollicular cells, which are of neural crest origin, produce the polypeptide hormone **calcitonin**, which acts to reduce the concentration of calcium in the blood. The

parathyroid glands produce **parathyroid hormone**, which increases blood calcium levels.

Thyroid

Development of the thyroid gland begins with local mesodermal inductive signals acting on the ventral endoderm of the foregut. This process results in the specification of a small number of endodermal cells (as few as 60 in the mouse) to be destined to a thyroid fate. These founder cells, the **thyroid anlage**, increase in number to form a thickened placode that soon begins to extend into the surrounding mesodermal mesenchyme, at which point it is called a **thyroid bud**. These cells are characterized by the expression of four transcription factors (**Hhex, Nkx2-1, Pax-8**, and **Foxe-1**), which operate together in a complex interacting pattern and all of which are required for further thyroid development.

The unpaired primordium (thyroid bud) of the **thyroid gland** appears in the ventral midline of the pharynx between the first and second pouches (see Fig. 14.37). Starting during the fourth week as an endodermal thickening just caudal to the median tongue bud (tuberculum impar), the thyroid primordium soon elongates to form a prominent downgrowth called the **thyroid diverticulum**. The pathway of caudal extension of the bilobed thyroid diverticulum is determined by the pattern of arteries in the neck and extension and continues during pharyngeal development. During its caudal migration, the tip of the thyroid diverticulum expands and bifurcates to form the thyroid gland itself, which consists of two main lobes connected by an isthmus. For some time, the gland remains connected to its original site of origin by a narrow **thyroglossal duct**.

By about the seventh week, when the thyroid has reached its final location at the level of the second and third tracheal cartilages, the thyroglossal duct has largely regressed. Nevertheless, in almost half the population, the distal portion of the thyroglossal duct persists as the **pyramidal lobe of the thyroid**. The original site of the thyroid primordium persists as the **foramen cecum**, a small blind pit at the base of the tongue.

The thyroid gland undergoes histodifferentiation and begins functioning early in embryonic development. By 10 weeks of gestation, follicles containing some colloid material are evident, and a few weeks thereafter, the gland begins to synthesize noniodinated **thyroglobulin**. Secretion of **triiodothyronine**, one of the forms of thyroid hormone, is detectable by late in the fourth month.

Hypophysis

The **hypophysis** (pituitary gland) develops from two initially separate ectodermal primordia that secondarily unite. One of the primordia, called the **infundibular process**, forms as a ventral downgrowth from the floor of the diencephalon. The other primordium is **Rathke's pouch**, a midline outpocketing from the stomodeal ectoderm that extends toward the floor of the diencephalon as early as the fourth week. An inductive event from the overlying diencephalon, mediated first by BMP-4 and then by FGF-8, stimulates the formation of a **Rathke's pouch primordium** in the dorsal stomodeal ectoderm and provides cues for molecular events that stimulate cell proliferation. Through the action of **Hesx-1** (homeobox gene expressed in embryonic stem cells, formerly called Rpx)

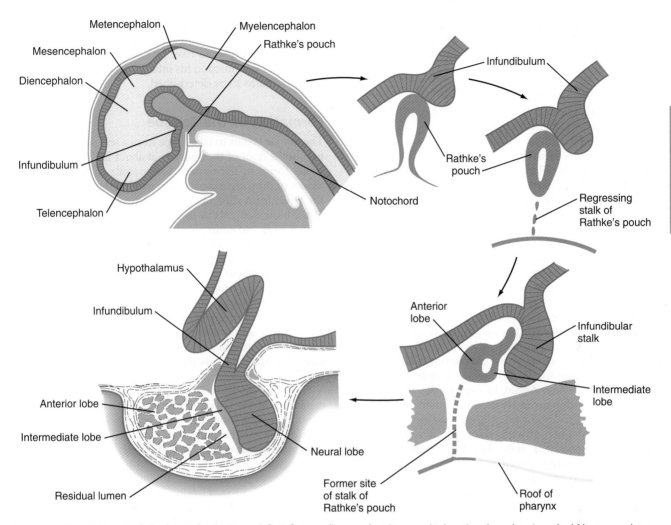

Fig. 14.38 **Development of the hypophysis.** *Upper left,* Reference diagram showing a sagittal section through a 4-week-old human embryo.

and the Lim-type homeobox-containing genes *Lhx3* and *Lhx4,* Rathke's pouch primordium forms the definitive Rathke's pouch (**Fig. 14.38**). In the early embryo, the cells in Rathke's pouch originate in the anterior ridge of the neural plate.

The infundibular process is intimately related to the hypothalamus (see Fig. 1.15), and certain hypothalamic neurosecretory neurons send their processes into the infundibular process, which ultimately becomes the **neural lobe of the hypophysis**. Throughout development, the histological structure of the infundibulum retains a neural character.

As development proceeds, Rathke's pouch elongates toward the infundibulum (see Fig. 14.38). While its blind end partially enfolds the infundibulum like a double-layered cup, the stalk of Rathke's pouch begins to regress. The outer wall of the cup thickens and assumes a glandular appearance in the course of its differentiation into the **pars distalis** (anterior lobe) of the hypophysis. The inner layer of the cup, which is closely adherent to the neural lobe, becomes the **pars intermedia** (intermediate lobe). It remains separated from the anterior lobe by a slitlike **residual lumen**, which represents all that remains of the original lumen of Rathke's pouch.

With the progression of pregnancy, the hypophysis undergoes a phase of cytodifferentiation. Late in the fetal period, specific cell types begin to produce small amounts of hormones. Molecular cascades underlying the differentiation

of specific cell types in the pituitary are being discovered (**Box 14.1**).

Although Rathke's pouch normally begins to lose its connections to the stomodeal epithelium by the end of the second month, portions of the tissue may occasionally persist along the pathway of the elongating stalk. If the tissue is normal, it is called a **pharyngeal hypophysis**. Sometimes, however, the residual tissue becomes neoplastic and forms hormone-secreting tumors called **craniopharyngiomas**.

Thymus and Lymphoid Organs

The paired endodermal thymic primordia begin to migrate from their third pharyngeal pouch origins during the sixth week. Their path of migration takes them through a substrate of mesenchymal cells until they reach the area of the future mediastinum behind the sternum. By the end of their migration, the two closely apposed thymic lobes are still epithelial structures. Soon, however, they become invested with a capsule of neural crest–derived connective tissue, which also forms septa among the endodermal epithelial cords and contributes to the thymic vasculature. In the absence of neural crest, the thymus fails to develop. An interaction between the neural crest and endodermal components of the thymic primordia conditions the latter for subsequent differentiation of thymic structure and function.

At about 9 to 10 weeks' gestation, blood-borne thymocyte precursors (**prothymocytes**), which originate in the hematopoietic tissue, begin to invade the epithelial thymus, in response to the secretion of the **chemokine CC121** by the thymus. Just before the prothymocytes invade the thymus, the thymic epithelium begins to express the transcription factor **WHN**, which is necessary for colonization of the thymic epithelium by the prothymocytes. In homozygous *WHN* mutants, the absence of such colonization results in the absence of functional T cells, thus leaving the individual severely immunocompromised and unable to reject foreign cells and tissue. Within the thymus, the prothymocytes force apart the epithelial cells and cause them to form a spongy **epithelial reticulum**. Responding to signals from the thymic epithelium, the prothymocytes proliferate and become redistributed, forming the cortical and medullary regions of the thymus.

By 14 to 15 weeks of gestation, blood vessels grow into the thymus, and a week later, some epithelial cells aggregate into small, spherical **Hassall's corpuscles**. At this point, the overall organization of the thymus is the same as that of adults. Functionally, the action of various **thymic hormones** causes the thymus to condition or instruct the prothymocytes migrating into it to become competent members of the **T-lymphocyte** family. The T lymphocytes leave the thymus and populate other lymphoid organs (e.g., lymph nodes, spleen) as fully functional immune cells.

The T lymphocytes are principally involved in **cellular immune responses**. Another population of lymphocytes that also originates in the bone marrow is instructed to become **B lymphocytes**, which are the mediators of **humoral immune responses**. B-lymphocyte precursors (**pro-B cells**) also must undergo conditioning to become fully functional, but their conditioning does not occur in the thymus. In birds, the pro-B cells pass through a cloacal lymphoid organ known as the **bursa of Fabricius**, where conditioning occurs. Humans do not possess a bursa, but its functional equivalent, although still undefined, is assumed to exist. B-lymphocyte conditioning is thought to occur in the bone marrow; in early embryos, conditioning possibly occurs in the liver.

The thymus and the bursa or mammalian equivalent are commonly referred to as **central lymphoid organs**. The lymphoid structures that are seeded by B and T lymphocytes are called **peripheral lymphoid organs**. (**Figure 14.39** shows the development and function of the lymphoid system.) Small **cervical thymus glands** have been discovered in the neck in mice. How their function fits into that of the overall lymphoid system remains to be determined.

Formation of the Tongue

The tongue begins to take shape from a series of ventral swellings in the floor of the pharynx at about the same time as the palate forms in the mouth. Major shifts in the positions of tissues of the tongue occur, thus making the characteristics of the adult form difficult to comprehend without knowledge of the basic elements of its embryonic development.

In 5-week-old embryos, the tongue is represented by a pair of **lateral lingual swellings** in the ventral regions of the first pharyngeal arches and two median unpaired swellings. The **tuberculum impar** is located between the first and second arches, and the **copula** (yoke) unites the second and third arches (**Figs. 14.40A** and **B** and **14.41**). The foramen cecum, which marks the original location of the thyroid primordium, serves as a convenient landmark delineating the border between the original tuberculum impar and the copula. Caudal to the copula is another swelling that represents the **epiglottis**.

Growth of the body of the tongue is accomplished by a great expansion of the lateral lingual swellings, with a minor contribution by the tuberculum impar (Fig. 14.40C and D). The root of the tongue is derived from the copula, along with additional ventromedial tissue between the third and fourth pharyngeal arches.

The dorsal surface of the tongue is covered with a large number of papillae. Development of the **filiform papillae**, which constitute the bulk of the papillae, follows a course that is remarkably similar to that of hair follicles. The ectoderm, which surrounds a mesenchymal core, expresses Hoxc-13 and the signaling factors shh, BMP-2 and BMP-4, and FGF-8. These molecular components characterize the inductive signaling system in almost all ectodermal derivatives, including hairs, feathers, and teeth.

Corresponding to its innervation by the hypoglossal nerve (cranial nerve XII), the musculature of the tongue migrates from a considerable distance (the **occipital [postotic] myotomes**). Similar to their counterparts in the limb, the myoblasts express Pax-3 during their migration to the tongue. The general sensory innervation of the tongue accurately reflects the pharyngeal arch origins of the epithelium. The lingual epithelium over the body of the tongue is innervated by the trigeminal nerve (cranial nerve V), in keeping with the first-arch origins of the lateral lingual swellings. Correspondingly, the root of the tongue is innervated by the glossopharyngeal nerve (cranial nerve IX—third arch) and the vagus nerve (cranial nerve X—fourth arch). The epithelium of the second arch is overgrown by that of the third arch; there is no general sensory innervation of the tongue by the seventh nerve.

Cranial nerves VII (facial) and IX innervate the taste buds. The contribution of cranial nerve VII is facilitated by its chorda tympani branch, which joins with the lingual branch of the trigeminal nerve and has access to the body of the tongue. Taste buds appear during the seventh week of gestation. In contrast to an earlier hypothesis that they are induced by fibers of the visceral afferent cranial nerves VII and IX,

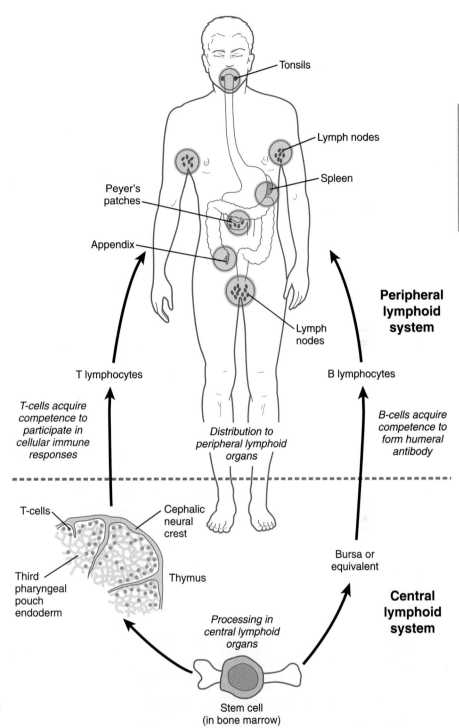

Fig. 14.39 Embryonic development of the lymphoid system.

which innervate taste buds postnatally, more recent evidence suggests that taste bud formation is independent of cranial nerves. An intrinsic signaling mechanism within the endodermal lingual epithelium, possibly mediated by a mechanism involving shh, Gli-1, and patched, seems to be the basis for the initial formation of taste buds. When formed, taste buds become strongly dependent on innervation for their maintenance. Considerable evidence indicates that the fetus is able to taste, and it has been postulated that the fetus uses the taste function to monitor its intra-amniotic environment.

Clinical Correlation 14.3 presents anomalies and syndromes involving the pharynx and pharyngeal arches.

Clinical Vignette

A nervous-looking 23-year-old woman complains to her physician of feeling too hot, losing too much weight, and sweating more than she had previously. On physical examination, her skin is warm and fine in texture, she has a fine tremor of her fingers, and her eyes bulge slightly. As part of the diagnostic follow-up, she is given a dose of radioactive iodine; then scanning is performed to localize the iodine. The scans show that most of the radioactive iodine is localized to a small mass of tissue at the base of the tongue. A neoplastic growth in the tissue mass is diagnosed, but the location of the mass is attributed to a congenital malformation.

Explain the embryological basis for the location of the neoplastic growth.

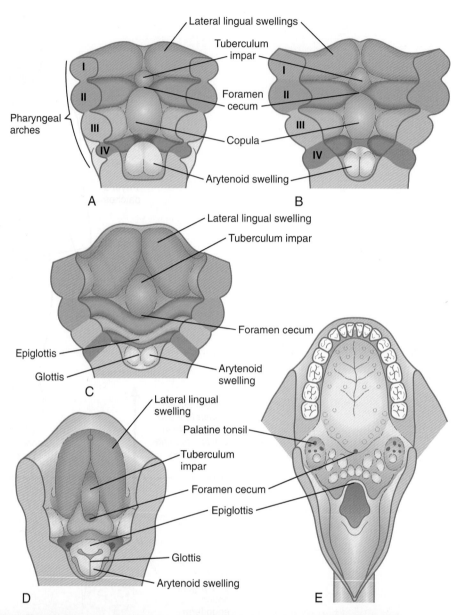

Fig. 14.40 Development of the tongue as seen from above. A, At 4 weeks. **B,** Late in the fifth week. **C,** Early in the sixth week. **D,** Middle of the seventh week. **E,** Adult.

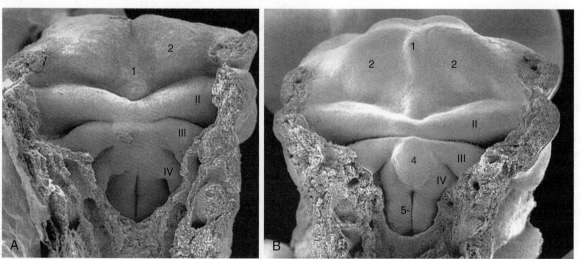

Fig. 14.41 Scanning electron micrographs looking down at the tongue-forming region of 5-week (**A**) and 6-week (**B**) human embryos. *1,* tuberculum impar; *2,* lateral lingual swellings (in arch I); *4,* copula; *5,* laryngotracheal groove; *II* to *IV,* pharyngeal arches. *(From Steding G: The anatomy of the human embryo, Basel, 2009, Karger; courtesy of Dr. J. Männer.)*

CLINICAL CORRELATION 14.3
Anomalies and Syndromes Involving the Pharynx and Pharyngeal Arches

Syndromes Involving the First Pharyngeal Arch

Several syndromes involve hypoplasia of the mandible and other structures arising from the first pharyngeal arch. Various mechanisms can account for these syndromes. Many have a genetic basis, and others result from exposure to environmental teratogens. The numerous studies on genetically manipulated mice are beginning to produce malformations that may have human counterparts. With the fine balance of signaling molecules and transcription factors that are involved in the development of the pharyngeal arches, it is not surprising that disturbances in the function of single genes, whether through mutation or the action of teratogens, can result in a visible morphological anomaly. Hypoplasia of the lower face has been associated with the ingestion of isotretinoin (a vitamin A derivative used for the treatment of acne) during early pregnancy.

Pierre Robin syndrome involves extreme **micrognathia** (small mandible), cleft palate, and associated defects of the ear. An imbalance often exists between the size of the tongue and the very hypoplastic jaw, which can lead to respiratory distress caused by mechanical interference of the pharyngeal airway by the large tongue. Although many cases of Pierre Robin syndrome are sporadic, others have a genetic basis.

Treacher Collins syndrome (**mandibulofacial dysostosis**) is typically inherited as an autosomal dominant condition. The responsible gene, called *TCOF1*, has been identified. Operating through the **Treacle** protein, it affects the survival and proliferation of cranial neural crest cells. In mutations of this gene, neural crest cell migration is normal, but increased apoptosis and decreased proliferation result in a much reduced population of neural crest cells in the first pharyngeal arch. This syndrome includes various anomalies, not all of which are found in all patients. Common components of the syndrome include hypoplasia of the mandible

and facial bones, malformations of the external and middle ears, high or cleft palate, faulty dentition, and coloboma-type defects of the lower eyelid (**Fig. 14.42**).

The most extreme form of first-arch hypoplasia is **agnathia**, in which the lower jaw basically fails to form (**Fig. 14.43**). In severe agnathia, the external ears remain in the ventral cervical region and may join in the ventral midline.

Lateral Cysts, Sinuses, and Fistulas

Structural malformations, such as lateral cysts, sinuses, and fistulas, can be related directly to the abnormal persistence of pharyngeal grooves, pharyngeal pouches, or both. A **cyst** is a completely enclosed, epithelially lined cavity that may be derived from the persistence of part of a pharyngeal pouch, a pharyngeal groove, or a cervical sinus. A **sinus** is closed on one end and open to the

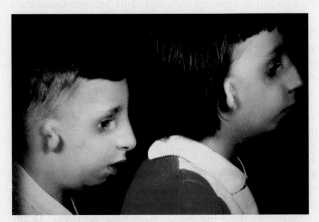

Fig. 14.42 **Siblings with Treacher Collins syndrome.** (Courtesy A. Burdi, Ann Arbor, Mich.)

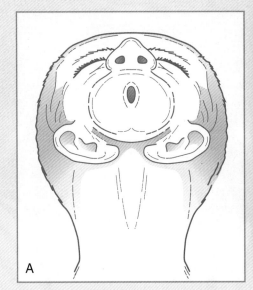
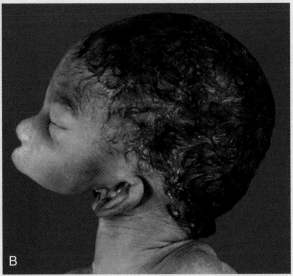

Fig. 14.43 **Agnathia.** A, Ventral view of the upturned face of an infant. B, Lateral view of the face of a fetus with agnathia. Note the cervical location of the external ears. (Courtesy of M. Barr, Ann Arbor, Mich.)

Continued

CLINICAL CORRELATION 14.3
Anomalies and Syndromes Involving the Pharynx and Pharyngeal Arches—cont'd

outside or to the pharynx. A **fistula** (Latin for "pipe") is an epithelially lined tube that is open at both ends—in this case, to the outside and to the pharynx.

The postnatal location of these structures accurately marks the location of their embryonic precursors. External openings of fistulas are typically found anterior to the sternocleidomastoid muscle in the neck (**Fig. 14.44**). Fistulas from remnants of pharyngeal grooves II or III may result from incomplete closure of the cervical sinus by tissue of the hyoid arch. Although present from birth, cervical cysts are often not manifested until after puberty. At that time, they expand because of increased amounts of secretions by the epithelium lining the inner surface of the cyst, changes that correspond to maturational changes in the normal epidermis.

Preauricular sinuses or **fistulas**, which are found in a triangular region in front of the ear, are also common. These structures are assumed to represent persistent clefts between preauricular hillocks on the first and second arches. True fistulas (**cervicoaural fistulas**) represent persisting ventral portions of the first pharyngeal groove. These extend from a pharyngeal opening to somewhere along the auditory tube or even the external auditory meatus.

Thyroglossal Duct Remnants

Various abnormal structures can persist along the pathway of the thyroglossal duct. **Ectopic thyroid tissue** can be found anywhere along the pathway of migration of the thyroid primordium from the foramen cecum in the tongue to the isthmus of the normal thyroid gland (**Fig. 14.45**). This fact must be considered in the clinical diagnosis or surgical treatment of carcinomas and other conditions affecting thyroid tissue. Less common are midline cysts or sinuses involving the former thyroglossal duct (**Fig. 14.46**).

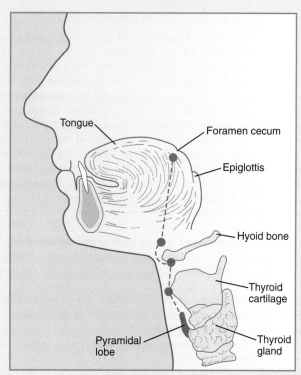

Fig. 14.45 Common locations *(red circles)* of thyroglossal duct remnants.

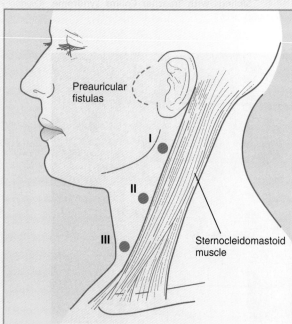

Fig. 14.44 Common locations of lateral cervical (branchial) cysts and sinuses *(red circles)* and preauricular fistulas. Roman numerals refer to the cervical cleft origin of the cysts.

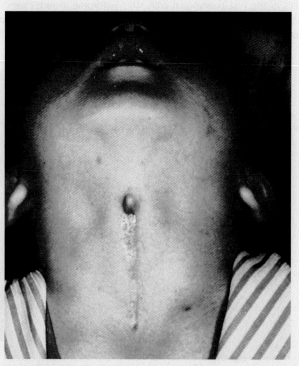

Fig. 14.46 **Individual with a thyroglossal duct sinus in the ventral midline of the neck.** *(Courtesy A. Burdi, Ann Arbor, Mich.)*

CLINICAL CORRELATION 14.3
Anomalies and Syndromes Involving the Pharynx and Pharyngeal Arches—cont'd

Because of their location, these entities can usually be easily distinguished from their lateral cervical counterparts.

Malformations of the Tongue

The most common malformation of the tongue is **ankyloglossia** (tongue-tie). This condition is caused by subnormal regression of the **frenulum**, the thin midline tissue that connects the ventral surface of the tongue to the floor of the mouth. Nonsyndromic ankyloglossia is caused by a mutation in the T-box transcription factor *TBX22*. Less common malformations of the tongue are **macroglossia** and **microglossia**, which are characterized by hyperplasia and hypoplasia of lingual tissue. Although sometimes associated with macroglossia, **furrowed tongue** (**Fig. 14.47**) is not normally associated with major functional disturbances.

Ectopic Parathyroid or Thymic Tissue

Because of their extensive migrations during early embryogenesis, parathyroid glands and components of the thymus gland are often found in abnormal sites (**Fig. 14.48**). Typically, this displacement is not accompanied by functional abnormalities, but awareness of the possibility of ectopic tissue or even supernumerary parathyroid glands is important for the surgeon.

DiGeorge's Syndrome

DiGeorge's syndrome is a cranial neural crest deficiency and is manifested by immunological defects and hypoparathyroidism (see Clinical Correlation 12.1). The underlying pathological condition is failure of differentiation of the thymus and parathyroid glands. Associated anomalies are malformations of first-arch structures and defects of the outflow tract of the heart, which contains an important cranial neural crest contribution as well. *Hoxa3* mutant mice exhibit many of the characteristics of human DiGeorge's syndrome.

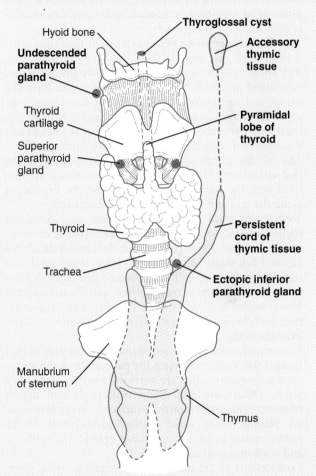

Fig. 14.48 Locations where abnormally positioned pharyngeal glands or portions of glands can be found.

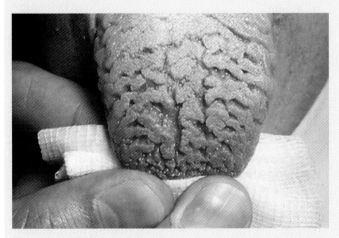

Fig. 14.47 **Furrowed tongue.** *(From Robert J: Gorlin Collection, Division of Oral and Maxillofacial Pathology, University of Minnesota Dental School; courtesy of Dr. Ioannis Koutlas.)*

Summary

■ The early craniofacial region arises from the rostral portions of the neural tube, the notochord, and the pharynx, which is surrounded by a series of paired aortic arches. Between the aortic arches and the overlying ectoderm are large masses of neural crest–derived and mesodermally derived mesenchyme. Some of these components show evidence of anatomical segmentation or segmental patterns of gene expression.

■ Massive migrations of segmental groups of neural crest cells provide the mesenchyme for much of the facial region. The musculature of the craniofacial region is derived from somitomeric mesoderm or the occipital somites. The connective tissue component of the facial musculature is of neural crest origin.

■ The pharyngeal (branchial) region is organized around paired mesenchymal pharyngeal arches, which alternate with endodermally lined pharyngeal pouches and ectodermally lined pharyngeal grooves.

- The face and lower jaw arise from an unpaired frontonasal prominence and paired nasomedial, maxillary, and mandibular processes. Through differential growth and fusion, the nasomedial processes form the upper jaw and lip, and the frontonasal prominence forms the upper part of the face. The expanding mandibular processes merge to form the lower jaw and lip. A nasolacrimal groove between the nasolateral and maxillary processes ultimately becomes canalized to form the nasolacrimal duct, which connects the orbit to the nasal cavity.

- The palate arises from the fusion of an unpaired median palatine process and paired lateral palatine processes. The former forms the primary palate, and the latter forms the secondary palate. Fusion of the lateral palatal shelves involves removal of the midline epithelial seam by a combination of apoptosis, migration, and transformation of epithelial cells into mesenchyme.

- The olfactory apparatus begins as a pair of thickened ectodermal nasal placodes. As these sink to form nasal pits, they are surrounded by horseshoe-shaped nasomedial and nasolateral processes. The nasomedial processes form the bridge and septum of the nose, and the nasolateral processes form the alae of the nose. The deepening nasal pits break into the oral cavity, and only later are the nasal cavities separated from the oral cavity by the palate.

- The salivary glands arise as epithelial outgrowths of the oral epithelium. Through a series of continuing interactions with the surrounding mesenchyme, the expanding glandular epithelium branches and differentiates.

- Teeth form from interactions between oral ectoderm (dental lamina) and neural crest mesenchyme. A developing tooth is first a tooth bud, which then passes through a cap and bell stage. Late in the bell stage, ectodermal cells (ameloblasts) of the epithelial enamel organ begin to form enamel, and the neural crest–derived epithelium (odontoblasts) begins to secrete dentin. Precursors of the permanent teeth form dental primordia along the more advanced primary teeth.

- Malformations of the face are common. Many, such as cleft lip and cleft palate, represent the persistence of the structural arrangements that are normal for earlier embryonic stages. Others, such as holoprosencephaly and hypertelorism, result from growth disturbances in the frontonasal process. Most facial malformations seem to be multifactorial in origin, involving genetic susceptibility and environmental causes.

- Components of the pharynx (pharyngeal grooves, pharyngeal arches, and pharyngeal pouches) give rise to a wide variety of structures. The first arch gives rise to the upper and lower jaws and associated structures. The first groove and pouch, along with associated mesenchyme from the first and second arches, form the many structures of the external and middle ear. The second, third, and fourth pharyngeal grooves become obliterated and form the outer surface of the neck, and the components of the second through fourth arches form the pharyngeal skeleton and much of the musculature and connective tissue of the pharyngeal part of the neck. The endoderm of the third and fourth pouches forms the thymus and parathyroid glands. The thyroid gland arises from an unpaired ventral endodermal outgrowth of the upper pharynx.

- The tongue originates from multiple ventral swellings in the floor of the pharynx. The bulk of the tongue comes from the paired lateral lingual swellings in the region of the first pharyngeal arches. The unpaired tuberculum impar and copula also contribute to the formation of the tongue. The tongue musculature, along with the hypoglossal nerve (cranial nerve XII), which supplies the muscles, arises from the occipital somites. General sensory innervation of the tongue (from cranial nerves V, IX, and X) corresponds with the embryological origin of the innervated part of the tongue. Cranial nerves VII and IX innervate the taste buds on the tongue.

- Many malformations of the lower face and jaw are related to hypoplasia of the first pharyngeal arches. Cysts, sinuses, and fistulas of the neck are commonly caused by abnormal persistence of pharyngeal grooves or pouches. Ectopic glandular tissue (thyroid, thymus, or parathyroid) is explained by the persistence of tissue rests along the pathway of migration of the glands. Certain syndromes (e.g., DiGeorge's syndrome), which seemingly affect disparate organs, can be attributed to neural crest defects.

Review Questions

1. The facial nerve (cranial nerve VII) supplies muscles derived from which pharyngeal arch?
A. First
B. Second
C. Third
D. Fourth
E. Sixth

2. Cleft lip results from lack of fusion of the:
A. Nasomedial and nasolateral processes
B. Nasomedial and maxillary processes
C. Nasolateral and maxillary processes
D. Nasolateral and mandibular processes
E. Nasomedial and mandibular processes

3. In cases of holoprosencephaly, defects of facial structures are typically secondary to defects of the:
A. Pharynx
B. Oral cavity
C. Forebrain
D. Eyes
E. Hindbrain

4. Meckel's cartilage is a prominent structure in the early formation of the:
A. Upper jaw
B. Hard palate
C. Nasal septum
D. Soft palate
E. Lower jaw

5. An early induction in tooth development consists of the ectoderm of the dental epithelium acting on the underlying neural crest mesenchyme. Which of the following molecules is an important mediator of the inductive stimulus?

A. BMP-4
B. Tenascin
C. Hoxb-13
D. Msx-1
E. Syndecan

6. A 15-year-old boy with mild acne developed a tender boil along the anterior border of the sternocleidomastoid muscle. What embryological condition would be included in a differential diagnosis?

7. The physician of the 15-year-old boy described in the previous question determined that the boy had a congenital cyst that needed to be removed surgically. What should the surgeon consider during removal of the cyst?

8. Why does a person sometimes get a runny nose when crying?

9. A woman who took an anticonvulsant drug during the tenth week of pregnancy gave birth to an infant with bilateral cleft lip and cleft palate. She sued the physician, blaming the facial malformations on the drug, and you are called in as an expert witness for the defense. What would be the basis for your case?

10. A woman who averaged three mixed drinks a day during pregnancy gave birth to an infant who was mildly retarded and who had a small notch in an upturned upper lip and a reduced olfactory sensitivity. What is the basis for this constellation of defects?

References

Allam KA and others: The spectrum of median craniofacial dysplasia, *Plast Reconstr Surg* 127:812-821, 2011.

Alt B and others: Arteries define the position of the thyroid gland during its developmental relocalisation, *Development* 133:3797-3804, 2006.

Barteczko K, Jacob M: A re-evaluation of the premaxillary bone in humans, *Anat Embryol* 207:417-437, 2004.

Bei M: Molecular genetics of tooth development, *Curr Opin Genet Dev* 19:504-510, 2009.

Brugmann SA and others: Wnt signaling mediates regional specification in the vertebrate face, *Development* 134:3283-3295, 2007.

Carstens MH: Neural tube programming and craniofacial cleft formation, I: the neuromeric organization of the head and neck, *Eur J Paediatr Neurol* 8:181-210, 2004.

Carstens MH: Development of the facial midline, *J Craniofac Surg* 13:129-187, 2002.

Chai Y, Maxson RE: Recent advances in craniofacial morphogenesis, *Dev Dyn* 235:2353-2375, 2006.

Clouthier DE, Garcia E, Schilling TF: Regulation of facial morphogenesis by endothelin signaling: insights from mice and fish, *Am J Med Genet A* 152A:2962-2973, 2010.

Cobourne MT, Sharpe PT: Making up the numbers: the molecular control of mammalian dental formula, *Semin Cell Dev Biol* 21:314-324, 2010.

Cohen MM, Shiota K: Teratogenesis of holoprosencephaly, *Am J Med Genet* 109:1-15, 2002.

Cordero DR and others: Cranial neural crest cells on the move: their roles in craniofacial development, *Am J Med Genet A* 155:270-279, 2011.

Couly G and others: Interactions between Hox-negative cephalic neural crest cells and the foregut endoderm in patterning the facial skeleton in the vertebrate head, *Development* 129:1061-1082, 2002.

Creuzet S, Couly G, Le Douarin NM: Patterning of the neural crest derivatives during development of the vertebrate head: insights from avian studies, *J Anat* 207:447-459, 2005.

de Felice M, Di Lauro R: Minireview: Intrinsic and extrinsic factors in thyroid gland development: an update, *Endocrinology* 152:2948-2956, 2011.

Depew MJ, Compagnucci C: Tweaking the hinge and caps: testing a model of the organization of jaws, *J Exp Zool B Mol Dev Evol* 310B:315-335, 2008.

Depew MJ and others: Reassessing the *Dlx* code: the genetic regulation of branchial arch skeletal pattern and development, *J Anat* 207:501-561, 2005.

Fagman H, Nilsson M: Morphogenetics of early thyroid development, *J Mol Endocrinol* 46:R33-R42, 2011.

Gitton Y and others: Evolving maps in craniofacial development, *Semin Cell Dev Biol* 21:301-308, 2010.

Gordon J, Manley NR: Mechanisms of thymus organogenesis and morphogenesis, *Development* 138:3865-3878, 2011.

Gorlin RJ, Cohen MM, Hennekam RCM: *Syndromes of the head and neck*, ed 4, Oxford, 2001, Oxford University Press.

Graham A: Deconstructing the pharyngeal metamere, *J Exp Zool B Mol Dev Evol* 310B:336-344, 2008.

Graham A, Okabe M, Quinlan R: The role of the endoderm in the development and evolution of the pharyngeal arches, *J Anat* 207:479-487, 2005.

Greene RM, Pisano MM: Palate morphogenesis: current understanding and future directions, *Birth Defects Res C Embryo Today* 90:133-154, 2010.

Gritli-Linde A: Molecular control of secondary palate development, *Dev Biol* 301:309-336, 2007.

Hanken J, Hall BK, eds: *The skull, vol 1, Development*, Chicago, 1993, University of Chicago Press.

Helms JA, Cordero D, Tapadia MD: New insights into craniofacial morphogenesis, *Development* 132:851-861, 2005.

Hinrichsen K: The early development of morphology and patterns of the face in the human embryo, *Adv Anat Embryol Cell Biol* 98:1-79, 1985.

Jheon AH, Schneider RA: The cells that fill the bill: neural crest and the evolution of craniofacial development, *J Dent Res* 88:12-21, 2009.

Jiang R, Bush JO, Lidral AC: Development of the upper lip: morphogenetic and molecular mechanisms, *Dev Dyn* 235:1152-1166, 2006.

Kantaputra PN: Dentinogenesis imperfecta-associated syndromes, *Am J Med Genet A* 104A:75-78, 2001.

Kantaputra PN, Gorlin RJ: Double dens invaginatus of molarized maxillary central incisors, premolarization of maxillary lateral incisors, multituberculism of the mandibular incisors, canines and first molar, and sensorineural hearing loss, *Clin Dysmorphol* 1:128-136, 1992.

Kantaputra PN, Matangkasombut O, Sripathomsawat W: Split hand-split-foot-ectodermal dysplasia and amelogenesis imperfecta with a *TP63* mutation, *Am J Med Genet A* 158A:188-192, 2012.

Kantaputra PN and others: Cleft lip and cleft palate, ankyloglossia, and hypodontia are associated with *TBX22* mutations, *J Dent Res* 90:450-455, 2011.

Kantaputra P and others: The smallest teeth in the world are caused by mutations in the *PCNT* gene, *Am J Med Genet A* 155A:1398-1403, 2011.

Kassai Y and others: Regulation of mammalian tooth cusp patterning by ectodin, *Science* 309:2067-2070, 2005.

Kawauchi S and others: *FGF8* expression defines a morphogenetic center required for olfactory neurogenesis and nasal cavity development in the mouse, *Development* 132:5211-5223, 2005.

Kelberman D and others: Genetic regulation of pituitary gland development in human and mouse, *Endocr Rev* 30:790-829, 2009.

Kjaer I: Orthodontics and foetal pathology: a personal view on craniofacial patterning, *Eur J Orthodont* 32:140-147, 2010.

Klingensmith J and others: Roles of bone morphogenetic protein signaling and its antagonism in holoprosencephaly, *Am J Med Genet C Semin Med Genet* 154C:43-51, 2010.

Knox SM and others: Parasympathetic innervation maintains epithelial progenitor cells during salivary organogenesis, *Science* 329:1645-1647, 2010.

Kuratani S: Craniofacial development and the evolution of vertebrates: the old problems on a new background, *Zool Sci* 22:1-19, 2005.

Lee S-H and others: A new origin for the maxillary jaw, *Dev Biol* 276:207-224, 2004.

Liu B, Rooker SM, Helms JA: Molecular control of facial morphology, *Semin Cell Dev Biol* 21:309-313, 2010.

Mangold E, Ludwig KU, Nöthen MM: Breakthroughs in the genetics of craniofacial clefting, *Trends Mol Med* 17:725-733, 2011.

Mehta A, Dattani MT: Developmental disorders of the hypothalamus and pituitary gland associated with congenital hypopituitarism, *Best Pract Clin Endocrinol Metabol* 22:191-206, 2008.

Meng L and others: Biological mechanisms in palatogenesis and cleft palate, *J Dent Res* 88:22-33, 2009.

Merida-Velasco JR and others: Development of the human temporomandibular joint, *Anat Rec* 255:20-33, 1999.

Minoux M, Rijli FM: Molecular mechanisms of cranial neural crest migration and patterning in craniofacial development, *Development* 137:2605-2621, 2010.

Mitsiadis TA, Graf D: Cell fate determination during tooth development and regeneration, *Birth Defects Res C Embryo Today* 87:199-211, 2009.

Müller F, O'Rahilly R: Olfactory structures in staged human embryos, *Cells Tissues Organs* 178:93-116, 2004.

Nie X, Luukko K, Kettunen P: BMP Signalling in craniofacial development, *Int J Dev Biol* 50:511-521, 2006.

Noden DM, Francis-West P: The differentiation and morphogenesis of craniofacial muscles, *Dev Dyn* 235:1194-1218, 2006.

Noden DM, Trainor PA: Relations and interactions between cranial mesoderm and neural crest populations, *J Anat* 207:575-601, 2005.

Patel VN, Rebustini IT, Hoffman MP: Salivary gland branching morphogenesis, *Differentiation* 74:349-364, 2006.

Raetzman LT, Cai JX, Camper SA: *Hes1* is required for pituitary growth and melanotrope specification, *Dev Biol* 304:455-466, 2007.

Rodríguez-Vázquez JF and others: Morphogenesis of the second pharyngeal arch cartilage (Reichert's cartilage) in human embryos, *J Anat* 208:179-189, 2006.

Smith TD, Bhatnagar KP: The human vomeronasal organ, part II: prenatal development, *J Anat* 197:421-436, 2000.

Song Y and others: Control of retinoic acid synthesis and FGF expression in the nasal pit is required to pattern the craniofacial skeleton, *Dev Biol* 276:313-329, 2004.

Szabo-Rogers HL and others: New directions in craniofacial morphogenesis, *Dev Biol* 341:84-94, 2010.

Tapadia MD, Cordero DR, Helms JA: It's all in your head: new insights into craniofacial development and deformation, *J Anat* 207:461-477, 2005.

Thesleff I, Mikkola M: The role of growth factors in tooth development, *Int Rev Cytol* 217:93-135, 2002.

Townsend G and others: Morphogenetic fields within the human dentition: a new, clinically relevant synthesis of an old concept, *Arch Oral Biol* 54S:S34-S44, 2009.

Trainor PA: Craniofacial birth defects: The role of neural crest cells in the etiology and pathogenesis of Treacher Collins syndrome and the potential for prevention, *Am J Med Genet A* 152A:2984-2994, 2010.

Tucker A, Matthews KL, Sharpe PT: Transformation of tooth type induced by inhibition of BMP signaling, *Science* 282:1136-1138, 1998.

Tucker A, Sharpe P: The cutting-edge of mammalian development: how the embryo makes teeth, *Nat Rev Genet* 5:499-508, 2004.

Tucker A and others: FGF-8 determines rostral-caudal polarity in the first branchial arch, *Development* 126:51-61, 1998.

Zhang Z and others: Antagonistic actions of Msx1 and Osr2 pattern mammalian teeth into a single row, *Science* 323:1232-1234, 2009.

Zhu X, Rosenfeld MG: Transcriptional control of precursor cell proliferation in the early phases of pituitary development, *Curr Opin Genet Dev* 14:567-574, 2004.

Digestive and Respiratory Systems and Body Cavities

The initial formation of the digestive system by the lateral folding of the endodermal germ layer into a tube is described in Chapter 6. From its beginnings as a simple tubular gut, development of the digestive system proceeds on several levels, including molecular patterning, elongation and morphogenesis of the digestive tube itself, inductions and tissue interactions leading to the formation of the digestive glands, and the biochemical maturation of the secretory and absorptive epithelia associated with the digestive tract. Clinical Correlations 15.1 to 15.3, later in the chapter, discuss malformations associated with the digestive system.

Formation of the respiratory system begins with a very unimposing ventral outpocketing of the foregut. Soon, however, this outpocketing embarks on a unique course of development while still following some of the basic patterns of epithelial-mesenchymal interactions characteristic of other gut-associated glands. Initially, the digestive and respiratory systems form in a common body cavity, but functional considerations later necessitate the subdivision of this primitive body cavity into thoracic and abdominal components. Later in the chapter, Clinical Correlation 15.4 presents malformations associated with the respiratory system, and Clinical Correlation 15.5 discusses malformations related to other body cavities.

Digestive System

Chapter 6 describes the formation of the primitive endodermal digestive tube, which is bounded at its cephalic end by the **oropharyngeal membrane** and at its caudal end by the **cloacal plate** (see Fig. 6.20). Because of its intimate relationship with the yolk sac through the **yolk stalk**, the gut can be divided into a **foregut**, an open-bottomed **midgut**, and a **hindgut**.

Patterning the Gut

Initial patterning of the gut endoderm begins during the period of late gastrulation, as the sheet of newly formed endoderm begins to form a gut tube. After initial broad patterning into anterior and posterior domains by **nodal** and **fibroblast growth factor-4** (FGF-4), respectively, the overall organization of the gut gradually takes shape. Much of the patterning and of early morphogenesis of the gut occurs in response to the actions of several sets of molecular signals. As

development and early organogenesis proceed, the same signaling molecules are reused. Paradoxically, the same molecule may play opposite roles in the same area, but at different times (i.e., it may first act as a stimulator, and then, within hours or days, it may function as an inhibitor).

Patterning of the broad foregut area occurs through the inhibition of Wnt signals (**Fig. 15.1**). The foregut domain is then marked by the expression of the transcription factors, **Sox-2**, **Hhex**, and **Foxa-2**. In contrast, a mix of activity by Wnts, FGFs, and bone morphogenetic proteins (BMPs), along with retinoic acid, represses foregut identity and maintains the regional identity of the hindgut. This is marked by the expression of the transcription factor **Cdx-2** throughout the broad hindgut and the later expression of **Pdx-1** in the midgut as this region emerges as a separate entity. Cdx-2 acts upstream of a broad range of **Hox** activity (**Fig. 15.2**) that is expressed throughout the gut. The activity of specific signaling molecules is associated with important transition points along the gut. **FGF-4** is strongly expressed near the foregut-midgut boundary (around the duodenal-jejunal junction), and **FGF-10** is associated with the establishment of the cecum.

Largely through the action of Cdx-2, the orderly expression of homeobox-containing genes then takes over in the regional patterning of the digestive system (see Fig. 15.2).

Mice bearing mutant copies of some of these genes develop several of the common structural malformations of the digestive tract that occur in humans. More dramatically, mice deficient in retinoic acid, a broad early patterning molecule, fail to form lungs and show severe defects of other posterior foregut derivatives, such as the stomach, duodenum, and liver. Development of the gut tube proper involves continuous elongation, herniation past the body wall, and rotation and folding for efficient packing into the body cavity, as well as histogenesis and later functional maturation.

By the end of the first month, small endodermal diverticula, which represent primordia of the major digestive glands, can be identified (**Fig. 15.3**). (Development of the pharynx and its glandular derivatives is discussed in Chapter 14.) The digestive glands and respiratory structures grow in complex branching patterns, resembling fractals, as the result of continuous epithelial-mesenchymal interactions. These interactions also occur in the developing digestive tube itself, with specific regional mesenchymal influences determining the character of the epithelium lining that part of the digestive tract.

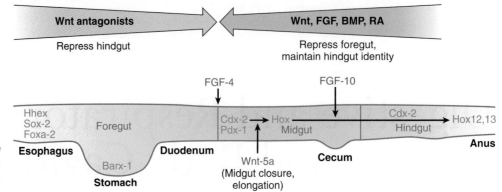

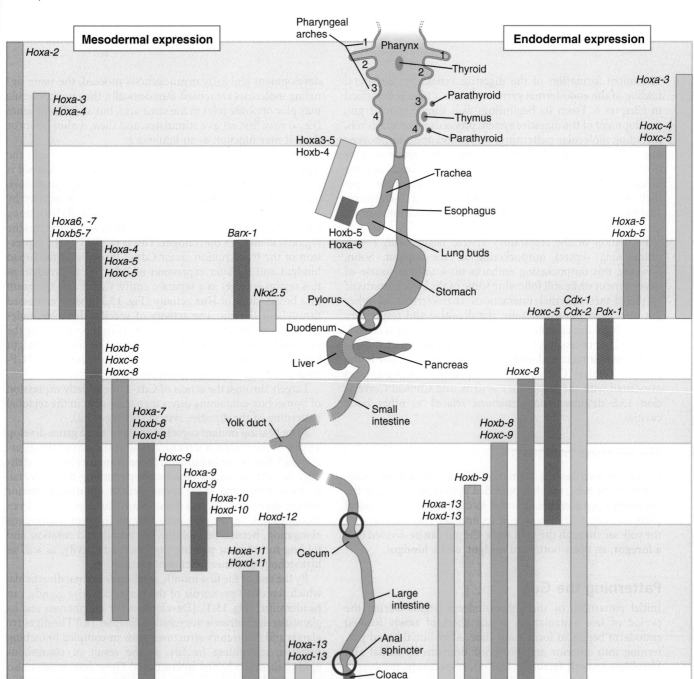

Fig. 15.1 **Early patterning of the gut.** See text for details. *Red letters,* signaling molecules; *blue letters,* transcription factors.

Fig. 15.2 *Hox* **gene expression along the developing digestive tract.** Expression along the gut endoderm *(right)* and in the gut-associated mesoderm *(left).* The *circles* represent areas where sphincters are located.

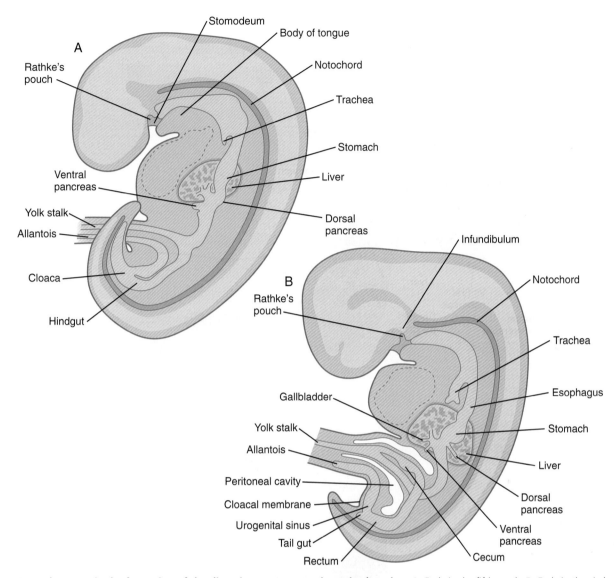

Fig. 15.3 **Early stages in the formation of the digestive tract as seen in sagittal section.** A, Early in the fifth week. B, Early in the sixth week.

Each of the glandular derivatives of the digestive tract, as well as the major regions along the gut, is the result of a specific response by a small population of founder cells for each organ to a set of environmental inductive signals. At a first level, certain regions of the gut must be prepared to be either responsive or refractive to these signals. For example, after the overall foregut is specified by the suppression of Wnt and FGF signals, transforming growth factor-β (TGF-β) signaling restricts the specification of foregut endoderm to allow the prehepatic and prepancreatic endoderm to remain receptive to inductive signals. By the same token, other influences on the dorsal side of the foregut repress the ability of these cells to become liver or pancreas.

During neurulation, as the head bends sharply to create the foregut, the ventral foregut endoderm is closely opposed to two mesodermal masses: the **cardiac mesoderm** and the primordium of the **septum transversum** (see **Fig. 15.4**). High levels of FGF, secreted by the cardiac mesoderm, and also retinoic acid induce the formation of the liver, lung bud, and thyroid (see Fig. 15.4). BMP-4 from the mesoderm of the septum transversum is also required for liver induction. By

contrast, endodermal movements carry the preventral pancreas cells far enough from the cardiac mesoderm to expose them to a low level of FGF, thus allowing the ventral pancreas to develop. For the dorsal pancreas to develop, locally produced sonic hedgehog (shh) must be inactivated by activin and FGF emanating from the notochord. In addition, retinoic acid from the somitic mesoderm is needed for induction of the dorsal pancreas. Meanwhile, in the hindgut, the actions of Wnt and other signaling molecules repress the expression of genes, such as **Hhex** and **Pdx1**, which are essential for formation of the liver and pancreas, respectively.

Induction of these organs is marked by the activation of transcription factors specific for the organ and stage of development of that organ. Some of these factors are schematically represented in Figure 15.4B.

Formation of the Esophagus

Just caudal to the most posterior pharyngeal pouches of a 4-week-old embryo, the pharynx becomes abruptly narrowed, and a small ventral outgrowth (lung bud) appears (see Fig.

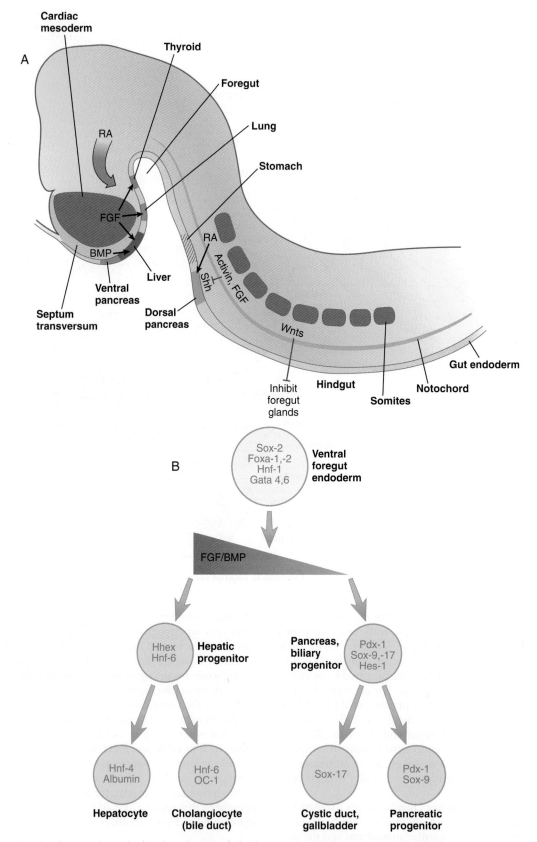

Fig. 15.4   A, Major signaling events involved in the induction of glands arising from gut endoderm. **B,** Some of the important transcription factors expressed at early stages of development of the liver and pancreas. The *red wedge* represents the gradient of fibroblast growth factor (FGF) and bone morphogenetic protein (BMP), which are required in high concentrations for formation of the liver and in low concentration for pancreatic development. RA, retinoic acid; shh, sonic hedgehog.

6.20). The region of foregut just caudal to the lung bud is the **esophagus**. This segment is initially very short, with the stomach seeming to reach almost to the pharynx. In the second month of development, during which the gut elongates considerably, the esophagus assumes nearly postnatal proportions in relation to the location of the stomach.

Although the esophagus grossly resembles a simple tube, it undergoes a series of striking differentiative changes at the tissue level. In its earliest stages, the endodermal lining epithelium of the esophagus is stratified columnar. By 8 weeks, the epithelium has partially occluded the lumen of the esophagus, and large vacuoles appear (**Fig. 15.5**). In succeeding weeks, the vacuoles coalesce, and the esophageal lumen recanalizes, but with multilayered ciliated epithelium. During the fourth month, this epithelium finally is replaced with the stratified squamous epithelium that characterizes the mature esophagus.

Deeper in the esophageal wall, layers of muscle also differentiate in response to inductive signals from the gut endoderm. Very early (5 weeks' gestation), the primordium of the inner circular muscular layer of esophagus is recognizable, and by 8 weeks, the outer longitudinal layer of muscle begins to take shape. The esophageal wall contains smooth and skeletal muscle. The smooth muscle cells differentiate from the local splanchnic mesoderm associated with the gut, and the skeletal musculature is derived from paraxial mesoderm. All esophageal musculature is innervated by the vagus nerve (cranial nerve X).

The cross-sectional structure of the esophagus, similar to that of the rest of the gut, is organized into discrete layers. The innermost layer (**mucosa**) consists of the epithelium, derived from endoderm, and an underlying layer of connective tissue, the **lamina propria** (see Fig. 15.5C and D). A thick layer of loose connective tissue (**submucosa**) separates the mucosa from the outer layers of muscle (usually smooth muscle, with the exception of the upper esophagus). This radial organization is regulated by the epithelial expression of **shh**, acting

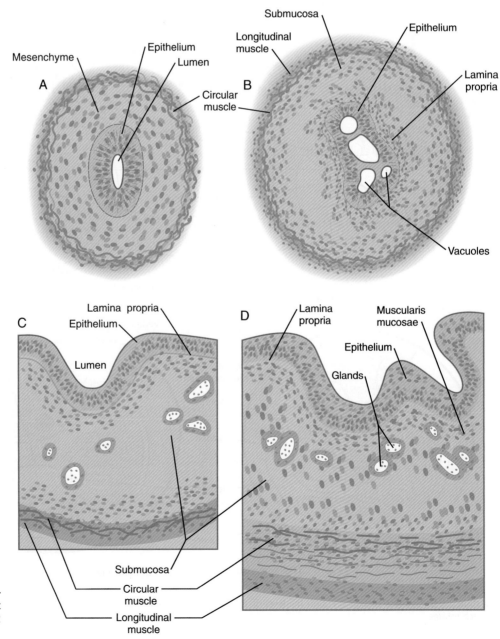

Fig. 15.5 **Stages in the histogenesis of the esophagus. A,** At 7 weeks. **B,** At 8 weeks. **C,** At 12 weeks. **D,** At 34 weeks.

through its receptor, **patched**, and **BMP-4**. Shh inhibits the formation of smooth muscle in the submucosal layer of the esophagus. Farther from the source of the endodermal shh, smooth muscle can differentiate in the outer wall of the intestine. How the developing smooth muscle layer of the mucosa (muscularis mucosae) escapes this inhibitory influence is also unclear. Intestinal mesenchyme can spontaneously differentiate into smooth muscle in the absence of an epithelium (which produces shh). Because in humans the muscularis mucosae differentiates considerably later than the outer muscular layers, it is possible that inhibitory levels of shh are reduced by that time.

Formation of the Stomach

Formation of the stomach within the foregut is first specified by the action of the transcription factors **Hoxa-5** and **Barx-1**, which inhibits the posteriorizing effects of Wnt signaling in the region of the future stomach. A second phase of specification follows, in which a descending posterior-to-anterior gradient of **FGF-10**, produced in the gastric mesoderm, begins the process of regional differentiation of the glandular character of the gastric epithelium.

Very early in the formation of the digestive tract, the **stomach** is recognizable as a dilated region with a shape remarkably similar to that of the adult stomach (see Fig. 15.3). The early stomach is suspended from the dorsal body wall by a portion of the dorsal mesentery called the **dorsal mesogastrium**. It is connected to the ventral body wall by a ventral mesentery that also encloses the developing liver (**Fig. 15.6**).

When the stomach first appears, its concave border faces ventrally, and its convex border faces dorsally. Two concomitant positional shifts bring the stomach to its adult configuration. The first is an approximately 90-degree rotation about its craniocaudal axis so that its originally dorsal convex border faces left, and its ventral concave border faces right. The other positional shift consists of a minor tipping of the caudal (pyloric) end of the stomach in a cranial direction so that the long axis of the stomach is positioned diagonally across the body (**Fig. 15.7**).

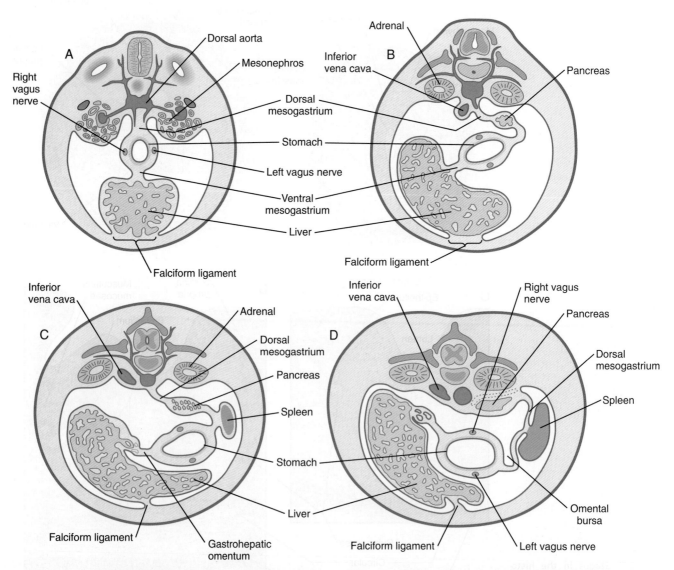

Fig. 15.6   A to **D,** Cross sections through the level of the developing stomach, showing changes in the relationships of the mesenteries as the stomach rotates.

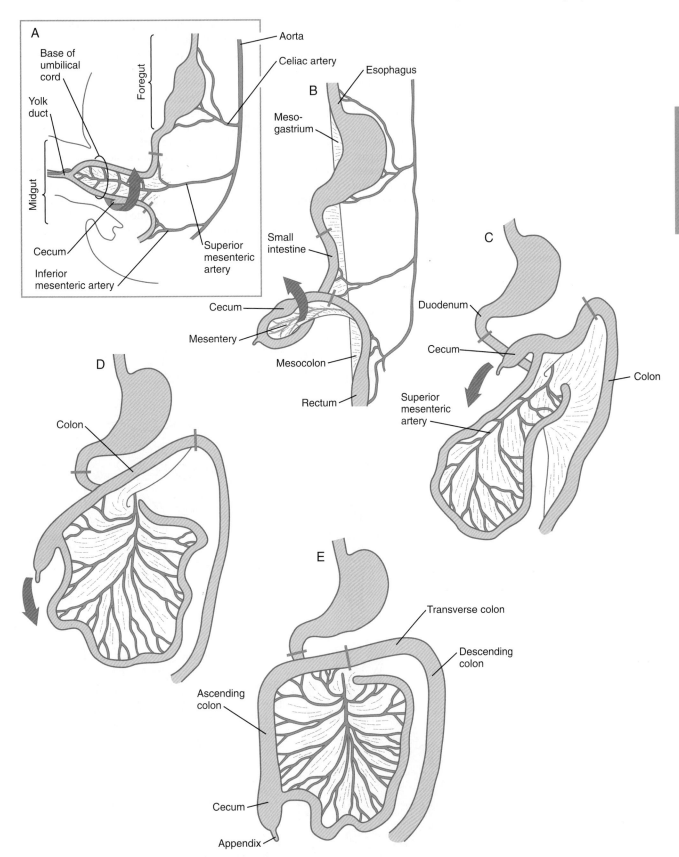

Fig. 15.7 **Stages in the development and rotation of the gut.** A, At 5 weeks. B, At 6 weeks. C, At 11 weeks. D, At 12 weeks. E, Fetal period. Areas between the *green lines* represent the midgut, which is supplied by the superior mesenteric artery.

During rotation of the stomach, the dorsal mesogastrium is carried with it, thus leading to the formation of a pouchlike structure called the **omental bursa** (bursa is a Latin word meaning "sac" or "pouch"). Both the spleen and the tail of the pancreas are embedded in the dorsal mesogastrium (see Fig. 15.6). Another viewpoint suggests that the right pneumatoenteric recess, a projection from the pleural cavity into the dorsal mesogastrium, persists as the omental bursa.

As the stomach rotates, the dorsal mesogastrium and the omental bursa that it encloses enlarge dramatically. Soon, part of the dorsal mesogastrium, which becomes the **greater omentum**, overhangs the transverse colon and portions of the small intestines as a large, double flap of fatty tissue (**Fig. 15.8**). The two sides of the greater omentum ultimately fuse and obliterate the omental bursa within the greater omentum. The rapidly enlarging liver occupies an increasingly large portion of the ventral mesentery.

At the histological level, the **gastric mucosa** begins to take shape late in the second month with the appearance of folds (**rugae**) and the first **gastric pits**. During the early fetal period, the individual cell types that characterize the gastric mucosa begin to differentiate. Biochemical and cytochemical studies have shown the gradual functional differentiation of specific cell types during the late fetal period. In most mammals, including humans, cells of the gastric mucosa begin to secrete hydrochloric acid shortly before birth.

The caudal end of the stomach is physiologically separated from the small intestine by the muscular **pyloric sphincter**. Formation of the pyloric sphincter is directed by the transcription factors **Sox-9** and **Nkx 2.5**, whose expression in the area of pyloric mesoderm is stimulated by BMP-4 signals. In addition, several *Hox* genes are needed to make each of the three major sphincters (pyloric, ileocecal, and anal) in the digestive tract.

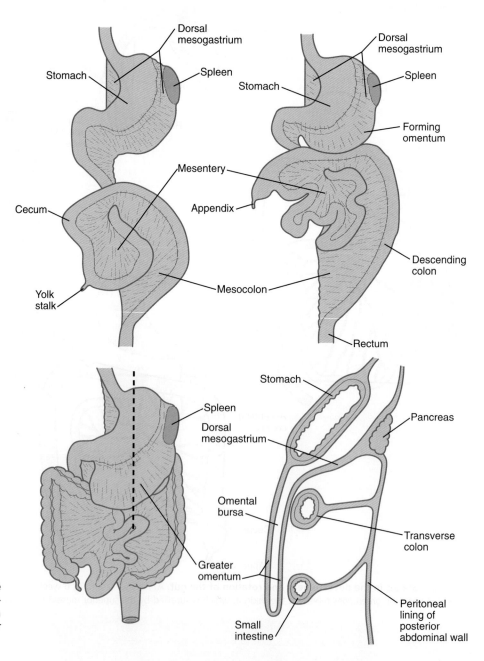

Fig. 15.8 **Stages in the rotation of the stomach and intestines and development of the greater omentum.** A section through the level of the *dashed line (lower left)* is shown *(lower right)*.

Clinical Correlation 15.1 presents malformations of the esophagus and stomach.

Development of the Spleen

Development of the spleen is not well understood. Initially, two bilaterally symmetrical organ fields are reduced by regression to one on the left. The spleen is first recognizable as a mesenchymal condensation in the dorsal mesogastrium at 4 weeks and is initially closely associated with the developing dorsal pancreas. Initiation of splenic development requires the cooperative action of a basic helix-loop-helix protein (**Pod-1**) and a homeobox-containing protein (**Bapx-1**), acting through another transcription factor, **Pbx-1**. Such a combination is emerging as a common theme in the initiation of development of several organs. These substances act on the downstream molecules **Nkx 2.5** and the oncogene *Hox-11* (T-cell leukemia homeobox-1) in early splenic development (**Fig. 15.9**).

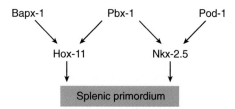

Fig. 15.9 **Molecular pathways in development of the spleen.**

CLINICAL CORRELATION 15.1
Malformations of the Esophagus and Stomach

Esophagus

The most common anomalies of the esophagus are associated with abnormalities of the developing respiratory tract (see p. 364). Other rare anomalies are **stenosis** and **atresia** of the esophagus. Stenosis is usually attributed to abnormal recanalization of the esophagus after epithelial occlusion of its lumen. Experimental evidence suggests that abnormal separation of the early notochord from the dorsal foregut endoderm is often associated with esophageal atresia, possibly through the incorporation of some of the dorsal foregut cells into an abnormal notochord. Atresia of the esophagus is most commonly associated with abnormal development of the respiratory tract. In both these conditions, impaired swallowing by the fetus can lead to an excessive accumulation of amniotic fluid (**polyhydramnios**). Just after birth, a newborn with these anomalies commonly has difficulty swallowing milk, and regurgitation and choking while drinking are indications for examination of the patency of the esophagus.

Stomach

Pyloric Stenosis

Pyloric stenosis, which seems to be more physiological than anatomical, consists of hypertrophy of the circular layer of smooth muscle that surrounds the pyloric (outlet) end of the stomach. The hypertrophy causes a narrowing (stenosis) of the pyloric opening and impedes the passage of food. Several hours after a meal, the infant violently vomits (**projectile vomiting**) the contents of the meal. The enlarged pyloric end of the stomach can often be palpated on physical examination. Although pyloric stenosis is commonly treated by a simple surgical incision through the layer of circular smooth muscle of the pylorus, the hypertrophy sometimes diminishes without treatment by several weeks after birth. The pathogenesis of this defect remains unknown, but it seems to have a genetic basis. Pyloric stenosis is more common in male than in female infants, and the incidence has been reported as 1 in 200 to 1 in 1000 infants.

Heterotopic Gastric Mucosa

Heterotopic gastric mucosa has been found in a variety of otherwise normal organs (**Fig. 15.10**). This condition is often clinically significant because if the heterotopic mucosa secretes hydrochloric acid, ulcers can form in unexpected locations. Many cases of heterotopic tissue within the gastrointestinal tract are now thought to be caused by the inappropriate expression of genes that are characteristic of other regions of the gut, but the question remains: What is the basis of the inappropriate gene expression, and why does it often occur within a very restricted area? Given the recognition of increasingly complex networks of genetic control in the gut, it is not difficult to imagine that on occasion normal developmental controls go awry, but understanding the genetic basis of specific instances of ectopic mucosa still remains elusive.

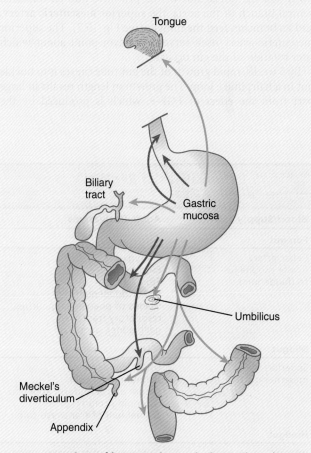

Fig. 15.10 **Locations of heterotopic gastric tissue.** The *red arrows* point to the most frequently occurring sites. The *pink arrows* indicate less common sites of occurrence. *(Based on Gray SW, Skandalakis JE: Embryology for surgeons, Philadelphia, 1972, Saunders.)*

The splenic primordium consists of a condensation of mesenchyme covered by the overlying mesothelium of the dorsal mesogastrium, both of which contribute to the stroma of the spleen. Its normal left-sided location is determined early by the mechanisms that dictate heart asymmetry; Nkx 2.5, an important determinant of early heart development, is also expressed in the splenic primordium. Hematopoietic cells move into the spleen in the late embryonic period, and from the third to fifth months, the spleen and the liver serve as major sites of hematopoiesis during the first trimester of pregnancy. Later, the splenic primordium becomes infiltrated by lymphoid cells, and by the fourth month, the complex vascular structure of the red pulp begins to take shape.

Formation of the Intestines

The intestines are formed from the posterior part of the foregut, the midgut, and the hindgut (**Table 15.1**). **Table 15.2** summarizes the chronology of major stages in the development of the digestive tract. Two points of reference are useful in understanding the gross transformation of the primitive gut tube from a relatively straight cylinder to the complex folded arrangement characteristic of the adult intestinal tract. The first is the yolk stalk, which extends from the floor of the midgut to the yolk sac. In the adult, the site of attachment of the yolk stalk is on the small intestine about 2 feet cranial to the junction between small and large intestine (**ileocecal junction**). On the dorsal side of the primitive gut, an unpaired ventral branch of the aorta, the **superior mesenteric artery**, and its branches feed the midgut (see Fig. 15.7). The superior mesenteric artery itself serves as a pivot point about which later rotation of the gut occurs.

By 5 weeks, rapid growth of the gut tube causes it to buckle out in a hairpinlike loop. The growth in length results in large part from the effect of **FGF-9**, which is produced by the

epithelium and stimulates proliferation of the fibroblasts in the intestinal walls. The major change that causes the intestines to assume their adult positions is a counterclockwise rotation of the caudal limb of the intestinal loop (with the yolk stalk attachment and superior mesenteric artery as reference points) around the cephalic limb from its ventral aspect. The main consequence of this rotation is to bring the future colon across the small intestine so that it can readily assume its C-shaped position along the ventral abdominal wall (see Fig. 15.7). Behind the colon, the small intestine undergoes great elongation and becomes packed in its characteristic position in the abdominal cavity.

The rotation and other positional changes of the gut occur partly because the length of the gut increases more than the length of the embryo. From almost the first stages, the volume of the expanded gut tract is greater than the body cavity can accommodate. Consequently, the developing intestines herniate into the body stalk (the umbilical cord after further development) (**Fig. 15.11**). Intestinal herniation begins by 6 or 7 weeks of embryogenesis. By 9 weeks, the abdominal cavity has enlarged sufficiently to accommodate the intestinal tract, and the herniated intestinal loops begin to move through the intestinal ring back into the abdominal cavity. Coils of small intestine return first. As they do, they force the distal part of

Table 15.1 Derivatives of Regions of the Primitive Gut

Blood Supply	Adult Derivatives
Foregut	
Celiac artery (lower esophagus to duodenum)	Pharynx Esophagus Stomach Upper duodenum Glands of pharyngeal pouches, respiratory tract, liver, gallbladder, pancreas
Midgut	
Superior mesenteric artery	Lower duodenum Jejunum and ileum Cecum and vermiform appendix Ascending colon Cranial half of transverse colon
Hindgut	
Inferior mesenteric artery	Caudal half of transverse colon Descending colon Rectum Superior part of anal canal

Table 15.2 Timelines in Development of the Digestive System

Normal Time (wk)	Developmental Events
3	Tubular gut beginning to form; early induction of major digestive glands
4	Most of gut tubular; primordia of liver, dorsal and ventral pancreas, and trachea visible; rupture of oropharyngeal membrane
5	Expansion and early rotation of stomach; intestinal loop beginning to form; cecum and bile duct evident
6	Rotation of stomach completed, prominent intestinal loop; appearance of allantois and appendix; urorectal septum beginning to subdivide cloaca into rectum and urogenital sinus
7	Herniation of intestinal loop; rapid growth of liver; fusion of dorsal and ventral pancreas; cloacal septation complete
8	Counterclockwise rotation of herniated intestinal loop; recanalization of intestine; early penetration of parasympathetic neuronal precursors from cranial neural crest into gut
9	Return of herniated gut into body cavity; differentiation of epithelial types in intestinal lining
11	Villi appearing in small intestine; differentiation of goblet cells
16	Villi lining entire intestine (including colon)
20	Peyer's patches seen in small intestine

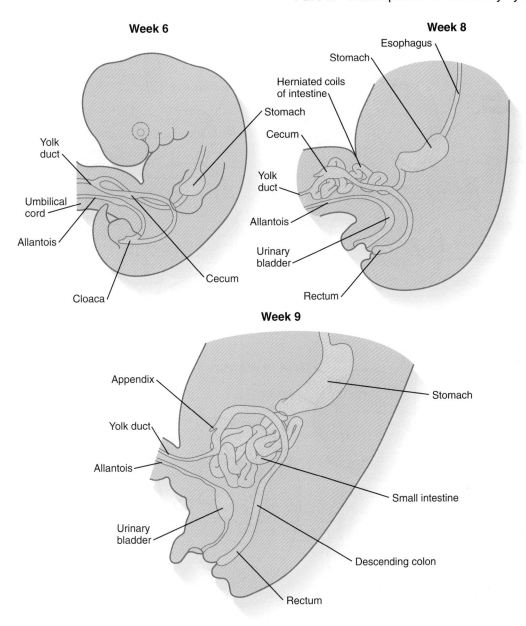

Week 6

Yolk
duct

Umbilical
cord

Allantois

Cloaca

Cecum

Stomach

Week 8

Esophagus

Stomach

Herniated coils
of intestine

Stomach

Cecum

Yolk
duct

Allantois

Urinary
bladder

Rectum

Week 9

Appendix

Yolk duct

Allantois

Urinary
bladder

Stomach

Small intestine

Descending colon

Rectum

Fig. 15.11　**Stages of herniation of the intestines into the body stalk and their return.**

the colon, which was never herniated, to the left side of the peritoneal cavity, thus establishing the definitive position of the descending colon. After the small intestine has assumed its intra-abdominal position, the herniated proximal part of the colon also returns, with its cecal end swinging to the right and downward (see Fig. 15.7).

During these coilings, herniations, and return movements, the intestines are suspended from the dorsal body wall by a mesentery (**Fig. 15.12**). Experimentation has shown that looping of the intestine is caused principally by tension-compression relationships between the intestine and its dorsal mesentery. When the intestine is separated from the mesentery, the normal looping does not occur. As the intestines assume their definitive positions within the body cavity, their mesenteries follow. Parts of the mesentery associated with the duodenum and colon (**mesoduodenum** and **mesocolon**) fuse with the peritoneal lining of the dorsal body wall.

Starting in the sixth week, the primordium of the **cecum** becomes apparent as a swelling in the caudal limb of the midgut (see Fig. 15.7). In succeeding weeks, the cecal enlargement becomes so prominent that the distal small intestine enters the colon at a right angle. The sphincterlike boundary at the cecum between the small and large intestines, similar to that in other regions of the gut, is regulated by a high concentration of Cdx-2 and a sequence of *Hox* gene expression. In mice, deletion of *Hoxd4, Hoxd8* to *Hoxd11,* and *Hoxd13* results in the absence of this region. When the overall pattern has been set by combinations of *Hox* genes, cecal development depends on an interaction between FGF-9 produced by the cecal epithelium and FGF-10 produced by the overlying mesoderm.

The tip of the cecum elongates, but its diameter does not increase in proportion to the rest of the cecum. This wormlike appendage is aptly called the **vermiform appendix.**

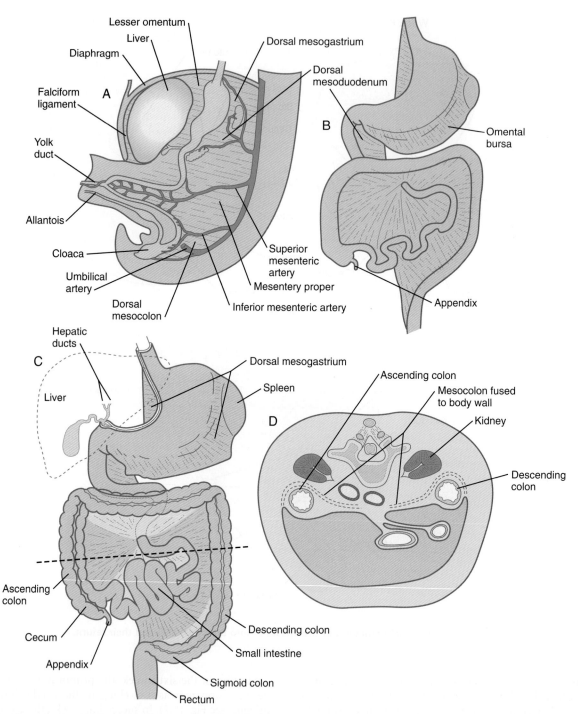

Fig. 15.12 **Stages in the development of the mesenteries. A,** At 5 weeks. **B,** In the third month. **C,** During the late fetal period. **D,** Cross section through the *dashed line* in **C.** In **C,** the *shaded areas* represent regions where the mesentery is fused to the dorsal body wall.

Partitioning of the Cloaca

In the early embryo, the caudal end of the hindgut terminates in the endodermally lined **cloaca**, which, in lower vertebrates, serves as a common termination for the digestive and urogenital systems. The cloaca also includes the base of the allantois, which later expands as a common **urogenital sinus** (see Chapter 16). A **cloacal (proctodeal) membrane** consisting of apposed layers of ectoderm and endoderm acts as a barrier between the cloaca and an ectodermal depression known as

the **proctodeum** (**Fig. 15.13**). A shelf of mesodermal tissue called the **urorectal septum** is situated between the hindgut and the base of the allantois. During weeks 6 and 7, the urorectal septum advances toward the cloacal membrane. At the same time, lateral mesodermal ridges extend into the cloaca.

The combined ingrowth of the lateral ridges and growth of the urorectal septum toward the cloacal membrane divide the cloaca into the **rectum** and **urogenital sinus** (see Fig. 15.13B). Double mutants of *Hoxa13* and *Hoxd13* result in the absence

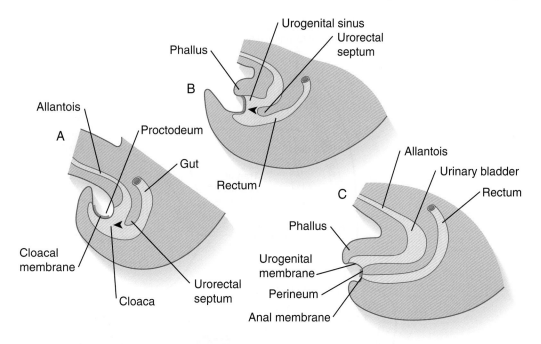

Fig. 15.13 **Stages in the subdivision of the common cloaca by the urorectal septum.** **A,** In the fifth week. **B,** In the sixth week. **C,** In the eighth week. The *arrowheads* indicate the direction of growth of the urorectal septum.

of cloacal partitioning, along with hypodevelopment of the phallus (genital tubercle). In addition, they lead to the absence of the smooth muscle component of the anal sphincter. According to classic embryology, the urorectal septum fuses with the cloacal membrane, thus dividing it into an **anal membrane** and a **urogenital membrane** before these membranes break down (see Fig. 15.13C). Other research suggests that the cloacal membrane undergoes apoptosis and breaks down without its fusion with the urorectal septum. The area where the urorectal septum and lateral mesodermal folds fuse with the cloacal membrane becomes the **perineal body**, which represents the partition between the digestive and urogenital systems.

The actual anal canal consists of a craniocaudal transition from columnar colonic (rectal) epithelium to a transitional region of cloacally derived endodermal epithelium leading into a zone of squamous epithelium that merges with the external perianal skin. These zones are surrounded by the smooth muscle internal anal sphincter.

Histogenesis of the Intestinal Tract

Shortly after its initial formation, the intestinal tract consists of a simple layer of columnar endodermal epithelium surrounded by a layer of splanchnopleural mesoderm. Three major phases are involved in the histogenesis of the intestinal epithelium: (1) an early phase of epithelial proliferation and morphogenesis, (2) an intermediate period of cellular differentiation in which the distinctive cell types characteristic of the intestinal epithelium appear, and (3) a final phase of biochemical and functional maturation of the different types of epithelial cells. The mesenchymal wall of the intestine also differentiates into several layers of highly innervated smooth muscle and connective tissue. An overall craniocaudal

gradient of differentiation is present within the developing intestine.

The endoderm of the early foregut is capable of producing cell types other than those of the gut tube itself, such as liver cells. A two-phase series of inhibitory influences from the gut mesoderm restricts the overlying endoderm to forming only the appropriate epithelial cell types through the activity of the transcription factors **Foxa-2** (formerly called hepatic nuclear factor-3) and **GATA-4**, which are essential for formation of anterior regions of endoderm.

Early in the second month, the epithelium of the small intestine begins a phase of rapid proliferation that causes the epithelium temporarily to occlude the lumen by 6 to 7 weeks' gestation. Within a couple of weeks, recanalization of the intestinal lumen has occurred. At about this time, small, crack-like secondary lumina appear beneath the surface of the multilayered epithelium, and aggregates of mesoderm push into the epithelium. A combination of coalescence of the secondary lumina with continued mesenchymal upgrowth beneath the epithelium results in the formation of numerous fingerlike **intestinal villi**, which greatly increase the absorptive surface of the intestinal surface. By this time, the epithelium has transformed from a stratified into a simple columnar type.

With the formation of villi, pitlike **intestinal crypts** form at the bases of the villi. Toward the bottom of the crypts are epithelial **intestinal stem cells**, which, in response to **Wnt** signaling, have a high rate of mitosis and serve as the source of epithelial cells for the entire intestinal surface (**Fig. 15.14**). Despite the presence of four to six stem cells per crypt, it has been shown that each crypt is monoclonal (i.e., all the existing cells are descendants of a single stem cell from earlier in development). Toward the top of a crypt, shh and Indian hedgehog (Ihh) signaling stimulates the activity of **BMP**. This BMP has two main functions. It counteracts the effects of Wnt and thus

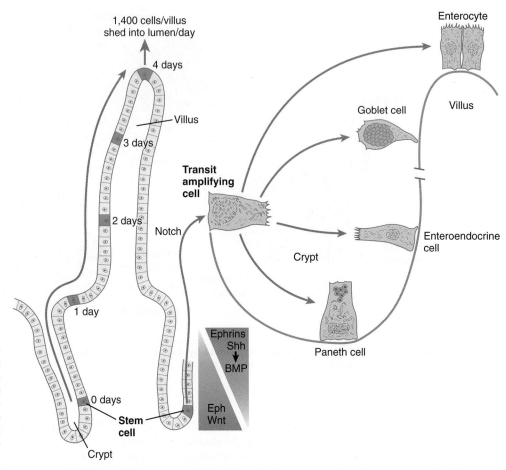

Fig. 15.14 Differentiation of intestinal epithelial cells from stem cells located in the crypts. The time scale shows the typical course of migration of daughter cells starting with their generation from the stem cell population to their being shed from the villus into the intestinal lumen.

keeps proliferation deep in the crypt, and it also facilitates cellular differentiation. Aided in part by an **ephrin-eph** gradient, progeny of the stem cells make their way up the wall of the crypt as multipotential **transit amplifying cells**, which, under the influence of the **Delta-Notch** system, begin to differentiate into the four main mature cell types of the intestinal epithelium (see Fig. 15.14). These cells then both differentiate and migrate toward the tip of the villus, until after about 4 days they die and are shed into the intestinal lumen as they are replaced below by new epithelial cells derived from the crypts. Human intestinal epithelial cells develop the intrinsic capacity for **apoptosis** by 18 to 20 weeks of gestation.

By the end of the second trimester of pregnancy, all cell types found in the adult intestinal lining have differentiated, but many of these cells do not possess adult functional patterns. Several specific biochemical patterns of differentiation are present by 12 weeks' gestation and mature during the fetal period. For example, **lactase**, an enzyme that breaks down the disaccharide **lactose** (milk sugar), is one of the digestive enzymes synthesized in the fetal period in anticipation of the early postnatal period during which the newborn subsists principally on the mother's milk. Further biochemical differentiation of the intestine occurs after birth, often in response to specific dietary patterns.

Histodifferentiation of the intestinal tract is not an isolated property of the individual tissue components of the intestinal wall. During the early embryonic period and sometimes into postnatal life, the epithelial and mesodermal components of the intestinal wall communicate by inductive interactions. In region-specific manners, these interactions involve hedgehog signaling (**shh** for the foregut and midgut and **Ihh** for the hindgut) from the endodermal epithelium. BMP signaling from the mesoderm is involved in positioning the crypts and villi in the small intestine and the glands of the colon. Interspecies recombination experiments show that the gut mesoderm exerts a regional influence on intestinal epithelial differentiation (e.g., whether the epithelium differentiates into a duodenal or colonic phenotype). When regional determination is set, however, the controls for biochemical differentiation of the epithelium are inherent. This pattern of inductive influence and the epithelial reaction are similar to those outlined earlier for dermal-epidermal interactions in the developing skin (see Chapter 9).

Final enzymatic differentiation of intestinal absorptive cells is strongly influenced by glucocorticoids, and the underlying mesoderm seems to mediate this hormonal effect. In a converse inductive influence, the intestinal endoderm, through the action of shh signaling, induces the differentiation of smooth muscle from mesenchymal cells in the wall of the intestine.

Although the intestine develops many functional capabilities during the fetal period, no major digestive function occurs until feeding begins after birth. The intestines of the fetus contain a greenish material called **meconium** (see Fig. 18.9), which is a mixture of lanugo hairs and vernix caseosa sloughed from the skin, desquamated cells from the gut, bile secretion, and other materials swallowed with the amniotic fluid.

Formation of Enteric Ganglia

As outlined in Chapter 12, the **enteric ganglia** of the gut are derived from neural crest. Pax-3–expressing cells from the vagal neural crest migrate into the foregut and spread in a wavelike fashion throughout the entire length of the gut. Slightly later, cells from the sacral crest enter the hindgut and intermingle with cells derived from the vagal neural crest. The migratory properties of vagal crest cells are much more pronounced than are those of cells from the sacral crest. Initially, the vagal crest cells migrate throughout the mesenchyme of the gut, but as the smooth muscle of the intestines begins to differentiate, the migrating vagal crest cells become

preferentially distributed between the smooth muscle and the serosa, where the myenteric plexuses form. They are absent from the connective tissue of the submucosa because of the inhibiting effects of **shh**, secreted by the epithelial cells. During migration through the gut, the population of neural crest cells undergoes a massive expansion until the number of enteric neurons ultimately exceeds the number of neurons present in the spinal cord. Glial cells also differentiate from neural crest precursors in the gut, but the environmental factors that contribute to the differentiation of neural crest cells in the gut wall remain poorly understood.

Clinical Correlation 15.2 presents malformations of the intestinal tract.

CLINICAL CORRELATION 15.2
Malformations of the Intestinal Tract

Duodenal Stenosis and Atresia

Duodenal stenosis and atresia typically result from absent or incomplete recanalization of the duodenal lumen after it is plugged by endothelium or from vascular obstruction during pregnancy. These malformations are rare.

Vitelline Duct Remnants

The most common family of anomalies of the intestinal tract is some form of persistence of the **vitelline (yolk) duct**. The most common member of this family is **Meckel's diverticulum**, which is present in 2% to 4% of the population. Typical Meckel's diverticulum is a blind pouch a few centimeters long located on the antimesenteric border of the ileum about 50 cm cranial from the

ileocecal junction (**Fig. 15.15A and E**). This structure represents the persistent proximal portion of the yolk stalk. Simple Meckel's diverticula are often asymptomatic, but they occasionally become inflamed or contain ectopic tissue (e.g., gastric, pancreatic, or even endometrial tissue), which can cause ulceration. It has been suggested that in the absence of intestinal mesodermal restriction (the wall of Meckel's diverticulum is derived from the yolk duct) of the endoderm lining Meckel's diverticulum, the endoderm retains the developmental capability to form various types of cellular phenotypes derived from endoderm.

In some cases, a ligament connects Meckel's diverticulum to the umbilicus (see Fig. 15.15B), or a simple vitelline ligament that may have an associated persisting **vitelline artery** can connect the

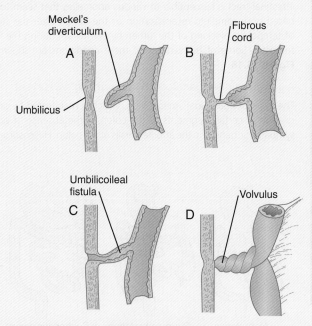

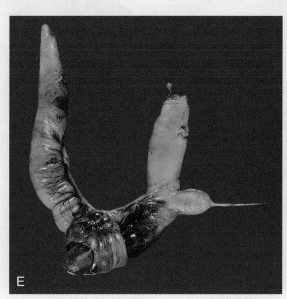

Fig. 15.15 Varieties of vitelline duct remnants. A, Meckel's diverticulum. **B,** Fibrous cord connecting a Meckel's diverticulum to the umbilicus. **C,** Umbilicoileal (vitelline) fistula. **D,** Volvulus caused by rotation of the intestine around a vitelline duct remnant. **E,** Meckel's diverticulum protruding at the right from a segment of ileum. The bowel below the diverticulum is reddish because of an associated intussusception just below the Meckel's diverticulum. *(Photo 2681 from the Arey-DaPeña Pediatric Pathology Photographic Collection, Human Development Anatomy Center, National Museum of Health and Medicine, Armed Forces Institute of Pathology, Washington, D.C.)*

Continued

CLINICAL CORRELATION 15.2
Malformations of the Intestinal Tract—cont'd

intestine to the umbilicus. Occasionally, the intestine rotates about such a ligament and causes a condition known as **volvulus** (see Fig. 15.15D). This disorder can lead to strangulation of the bowel.

A persistent vitelline duct can take the form of a **vitelline fistula** (see Fig. 15.15C), which constitutes a direct connection between the intestinal lumen and the outside of the body via the umbilicus. Rarely, a **vitelline duct cyst** is present along the length of a vitelline ligament.

Omphalocele

Omphalocele represents the failure of return of the intestinal loops into the body cavity during the tenth week. The primary defect in omphalocele is most likely a reduced prominence of the lateral body wall that does not provide sufficient space for the complete return of the intestines to the body cavity. After birth, the intestinal loops can be easily seen within an almost transparent sac consisting of amnion on the outside and peritoneal membrane on the inside (**Fig. 15.16**). The incidence of omphalocele is approximately 1 in 3500 births, but half of the infants with this condition are stillborn.

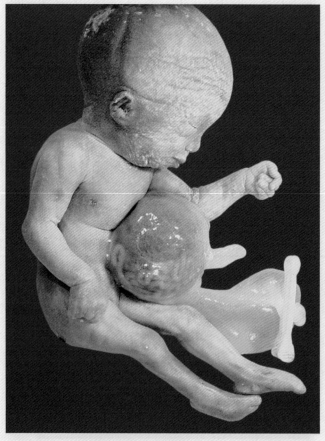

Fig. 15.16 **Omphalocele in a stillborn.** Loops of small intestine can be clearly seen through the nearly transparent amniotic membrane that covers the omphalocele. *(Courtesy of M. Barr, Ann Arbor, Mich.)*

Congenital Umbilical Hernia

In congenital umbilical hernia, which is especially common in premature infants, the intestines return normally into the body cavity, but the musculature (rectus abdominis) of the ventral abdominal wall fails to close the umbilical ring, thus allowing a varying amount of omentum or bowel to protrude through the umbilicus. In contrast to omphalocele, the protruding tissue in an umbilical hernia is covered by skin, rather than amniotic membrane.

Omphalocele and congenital umbilical hernia are associated with closure defects in the ventral abdominal wall. If these defects are large, they may be accompanied by massive protrusion of abdominal contents or with other closure defects, such as exstrophy of the bladder (see Chapter 16).

Abnormal Rotation of the Gut

Sometimes the intestines undergo no or abnormal rotation as they return to the abdominal cavity; this can result in a wide spectrum of anatomical anomalies (**Fig. 15.17**). In most cases, these anomalies are asymptomatic, but occasionally, they can lead to volvulus or another form of strangulation of the gut. The major rotation of the gut occurs after smooth muscle has formed in the intestinal walls. Mice mutant for both sonic hedgehog (shh) and Indian hedgehog have greatly reduced smooth muscle in the gut wall and frequently exhibit malrotation of the intestines. Research has placed much more emphasis on mechanical relationships between the dorsal mesentery and the intestine for patterns of gut rotation.

Intestinal Duplications, Diverticula, and Atresia

As with the esophagus and duodenum, the remainder of the intestinal tract is susceptible to various anomalies that seem to be based on incomplete recanalization of the lumen after the stage of temporary blocking of the lumen by epithelium during the first trimester. Some of the variants of these conditions are shown in **Figure 15.18**.

Aganglionic Megacolon (Hirschsprung's Disease)

The basis of aganglionic megacolon, which is manifested by great dilation of certain segments of the colon, is the absence of parasympathetic ganglia in the affected walls of the colon. Hirschsprung's

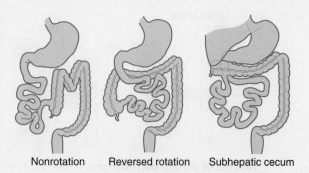

Nonrotation   Reversed rotation   Subhepatic cecum

Fig. 15.17 **Types of abnormal rotations of the gut.**

CLINICAL CORRELATION 15.2
Malformations of the Intestinal Tract—cont'd

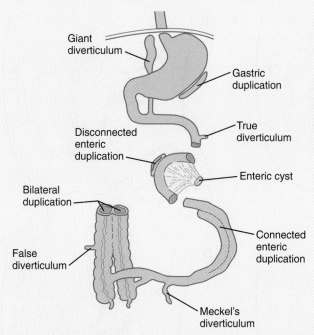

Fig. 15.18   **Types of diverticula and duplications that can occur in the digestive tract.** *(Based on Gray SW, Skandalakis JE:* Embryology for surgeons, *Philadelphia, 1972, Saunders.)*

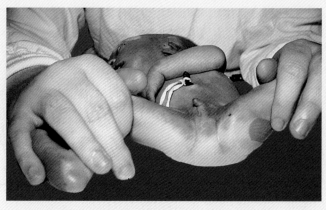

Fig. 15.19   **Anal atresia in a newborn.** No trace of an anal opening is seen. *(Courtesy of M. Barr, Ann Arbor, Mich.)*

disease seems to be truly multifactorial, with dominant and recessive mutations resulting in the condition. Many patients with Hirschsprung's disease do not express the **c-RET** oncogene. C-RET, along with a coreceptor, Gfra-1, is a receptor for glia-derived neurotrophic factor (GDNF). This gene is activated by the combination of Pax-3 with SOX-10, both of which are necessary for the formation of enteric ganglia. Similarly, mutations of *SOX10*, which probably interferes with Pax-3 function, can result in a combined Waardenburg-Hirschsprung syndrome. Mutants of *Ret, Gfra1*, and *GDNF* interfere with the migration of vagal neural crest cells into the gut.

Known mutations currently account for only about half of the cases of Hirschsprung's disease. Other mutations underlying aganglionic megacolon could involve defects in the migration or proliferation of neural crest precursor cells. Death of precursor cells before or after they reach their final destination in the hindgut could reduce the number of enteric ganglia. Alternatively, the local environment could prevent the successful migration of neural crest cells into the colon. Evidence from mutant mice developing aganglionic segments of the bowel strongly suggests that the environment of the gut wall inhibits the migration of neural crest cells into the affected segment of gut. This was shown by experiments in which crest cells from mutant mice were capable of colonizing normal gut, but normal crest cells could not migrate into gut segments of mutant mice. The accumulation of laminin in the gut wall

as the result of overproduction of **endothelin-3** serves as a stop signal for neural crest migration.

The distal colon is the most commonly affected region for aganglionosis, but in a few cases, aganglionic segments extend as far cranially as the ascending colon. Estimates of the frequency of megacolon vary widely, from 1 in 1000 to 1 in 30,000 births.

More recent experiments have shown that cells derived from the vagal neural crest exhibit much stronger migratory properties in the hindgut than do cells from the sacral crest. This finding has led to the suggestion that vagal crest cells be transplanted into aganglionic segments of the colon in an attempt to correct the deficit of enteric neurons in this area.

Imperforate Anus

Imperforate anus, which occurs in 1 of every 4000 to 5000 births, includes a spectrum of anal defects that can range from a simple membrane covering the anal opening (persistence of the cloacal membrane) to atresia of various lengths of the anal canal, rectum, or both. Grossly, all defects are characterized by the absence of an anal opening (**Fig. 15.19**). Deletions of *Hoxa13* and *Hoxd13* in mice result in defects in morphogenesis of the anal sphincter, and mutants of shh and its downstream molecules *Gli2* and *Gli3* cause the colon to end in a blind sac, with no anus forming. Any examination of a newborn must include a determination of the presence of an anal opening. The extent of the atretic segment is important when considering the surgical treatment of imperforate anus. Treatment of a persistent anal membrane can be trivial, whereas more extensive defects, especially defects involving the anal musculature, constitute very challenging surgical problems.

Hindgut Fistulas

In many cases, anal atresia is accompanied by a fistula linking the patent portion of the hindgut to another structure in the region of the original urogenital sinus region. Common types of fistulas connect the hindgut with the vagina, the urethra, or the bladder, and others may lead to the surface in the perineal area (**Fig. 15.20**).

Continued

CLINICAL CORRELATION 15.2
Malformations of the Intestinal Tract—cont'd

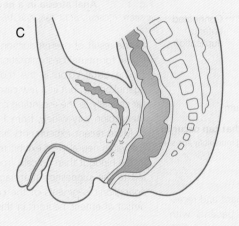
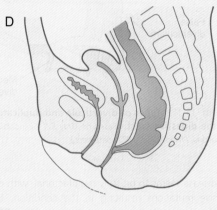
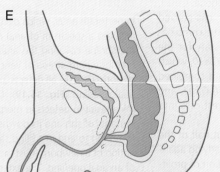
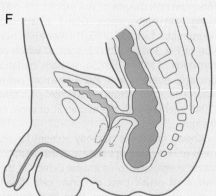

Fig. 15.20 **Varieties of hindgut fistulas and atresias.** A, Persistent anal membrane. B, Anal atresia. C, Anoperineal fistula. D, Rectovaginal fistula. E, Rectourethral fistula. F, Rectovesical fistula.

Glands of the Digestive System

The glands of the digestive system arise through inductive processes between the early gut epithelium and the surrounding mesenchyme. The various glandular epithelia have considerably different requirements in the types of mesenchyme that can support their development. In tissue recombination experiments, pancreatic epithelium undergoes typical development when it is juxtaposed with mesenchyme from almost any source. The development of salivary gland epithelium is supported by mesenchyme from lung or accessory sexual glands, but not by many other types of mesenchyme. Inductive support of hepatic (liver) epithelium follows a distinctive pattern. Normal epithelial development is supported by mesenchyme derived from lateral plate or intermediate mesoderm, but axial mesenchyme (either somitic or neural crest) fails to support hepatic differentiation. The inductive properties of certain glandular mesenchymes may be correlated with different modes of vascularization of these mesenchymes (see p. 414).

Formation of the Liver

After initial induction by the cardiac mesoderm and the septum transversum (see Fig. 15.4), the gut-derived hepatic endoderm thickens to form a pseudostratified epithelium (**Fig. 15.21**). The nuclear dynamics within the pseudostratified epithelium parallel those in the early neural tube (see Fig. 11.4). The nuclei undergo DNA synthesis (S phase of the cell cycle) in the basal position near the basal lamina surrounding the liver bud. Then the nuclei migrate to the apical (luminal) position, where they undergo mitotic division. The pathways and fates of the daughter cells have not yet been clearly determined. The transition to the pseudostratified stage requires the activity of the homeobox gene **Hhex**, without which the liver does not form (see Fig. 15.21).

Early in the third week, through the actions of Hhex and other transcription factors, cells within the hepatic epithelium lose their epithelial characteristics by the downregulation of **E-cadherin** and migrate through the underlying basal lamina, which has been degraded by **matrix metalloproteinases** (**MMPs**). These migrating cells make their way into the underlying mesenchyme of the septum transversum and form **hepatic cords**. Early in the formation of the liver, the future hepatic cells already express the *albumin* gene, one of the major characteristics of mature hepatocytes.

The original hepatic diverticulum branches into many hepatic cords, which are closely associated with splanchnic mesoderm of the septum transversum. The mesoderm supports continued growth and proliferation of the hepatic endoderm. This occurs partly through the actions of **hepatic growth factor** (**HGF**), which is bound by the receptor molecule, **c-met**, located on the surface of the endodermal hepatocytes. Experimental studies have shown that mesoderm from either the splanchnopleural or the somatopleural components of the lateral plate mesoderm can support further hepatic growth and differentiation, whereas paraxial mesoderm has only a limited capacity to support hepatic development.

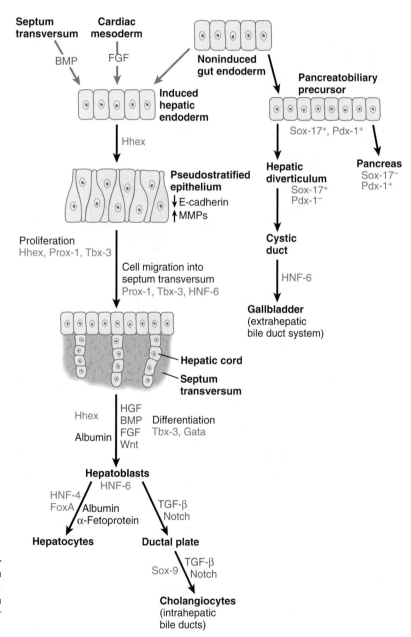

Fig. 15.21 **Major developmental events in the formation of the liver and bile duct system.** Growth factors are printed in *red,* and transcription factors in *blue.* BMP, bone morphogenetic protein; FGF, fibroblast growth factor; HGF, hepatic growth factor; HNF, hepatic nuclear factor; TGF, transforming growth factor.

The cells of the hepatic cords (**hepatoblasts**) are bipotential: they can form either hepatic parenchymal cells (**hepatocytes**) or intrahepatic bile duct cells (**cholangiocytes**). Guided by the transcription factors **hepatic nuclear factor-4** (**HNF-4**) and **FoxA**, some hepatoblasts differentiate into hepatocytes, which begin to express molecules (e.g., albumin and α-fetoprotein) characteristic of mature hepatic parenchymal cells. Other hepatoblasts, under the influence of **TGF-β** and **Notch**, gather as a single layer around branches of the portal vein as a **ductal plate** (**Fig. 15.22A**). Through mechanisms still not well understood, two ductlike structures (future bile ducts) begin to form around each vein. Initially, the cells comprising the walls of the ducts are hybrid in character (Fig. 15.22B). Cells closest to the vein have the characteristics of cholangiocytes, whereas those on the opposite face more closely resemble hepatocytes. Eventually, all the cells lining the bile duct become full-fledged cholangiocytes. These bile ducts branch to form networks leading toward the edges of the

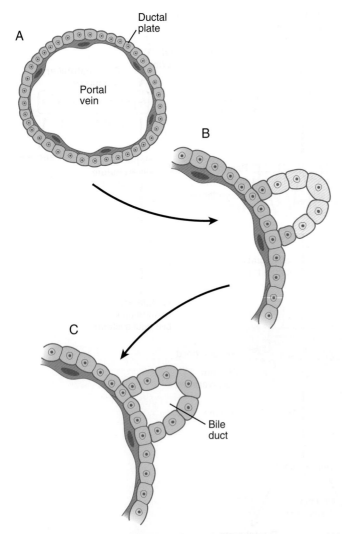

Fig. 15.22   Stages in the formation of intrahepatic bile ducts.
A, A layer of hepatic endodermal cells forms a ductal plate around a portal vein. **B,** In early stages in the formation of a bile duct, the cells at the outer perimeter have the characteristics of hepatocytes, whereas the inner cells exhibit a cholangiocyte phenotype. **C,** A differentiated bile duct, with all cells having the cholangiocyte phenotype. *(Based on Lemaigre FP: Prog Mol Biol Transl Sci 97:103-126, 2010.)*

hepatic lobes and constitute the intrahepatic component of the bile duct system.

The other major component of the bile duct system, consisting of the larger hepatic ducts, the cystic duct, the gallbladder, and the common bile duct, arises outside the main body of the liver and is called the **extrahepatic biliary tree**. Its precursor cells arise as a component of a common **pancreatobiliary precursor** (see Fig. 15.21), which is located caudal to the prehepatic endoderm. Expressing **Sox-17** and **Pdx-1**, these cells are bipotential, like those of the hepatic diverticulum. Some of these cells cease expressing Sox-17, but continue to express Pdx-1 and go on to form the ventral pancreas. Others, which lose Pdx-1 expression but continue to express Sox-17, become precursors of the extrahepatic biliary tree. They extend to form the **cystic duct**, and a dilatation from that foreshadows the further development of the **gallbladder** (**Fig. 15.23**). How the intrahepatic and extrahepatic bile ducts become connected remains unclear.

Within the substance of the liver, the hepatic cords form a series of loosely packed and highly irregular sheets that alternate with mesodermally lined **sinusoids**, through which blood circulates and exchanges nutrients with the hepatocytes. The sinusoids are the first vessels to form in the liver, and they arise from the mesenchyme of the proepicardium and the septum transversum. This same source also gives rise to **stellate cells**, cells residing in the space (of Dissé) between the hepatocytes and the sinusoidal endothelium. These cells store vitamin A and can also modulate the sinusoidal circulation, but when chronically activated postnatally, they form the basis for fibrosis of the liver.

The entire liver soon becomes too large to be contained in the septum transversum, and it protrudes into the ventral mesentery within the abdominal cavity. As it continues to expand, the rapidly growing liver remains covered by a glistening, translucent layer of mesenteric tissue that now serves as the connective tissue capsule of the liver. Between the liver and the ventral body wall is a thin, sickle-shaped piece of ventral mesentery: the **falciform ligament**. The ventral mesentery between the liver and the stomach is the **lesser omentum** (see Fig. 15.6).

Development of Hepatic Function

Development of the liver is not only a matter of increasing its mass and structural complexity. As the liver develops, its cells gradually acquire the capacity to perform the many biochemical functions characterizing the mature, functioning liver. One characteristic major function of the liver is to produce the plasma protein **serum albumin**. The mRNA for albumin has been detected in mammalian hepatocytes during the earliest stages of their ingrowth into the hepatic mesoderm, and it seems to depend on the earlier expression of the transcription factor HNF-3 (Foxa-3).

Major functions of the adult liver include the synthesis and storage of **glycogen**, which serves as a carbohydrate reserve. As the fetal period progresses, the liver actively stores glycogen. This function is strongly stimulated by adrenocortical hormones and is indirectly stimulated by the anterior pituitary. Similarly, the fetal period includes the functional development of the system of enzymes involved in the synthesis of urea from nitrogenous metabolites. By birth, these enzymes have attained full functional capacity.

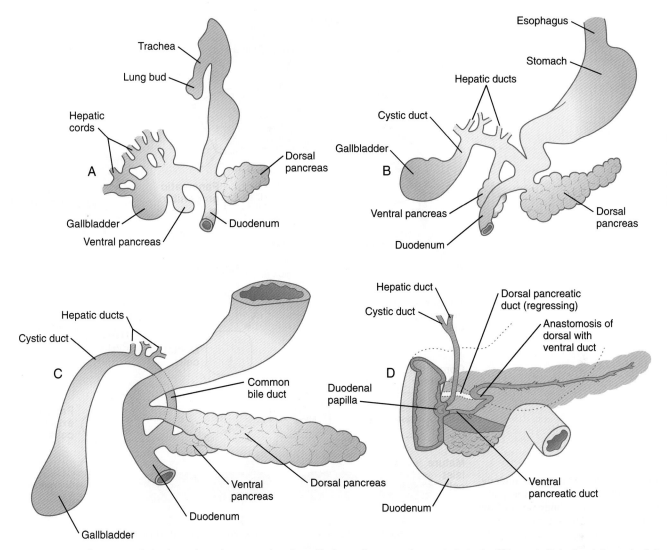

Fig. 15.23   Development of the hepatic and pancreatic primordia from the ventral aspect. **A,** In the fifth week. **B,** In the sixth week. **C,** In the seventh week. **D,** In the late fetus, showing fusion of the dorsal and ventral pancreatic ducts and regression of the distal portion of the dorsal duct.

A major function of the embryonic liver is the production of blood cells. After yolk sac hematopoiesis, the liver is one of the chief sites of intraembryonic blood formation (see Fig. 17.2). Hematopoietic cells, seeding the liver from origins in other sites, appear in small clusters among the hepatic parenchymal cells. As the liver matures, the intrahepatic environment no longer supports blood cell development, and hematopoiesis migrates to other sites in the fetus.

At approximately 12 weeks' gestation, the hepatocytes begin to produce **bile**, largely through the breakdown of hemoglobin. The bile drains down the newly formed bile duct system and is stored in the gallbladder. As bile is released into the intestines, it stains the other intestinal contents a dark green, which is one of the characteristics of meconium.

Formation of the Pancreas

The pancreas begins as separate dorsal and ventral primordia within the duodenal endoderm (see Fig. 6.20D). Early development of each of these two primordia is under different molecular controls. As discussed previously, ventral endoderm of the hepatic diverticulum is patterned to differentiate into ventral pancreatic tissue by a default mechanism in areas where liver induction does not occur. In primitive vertebrates, pancreatic function is distributed in cells within the foregut, rather than in a discrete gland, and some investigators have speculated that the ventral pancreatic bud represents the extension of this system, whereas the discrete dorsal pancreas is an evolutionary newcomer.

The dorsal pancreas is induced from the dorsal gut endoderm by **activin** and **FGF** signals emanating from the notochord, which early in development is directly opposed to the endoderm (**Fig. 15.24**). Shh activity in the dorsal endoderm must be repressed, or pancreatic differentiation does not occur. Initiation of development of the ventral pancreas is under a different set of developmental controls and heavily depends on the activity of the transcription factor **Ptf-1a**. During the earliest stages of dorsal pancreatic bud formation, the **pancreatic progenitor cells** express the transcription factors **Pdx-1** and **Hoxb-9**. If Pdx-1 expression is eliminated

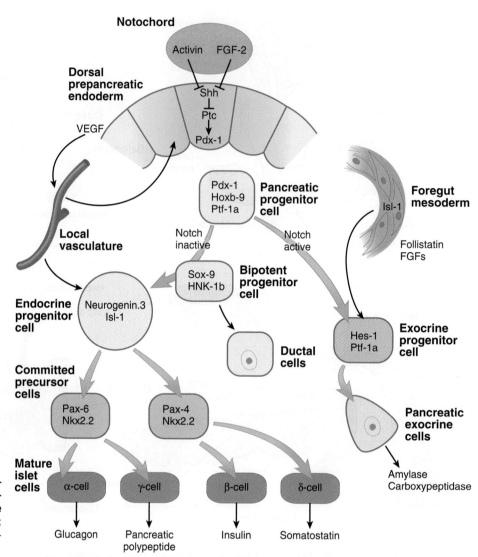

Fig. 15.24 **Molecular events underlying the differentiation of the endocrine and exocrine components of the pancreas.** FGF, fibroblast growth factor; Shh, sonic hedgehog; VEGF, vascular endothelial growth factor.

by targeted mutagenesis, development of the pancreatic bud ceases. From this point, different environmental signals and intracellular responses result in the differentiation of two lineages of cells: the **pancreatic exocrine cells** and the **pancreatic endocrine cells** (see Fig. 15.24).

During the phase of early growth, the dorsal pancreas becomes considerably larger than the ventral pancreas. At about the same time, the duodenum rotates to the right and forms a C-shaped loop that carries the ventral pancreas and common bile duct behind it and into the dorsal mesentery. The ventral pancreas soon makes contact and fuses with the dorsal pancreas.

Both the dorsal pancreas and the ventral pancreas possess a large duct. After fusion of the two pancreatic primordia, the main duct of the ventral pancreas makes an anastomotic connection with the duct of the dorsal pancreas. The portion of the dorsal pancreatic duct between the anastomotic connection and the duodenum normally regresses, leaving the main duct of the ventral pancreas (**duct of Wirsung**) the definitive outlet from the pancreas into the duodenum (see Fig. 15.23).

The pancreas is a dual organ with endocrine and exocrine functions. The exocrine portion consists of large numbers of **acini**, which are connected to a secretory duct system. The

endocrine component consists of roughly 1 million richly vascularized **islets of Langerhans**, which are scattered among the acini.

In certain pancreatic progenitor cells, the action of the signaling molecules, **follistatin**, and several **FGFs** emanating from the surrounding mesoderm, in combination with the activation of the **Notch** receptor system (see p. 69), results in the differentiation of many of the pancreatic cells along the exocrine pathway. These cells, which secrete digestive hormones, such as **amylase** and **carboxypeptidase**, are ultimately responsible for the gross morphogenesis of the pancreas. During outgrowth of the pancreatic primordia, the exocrine cells assume the form of sequentially budding cords. From these cellular cords, the acini and their ducts differentiate. Tissue recombination experiments, conducted in vitro and in vivo, have shown that the presence of mesenchyme is necessary for the formation of acini, but that ducts can form in the absence of mesenchyme if the endodermal precursor cells are exposed to a gel rich in basement membrane material.

Although the presence of mesenchyme is required for the differentiation of acini, the mesenchyme need not be of pancreatic origin. In vitro, pancreatic endoderm combined with salivary gland mesoderm differentiates even better than that exposed to pancreatic mesenchyme. This finding shows that in the case of the pancreas, the inductive influence of the mesenchyme is permissive, rather than instructive.

Differentiation of the acini is divided into three phases (**Fig. 15.25**). The first, called the **predifferentiated state**, occurs while the pancreatic primordia are first taking shape. A population of pancreatic progenitor cells that exhibits virtually undetectable levels of digestive enzyme activity is established. As the pancreatic buds begin to grow outward, the epithelium undergoes a transition into a second, **protodifferentiated state**. During this phase, the exocrine cells synthesize low levels of many hydrolytic enzymes that they will ultimately produce. After the main period of outgrowth, the pancreatic acinar cells pass through another transition before attaining a

third, **differentiated state**. By this time, these cells have acquired an elaborate protein-synthesizing apparatus, and the inactive forms of the polypeptide digestive enzymes are stored in the cytoplasm as **zymogen granules**. Glucocorticoid hormones from the fetal adrenal cortex stimulate increased production of certain digestive enzymes.

Development of the islets of Langerhans follows a different course from that of the acini. The islets of Langerhans are formed from groups of epithelial cells that break away from the acinar epithelial cells during the second (protodifferentiated) phase of acinar cell development. In a pathway not involving activation of the Notch system, but requiring signals from the local vasculature, a bipotential precursor cell has the potential to become either an endocrine cell or a ductal cell

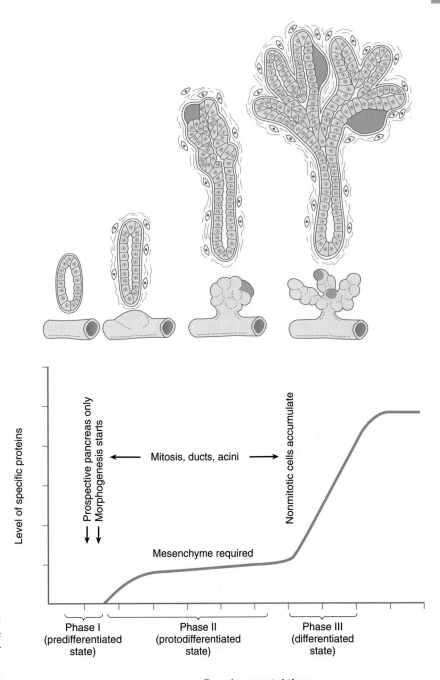

Fig. 15.25 **Stages of structural and functional differentiation of the pancreas.** The *green areas* represent primitive islets. *(Adapted from Pictet R, Rutter W: Handbook of physiology, section 7: Endocrinology, vol 1, Washington, D.C., 1972, American Physiological Society, pp 25-66.)*

(see Fig. 15.24). Those cells entering the endocrine lineage as **endocrine progenitor cells** express the transcription factors **neurogenin-3** and **Isl-1**. The endocrine progenitor cells give rise to two types of progenies (**committed precursor cells**), each of which is characterized by the expression of a different *Pax* gene. One type, which differentiates at 8 to 9 weeks, gives rise to α and γ **cells**, which produce **glucagon** and **pancreatic polypeptide**. The other type, which differentiates later, gives rise to β and δ **cells**, which produce **insulin** and **somatostatin**. During the second phase of pancreatic differentiation (protodifferentiated state), the levels of glucagon synthesis considerably exceed the levels of insulin. By the third phase of pancreatic development, secretory granules are evident in the cytoplasm of most islet cells. Insulin and glucagon are present in the fetal circulation by the end of the fifth month of gestation.

Clinical Correlation 15.3 presents anomalies of the liver and pancreas.

CLINICAL CORRELATION 15.3
Anomalies of the Liver and Pancreas

Many minor variations in the shape of the liver or bile ducts occur, but these variations normally have no functional significance. One of the most serious malformations involving the liver is **biliary atresia**. This malformation can involve any level ranging from the tiny bile canaliculi to the major bile-carrying ducts. **Alagille's syndrome**, which is characterized by biliary atresia and cardiac defects, is caused by mutations in **Jagged-1**, a ligand for the Notch receptor. Newborns with this condition typically develop severe **jaundice** shortly after birth. Some patients can be treated surgically; for others, a liver transplant is necessary.

Rarely, a ring of pancreatic tissue completely encircles the duodenum and forms an **annular pancreas (Fig. 15.26)**. This anomaly can sometimes cause obstruction of the duodenum after birth. The cause of annular pancreas is not established, but a commonly accepted explanation is that outgrowths from a bifid ventral pancreas may encircle the duodenum from both sides. Studies in mice suggest that locally reduced sonic hedgehog (shh) signaling may permit overgrowth of the ventral pancreatic bud tissue.

Heterotopic pancreatic tissue can occasionally be found along the digestive tract, and it occurs most frequently in the duodenum or mucosa of the stomach (**Fig. 15.27**). About 6% of Meckel's diverticula contain heterotopic pancreatic tissue.

A serious genetic condition affecting both the liver and the pancreas, as well as the kidneys and other organs, is **polycystic disease**. This condition results from malfunction of **primary cilia**, and it can be caused by the defective formation of any of several proteins (e.g., **Polycystin-1** or **-2** in autosomal primary polycystic kidney disease) involved in the function of the primary cilia. Mutations of the gene coding for the transmembrane protein **polyductin** in cholangiocytes are responsible for some cases of polycystic disease in the liver. Symptoms of polycystic liver disease include swelling of the liver and abdominal discomfort, especially after eating.

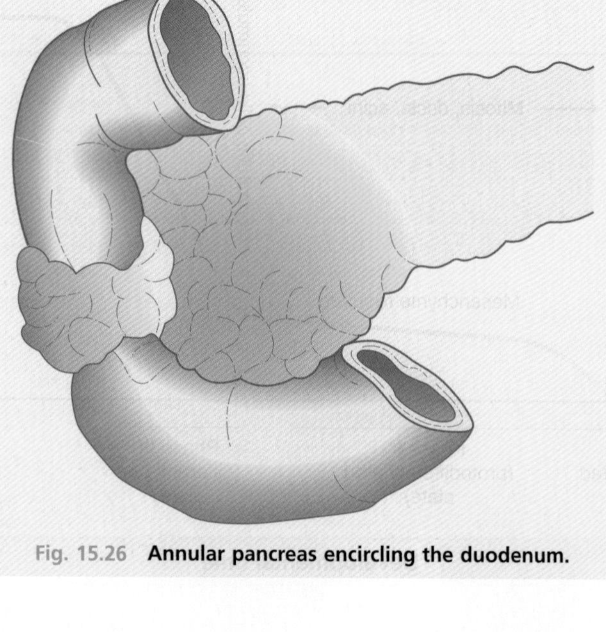

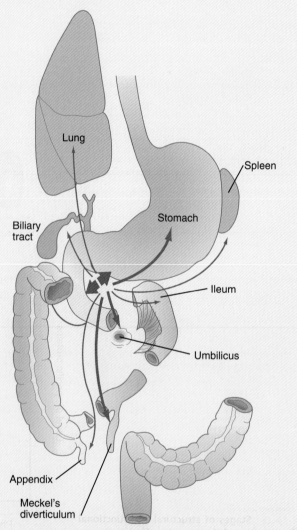

Fig. 15.27 **Most common locations in which heterotopic pancreatic tissue can be found.** The thickness of the *arrows* corresponds to the frequency of heterotopic tissue in that location. *(Based on Gray SW, Skandalakis JE: Embryology for surgeons, Philadelphia, 1972, Saunders.)*

Fig. 15.26 **Annular pancreas encircling the duodenum.**

Respiratory System

The location of the future respiratory system in the ventral foregut is indicated by a zone of expression of the transcription factor **Nkx 2.1**, which also marks the site of formation of the thyroid gland. The dorsal wall of the foregut in this area is characterized by **Sox-2** expression. Specification (induction) of the respiratory region is mediated by Wnt and FGF signals from the adjacent mesoderm. Late in the fourth week, paired **lung buds** begin to protrude from the posterior part of the respiratory endoderm (**Fig. 15.28A**). Dorsal to the lung buds, a pair of lateral mesodermal ridges begins to grow inward. Under the influence of **Wnt** signaling, these ridges fuse in a posterior-to-anterior direction. As they do so, they create a septum that separates the newly forming trachea from the esophagus. Through a series of interactions with the surrounding mesoderm, the early respiratory diverticulum (trachea plus lung buds) elongates, and the lung buds begin a set of 23 bifurcations that continue into postnatal life.

Formation of the Larynx

During weeks 4 and 5 of gestation, a rapid proliferation of the fourth and sixth pharyngeal arch mesenchyme around the site of origin of the respiratory bud converts the opening slit from the esophagus into a T-shaped **glottis** bounded by two lateral **arytenoid swellings** and a cranial **epiglottis**. The mesenchyme surrounding the laryngeal orifice ultimately differentiates into the **thyroid, cricoid**, and **arytenoid cartilages**, which form the skeletal supports of the **larynx**. Similar to the esophagus, the lumen of the larynx undergoes a temporary epithelial occlusion. In the process of recanalization during weeks 9 and 10, a pair of lateral folds and recesses forms the structural basis for the **vocal cords** and adjacent **laryngeal ventricles**. The somitomere-derived musculature of the larynx is innervated by branches of the vagus nerve (cranial nerve X), the musculature associated with the fourth arch is innervated by the **superior laryngeal nerve**, and the musculature of the sixth arch is innervated by the **recurrent laryngeal nerve**.

Formation of the Trachea and Bronchial Tree

At the initial appearance of the respiratory diverticulum, a pair of bronchial buds appears at its end (Fig. 15.28B). It now seems that the precursors of the trachea and lung buds are derived from separate sources of cells, and that the lung buds give rise to the bronchi and distal respiratory tree. The straight portion of the respiratory diverticulum is the primordium of the trachea. The bronchial buds, which ultimately become the primary bronchi, give rise to additional buds—three on the right and two on the left. These buds become the **secondary**, or **stem**, **bronchi**, and their numbers presage the formation of the three lobes of the right lung and the two lobes of the left (see Fig. 15.28). From this point, each secondary bronchial bud undergoes a long series of branchings throughout embryonic and fetal life.

Morphogenesis of the lung continues after birth. Stabilization of the morphological pattern of the lungs does not occur until about 8 years of age. An array of *Hox* genes (*Hoxa3* to *Hoxa5* and *Hoxb3* to *Hoxb6*) is expressed early in the developing respiratory tract. Combinatorial patterns of expression of *Hox* genes are involved in regional specification of the respiratory tract.

The mesoderm surrounding the endoderm controls the extent of branching within the respiratory tract. Numerous tissue recombination experiments have shown that the mesoderm surrounding the trachea inhibits branching, whereas the mesoderm surrounding the bronchial buds promotes branching. If tracheal endoderm is combined with bronchial mesoderm, abnormal budding is induced. Conversely, tracheal mesoderm placed around bronchial endoderm inhibits bronchial budding. Mesoderm of certain other organs, such as salivary glands, can promote budding of the bronchial endoderm, but a pattern of branching characteristic of the mesoderm is induced. A mesoderm capable of promoting or sustaining budding must maintain a high rate of proliferation of the epithelial cells. Generally, the pattern of the epithelial organ is largely determined by the mesoderm. Structural and functional differentiation of the epithelium is a specific property of the epithelial cells, but the epithelial phenotype corresponds to the region dictated by the mesoderm.

The basic principles underlying pulmonary branching are similar to those operating in the development of the salivary glands and pancreas. At points of branching, epithelial cell proliferation is reduced, and the deposition of types I, III, and IV collagen, fibronectin, and proteoglycans stabilizes the morphology of the branching point and more proximal ductal regions. Heightened epithelial cell proliferation characterizes the rapidly expanding portions of the epithelial buds (**Fig. 15.29**).

The activities of many molecules contribute to lung morphogenesis. More than 50 genes are involved in morphogenesis of the tracheal system in *Drosophila*, which shows remarkable parallels to the mammalian respiratory system. A prime mover in generating branching is **FGF-10**, which, in response to the action of **retinoic acid*** and **Tbx-4** and **Tbx-5**, is produced by the mesenchyme off the tip of a growing bud in the respiratory system. In FGF-10 knockout mice, budding of the developing lungs fails to occur. FGF-10 acts as a signaling center, by stimulating cell proliferation in the epithelium at the tip of the bud and causing the epithelium to grow out toward the source of the FGF-10 (see Fig. 15.29A). Apical epithelial proliferation is also promoted by the expression of the transcription factor Nkx 2.1 in these cells.

Branching is initiated with the stimulation of secretion of BMP-4 in the apical epithelial cells; this inhibits their proliferation. Simultaneously, shh, which is also produced by the epithelium, stimulates proliferation of the mesenchymal cells off the tip and inhibits the formation of FGF-10 (see Fig. 15.29B). These mesenchymal cells begin to produce **TGF-β1**, which, in addition to inhibiting FGF-10 production along with shh, promotes the synthesis of extracellular matrix molecules just distal to the apical epithelial cells. These molecules, including **fibronectin** and **collagen types I, III,** and **IV**, stabilize the formerly growing epithelial tip.

While the cell proliferation in the epithelial tip is reduced and the cells become bound by newly secreted extracellular

---

*Retinoic acid has more recently been shown to stimulate the regeneration of alveoli in damaged adult lungs.

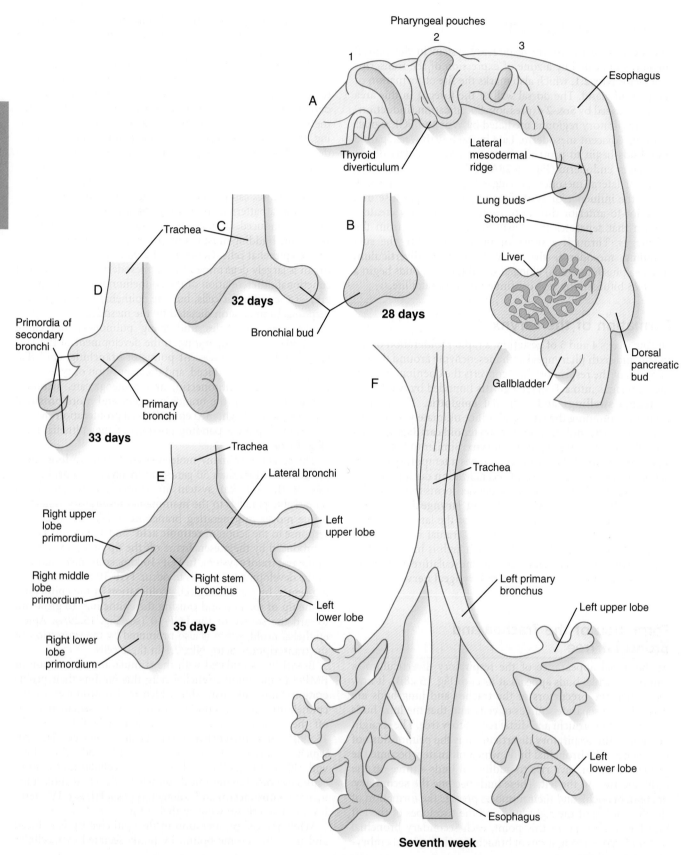

Fig. 15.28 **Development of the major branching patterns of the lungs.** **A,** Lateral view of the pharynx, showing the respiratory diverticulum in a 4-week-old embryo. **B,** At 4 weeks. **C,** At 32 days. **D,** At 33 days. **E,** At the end of the fifth week. **F,** Early in the seventh week.

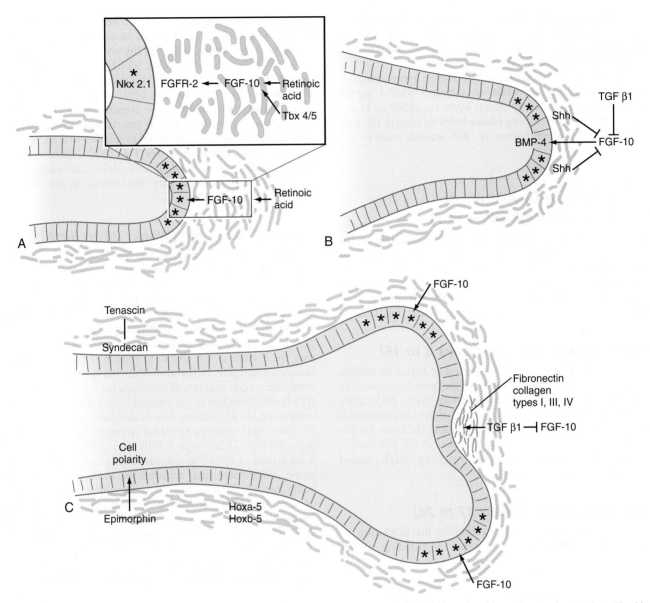

Fig. 15.29 Molecular aspects of outgrowth and branching of the respiratory tree. A, The tip of an elongating respiratory duct. Fibroblast growth factor-10 (FGF-10) secretion in the mesenchyme stimulates the growth of the tip of the epithelial duct toward it. **B,** The prelude to branching. Inhibition of FGF-10 signaling at the tip of the duct leads to stabilization of that area. **C,** Cleft formation. Extracellular matrix molecules are deposited in the newly forming cleft, and two new centers of outgrowth, stimulated by FGF-10 signaling, indicate the start of branching. *Asterisk* indicates dividing cells. BMP, bone morphogenetic protein; FGFR, fibroblast growth factor receptor; Shh, sonic hedgehog; TGF, transforming growth factor.

matrix molecules, FGF-10 is secreted by the mesenchyme lateral to the old apex, where the concentrations of shh and TGF-β1 are reduced to less than the inhibitory level (see Fig. 15.29C). This activity sets up two new signaling centers on either side of the original one, and the cycle of apical epithelial proliferation begins anew. As the new apical growth centers mature, FGF-10 signaling is again inhibited, and each of the two existing tips begins its own branching cycle. The concurrent presence of the epithelial cell–associated proteoglycan **syndecan** is important for maintaining the stability of epithelial sheets along the ducts. Interacting with the extracellular matrix protein **tenascin**, syndecan is found along

already formed ducts, but not in areas where branching is occurring in terminal saccular regions of the developing airway (see Fig. 15.29).

As with branching morphogenesis, the formation and maintenance of epithelially lined ducts involve special sets of molecular components. Hoxb-5 is expressed during the early development of smaller bronchioles (e.g., terminal bronchioles), but not in the components of the lung that are involved in actual respiratory exchange (i.e., respiratory bronchioles, alveolar ducts, alveoli). The protein **epimorphin** is important in the later formation of epithelial tubes. Epimorphin is located in the mesenchyme and seems to provide a signal that allows overlying epithelial cells to establish proper polarity or cell arrangements. In the embryonic lung, the developing epithelial ducts become disorganized and do not form lumina if epimorphin is blocked by specific antibodies.

Smooth muscle formation in the mesenchyme alongside the respiratory tract depends on shh and BMP-4 signals coming from the distal epithelial buds. In addition, FGF-9 secreted by the surrounding pleura helps to control the proliferation and differentiation of the smooth muscle cell precursors.

Stages in Lung Development
Embryonic Stage (Weeks 4 to 7)

The embryonic stage includes the initial formation of the respiratory diverticulum up to the formation of all major bronchopulmonary segments. During this period, the developing lungs grow into and begin to fill the bilateral **pleural cavities**. These structures represent the major components of the thoracic body cavity above the pericardium (**Fig. 15.30**).

Pseudoglandular Stage (Weeks 8 to 16)

The pseudoglandular stage is the period of major formation and growth of the duct systems within the bronchopulmonary segments before their terminal portions form respiratory components. The histological structure of the lung resembles that of a gland (**Fig. 15.31**), thus providing the basis for the designation of this stage. During this period, the pulmonary arterial system begins to form. The elongating vessels parallel the major developing ducts.

Canalicular Stage (Weeks 17 to 26)

The canalicular stage is characterized by the formation of **respiratory bronchioles** as the result of budding of the

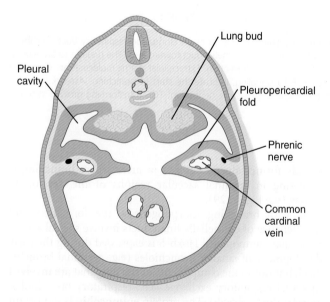

Fig. 15.30 **Cross section through the thorax showing the lung buds growing into the pleural cavities.** The pleuropericardial folds separate the future pleural from the pericardial cavities.

terminal components of the system of bronchioles that formed during the pseudoglandular stage. An array of many different cell types forms along the developing respiratory tree. A gradient of BMP-4 and Wnt signaling, which is highest at the distal tips of the branches, prevents the distal cells from forming phenotypes more characteristic of the larger branches of the bronchial tree. The other major events during this stage are the intense ingrowth of blood vessels into the developing lungs and the close association of capillaries with the walls of the respiratory bronchioles (see Fig. 15.31). Occasionally, a fetus born toward the end of this period can survive with intensive care, but respiratory immaturity is the principal reason for poor viability.

Terminal Sac Stage (Weeks 26 to Birth)

During the terminal sac stage, the terminal air sacs (**alveoli**) bud off the respiratory bronchioles that largely formed during the canalicular stage. The epithelium lining the alveoli differentiates into two types of cells: **type I alveolar cells** (pneumocytes), across which gas exchange occurs after birth; and **type II alveolar (secretory epithelial) cells**. Type II alveolar cells form **pulmonary surfactant**, the material that spreads over the surface of the alveoli to reduce surface tension and facilitate expansion of the alveoli during breathing. Research involving specific markers of the epithelial cells has shown that type II cells form first in the alveolar lining. After proliferation, some type II cells flatten, lose their characteristic secretory function, and undergo terminal differentiation into type I pneumocytes. Other type I cells may differentiate directly from a pool of epithelial precursor cells in the early alveolar lining. With increasing amounts of pulmonary surfactant being formed, the fetus has a correspondingly greater chance of survival if born prematurely. In the fetus, the respiratory passageways in the lungs are filled with fluid (see Chapter 18). During the last 4 weeks of pregnancy, greatly increased formation of alveoli results in an exponential increase in the respiratory surface area of the lung. These weeks are sometimes referred to as the **alveolar period of lung development**.

Postnatal Stage

At birth, the mammalian lung is far from mature. An estimated 90% or more of the roughly 300 million alveoli found in the mature human lung are formed after birth. The major mechanism for this increase is the formation of secondary connective tissue septa that divide existing alveolar sacs. When they first appear, the secondary septa are relatively thick. In time, they transform into thinner mature septa capable of full respiratory exchange function.

Clinical Correlation 15.4 presents malformations of the respiratory system.

Body Cavities

Formation of the Common Coelom and Mesentery

As the lateral plate mesoderm of the early embryo splits and then folds laterally, the space between the somatic and splanchnic layers of mesoderm becomes the common

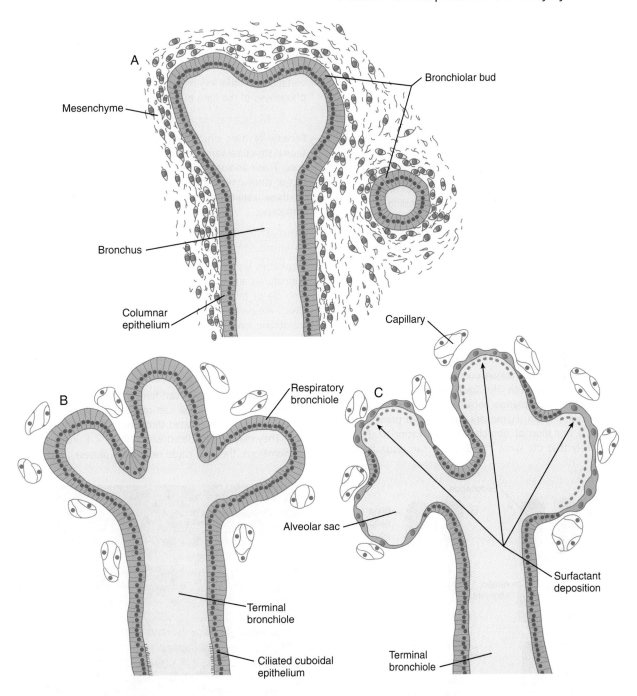

Fig. 15.31 Stages in histogenesis of the lungs. A, Pseudoglandular phase (up to 17 weeks). **B,** Canalicular phase (17 to 26 weeks). **C,** Terminal sac phase (26 weeks to birth).

intraembryonic coelom (Fig. 15.34). The same folding process that results in the completion of the ventral body wall and the separation of the intraembryonic from the extraembryonic coelom also brings the two layers of splanchnic mesoderm around the newly formed gut as the **primary (common) mesentery.** The primary mesentery suspends the gut from the dorsal body wall as the **dorsal mesentery** and attaches it to the ventral body wall as the **ventral mesentery.** This placement effectively divides the coelom into right

and left components. Soon, however, most of the ventral mesentery breaks down and causes a confluence of the right and left halves of the coelom. In the region of the developing stomach and liver, the ventral mesentery persists, thus forming the **ventral mesogastrium** and the falciform ligament of the liver (see Fig. 15.6). Further cranially, the tubular primordium of the heart is similarly supported by a **dorsal mesocardium** and briefly by a **ventral mesocardium,** which soon breaks down.

CLINICAL CORRELATION 15.4
Malformations of the Respiratory System

Tracheoesophageal Fistulas

The most common family of malformations of the respiratory tract is related to abnormal separation of the tracheal bud from the esophagus during early development of the respiratory system. Many common anatomical varieties of tracheoesophageal fistulas exist (**Fig. 15.32**), but virtually all involve the stenosis or atresia of a segment of trachea or esophagus and an abnormal connection between them. These are manifested early after birth by the newborn's choking or regurgitation of milk when feeding.

Expression of certain genes is important in the normal formation of a mesenchymal partition between the esophagus and the developing trachea. **Nkx 2.1** and **bone morphogenetic protein-4** (**BMP-4**) are expressed in the ventral foregut mesoderm in the area where the trachea forms. Mutants of these genes are characterized by a high incidence of tracheoesophageal fistulas. The loss of Wnt signaling, leading to the downregulation of Nkx 2.1 ventrally, and reduced Sox-2 activity in the dorsal foregut have both been linked to tracheoesophageal fistulas.

Tracheal or Pulmonary Agenesis

Tracheal agenesis and pulmonary agenesis are rare malformations that are incompatible with life. Tracheal agenesis apparently is caused by defective septation between the esophagus and the respiratory diverticulum. Pulmonary agenesis is a primary consequence of a mutation of fibroblast growth factor-10 (FGF-10), but it is likely that the same malformation can result from null mutations of other key molecules involved in early branching morphogenesis of the lung buds.

Gross Malformations of the Lungs

Because of their structural complexity, the lungs are subject to several structural variations or malformations (e.g., abnormal lobation). These anomalies are usually asymptomatic, but they can be foci of chronic respiratory infection. Recognition of the possibility of these variations from normal is important for pulmonary surgeons.

Respiratory Distress Syndrome (Hyaline Membrane Disease)

Respiratory distress syndrome is often manifested in infants born prematurely and is characterized by labored breathing. In infants who die of this condition, the lungs are underinflated, and the alveoli are partially filled with a proteinaceous fluid that forms a membrane over the respiratory surfaces (**Fig. 15.33**). This syndrome is related to insufficiencies in the formation of surfactant by the type II alveolar cells.

Congenital Cysts in the Lung

Abnormal cystic structures can form in the lung or other parts of the respiratory tract. These can range from large single cysts to numerous small cysts located throughout the parenchyma of the lung. They may be associated with polycystic kidneys. If the cysts are numerous, they can cause respiratory distress.

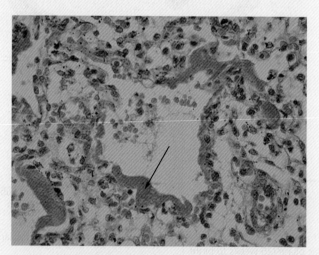

Fig. 15.33 **Photomicrograph from the lung of a newborn who died of hyaline membrane disease.** The *arrow* points to the "membrane" that interferes with gas exchange. *(Slide 427 from the Arey-DaPeña Pediatric Pathology Photographic Collection, Human Developmental Anatomy Center, National Museum of Health and Medicine, Armed Forces Institute of Pathology, Washington, D.C.)*

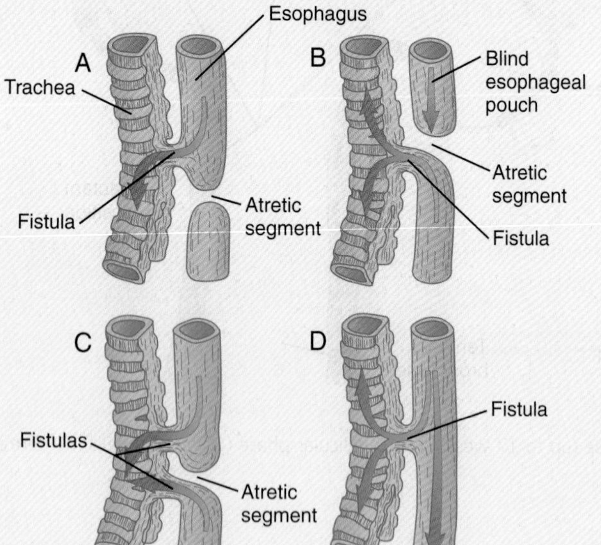

Fig. 15.32 **Varieties of tracheoesophageal fistulas.** A, Fistula above the atretic esophageal segment. B, Fistula below the atretic esophageal segment. C, Fistulas above and below the atretic esophageal segment. D, Fistula between the patent esophagus and the trachea.

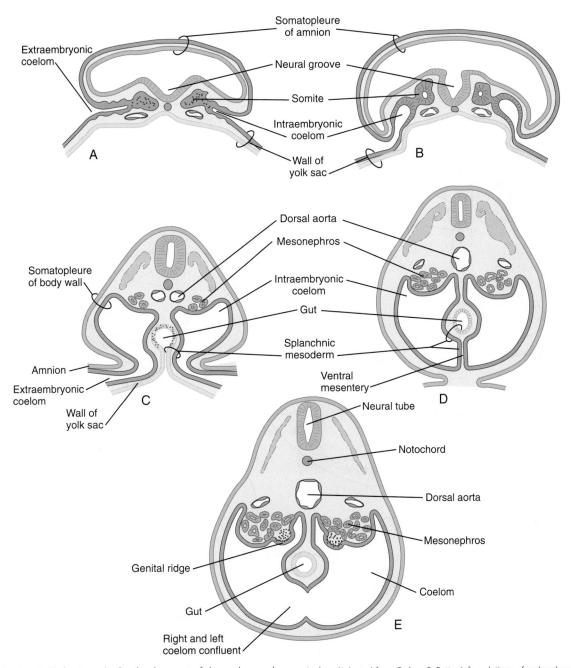

Fig. 15.34 A to E, Early stages in the development of the coelom and mesenteries. *(Adapted from Carlson B: Patten's foundations of embryology, ed 6, New York, 1996, McGraw-Hill.)*

Formation of the Septum Transversum and Pleural Canals

A major factor in division of the common coelom into thoracic and abdominal components is the **septum transversum**. This septum grows from the ventral body wall as a semicircular shelf, which separates the heart from the developing liver (**Fig. 15.35**). During its early development, a major portion of the liver is embedded in the septum transversum. Ultimately, the septum transversum constitutes a significant component of the diaphragm (see p. 367).

The expanding septum transversum serves as a partial partition between the pericardial and the peritoneal portions of the coelom. By the time the expanding edge of the septum transversum reaches the floor of the foregut, it has almost cut the common coelom into two parts. Two short channels located on either side of the foregut connect the two major parts (**Fig. 15.36**). Initially known as the **pleural (pericardioperitoneal) canals**, these channels represent the spaces into which the developing lungs grow. The pleural canals enlarge greatly as the lungs increase in size and ultimately form the **pleural cavities**.

The pleural canals are partially delimited by two paired folds of tissue: the pleuropericardial and pleuroperitoneal folds. The **pleuropericardial folds** (see Fig. 15.30) are ridges of tissue associated with the common cardinal veins, which

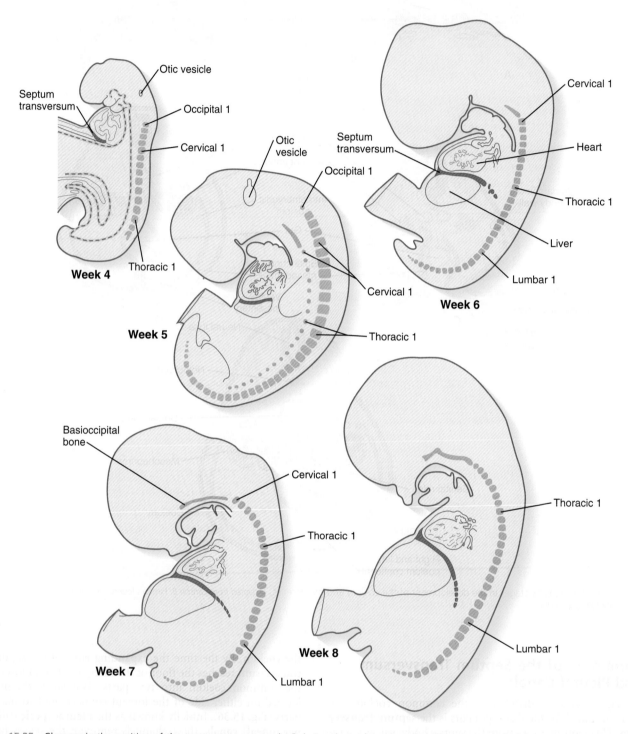

Fig. 15.35 Changes in the position of the septum transversum *(red)* during the embryonic period. The *gray* repeating structures are somites. The *orange* repeating structures are elements of the axial skeleton.

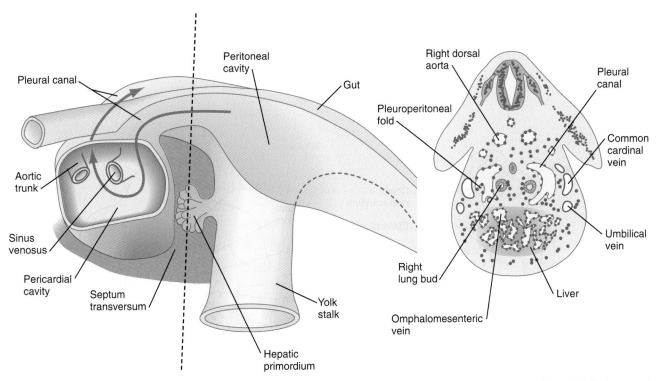

Fig. 15.36 **Relationships among the pericardial cavity, pleural canals, and peritoneal cavity.** The *red arrow* passes from the left pleural cavity into the pericardial cavity and then into the right pleural canal. The *dashed line* on the left represents the level of the cross section on the right.

bulge into the dorsolateral wall of the coelom as they arch toward the midline of the thoracic portion of the coelom and enter the sinus venosus of the heart (**Fig. 15.37**). Initially, the pleuropericardial folds are not large and cause only a narrowing at the junction of the pericardial cavity and pleural canals. As the lungs expand, however, the folds form prominent shelves that meet at the midline and form the fibrous (parietal) layer of the pericardium.

The paired **phrenic nerves** are associated with the pleuropericardial folds. These nerves arise from joined branches of cervical roots 3, 4, and 5 and supply the muscle fibers of the diaphragm. With the shifts in positions of various components of the body during growth, the diaphragm ultimately descends to the level of the lower thoracic vertebrae. As it does, it carries the phrenic nerves with it. Even in adults, the pathway of the phrenic nerves through the fibrous pericardium is a reminder of their early association with the pleuropericardial folds.

At the caudal ends of the pleural canals, another pair of folds, the **pleuroperitoneal folds**, becomes prominent as the expanding lungs push into the mesoderm of the body wall. The pleuroperitoneal folds occupy successively greater portions of the pleural canal until they fuse with the septum transversum and the mesentery of the esophagus, thereby effectively obliterating the pleural canal (**Fig. 15.38**). Cells from the pleuroperitoneal folds continue into the abdominal cavity and contribute to the connective tissue that connects the liver and the right adrenal gland. All connections between the abdominal cavity and the thoracic cavity are eliminated.

Formation of the Diaphragm

The **diaphragm**, which separates the thoracic from the abdominal cavity in adults, is a composite structure derived from several embryonic components (see Fig. 15.38). The large ventral component of the diaphragm arises from the septum transversum, which fuses with the ventral part of the esophageal mesentery. Converging on the esophageal mesentery from the dorsolateral sides are the pleuroperitoneal folds. These components form the bulk of the diaphragm. As the lungs continue to grow, their caudal tips excavate additional space in the body wall. The body wall mesenchyme separated from the body wall proper becomes a third component of the diaphragm by forming a thin rim of tissue along its dorsolateral borders. In keeping with their motor innervation by the vagus nerve (cranial nerve X), the cellular precursors of the diaphragmatic musculature migrate caudally into the body cavity from their site of origin in the occipital somites.

Clinical Correlation 15.5 presents malformations of the body cavities, diaphragm, and body wall.

Clinical Vignette

A 14-year-old girl has been troubled for several years with moderately severe upper abdominal pains that recur on a fairly regular basis at roughly monthly intervals. After passing through several pediatric and medical clinics without obtaining relief, she is sent to a psychiatrist, who also cannot resolve her symptoms. Finally, an astute physician suspects that her symptoms may be caused by a congenital anomaly. Further testing and ultimately surgery proved this suspicion to be correct.

What was the diagnosis?

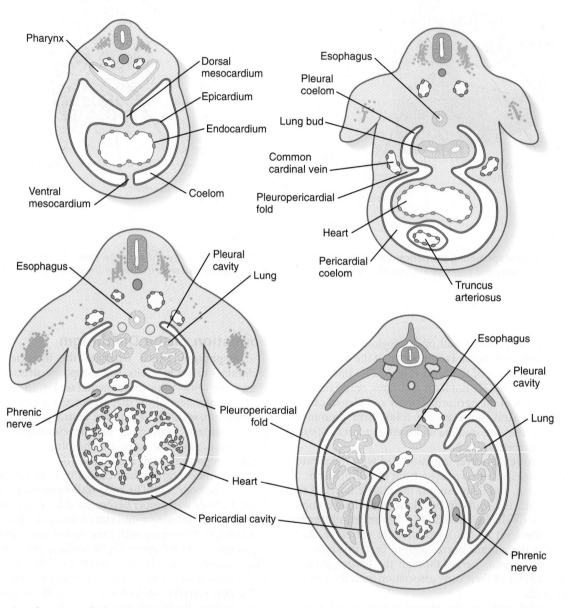

Fig. 15.37 **Development of the pleuropericardial folds.** *(Adapted from Carlson B:* Patten's foundations of embryology, *ed 6, New York, 1996, McGraw-Hill.)*

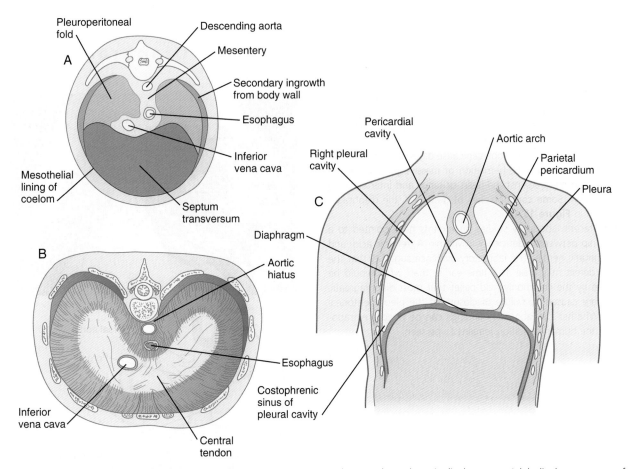

Fig. 15.38 **Stages in the formation of the diaphragm.** **A,** Components making up the embryonic diaphragm. **B,** Adult diaphragm as seen from the thoracic side. **C,** Frontal section showing relations of the diaphragm to the pleural and pericardial cavities.

CLINICAL CORRELATION 15.5
Malformations of the Body Cavities, Diaphragm, and Body Wall

Ventral Body Wall Defects, Ectopia Cordis, Gastroschisis, and Omphalocele

The opposing sides of the body wall occasionally fail to fuse as the embryo assumes its cylindrical shape late in the first month. Several defective mechanisms, such as hypoplasia of the tissues, can account for these defects. A quantitatively minor defect in closure of the thoracic wall is manifested as **failure of sternal fusion** (**Fig. 15.39**). If growth of the two sides of the thoracic wall is severely defective, the heart can form outside the thoracic cavity, thus resulting in **ectopia cordis** (**Fig. 15.40**).

Closure defects of the ventral abdominal wall can lead to similar gross malformations. In many cases of omphalocele (see Fig. 15.16), hypoplasia of the abdominal wall itself and deficiencies of abdominal musculature are evident. More serious cases involve evisceration of the abdominal contents through a fissure between the umbilicus and sternum (**gastroschisis**) (**Fig. 15.41**). Caudal to the umbilicus, an associated closure defect of the urinary bladder (exstrophy of the bladder [see Fig. 16.20]) is common.

Diaphragmatic Hernias

Incomplete fusion or hypoplasia of one or more of the components of the diaphragm can lead to an open connection between the

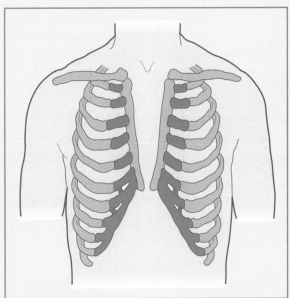

Fig. 15.39 **Failure of fusion of the paired components of the embryonic sternum.**

Continued

CLINICAL CORRELATION 15.5
Malformations of the Body Cavities, Diaphragm, and Body Wall—cont'd

abdominal and thoracic cavities. If the defect is large enough, various structures in the abdominal cavity (usually part of the stomach or intestines) can herniate into the thoracic cavity, or, more rarely, thoracic structures can penetrate into the abdominal cavity. Minor cases of herniation can cause digestive symptoms. In the case of major defects, herniation of massive portions of the intestines can press against the heart or lungs and interfere with their function. Some common sites of defects in the diaphragm are shown in **Figure 15.42**.

More recent laboratory studies on rodents have pointed to a relationship between a deficiency in vitamin A (retinoic acid) and diaphragmatic hernia. The laboratory evidence suggests that the primary defect may arise at a time earlier than what would be predicted by the commonly held belief that most diaphragmatic hernias are caused by failure of closure of the pleuroperitoneal canals. Whether the laboratory findings can be directly extrapolated to the human condition remains to be seen.

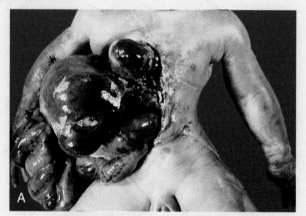

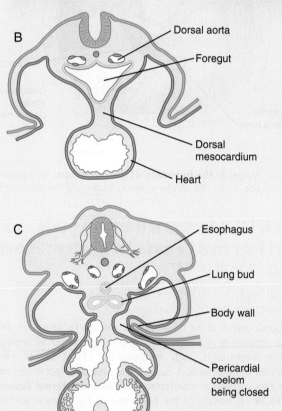

Fig. 15.40 **Ectopia cordis.** A, Fetus with a major ventral abdominal wall defect that combines gastroschisis and ectopia cordis. B and C, Cross sections illustrating the inability of the folding sides of the body wall to encompass the developing heart, resulting in ectopia cordis. *(Courtesy of M. Barr, Ann Arbor, Mich.)*

CLINICAL CORRELATION 15.5
Malformations of the Body Cavities, Diaphragm, and Body Wall—cont'd

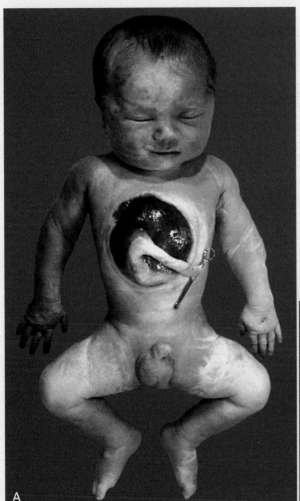

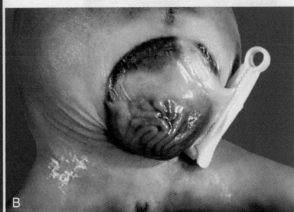

Fig. 15.41 Defective closure of the ventral abdominal wall cranial (**A**) and caudal (**B**) to the umbilicus. *(Courtesy of M. Barr, Ann Arbor, Mich.)*

Continued

CLINICAL CORRELATION 15.5
Malformations of the Body Cavities, Diaphragm, and Body Wall—cont'd

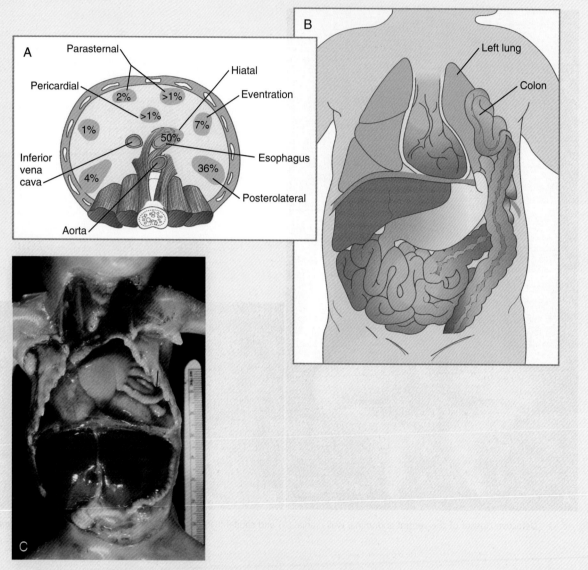

Fig. 15.42 **A,** Common sites of diaphragmatic hernias. Percentages of occurrence are indicated. **B,** Diaphragmatic hernia with intestines entering the left pleural cavity and compressing the left lung. **C,** Fetus with diaphragmatic hernia. Loops of intestine in the left pleural cavity *(arrow)* are compressing the left lung. *(Courtesy of M. Barr, Ann Arbor, Mich.)*

Summary

- The digestive system arises from the primitive endodermally lined gut tube, which is bounded cranially by the oropharyngeal membrane and caudally by the cloacal membrane. The gut is divided into foregut, midgut, and hindgut segments, with the midgut opening into the yolk sac. Specification of the various regions of the gut tract depends on a pattern set up by tightly regulated combinations of *Hox* genes. Development of virtually all parts of the gut depends on epithelial-mesenchymal interactions. Responding to such interactions, primordia of the

respiratory system, the liver, the pancreas, and other digestive glands bud out from the original gut tube.

- The esophagus takes shape as a simple tubular structure between the pharynx and stomach. At one stage, the epithelium occludes the lumen of the esophagus; the lumen later recanalizes. The developing stomach is suspended from a dorsal and ventral mesogastrium. Through two types of rotation, the stomach attains its adult position. Common malformations of the stomach include the following: pyloric stenosis, which interferes with emptying of the stomach; and ectopic gastric mucosa, which can produce ulcers in unexpected locations.

- As they grow, the intestines form a hairpinlike loop that herniates into the body stalk. Further growth of the small intestine causes small intestinal loops to accumulate in the body stalk. As the intestines retract into the body cavity, they rotate around the superior mesenteric artery. This rotation results in the characteristic positioning of the colon around the small intestine in the abdominal cavity. During these changes in position, parts of the dorsal mesentery fuse with the peritoneal lining of the dorsal body wall. In the posterior part of the gut, the urorectal septum partitions the cloaca into the rectum and urogenital sinus.

- During its differentiation, the lining of the intestinal tract passes through phases of (1) epithelial proliferation, (2) cellular differentiation, and (3) biochemical and functional maturation. Similar to the esophagus, the small intestine goes through a period of occlusion of the lumen by the epithelium. At later stages, intestinal crypts located at the base of villi contain epithelial stem cells, which supply the entire intestinal epithelial surface with various epithelial cells.

- The intestinal tract is subject to a variety of malformations, including local stenosis, atresia, duplications, diverticula, and abnormal rotation. Incomplete resorption of the vitelline duct can give rise to Meckel's diverticulum, vitelline duct ligaments, cysts, or fistulas. Omphalocele is the failure of the intestines to return to the body cavity from the body stalk. Aganglionic megacolon is caused by the failure of parasympathetic neurons to populate the distal part of the colon. Failure of the anal membrane to break down (imperforate anus) may be associated with fistulas connecting the digestive tract to various regions of the urogenital system.

- Digestive glands arise as epithelial diverticula from the gut. Their formation and further outgrowth are based on inductive interactions with the surrounding mesenchyme. The primordium of the liver arises in the septum transversum, but as it expands, it protrudes into the ventral mesentery. As it develops, the liver acquires the capacity to synthesize and secrete serum albumin and to store glycogen, among other biochemical functions. The pancreas grows out as dorsal and ventral pancreatic buds, which ultimately fuse to form a single pancreas. Within the pancreas, the epithelium forms exocrine components, which secrete digestive enzymes, and endocrine components (islets of Langerhans), which secrete insulin and glucagon.

- The respiratory system arises as a ventral outgrowth from the gut just caudal to the pharynx. Through epithelial-mesenchymal interactions, the tip of the respiratory diverticulum undergoes up to 23 sets of dichotomous branchings. Other interactions with the surrounding mesenchyme stabilize the tubular parts of the respiratory tract by inhibiting branching. Lung development goes through several stages: (1) the embryonic stage, (2) the pseudoglandular stage, (3) the canalicular stage, (4) the terminal sac stage, and (5) the postnatal stage.

- Important malformations of the respiratory tract include tracheoesophageal fistulas, which result in abnormal connections between the trachea and esophagus. Atresia of components of the respiratory system is rare, but anatomical variations in the morphology of the lungs are common. Respiratory distress syndrome, commonly seen in premature infants, is related to insufficiencies in the formation of pulmonary surfactant by type II alveolar cells.

- In its most basic condition, the intraembryonic coelom is separated into right and left components by the dorsal and ventral mesenteries, which suspend the gut. Except for the region of the stomach and liver, the ventral mesentery disappears. In the region of the heart, the dorsal mesocardium persists, and the ventral mesocardium disappears.

- The septum transversum divides the coelom into thoracic and abdominal regions, which are connected by pleural canals. The developing lungs grow into the pleural canals, which are partially delimited by paired pleuropericardial and pleuroperitoneal folds. The definitive diaphragm is formed from (1) the septum transversum, (2) the pleuroperitoneal folds, and (3) ingrowths from body wall mesenchyme.

- Quantitative deficiencies in ventral body wall tissue can result in abnormalities ranging from failure of sternal fusion to ectopia cordis in the thorax and omphalocele to gastroschisis or exstrophy of the bladder or both in the abdomen. Defects in the diaphragm are diaphragmatic hernias and can result in herniation of the intestines into the thoracic cavity.

Review Questions

1. Which condition is most closely associated with a disturbance of neural crest?
A. Anal atresia
B. Meckel's diverticulum
C. Omphalocele
D. Volvulus
E. Aganglionic megacolon

2. Meckel's diverticulum is most commonly located in the:
A. Ileum
B. Ascending colon
C. Jejunum
D. Transverse colon
E. Duodenum

3. The primordium of which structure is located in the septum transversum?
A. Dorsal pancreas
B. Lung
C. Liver
D. Thymus
E. Spleen

4. The yolk stalk is most closely associated with which artery?
A. Celiac
B. Umbilical
C. Superior mesenteric
D. Aorta
E. Inferior mesenteric

5. The dorsal pancreatic bud is initially induced from the gut endoderm by the:
A. Liver
B. Notochord

C. Lung bud
D. Yolk sac
E. None of the above

6. Splanchnic mesoderm acts as an inducer of all of the following tissues or organs except:
A. Teeth
B. Trachea
C. Liver
D. Lungs
E. Pancreas

7. During the first feeding, a newborn begins to choke. What congenital anomalies should be included in the differential diagnosis?

8. A newborn took the first feeding of milk without incident, but an hour later was crying in pain and vomited the milk with considerable force. Examination revealed a hard mass near the midline in the upper region of the abdomen. What was the diagnosis?

9. An infant was noted to extrude a small amount of mucus and fluid from the umbilicus when crying or straining. This should make the physician think of what congenital anomaly in the differential diagnosis?

10. A newborn was given a cursory physical examination and was taken home by the mother 1 day after delivery. Several days later, the mother brought the infant to the clinic. The infant was in obvious severe discomfort with a swollen abdomen. Physical examination revealed that an important congenital anomaly had been overlooked at the original examination. What was that anomaly?

References

Asayesh A and others: Spleen versus pancreas: strict control of organ interrelationship revealed by analysis of *Bapx1*$^{-/-}$ mice, *Genes Dev* 20:2208-2213, 2006.

Bort R and others: Hex homeobox gene controls the transition of the endoderm to a pseudostratified, cell emergent epithelium for liver bud development, *Dev Biol* 290:44-56, 2006.

Brewer S, Williams T: Finally, a sense of closure? Animal models of human ventral body wall defects, *Bioessays* 26:1307-1321, 2004.

Cardoso WV, Lü J: Regulation of early lung morphogenesis: questions, facts and controversies, *Development* 133:1611-1624, 2006.

Chia LA, Kuo CJ: The intestinal stem cell, *Prog Mol Biol Transl Sci* 96:157-173, 2010.

Dessimoz J and others: FGF signaling is necessary for establishing gut tube domains along the anterior-posterior axis in vivo, *Mech Dev* 123:42-55, 2006.

Felix JF and others: Genetic and environmental factors in the etiology of esophageal atresia and/or tracheoesophageal fistula: an overview of the current concepts, *Birth Def Res A Clin Mol Teratol* 85:747-754, 2009.

Fukuda K, Yasugi S: The molecular mechanisms of stomach development in vertebrates, *Dev Growth Differ* 47:375-382, 2005.

Gallot D and others: Congenital diaphragmatic hernia: a retinoid-signaling pathway disruption during lung development? *Birth Def Res A Clin Mol Teratol* 73:523-531, 2005.

Gittes GK: Developmental biology of the pancreas: a comprehensive review, *Dev Biol* 326:4-35, 2009.

Gray SW, Skandalakis JE: *Embryology for surgeons*, Philadelphia, 1972, Saunders.

Grigorieff A, Clevers H: Wnt signaling in the intestinal epithelium: from endoderm to cancer, *Genes Dev* 19:877-890, 2005.

Groenman F, Unger S, Post M: The molecular basis for abnormal human lung development, *Biol Neonate* 87:164-177, 2005.

Hayashi S and others: Pleuroperitoneal canal closure and the fetal adrenal gland, *Anat Rec* 294:633-644, 2011.

Heath JK: Transcriptional networks and signaling pathways that govern vertebrate intestinal development, *Curr Top Dev Biol* 90:159-192, 2010.

Jacquemin P and others: An endothelial-mesenchymal relay pathway regulates early phases of pancreas development, *Dev Biol* 290:189-199, 2006.

Jensen J: Gene regulatory factors in pancreatic development, *Dev Dyn* 229:176-200, 2004.

Kawazoe Y and others: Region-specific gastrointestinal *Hox* code during murine embryonal gut development, *Dev Growth Differ* 44:77-84, 2002.

Khurana S, Mills JC: The gastric mucosa: development and differentiation, *Prog Mol Biol Transl Sci* 96:93-115, 2010.

Lau J, Kawahira H, Hebrok M: Hedgehog signaling in pancreas development and disease, *Cell Mol Life Sci* 63:642-652, 2006.

Lemaigre FP: Molecular mechanisms of biliary development, *Prog Mol Biol Transl Sci* 97:103-126, 2010.

Li Y and others: Aberrant Bmp signaling and notochord delamination in the pathogenesis of esophageal atresia, *Dev Dyn* 236:746-754, 2007.

MacDonald RJ, Swift GH, Real FX: Transcriptional control of acinar development and homeostasis, *Prog Mol Biol Transl Sci* 97:1-40, 2010.

Maeda Y, Davé V, Whitsett JA: Transcriptional control of lung morphogenesis, *Physiol Rev* 87:219-244, 2007.

Marshman E, Booth C, Potten CS: The intestinal epithelial stem cell, *Bioessays* 24:91-98, 2002.

McLin VA, Henning SJ, Jamrich M: The role of the visceral mesoderm in the development of the gastrointestinal tract, *Gastroenterology* 136:2074-2091, 2009.

Moniot B and others: SOX9 specifies the pyloric sphincter epithelium through mesenchymal-epithelial signals, *Development* 131:3795-3804, 2004.

Morrisey EE, Hogan BLM: Preparing for the first breath: genetic and cellular mechanisms in lung development, *Dev Cell* 18:8-23, 2010.

Murtaugh LC: Pancreas and beta-cell development: from the actual to the possible, *Development* 134:427-438, 2007.

Nagaoka M, Duncan SA: Transcriptional control of hepatocyte differentiation, *Prog Mol Biol Transl Sci* 97:79-101, 2010.

Nievelstein RAJ and others: Normal and abnormal embryonic development of the anorectum in human embryos, *Teratology* 57:70-78, 1998.

O'Rahilly R: The timing and sequence of events in the development of the human digestive system and associated structures during the embryonic period proper, *Anat Embryol* 153:123-136, 1978.

O'Rahilly R, Boyden EA: The timing and sequence of events in the development of the human respiratory system during the embryonic period proper, *Z Anat Entwicklungsgesch* 141:237-250, 1973.

O'Rahilly R, Tucker JA: The early development of the larynx in staged human embryos, *Ann Otolaryngol Rhinol Laryngol* 82(Suppl 7):1-27, 1973.

Park WY and others: FGF-10 is a chemotactic factor for distal epithelial buds during lung development, *Dev Biol* 201:125-134, 1998.

Que J and others: Morphogenesis of the trachea and esophagus: current players and new roles for noggin and BMPs, *Differentiation* 74:422-437, 2006.

Rishniw M and others: Skeletal myogenesis in the mouse esophagus does not occur through transdifferentiation, *Genesis* 36:81-82, 2003.

Sala FG and others: Fibroblast growth factor 10 is required for survival and proliferation but not differentiation of intestinal epithelial progenitor cells during murine colon development, *Dev Biol* 299:373-385, 2006.

Savin T and others: On the growth and form of the gut, *Nature* 476:57-62, 2011.

Schäfer K-H, van Ginneken C, Copray S: Plasticity and neural stem cells in the enteric nervous system, *Anat Rec* 292:1940-1952, 2009.

Si-Tayeb K, Lemaigre FP, Duncan SA: Organogenesis and development of the liver, *Dev Cell* 18:175-189, 2010.

Spence JR, Lauf R, Shroyer NF: Vertebrate intestinal endoderm development, *Dev Dyn* 240:501-520, 2011.

Stephens FD: Embryology of the cloaca and embryogenesis of anorectal malformations, *Birth Defects Orig Artic Ser* 24:177-209, 1988.

Udager A, Prakash A, Gumucio DL: Dividing the tubular gut: generation of organ boundaries at the pylorus, *Prog Mol Biol Transl Sci* 96:35-62, 2010.

Vachon PH and others: Early establishment of epithelial apoptosis in the developing human small intestine, *Int J Dev Biol* 44:891-898, 2000.

van der Putte SCJ: The development of the human anorectum, *Anat Rec* 292:951-954, 2009.

Van Veenendaal MB, Liem KD, Marres HAM: Congenital absence of the trachea, *Eur J Pediatr* 159:8-13, 2000.

Varga I and others: Congenital anomalies of the spleen from an embryological point of view, *Med Sci Monit* 15:RA269-276, 2009.

Wang Z and others: Retinoic acid regulates morphogenesis and patterning of posterior foregut derivatives, *Dev Biol* 297:433-445, 2006.

Warburton D and others: Lung organogenesis, *Curr Top Dev Biol* 90:73-158, 2010.

Warburton D and others: Molecular mechanisms of early lung specification and branching morphogenesis, *Pediatr Res* 57:26R-37R, 2005.

Wartiovaara K, Salo M, Sariola H: Hirschsprung's disease genes and the development of the enteric nervous system, *Ann Med* 30:66-74, 1998.

Zákány J, Duboule D: *Hox* genes and the making of sphincters, *Nature* 401:761-762, 1999.

Zaret KS, Grompe M: Generation and regeneration of cells of the liver and pancreas, *Science* 322:1490-1494, 2008.

Zhang X and others: Reciprocal epithelial-mesenchymal FGF signaling is required for cecal development, *Development* 133:173-180, 2005.

Zorn AM, Wells JM: Vertebrate endoderm development and organ formation, *Annu Rev Cell Dev Biol* 25:221-251, 2009.

Chapter **16**

Urogenital System

The urogenital system arises from the intermediate mesoderm of the early embryo (see Fig. 6.7). Several major themes underlie the development of urinary and genital structures from this common precursor. The first is the interconnectedness of urinary and genital development, in which early components of one system are taken over by another during its later development. A second is the recapitulation during human ontogeny of kidney types (the equivalent of organ isoforms) that are terminal forms of the kidney in lower vertebrates. A third theme comprises the dependence of differentiation and the maintenance of many structures in the urogenital system on epithelial-mesenchymal interactions. Finally, the sexual differentiation of many structures passes from an indifferent stage, in which male and female differences are not readily apparent, to a male or female pathway, depending on the presence of specific promoting or inhibiting factors acting on the structure. Although phenotypic sex is genetically determined, genetic sex can be overridden by environmental factors, thus leading to a discordance between the two. Clinical Correlations 16.1 and 16.2, later in this chapter, discuss abnormalities of the urinary and genital systems, respectively.

Urinary System

The urinary system begins to take shape before any gonadal development is evident. Embryogenesis of the kidney begins with the formation of an elongated pair of excretory organs similar in structure and function to the kidneys of lower vertebrates. These early forms of the kidney are later supplanted by the definitive metanephric kidneys, but as they regress, certain components are retained to be reused by other components of the urogenital system.

Early Forms of the Kidney

The common representation of mammalian kidney development includes three successive phases beginning with the appearance of the **pronephros**, the developmental homologue of the type of kidney found in only the lowest vertebrates. In human embryos, the first evidence of a urinary system consists of the appearance of a few segmentally arranged sets of epithelial cords that differentiate from the anterior intermediate mesoderm at about 22 days' gestation. These structures are more appropriately called **nephrotomes**. The nephrotomes connect laterally with a pair of **primary nephric (pronephric) ducts**, which grow toward the cloaca (**Fig. 16.1**). The earliest stages in the development of the urinary system depend on

the action of retinoic acid, which sets the expression limits of *Hox 4-11* genes that determine the craniocaudal limits of the early urinary system. The molecular response by the intermediate mesoderm is the expression of the transcription factors **Pax-2** and **Pax-8**, which then induce **Lim-1** (Lhx-1) in the intermediate mesoderm. Lim-1 is required for the aggregation of the mesenchymal cells of the intermediate mesoderm into the primary nephric ducts.

As the primary nephric ducts extend caudally, they stimulate the intermediate mesoderm to form additional segmental sets of tubules. The conversion of the mesenchymal cells of the intermediate mesoderm into epithelial tubules depends on the expression of Pax-2, and in the absence of this molecule, further development of kidney tubules does not occur. These tubules are structurally equivalent to the **mesonephric tubules** of fishes and amphibians. A typical mesonephric unit consists of a vascular **glomerulus**, which is partially surrounded by an epithelial glomerular capsule. The glomerular capsule is continuous with a contorted mesonephric tubule, which is surrounded by a mesh of capillaries (see Fig. 16.1B). Each mesonephric tubule empties separately into the continuation of the primary nephric duct, which becomes known as the **mesonephric (wolffian) duct**.

The formation of pairs of mesonephric tubules occurs along a craniocaudal gradient. The first 4 to 6 pairs of mesonephric tubules (and the pronephric tubules) arise as outgrowths from the primary nephric ducts. Farther caudally, mesonephric tubules, up to a total of 36 to 40, take shape separately in the intermediate mesoderm slightly behind the caudal extension of the mesonephric ducts. By the end of the fourth week of gestation, the mesonephric ducts attach to the cloaca, and a continuous lumen is present throughout each. There is a difference in the developmental controls between the most cranial 4 to 6 pairs of mesonephric tubules and the remaining caudal tubules. Knockouts for the *WT-1* (Wilms' tumor suppressor) gene result in the absence of posterior mesonephric tubules, whereas the cranial tubules that bud off the pronephric duct form normally. As is the case in the formation of the metanephros (see later), WT-1 regulates the transformation from mesenchyme to epithelium during the early formation of renal (mesonephric) tubules. Very near its attachment site to the cloaca, the mesonephric duct develops an epithelial outgrowth called the **ureteric bud** (see Fig. 16.1A).

Early in the fifth week of gestation, the ureteric bud begins to grow into the most posterior region of the intermediate mesoderm. It then sets up a series of continuous inductive interactions leading to the formation of the definitive kidney, the **metanephros**.

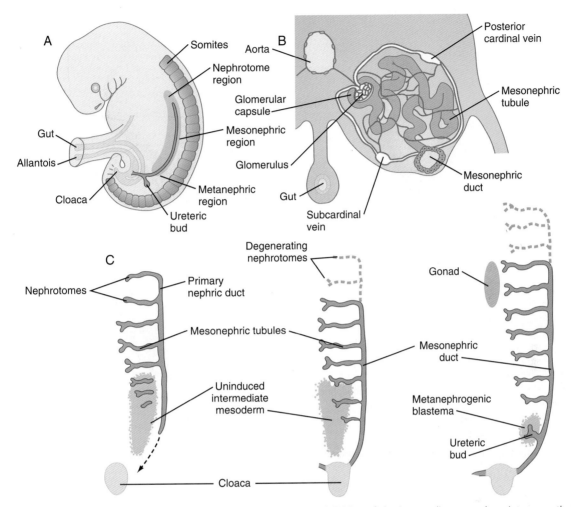

Fig. 16.1 **Early stages in the establishment of the urinary system.** **A,** Subdivision of the intermediate mesoderm into areas that will form nephrotomes, mesonephros, and metanephros. **B,** Cross section through mesonephros showing a well-developed mesonephric tubule and its associated vasculature. **C,** Caudal progression of formation of the mesonephros and degeneration of the most cranial segments of the primitive kidney.

Although there is evidence of urinary function in the mammalian mesonephric kidney, the physiology of the mesonephros has not been extensively investigated. Urine formation in the mesonephros begins with a filtrate of blood from the glomerulus into the glomerular capsule. This filtrate flows into the tubular portion of the mesonephros, where the selective resorption of ions and other substances occurs. The return of resorbed materials to the blood is facilitated by the presence of a dense plexus of capillaries around the mesonephric tubules.

The structure of the human embryonic mesonephros is very similar to that of adult fishes and aquatic amphibians, and it functions principally to filter and remove body wastes. Because these species and the amniote embryo exist in an aquatic environment, there is little need to conserve water. The mesonephros does not develop a medullary region or an elaborate system for concentrating urine as the adult human kidney must.

The mesonephros is most prominent while the definitive metanephros is beginning to take shape. Although it rapidly regresses as a urinary unit after the metanephric kidneys become functional, the mesonephric ducts and some of the mesonephric tubules persist in the male and become incorporated as integral components of the genital duct system (**Fig. 16.2**).

Metanephros

Development of the metanephros begins early in week 5 of gestation, when the ureteric bud (**metanephric diverticulum**) grows into the posterior portion of the intermediate mesoderm. Mesenchymal cells of the intermediate mesoderm condense around the metanephric diverticulum to form the **metanephrogenic blastema** (see Fig. 16.1C). Outgrowth of the ureteric bud from the mesonephric duct is a response to the secretion of **glial cell line–derived neurotrophic factor** (**GDNF**) by the undifferentiated mesenchyme of the metanephrogenic blastema (**Fig. 16.3A**). This inductive signal is bound by **c-Ret**, a member of the tyrosine kinase receptor superfamily, and the coreceptor **Gfra-1**, which are located in the plasma membranes of the epithelial cells of the early ureteric bud. The formation of GDNF in the metanephric mesenchyme is regulated by **WT-1**. The posterior location of the ureteric bud results from a combination of repression of GDNF expression in the more anterior regions by the actions of **Slit-2/Robo-2** in the mesenchyme and **Sprouty**, which reduces the sensitivity of the anterior mesonephric duct to the action of GDNF. Bone morphogenetic protein (**BMP**) signaling in the surrounding mesoderm is also inhibitory to outgrowth of the

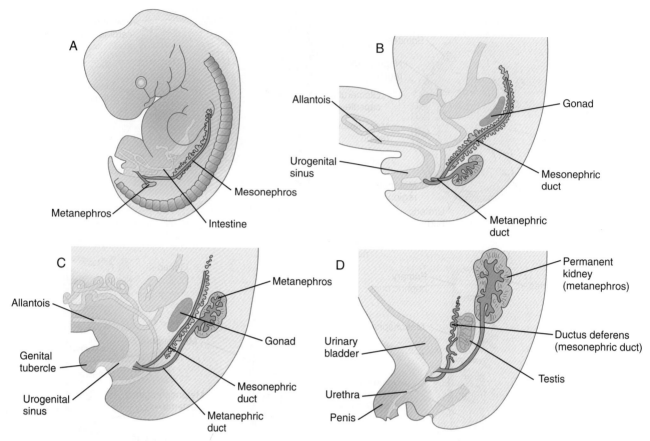

Fig. 16.2   **Stages in the formation of the metanephros. A, At 6 weeks. B, At 7 weeks. C, At 8 weeks. D, At 3 months (male).**

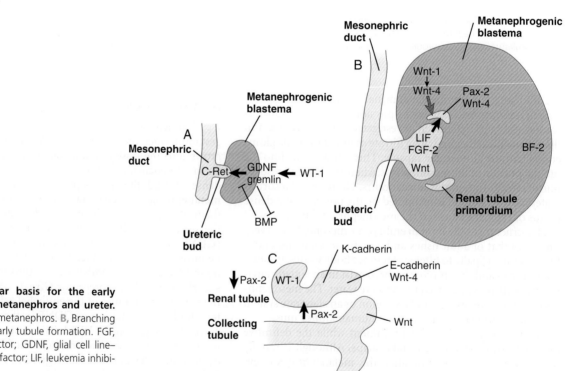

Fig. 16.3 **Molecular basis for the early formation of the metanephros and ureter.** A, Early induction of metanephros. B, Branching of ureteric bud. C, Early tubule formation. FGF, fibroblast growth factor; GDNF, glial cell line–derived neurotrophic factor; LIF, leukemia inhibitory factor.

ureteric bud, but within the metanephrogenic blastema its action is counteracted by the BMP-inhibitory actions of **gremlin**, which is produced within the blastema itself.

The outgrowing ureteric bud is associated with two types of mesenchyme: intermediate mesoderm and tailbud mesenchyme. These two types of mesenchyme create a sharp border between the forming ureter (associated with tailbud mesenchyme) and the intrarenal collecting duct system (associated with intermediate mesoderm). **BMP-4**, secreted by the surrounding tailbud mesenchyme, causes the ureteric epithelium to form **uroplakins**, proteins that render the epithelium of the ureter impermeable to water. The renal pelvis of the adult kidney shares properties with the ureter and the collecting system, and its cellular origins are unclear.

The morphological foundations for the development of the metanephric kidney are the elongation and branching (up to 14 or 15 times) of the ureteric bud, which becomes the **collecting (metanephric) duct** system of the metanephros, and the formation of renal tubules from mesenchymal condensations (metanephrogenic blastema) located around the tips of the branches. The mechanism underlying these events is a series of reciprocal inductive interactions between the tips of the branches of the metanephric ducts and the surrounding metanephrogenic blastemal cells. Without the metanephric duct system, tubules do not form; conversely, the metanephrogenic mesoderm acts on the metanephric duct system and induces its characteristic branching. The pattern of branching is largely determined by the surrounding mesenchyme. If lung bud mesenchyme is substituted for metanephric mesenchyme, the pattern of branching of the ureteric bud closely resembles that of the lung.

The mechanism of branching of the ureteric bud is similar to that which occurs in the initial induction of the metanephros. At the tip of each branch point, a highly localized reciprocal induction system operates. In response to the GDNF signal from the metanephrogenic mesenchyme, the branching epithelial tips of the ureteric buds produce the signaling molecules **fibroblast growth factor-2 (FGF-2)** and **leukemia inhibitory factor**, which induce the surrounding metanephrogenic mesenchyme to begin to form the epithelial precursors of renal tubules. **BMP-7**, which is produced in the same area, prevents mesenchymal cell death and maintains the mesenchymal cells in a developmentally labile state. Expression of the transcription factor **Wnt-9b** in the tips of the ureteric buds is important for the branching response.

Tubule formation also requires a sequential inductive signal, **Wnt-4**, which is produced by the metanephric mesenchyme itself. This early induction patterns the metanephric mesenchyme into a tubular epithelial domain, in which the cells express Wnt-4 and Pax-2, and a stromal region, in which the mesenchymal cells express a winged helix transcription factor, **BF-2**, which may regulate stromal inductive signals.

The formation of individual functional tubules (**nephrons**) in the developing metanephros involves three mesodermal cell lineages: epithelial cells derived from the ureteric bud, mesenchymal cells of the metanephrogenic blastema, and ingrowing vascular endothelial cells. The earliest stage is the condensation of mesenchymal blastemal cells around the terminal bud of the ureteric bud (later to become the metanephric duct). The preinduced mesenchyme contains several interstitial proteins, such as types I and III collagen and fibronectin. As the mesenchymal cells condense after local induction by the branching tips of the ureteric bud, these proteins are lost and are replaced with epithelial-type proteins (type IV collagen, syndecan-1, laminin, and heparin sulfate proteoglycan), which are ultimately localized to the basement membranes (**Fig. 16.4**).

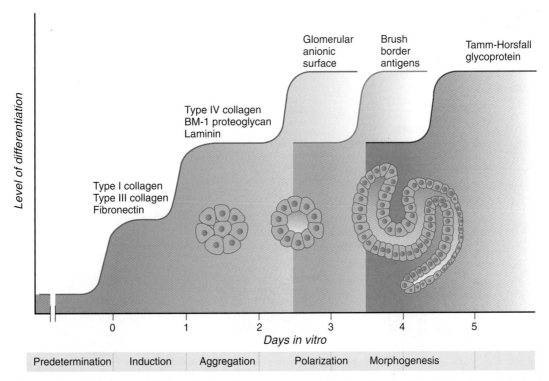

Fig. 16.4   **Multiphasic determination and differentiation of mouse metanephric mesoderm in vitro.** *(Adapted from Saxén L and others:* Biology of human growth, *New York, 1981, Raven Press.)*

As the terminal bud of the metanephric duct branches, each tip is surrounded by a cap of condensed mesenchyme. Soon, this cap becomes subdivided into a persisting mesenchymal **cap** and, at its end, a region where the mesenchyme is transforming into an epithelial **nephrogenic vesicle** (**Fig. 16.5A**). A single condensation of mesenchymal cells undergoes a defined series of stages to form a renal tubule. After a growth phase, mitotic activity within the rounded blastemal mesenchyme decreases, and the primordium of the tubule assumes a comma shape. Within the comma, a group of cells farthest from the end of the metanephric duct becomes polarized and forms a central lumen and a basal lamina on the outer surface. This marks the transformation of the induced mesenchymal cells into an epithelium—the specialized **podocytes**, which ultimately surround the vascular endothelium of the glomerulus.

A consequence of this epithelial transformation is the formation of a slit just beneath the transforming podocyte precursors in the tubular primordium (Fig. 16.5B). Precursors of vascular endothelial cells grow into this slit, which ultimately forms the glomerulus. Induced metanephric mesenchyme stimulates the ingrowth of endothelial cells, possibly by the release of a factor similar to FGF. Uninduced mesenchyme does not possess this capability. The endothelial cells are connected with branches from the dorsal aorta, and they form a complex looping structure that ultimately becomes the renal glomerulus. Cells of the glomerular endothelium and the adjoining podocyte epithelium form a thick basement membrane between them. This basement membrane later serves as an important component of the renal filtration apparatus.

As the glomerular apparatus of the nephron takes shape, another slit forms in the comma-shaped tubular primordium, thus transforming it into an S-shaped structure (Fig. 16.5C). Cells in the rest of the tubule primordium also undergo an epithelial transformation to form the remainder of the renal

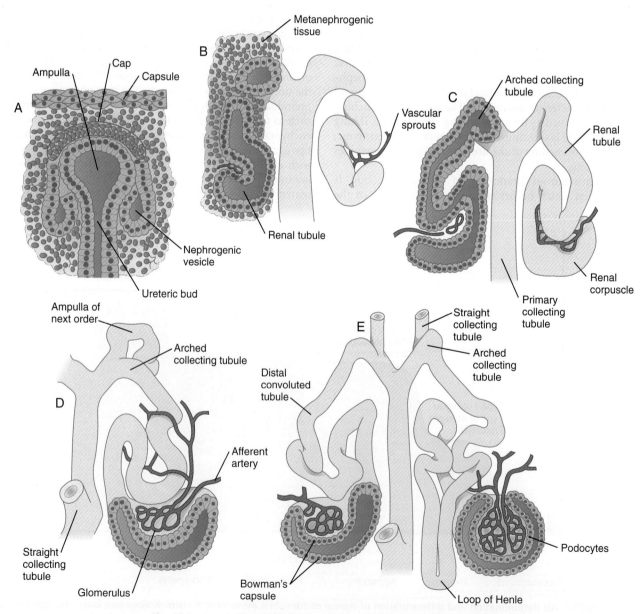

Fig. 16.5 A to E, **Stages in the development of a metanephric tubule.**

tubule. This transformation involves the acquisition of polarity by the differentiating epithelial cells. It is correlated with the deposition of laminin in the extracellular matrix along the basal surface of the cells and the concentration of the integral membrane glycoprotein **uvomorulin** (**E-cadherin**), which seals the lateral borders of the cells (**Fig. 16.6**). As the differentiating tubule assumes an S shape, differing patterns of gene expression are seen along its length. Near the future glomerular end, levels of Pax-2 expression decrease as WT-1 becomes strongly expressed (see Fig. 16.3). Lim-1 expression and the downstream **Delta/Notch** system are now known to play a prominent role in generating the proximal convoluted tubule. At the other end of the tubule (future distal convoluted tubule), Wnt-4 and E-cadherin remain prominent, whereas in the middle (future proximal convoluted tubule), K-cadherin is a prominent cellular marker. Many of the uninduced mesenchymal cells between tubules undergo apoptosis.

Differentiation of the renal tubule progresses from the glomerulus to the proximal and then distal convoluted tubule. During differentiation of the nephron, a portion of the tubule develops into an elongated hairpin loop that extends into the medulla of the kidney as the **loop of Henle**. As they differentiate, the tubular epithelial cells develop molecular features characteristic of the mature kidney (e.g., brush border antigens or the Tamm-Horsfall glycoprotein [see Fig. 16.4]).

Growth of the kidney involves the formation of approximately 15 successive generations of nephrons in its peripheral zone, with the outermost nephrons less mature than the nephrons farther inward. Development of the internal architecture of the kidney is complex, involving the formation of highly ordered arcades of nephrons (**Fig. 16.7**). Details are beyond the scope of this text.

Later Changes in Kidney Development

While the many sets of nephrons are differentiating, the kidney becomes progressively larger. The branched system of ducts also becomes much larger and more complex, and it forms the pelvis and system of **calyces** of the kidney (**Fig. 16.8**). These structures collect the urine and funnel it into the ureters. During much of the fetal period, the kidneys are divided into grossly visible lobes. By birth, the lobation is already much less evident, and it disappears during the neonatal period.

When they first take shape, the metanephric kidneys are located deep in the pelvic region. During the late embryonic and early fetal period, they undergo a pronounced shift in position that moves them into the abdominal region. This shift results partly from actual migration and partly from a marked expansion of the caudal region of the embryo. Two concurrent components to the migration occur. One is a caudocranial shift from the level of L4 to L1 or even the T12 vertebra (**Fig. 16.9**). The other is a lateral displacement.

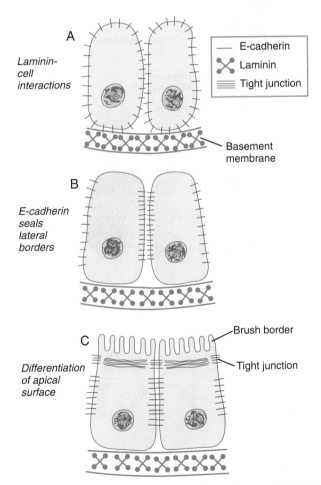

Fig. 16.6 **Stages in the transformation of renal mesenchyme into epithelium, with emphasis on the role of laminin and E-cadherin (uvomorulin).** A, Development of polarity is triggered by interactions between laminin and the cell surface, but E-cadherin is still distributed in a nonpolar manner. B, E-cadherin redistribution occurs, and E-cadherin interactions seal the lateral borders of the cells. C, Apical border of epithelial cells differentiates, as seen by formation of a brush border. *(Based on Ekblom P: FASEB J 3:2141-2150, 1989.)*

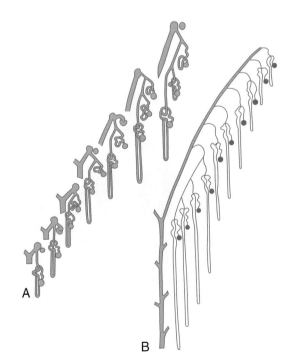

Fig. 16.7 **Formation of arcades of nephrons in the developing human metanephros.** A, Early stages. B, Arrangement of nephrons at the time of birth. *(Based on Osathanondh V, Potter EL: Arch Pathol 76:271-302, 1963.)*

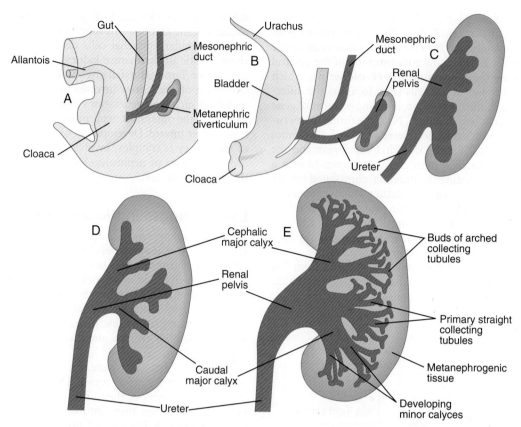

Fig. 16.8　A to E, **Later changes in the development of the metanephros.**

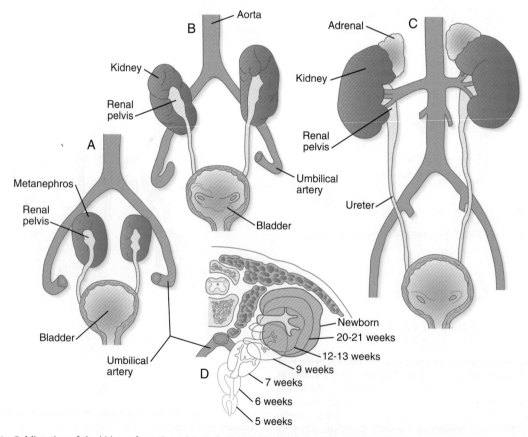

Fig. 16.9　A to C, Migration of the kidneys from the pelvis to their definitive adult level. D, Cross section of the pathway of migration of the kidneys out of the pelvis.

These changes bring the kidneys into contact with the adrenal glands, which form a cap of glandular tissue on the cranial pole of each kidney. During their migration, the kidneys also undergo a 90-degree rotation, with the pelvis ultimately facing the midline. As they are migrating out of the pelvic cavity, the kidneys slide over the large umbilical arteries, which branch from the caudal end of the aorta. All these changes occur behind the peritoneum because the kidneys are retroperitoneal organs. During the early phases of migration of the metanephric kidneys, the mesonephric kidneys regress. The mesonephric ducts are retained, however, as they become closely associated with the developing gonads.

Although normally supplied by one large renal artery branching directly from the aorta, the adult kidney consists of five vascular lobes. The arteries feeding each of these lobes were originally segmental vessels that supplied the mesonephros and were taken over by the migrating metanephros. Their aortic origins are typically reduced to the single pair of renal arteries, but anatomical variations are common.

Formation of the Urinary Bladder

The division of the cloaca into the rectum and urogenital sinus region was introduced in Chapter 15 (see Fig. 15.13). The urogenital sinus is continuous with the allantois, which has an expanded base continuous with the urogenital sinus and an attenuated tubular process that extends into the body stalk on the other end. Along with part of the urogenital sinus, the dilated base of the allantois continues to expand to form the **urinary bladder**, and its attenuated distal end solidifies into the cordlike urachus, which ultimately forms the **median umbilical ligament** that leads from the bladder to the umbilical region (see Fig. 16.19).

As the bladder grows, its expanding wall, which is derived from tailbud mesenchyme, incorporates the mesonephric ducts and the ureteric buds (**Fig. 16.10**). The result is that these structures open separately into the posterior wall of the bladder. Through a poorly defined mechanism possibly involving mechanical tension exerted by the migrating kidneys,

the ends of the ureters open into the bladder laterally and cephalically to the mesonephric ducts. The region bounded by these structures is called the **trigone** of the bladder, but much of the substance of the trigone itself is composed of musculature from the bladder. Only small strips of smooth muscle along the edges of the trigone may arise from ureteral smooth muscle. At the entrance of the mesonephric ducts, the bladder becomes sharply attenuated. This region, originally part of the urogenital sinus, forms the **urethra**, which serves as the outlet of the bladder (see p. 399).

Clinical Correlation 16.1 presents congenital anomalies of the urinary system.

Genital System

Development of the genital system is one phase in the overall sexual differentiation of an individual (**Fig. 16.21**). Sexual determination begins at fertilization, when a Y chromosome or an additional X chromosome is joined to the X chromosome already in the egg. This phase represents the genetic determination of gender. Although the genetic gender of the embryo is fixed at fertilization, the gross phenotypic gender of the embryo is not manifested until the seventh week of development. Before that time, the principal morphological indicator of the embryo's gender is the presence or absence of the **sex chromatin** (**Barr body**) in the female. The Barr body is the result of inactivation of one of the X chromosomes. During this morphologically **indifferent stage** of sexual development, the gametes migrate into the gonadal primordia from the yolk sac.

The phenotypic differentiation of gender is traditionally considered to begin with the gonads* and progresses with gonadal influences on the sexual duct systems. Similar

---

*More recent research has shown gender differences as early as the preimplantation embryo. The Sry genes (see later section) are already transcribed before implantation. In addition, the XY preimplantation embryo develops more rapidly than the XX embryo. Male and female preimplantation embryos are antigenically distinguishable. This suggests differences in gene expression.

Text continued on p. 389

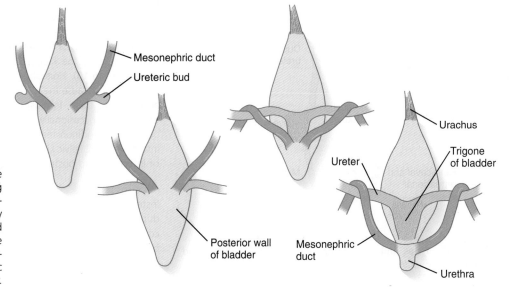

Fig. 16.10 Dorsal views of the developing urinary bladder showing changing relationships of the mesonephric ducts and the ureters as they approach and become incorporated into the bladder. In the two on the *right,* note the incorporation of portions of the walls of the mesonephric ducts into the trigone of the bladder.

Mesonephric duct
Ureteric bud
Urachus
Trigone of bladder
Ureter
Posterior wall of bladder
Mesonephric duct
Urethra

CLINICAL CORRELATION 16.1
Congenital Anomalies of the Urinary System

Anomalies of the urinary system are common (3% to 4% of live births). Many are asymptomatic, and others are manifest only later in life. **Figure 16.11** summarizes the locations of many frequently encountered malformations of the urinary system.

Renal Agenesis

Renal agenesis is the unilateral or bilateral absence of any trace of kidney tissue (**Fig. 16.12A**). Unilateral renal agenesis is seen in roughly 0.1% of adults, whereas bilateral renal agenesis occurs in 1 in 3000 to 4000 newborns. The ureter may be present. This anomaly is usually ascribed to a faulty inductive interaction between the ureteric bud and the metanephrogenic mesenchyme. As many as 50% of cases of renal agenesis in humans have been attributed to mutations of RET or glial cell line–derived neurotrophic factor (GDNF), which are key players in the earliest induction of the ureteric bud. Individuals with unilateral renal agenesis are often asymptomatic, but typically the single kidney undergoes **compensatory hypertrophy** to maintain a normal functional mass of renal tissue.

An infant born with bilateral renal agenesis dies within a few days after birth. Because of the lack of urine output, reduction in the volume of amniotic fluid (**oligohydramnios**) during pregnancy is often an associated feature. Infants born with bilateral renal agenesis characteristically exhibit the **Potter sequence**, consisting of a flattened nose, wide interpupillary space, a receding chin, tapering fingers, low-set ears, hip dislocation, and pulmonary hypoplasia (**Fig. 16.13**). Respiratory failure from pulmonary hypo-

plasia is a common cause of neonatal death in this condition, especially when pulmonary hypoplasia is caused by disorders other than renal agenesis. A sequence is classified as a set of malformations secondary to a primary disturbance in development. In the Potter sequence, reduced urinary output secondary to renal agenesis or a urinary blockage is the factor that sets in motion the other disorders seen in this constellation (**Fig. 16.14**). The actual mechanical effects result from the lack of mechanical buffering by the greatly reduced amount of amniotic fluid.

Renal Hypoplasia

An intermediate condition between renal agenesis and a normal kidney is **renal hypoplasia** (see Fig. 16.12B), in which one kidney or, more rarely, both kidneys will be substantially smaller than normal even though a certain degree of function may be retained. Although a specific cause for renal hypoplasia has not been identified, some cases may be related to deficiencies in growth factors or their receptors that are active during later critical phases of metanephrogenesis. As with renal agenesis, the normal counterpart to a hypoplastic kidney is likely to undergo compensatory hypertrophy.

Renal Duplications

Renal duplications range from a simple duplication of the renal pelvis to a completely separate **supernumerary kidney**. Similar to hypoplastic kidneys, renal duplications may be asymptomatic, although the incidence of renal infections may be increased. Many variants of **duplications of the ureter** have also been described (see Fig. 16.12). Duplication anomalies are commonly attributed to splitting or wide separation of branches of the ureteric bud, the latter resulting from ectopic expression of GDNF more proximally along the mesonephric duct.

Anomalies of Renal Migration and Rotation

The most common disturbance of renal migration leaves a kidney in the pelvic cavity (**Fig. 16.15A**). This disturbance is usually associated with malrotation of the kidney, so that the hilus of the **pelvic kidney** faces anteriorly instead of toward the midline. Another category of migratory malformation is **crossed ectopia**, in which one kidney and its associated ureter are found on the same side of the body as the other kidney (Fig. 16.15B). In this condition, the ectopic kidney may be fused with the normal kidney.

In the condition of **horseshoe kidney**, which can occur in 1 in 400 individuals, the kidneys are typically fused at their inferior poles (**Fig. 16.16**). Horseshoe kidneys cannot migrate out of the pelvic cavity because the inferior mesenteric artery, coming off the aorta, blocks them. In most cases, horseshoe kidneys are asymptomatic, but occasionally pain or obstruction of the ureters may occur. This condition may be associated with anomalies of other internal organs. Pelvic kidneys are subject to an increased incidence of infections and obstructions of the ureters.

Anomalies of the Renal Arteries

Instead of a single renal artery branching off each side of the aorta, duplications or major extrarenal branches of the renal artery are common. Because of the appropriation of segmental arterial branches to the mesonephros by the metanephros, consolidation of the major external arterial supply to the kidney occasionally does not occur.

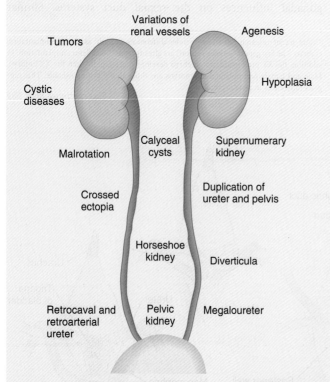

Variations of renal vessels

Tumors

Agenesis

Cystic diseases

Hypoplasia

Malrotation

Calyceal cysts

Supernumerary kidney

Crossed ectopia

Duplication of ureter and pelvis

Horseshoe kidney

Diverticula

Retrocaval and retroarterial ureter

Pelvic kidney

Megaloureter

Fig. 16.11 **Types and sites of anomalies of the kidneys and ureters.** (*Adapted from Gray SW, Skandalakis JE: Embryology for surgeons, Philadelphia, 1972, Saunders.*)

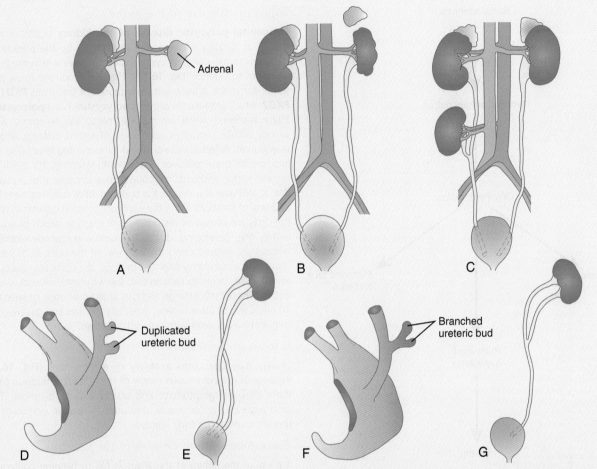

Adrenal

Duplicated ureteric bud

Branched ureteric bud

Fig. 16.12　Common renal anomalies. A, Unilateral renal agenesis. The ureter is also missing. **B,** Unilateral renal hypoplasia. **C,** Supernumerary kidney. **D** and **E,** Complete duplication of ureter, presumably arising from two separate ureteric buds. **F** and **G,** Partial duplication of ureter, presumably arising from a bifurcated ureteric bud.

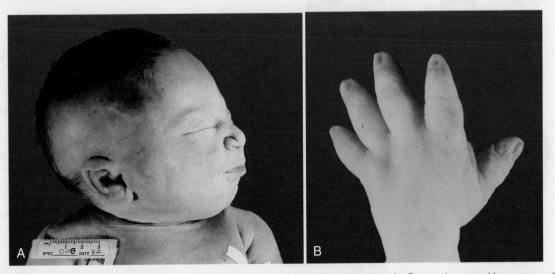

Fig. 16.13　A, Potter's facies, which is characteristic of a fetus exposed to oligohydramnios. Note the flattened nose and low-set ears. **B,** Potter's hand with thickened, tapering fingers. *(Courtesy of M. Barr, Ann Arbor, Mich.)*

Continued

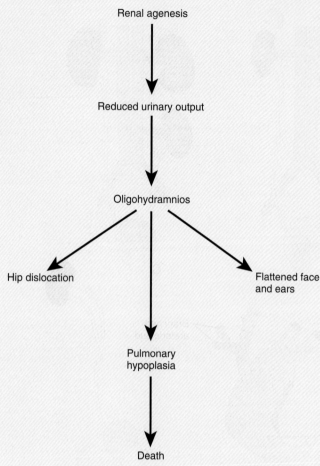

Fig. 16.14 **Major steps in development of the Potter sequence.**

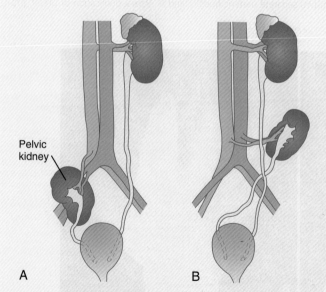

Fig. 16.15 **Migration defects of the kidney.** A, Pelvic kidney. B, Crossed ectopia. The right kidney has crossed the left ureter and has migrated only part of the normal distance.

Polycystic Disease of the Kidney

Congenital polycystic disease of the kidney occurs in more than 1 in 800 live births and is manifested by the presence of hundreds to thousands of cysts of different sizes within the parenchyma of the kidney (**Fig. 16.17**). The most common form, autosomal dominant, is the result of mutations of the genes *PKD1* and *PKD2*, which produce the proteins **polycystin-1** and **polycystin-2**. These proteins, which are surface membrane receptors, affect various cellular processes, such as proliferation, polarity, and differentiation. Affected individuals exhibit persisting fetal patterns of location of these proteins, along with receptors for epidermal growth factor and sodium, potassium–adenosine triphosphatase (Na^+,K^+-ATPase); the result is the budding off of spherical cysts from a variety of locations along the nephron. In some genetic mutants, the cysts are caused by disturbances in the orientation of mitoses within the developing ducts. In normal ductal development, mitoses are aligned along the long axis of the duct. In mutations that result in randomly oriented mitoses, the collecting ducts and even tubules begin to balloon out, forming cysts, instead of elongating. These cysts enlarge and can attain diameters greater than 10 cm. Cysts of other organs, especially the liver and pancreas, are frequently associated with polycystic kidneys.

Ectopic Ureteral Orifices

Ureters may open into a variety of ectopic sites (**Fig. 16.18**). Because of the continuous supply of urine flowing through them, these sites are symptomatic and usually easy to diagnose. Their embryogenesis is commonly attributed to ectopic origins of the ureteric buds in the early embryo.

Cysts, Sinuses, and Fistulas of the Urachus

If parts of the lumen of the allantois fail to become obliterated, **urachal cysts, sinuses**, or **fistulas** can form (**Fig. 16.19**). In the case of urachal fistula, urine seeps from the umbilicus. Urachal sinuses or cysts may swell in later life if they are not evident in an infant.

Exstrophy of the Bladder

Exstrophy of the bladder is a major defect in which the urinary bladder opens broadly onto the abdominal wall (**Fig. 16.20**). Rather than being a primary defect of the urinary system, it is most commonly attributed to an insufficiency of mesodermal tissue of the ventral abdominal wall. Although initially the ventral body wall may be closed with ectoderm, it breaks down in the absence of mesoderm, and degeneration of the anterior wall of the bladder typically follows. In male infants, exstrophy of the bladder commonly involves the phallus, and a condition called **epispadias** results (see p. 404). A reduction in the expression of sonic hedgehog (shh) signaling in the pericloacal epithelia may contribute to a deficiency of tissue in the bladder and the external genitalia. According to a different hypothesis, aneurysmic swellings of the dorsal aortae in the area may prevent tissues from fusing along the ventral midline and keep the walls of the cloaca from closing around the future bladder.

CLINICAL CORRELATION 16.1
Congenital Anomalies of the Urinary System—cont'd

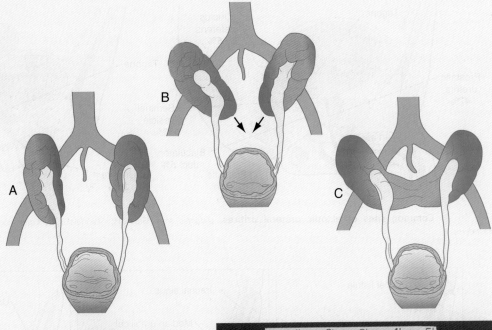

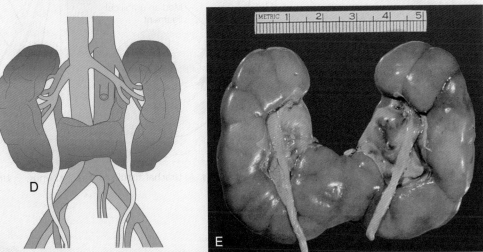

Fig. 16.16 **Stages in the formation of a horseshoe kidney.** **A** to **C,** As the kidneys migrate out of the pelvis, their caudal poles touch and fuse. **D,** Pelvic kidney in an adult. Note the lack of rotation of the kidneys so that the ureters face ventrally instead of medially. **E,** Horseshoe kidney. (*E, Photo 914E from the Arey-DaPeña Pediatric Pathology Photographic Collection, Human Developmental Anatomy Center, National Museum of Health and Medicine, Armed Forces Institute of Pathology, Washington, D.C.*)

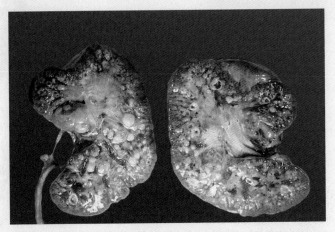

Fig. 16.17 **Polycystic kidneys.** (*Courtesy of M. Barr, Ann Arbor, Mich.*)

Continued

CLINICAL CORRELATION 16.1
Congenital Anomalies of the Urinary System—cont'd

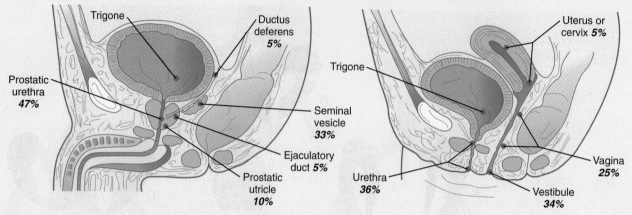

Fig. 16.18 **Common sites of ectopic ureteral orifices.** *(Adapted from Gray SW, Skandalakis JE: Embryology for surgeons, Philadelphia, 1972, Saunders.)*

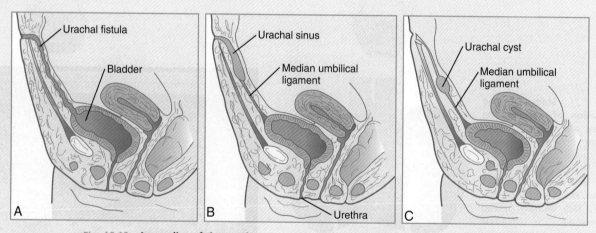

Fig. 16.19 **Anomalies of the urachus.** A, Urachal fistula. B, Urachal sinus. C, Urachal cyst.

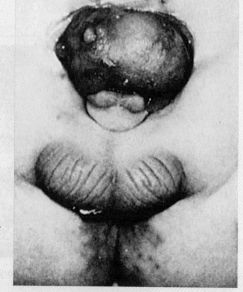

Fig. 16.20 Exstrophy of the bladder in a male infant, showing protrusion of the posterior wall of the bladder through a defect in the lower abdominal wall. At the base of the open bladder is an abnormal, partially bifid penis with an open urethra (not seen) on its dorsal surface. A wide, shallow scrotum is separated from the penis. *(From Crowley LV: An introduction to clinical embryology, St. Louis, 1974, Mosby.)*

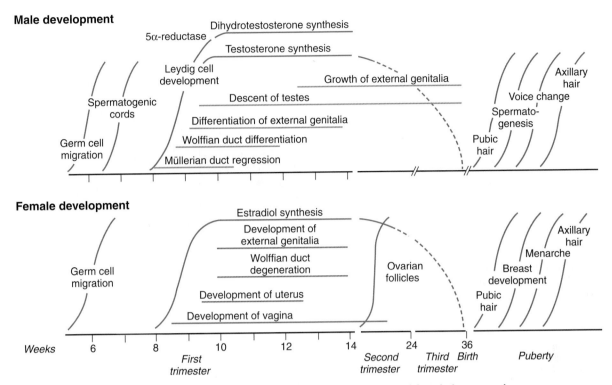

Fig. 16.21 **Major events in the sexual differentiation of male and female human embryos.**

influences on the differentiation of the external genitalia and finally on the development of the secondary sexual characteristics (e.g., body configuration, breasts, hair patterns) complete the events that constitute the overall process of sexual differentiation. Sexual differentiation of the brain, which has an influence on behavior, also occurs.

Under certain circumstances, an individual's genetic gender can be overridden by environmental factors so that the genotypic sex and the phenotypic sex do not correspond. An important general principle is that the development of phenotypic maleness requires the action of substances produced by the testis. In the absence of specific testicular influences or the ability to respond to them, a female phenotype results. Based on present information, the female phenotype is considered to be a baseline, or default, condition, which must be acted on by male influences to produce a male phenotype.

Genetic Determination of Gender

Since 1923, scientists have recognized that the XX and XY chromosomal pairings represent the genetic basis for human femaleness and maleness. For many decades, scientists believed that the presence of two X chromosomes was the sex-determining factor, but in 1959, it was established that the differentiation between maleness and femaleness in humans depends on the presence of a Y chromosome. Nevertheless, the link between the Y chromosome and the determination of the testis remained obscure. During more recent decades, three candidates for the **testis-determining factor** were proposed.

The first was the **H-Y antigen**, a minor histocompatibility antigen present on the cells of males but not females. The H-Y antigen was mapped to the long arm of the human Y chromosome. It had been considered to be the product of the mammalian testis-determining gene. Then a strain of mice *(Sxr)* was found to produce males in the absence of the H-Y antigen. *Sxr* mice were found to have a transposition of a region of the Y chromosome onto the X chromosome, but the locus coding for the H-Y antigen was not included. In addition, certain phenotypic human males with an XX genotype were shown to be missing the genetic material for the H-Y antigen.

The next candidate was a locus on the short arm of the Y chromosome called the *zinc finger Y (ZFY)* gene. With DNA hybridization techniques, this gene has been found in XX male humans and in mice in which small pieces of the X and Y chromosomes were swapped during crossing-over in meiosis. Conversely, this gene was missing in some rare XY human females. Certain XX males were found to lack the gene, however, and other rare cases of anomalies of sexual differentiation did not show a correspondence between sexual phenotype and the expected presence or absence of the *ZFY* gene.

The most recent candidate for the testis-determining gene is one called *Sry,* a member of the *Sox* family of transcription factors and probably an evolutionary derivative of *Sox-3*; it is also located within a 35-kb region on the short arm of the Y chromosome (**Fig. 16.22**). The *Sry* gene encodes a 223-amino acid nonhistone protein belonging to a family of proteins that contain a highly conserved 79-amino acid DNA-binding region called a **high mobility group box.** After the gene was cloned, it was detected in many cases of gender reversal, including XX males with no *ZFY* genes. The *SRY* gene on the human Y chromosome is located near the homologous region, thus making it susceptible to translocation to the X chromosome.

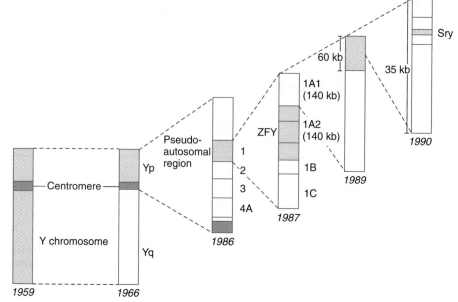

Fig. 16.22 **A history in the progress of localization of the sex-determining gene on the Y chromosome.** *(Adapted from Sultan C and others: SRY and male sex determination, Horm Res 36:1-3, 1991.)*

The *Sry* gene is also absent in a strain of XY mice that are phenotypically female. Further experimental evidence consisted of producing transgenic mice with the insertion of a 14-kb fragment of the Y chromosome containing the *Sry* gene. Many of the transgenic XX mice developed into phenotypic males with normal testes and male behavior. In situ hybridization studies in mice have shown that expression of the *Sry* gene product occurs in male gonadal tissue at the time of sex determination, but it is not expressed in the gonads of female embryos.

Specification of Germ Cells, Migration into the Gonads, and Entry into Meiosis

The early appearance of primordial germ cells (PGCs) in the lining of the yolk sac and their migration into the gonads in human embryos is briefly described in Chapter 1. Both descriptive and experimental studies in the mouse have shown that PGCs originate in the epiblast-derived extraembryonic mesoderm at the posterior end of the primitive streak. In the mouse, as few as six precursor cells become specified to become PGCs in response to BMP-2, BMP-4, and BMP-8b, which are secreted by nearby extraembryonic ectoderm. These cells maintain pluripotency by expressing Sox, Nanog, and Oct-4, much as these genes maintain the undifferentiated condition of blastomeres of cleaving embryos (see p. 42). They are protected by the transcriptional repressor **Blimp-1** from entering the default transcriptional program that directs cells of the epiblast to become somatic cells.

Once specified, the PGCs begin a phase of active migration (see Fig. 1.1), which takes them first from the base of the allantois to the future hindgut and, in a second phase, from the hindgut up the dorsal mesentery into the **genital ridges** (future gonads). During the migratory phase, PGCs are protected from undergoing apoptosis by the actions of **Nanos-3**, an evolutionarily conserved protein involved in germ cell

maintenance. The initial stages of migration of PGCs at some distance from the gonads are accomplished by active ameboid movement of the cells in response to a permissive extracellular matrix substrate. Tissue displacements through differential growth of the posterior region of the embryo may also contribute. During their migration, many PGCs are linked to one another through long cytoplasmic processes. How these interconnections control either migration or settling down in the gonads remains to be determined. As they migrate through the dorsal mesentery, the PGCs proliferate in response to mitogenic factors such as leukemia inhibitory factor and **Steel factor** (Kit-ligand).

As the germ cells approach the genital ridges late in the fifth week of development, they may be influenced by chemotactic factors secreted by the newly forming gonads. Such influences have been shown by grafting embryonic tissues (e.g., hindgut, which contains dispersed germ cells) into the body cavity of a host embryo. The PGCs of the graft typically concentrate on the side of the graft nearest the genital ridges of the host or sometimes migrate into the genital ridges from the graft. Approximately 1000 to 2000 PGCs enter the genital ridges. When the PGCs have penetrated the genital ridges, their migratory behavior ceases, and a new set of genes is activated.

After entry into the genital ridges, PGCs in females enter meiosis, whereas those in males undergo mitotic arrest. Initially, male and female PGCs are equivalent, and under the influence of **Dazl** (Deleted in azoospermia-like), they both progress to a meiosis-competent stage (**Fig. 16.23**). At this point, differing environments in male and female gonads exert a profound effect on the PGCs. In the female, **retinoic acid**, produced in the tubules of the adjoining mesonephros, is found in the gonad. Working through **Stra-8**, which is required for premeiotic DNA replication, retinoic acid stimulates the PGCs to enter the meiotic cycle. In the male gonad, the action of the cytochrome P450 enzyme **Cyp26b1** catabolizes the mesonephrically derived retinoic acid into inactive

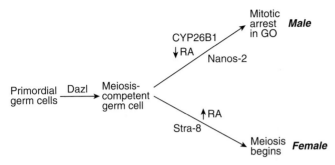

Fig. 16.23 Scheme showing the effect of exposure to different concentrations of retinoic acid (RA) on the fate of primordial germ cells in males and females.

metabolites. This action, along with the antimeiotic activity of **Nanos-2** within the germ cells, prevents entrance of the PGCs into meiosis. Instead, they become arrested at the G_0 phase of the mitotic cycle, where they remain until after birth. If Cyp26b1 is inactivated, male PGCs, like their female counterparts, are also propelled into the meiotic cycle. The female gonad suppresses the formation of both these meiosis-inhibiting factors (Cyp26b1 and Nanos-2).

Whether the PGCs, when in the gonad, develop into male or female gametes depends on the environment of the somatic cells in the gonad, not on their own genetic endowment. XY PGCs begin to develop into eggs if they are transplanted into a female gonad, and XX PGCs begin to develop into spermatogonia when they are transplanted into a male gonad.

Some PGCs follow inappropriate migratory pathways that lead the cells to settle into extragonadal sites. These cells normally start to develop as oogonia, regardless of genotype; they then degenerate. Rarely, however, the PGCs persist in ectopic sites, such as the mediastinum or the sacrococcygeal region, and ultimately may give rise to **teratomas** (see Chapter 1).

Establishment of Gonadal Gender
Origin of the Gonads and Adrenal Cortex

The gonads arise from an elongated region of steroidogenic mesoderm along the ventromedial border of the mesonephros. Cells in the cranial part of this region condense to form the **adrenocortical primordia**, and cells in the caudal part become the **genital ridges**, which are identifiable midway through the fifth week of gestation. The early genital ridges consist of two major populations of cells: one derived from the **coelomic epithelium** and the other arising from the **mesonephric ridge**.

One of the earliest genes required for development of the gonads is **WT-1**, which is expressed throughout the intermediate mesoderm and, as discussed earlier (see p. 376), is important in early kidney formation (see Fig. 16.3). **Steroidogenic factor-1** (**SF-1**) is expressed in the early indifferent gonad and in the developing adrenal cortex. SF-1 apparently is involved in the somatic cells of the early gonad, rather than the PGCs. In keeping with its association with endocrine organs, SF-1 is also expressed in cells of the pituitary and hypothalamus. The other major gene involved in the earliest phase of gonadal development is **Lim1**. As noted earlier in this text (see Fig. 5.9), the absence of **Lim1** expression results in the lack of formation of the anterior head; in addition, neither kidneys nor gonads form.

In humans, **adrenocortical primordia** appear during the fourth week as thickenings of the coelomic mesothelium at the cranial end of the genital ridge. By 8 weeks of gestation, the primordia are located at the cranial pole of the mesonephros. As neural crest cells migrate into the interior of an adrenal gland, a capsule forms around the gland. The cortex begins to differentiate into an inner zone that secretes steroidal precursors for placenta-derived estradiol, which is essential for the maintenance of pregnancy. An outer zone forms beneath the capsule, and in the center the neural crest cells differentiate into adrenal medullary cells, perhaps under the influence of the cortical environment.

Differentiation of the Testes

When the genital ridges first appear, those of males and females are morphologically indistinguishable (indifferent stage). The general principle underlying gonadal differentiation is that under the influence of the **Sry** gene (testis-determining factor) on the Y chromosome, the indifferent gonad differentiates into a testis (**Fig. 16.24**). In the absence of expression of products of this gene, the gonad later differentiates into an ovary.

In males, transcripts of the **Sry** gene are detected only in the genital ridge just at the onset of differentiation of the testis. Neither expression of the **Sry** gene nor later differentiation of the testis depends on the presence of germ cells. The sex-determining genes act on the somatic portion of the testis and not on the germ cells. In males, Sry stimulates the expression of **Sox-9**, which initiates the pathway of differentiation of undifferentiated stromal cells into Sertoli cells. Sox-9 stimulates FGF-9 activity, which reinforces Sox-9 activity (see Fig. 16.27).

Timing is important in differentiation of the testis. The testis develops more rapidly than the ovary. Precursors of the Sertoli cells must be prepared to receive the genetic signals (**Sox-9**) for testicular differentiation by a certain time. If not, the PGCs begin to undergo meiosis, and the gonad differentiates into an ovary. Early differentiation of the testis also seems to depend on a signal from the mesonephros, possibly WT-1. In the absence of the mesonephros, internal structures (testis cords) differentiate poorly.

The morphology of early gonadal differentiation has been controversial, with several proposed scenarios of cell lineage and interactions. According to more recent morphological evidence, the genital ridges first appear midway in the fifth week through the proliferation of **coelomic epithelial cells** along the medial border of the mesonephros (**Fig. 16.25**). Later in the fifth week, the PGCs enter the early genital ridge, and the coelomic epithelium sends short epithelial pillars toward the interior of the gonad. Early in the sixth week, under the influence of the transcription factor Sox-9, a set of **primitive sex cords** takes shape in the genital ridge, and the PGCs migrate into the primitive sex cords. The primitive sex cords are partitioned from one another by endothelial cells that grow into the genital ridge, probably from the mesonephros. The cords are then surrounded by a thin layer of myoid cells of local origin. The myoid cells, which in the adult contract and help to move developing sperm cells along the seminiferous tubules, have no equivalent in the ovary. It is still unclear how many types of cells migrate into the gonad from the mesonephros, and if so, what is the nature of the stimulus for their migration. Ovarian tissue does not attract these cells.

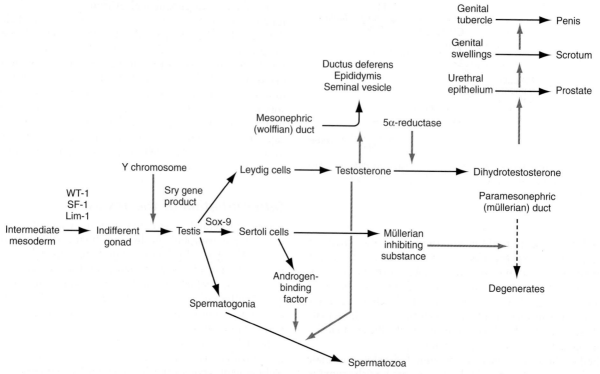

Fig. 16.24 **Differentiation of the male phenotype.**

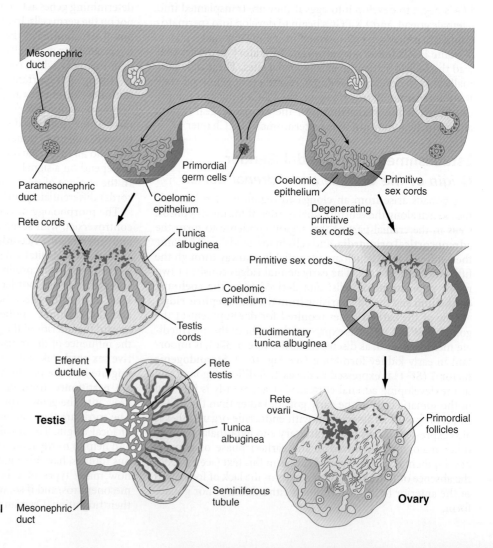

Fig. 16.25 **Morphology of gonadal differentiation.**

1. Induce migration of mesenchymal cells (endothelial cells) from mesonephros into testis
2. Give off signals that inhibit the entry of male germ cells into meiosis
3. Emit signals that induce the differentiation of Leydig cells, which secrete testosterone
4. Secrete müllerian inhibiting substance
5. Secrete androgen-binding factor

Table 16.1 **Origins of Cellular Components of Embryonic Gonads**

Site of Origin	Testis	Ovary
Primordial germ cells	Spermatogonia	Oogonia
Genital ridge	Leydig cells* Peritubular myoid cells	Theca cells —
	Sertoli cells*	Follicular granulosa cells
	Leydig cells*	Theca cells
Coelomic epithelium	Sertoli cells*	Follicular granulosa cells
	Leydig cells*	Theca cells
Mesonephros	Sertoli cells*	Follicular granulosa cells
	Leydig cells* Vascular endothelial cells	Theca cells ??
	Rete testis	Rete ovarii

*Origin still uncertain.

Late in the sixth week of gestation, the testis shows evidence of differentiation. The primitive sex cords enlarge and are better defined, and their cells are thought to represent the precursors of the Sertoli cells (**Box 16.1**). As the sex cords differentiate, they are separated from the surface epithelium (germinal epithelium) by a dense layer of connective tissue called the **tunica albuginea**. The deepest portions of the testicular sex cords are in contact with the fifth to twelfth sets of mesonephric tubules. The outer portions of the testicular sex cords form the **seminiferous tubules**, and the inner portions become meshlike and ultimately form the **rete testis**. The rete testis ultimately joins the **efferent ductules**, which are derived from mesonephric tubules.

For the first 2 months of development, **Leydig cells** are not identifiable in the embryonic testis. In addition to local steroidogenic cells arising from the genital ridge, Leydig cell precursors may migrate into the testis from the mesonephros (**Table 16.1**). These become recognizable during the eighth week and soon begin to synthesize androgenic hormones (testosterone and androstenedione). This endocrine activity is important because differentiation of the male sexual duct system and the external genitalia depends on the sex hormones secreted by the fetal testis. Fetal Leydig cells secrete their hormonal products at just the period when differentiation of the hormonally sensitive genital ducts occurs (9 to 14 weeks). After weeks 17 and 18, the fetal Leydig cells gradually involute and do not reappear until puberty, when they stimulate spermatogenesis. The fetal Leydig cells can be viewed as a cellular isoform that is later replaced by the definitive adult form of the cells. By 8 weeks, the embryonic Sertoli cells produce **müllerian inhibiting substance** (see p. 394), which also plays an important role in shaping the sexual duct system by causing involution of the precursors of the female genital ducts.

During the late embryonic and fetal periods and after birth, the PGCs in the testis divide slowly by mitosis, but the fetal Sertoli cells are insensitive to androgens and fail to mature. Sertoli cells are thought to secrete a **meiosis-inhibiting factor**, but such a factor has not been isolated or characterized. Embryonic male germ cells are also protected from the meiosis-inducing effects of retinoic acid by their location deep within the testis cords. The environment of the testis does not become favorable for meiosis and spermatogenesis until puberty.

Differentiation of the Ovaries

Despite considerable research, the factors leading to the differentiation of ovaries remain incompletely understood.

According to one hypothesis, **Wnt-4** and **Rspo-1** signals repress FGF-9 expression, thus leading to a reduction of Sox-9. This reduction inhibits testis development and leads to formation of an ovary. In contrast to the testes, the presence of viable germ cells is essential for ovarian differentiation. If PGCs fail to reach the genital ridges, or if they are abnormal (e.g., XO) and degenerate, the gonad regresses, and **streak ovaries** (vestigial ovaries) result.

After the PGCs have entered the future ovary, they remain concentrated in the outer cortical region or near the corticomedullary border. Similar to the testis, the ovary contains primitive sex cords in the medullary region, but these are not as well developed as the sex cords in the testis. The origin of the cells that form the ovarian follicles has not been established. Three sites of origin have been proposed for the follicular epithelial cells: (1) the coelomic epithelium (secondary sex cords), (2) the primitive sex cords of mesonephric origin, and (3) the stroma of the genital ridge itself. Shortly after they arrive in the gonad, clumps of PGCs in the medullary region become partially surrounded by follicular cells and take the form of **cell nests**. The PGCs, now properly called **oogonia**, briefly proliferate by mitosis until, under the influence of **retinoic acid**, they enter prophase of the first meiotic division. The retinoic acid, which is produced in the mesonephros (see earlier) is probably associated with the mesonephrically derived **rete ovarii**, which is located in the medulla of the ovary. By week 22, oogonia in the cortical region also enter meiosis. By the fetal period, the oogonia, now called **oocytes**, separate from the cell nests and become individually associated with follicular cells to form **primordial follicles** (see Fig. 1.5). The oocytes continue in meiosis until they reach the diplotene stage of prophase of the first meiotic division. Meiosis is then arrested, and the oocytes remain in this stage until the block is removed. In adults, this occurs in individual oocytes just days before ovulation. In premenopausal women, 50 years may have elapsed since these oocytes entered the meiotic block in embryonic life.

Table 16.2 Homologies in the Male and Female Urogenital Systems

Indifferent Structure	Male Derivative	Female Derivative
Genital ridge	Testis	Ovary
Primordial germ cells	Spermatozoa	Ova
Sex cords	Seminiferous tubules (Sertoli cells)	Follicular (granulosa) cells
Mesonephric tubules	Efferent ductules Paradidymis	Oöphoron Paroöphoron
Mesonephric (wolffian) ducts	Appendix of epididymis Epididymal duct Ductus deferens Ejaculatory duct Seminal vesicles	Appendix of ovary Gartner's duct
Mesonephric ligaments	Gubernaculum testis	Round ligament of ovary Round ligament of uterus
Paramesonephric (müllerian) ducts	Appendix of testis Prostate utricle	Uterine tubes Uterus Upper vagina
Definitive urogenital sinus (lower part)	Penile urethra Bulbourethral glands	Lower vagina Vaginal vestibule
Early urogenital sinus (upper part)	Urinary bladder Prostatic urethra Prostate gland	Urinary bladder Urethra Glands of Skene
Genital tubercle	Penis	Clitoris
Genital folds	Floor of penile urethra	Labia minora
Genital swellings	Scrotum	Labia majora

In the fetal ovary, an inconspicuous tunica albuginea forms at the corticomedullary junction. The cortex of the ovary is the dominant component, and it contains most of the oocytes. The medulla fills with connective tissue and blood vessels that are derived from the mesonephros. The testis is characterized by a dominance of the medullary component located inside a prominent tunica albuginea.

The developing ovary does not maintain a relationship with the mesonephros. Normally, the mesonephric tubules in the female embryo degenerate, leaving only a few remnants (**Table 16.2**).

Sexual Duct System

Similar to the gonads, the sexual ducts pass through an early indifferent stage. As the fetal testes begin to function in the male, their secretion products act on the indifferent ducts and cause some components of the duct system to develop further and others to regress. In females, the absence of testicular secretory products results in the preservation of structures

that regress and the regression of structures that persist in males.

Indifferent Sexual Duct System

The indifferent sexual duct system consists of the **mesonephric (wolffian) ducts** and the **paramesonephric (müllerian) ducts** (**Fig. 16.26**). The paramesonephric ducts appear between 44 and 48 days of gestation as longitudinal invaginations of the coelomic mesothelium along the mesonephric ridge lateral to the mesonephric ducts. Arising from thickened placodelike structures, the invaginations, which take on the form of epitheliumlike cords, extend toward the mesonephric ducts under the influence of Wnt-4 produced by the mesonephros. When associated with the mesonephric ducts, the tips of the paramesonephric ducts form a proliferative center and depend on a **Wnt-9b** signal from the mesonephric ducts for their continued caudal advancement toward the urogenital sinus. If the mesonephric ducts are interrupted, the caudally elongating paramesonephric ducts do not extend past the cut ends. The paramesonephric ducts do not develop a true lumen until they have contacted the urogenital sinus. The cranial end of each paramesonephric duct opens into the coelomic cavity as a funnel-shaped structure. The fate of the indifferent genital ducts depends on the gender of the gonad.

Sexual Duct System of Males

Development of the sexual duct system in the male depends on secretions from the testis. Under the influence of **müllerian inhibiting substance** (sometimes called antimüllerian hormone), a glycoprotein of the transforming growth factor-β family secreted by the Sertoli cells of the testes at 8 weeks' gestation, the paramesonephric ducts degenerate, leaving only remnants at their cranial and caudal ends (**Figs. 16.27** and **16.28**; see Table 16.2). Müllerian inhibiting substance apparently does not directly affect the epithelium of the paramesonephric ducts, but rather affects the surrounding mesenchyme. These mesenchymal cells express a gene that encodes a serine-threonine kinase membrane-bound receptor, which binds the müllerian inhibiting substance. Then the surrounding mesenchymal cells instruct the epithelial cells of the müllerian duct to regress through apoptosis and transformation of the epithelial cells into mesenchyme.

Two signals from the Sertoli cells, **desert hedgehog** and **platelet-derived growth factor**, stimulate the differentiation of fetal Leydig cells, which then begin to secrete testosterone. Under the influence of testosterone, the mesonephric ducts continue to develop even though the mesonephric kidneys are degenerating. The mesonephric ducts differentiate into the paired **ductus deferens**, which constitutes the path of sperm transport from the testis to the urethra. Portions of degenerating mesonephric tubules may persist near the testis as the **paradidymis**. *Hox* genes play a role in the specification of the various regions of the male reproductive tract. **Hoxa-10** is expressed along the mesonephric duct from the caudal epididymis to the point where the ductus deferens inserts into the urethra. Mutants of *Hoxa10* and *Hoxa11* exhibit a homeotic transformation that results in the partial transformation of ductus deferens to epididymis.

Associated with development of the male genital duct system (both the ductus deferens and the urethra) is the

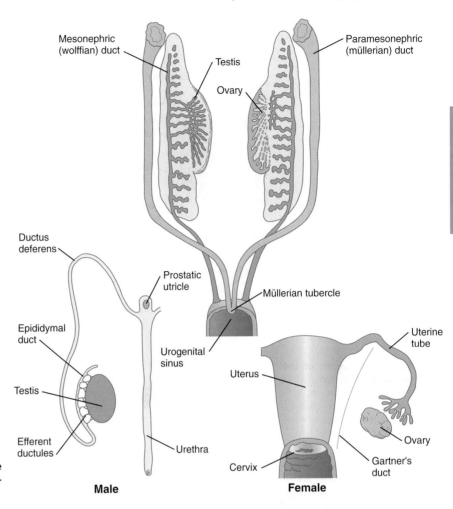

Fig. 16.26 **Indifferent condition of the genital ducts in the embryo at approximately 6 weeks.**

Male

Female

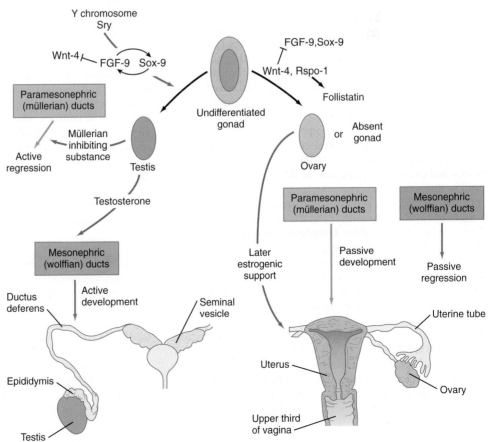

Fig. 16.27 **Factors involved in sexual differentiation of the genital tract.** FGF, fibroblast growth factor. *(After Hutson JM and others: In Burger H, deKrester D, eds: The testis, ed 2, New York, 1989, Raven Press, pp 143-179.)*

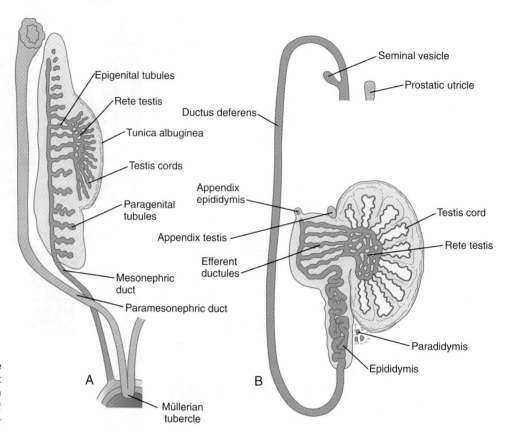

Fig. 16.28 Development of the male genital duct system. A, At the end of the second month. **B,** In the late fetus. *(Adapted from Sadler T: Langman's medical embryology, ed 6, Baltimore, 1990, Williams & Wilkins.)*

formation of the male accessory sex glands: the **seminal vesicles**, the **prostate**, and the **bulbourethral glands** (**Fig. 16.29**). These glands arise as epithelial outgrowths from their associated duct systems (seminal vesicles from the ductus deferens and the others from the urogenital sinus, the precursor of the urethra), and their formation involves epithelial-mesenchymal interactions similar to those of other glands. In addition, these glands depend on androgenic stimulation for their development. Specifically, the mesenchymal cells develop androgen receptors and seem to be the primary targets of the circulating androgenic hormones. (At this stage, the epithelial cells do not contain androgen receptors.) After stimulation by the androgens, the mesenchymal cells act on the associated epithelium through the local paracrine effects of growth factors and cause this epithelium to differentiate with gland-specific characteristics.

In the developing prostate, the urogenital mesenchyme induces epithelial outgrowths from the urogenital sinus endoderm just below the bladder. **Dihydrotestosterone** (see later), acting through receptors in the mesenchyme, and the resultant secretion of **FGF-10** and **transforming growth factor-β1** by the mesenchyme, regulates the production of **sonic hedgehog** (**shh**) in the epithelium of the urogenital sinus. In response to shh signaling and the involvement of retinoic acid, the prostatic ducts begin to bud off the epithelium of the urogenital sinus. The extent of budding is regulated by the inhibitory action of **BMP-4**, which is most strongly expressed lateral to the area where the prostatic ducts begin to bud off. Underlying all these molecular interactions is the action of the transcription factors, **Hoxa-13** and **Hoxd-13**, which determine that the organ that will form in this site is the prostate. Null

mutants of these *Hox* genes show a reduced number of prostatic ducts. The developing prostatic epithelium also induces the surrounding mesenchyme to differentiate into smooth muscle cells.

Tissue recombination experiments in which glandular mesoderm from mice with **testicular feminization syndrome** (lack of testosterone receptors resulting in no response to testosterone) was combined with normal epithelium showed that the mesodermal component of the glandular primordia is the hormonal target. Differentiation of the epithelium did not occur. In contrast, when normal glandular mesoderm was combined with epithelium from animals with testicular feminization syndrome, normal development occurred.

In the embryo, the tissues around the urogenital sinus synthesize an enzyme (**5α-reductase**) that converts testosterone to dihydrotestosterone. Through the action of appropriate receptors of either form of testosterone, crucial tissues of the male reproductive tract are maintained and grow (**Fig. 16.30**).

Sexual Duct System of Females

If ovaries are present, or if the gonads are absent or dysgenic, the sexual duct system differentiates into a female phenotype. In the absence of testosterone secreted by the testes, the mesonephric ducts regress, leaving only rudimentary structures (see Table 16.2). In contrast, the absence of müllerian inhibitory substance allows the paramesonephric (müllerian) ducts to continue to develop into the major structures of the female genital tract (**Fig. 16.31**).

Early formation of the paramesonephric ducts depends on Wnt signaling. In the absence of Wnt-4, the paramesonephric

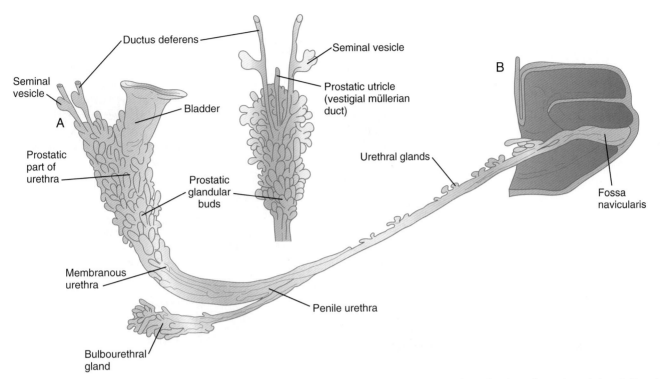

Fig. 16.29 **Development of the male urethra and accessory sex glands in an embryo at approximately 16 weeks.** A, Lateral view. B, Dorsal view of prostatic region. *(After Didusch. From Johnson FP: J Urol 4:447-502, 1920.)*

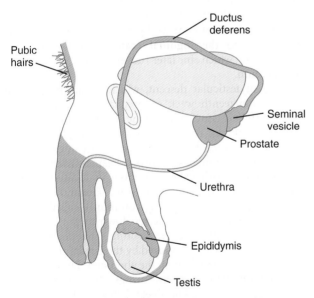

Fig. 16.30 Regions of the male reproductive tract sensitive to testosterone *(brown)* and dihydrotestosterone *(blue)*.

ducts fail to form. Wnt-7a, which is also involved in setting up the dorsoventral axis of the developing limb, is expressed in the epithelium throughout the entire paramesonephric ducts and is required for their normal development. Later in development, Wnt-7a expression becomes restricted to the uterine epithelium. In some manner, Wnt-7a seems to be involved in maintaining the expression of a sequence of *Hox* genes (**Hoxd-10** through **Hoxd-13** and the Hoxa paralogues) that are spread along the female reproductive tract (**Fig. 16.32**): Hoxa-9 is expressed in the uterine tubes; Hoxa-10, in the uterus; Hoxa-11, in the uterus and cervix; and Hoxa-12, in the upper vagina.

As is the case in the male reproductive tract, mutations of *Hox* genes result in homeotic transformations. In the absence of Hoxa-10, the cranial part of the uterus becomes transformed into uterine tube. In contrast to other areas of the body, *Hox* gene expression throughout the female reproductive tract, at least in the mouse, continues into adult life. This continued expression may be related to the developmental plasticity required of the female reproductive tract throughout the reproductive cycle.

The cranial portions of the paramesonephric ducts become the **uterine tubes**, with the cranial openings into the coelomic cavity persisting as the fimbriated ends. Toward their caudal ends, the paramesonephric ducts begin to approach the midline and cross the mesonephric ducts ventrally. This crossing and ultimate meeting in the midline are caused by the medial swinging of the entire urogenital ridge (**Fig. 16.33**). The region of midline fusion of the paramesonephric ducts ultimately becomes the uterus, and the ridge tissue that is carried along with the paramesonephric ducts forms the **broad ligament of the uterus**.

The formation of the vagina remains poorly understood, and several explanations for its origin have been posited. According to one commonly held hypothesis, the fused paramesonephric ducts form the upper part of the vagina, and epithelial tissue from the **müllerian tubercle** (**uterovaginal plate**) hollows out to form the lower part (**Fig. 16.34**). More recently, several investigators have suggested that the most caudal portions of the mesonephric ducts participate in the formation of the vagina either by directly contributing cells to its wall or by inductively acting on the paramesonephric tissue, which appears to diverge toward the mesonephric ducts at the very tip of the fused portion. Full development of the female reproductive tract depends on estrogenic hormones secreted by the fetal ovaries.

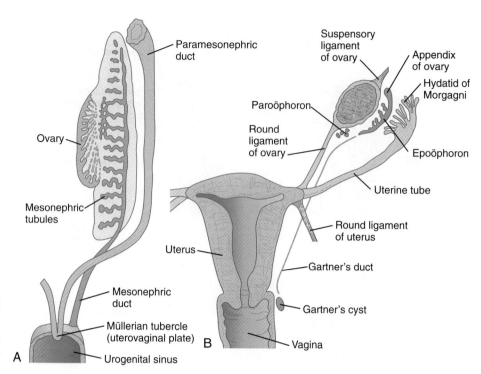

Fig. 16.31 **Development of the female genital duct system. A,** At the end of the second month. **B,** Mature condition. *(Adapted from Sadler T: Langman's medical embryology, ed 6, Baltimore, 1990, Williams & Wilkins.)*

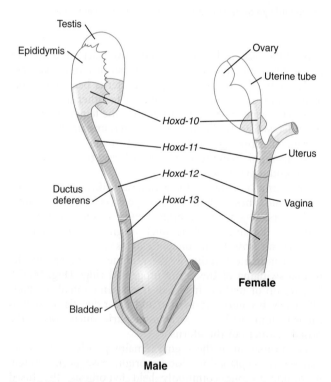

Fig. 16.32 **Gradients of *Hoxd* gene expression in the male and female internal genitalia of mouse embryos.** *(Based on studies by Dollé and others [1991].)*

Descent of the Gonads
Descent of the Testes

The testes do not remain in their original site of development; they migrate from their intra-abdominal location into the scrotum (**Fig. 16.35**). Similar to the kidneys, the testes are retroperitoneal structures, and their descent occurs behind the peritoneal epithelium. Before their descent, the testes are anchored cranially to the **cranial suspensory ligament**, derived from the diaphragmatic ligament of the mesonephros, and caudally to the **inguinal (caudal) ligament of the mesonephros**, which in later development is called the **gubernaculum**.

Control of testicular descent, which occurs between the tenth and fourteenth week of pregnancy, occurs in three phases. The first is associated with the enlargement of the testes and the concomitant regression of the mesonephric kidneys. Under the influence of androgens, acting through androgen receptors in the cranial suspensory ligament, the ligament regresses, thus releasing the testis from its location near the diaphragm. This regression causes some caudal displacement of the testes. The second phase, commonly called **transabdominal descent**, brings the testes down to the level of the inguinal ring, but not into the scrotum. This phase depends on the activity of **Insl-3**, produced by the Leydig cells, without which the testes remain high in the abdomen. The third phase, called **transinguinal descent**, brings the testes into the scrotum, usually just a few weeks before birth. This phase involves the action of testosterone and the guidance of the **inguinal ligament of the mesonephros**, which in later development is called the **gubernaculum**. Whether the gubernaculum actively pulls the testis into the scrotum or just acts as a fixed point while the other tissues grow has not been resolved.

Testicular descent begins during the seventh month and may not be completed until birth. As it descends into the scrotum, the testis slides behind an extension of the peritoneal cavity known as the vaginal process (see Fig. 16.35C). Although this cavity largely closes off with maturation of the testis, it remains as a potential mechanical weak point. With straining, it can open and permit the herniation of intestine into the scrotum.

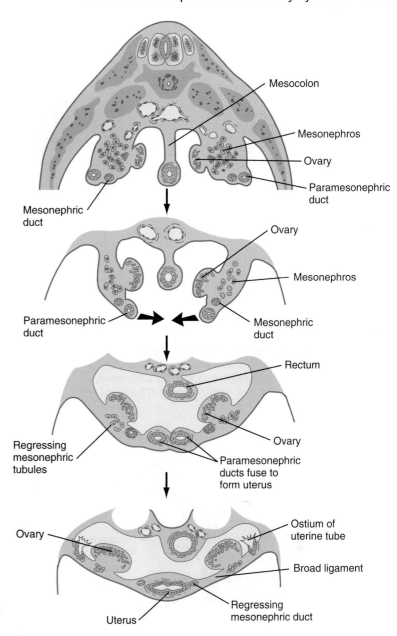

Fig. 16.33 **Formation of the broad ligament in the female embryo.**

Descent of the Ovaries

Although not so dramatically as the testes, the ovaries also undergo a distinct caudal shift in position. In conjunction with their growth and the crossing over of the paramesonephric ducts, the ovaries move caudally and laterally. Their position is stabilized by two ligaments, both of which are remnants of structures associated with the mesonephros. Cranially, the **diaphragmatic ligament of the mesonephros** becomes the **suspensory ligament of the ovary**. The superior portion of the inguinal ligament (called the **caudal gonadal ligament** by some authors) develops into the **round ligament of the ovary**, and the inferior portion of the inguinal ligament becomes the **round ligament of the uterus** (see Fig. 16.31). The most caudal ends of the round ligaments of the uterus become embedded in the dense fascial connective tissue of the labia majora.

External Genitalia

Indifferent Stage

The external genitalia are derived from a complex of mesodermal tissue located around the cloaca. A very early midline elevation called the **genital eminence** is situated just cephalic to the proctodeal depression. This structure soon develops into a prominent **genital tubercle** (**Fig. 16.36**), which is flanked by a pair of **genital folds** extending toward the proctodeum. Lateral to these are paired **genital swellings** (**Fig. 16.37**). When the original cloacal membrane breaks down during the eighth week, the urogenital sinus opens directly to the outside between the genital folds. An endodermal **urethral plate** lines much of the open urogenital sinus. These structures, which are virtually identical in male and female embryos

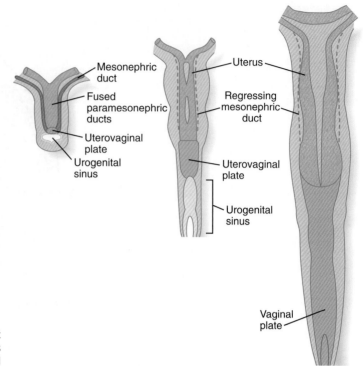

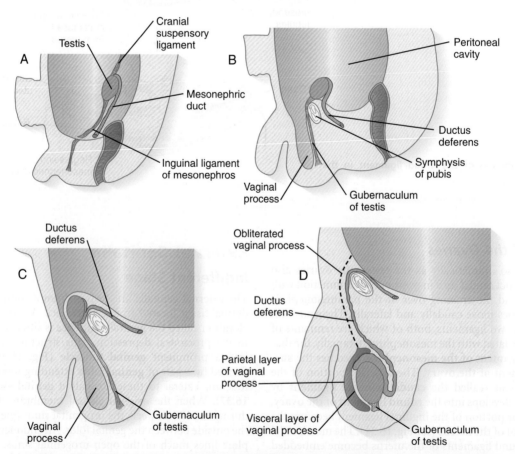

Fig. 16.34 **Development of the uterus and vagina.** Contact between the fused müllerian ducts and urogenital sinus stimulates proliferation of the junctional endoderm to form the uterovaginal plate. Later canalization of the plate forms the lumen of the vagina.

Fig. 16.35 **Descent of the testis in the male fetus. A,** In the second month. **B,** In the third month. **C,** In the seventh month. **D,** At term.

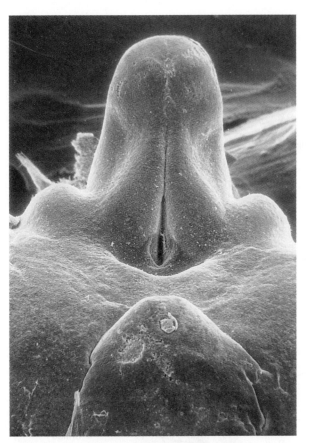

Fig. 16.36 Scanning electron micrograph of the inferior aspect of the indifferent external genitalia of a human embryo at the end of the eighth week of development. *(From Jirásek JE: Atlas of human prenatal morphogenesis, Amsterdam, 1983, Martinus Nijhoff.)*

during the indifferent stage, form the basis for development of the external genitalia.

Development of the indifferent stage of the genital tubercle is essentially independent of hormonal influences, whereas the differences in later development between males and females are highly dependent on the effects of sex hormones.

There are some parallels between development of the genital tubercle and that of the limbs, but significant differences also exist. Both use many of the same molecular players, starting with an underlying basis of *Hox* gene expression. Being at the terminal part of the urogenital system, the genital tubercle expresses the 5' elements along the *Hox* gene clusters, specifically Hoxa-13 and Hoxd-13. The signal initiating the development of the genital tubercle is still not known. Shh, which is expressed in the endodermal urethral epithelial plate, is the principal molecule that acts on the mesenchyme and outer ectoderm to cause outgrowth of the genital tubercle. Many members of the FGF and Wnt families of signaling molecules are active in the genital tubercle, but their exact functions remain incompletely understood. Small pieces of genital tubercle exhibit polarizing activity when they are grafted into anterior regions of the limb bud (see Chapter 10). This activity is probably caused by the presence of shh in the endoderm of the tubercle.

External Genitalia of Males

Under the influence of dihydrotestosterone (see Fig. 16.30), the genital tubercle in the male undergoes a second phase of elongation to form the penis, and the genital swellings enlarge to form scrotal pouches (see Fig. 16.37). As this growth is occurring, the urethra takes shape. The male urethra forms in a proximodistal direction by the ventral folding and midline fusion of the genital folds. This results in the formation of a midline epithelial seam on the ventral face of the elongating genital tubercle (**Fig. 16.38B**). The midline seam undergoes proximodistal remodeling by secondary canalization and detachment from the ventral surface epithelium to form the urethra proper (see Fig. 16.38C). Various signaling systems, including BMP-7, Eph-ephrin, and FGF, are involved in ventral closure of the urethra. The entire length of the urethra is formed from the endodermal lining of the urogenital sinus, and the histological characteristic of the distal epithelial lining (a stratified squamous epithelium) can be accounted for by the specific inductive effect of the glandular mesenchyme acting on the urethral epithelium. After the urethra has formed and has become detached from the ventral epithelial seam, the line of fusion of the urethral folds is marked by the persistence of a ventral **raphe**, which is continuous with the midline raphe that passes between the scrotal swellings.

Outgrowth of the male phallus is highly testosterone dependent. In the absence of testosterone or the absence of functional testosterone receptors in the testicular feminization syndrome (see Clinical Correlation 16.2), significant outgrowth of the penis does not occur.

External Genitalia of Females

In females, the pattern of external genitalia is similar to the pattern of the indifferent stage (see Fig. 16.37). The genital tubercle becomes the **clitoris**, the genital folds become the **labia minora**, and the genital swellings develop into the **labia majora**. The urogenital sinus remains open as the vestibule, into which the urethra and the vagina open. The female **urethra**, developing from the more cranial part of the urogenital sinus, is equivalent to the prostatic urethra of the male, which has a similar origin. The lack of outgrowth of the clitoris was traditionally considered to result solely from the absence of a dihydrotestosterone influence, but more recent research has also implicated an inhibitory influence of estrogen receptors. In mice, if estrogen receptors are inactivated, the clitoris undergoes elongation, and partial masculinization of the external genitalia occurs. This may be caused by the masculinizing influence of basal levels of androgens, which, in normal development, is repressed by estrogens.

Clinical Correlation 16.2 presents malformations of the genital system.

Clinical Vignette

A female athlete with amenorrhea is subjected to a routine sex chromatin test and is told that she cannot compete because she is a male.
1. What was the appearance of the cells that were tested?
2. What was the most likely basis for her female phenotype?
3. What is the likely anatomy of her gonads and genital duct systems, and why?

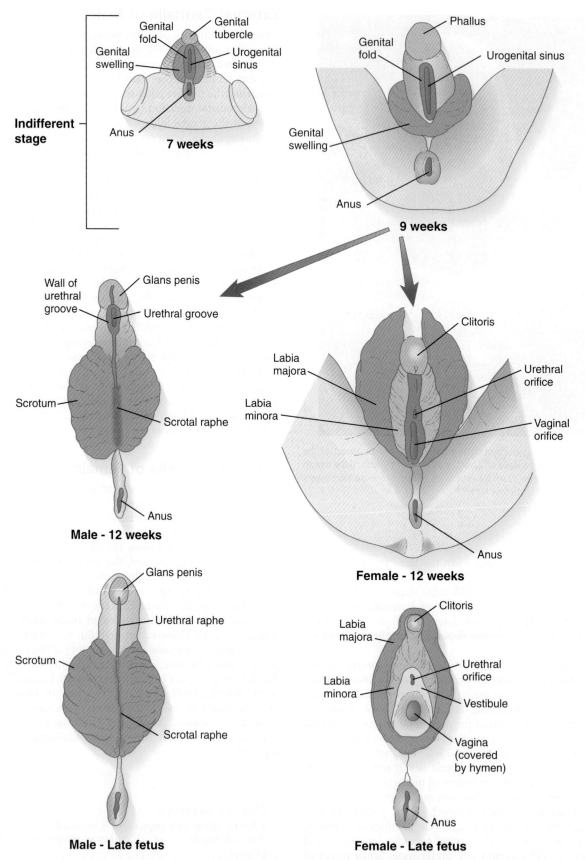

Fig. 16.37   Differentiation of the external genitalia of embryos.

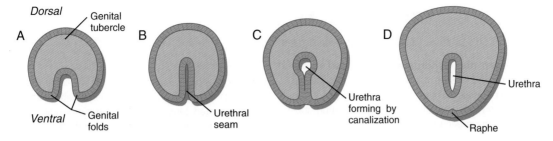

Fig. 16.38   **A** to **D, Cross-sectional views through the developing penis that show the mode of formation of the male penile urethra.**

CLINICAL CORRELATION 16.2
Malformations of the Genital System

Abnormalities of Sexual Differentiation

Turner's Syndrome (Gonadal Dysgenesis)

Turner's syndrome results from a chromosomal anomaly (45,XO) (see p. 142). Individuals with this syndrome possess primordial germ cells that degenerate shortly after reaching the gonads. Differentiation of the gonad fails to occur, thus leading to the formation of a **streak gonad**. In the absence of gonadal hormones, the genitalia develop along female lines, but they remain infantile. The mesonephric duct system regresses for lack of androgenic hormonal stimulation.

True Hermaphroditism

Individuals with true hermaphroditism, which is an extremely rare condition, possess both testicular and ovarian tissues. In cases of genetic mosaicism, an ovary and a testis may be present; in other cases, ovarian and testicular tissues are present in the same gonad (**ovotestis**). Most true hermaphrodites have a 46,XX chromosome constitution, and the external genitalia are basically female, although typically the clitoris is hypertrophied. Such individuals are usually reared as girls.

Female Pseudohermaphroditism

Female pseudohermaphrodites are genetically female (46,XX) and are sex chromatin positive. The internal genitalia are typically female, but the external genitalia are masculinized, either from excessive production of androgenic hormones by the adrenal cortex (**congenital virilizing adrenal hyperplasia**) or from inappropriate hormonal treatment of pregnant women. The degree of external masculinization can vary from simple clitoral enlargement to partial fusion of the labia majora into a scrotumlike structure.

Male Pseudohermaphroditism

Male pseudohermaphrodites are sex chromatin negative (46,XY). Because this condition commonly results from inadequate hormone production by the fetal testes, the phenotype can vary. It is often associated with hypoplasia of the phallus, and there may be various degrees of persistence of paramesonephric duct structures.

Testicular Feminization (Androgen Insensitivity) Syndrome

Individuals with testicular feminization syndrome are genetic males (46,XY) and possess internal testes, but they typically have a normal female external phenotype and are raised as girls (see Fig. 9.13). Often, testicular feminization is not discovered until the individual seeks treatment for amenorrhea or is tested for sex chromatin before athletic events. The testes typically produce testosterone, but because of a deficiency in receptors caused by a mutation on the X chromosome, the testosterone is unable to act on the appropriate tissues. Because müllerian inhibitory substance is produced by the testes, the uterus and upper part of the vagina are absent.

Vestigial Structures from the Embryonic Genital Ducts

Vestigial structures are remnants from the regression of embryonic genital ducts, which is rarely complete. These structures are so common that they are not always considered to be malformations, although they can become cystic.

Mesonephric Duct Remnants

In males, a persisting blind cranial end of the mesonephric duct can appear as the **appendix of the epididymis** (see Fig. 16.28). Remnants of a few mesonephric tubules caudal to the efferent ductules occasionally appear as the **paradidymis**.

In females, the remains of the cranial parts of the mesonephros may persist as the **epoöphoron**, or **paroöphoron** (see Fig. 16.31). The caudal part of the mesonephric ducts is often seen in histological sections along the uterus or upper vagina as Gartner's ducts. Portions of these duct remnants sometimes enlarge to form cysts.

Paramesonephric Duct Remnants

The cranial tip of the paramesonephric duct may remain as the small **appendix of the testis** (see Fig. 16.28). The fused caudal ends of the paramesonephric ducts are commonly seen in the prostate gland as a small midline **prostatic utricle**, which represents the rudimentary uterine primordium. In male newborns, the prostatic utricle is typically slightly enlarged because of the influence of maternal estrogenic hormones during pregnancy, but it regresses soon after birth. This structure can enlarge to form a uteruslike structure in some cases of male pseudohermaphroditism. In females, a small part of the cranial tip of the paramesonephric duct may persist at the fimbriated end of the uterine tube as the **hydatid of Morgagni** (see Fig. 16.31).

Other Abnormalities of the Genital Duct System

Males

Abnormalities of the mesonephric duct system are rare, but duplications or diverticula of the ductus deferens or urethra can occur. There is a correlation of absent or rudimentary ductus deferens in boys with cystic fibrosis. It may be the result of a defect in a gene situated alongside the gene causing cystic fibrosis.

Continued

CLINICAL CORRELATION 16.2
Malformations of the Genital System—cont'd

Persistent müllerian duct syndrome, characterized by the formation of a uterus and uterine tubes, has been described in some 46,XY phenotypic males. There is no single cause for this condition, and mutations of genes for müllerian inhibiting substance and its receptor have been documented.

Females

Malformations of the uterus or vagina are attributed to abnormalities of fusion or regression of the caudal ends of the paramesonephric ducts (**Fig. 16.39**). Uterine anomalies range from a small septum extending from the dorsal wall of the uterus to complete duplication of the uterus and cervix. Numerous successful pregnancies have been recorded in women with uterine malformations. **Agenesis of the vagina** has been attributed to a failure of formation of the epithelial vaginal plate from the site of joining of the müllerian tubercle with the urogenital sinus.

Abnormalities of Testicular Descent

Cryptorchidism

Undescended testes are common in premature male infants and are seen in about 3% of term male infants. Normally, the testes of these individuals descend into the scrotum within the first few months after birth. If they do not, the condition of **cryptorchidism** results. Descent of the testes requires the activity of **Insl-3** and androgens, but how disturbances in these molecules result in cryptorchidism remains obscure. Cryptorchidism results in sterility because spermatogenesis does not normally occur at the temperature of the body cavity. There is also a 50-fold greater incidence of malignant disease in undescended testes.

Ectopic Testes

A testis occasionally migrates to some site other than the scrotum, including the thigh, perineum, and ventral abdominal wall. Because of the elevated temperature of the surrounding tissues, ectopic testes produce reduced numbers of viable spermatozoa.

Congenital Inguinal Hernia

If the peritoneal canal that leads into the fetal scrotum fails to close, a condition called **persistent vaginal process** occurs. This space may become occupied by loops of bowel that herniate into the scrotum.

Malformations of the External Genitalia

Males

The most common malformation of the penis is hypospadias, in which the urethra opens onto the ventral surface of the penis, rather than at the end of the glans (**Fig. 16.40**). The degree of hypospadias can range from a mild ventral deviation of the urethral opening to an elongated opening representing an unfused portion of the urogenital sinus. In mice, lack of local expression of *Hoxa-13*, disturbances in the Eph-ephrin system, or faulty fibroblast growth factor (FGF) or bone morphogenetic protein (BMP) signaling can all result in the same varieties of hypospadias as those seen in humans. In the more severe varieties, the penis is often bowed ventrally (**chordee**).

Isolated **epispadias**, with the urethra opening on the dorsal surface of the penis, is very rare. A dorsal groove on the penis is commonly associated with exstrophy of the bladder (see Fig. 16.20).

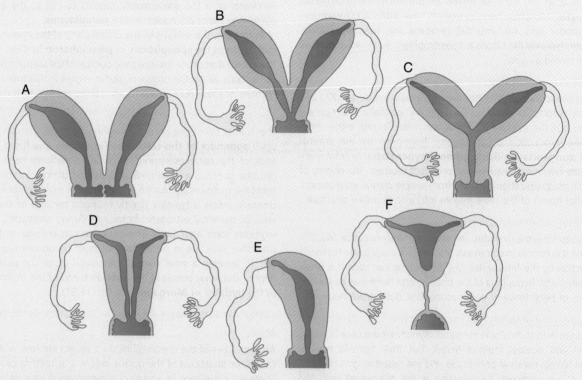

Fig. 16.39   **Abnormalities of the uterus and vagina. A,** Double uterus and double vagina. **B,** Double uterus and single vagina. **C,** Bicornuate uterus. **D,** Septate uterus. **E,** Unicornuate uterus. **F,** Atresia of the cervix.

CLINICAL CORRELATION 16.2
Malformations of the Genital System—cont'd

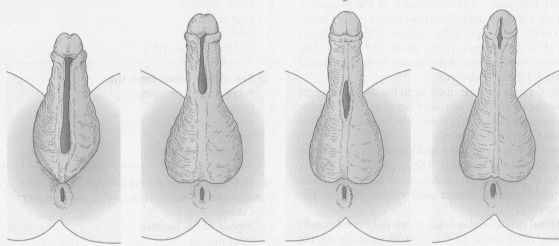

Fig. 16.40   Variations in the extent of hypospadias.

Duplication of the penis occurs most commonly in association with exstrophy of the bladder and seems to result from the early separation of the tissues destined to form the genital tubercle. Duplication of the penis very rarely occurs separately from exstrophy of the bladder.

Congenital absence of the penis (or clitoris in females) is rare. This condition most likely results from mutations of the distalmost *Hox* genes, specifically *Hoxa-13* or *Hoxd-13,* or a lack of sonic hedgehog (shh) function in the area of the genital tubercle.

Females

Anomalies of the female external genitalia can range from hormonally induced enlargement of the clitoris to duplications. Exposure to androgens may also masculinize the genital swellings, with resulting scrotalization of the labia majora. Depending on the degree of severity, wrinkling of the skin and partial fusion may occur.

Summary

- The urogenital system arises from the intermediate mesoderm. The urinary system arises before gonadal development begins.

- Kidney development begins with the formation of pairs of nephrotomes that connect with a pair of primary nephric ducts. Farther caudally to the nephrotomes, pairs of mesonephric tubules form in a craniocaudal sequence and connect to the primary nephric ducts, which become known as mesonephric ducts. In the caudal part of each mesonephric duct, a ureteric bud grows out and induces the surrounding mesoderm to form the metanephros.

- Outgrowth of the ureteric bud is stimulated by GDNF, produced by the metanephrogenic mesenchyme. This inductive signal is bound by c-Ret on the ureteric bud. FGF-2, BMP-7, and leukemia inhibitory factor, secreted by the ureteric bud, stimulate the formation of renal tubules in the metanephrogenic mesenchyme.

- Within the developing metanephros, nephrons (functional units of the kidney) form from three sources: the metanephrogenic blastema, the metanephrogenic diverticulum, and ingrowing vascular endothelial cells. Nephrons continue to form throughout fetal life. The induction of nephrons involves reciprocal inductions between terminal branches of the collecting duct system (ureteric bud) and the metanephrogenic mesoderm. Many molecular interactions mediate these inductions.

- The kidneys arise in the pelvic basin. During the late embryonic and early fetal period, the kidneys shift into the abdominal region, where they become associated with the adrenal glands. The urinary bladder arises from the base of the allantois.

- The urinary system is subject to various malformations. The most severe is renal agenesis, which is probably caused by faulty induction in the early embryo. Abnormal migration can result in pelvic kidneys, other ectopic kidneys, or horseshoe kidney. Polycystic disease of the kidney is associated with cysts in other internal organs. Faulty closure of the allantois results in urachal cysts, sinuses, or fistulas.

- Sex determination begins at fertilization by the contribution of an X or a Y chromosome to the egg by the sperm. The early embryo is sexually indifferent. Through the action of the *Sry* gene, the indifferent gonad in the male develops into a testis. In the absence of this gene, the gonad becomes an ovary.

- Gonadal differentiation begins after migration of the PGCs into the indifferent gonads. Under the influence of the *Sry* gene product (testis-determining factor), the testis begins to differentiate. The presence of germ cells is not required for differentiation of testis cords. In the embryonic testis, Leydig cells secrete testosterone, and Sertoli cells produce müllerian inhibitory substance. In the absence of *Sry* expression, the gonad differentiates into an ovary and contains follicles. Ovarian follicular differentiation does not occur in the absence of germ cells.

■ The sexual duct system consists of the mesonephric (wolffian) and paramesonephric (müllerian) ducts. The duct system is originally indifferent. In the male, müllerian inhibitory substance causes regression of the paramesonephric duct system, and testosterone causes further development of the mesonephric duct system. In the female, the mesonephric ducts regress in the absence of testosterone, and the paramesonephric ducts persist in the absence of müllerian inhibitory substance.

■ In males, the mesonephric ducts form the ductus deferens and give rise to the male accessory sex glands. In females, the paramesonephric ducts form the uterine tubes, the uterus, and part of the vagina.

■ The testes descend from the abdominal cavity into the scrotum later in development. The ovaries also shift to a more caudal position. Faulty descent of the testes results in cryptorchidism and is associated with sterility and testicular tumors.

■ The external genitalia also begin in an indifferent condition. Basic components of the external genitalia are the genital tubercle, genital folds, and genital swellings. Under the influence of dihydrotestosterone, the genital tubercle elongates into a phallus, and the genital folds fuse to form the penile urethra. The genital swellings form the scrotum. In the female, the genital tubercle forms the clitoris, the genital folds form the labia minora, and the genital swellings form the labia majora.

■ If an individual possesses only one X chromosome (XO), Turner's syndrome results. Such individuals have a female phenotype with streak gonads. True hermaphroditism or pseudohermaphroditism can result from various causes. Testicular feminization is found in genetic males lacking testosterone receptors. Such individuals are phenotypic females. Major abnormalities of the sexual ducts are rare, but they can lead to duplications or the absence of the uterus in females.

Review Questions

1. Which of the following does not connect directly with the primary nephric (mesonephric) duct?
A. Metanephros
B. Cloaca
C. Nephrotomes
D. Mesonephric tubules
E. Ureteric bud

2. Which association is correct?
A. Potter's facies and hydramnios
B. Urachal fistula and hydramnios
C. Horseshoe kidney and superior mesenteric artery
D. GDNF and metanephrogenic blastema
E. Bilateral renal agenesis and compensatory hypertrophy

3. Which defect is strongly associated with oligohydramnios?
A. Pelvic kidney
B. Renal agenesis
C. Horseshoe kidney
D. Crossed ectopia
E. Polycystic kidney

4. Which anomaly is most closely associated with exstrophy of the bladder?
A. Epispadias
B. Renal agenesis
C. Anal atresia
D. Pelvic kidney
E. Ectopic ureteral orifice

5. The uterus arises from the:
A. Paramesonephric ducts
B. Urogenital sinus
C. Mesonephric tubules
D. Pronephric ducts
E. Mesonephric ducts

6. The floor of the penile urethra in the male is homologous to what structure in the female?
A. Clitoris
B. Trigone of the bladder
C. Labia majora
D. Labia minora
E. Perineum

7. The metanephrogenic blastema is induced by the:
A. Pronephric duct
B. Ureteric bud
C. Mesonephric tubules
D. Allantois
E. Mesonephric duct

8. Drops of a yellowish fluid were observed around the umbilicus of a young infant. What is a likely diagnosis, and what is the embryological basis?

9. A woman who gained relatively little weight during pregnancy gives birth to an infant with large, low-set ears, a flattened nose, and a wide interpupillary space. Within hours after birth, the infant is obviously in great distress and dies after 2 days. What is the diagnosis?

10. A seemingly normal woman experiences pelvic pain during the later stages of pregnancy. An ultrasound examination reveals that she has a bicornuate uterus. What is the embryological basis for this condition?

References

Barsoum I, Yao HH-C: The road to maleness: from testis to wolffian duct, *Trends Endocrinol Metab* 17:223-228, 2006.

Baskin LS and others: Urethral seam formation and hypospadias, *Cell Tissue Res* 305:379-387, 2001.

Basson MA and others: Sprouty1 is a critical regulator of GDNF/RET-mediated kidney induction, *Dev Cell* 8:229-239, 2005.

Bowles J, Koopman P: Retinoic acid, meiosis and germ cell fate in mammals, *Development* 134:3401-3411, 2007.

Boyle S, de Caestecker M: Role of transcriptional networks in coordinating early events during kidney development, *Am J Physiol Renal Physiol* 291:F1-F8, 2006.

Brennan J, Capel B: One tissue, two fates: molecular genetic events that underlie testis versus ovary development, *Nat Rev Genet* 5:509-521, 2004.

Brenner-Anantharam A and others: Tailbud-derived mesenchyme promotes urinary tract segmentation via BMP-4 signaling, *Development* 134:1967-1975, 2007.

Cartry J and others: Retinoic acid signalling is required for specification of pronephric cell fate, *Dev Biol* 299:35-51, 2006.

Cohn MJ: Development of the external genitalia: conserved and divergent mechanisms of appendage patterning, *Dev Dyn* 240:1108-1115, 2011.

Combes AN and others: Endothelial cell migration directs testis cord formation, *Dev Biol* 326:112-120, 2009.

Costantini F: Renal branching morphogenesis: concepts, questions, and recent advances, *Differentiation* 74:402-421, 2006.

Costantini F, Kopan R: Patterning a complex organ: branching morphogenesis and nephron segmentation in kidney development, *Dev Cell* 18:698-712, 2010.

Costantini F, Shakya R: GDNF/Ret signaling and the development of the kidney, *Bioessays* 28:117-127, 2006.

DeFalco T, Capel B: Gonad morphogenesis: divergent means to a convergent end, *Annu Rev Cell Dev Biol* 25:457-482, 2009.

Dressler GR: Advances in early kidney specification, development and patterning, *Development* 136:3863-3874, 2009.

Durcova-Hills G, Capel B: Development of germ cells in the mouse, *Curr Top Dev Biol* 83:185-212, 2008.

Ewen KA, Koopman P: Mouse germ cell development: from specification to sex determination, *Mol Cell Endocrinol* 323:76-93, 2010.

Guigon CJ, Magre S: Contribution of germ cells to the differentiation and maturation of the ovary: insights from models of germ cell depletion, *Biol Reprod* 74:450-458, 2005.

Guioli S, Sekido R, Lovell-Badge R: The origin of the mullerian duct in chick and mouse, *Dev Biol* 302:389-398, 2007.

Hammer GD, Parker KL, Schimmer BP: Minireview: transcriptional regulation of adrenocortical development, *Endocrinology* 146:1018-1024, 2005.

Haraguchi R and others: Molecular analysis of coordinated bladder and urogenital organ formation by hedgehog signaling, *Development* 134:525-533, 2007.

Joseph A, Yao H, Hinton BT: Development and morphogenesis of the wolffian/epididymal duct, more twists and turns, *Dev Biol* 325:6-14, 2009.

Kashimada K, Koopman P: *Sry*: The master switch in mammalian sex determination, *Development* 137:3921-3930, 2010.

Kim Y, Capel B: Balancing the bipotential gonad between alternative organ fates: a new perspective on an old problem, *Dev Dyn* 235:2292-2300, 2006.

Kimble J, Page DC: The mysteries of sexual identity: the germ cell's perspective, *Science* 316:400-401, 2007.

Klonisch T, Fowler PA, Hombach-Klonisch S: Molecular and genetic regulation of testis descent and external genitalia development, *Dev Biol* 270:1018, 2004.

Lin Y and others: Germ cell-intrinsic and -extrinsic factors govern meiotic initiation in mouse embryos, *Science* 322:1685-1687, 2008.

Little M and others: Kidney development: two tales of tubulogenesis, *Curr Top Dev Biol* 90:193-229, 2010.

Liu C-F, Liu C, Yao HH-C: Building pathways for ovary organogenesis in the mouse embryo, *Curr Top Dev Biol* 90:263-290, 2010.

Ludbrook LM, Harley VR: Sex determination: a "window" of DAX1 activity, *Trends Endocrinol Metab* 15:116-121, 2004.

Ludwig KS, Landmann L: Early development of the human mesonephros, *Anat Embryol* 209:439-447, 2005.

Maatouk DM, Capel B: Sexual development of the soma in the mouse, *Curr Top Dev Biol* 83:151-183, 2008.

Männer J, Kluth D: The morphogenesis of the exstrophy-epispadias complex: a new concept based on observations made in early embryonic cases of cloacal exstrophy, *Anat Embryol* 210:51-57, 2005.

Mathiot A and others: Uncommon ureteric ectopias: embryological implications, *Anat Embryol* 207:4889-4893, 2004.

O'Rahilly RO, Muecke EC: The timing and sequence of events in the development of the human urinary system during the embryonic period proper, *Z Anat Entwicklungsgesch* 138:99-109, 1972.

Orvis GD, Behringer RB: Cellular mechanisms of müllerian duct formation in the mouse, *Dev Biol* 306:493-504, 2007.

Richards JS, Pangas SA: The ovary: basic biology and clinical implications, *J Clin Invest* 120:963-972, 2010.

Saga Y: Mouse germ cell development during embryogenesis, *Curr Opin Genet Dev* 18:337-341, 2008.

Saxén L: *Organogenesis of the kidney*, Cambridge, 1987, Cambridge University Press.

Sebastian CJ, van der Putte SC, Sie-Go DM: Development and structure of the glandopreputial sulcus of the human clitoris with a special reference to glandopreputial glands, *Anat Rec* 294:156-164, 2011.

Sekido R, Lovell-Badge R: Mechanisms of gonadal morphogenesis are not conserved between chick and mouse, *Dev Biol* 302:132-142, 2007.

Swain A, Lovell-Badge R: Sex determination and differentiation. In Rossant J, Tam PPL, eds: *Mouse development*, San Diego, 2002, Academic Press, pp 371-393.

Thompson AA, Marker PC: Branching morphogenesis in the prostate gland and seminal vesicles, *Differentiation* 74:382-392, 2006.

Van der Werff JFA and others: Normal development of the male anterior urethra, *Teratology* 61:172-183, 2000.

Vezina CM and others: Retinoic acid induces prostatic bud formation, *Dev Dyn* 237:1321-1333, 2008.

Viana R and others: The development of the bladder trigone, the center of the anti-reflux mechanism, *Development* 134:3763-3769, 2007.

Wainwright EN, Wilhelm D: The game plan: cellular and molecular mechanisms of mammalian testis development, *Curr Top Dev Biol* 90:231-262, 2010.

Wilhelm D, Palmer S, Koopman P: Sex determination and gonadal development in mammals, *Physiol Rev* 87:1-28, 2007.

Wilson PD: Polycystic kidney disease, *N Engl J Med* 350:151-164, 2004.

Yamada G and others: Molecular genetic cascades for external genitalia formation: an emerging organogenesis program, *Dev Dyn* 235:1738-1752, 2006.

Yin Y, Ma L: Development of the mammalian female reproductive tract, *J Biochem* 137:677-683, 2005.

Cardiovascular System

This chapter follows the development of the heart from a simple tubular structure to the four-chambered organ that can assume the full burden of maintaining an independent circulation at birth. Similarly, the pattern of blood vessels is traced from their first appearance to an integrated system that carries blood to all parts of the embryo and the placenta. (The early stages in the establishment of the heart and blood vessels are described in Chapter 6 [see Figs. 6.14 to 6.19], and the general plan of the embryonic circulation is summarized in Figure 6.26.) Cellular aspects of blood formation are also briefly described. Clinical Correlations 17.1 and 17.2 at the end of the chapter discuss malformations of the heart and the blood vessels. Table 17.6, also at the end of the chapter, summarizes the timelines in cardiac development.

Functionally, the embryonic heart needs only to act like a simple pump that maintains the flow of blood through the body of the embryo and into the placenta, where fetal wastes are exchanged for oxygen and nutrients. An equally important function, however, is to anticipate the radical changes in the circulation that occur at birth as a consequence of the abrupt cutting off of the placental circulation and the initiation of breathing. To meet the complex requirements of the postnatal circulatory system, the embryonic heart must develop four chambers that can receive or pump the full flow of blood circulating throughout the body. The heart must also adapt to the condition of the fetal lungs, which are poorly developed and for much of the fetal period do not possess a vasculature that can accommodate a large flow of blood. This physiological dilemma is resolved by the presence of two shunts that allow each chamber of the heart to handle large amounts of blood while sparing the underdeveloped pulmonary vascular channels.

Cardiac morphogenesis involves intrinsic cellular and molecular interactions, but these must occur against a background of ongoing mechanical function. Some of these mechanisms remain elusive, but others are becoming better defined through research on normal and abnormal cardiac development.

Development of the vasculature at the level of gross patterns of arteries and veins has been well understood for many years. More recently, new cellular and molecular markers have enabled investigators to outline the cellular origins and factors controlling differentiation of the arteries and veins in specific organs or regions of the body.

Development of Blood and the Vascular System

Development of the vascular system begins in the wall of the yolk sac during the third week of gestation (18 days) with the formation of blood islands (see Fig. 6.19). At this time, the embryo has attained a size that is too large for the distribution of oxygen to all tissues by diffusion alone. This situation necessitates the very early development of the heart and the vascular system. Because the tissues that normally produce blood cells in an adult have not yet begun to form, yolk sac hematopoiesis serves as a temporary adaptation for accommodating the immediate needs of the embryo.

The origin and nature of the cells constituting the blood islands are the subjects of two major hypotheses. According to a long-standing hypothesis, founder cells of the blood islands, called **hemangioblasts**, have a bipotential developmental capacity and can give rise to either endothelial cells or hematopoietic stem cells. When a commitment has been made to one of these two lineages, daughter cells lose the capacity to form the other type of cell. A more recent hypothesis posits that by the time the yolk sac is colonized by hemangioblasts, these cells have already been segregated into hematopoietic and endothelial lineages. Research on mouse embryos suggests that instead of blood islands, the hematopoietic cells aggregate into a blood band that surrounds the yolk sac.

Embryonic Hematopoiesis

Hemangiogenic precursor cells first arise in the posterolateral mesoderm during gastrulation and from there migrate to the earliest blood-forming organs (**Fig. 17.1**). Under the influence of **Runx-1**, some of their progeny follow the hematopoietic lineage, whereas others, responding to **Hoxa3** enter the endothelial lineage. Still other progeny will enter a third lineage and eventually form vascular smooth muscle cells. Although blood cell formation (**hematopoiesis**) begins in the yolk sac, the yolk sac–derived cells are soon replaced by blood cells that are independently derived from other sites of hematopoiesis (**Fig. 17.2**).

The blood islands contain pluripotential **hematopoietic stem cells**, which can give rise to most types of cells found in the embryonic blood. The erythrocytes produced in the yolk sac are large nucleated cells that enter the bloodstream just

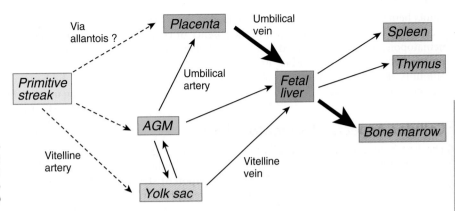

Fig. 17.1 **Sites of embryonic hematopoiesis and routes by which organs are seeded by embryonic blood cells.** AGM, aorta/genital ridge/mesonephros region. *(Adapted from Mikkola HKA, Orkin SH:* Development *133:3733-3744, 2006.)*

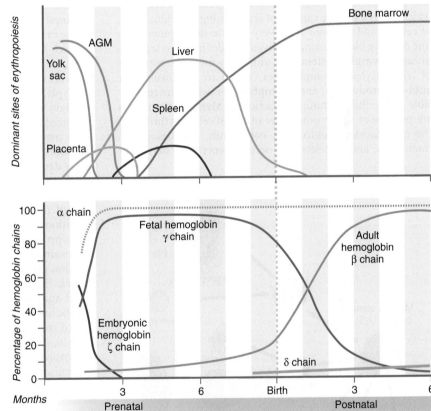

Fig. 17.2 **Sites of hematopoiesis** *(above)* **and phases of hemoglobin synthesis** *(below)* **in the human embryo.** The *upper graph* highlights the relative importance of the various sites of hematopoiesis. The *lower graph* shows the percentages of the various hemoglobin polypeptide chains present in the blood at any given time. The α chain is treated separately from the others. AGM, aorta/genital ridge/mesonephros region. *(Based on Carlson B:* Patten's foundations of embryology, *ed 6, New York, 1996, McGraw-Hill).*

before the heart tube begins to beat at about 22 days' gestation. For the first 6 weeks, the circulating erythrocytes are largely yolk sac derived, but during that time, preparations for the next stages of hematopoiesis are taking place.

Analysis of human embryos has shown that, starting at 28 days, definitive **intraembryonic hematopoiesis** begins in small clusters of cells (**para-aortic clusters**) in the splanchnopleuric mesoderm associated with the ventral wall of the dorsal aorta and shortly thereafter in the **aorta/genital ridge/mesonephros (AGM) region**. Precursor cells from the AGM region make their way via the blood to blood-forming sites in the liver, the yolk sac, and the placenta. Hematopoietic stem cells formed in the AGM, the yolk sac, and the placenta become transported to the liver via the circulation to the liver (see Fig. 17.1). By 5 to 6 weeks of gestation, sites of hematopoiesis become prominent in the liver. In both the yolk sac and the early sites of embryonic hematopoiesis, the endothelial cells themselves briefly retain the capacity for producing blood-forming cells. There is now evidence that in the AGM region, nitric oxide gas signaling, resulting from shear stress caused by blood flow on the endothelial cells, can induce their transformation into hematopoietic stem cells.

The erythrocytes produced by the liver are quite different from the erythrocytes derived from the yolk sac. Although still considerably larger than normal adult red blood cells, liver-derived erythrocytes are non-nucleated and contain different types of hemoglobin. By 6 to 8 weeks of gestation in humans, the liver replaces the yolk sac as the main source of blood cells. Although the liver continues to produce red blood cells until the early neonatal period, its contribution begins to decline in the sixth month of pregnancy. At this time, the formation of blood cells shifts to the bone marrow, the definitive site of adult hematopoiesis. This shift is controlled by cortisol secreted by the fetal adrenal cortex. In the absence of cortisol, hematopoiesis remains confined to the liver. Before hematopoiesis becomes well established in the bone marrow, small amounts of blood formation may also occur in the omentum and possibly the spleen.

Cellular Aspects of Hematopoiesis

The first **hematopoietic stem cells** that arise in the embryo are truly pluripotential in that they can give rise to all the cell types found in the blood (**Fig. 17.3**). These **pluripotent stem cells**, sometimes called **hemocytoblasts**, have great proliferative ability. They produce vast numbers of progeny, most of which are cells at the next stage of differentiation, but they also produce small numbers of their original stem cell type, which act as a reserve capable of replenishing individual lines of cells should the need arise. Very early in development, the line of active blood-forming cells subdivides into two separate lineages. **Lymphoid stem cells** ultimately form the two lines of lymphocytes: **B lymphocytes** (which are responsible for antibody production) and **T lymphocytes** (which are responsible for cellular immune reactions). **Myeloid stem cells** are precursors to the other lines of blood cells: erythrocytes, the granulocytes (neutrophils, eosinophils, and basophils), monocytes, and platelets. The second-generation stem cells

(lymphoid and myeloid) are still pluripotent, although their developmental potency is restricted because neither lymphoid cells nor myeloid cells can form the progeny of the other type.

Stemming from their behavior in certain experimental situations, the hematopoietic stem cells are often called **colony-forming units** (**CFUs**). The first-generation stem cell is called the CFU-ML because it can give rise to myeloid and lymphoid lines of cells. Stem cells of the second generation are called CFU-L (lymphocytes) and CFU-S (spleen) (determined from experiments in which stem cell differentiation was studied in irradiated spleens). In some cases, the progeny of CFU-ML and CFU-S are **committed stem cells**, which are capable of forming only one type of mature blood cell. For each lineage, the forming cell types must pass through several stages of differentiation before they attain their mature phenotype.

What controls the diversification of stem cells into specific cell lines? Experiments begun in the 1970s provided evidence for the existence of specific **colony-stimulating factors** (**CSFs**) for each line of blood cell. CSFs are diffusible proteins that stimulate the proliferation of hematopoietic stem cells. Some CSFs act on several types of stem cells; others stimulate only one type. Although much remains to be learned about the sites of origin and modes of action of CSFs, many CSFs seem to be produced locally in stromal cells of the bone marrow, and some may be stored on the local extracellular matrix. CSFs are bound by small numbers of surface receptors on their target stem cells. Functionally, CSFs represent mechanisms for stimulating the expansion of specific types of blood cells when the need arises. Recognition of the existence of CSFs has prompted considerable interest in their clinical application to conditions characterized by a deficiency of white blood cells (leukopenia).

Certain *Hox* genes, especially those of the *Hoxa* and *Hoxb* families, play an important role in some aspects of hematopoiesis. Exposure of bone marrow to antisense oligonucleotides against specific *Hox* genes results in the suppression of specific lines of differentiation of blood cells. Conversely, engineered overexpression of genes, such as *Hoxb8*, *Hoxa9*, and *Hoxa10*, causes leukemia in mice. Evidence is increasing for the involvement of *Hox* genes in the pathogenesis of human leukemias. One important function of the *Hox* genes in hematopoiesis is the regulation of proliferation. Several growth factors, especially bone morphogenetic protein-4 (BMP-4), Indian hedgehog, and Wnt proteins, are important in stimulating and maintaining hematopoietic stem cell activity.

Erythropoiesis

Red blood cell formation (**erythropoiesis**) occurs in three waves during the embryonic period. The first wave begins with precursors within the yolk sac, which produce primitive nucleated erythrocytes that mature within the bloodstream. The second wave also begins in the yolk sac, but the precursor cells then colonize the embryonic liver and produce the first of a generation of definitive fetal erythrocytes that are dominant during the prenatal period. The third wave consists of precursor cells that enter the liver from the AGM mesoderm and the placenta. Some of these definitive erythroid progenitor cells send progeny directly from the liver into the bloodstream as definitive fetal erythrocytes. Others seed the bone marrow and produce adult-type erythrocytes later in the fetal period.

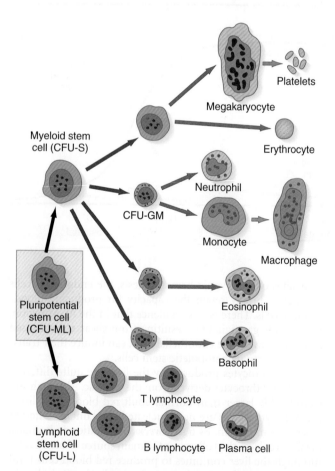

Fig. 17.3   Major cell lineages during hematopoiesis. Mature blood cells are shown on the *right*. CFU, colony-forming unit; GM, granulocyte and monocyte; L, lymphocyte; ML, myeloid and lymphoid; S, spleen.

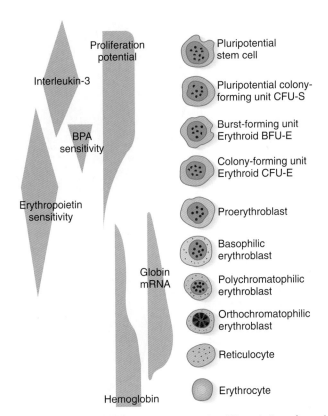

Fig. 17.4 *Right,* Morphological stages in the differentiation of a red blood cell from a pluripotential stem cell. *Left,* Molecular correlates of differentiation. The thickness of the *tan background* is proportional to the amount at the corresponding stages of erythropoiesis. BFU, burst-forming units; BPA, burst-promoting activity; CFU, colony-forming unit; E, erythroid; S, spleen.

The erythrocyte lineage represents one line of descent from the CFU-S cells. Although the erythroid progenitor cells are restricted to forming only red blood cells, there are many generations of precursor cells (**Fig. 17.4**). The earliest stages of erythropoiesis are recognized by the behavior of the precursor cells in culture, rather than by morphological or biochemical differences. These are called **erythroid burst-forming units** (**BFU-E**) and **erythroid CFUs** (**CFU-E**). Each responds to different stimulatory factors. The pluripotent CFU-S precursors (see Fig. 17.3) respond to **interleukin-3**, a product of macrophages in adult bone marrow. A hormone designated as **burst-promoting activity** stimulates mitosis of the BFU-E precursors (see Fig. 17.4). A CFU-E cell, which has a lesser proliferative capacity than a BFU-E cell, requires the presence of **erythropoietin** as a stimulatory factor.

Erythropoietin is a glycoprotein that stimulates the synthesis of the mRNA for globin and is first produced in the fetal liver. Later in development, synthesis shifts to the kidney, which remains the site of erythropoietin production in adults. Under conditions of hypoxia (e.g., from blood loss or high altitudes), the production of erythropoietin by the kidneys increases, thereby stimulating the production of more red blood cells to compensate for the increased need. In adult erythropoiesis, the CFU-E stage seems to be the one most responsive to environmental influences. The placenta is apparently impervious to erythropoietin, and this property insulates the embryo from changes in erythropoietin levels of the mother and eliminates the influence of fetal erythropoietin on the blood-forming apparatus of the mother.

One or two generations after the CFU-E stage, successive generations of erythrocyte precursor cells can be recognized by their morphology. The first recognizable stage is the **proerythroblast** (**Fig. 17.5**), a large, highly basophilic cell that has not yet produced sufficient hemoglobin to be detected by cytochemical analysis. Such a cell has a large nucleolus, much uncondensed nuclear chromatin, numerous ribosomes, and a high concentration of globin mRNAs. These are classic cytological characteristics of an undifferentiated cell.

Succeeding stages of erythroid differentiation (**basophilic, polychromatophilic,** and **orthochromatic erythroblasts**) are characterized by a progressive change in the balance between the accumulation of newly synthesized hemoglobin and the decline of first the RNA-producing machinery and later the protein-synthesizing apparatus. The overall size of the cell decreases, and the nucleus becomes increasingly **pyknotic** (smaller with more condensed chromatin) until it is finally extruded at the stage of the orthochromatic erythrocyte. After the loss of the nucleus and most cytoplasmic organelles, the immature red blood cell, which still contains a small number of polysomes, is a **reticulocyte**. Reticulocytes are released into the bloodstream, where they continue to produce small amounts of hemoglobin for 1 or 2 days.

The final stage of hematopoiesis is the mature **erythrocyte**, which is a terminally differentiated cell because of the loss of its nucleus and most of its cytoplasmic organelles. Erythrocytes in embryos are larger than their adult counterparts and have a shorter life span (50 to 70 days in the fetus versus 120 days in adults).

Hemoglobin Synthesis and its Control

Both the red blood cells and the hemoglobin within them undergo isoform transitions during embryonic development. The adult hemoglobin molecule is a complex composed of heme and four globin chains: two α and two β chains. The α and β subunits are products of genes located on chromosomes 16 and 11 (**Fig. 17.6**). Different isoforms of the subunits are encoded linearly on these chromosomes.

During the period of yolk sac hematopoiesis, embryonic globin isoforms are produced. The earliest embryonic hemoglobin, sometimes called **Gower 1**, is composed of two ζ (α-type) and two ε (β-type) chains. After passing through a couple of transitional forms (**Table 17.1**), hemoglobin synthesis enters a fetal stage by 12 weeks, which corresponds to the shift in the site of erythropoiesis from the yolk sac to the liver. Fetal hemoglobin consists of two adult-type α chains, which form very early in embryogenesis, and two γ chains, the major fetal isoform of the β chain. Fetal hemoglobin is the predominant form during the remainder of pregnancy. The main adaptive value of the fetal isoform of hemoglobin is that it has a higher affinity for oxygen than the adult form. This is advantageous to the fetus, which depends on the oxygen concentration of the maternal blood. Starting at about 30 weeks' gestation, there is a gradual switch from the fetal to the adult type of hemoglobin, with $\alpha_2\beta_2$ being the predominant type. A minor but functionally similar variant is $\alpha_2\delta_2$.

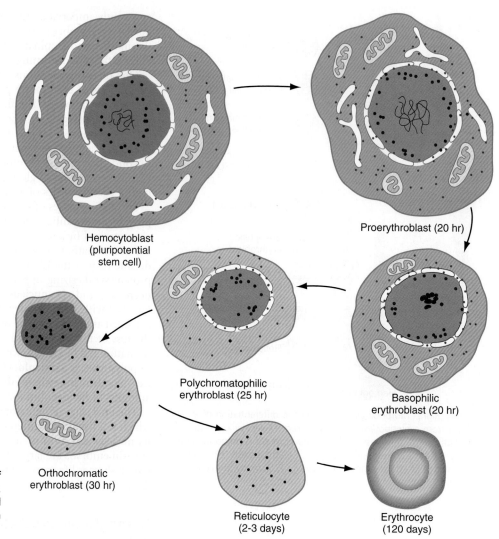

Fig. 17.5 **Structural features of erythropoiesis.** In successive stages, cytoplasmic basophilia decreases and the concentration of hemoglobin increases in the cells.

Hemocytoblast (pluripotential stem cell)

Proerythroblast (20 hr)

Basophilic erythroblast (20 hr)

Polychromatophilic erythroblast (25 hr)

Orthochromatic erythroblast (30 hr)

Reticulocyte (2-3 days)

Erythrocyte (120 days)

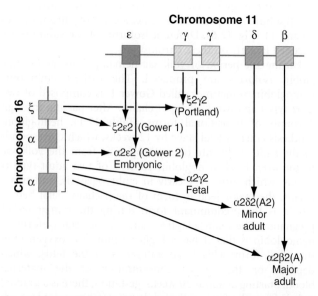

Chromosome 11

ε   γ   γ   δ   β

ξ2γ2 (Portland)

ξ2ε2 (Gower 1)

α2ε2 (Gower 2) Embryonic

α2γ2 Fetal

α2δ2(A2) Minor adult

α2β2(A) Major adult

Chromosome 16

ξ

α

α

Fig. 17.6 **Organization of hemoglobin genes along chromosomes 11 and 16 and their sequential activation during embryonic development.**

Table 17.1 Developmental Isoforms of Human Hemoglobin		
Developmental Stage	**Hemoglobin Type**	**Globin Chain Composition**
Embryo	Gower 1	$\zeta_2\varepsilon_2$
Embryo	Gower 2	$\alpha_2\varepsilon_2$
Embryo	Portland	$\zeta_2\gamma_2$
Embryo to fetus	Fetal	$\alpha_2\gamma_2$
Fetus to adult	A (adult)	$\alpha_2\beta_2$
Adult	A_2	$\alpha_2\delta_2$
Adult	Fetal	$\alpha_2\gamma_2$*

*The fetal hemoglobin expressed in adults differs from true fetal hemoglobin by an amino acid substitution at the 136 position of the γ chain.

Adapted from Brown MS: In Stockman J, Pochedly C, eds: *Developmental and neonatal hematology*, New York, 1988, Raven Press.

Formation of Embryonic Blood Vessels

The early embryo is devoid of blood vessels. Although blood islands appear in the wall of the yolk sac, and extraembryonic vascular channels form in association with them (see Fig. 6.19), much of the vasculature of the embryonic body is derived from intraembryonic sources. During the early period of somite formation, networks of small vessels rapidly appear in many regions of the embryonic body.

The formation of blood vessels in the embryo consists of several phases (**Fig. 17.7**). The first is the specification of a population of vascular precursors, called **angioblasts**. These cells become organized into a **primary capillary plexus** through a process known as **vasculogenesis**. To keep pace with the rapidly growing embryo, the primary capillary plexus must rapidly undergo reorganization through the resorption of existing vessels and the sprouting of new branches to support the expanding vascular network. This latter process is called **angiogenesis**. Angiogenesis continues not only in the prenatal period, but also throughout adult life, as tissues and organs continually adapt to changing conditions of life, whether normal or pathological.

Detailed descriptive studies and transplantation experiments involving intrinsic cellular labels or graft-specific monoclonal antibody labels have shown that **angioblasts** arise from most mesodermal tissues of the body, except notochord and prechordal mesoderm (**Table 17.2**). Embryonic blood vessels form from angioblasts by three main mechanisms. Many of the larger blood vessels, such as the dorsal aortae, are formed by the coalescence of angioblasts in situ. Other equally large channels, such as the endocardium, are formed by angioblasts migrating into the region from other sites. Other vessels, especially the intersegmental vessels of the main body axis and vessels of the central nervous system, arise as vascular sprouts from existing larger vessels. Many of the angioblasts of the trunk are originally associated with the splanchnic mesoderm.

The developmental processes leading to the initial formation of the aorta are beginning to be understood. The endothelium of the early paired aortae is derived from splanchnopleure and requires an interaction with the underlying endoderm for its development. While the aortae are still in the paired stage, somite-derived cells contribute to their dorsal walls. Concomitantly, the ventral splanchnopleure-derived endothelium begins to give rise to clusters of hematopoietic stem cells. Then the dorsal somite-derived endothelial

Table 17.2 Distribution of Endogenous Angioblasts in Embryonic Tissues

Tissues	Angioblasts
Cephalic	
Paraxial mesoderm	+
Lateral mesoderm	+
Prechordal mesoderm	−
Notochord	−
Brain	−
Neural crest	−
Trunk	
Whole somites	+
Dorsal half somites	+
Segmental plate mesoderm	+
Lateral somatic mesoderm	+
Lateral splanchnic mesoderm	+
Spinal cord	−

From Noden DM: *Ann NY Acad Sci* 588:236-249, 1990.

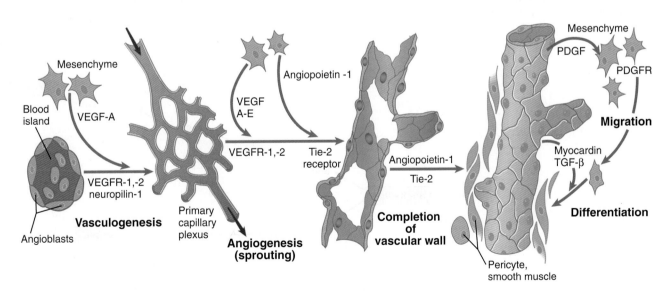

Fig. 17.7 Scheme illustrating vasculogenesis, angiogenesis, and assembly of the vascular wall. Angioblasts, initially expressing vascular endothelial growth factor receptor (VEGFR-2), are stimulated by vascular endothelial growth factor (VEGF-A), secreted by the surrounding mesenchyme, to form the primary capillary plexus by the process of vasculogenesis. Under additional stimulation by growth factors, competent endothelial cells of the primary capillary plexus form vascular sprouts in the earliest stages of angiogenesis. This is followed by the recruitment of surrounding mesenchymal cells to form the cellular elements of the vascular wall. PDGF, platelet-derived growth factor; PDGFR, platelet-derived growth factor receptor; TGF-β, transforming growth factor-β.

cells overgrow the ventral splanchnopleure-derived endothelial cells. When this is completed, hematopoiesis in the aorta ceases.

All stages in the formation of the vascular system occur in response to the influence of powerful growth factors and their receptors. The initial phase of recruitment of a population of angioblasts from the mesoderm is characterized by the appearance of a transmembrane **vascular endothelial growth factor receptor** (**VEGFR-2**) on their surfaces (see Fig. 17.7). Soon, in response to the production of **vascular endothelial growth factor** (**VEGF-A**) by the surrounding mesenchyme, the phase of vasculogenesis occurs, and the angioblasts form the cellular tubes that become the basis for the primary capillary plexus.

The formation of vascular endothelial sprouts, the cellular basis for angiogenesis, occurs against a background of VEGF/VEGFR-1 and VEGF/VEGFR-2 interactions, but with a new set of players added. A sprouting factor, **angiopoietin-1**, interacts with its receptor, **Tie-2**, on the endothelial cells at sites where endothelial sprouts will occur. The **Notch** signaling pathway is also strongly tied to the formation of vascular sprouts (a common denominator with other organ systems that display branching morphogenesis), but its connection to the angiopoietin-1/Tie-2 mechanism remains unclear.

The next step in building a blood vessel is formation of the vascular wall, which in the trunk and extremities is derived from local mesoderm that becomes associated with the endothelial lining of the vessel. In the head and many areas of the aortic arch system, mesenchyme derived from neural crest ectoderm is a major contributor to the connective tissue and smooth muscle of the vascular wall. The neural crest, however, does not give rise to endothelial cells.

Two-way molecular signaling is involved in building up the walls of blood vessels. In response to the angiopoietin-1/Tie-2 interaction that occurs during angiogenesis, the endothelial cells release their own signaling molecule, **platelet-derived growth factor**, which stimulates the migration of mesenchymal cells toward the vascular endothelium. The release of other growth factors (transforming growth factor-β [TGF-β]

and **myocardin**, a master regulator of smooth muscle formation) by the endothelial cells stimulates the differentiation of the mesenchymal cells into vascular smooth muscle or pericytes.

Research has shed considerable light on the differentiation of the arterial versus venous system. The arterial or venous identity of endothelial cells is established very early in their development, before angiogenesis and before the onset of circulation. The endothelial cells of developing arteries express the membrane-bound ligand **Ephrin-B2**, whereas the endothelial cells of developing veins express the receptor **Eph-B4** on their surface membranes. These characteristic phenotypes are the results of different signaling cascades. Arterial vessels are the first to differentiate from generic endothelial background, and a set of signals, starting with **sonic hedgehog** (**shh**), finally leads to the acquisition of an arterial phenotype (**Fig. 17.8**). **Notch**, one of the links in that cascade, not only causes progression of the sequence of arterial differentiation leading to Ephrin-B2 expression, but it inhibits the expression of Eph-4 and the pathway leading to the venous phenotype. In what was earlier assumed to be a purely default mode, venous differentiation occurs under the influence of **COUP-TFII** (Chicken *o*valbumin *u*pstream *p*romoter-transcription factor II), which suppresses the arterial pathway by inhibiting Notch signaling, but is a determinant of venous differentiation by acting upstream of Eph-4. Under the influence of **Sox-18** and **Prox-1**, lymphatic vessels form and branch off from the veins. Prox-1 is a master regulator of venous identity. Later, physiological and local factors play a role in the differentiation of blood vessels. When the flow of blood to the yolk sac is greatly reduced, vessels slated to become arteries develop venous characteristics; correspondingly, developing veins exposed to high blood pressure transform into arteries.

As with myoblasts, angioblasts seem to react to local environmental cues that determine the specific morphological pattern of a blood vessel. An unexpected finding is that the pattern of the peripheral innervation often determines the pattern of the smaller arteries. The growing tips of endothelial

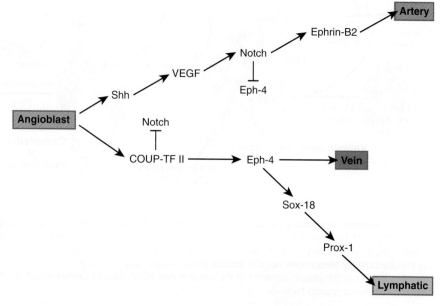

Fig. 17.8 Diagrammatic representation of major pathways leading to the differentiation of arterial, venous and lymphatic endothelium. COUP-TF II, chicken *o*valbumin *u*pstream *p*romoter-transcription factor II; Shh, sonic hedgehog; VEGF, vascular endothelial growth factor.

Table 17.3 Responsiveness of Axonal Growth Cones and Endothelial Cell Tips to Environmental Ligands		
Ligand	**Growth Cones**	**Endothelial Tips**
Attractants		
Semaphorin	+	−
VEGF	−	+
Netrin	+	+
Slit	+	+
Repellants		
Semaphorin	+	+
Ephrin	+	+
Netrin	+	+
Slit	+	+

VEGF, vascular endothelial growth factor.

cell buds and axonal growth cones of outgrowing nerve fibers contain receptors that respond remarkably similarly to the major families of environmental ligands (**Table 17.3**). VEGF, secreted by the nerve fibers, acts as an effective patterning agent for the blood vessels. The smooth muscle cells of the developing arteries secrete a factor, **artemin**, that guides the extension of sympathetic nerve fibers along the vessel wall.

Tracing studies of transplanted angioblasts have shown that some can of these cells migrate long distances. Angioblasts that have migrated far from the place into which they were grafted become integrated into morphologically normal blood vessels in the areas where they settle.

Local factors also influence the initiation of vasculogenesis. In some organs (e.g., the liver) or parts of organs (e.g., the bronchi of the respiratory system), the blood vessels supplying the regions arise from local mesoderm, whereas other organs (e.g., the metanephric kidneys) or parts of organs (e.g., the alveoli of the lungs) are supplied by blood vessels that grow into the mesenchyme from other tissues. In the latter type of vascularization mechanism, evidence is increasing that these organ primordia produce their own **angiogenesis factors** that stimulate the growth of vascular sprouts (by promoting mitosis of endothelial cells) into the glandular mesenchyme. Nearby blood vessels, in turn, influence the morphogenesis and differentiation of many structures (e.g., pancreas, glomerulus, liver) with which they are associated.

Development of the Arteries
Aorta, Aortic Arches, and Their Derivatives

The dorsal aorta forms from the direct aggregation of endothelial precursor cells derived from the lateral plate mesoderm. These cells form a vessel directly by a vasculogenesis mechanism. Vasculogenesis is stimulated by VEGF and other factors produced by the endoderm and BMP in the lateral mesoderm. When first formed, the cranial part of the dorsal aorta is a paired, with each member located lateral to the midline. The reason is that in the midline the notochord secretes the BMP antagonists **noggin** and **chordin**, which inhibit the activity of

BMP and also inactivate the vasculogenic influences from the endoderm. Late in the fourth week, hematopoietic stem cells form in the lining of the ventral part of the aorta (see p. 409).

The system of aortic arches in early human embryos is organized along the same principles as the system of arteries supplying blood to the gills of many aquatic lower vertebrates. Blood exits from a common ventricle in the heart into a ventral aortic root, from which it is distributed through the branchial arches by pairs of aortic arches (**Fig. 17.9A**). In gilled vertebrates, the aortic arch arteries branch into capillary beds, where the blood becomes reoxygenated as it passes through the gills. In mammalian embryos, the aortic arches remain continuous vessels because gas exchange occurs in the placenta and not in the pharyngeal arches. The aortic arches empty into paired dorsal aortae where the blood enters the regular systemic circulation. In human embryos, all aortic arches are never present at the same time. Their formation and remodeling show a pronounced craniocaudal gradient. Blood from the outflow tract of the heart (the truncoconal region) flows into an **aortic sac**, which differs from the truncoconal region in the construction of its wall. The aortic arches branch off from the aortic sac.

The developmental anatomy of the aortic arch system illustrates well the principle of morphological adaptation of the vascular bed during different stages of embryogenesis (**Table 17.4**). Continued development of the cranial and cervical regions causes components of the first three arches and associated aortic roots to be remodeled into the carotid artery system (see Fig. 17.9). With the remodeling of the heart tube and the internal division of the outflow tract into aortic and pulmonary components, the fourth arches undergo an asymmetrical adaptation to the early asymmetry of the heart. The left fourth aortic arch is retained as a major channel (arch of the aorta), which carries the entire output from the left ventricle of the heart. The right fourth arch is incorporated into the right subclavian artery.

Embryology textbooks traditionally depict the aortic arch system as consisting of six pairs of vascular arches, but the fifth and sixth arches never appear as discrete vascular channels similar to the first through fourth arches. The fifth aortic arch, if it exists at all, is represented by no more than a few capillary loops. The sixth (**pulmonary arch**) arises as a capillary plexus associated with the early trachea and lung buds. The capillary plexus is supplied by ventral segmental arteries arising from the paired dorsal aortae in that region (**Fig. 17.10**). The equivalent of the sixth arch is represented by a discrete distal segment (ventral segmental artery) connected to the dorsal aorta and a plexuslike proximal segment that establishes a connection between the aortic sac at the base of the fourth arch and the distal segmental component. As the respiratory diverticulum and early lung buds elongate, parts of the pulmonary capillary network consolidate to form a pair of discrete **pulmonary arteries** that connect to the putative sixth arch. Although the term **sixth aortic arch** is frequently used in anatomical and clinical literature, **pulmonary arch** is a more appropriate term because it does not imply equivalence to the other aortic arches.

Similar to the fourth aortic arch, the pulmonary arch develops asymmetrically. On the left side, it becomes a large channel. Its distal segment, which was derived from a ventral segmental artery, persists as a major channel (**ductus arteriosus**) that shunts blood from the left pulmonary artery to the aorta (see

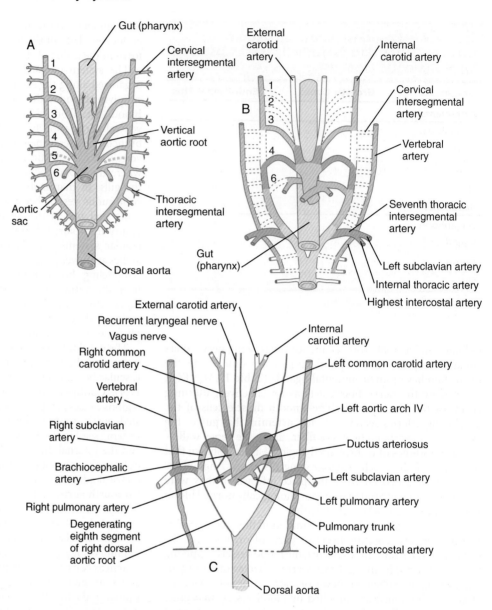

Fig. 17.9 **A,** Schematic representation of the embryonic aortic arch system. **B** and **C,** Later steps in the transformation of the aortic arch system in a human. Disposition of the recurrent laryngeal nerve in relation to the right fourth and left sixth arch is also shown in **C.**

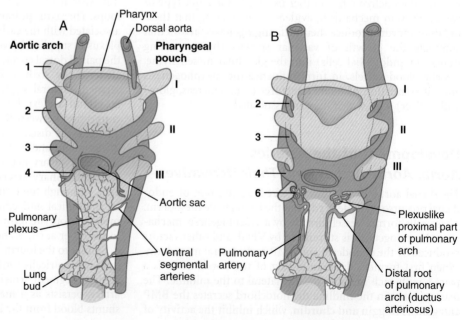

Fig. 17.10 **A** and **B,** Development of the pulmonary arch, showing the early pulmonary plexus in relation to several ventral segmental arteries associated with the early respiratory diverticulum **(A)** and their consolidation into discrete vessels that establish a connection with the bases of the fourth aortic arches **(B)**. *(Based on DeRuiter MC and others:* Anat Embryol *179:309-325, 1989.)*

Table 17.4 Adult Derivatives of the Aortic Arch System

	Right Side	Left Side
Aortic Arches		
1	Disappearance of most of structure Part of maxillary artery	Disappearance of most of structure Part of maxillary artery
2	Disappearance of most of structure Hyoid and stapedial arteries	Disappearance of most of structure Hyoid and stapedial arteries
3	Ventral part—common carotid artery Dorsal part—internal carotid artery	Ventral part—common carotid artery Dorsal part—internal carotid artery
4	Proximal part of right subclavian artery	Part of arch of aorta
5	Rarely recognizable, even in early embryo	Rarely recognizable, even in early embryo
6 (pulmonary)	Part of right pulmonary artery	Ductus arteriosus Part of left pulmonary artery
Ventral Aortic Roots		
Cranial to third arch	External carotid artery	External carotid artery
Between third and fourth arches	Common carotid artery	Common carotid artery
Between fourth and sixth arches	Right brachiocephalic artery	Ascending part of aorta
Dorsal Aortic Roots		
Cranial to third arch	Internal carotid artery	Internal carotid artery
Between third and fourth arches	Disappearance of structure	Disappearance of structure
Between fourth and pulmonary arches	Central part of right subclavian artery	Descending aorta
Caudal to pulmonary arch	Disappearance of structure	Descending aorta

Fig. 17.9C). The lungs are protected by this shunt from a flow of blood that is greater than what their vasculature can handle during most of the intrauterine period. On the right side, the distal segment of the pulmonary arch regresses, and the proximal segment (the base of the right pulmonary artery) branches off from the pulmonary trunk.

The asymmetry of the derivatives of the pulmonary arch accounts for the difference between the course of the right and left **recurrent laryngeal nerves**, which are branches of the vagus nerve (cranial nerve X). These nerves, which supply the larynx, hook around the pulmonary arches. As the heart descends into the thoracic cavity from the cervical region, the branch point from the vagus of each recurrent laryngeal nerve is correspondingly moved. On the left side, the nerve is associated with the ductus arteriosus (see Fig. 17.9C), which persists throughout the fetal period, so it is pulled deep into the thoracic cavity. On the right side, with the regression of much of the right pulmonary arch, the nerve moves to the level of the fourth arch, which constitutes an anatomical barrier. The positions of the right and left recurrent laryngeal nerves in an adult reflect this asymmetry, with the right nerve curving under the right subclavian artery (fourth arch) and the left nerve hooking around the **ligamentum arteriosum** (the adult derivative of the ductus arteriosus, the distal segment of the left pulmonary arch).

As the aortic arches become modulated into their adult configuration, specialized **baroreceptors** begin to form. One of these is the carotid sinus, which is located in the proximal portion of each internal carotid artery and is innervated by

fibers of the glossopharyngeal nerve. Similarly, baroreceptors located in the proximal part of the right subclavian artery and in the aortic arch between the left common carotid and subclavian arteries are innervated by branches of the vagus and recurrent laryngeal nerves. Chemosensory structures associated with the aortic arches are the paired **carotid bodies**, which are located at the bifurcation of the internal and external carotid arteries and are innervated by a sensory branch of the glossopharyngeal nerve and sympathetic nerve fibers from the superior cervical ganglion.

Major Branches of the Aorta

In the early embryo, when the dorsal aortae are still paired vessels, three sets of arterial branches arise from them—**dorsal intersegmental, lateral segmental**, and **ventral segmental** (Fig. 17.11). These branches undergo a variety of modifications in form before assuming their adult configurations (Table 17.5). The ventral segmental arteries arise as paired vessels that course over the dorsal and lateral walls of the gut and yolk sac. With the closure of the gut and the narrowing of the dorsal mesentery, certain branches fuse in the midline to form the celiac, superior, and inferior mesenteric arteries.

The **umbilical arteries** begin as pure ventral segmental branches supplying allantoic mesoderm, but their bases later connect with lumbar intersegmental vessels. The most proximal umbilical channels then regress, and the intersegmental branches become their main branches off the aorta. Like their

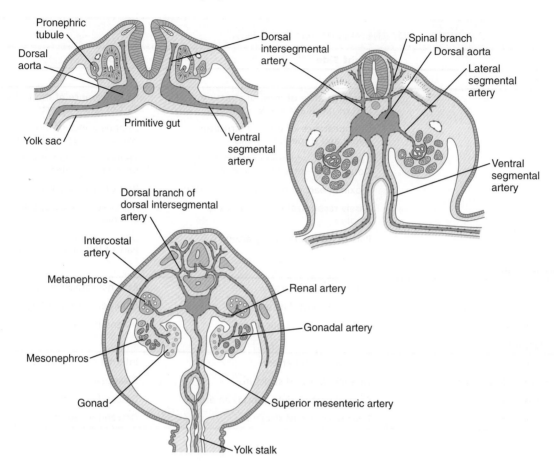

Fig. 17.11 **Types of segmental branches coming off the abdominal aorta at different stages of development.**

Table 17.5 Major Arterial Branches of the Aorta

Embryonic Vessels	Adult Derivatives
Dorsal Intersegmental Branches (Paired)	
Cervical intersegmental (1-16)	Lateral branches joining to become vertebral arteries
Seventh intersegmentals	Subclavian arteries
Thoracic intersegmentals	Intercostal arteries
Lumbar intersegmentals	Iliac arteries
Lateral Segmental Branches	
Up to 20 pairs of vessels supplying the mesonephros	Adrenal arteries, renal arteries, gonadal (ovarian or spermatic) arteries
Ventral Segmental Branches*	
Vitelline vessels	Celiac artery, superior and inferior mesenteric arteries
Allantoic vessels	Umbilical arteries

*Originally paired in areas where the embryonic aorta itself consists of paired components.

subclavian counterparts in the arms, the initially small arterial branches (**iliac arteries**) supplying the leg buds appear as components of the dorsal intersegmental (lumbar) branches of the aorta. After the umbilical arteries incorporate the proximal segments of the intersegmental vessels, the iliac arteries arise as branches off the umbilical arteries.

Arteries of the Head

The arteries supplying the head arise from two sources. Ventrally, the aortic arch system (first to third arches and corresponding roots) gives rise to the arteries supplying the face (**external carotid arteries**) and the frontal part of the base of the brain (**internal carotid arteries**) (see Fig. 17.9).

At the level of the spinal cord, the **vertebral arteries**, which form through connections of lateral branches of dorsal intersegmental arteries, grow toward the brain. Soon, they veer toward the midline and fuse with the **basilar artery**, which has already formed earlier from the fusion of the bilateral **longitudinal neural arteries** (**Fig. 17.12**). The longitudinal neural arteries, which run along the ventral sides of the brain from the midbrain through the caudal end of the hindbrain, represent the major arterial supply to the brain during the fourth week of development. The basilar artery runs along the ventral surface of the brainstem and supplies the brainstem with a series of paired arteries. As the basilar artery approaches the level of the diencephalon and the internal carotid arteries, sets

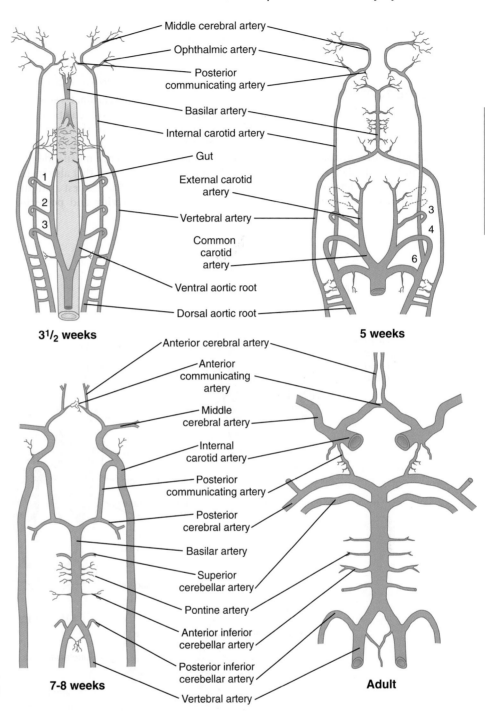

Middle cerebral artery

Ophthalmic artery

Posterior communicating artery

Basilar artery

Internal carotid artery

Gut

External carotid artery

Vertebral artery

Common carotid artery

Ventral aortic root

Dorsal aortic root

3½ weeks

5 weeks

Anterior cerebral artery

Anterior communicating artery

Middle cerebral artery

Internal carotid artery

Posterior communicating artery

Posterior cerebral artery

Basilar artery

Superior cerebellar artery

Pontine artery

Anterior inferior cerebellar artery

Posterior inferior cerebellar artery

Vertebral artery

7-8 weeks

Adult

Fig. 17.12   Stages in the development of the major arteries supplying the brain. Ventral views.

of branches from each of these major vessels grow out and fuse, to form **posterior communicating arteries**, which join the circulations of the basilar and internal carotid arteries. Two other small branches off the internal carotid system fuse in the midline to complete a vascular ring (**circle of Willis**), which underlies the base of the diencephalon and encircles the optic chiasm and pituitary stalk. The circle of Willis is a structural adaptation that ensures a continuous blood supply in the event of occlusion of some major arteries supplying the brain. In addition, it is a structural landmark that marks the transition between mesodermally derived blood vessels of the trunk and those of the head, which are primarily of neural crest origin.

Coronary Arteries

Although intuitively one would expect the coronary arteries to arise as branches growing out from the aorta, experimental studies have shown instead that cellular precursors of the coronary arteries, arising in the same cellular primordium as the future epicardium, migrate toward the aorta and invade its wall (see Fig. 6.18B). The smooth muscle cells of the coronary vessels have a purely mesodermal origin, instead of the mixed neural crest and mesodermal origin seen in the aortic arch derivatives. Studies involving the tagging of cells with retroviral markers confirm that progenitor cells of the coronary vessels penetrate the wall of the already beating heart as the epicardial layer envelops the myocardium, and that after forming in situ, the coronary arteries secondarily enter the aorta.

Development of Veins

Veins follow a morphologically complex pattern of development characterized by the formation of highly irregular networks of capillaries and the ultimate expansion of certain channels into definitive veins. Because of the multichanneled beginnings and the number of options (**Fig. 17.13**), the adult venous system is characterized by a higher incidence of anatomical variations than the arterial system. A detailed description of the development of venous channels is beyond the scope of this text.

Cardinal Veins

The cardinal veins form the basis for the intraembryonic venous circulation. Several sets of cardinal veins appear at different times and in different locations. Within any set of cardinal veins, some segments regress, and others persist, either as independent channels or as components of composite veins that also include portions of other cardinal veins.

The earliest pattern of cardinal veins consists of paired **anterior** and **posterior cardinal veins**, which drain blood

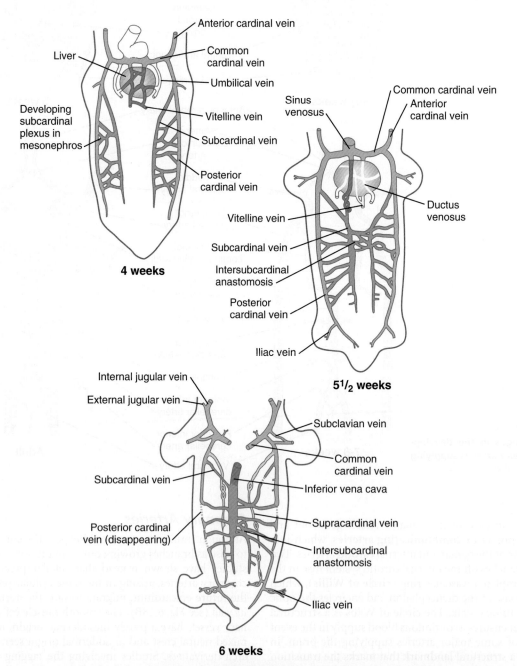

Fig. 17.13 Development of the cardinal vein system in a human embryo. The colors of the original embryonic cardinal veins are carried through in all drawings to facilitate an understanding of the derivations of the adult veins. *(Based on McClure CFW, Butler EG: Am J Anat 35:331-383, 1925.)*

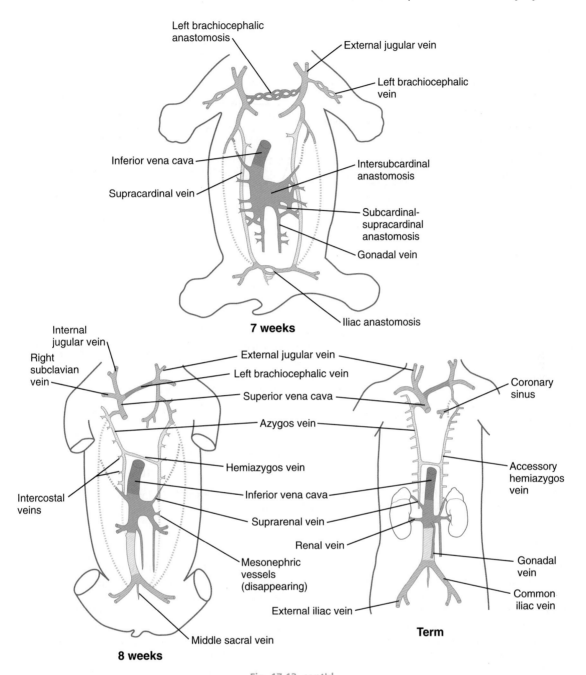

Left brachiocephalic anastomosis

External jugular vein

Left brachiocephalic vein

Inferior vena cava

Supracardinal vein

Intersubcardinal anastomosis

Subcardinal-supracardinal anastomosis

Gonadal vein

Iliac anastomosis

7 weeks

Internal jugular vein

Right subclavian vein

External jugular vein

Left brachiocephalic vein

Superior vena cava

Azygos vein

Coronary sinus

Hemiazygos vein

Inferior vena cava

Accessory hemiazygos vein

Intercostal veins

Suprarenal vein

Renal vein

Mesonephric vessels (disappearing)

Gonadal vein

Common iliac vein

External iliac vein

Term

Middle sacral vein

8 weeks

Fig. 17.13, cont'd

from the head and body into a pair of short **common cardinal veins** (see Fig. 17.13). The common cardinal veins, in turn, empty their blood into the **sinus venosus** of the primitive heart (see Fig. 17.15).

In the cranial region, the originally symmetrical anterior cardinal veins are transformed into the **internal jugular veins** (**Fig. 17.14**). As the heart rotates to the right, the base of the left internal jugular vein is attenuated. At the same time, a new anastomotic channel, which ultimately forms the **left brachiocephalic vein**, connects the left internal jugular vein to the right one. Through this anastomosis, the blood from the left side of the head is drained into the original right anterior cardinal vein, which ultimately becomes the **superior vena**

cava, emptying into the right atrium of the heart. Meanwhile, the proximal part of the left common cardinal vein persists as a small channel, the **coronary sinus**, which is the final drainage pathway of many of the coronary veins, also into the right atrium of the heart.

In the trunk, a pair of **subcardinal veins** arises in association with the developing mesonephros. The subcardinal veins are connected with the posterior cardinal veins and to each other through numerous anastomoses. Both the postcardinal and the subcardinal veins drain the mesonephric kidneys through numerous small side branches. As the mesonephric kidneys start to regress, the veins draining them also begin to break up. At this point, a pair of **supracardinal veins** appears

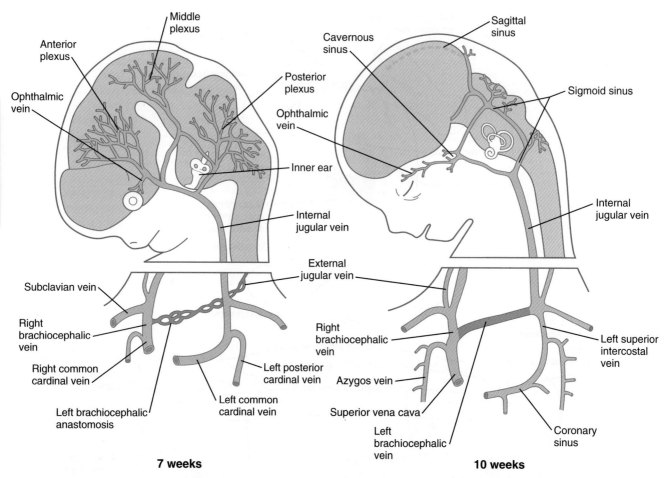

Fig. 17.14 **Stages in the formation of the major veins that drain the head and upper trunk.** At 7 weeks, an anastomosis *(purple)* between the left and right jugular vein forms the basis for establishment of the left brachiocephalic vein.

in the body wall dorsal to the subcardinal veins. Over time, all three sets of cardinal veins in the body break up to varying degrees, with surviving remnants incorporated into the **inferior vena cava**. The inferior vena cava forms as a single asymmetrical vessel that runs parallel to the aorta on the right (see Fig. 17.13). Most of the named veins of the thoracic and abdominal cavities are derived from persisting segments of the cardinal vein system.

Vitelline and Umbilical Veins

The extraembryonic **vitelline** and **umbilical veins** begin as pairs of symmetrical vessels that drain separately into the sinus venosus of the heart (**Fig. 17.15**). Over time, these vessels become intimately associated with the rapidly growing liver. The vitelline veins, which drain the yolk sac, develop sets of anastomosing channels within and outside the liver. Outside the liver, the two vitelline veins and their side-to-side anastomotic channels become closely associated with the duodenum. Through the persistence of some channels and the disappearance of others, the **hepatic portal vein**, which drains the intestines, takes shape. Within the liver, the vitelline plexus becomes transformed into a capillary bed that allows the broad distribution of food materials absorbed from the gut through the functional parts of the liver. From the hepatic

capillary bed, the blood that arrives from the hepatic portal vein passes into a set of **hepatic veins**, which empty the blood into the sinus venosus.

The originally symmetrical umbilical veins soon lose their own hepatic segments and drain directly into the liver by combining with the intrahepatic vascular plexus of the vitelline veins. Soon, a major channel, the **ductus venosus**, forms and shunts much of the blood entering from the left umbilical vein directly through the liver and into the inferior vena cava. The ductus venosus is an important adaptation for maintaining a functional embryonic pattern of blood circulation. Shortly thereafter, the right umbilical vein degenerates, leaving the left umbilical vein the sole channel for bringing blood that has been reoxygenated and purified in the placenta back to the embryonic body. The ductus venosus permits the incoming oxygenated placental blood to bypass the capillary networks of the liver and to distribute it to the organs (e.g., brain, heart) that need it most.

Pulmonary Veins

The pulmonary veins are phylogenetically recent structures that form independently, rather than taking over portions of the older cardinal vein systems. From each lung, venous drainage channels converge until they ultimately form a single large

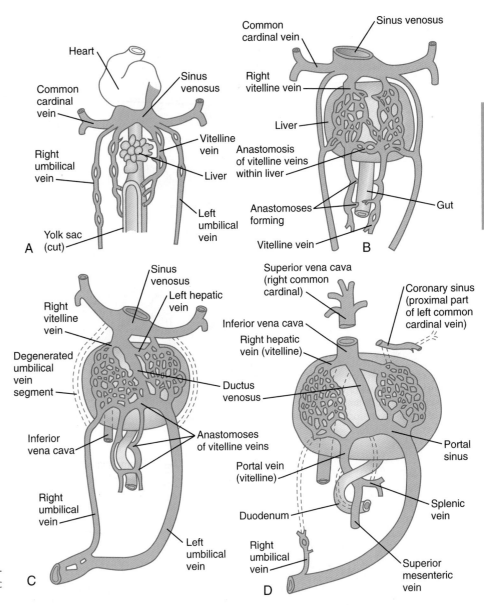

Fig. 17.15 **A** to **D,** Stages in the development of the umbilical and hepatic portal veins and intrahepatic circulation.

common pulmonary vein, which empties into the left atrium of the heart. As the atrium expands, the common pulmonary vein becomes incorporated into its wall (**Fig. 17.16**). Ultimately, the absorption passes the first and second branch points of the original pulmonary veins, with resulting entrance of four independent pulmonary veins into the left atrium.

Development of Lymphatic Channels

The lymphatic system arises by cells budding off from the lateral surfaces of the anterior and posterior cardinal veins (see Fig. 17.8). As they are preparing to bud off from the veins, the prelymphatic endothelial cells express the homeobox genes *Sox18* and *Prox-1*. This is followed by the expression of **VEGFR-3** and their specification to become lymphatic endothelial cells. These cells respond to two specific isoforms of VEGF: VEGF-C and VEGF-D. The emigrating lymphatic endothelial cells aggregate in areas where the lateral plate mesoderm produces VEGF-C to form **primary lymph sacs**.

The first of these are the two **jugular lymph sacs** arising from the anterior cardinal veins and abdominal lymph sacs, which arise from the endothelium of the posterior cardinal veins (**Fig. 17.17**). In the abdomen, a **retroperitoneal lymph sac** forms on the posterior body wall at the root of the mesentery during the eighth week. Later, a **cisterna chyli** forms at the same level, but dorsal to the aorta. At about the same time, a pair of **posterior lymph sacs** arises at the bifurcation of the femoral and sciatic veins. By the end of the ninth week, lymphatic vessels connect these lymph sacs.

Two major lymphatic vessels connect the cisterna chyli with the jugular lymph sacs. An anastomosis forms between these two channels. A single lymphatic vessel consisting of the caudal part of the right channel, the anastomotic segment, and the cranial part of the left channel ultimately becomes the definitive **thoracic duct** of an adult. The thoracic duct drains lymph from most of the body and the left side of the head into the venous system at the junction of the left internal jugular and subclavian veins. The right lymphatic duct, which drains

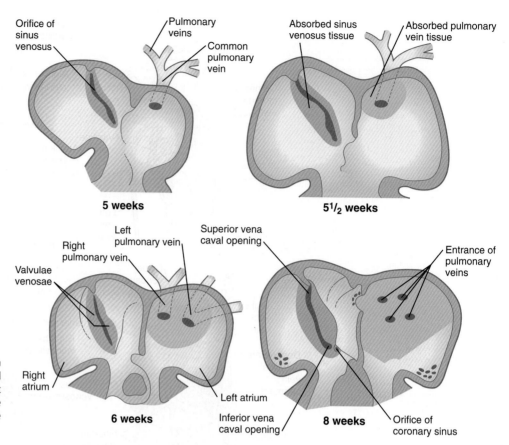

Fig. 17.16 Stages in the absorption of the common pulmonary vein and its branches into the wall of the left atrium and changes over time in the opening of the sinus venosus into the right atrium.

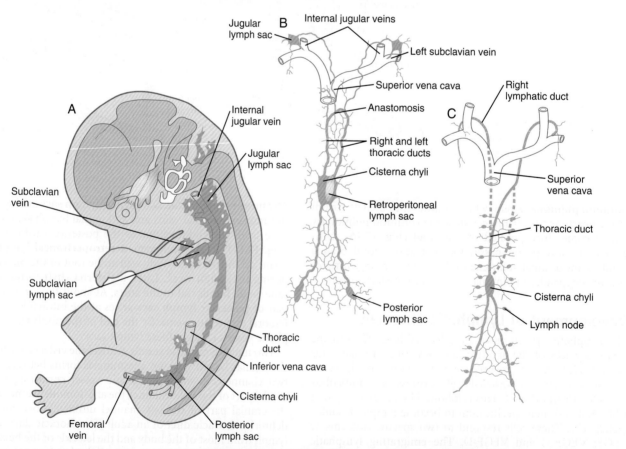

Fig. 17.17 **Stages in the development of the major lymphatic channels. A** and **B** show 9-week-old embryos. **C** shows the fetal period. Between **B** and **C**, the transformation between the reasonably symmetrical disposition of the main lymphatic channels and the asymmetrical condition that is characteristic of an adult can be seen.

the right side of the head and upper part of the thorax and right arm, also empties into the venous system at the original location of the right jugular sac. In contrast to the lymph sacs, many of the peripheral lymphatic vessels originate from local **lymphangioblasts**.

Development and Partitioning of the Heart

Early Development of the Heart
Cellular Origins

The heart arises from splanchnic mesoderm. The first part of the heart to develop is the **primary heart field** (cardiac crescent; see Fig. 6.14B), which forms the left ventricle and the atria, the phylogenetically most primitive components of the mammalian heart. Exposure of the most posterior cells of the cardiac crescent to a **retinoic acid** gradient arising in the posterior mesoderm conditions these cells to adopt an atrial identity, whereas the more anterior cells not exposed

to retinoic acid by default assume a ventricular identity (**Fig. 17.18**). The appearance of the **secondary heart field** provides the cellular material for formation of the evolutionarily more recent right ventricle and outflow tract. Cells of the secondary heart field also form the proepicardium (see p. 105) and contribute myocardial cells to the inflow (atrial) areas of the heart.

Cells of the secondary heart field arise from multipotent precursors within the pharyngeal mesoderm. These precursor cells can form either skeletal or cardiac muscle. Those from the first pharyngeal arch can either migrate to the head and form the masticatory muscles or migrate caudally into the secondary heart field, where they later become incorporated into the right ventricle. Similarly, precursor cells from the second pharyngeal arch can give rise to the muscles of facial expression or contribute to the outflow tract of the heart.

Several sets of molecules (MEF2, NKX2, GATA, Tbx, and Hand) form what has been called the core regulatory network that guides the differentiation of cardiac tissue. These sets of molecules are differentially regulated, however, by upstream activators that are specific to either the primary or

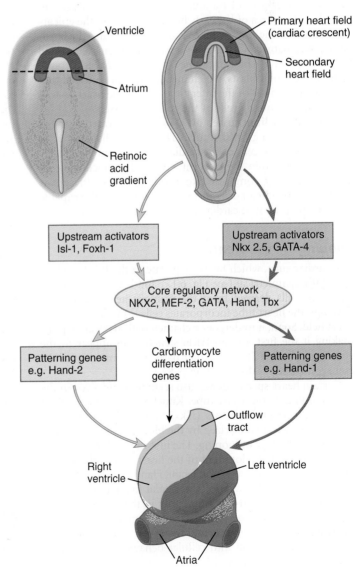

Fig. 17.18 **Steps involved in early establishment of the heart.**

the secondary heart fields (see Fig. 17.18). Even within the core regulatory network, molecular variants are unique to derivatives of the primary or secondary heart fields. **Hand-1** is expressed in cells derived from the primary heart field, and in its absence left ventricular anomalies are seen. **Hand-2** is expressed in cells derived from the secondary heart field, and if it is not expressed, the right ventricle does not form. Research has shown that the actions of **microRNAs** are tightly integrated into the overall regulatory programs that guide development of the heart.

As the outflow tract develops, it receives additional cellular contributions from various sources. Endothelial components arise from cephalic paraxial and lateral mesoderm in the region of the otic placode. A major cellular component of the wall of the outflow tract is derived from cardiac neural crest, specifically from the level between the midotic placode to the caudal end of the third somite (see Fig. 12.11). These cranial components integrate with the bilateral cardiac primordia while in the cervical region. As the heart descends into the thoracic cavity, the cranially derived cells of the outflow tract accompany it.

Another source of cells for the developing heart is the **proepicardium** (see Fig. 6.18). The proepicardium, which consists of an aggregate of mesothelial cells in the pericardium near the inflow area of the early heart, gives rise to the epicardium, most of the interstitial cells of the heart, and the coronary vasculature. Through an epitheliomesenchymal transformation mechanism, some of the epicardial cells transform into the fibroblasts, which invade the heart musculature and constitute most of the interstitial cells of the heart, as well as the smooth muscle cells of the coronary vessels.

The **endocardium** arises by a process of vasculogenesis within the forming cardiac tube. The endocardial cells arise from mesodermal cells of the cardiac crescent, but it is uncertain whether, within the crescent, endothelial precursor cells are already fully committed to the endothelial lineage or whether the precursor cells are bipotential and are also capable of differentiating into cardiac muscle cells.

Looping of the Heart

The cardiac tube, which takes shape late in the third week, is bilaterally symmetrical (see Fig. 6.17) and is composed principally of cells derived from the primary heart field. As it develops, the heart tube incorporates cells from the secondary heart field. Soon, it undergoes a characteristic dextral looping, making it the first asymmetrical structure to appear in the embryonic body. (The molecular basis for asymmetry of the embryonic body is discussed on p. 87.)

Certain heart-specific genes also seem to be involved in the early looping of the heart tube. Knockouts of four types of cardiac transcription factors (Nkx2.5, MEF-2, Hand-1, and Hand-2) all are characterized by the blocking of heart development at the stage of looping. The first molecular indication of asymmetrical development of the heart tube is the shift in expression of the transcription factor, Hand-1 (**e-Hand**), from both sides to the left side of the caudal heart tube. Hand-2 (**d-Hand**) is expressed predominantly in the primordium of the right ventricle. The Hand molecules may play a role in interpreting earlier asymmetrical molecular signals and translating this information into cellular behavior that results in looping.

As the straight heart tube begins to undergo looping, the originally ventral surface of the heart tube becomes the outer border of the loop, and the originally dorsal surface becomes the inner border of the loop. When the individual heart chambers begin to form, they arise as outpocketings from the outer border of the heart loop. Characterization of the cellular basis for cardiac looping has proven very difficult. Although it has been clearly shown that looping is an intrinsic property of the developing heart tube, the mechanisms by which molecular instructions are translated into structural changes remain obscure.

Research has revealed that the outpocketing (ballooning) of the future ventricular and atrial chambers is accompanied by the development of important structural and functional properties. The early symmetrical heart tube (called the **primary myocardium**) is characterized by slow growth, slow conduction of impulses, slow contraction, and the ability to undergo spontaneous depolarization. As the heart begins to loop, bulges that represent the incipient ventricular and atrial chambers appear on the outer surfaces of the loop and at the inflow end of the heart tube (**Fig. 17.19**).

In contrast to the primary myocardium, the cells of the **chamber myocardium** are characterized by a high proliferative capacity, strong contractility, high conduction velocity, and a low capacity to generate spontaneous impulses. Development of the primary heart tube is guided by the transcription factor Tbx-2, whereas development of the ventricular chambers is controlled by Tbx-5. As the heart undergoes developmental remodeling, the chamber myocardium constitutes the bulk of the atrial and ventricular chambers, whereas the primary myocardium is retained in the outflow tract (where it is later joined by cells from the neural crest), the atrioventricular canal and atrioventricular valves (where it is joined by cells originating from the proepicardium), and parts of the conducting system (the sinoatrial and atrioventricular nodes and the atrioventricular bundle).

The result of cardiac looping is an S-shaped heart in which the originally caudal inflow part of the heart (**atrium**) becomes positioned dorsal to the outflow tract. In the early heart, the outflow tract is commonly called the **bulbus cordis**, and it leads directly into the aortic sac and the incipient aortic arch system (**Fig. 17.20**). In early looping, the cranial limb of the S represents the bulbus cordis, and the middle limb represents the ventricular part of the heart, with the primordium of the right ventricle closest to the outflow tract, and that of the left ventricle next to the caudal limb. The caudal limb represents a common atrium. Later, the common atrium bulges on either side (**Fig. 17.21**), and an internal septum begins to divide the common ventricle into right and left chambers. The outflow tract (bulbus cordis of the early heart) retains its gross tubular appearance. Its distal part, which leads directly into the aortic arch system, is called the **truncus arteriosus**. The shorter transitional segment between the truncus and the ventricle is called the **conus arteriosus**. The conus is separated from the ventricles by faint grooves.

Early Atrioventricular Partitioning of the Heart

Early in heart development, the atrium becomes partially separated from the ventricle by the formation of thickened

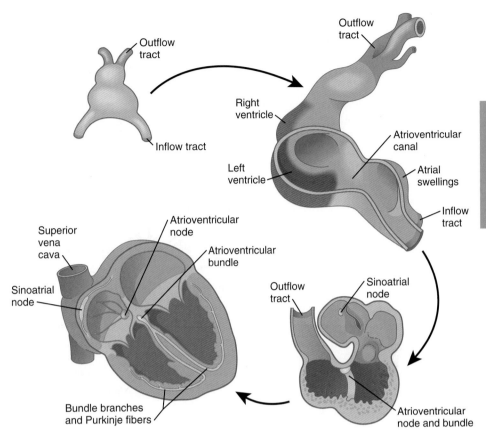

Fig. 17.19 Early development of the heart that shows the ballooning of the four chambers from the primary myocardium *(tan). (Adapted from Christoffels AM, Burch JBE, Moorman AFM: Trends Cardiovasc Med 14:301-307, 2004.)*

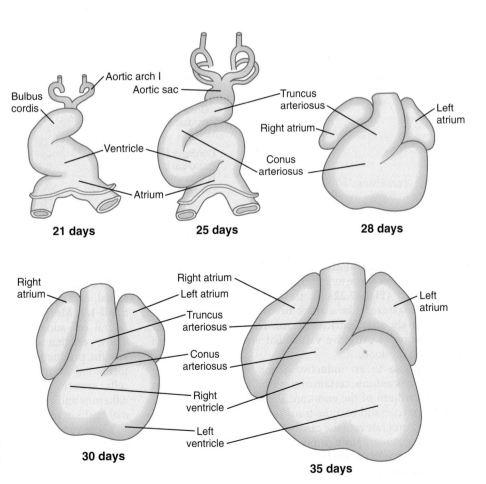

Fig. 17.20 Ventral views of human embryonic hearts that illustrate bending of the cardiac tube and the establishment of its regional divisions.

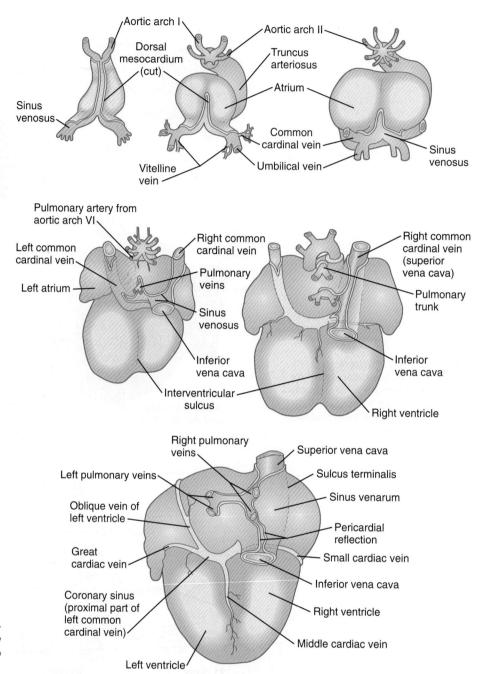

Fig. 17.21 Dorsal views of the development of the human heart that show changes in the venous input channels into the heart.

atrioventricular cushions. A similar but less pronounced thickening forms at the junction between the ventricle and the outflow tract (**Fig. 17.22**). In these areas, the cardiac jelly, which is organized like a thick basement membrane, protrudes into the atrioventricular canal. The **endocardial cushions** function as primitive valves that assist in the forward propulsion of blood.

In response to an inductive action by the underlying primary myocardium, certain antigenically distinct cells from the endocardium of the endocardial cushions lose their epithelial character and become transformed into mesenchymal cells, which migrate into the cardiac jelly. The inductive stimulus, which emanates from primary, but not chamber, myocardial cells, was initially described as being packaged as 20- to 50-nm particles, called **adherons** (see Fig. 17.22). Current data

suggest that **TGF-β3** and **BMP-2**, which act upstream of **Notch** and **Tbx-2**, are important components of the overall inductive signal, as a result of which certain endothelial cells express **Snail-1/2**, **Msx-1**, and **Twist-1**. Endocardial cells express the neural cell adhesion molecule (N-CAM) on their surfaces. Those cells that transform into mesenchymal cells downregulate the production of N-CAM, which facilitates their transformation into motile cells. The transformed mesenchymal cells secrete proteases, which destroy inductively active adherons and restore morphogenetic stability to the endocardial cushion regions.

These cellular and molecular events form the basis for the early formation of the major heart valves. Disturbances in these processes can account for many malformations of the heart.

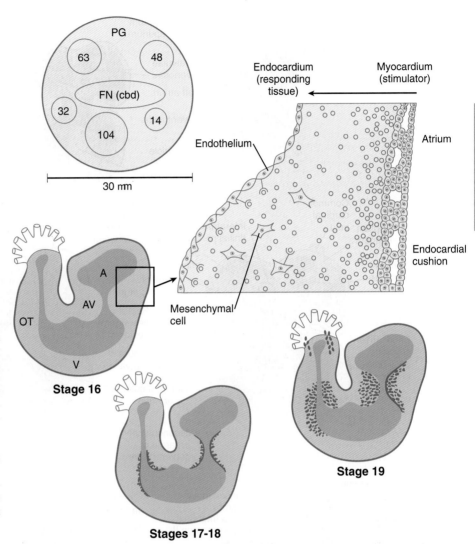

Fig. 17.22 Seeding of the endocardial cushions by mesenchymal cells in the developing avian heart. The higher power drawing in the *upper right* shows details of the transformation of endocardial to mesenchymal cells. Also shown is a hypothetical model of the 30-nm adheron-like particle that seems to induce this transformation in the endocardial cells. *Numbered circles* indicate other proteins. A, atrium; AV, atrioventricular canal; FN (cbd), fibronectin (cell-binding domain); OT, outlet tract; PG, proteoglycan; V, ventricle. *(Based on Bolender DL, Markwald RR: In Feinberg RN, Sherer GK, eds:* The development of the vascular system, *Basel, 1991, Karger, pp 109-124.)*

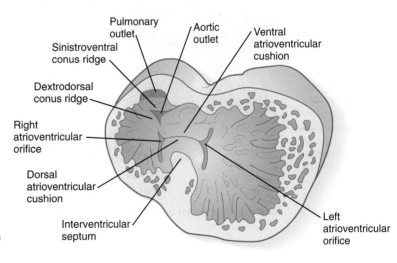

Fig. 17.23 Partitioning the atrioventricular canal from the ventricular aspect.

Later Partitioning of the Heart
Separation of the Atria from the Ventricles

The endocardial cushions (see Fig. 17.22), which ultimately become transformed into dense connective tissue, form on the dorsal and ventral walls of the atrioventricular canal. As they

grow into the canal, the two cushions meet and separate the atrioventricular canal into right and left channels (**Figs. 17.23 and 17.24**). The early endocardial cushions serve as primitive valves that assist in the forward propulsion of blood through the heart. Later in development, thin leaflets of anatomical valves take shape in the atrioventricular canal. The definitive

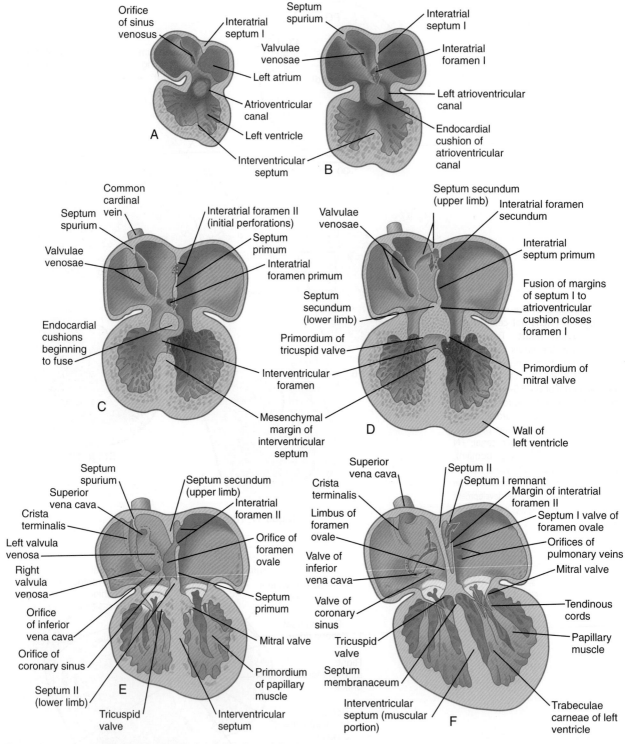

Fig. 17.24   A to F, Stages in the internal partitioning of the heart. *(After Patten BM: Human embryology, ed 3, New York, 1968, McGraw-Hill.)*

valve leaflets do not seem to come from endocardial cushion tissue as much as from invagination of superficial epicardially derived tissues of the atrioventricular groove. The valve that protects the right atrioventricular canal develops three leaflets (**tricuspid valve**), but the valve in the left canal (**mitral, or bicuspid, valve**) develops only two. At the molecular level, inhibition of expression of **Tbx-2** on the atrial and the ven-

tricular sides of the atrioventricular canal effects the separation between the atria and ventricles.

Partitioning of the Atria

While the atrioventricular canals are forming, a series of structural changes divides the common atrium into separate left and right chambers. Partitioning begins in the fifth week with the downgrowth of a crescentic **interatrial septum primum** from the cephalic wall between the bulging atrial chambers (see Fig. 17.24). The apices of the crescent of the septum primum extend toward the atrioventricular canal and merge with the endocardial cushions. The space between the leading edge of the septum primum and the endocardial cushions is called the **interatrial foramen primum**. This space serves as a shunt, permitting blood to pass directly from the right to the left atrium.

Circulatory shunts in the developing heart satisfy a very practical need. All incoming blood enters to the right side of the interatrial septum primum. Because of the late development of the lungs, however, and the poor carrying capacity of the pulmonary vessels during most of the fetal period, the pulmonary circulation cannot handle a full load of blood. If the heart were to form four totally separate chambers from the beginning, the pulmonary circulation would be overstressed, and the left side of the heart would not be pumping enough blood to foster normal development, especially in the early weeks.

The problem of maintaining a balanced circulatory load on all chambers of the heart is met by the existence of two shunts that allow most of the circulating blood to bypass the lungs. One shunt is a direct connection between the right and left atria that allows blood entering the right atrium to bypass the pulmonary circulation completely by passing directly into the left atrium. This shunt permits the normal functional development of the left atrium. If all the blood entering the right atrium passed directly into the left atrium, however, the right ventricle would have nothing to pump against and would become hypoplastic. In midpregnancy, more than 30% of the blood entering the right atrium is shunted directly into the left atrium; near term, the percentage is reduced to less than 20%. With the arrangement of the openings of vascular channels into the right atrium, a significant amount of blood also enters the right ventricle and leaves that chamber through the pulmonary outflow tract. Most of the blood leaving the right ventricle, which is still far too much to be accommodated by the vasculature of the lungs, bypasses the lungs via the ductus arteriosus and empties directly into the descending aorta. By these two mechanisms, the heart is evenly exercised, and the pulmonary circulation is protected.

When the interatrial septum primum is almost ready to fuse with the endocardial cushions, an area of genetically programmed cell death causes the appearance of multiple perforations near its cephalic end (see Fig. 17.24C). As the leading edge of the septum primum fuses with the endocardial cushions, thus obliterating the foramen primum, the cephalic perforations in the septum primum coalesce and give rise to the **interatrial foramen secundum**. This new foramen preserves the direct connection between the right and left atria.

Shortly after the appearance of the foramen secundum, a crescentic **septum secundum** begins to form just to the right of the septum primum. This structure, which grows out from the dorsal to the ventral part of the atrium, forms a **foramen ovale**. The position of the foramen ovale allows most of the blood that enters the right atrium through the inferior vena cava to pass directly through it and the foramen secundum into the left atrium. The arrangement of the two interatrial septa allows them to act like a one-way valve, however, and permits blood to flow from the right to the left atrium, but not in the reverse direction.

Repositioning of the Sinus Venosus and the Venous Inflow into the Right Atrium

During the stage of the straight tubular heart, the sinus venosus is a bilaterally symmetrical chamber into which the major veins of the body empty (see Fig. 17.15). As the heart undergoes looping and the interatrial septa form, the entrance of the sinus venosus shifts completely to the right atrium (see Figs. 17.21 and 17.24). As this occurs, the right horn of the sinus venosus becomes increasingly incorporated into the wall of the right atrium, so the much reduced left horn, the **coronary sinus** (which is the common drainage channel for the coronary veins), opens directly into the right atrium (see Fig. 17.16). Also in the right atrium, valvelike flaps of tissue (**valvulae venosae**) form around the entrances of the superior and inferior venae cavae. Because of the orientation of the orifice and its pressure, blood entering the right atrium from the inferior vena cava passes mostly through the interatrial shunt and into the left atrium, whereas blood entering from the superior vena cava and the coronary sinus flows through the tricuspid valve into the right ventricle.

Partitioning of the Ventricles

When the interatrial septa are first forming, a muscular **interventricular septum** begins to grow from the apex of the ventricular loop between the ballooning right and left ventricular chambers toward the atrioventricular endocardial cushions (see Fig. 17.24C). The early division of the common ventricle is also reflected by the presence of a groove on the outer surface of the heart (**Fig. 17.25**). Although an **interventricular foramen** is initially present, it is ultimately obliterated by (1) further growth of the muscular interventricular septum, (2) a contribution by truncoconal ridge tissue that divides the outflow tract of the heart, and (3) a membranous component derived from endocardial cushion connective tissue.

Partitioning of the Outflow Tract of the Heart

In the very early tubular heart, the outflow tract is a single tube, the bulbus cordis. By the time the interventricular septum begins to form, the bulbus has elongated and can be divided into a proximal conus arteriosus and a distal truncus arteriosus (see Fig. 17.20). Closest to the heart, the wall of the outflow tract is composed largely of cells derived from the secondary heart field; more distally, cells derived from the neural crest predominate. Although initially a single channel, the outflow tract is partitioned into separate aortic and pulmonary channels through the appearance of two spiral **truncoconal ridges**, which are derived largely from neural crest mesenchyme. These ridges bulge into the lumen and finally meet, thus separating the lumen into two channels. The aortic sac, which is located distal to the truncoconal region, does not contain ridges. Partitioning of the outflow tract begins near the ventral aortic root between the fourth and sixth arches and extends toward the ventricles, spiraling as it goes (**Fig. 17.26**). This accounts for the partial spiraling of the aorta and the pulmonary artery in the adult heart.

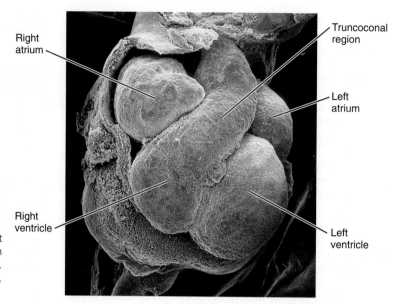

Fig. 17.25 Scanning electron micrograph showing a right oblique view of the heart of a human embryo early in the sixth week. The pericardium has been dissected free from the heart. *(From Jirásek J:* Atlas of human prenatal morphogenesis, *Boston, 1983, Martinus Nijhoff.)*

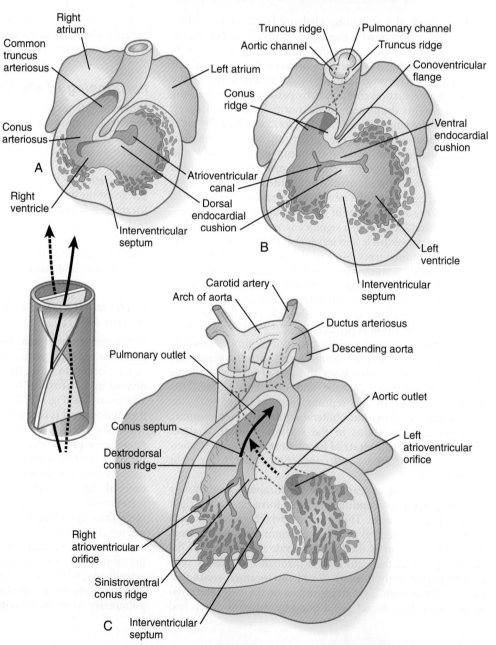

Fig. 17.26 A to **C**, Partitioning of the outflow tract of the developing heart. The truncoconal ridges undergo a 180-degree spiraling. *(After Kramer TC:* Am J Anat *71:343-370, 1942.)*

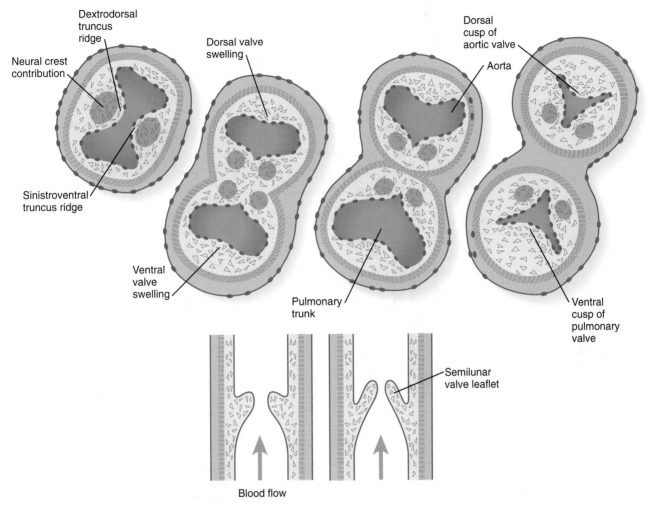

Fig. 17.27 **Formation of the semilunar valves in the outflow tract of the heart.** Neural crest cells *(green)* contribute to the formation of the valvular leaflets.

Before and during the partitioning process, the neural crest–derived cells of the wall of the outflow tract begin to produce elastic fibers, which provide the resiliency required of the aorta and other great vessels. Elastogenesis follows a gradient, first through the outflow tract, then into the aorta itself, and ultimately into the smaller arterial branches off the aorta.

At the base of the conus, where endocardial cushion tissue is formed in the same manner as in the atrioventricular canal, two new sets of **semilunar valves** form (**Fig. 17.27**). These valves, each of which has three leaflets, prevent ejected blood from washing back into the ventricles. Cranial neural crest cells and cardiac mesoderm contribute to the formation of the semilunar valves. As previously stated, the most proximal extensions of the truncoconal ridges contribute to the formation of the interventricular septum. Just past the aortic side of the aortic semilunar valve, the two coronary arteries join the aorta to supply the heart with blood.

Innervation of the Heart

Although initial heart development occurs independently of nerves, three sets of nerve fibers ultimately innervate the heart (**Fig. 17.28**). Sympathetic (adrenergic) nerve fibers, which act to speed up the heart beat, arrive as outgrowths from

sympathetic ganglia of the trunk. These nerve fibers are derived from trunk neural crest. Parasympathetic (cholinergic) innervation is derived from the cardiac neural crest. Neurons of the cardiac ganglia, which are the second-order parasympathetic neurons, migrate directly to the heart from the cardiac neural crest. These synapse with axons of first-order parasympathetic neurons that gain access to the heart via the vagus nerve. Sensory innervation of the heart is also supplied via the vagus nerve, but the sensory neurons originate from placodal ectoderm (nodose placode) (see Fig. 13.1). The direct innervation of the heart thus has three separate origins.

If the cardiac neural crest is removed in the early chick embryo, cholinergic cardiac ganglia still form. Experiments have determined that the nodose placodes compensate for the loss of neural crest by supplying neurons that replace the normal parasympathetic ones.

Initiation of Cardiac Function and the Conducting System of the Heart

The human heart begins to beat 21 to 23 days after fertilization, when it is still in the stage represented by the primary

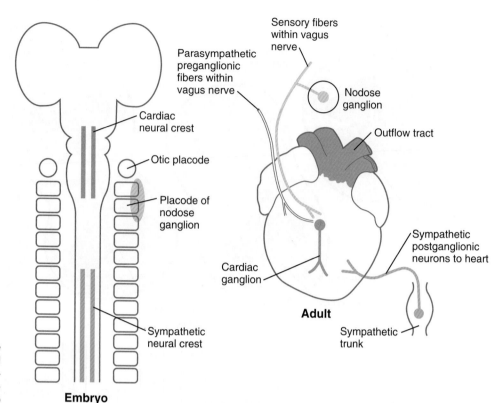

Fig. 17.28 Contributions of the cranial and trunk neural crest and the nodose placode to the innervation of the avian heart. *(Based on Kirby ML: Cell Tissue Res 252:17-22, 1988.)*

heart myocardium (see p. 107). The beat is slow (<40 beats/minute), and pacemaking activity begins near the inflow region of the heart and spreads toward the outflow tract through spontaneous depolarization of the cells. At this stage, the heart functions like a simple peristaltic pump.

As the atrial and ventricular chambers take shape, the differentiating cardiomyocytes are unable to generate or propagate beats in the same way as the cells of the primary myocardium. To coordinate the beat of the expanding chambers, it is necessary for the mammalian heart to develop a specialized conducting system, which takes advantage of some elements from the primary myocardium in the atrial region and adds to them a phylogenetically newer conducting system within the ventricular myocardium.

The **sinoatrial node** is the pacemaker of the mature heart (**Fig. 17.29**), and it is a direct descendant of the cells in the primary myocardium that initiate the first coordinated heart beats. Terminals of sympathetic and parasympathetic nerve fibers grow into the area to modulate the heart beat. The contractile stimulus then passes to the atrioventricular node through mechanisms still not well understood. From early development, the **atrioventricular node**, which is also a direct derivative of the primary myocardium, functions to slow down the conductive impulse to separate the contractions of the atrial and ventricular chambers. Activity of the transcriptional repressor **Tbx-3** prevents the primary myocardial cells destined to form the sinoatrial and atrioventricular nodes from differentiating into the more highly contractile and more poorly conducting cells that characterize the ventricular chambers. From the atrioventricular node, the pacemaking impulse then passes with increasing velocity down the

atrioventricular bundle and into left and right **bundle branches** before spreading out over the ventricular myocardium as the **Purkinje fibers**.

The atrioventricular node and bundle arise from a segment of a ring of myocardial cells that initially surround the interventricular foramen and later in some species become translocated to form a figure eight–shaped ring at the atrioventricular junction (see Fig. 17.29). Branches from this ring ultimately pass along either side of the interventricular septum and then arborize along the ventricular walls as Purkinje fibers.

The part of the conducting system consisting of bundle and Purkinje fibers represents a network of highly modified cardiac muscle fibers, whose structural and functional characteristics have been highly modified during development by paracrine influences. Stimulated by hemodynamic forces from the beating ventricles, endothelial cells produce an enzyme that activates the peptide **endothelin-1**, which along with other factors, such as **neuregulin** from the endocardium, stimulates the transformation of early cardiomyocytes into conducting cells of the Purkinje system. Purkinje cells elaborate **connexins**, which facilitate rapid conduction from one cell to another. Very rapid conduction is necessary to ensure a nearly simultaneous beat throughout the ventricle.

Fetal Circulation

In many respects, the overall plan of the embryonic circulation seems to be inefficient and more complex than needed to maintain the growth and development of the fetus. The

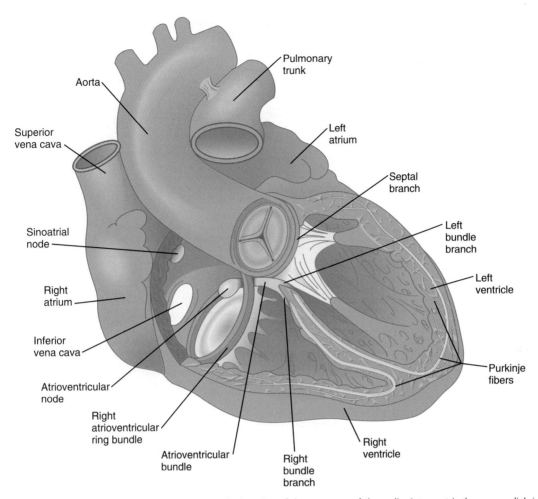

Fig. 17.29   Representation of the adult human heart showing the location of the remnants of the earlier interventricular myocardial ring *(green)* and the portion *(yellow)* that gives rise to the atrioventricular node and bundle. The left and right bundle branches emanate from this structure and ultimately spread out along the ventricular walls as the Purkinje fibers. *(Based on Moorman A, Lamers WH: In Harvey RP, Rosenthal N eds: Heart development, New York, 1999, Academic Press.)*

embryo must prepare for the moment, however, when it suddenly shifts to a totally different pattern of oxygenation of blood through the lungs, rather than the placenta, thus making the modifications of the fetal plan of circulation essential.

Highly oxygenated blood from the placenta enters the umbilical vein in a large stream that is sometimes under increased pressure because of uterine contractions. Within the substance of the liver, blood from the umbilical vein under higher pressure passes directly into the ductus venosus, which allows it to bypass the small circulatory channels of the liver and flow directly into the inferior vena cava (**Fig. 17.30**). When in the vena cava, blood has immediate access to the heart. Poorly oxygenated blood flowing in the inferior vena cava can be backed up because of the strength of the umbilical blood flow.

Functional evidence exists for a physiological sphincter in the ductus venosus, which forces much of the umbilical blood to pass through hepatic capillary channels and enter the inferior vena cava through hepatic veins when it tightens. This physiological sphincter considerably reduces the pressure of the umbilical blood and allows poorly oxygenated systemic blood from the inferior vena cava to enter the right atrium at a lower pressure. Higher-pressure blood entering the

umbilical vein from the placenta also tends to prevent blood from the hepatic portal vein from entering the ductus venosus. When the uterus is relaxed, and the umbilical venous blood is under low pressure, poorly oxygenated portal blood mixes with the umbilical blood in the ductus venosus. More mixing of umbilical and systemic blood occurs in the inferior vena cava as well.

In the right atrium, the orientation of the entrance of the inferior vena cava allows a stream of blood under slightly increased pressure to pass directly through the foramen ovale and foramen secundum into the left atrium (see Fig. 17.30). This is the route normally taken by highly oxygenated umbilical blood entering the body under increased pressure. Because the interatrial shunt of the fetus is smaller than the opening of the inferior vena cava, some of the highly oxygenated caval blood eddies in the right atrium and enters the right ventricle. When low-pressure blood (typically poorly oxygenated systemic blood) enters the right atrium, it joins with the venous blood draining the head through the superior vena cava and the heart through the coronary sinus and is mostly directed through the tricuspid valve into the right ventricle.

All blood entering the fetal right ventricle leaves through the pulmonary artery and passes toward the lungs. Even in the

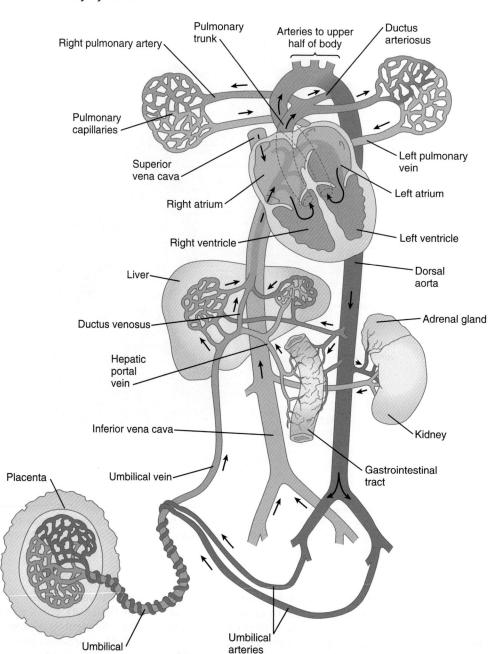

Pulmonary trunk

Right pulmonary artery

Arteries to upper half of body

Ductus arteriosus

Pulmonary capillaries

Superior vena cava

Right atrium

Right ventricle

Liver

Ductus venosus

Hepatic portal vein

Inferior vena cava

Placenta

Umbilical vein

Umbilical cord

Umbilical arteries

Left pulmonary vein

Left atrium

Left ventricle

Dorsal aorta

Adrenal gland

Kidney

Gastrointestinal tract

Fig. 17.30   Fetal circulation at term.

relatively late fetus, the pulmonary vasculature is not able to handle the full volume of blood that enters the pulmonary artery. A major reason for this is that the blood entering the lungs from the right ventricle is relatively poorly saturated (~50%) with oxygen. Especially later in fetal life, low oxygen saturation results in increasingly greater pulmonary vascular resistance. The blood that cannot be accommodated by the pulmonary arteries is shunted to the aorta via the ductus arteriosus. This structure protects the lungs from circulatory overload, yet it allows the right ventricle to exercise in preparation for its functioning at full capacity at birth. Only 12% of the right ventricular output passes through the lungs of the fetus. The control of patency of the ductus arteriosus has been subject to considerable speculation. Patency of the ductus

arteriosus and the ductus venosus in the fetus is maintained actively through the action of prostaglandin E_2 (ductus arteriosus) and prostaglandin I_2 (ductus venosus). Some of the effect of the prostaglandins is mediated by nitric oxide.

The left atrium receives a stream of highly oxygenated umbilical blood through the interatrial shunt and a small amount of poorly oxygenated blood from the pulmonary veins. This blood, which in aggregate is highly oxygenated, passes into the left ventricle and leaves the heart through the aorta. Some of the first arterial branches leaving the aorta supply the heart and brain, organs that require a high concentration of oxygen for normal development.

Where the aortic arch begins to descend, the ductus arteriosus empties poorly oxygenated blood into it. This mixture of well-oxygenated and poorly oxygenated blood is distributed to the tissues and organs that are supplied by the thoracic and abdominal branches of the aorta. Near its caudal end, the aorta gives off two large umbilical arteries, which carry blood to the placenta for renewal.

Clinical Correlation 17.1 presents malformations of the heart, and **Clinical Correlation 17.2** presents malformations of the blood vessels. **Table 17.6** summarizes the timelines in cardiac development.

Clinical Vignette

An 8-year-old boy is brought to the physician with a complaint of excessive fatigue and uncomfortable legs when walking or running. On physical examination, the physician notes a reduced dorsalis pedis pulse and some signs of cyanosis of the toes. The boy's hands show no signs of cyanosis.

On the basis of the physical examination, the physician suspects the presence of what malformation? Why?

Table 17.6 Timelines in Normal and Abnormal Cardiac Development

Normal Time	Developmental Events	Malformations Arising during Period
18 days	Horseshoe-shaped cardiac primordium appears	Lethal mutants
20 days	Bilateral cardiac primordia fuse Cardiac jelly appears Aortic arch is forming	Cardia bifida (experimental) — —
22 days	Heart is looping into S shape Heart begins to beat Dorsal mesocardium is breaking down Aortic arches I and II are forming	Dextrocardia — — —
24 days	Atria are beginning to bulge Right and left ventricles act like two pumps in series Outflow tract is distinguishable from right ventricle	— — —
Late fourth week	Sinus venosus is becoming incorporated into right atrium Endocardial cushions appear Early septum I appears between left and right atria Muscular interventricular septum is forming Truncoconal ridges are forming Aortic arch I is regressing Aortic arch III is formed Aortic arch IV is forming	Venous inflow malformations Persistent atrioventricular canal Common atrium Common ventricle Persistent truncus arteriosus — — Missing fourth aortic arch segment
Early fifth week	Endocardial cushions are coming together, forming right and left atrioventricular canals Further growth of interatrial septum I and muscular interventricular septum occurs Truncus arteriosus is dividing into aorta and pulmonary artery Atrioventricular bundle is forming; possible neurogenic control of heart beat Pulmonary veins are becoming incorporated into left atrium Aortic arches I and II have regressed Aortic arches III and IV have formed Aortic arch VI is forming Conduction system forms	Persistent atrioventricular canal Muscular ventricular septal defects Transposition of great vessels; aortic and pulmonary stenosis or atresia — Aberrant pulmonary drainage — — — —
Late fifth to early sixth week	Endocardial cushions fuse Interatrial foramen II is forming Interatrial septum I is almost contacting endocardial cushion Membranous part of interventricular septum starts to form Semilunar valves begin to form	— — Low atrial septal defects Membranous interventricular septal defects Aortic and pulmonary valvular stenosis
Late sixth week	Interatrial foramen II is large Interatrial septum II starts to form Atrioventricular valves and papillary muscles are forming Interventricular septum is almost complete Coronary circulation is becoming established	High atrial septal defects — Tricuspid or mitral valvular stenosis or atresia Membranous interventricular septal defects —
Eighth to ninth week	Membranous part of interventricular septum is completed	Membranous interventricular septal defects

CLINICAL CORRELATION 17.1
Malformations of the Heart

With an incidence of almost 1 per 100 live births, heart defects are the most common class of congenital malformations. Because of the close physiological balance of the circulation, most malformations produce symptoms. Clinically, heart malformations are typically classified as malformations that are associated with cyanosis (**cyanotic defects**) in postnatal life and those that are not (**acyanotic defects**).

Cyanosis results when the blood contains more than 5 g/dL of reduced hemoglobin. Cyanosis is readily recognizable by a purplish to bluish tinge to the skin in areas with a dense superficial capillary circulation. It is associated with **polycythemia**, an increased concentration of erythrocytes in the blood resulting from the overall decreased oxygen saturation of the blood. Long-term cyanosis is associated with a prominent clubbing of the ends of the fingers (see Fig. 17.42) and decreased growth. In severe cases of cyanosis, children often assume a squatting posture that may facilitate reoxygenation of the blood.

Postnatally, cyanosis is associated with the presence of a right-to-left shunt in which venous blood mixes with systemic blood. Some heart defects are acyanotic for many years, but then become cyanotic. These defects are initially characterized by a left-to-right shunt in which oxygenated systemic blood refluxes into the right atrium or ventricle. The net result is an increased pumping load on the right ventricle that ultimately leads to right ventricular hypertrophy. Over a long period, the increased blood flow through the lungs provokes a hypertensive reaction in the pulmonary vasculature that effectively increases the pressure in the right ventricle and atrium. When the blood pressure on the right side of the heart

exceeds that in the corresponding left chamber, the shunt reverses, and poorly oxygenated blood passes to the systemic circulation, thus leading to cyanosis. At this point, the condition of the patient who has the cardiac lesion often rapidly worsens.

Analysis of the numerous available lines of genetically modified mice has shown that interference with the function of many genes results in the appearance of a wide variety of heart and vascular defects. A given type of heart defect may be produced by interfering with any of a wide array of molecules (both signaling molecules and transcription factors) that are components of the cascade leading to the normal formation of a given part of the heart. The following treatment of heart defects is based on clinical and anatomical features. Their molecular underpinnings, when known, also are discussed, but for any given defect only major disrupted pathways are mentioned. For the sake of overall perspective, rather than focusing on specific details, **Figure 17.31** summarizes the current state of knowledge of the role of major transcription factors in the genesis of important congenital heart defects.

Chamber-to-Chamber Shunts

Atrial and ventricular septal defects are common, accounting for almost 50% of cases of congenital heart disease. Because of their simple nature, they were among the first heart defects to be treated with open heart surgery.

Interatrial Septal Defects

Several types of anatomical defects in the interatrial septum can result in a persisting shunt between the two atria. The most

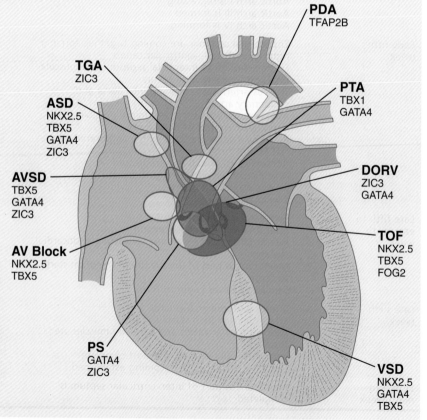

Fig. 17.31 **Scheme showing the association between structural anomalies of the heart and mutations of transcription factors.** ASD, atrial septal defect; AV, atrioventricular; AVSD, atrioventricular septal defect; DORV, double outlet right ventricle; PDA, patent ductus arteriosus; PS, pulmonary stenosis; PTA, persistent truncus arteriosus; TGA, transposition of great arteries; TOF, tetralogy of Fallot; VSD, ventricular septal defect. *(Adapted from Clark KL, Yutzey KE, Benson DW:* Annu Rev Physiol 68:97-121, 2006.)

CLINICAL CORRELATION 17.1
Malformations of the Heart—cont'd

common varieties are caused by excessive resorption of tissue around the foramen secundum or hypoplastic growth of the septum secundum (**Fig. 17.32A**). A less common variety is a low septal defect, which is usually caused by the lack of union between the leading edge of the septum primum and the endocardial cushions (Fig. 17.32B). If the defect is the result of a deficiency of endocardial cushion tissue, associated defects of the atrioventricular valves can considerably complicate the lesion. Lack of septation of the atrium results in a **common atrium**, a serious defect that is usually associated with other heart defects. Atrial septal defects are common heart malformations. Increasingly, atrial septal defects are being associated with chromosome 21. Individuals with Down syndrome (trisomy of chromosome 21) have a high incidence of defects of the atrial and ventricular septa.

Of the many genes whose mutations are associated with atrial septal defects, *Nkx 2-5*, *GATA4*, and *Tbx5* are most prominently represented. Individuals with autosomal dominant mutations of the *Nkx2-5* gene (see p. 426) have a high incidence of abnormalities of the septum secundum, with resulting atrial septal defects. Associated with the atrial septal defects is an equally high incidence of atrioventricular block, which can lead to sudden death in affected individuals whose hearts are not assisted by pacemakers. Before the discovery of this mutation, it was suspected that many of the cases of atrioventricular block resulted from disruption of the atrioventricular bundle by the repair procedure. During the early days of cardiac surgery, before the anatomy of the atrioven-tricular bundle was precisely determined, surgically induced bundle branch block was a problem in the repair of low atrial septal defects.

Another condition that is strongly associated with atrial (and ventricular) septal defects, as well as limb anomalies, is **Holt-Oram syndrome**. This syndrome is caused by a mutation in the *T-box* gene *Tbx-5*, the gene that is expressed in the upper limb, but not the lower limb (see p. 200).

Uncomplicated atrial septal defects are usually compatible with many years of symptom-free life. Even during the symptom-free period, blood from the left atrium, which is under slightly higher pressure than that in the right atrium, passes into the right atrium. This additional blood causes right atrial hypertrophy and results in increased blood flow into the lungs. Over many years, pulmonary hypertension can develop. This increases the blood pressure of the right ventricle and ultimately that of the right atrium. Only a few millimeters of increased right atrial pressure reverses the blood flow in the interatrial shunt and causes cyanosis.

A more serious condition is **premature closure of the foramen ovale**. In this situation, the entire input of blood into the right atrium passes into the right ventricle and causes massive hypertrophy of the right side of the heart. The left side is severely hypoplastic because of the reduced blood that the left chambers carry. Although this defect is usually compatible with intrauterine life, infants typically die shortly after birth because the hypoplastic left heart cannot handle a normal circulatory load.

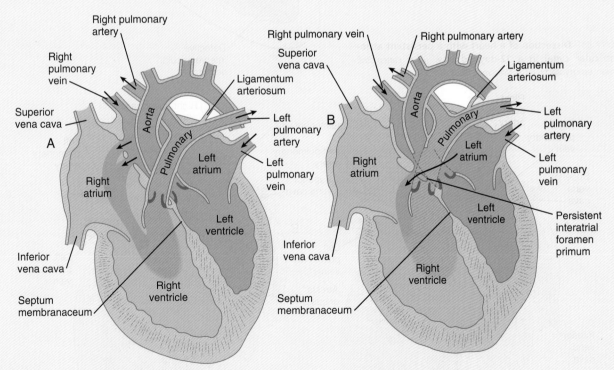

Fig. 17.32   High (**A**) and low (**B**) atrial septal defects in the heart. *Red* denotes well-oxygenated arterial blood, *blue* denotes poorly oxygenated venous blood, and *purple* denotes a mixture of arterial and venous blood.

Continued

CLINICAL CORRELATION 17.1
Malformations of the Heart—cont'd

Persistent Atrioventricular Canal

The usual basis for persistent atrioventricular canal is underdevelopment of the endocardial cushions that results in a lack of division of the early atrioventricular canal into right and left channels. Because of the large number of molecules involved in the normal formation of the endocardial cushions and the atrioventricular valves, defects in the valves have been attributed to mutations of many genes, some of which are involved in patterning and others in effecting epitheliomesenchymal transformation at the sites of the cushions.

A persistent atrioventricular canal is often associated with major interatrial and interventricular septal defects (**Fig. 17.33**). This severe defect leads to poor growth and a considerably shortened life. Despite the potential for mixing of blood, the predominant shunt direction is from left to right, and some patients have little cyanosis.

Tricuspid Atresia

In tricuspid atresia, the etiology of which is poorly understood, the normal valvular opening between the right atrium and the right ventricle is completely occluded (**Fig. 17.34B**). Such a defect alone causes death because the blood cannot gain access to the lungs for oxygenation. Children can survive with this malformation,

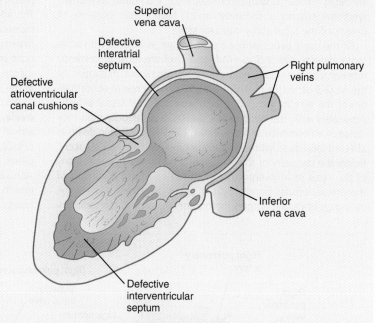

Fig. 17.33 **Dissection of a heart with a persistent atrioventricular canal in a 12-day-old boy.** (After Patten BM: Human embryology, ed 3, New York, 1968, McGraw-Hill.)

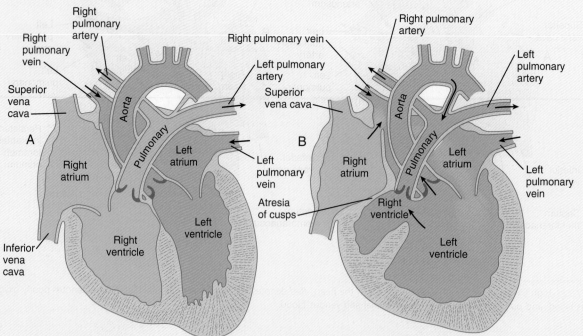

Fig. 17.34 A, Normal postnatal heart. B, Tricuspid atresia, with compensating defects in the interatrial septum and interventricular septum (arrows), which allow this patient to survive.

however, and this illustrates an important point in cardiac embryology. Often a primary lesion is accompanied by one or more secondary lesions (usually shunts) that permit survival, although frequently at a poor functional level.

In this condition, secondary shunts must accomplish two things. First, a persisting atrial septal defect must shunt the blood that cannot pass through the atretic tricuspid valve into the left atrium. The left atrial blood then flows into the left ventricle. Second, one or more secondary shunts must allow blood to gain access to the lungs so that it can become oxygenated. Left ventricular blood could enter the right ventricle and pulmonary arterial system if a defect is present in the interventricular septum. Another possibility is for the blood in the left ventricle to pass into the systemic circulation, where it can gain access to the lungs by passing from the aorta through a patent ductus arteriosus into the pulmonary arteries. From the lungs, the oxygenated blood enters the left atrium, perhaps to be recycled through the lungs again before entering the systemic circulation.

Mitral atresia can also occur, but it is much rarer than tricuspid atresia. Secondary compensating defects again have to be present for survival. Infants with these lesions typically survive only a few months or years.

Interventricular Septal Defect

Defects in the interventricular septum are the most common congenital cardiac defect in infants, but most of the defects close spontaneously before these children are 10 years old. In adults, these defects are not as common as atrial septal defects. Almost 70% of ventricular septal defects occur in the membranous part of the septum, where several embryonic tissues converge (**Fig. 17.35**). Because the pressure of the blood in the left ventricle is higher than that in the right, this lesion is initially associated with

left-to-right acyanotic shunting of blood flow (**Fig. 17.36**). The increased blood flow into the right ventricle produces right ventricular hypertrophy, however, and can lead to pulmonary hypertension, ultimately causing reversal of the shunt. The basic pathological dynamics are similar to those for atrial septal defects. Many of the mutations that cause atrial septal defects can also result in ventricular septal defects, but in addition ventricular septal defects are often seen in conjunction with malformations of the outflow tract.

Malformations of the Outflow Tract

The outflow tract of the heart (truncoconal region) is subject to various malformations. Such malformations are responsible for about 20% to 30% of all cases of congenital heart disease. Experimental studies have shown that defects of the outflow tract can generally be attributed to disturbances in fundamental aspects of early heart development: formation of the secondary heart field or the cardiac neural crest. Extirpation and transplantation experiments have shown specific requirements for cardiac neural crest cells in the normal development of the cardiac outflow tract (**Fig. 17.37**). If the cardiac neural crest is removed, ectodermal cells from the nodose placode populate the outflow tract, but septation of the outflow tract does not occur, thus leading to a persistent truncus arteriosus.

Although all malformations of this area cannot be attributed to defective neural crest development, circumstantial evidence suggests that this may be a significant factor. Some defects of the outflow tract are associated with translocations or deletions in chromosome 22, and many of these can involve both the neural crest and the secondary heart field. Lesions of the outflow tract can be produced experimentally by interfering with the function of specific genes, often genes that in the human are located on chromosome 22, a prominent example being *Tbx1*, and genes that affect properties of cranial neural crest cells. Outflow tract abnormalities are seen in mice deficient in **neurotrophin-3**, a member

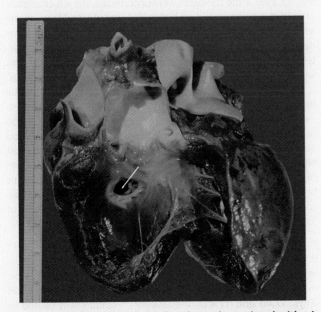

Fig. 17.35 **Ventricular septal defect (arrow) associated with tricuspid atresia.** *(Photo 147 from the Arey-DaPeña Pediatric Pathology Photographic Collection, Human Developmental Anatomy Center, National Museum of Health and Medicine, Armed Forces Institute of Pathology, Washington, D.C.)*

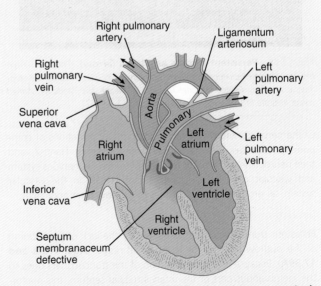

Fig. 17.36 **Interventricular septal defect (membranous portion).** Mixing of arterial and venous blood occurs in both outflow tracts, but especially in the pulmonary artery.

Continued

CLINICAL CORRELATION 17.1
Malformations of the Heart—cont'd

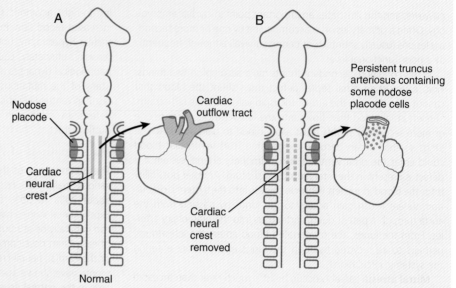

Fig. 17.37 Neural crest and morphogenesis of the outflow tract of the heart. **A,** Normal structure, showing cardiac neural crest contributing to the formation of the outflow tract in the avian heart. **B,** Removal of the cardiac neural crest leads to the formation of a persistent truncus arteriosus containing cells derived from the nodose placode. *(Based on Kirby ML, Waldo KL: Circulation 82:332-340, 1990.)*

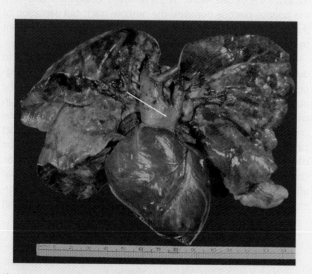

Fig. 17.38 Persistent truncus arteriosus (arrow). *(Photo 117 from the Arey-DaPeña Pediatric Pathology Photographic Collection, Human Developmental Anatomy Center, National Museum of Health and Medicine, Armed Forces Institute of Pathology, Washington, D.C.)*

of the nerve growth factor family. In addition, mutations of components of a cascade, starting with endothelin-1, Hand-2, and then neuropilin-1, a receptor for semaphorin in the nervous system and vascular endothelial growth factor (VEGF) in the vascular system, all produce various degrees of outflow tract anomalies.

Persistent Truncus Arteriosus

Persistent truncus arteriosus is caused by the lack of partitioning of the outflow tract by the truncoconal ridges (**Figs. 17.38 and 17.39A**). Because of the contribution of the truncoconal ridges to the membranous part of the interventricular septum, this malformation is almost always accompanied by a ventricular septal defect. A large arterial outflow vessel overrides the ventricular septum and receives blood that exits from each ventricle. As may be predicted, individuals with a persistent truncus arteriosus are highly cyanotic. Without treatment, 60% to 70% of infants born with this defect die within 6 months.

Transposition of the Great Vessels

Rarely, the truncoconal ridges fail to spiral as they divide the outflow tract into two channels. This defect results in two totally independent circulatory arcs, with the right ventricle emptying into the aorta and the left ventricle emptying into the pulmonary artery (Fig. 17.39B). If the condition were uncorrected, the left circulatory arc would continue pumping highly oxygenated blood through the left side of the heart and the lungs, whereas the right side of the heart would pump venous blood through the aorta into the systemic circulatory channels and back into the right atrium. This lesion, which is the most common cause of cyanosis in newborns, is compatible with life only if an atrial and a ventricular septal defect and an associated patent ductus arteriosus accompany it. Even with these anatomical compensations, the quality of blood reaching the body is poor. During fetal life, the pattern of blood flow is such that the posterior part of the body receives the most highly oxygenated blood, whereas the head receives the lesser oxygenated blood that would have ordinarily gone to the posterior body. As a result, the brain develops under somewhat unfavorable conditions and is frequently underweight at birth. Any level of functional impairment in the brain remains poorly documented.

Aortic and Pulmonary Stenosis

If the septation of the outflow tract by the truncoconal ridges is asymmetrical, either the aorta or the pulmonary artery is abnormally narrowed, resulting in **aortic** and **pulmonary stenosis** (**Figs. 17.40 and 17.41**). The severity of symptoms is related to the degree of stenosis. In the most extreme case, the stenosis is so severe that the lumen of the vessel is essentially obliterated. This

CLINICAL CORRELATION 17.1
Malformations of the Heart—cont'd

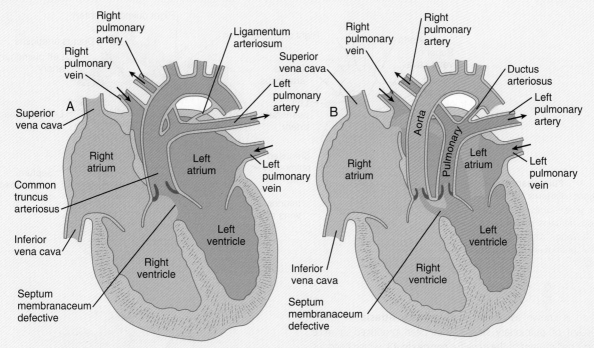

Fig. 17.39 **A,** Persistent truncus arteriosus. A single outflow tract is fed by blood entering from the right and left ventricles. The membranous part of the interventricular septum is commonly defective. **B,** Transposition of the great vessels caused by lack of spiraling of the truncoconal ridges in the early embryo. The aorta arises from the right ventricle, and the pulmonary artery arises from the left ventricle.

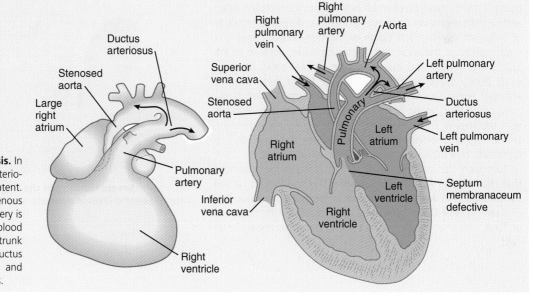

Fig. 17.40 **Aortic stenosis.** In severe cases, the ductus arteriosus commonly remains patent. *Right,* Mixed arterial and venous blood in the pulmonary artery is shown in *purple*. Initially, blood from the pulmonary trunk *(purple)* goes through the ductus arteriosus into the aorta, and this often leads to cyanosis.

Continued

CLINICAL CORRELATION 17.1
Malformations of the Heart—cont'd

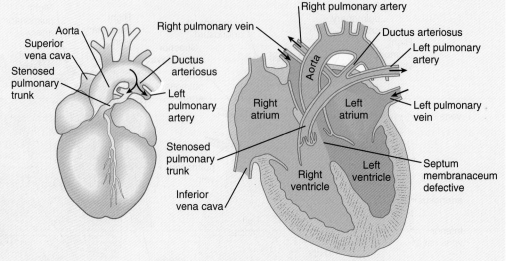

Fig. 17.41 **Pulmonary stenosis.** *Right,* Patterns of blood flow. In severe cases, the ductus arteriosus remains patent, with blood flowing from the aorta into the pulmonary circulation *(arrows).*

condition is known as **aortic** or **pulmonary atresia**. A lesion reminiscent of pulmonary stenosis has been produced in mice bearing a null mutant of the gene for **connexin 43**, which encodes a protein component of the gap junction channel. Why such a genetic lesion would affect principally the pulmonary outlet of the heart is unknown.

One of the best-known lesions of this type is **tetralogy of Fallot**, which is characterized by (1) pulmonary stenosis, (2) a membranous interventricular septal defect, (3) a large aorta (overriding aorta, the opening of which extends into the right ventricle), and (4) right ventricular hypertrophy. The basic defects in tetralogy of Fallot are asymmetrical fusion of the truncoconal ridges and malalignment of the aortic and pulmonary valves. Because of the pulmonary stenosis and the wider than normal aortic opening, some poorly oxygenated right ventricular blood leaves via the enlarged aorta, thus causing cyanosis. Tetralogy of Fallot is the most common cyanotic heart lesion in young children. Patients with tetralogy of Fallot are highly cyanotic from birth and exhibit severe digital clubbing (**Fig. 17.42**). When this condition is untreated, only 50% of patients survive past 2½ years.

Pulmonary stenosis or tetralogy of Fallot is one of the conditions that characterizes **Alagille's syndrome**. This condition is caused by a mutation in **Jagged-1**, a ligand of the Notch receptor. It is also seen in **Holt-Oram syndrome**, in association with mutations of *Tbx5*, and after experimental deletions of the secondary heart field.

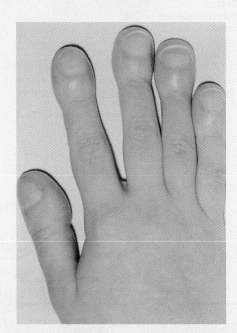

Fig. 17.42 **Severe clubbing of the fingers.** *(From Zitelli JB, Davis HW: Atlas of pediatric physical diagnosis, ed 4, St. Louis, 2002, Mosby.)*

CLINICAL CORRELATION 17.2
Malformations of the Blood Vessels

Because of their mode of formation, in which one vascular channel is favored within a dense network, blood vessels (especially veins) are subject to numerous variations from normal. Most variations seen in the dissecting laboratory are of little functional significance. Animal experiments suggest that disturbances in the neural crest are involved in the genesis of certain anomalies of the major arteries. When the cardiac neural crest is removed from early avian embryos, malformations involving the carotid arteries and arch of the aorta result. Malformations of the larger vessels can cause serious symptoms or may be significant during surgery.

Double Aortic Arch

Rarely, the segment of the right dorsal aortic arch between the exit of the right subclavian artery and its point of joining with the left aortic arch persists, instead of degenerating. This condition results in a complete vascular ring surrounding the trachea and esophagus (**Figs. 17.43A and 17.44**). A double aortic arch can cause **dyspnea** (difficulty breathing) in infants while they feed. Even if the condition is asymptomatic early in life, later growth typically narrows the diameter of the ring in relation to the size of the trachea and esophagus and causes symptoms in later years.

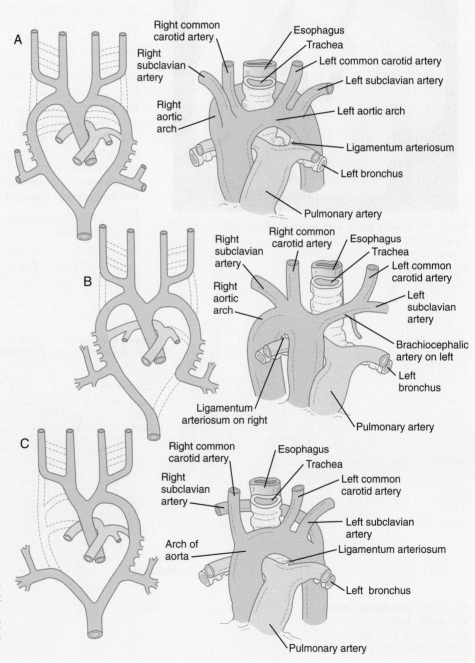

Fig. 17.43 **Aortic arch anomalies.** *Left,* Configuration of embryonic vessels. *Right,* Postnatal appearance. A, Double aortic arch. B, Right aortic arch. C, Right subclavian artery from the arch of the aorta.

Continued

CLINICAL CORRELATION 17.2
Malformations of the Blood Vessels—cont'd

Right Aortic Arch

A right aortic arch arises from the persistence of the complete embryonic right aortic arch and the disappearance in the left arch of the segment caudal to the exit of the left subclavian artery (Fig. 17.43B). This condition is essentially a mirror image of normal development of the aortic arch, and it can occur as an isolated anomaly or as part of complete situs inversus of the individual. Symptoms are typically mild or absent, unless an aberrant left subclavian artery presses against the esophagus or trachea.

Right Subclavian Artery Arising from the Arch of the Aorta

If the right fourth aortic arch degenerates between the common carotid artery and the exit of the right seventh thoracic intersegmental artery (see Fig. 17.8B and C), and if the segment between the exit of the right subclavian artery and the more distal segment of the right aortic arch (which normally disappears) persists, the right subclavian artery arises from the left aortic arch and passes behind the esophagus and trachea to reach the right arm (see Fig. 17.43C). As with a double aortic arch, this condition can cause difficulties in breathing and swallowing.

Interruption of the Left Aortic Arch

Interruption of the left aortic arch is an uncommon vascular malformation that usually results in a break proximal to the exit of the left subclavian artery (**Fig. 17.45**). To be compatible with life, this lesion is usually accompanied by a patent ductus arteriosus, which allows blood flow to the lower part of the body. This lesion has been produced in mice that are lacking in the winged helix transcription factor (**mesenchyme fork head-1 (MFH-1)** and **transforming growth factor-β2 (TGF-β2)**).

Patent Ductus Arteriosus and Premature Closure of the Ductus

One common vascular anomaly is failure of the ductus arteriosus to close after birth (**Fig. 17.46**). This malformation occurs in a

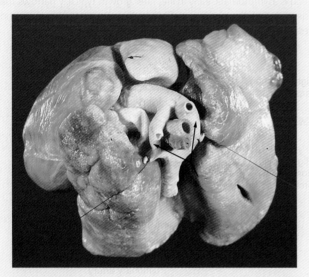

Fig. 17.44   Double aortic arch *(arrows).* *(Photo 5992 from the Arey-DaPeña Pediatric Pathology Photographic Collection, Human Developmental Anatomy Center, National Museum of Health and Medicine, Armed Forces Institute of Pathology, Washington, DC.)*

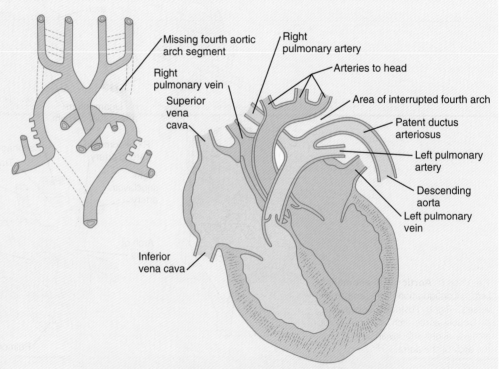

Fig. 17.45   Interruption of the left aortic arch. This lesion is associated with a deficiency in the MFH-1 winged helix transcription factor.

Missing fourth aortic arch segment

Right pulmonary vein

Superior vena cava

Inferior vena cava

Right pulmonary artery

Arteries to head

Area of interrupted fourth arch

Patent ductus arteriosus

Left pulmonary artery

Descending aorta

Left pulmonary vein

CLINICAL CORRELATION 17.2
Malformations of the Blood Vessels—cont'd

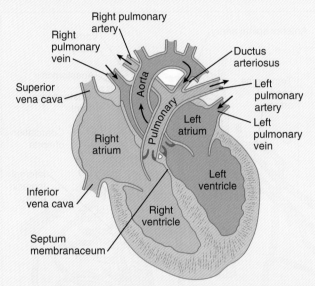

Fig. 17.46　**Patent ductus arteriosus showing the flow of blood from the aorta into the pulmonary circulation.** Later in life, pulmonary hypertension may result, causing the reversal of blood flow through the shunt and cyanosis.

higher than normal incidence in pregnancies complicated by rubella or hypoxia. At least half of infants with this condition experience no symptoms, but over many years the strong flow of blood from the higher pressure systemic (aortic) circulation into the pulmonary circulation overloads the vasculature of the lungs and results in pulmonary hypertension and ultimately heart failure. This causes a reversal of direction of blood flow through the persistent ductus and can result in lower body cyanosis.

Occasionally, the ductus arteriosus closes during fetal life. This causes major imbalances in blood flow and certain secondary structural abnormalities. Most significant clinically is hypertrophy of the right ventricle secondary to resistance caused by the closure of the ductus and to increased blood flow into the right ventricle from the left through interventricular septum.

Coarctation of the Aorta

Another common, nonlethal malformation of the vascular system is coarctation of the aorta, which occurs in two main variants. One consists of an abrupt narrowing of the descending aorta caudal to the entrance of the ductus arteriosus (**Fig. 17.47B**). The other variant, called **preductal coarctation**, occurs upstream from the ductus (see Fig. 17.47A). The former variety (**postductal coarctation**) is more common, accounting for more than 95% of all cases.

The embryogenesis of coarctation is still unclear. Several underlying causes may lead to the same condition. In patients with Down syndrome and Turner's syndrome, the incidence of coarctation of the aorta is increased.

In preductal coarctation of the aorta, which may be related to inadequate expression of MFH-1, the ductus arteriosus typically remains patent after birth. The blood supplying the trunk and limbs reaches the descending aorta through the ductus. This can lead to differential cyanosis, in which the head and upper trunk and arms have a normal color, but the lower trunk and limbs are cyanotic because of the flow of venous blood into the aorta through the patent ductus arteriosus.

The vasculature must compensate for a postductal coarctation in a different manner because the location of the narrowing in this case effectively cuts off the arterial circulation of the head and arms from that of the trunk and legs. The body responds by opening up collateral circulatory channels and connections through normally small arteries that lead from the upper to the lower body (Fig. 17.47C). Such channels are the internal thoracic arteries, the arteries associated with the scapula, and the anterior spinal artery. The unusually large flow of blood through these arteries passes through segmental branches (e.g., intercostal arteries) into the descending aorta caudal to the coarctation. The increased blood flow in the intercostal arteries causes a distinct notching in the posterior third of the third through eighth ribs that can be readily seen in radiological images. Despite these compensatory circulatory adaptations, the blood pressure in patients with a postductal coarctation is much higher in the arms than in the legs.

Malformations of the Venae Cavae

As could be expected from their complex mode of formation (see Fig. 17.13), the superior and inferior venae cavae are subject to a wide range of malformations. Common variants are duplications of the superior and inferior venae cavae or persistence of the left instead of the right segments of these vessels, along with the absence of the normal vessel. In most cases, these malformations are asymptomatic.

Anomalous Pulmonary Return

Because of the way the individual pulmonary veins are joined and the later absorption of the distal part of the pulmonary venous system into the left atrial wall, inappropriate connections of pulmonary veins to the heart can occur (**Fig. 17.48**). One common condition is for one or more branches of the pulmonary vein to enter the right instead of the left atrium. In other cases (**total anomalous pulmonary return**), all pulmonary veins empty into the right atrium or superior vena cava. Such a case must be accompanied by an associated shunt (e.g., interatrial shunt) to bring oxygenated blood into the systemic circulation.

Continued

CLINICAL CORRELATION 17.2
Malformations of the Blood Vessels—cont'd

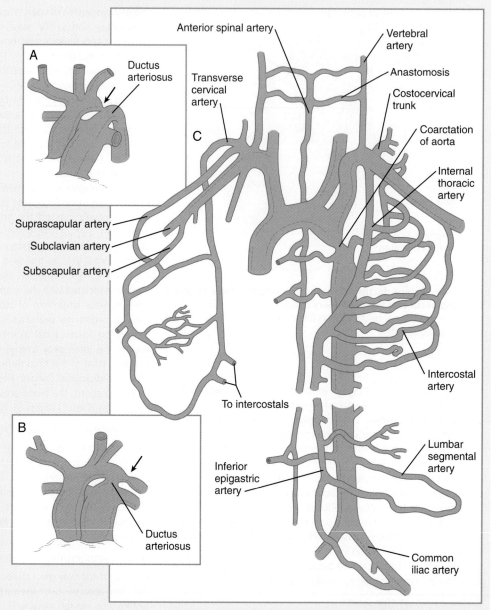

Fig. 17.47 **Coarctation of the aorta. A,** Preductal coarctation (*arrow*) with accompanying patent ductus arteriosus. **B,** Postductal coarctation (*arrow*) with an accompanying patent ductus arteriosus. **C,** Collateral circulation in postductal coarctation, with enlarged peripheral vessels carrying blood to the lower part of the body.

Vascular Malformations and Hemangiomas

Localized abnormalities of the vasculature exist in several forms. One of the most dramatic is a **hemangioma (Fig. 17.49),** which is actually a vascular tumor that typically appears within a few weeks of birth, rapidly expands, and then spontaneously regresses, usually before the age of 10 years. Hemangiomas are seen in 10% to 12% of all newborns, with a three to four times greater incidence in girls. The endothelium of a hemangioma is mitotically very active. In contrast, **vascular malformations** are typically purplish, with a raised and irregular surface. Malformations of larger vessels consist of tangles of vessels that are inactive mitoti-

cally, and their growth typically keeps pace with that of the rest of the body. They do not regress. Some familial cases occur because of mutations of Tie-2, the angiopoietin receptor. These mutations paradoxically result in a more active sprouting stimulus. **Capillary malformations,** often called **port wine stains,** constitute the most common vascular anomaly of the skin. Typically reddish and later changing to purple, these are harmless, but do not disappear.

Malformations of the Lymphatic System

Although minor anatomical variations of lymphatic channels are common, anomalies that cause symptoms are rare. These typically

CLINICAL CORRELATION 17.2
Malformations of the Blood Vessels—cont'd

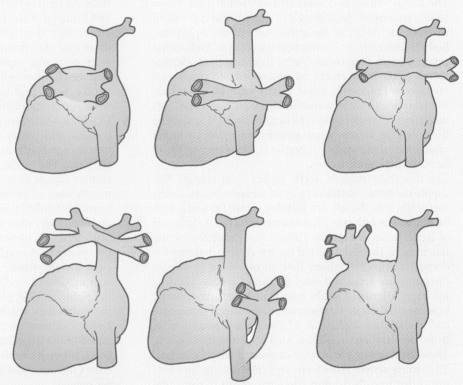

Fig. 17.48 Variants of anomalous pulmonary drainage.

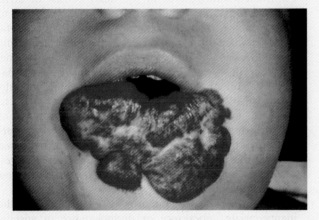

Fig. 17.49 **An oral hemangioma in a child.** (*Courtesy of A. Burdi, Ann Arbor, Mich.*)

manifest as swelling caused by dilation of major lymphatic vessels. The most common major lymphatic anomaly seen in fetuses is **cystic hygroma**, which manifests as large swellings, sometimes even collarlike, in the region of the neck (see Fig. 8.1A). Although the embryological basis for cystic hygroma is uncertain, excessive local production and growth of lymphatic tissue, possibly originating as pinched-off buds from the jugular lymph sacs, are the probable causes.

Congenital **lymphedema** is caused by maldeveloped or dysfunctional lymphatic vessels and is characterized by nonpitting localized swelling (lymphedema), along with increased local susceptibility to infections. This condition can be caused by aplasia or hypoplasia of lymphatic capillaries, absent or nonfunctional lymphatic valves, or nonfunctional smooth muscle cells in the walls of the lymphatic channels.

Summary

- The vascular system arises from mesodermal blood islands in the wall of the yolk sac. Hemangioblasts give rise to either blood cells or vascular endothelial cells. Nucleated red blood cells produced in the blood islands are the first blood cells found in the embryo. Later, hematopoiesis shifts to the embryonic body, beginning in the para-aortic bodies, and then to the liver, and finally to the bone marrow.

- During hematopoiesis, hemocytoblasts give rise to lymphoid and myeloid stem cells. Each of these stem cells further differentiates into the definitive lines of blood cells. Erythropoiesis involves the passage of precursor cells of red blood cells through several stages. The earliest stages are defined by behavioral, rather than morphological, characteristics. During later stages of differentiation, the precursor cells of erythrocytes gradually lose their RNA-producing machinery and accumulate increasing amounts of hemoglobin in their cytoplasm; at the same time, the nucleus becomes more condensed and is eventually lost. Hemoglobin also undergoes isoform transitions during embryonic development.

- The earliest blood and associated extraembryonic blood vessels arise from blood islands in the mesodermal wall of the yolk sac. Much of the vasculature of the embryonic body is derived from intraembryonic sources. Endothelial cell precursors (angioblasts) arise from most mesodermal tissues of the body except the notochord and prechordal mesoderm. Embryonic blood vessels form by three main mechanisms: (1) coalescence in situ (vasculogenesis), (2) migration of angioblasts into organs, and (3) sprouting from existing vessels (angiogenesis). Ingrowth of blood vessels into some organ primordia is stimulated by VEGF or other angiogenic factors.

- The first three pairs of aortic arches form arteries that supply the head. The fourth pair of arches develops asymmetrically, with the left arch forming part of the aortic arch of adults. The fifth pair of arches never forms. A sixth pair of arches arises as a capillary plexus that connects with the fourth arch. The distal part of the left sixth arch forms the ductus arteriosus, a shunt that allows blood to bypass the immature lungs and enter the aorta directly. Many of the larger arteries of adults arise from three sets of aortic branches: the dorsal intersegmental, the lateral segmental, and the ventral segmental. The coronary arteries arise from capillary plexuses associated with the epicardium. These plexuses are secondarily connected with the aorta.

- The venous system arises from very complex capillary networks that initially develop into components of the cardinal vein system. Anterior and posterior cardinal veins drain the head and trunk. They then empty into the paired common cardinal veins and ultimately into the sinus venosus of the heart. Paired subcardinal veins are associated with the developing mesonephros. Paired extraembryonic umbilical and vitelline veins pass through the developing liver and directly into the sinus venosus. The pulmonary veins arise as separate structures and empty into the left atrium. The lymphatic system first appears as six primary lymph sacs. These become connected by lymphatic channels. Lymphatics from most of the body collect into the thoracic duct, which empties into the venous system at the base of the left internal jugular vein.

- The heart arises from splanchnic mesoderm as a horseshoe-shaped primordium consisting of primary and secondary heart fields. Originally, bilateral endocardial tubes fuse in the midline. The fused cardiac tube then undergoes an S-shaped looping, and, soon, specific regions of the heart can be identified. Starting with the inflow tract, these regions are the sinus venosus, the atria, the ventricles, and the outflow tract (bulbus cordis). The outflow tract later divides into the conus arteriosus and the truncus arteriosus.

- Atrial endocardial cushions are thickenings between the atria and ventricles. The underlying myocardium induces cells from the endothelial lining of the endocardial cushion to leave the endocardial layer and transform into mesenchymal cells that invade the cardiac jelly. These events serve as the basis for the formation of the atrioventricular valves.

- Internal partitioning of the heart begins with the separation of the atria from the ventricles and formation of the mitral and tricuspid valves. The left and right atria become separated by growth of the septum primum and septum secundum, but throughout embryonic life a shunt remains from the right to the left atrium via the foramen secundum and foramen ovale. The sinus venosus and the venae cavae empty into the right atrium, and the pulmonary veins drain into the left atrium. The ventricles are divided by the interventricular septum. Spiral truncoconal ridges partition the common outflow tract into pulmonary and aortic trunks. Semilunar valves prevent the reflux of blood in these vessels into the heart.

- In addition to sensory innervation, the heart receives sympathetic and parasympathetic innervation. The conduction system distributes the contractile stimulus throughout the heart. The conduction system is derived from modified cardiac muscle cells. The heart begins to beat early in the fourth week of gestation. Physiological maturation of the heart beat follows maturation of the pacemaker system and the innervation of the heart.

- The fetal circulation brings oxygenated blood from the placenta through the umbilical vein and into the right atrium, where much of it is shunted into the left atrium. Other blood entering the right atrium passes into the right ventricle. Blood leaving the right ventricle enters the pulmonary trunk, which supplies some blood to the lungs and most to the aorta via the ductus arteriosus. Blood in the left atrium empties into the left ventricle and aorta, where it supplies the body. Poorly oxygenated blood enters the umbilical arteries and is carried to the placenta for renewal.

- Common malformations of the heart consist of atrial septal defects, which in postnatal life allow blood to pass from the left to the right atrium. Ventricular septal defects, which also result in a left-to-right shunting of blood, are more serious. Defects that block a channel for blood flow (e.g., tricuspid atresia) must be accompanied by secondary shunt defects to be compatible with life. A persistent atrioventricular canal can be attributed to a defect in the formation or further development of the atrioventricular endocardial cushions. Most malformations of the outflow tract of the heart seem to be related to inappropriate partitioning by the truncoconal ridges. The basis for this is frequently found in neural crest abnormalities.

- Malformations of the major arteries often result from the inappropriate appearance or disappearance of specific components of the aortic arch system. Some malformations, such as double aortic arch or right aortic arch, can interfere with swallowing or breathing because of pressure. Patent ductus arteriosus is caused by the failure of the ductus arteriosus to close properly after birth. Coarctation of the aorta must be compensated by either a patent ductus arteriosus or the opening of collateral vascular channels that allows blood to bypass the site of coarctation.

- Because of their complex mode of origin, veins are commonly subject to considerable variation, but these malformations are frequently asymptomatic. Anomalous pulmonary return, which brings oxygenated blood into the right atrium, must be accompanied by a right-to-left shunt to be compatible with life. Malformations of the lymphatic system can cause local swellings such as cystic hygroma, which results in a collarlike swelling in the neck.

Review Questions

1. Nucleated erythrocytes found circulating in the embryo are produced in the:
A. Yolk sac
B. Para-aortic clusters
C. Liver
D. Bone marrow
E. None of the above

2. In a 7-month fetus, blood draining the left temporalis muscle enters the heart via the:
A. Left anterior cardinal vein
B. Coronary sinus
C. Left common cardinal vein
D. Superior vena cava
E. None of the above

3. Adherons are inductive particles released by what structure in the endocardial cushion area?
A. Endocardium
B. Cardiac jelly
C. Myocardium
D. Epicardium
E. None of the above

4. Neural crest contributes to the structure of which of the following?
A. Truncus arteriosus
B. Ascending aorta
C. Pulmonary trunk
D. All of the above
E. None of the above

5. For which of these cardiovascular malformations is a patent ductus arteriosus necessary for survival of the individual?
A. Atrial septal defect
B. Ventricular septal defect
C. Double aortic arch
D. Right subclavian artery from arch of aorta
E. None of the above

6. Five days after birth, an infant becomes cyanotic during a prolonged crying spell. The cyanosis is most likely caused by venous blood entering the systemic circulation through the:
A. Interatrial septum
B. Ductus arteriosus
C. Ductus venosus
D. Umbilical vein
E. Interventricular septum

7. The internal carotid artery arises from aortic arch number:
A. 1
B. 2
C. 3
D. 4
E. 5

8. A 12-year-old boy tells his physician that over the past few months he has noticed some difficulty in swallowing when eating meat. The physician performs a physical examination and orders an upper gastrointestinal x-ray series. After examining the films, the physician refers the boy for some vascular studies. What is the reasoning behind this decision?

9. An individual with atresia of the mitral valve could not survive after birth without other defects of the cardiovascular system that could compensate for the primary defect, in this case a complete blockage between the left atrium and ventricle. Construct at least one set of associated defects that could physiologically compensate for the disruption caused by the mitral atresia.

10. What is the embryological basis for a duplication of the inferior vena cava caudal to the kidneys?

References

Anderson RH, Brown NA, Moorman FM: Development and structures of the venous pole of the heart, *Dev Dyn* 235:2-9, 2006.

Arthur HM, Bamforth SD: TGFβ signaling and congenital heart disease: insights from mouse studies, *Birth Def Res A Clin Mol Teratol* 91:423-4334, 2011.

Bernanke DH, Velkey JM: Development of the coronary blood supply: changing concepts and current ideas, *Anat Rec* 269:198-208, 2002.

Black BL: Transcriptional pathways in second heart field development, *Semin Cell Dev Biol* 18:67-76, 2007.

Bressan M and others: Notochord-derived BMP antagonists inhibit endothelial cell generation and network formation, *Dev Biol* 326:101-111, 2009.

Brickner ME, Hillis LD, Lange RA: Congenital heart disease in adults, parts I and II, *N Engl J Med* 342:256-263, 334-342, 2000.

Butler MG, Isogai S, Weinstein BM: Lymphatic development, *Birth Def Res C Embryo Today* 87:222-231, 2009.

Cao N, Yao Z-X: The hemangioblast: from concept to authentication, *Anat Rec* 294:580-588, 2011.

Chappell JC, Bautch VL: Vascular development: genetic mechanisms and links to vascular disease, *Curr Top Dev Biol* 90:43-72, 2010.

Christoffels VM, Burch JBE, Moorman AFM: Architectural plan for the heart: early patterning and delineation of the chambers and the nodes, *Trends Cardiovasc Med* 14:301-307, 2004.

Clark KL, Yutzey KE, Benson DW: Transcription factors and congenital heart defects, *Annu Rev Physiol* 68:97-121, 2006.

Congdon ED: Transformation of the aortic-arch system during the development of the human embryo, *Carnegie Contr Embryol* 14:47-110, 1922.

De Val S: Key transcriptional regulators of early vascular development, *Arterioscler Thromb Vasc Biol* 31:1469-1475, 2011.

Dunwoodie SL: Combinatorial signaling in the heart orchestrates cardiac induction, lineage specification and chamber formation, *Semin Cell Dev Biol* 18:54-66, 2007.

Dyer LA, Kirby ML: The role of secondary heart fields in cardiac development, *Dev Biol* 336:137-144, 2009.

Dzierzak E, Robin C: Placenta as a source of hematopoietic stem cells, *Trends Mol Med* 16:361-367, 2010.

Firulli AB, Thattaliyath BD: Transcription factors in cardiogenesis: the combinations that unlock the mysteries of the heart, *Int Rev Cytol* 214:1-62, 2002.

Garg V: Insights into the genetic basis of congenital heart disease, *Cell Mol Life Sci* 63:1141-1148, 2006.

Gittenberger-De Groot AC and others: Basics of cardiac development for the understanding of congenital heart malformations, *Pediatr Res* 57:169-176, 2005.

Gruber PJ, Epstein JA: Development gone awry: congenital heart disease, *Circ Res* 94:273-283, 2004.

Geudens I, Gerhardt H: Coordinating cell behaviour during blood vessel formation, *Development* 138:4569-4583, 2011.

Harris IS, Black BL: Development of the endocardium, *Pediatr Cardiol* 31:391-399, 2010.

Harvey RP, Rosenthal N, eds: *Heart development*, San Diego, 1999, Academic Press.

Heuser CH: The branchial vessels and their derivatives in the pig, *Carnegie Contr Embryol* 15:121-139, 1923.

Hutson MR, Kirby ML: Model systems for the study of heart development and disease: cardiac neural crest and conotruncal malformations, *Semin Cell Dev Biol* 18:101-110, 2007.

Iruela-Arispe ML, Davis GE: Cellular and molecular mechanisms of vascular lumen formation, *Dev Cell* 16:222-231, 2009.

Ishii Y and others: Endothelial cell lineages of the heart, *Cell Tissue Res* 335:67-73, 2009.

Kameda Y: *Hoxa3* and signaling molecules involved in aortic arch patterning and remodeling, *Cell Tissue Res* 336:165-178, 2009.

Kanjuh VI, Edwards JE: A review of congenital anomalies of the heart and great vessels according to functional categories, *Pediatr Clin North Am* 11:55-105, 1964.

Kume T: Specification of arterial, venous, and lymphatic endothelial cells during embryonic development, *Histol Histopathol* 25:637-646, 2010.

Lincoln J, Lange AW, Yutzey KE: Hearts and bones: shared regulatory mechanisms in heart valve, cartilage, tendon and bone development, *Dev Biol* 294:292-302, 2006.

Lincoln J, Yutzey KE: Molecular and developmental mechanisms of congenital heart valve disease, *Birth Def Res A Clin Mol Teratol* 91:526-534, 2011.

Liu N, Olson EN: MicroRNA regulatory networks in cardiovascular development, *Dev Cell* 18:510-525, 2010.

Ma L and others: BMP2 is essential for cardiac cushion epithelial-mesenchymal transition and myocardial patterning, *Development* 132:5601-5611, 2005.

MacGrogan D, Luna-Zurita L, de la Pompa JL: Notch signaling in cardiac valve development and disease, *Birth Def Res A Clin Mol Teratol* 91:449-459, 2011.

McGrath K, Palis J: Ontogeny of erythropoiesis in the mammalian embryo, *Curr Top Dev Biol* 82:1-22, 2008.

Medvinsky A, Rubtsov S, Taoudi S: Embryonic origin of the adult hematopoietic system: advances and questions, *Development* 138:1017-1031, 2011.

Melani M, Weinstein BM: Common factors regulating patterning of the nervous and vascular systems, *Annu Rev Cell Dev Biol* 26:639-665, 2010.

Mikawa T, Hurtado R: Development of the cardiac conduction system, *Semin Cell Dev Biol* 18:90-100, 2007.

Mikkola HKA, Orkin SH: The journey of developing hematopoietic stem cells, *Development* 133:3733-3744, 2006.

Miquerol L, Beyer S, Kelly RG: Establishment of the mouse ventricular conduction system, *Cardiovasc Res* 91:232-242, 2011.

Mukouyama Y-S and others: Sensory nerves determine the pattern of arterial differentiation and blood vessel branching in the skin, *Cell* 109:693-705, 2002.

Nakajima Y: Second lineage of heart forming region provides new understanding of conotruncal heart defects, *Congenit Anom (Kyoto)* 50:8-14, 2010.

Okamoto N and others: Formal genesis of the outflow tracts of the heart revisited: previous works in the light of recent observations, *Congenit Anom (Kyoto)* 50:141-158, 2010.

Oliver G, Srinivasan RS: Endothelial cell plasticity: how to become and remain a lymphatic endothelial cell, *Development* 137:363-372, 2010.

Olson EN: Gene regulatory networks in the evolution and development of the heart, *Science* 313:1922-1927, 2006.

Padget DH: The development of the cranial arteries in the human embryo, *Carnegie Contr Embryol* 212:207-261, 1948.

Parisot P and others: *Tbx1*, subpulmonary myocardium and conotruncal congenital heart defects, *Birth Def Res A Clin Mol Teratol* 91:477-484, 2011.

Petrenko VM, Gashev AA: Observations on the prenatal development of human lymphatic vessels with focus on basic structural elements of lymph flow, *Lymphat Res Biol* 6:89-95, 2008.

Poelmann RE, Gittenberger-de Groot AC: Apoptosis as an instrument in cardiovascular development, *Birth Def Res A Clin Mol Teratol* 75:305-313, 2005.

Pouget C and others: Somite-derived cells replace ventral aortic hemangioblasts and provide aortic smooth muscle cells of the trunk, *Development* 133:1013-1022, 2006.

Ransom J, Srivastava D: The genetics of cardiac birth defects, *Semin Cell Dev Biol* 18:132-139, 2007.

Rentschler S, Jain R, Epstein JA: Tissue-tissue interactions during morphogenesis of the outflow tract, *Pediatr Cardiol* 31:408-413, 2010.

Restivo A and others: Cardiac outflow tract: a review of some embryologic aspects of the conotruncal region of the heart, *Anat Rec A Discov Mol Cell Evol Biol* 288:936-943, 2006.

Rossant J, Howard L: Signaling pathways in vascular development, *Annu Rev Cell Dev Biol* 18:541-573, 2002.

Rudolph AM: Congenital cardiovascular malformations and the fetal circulation, *Arch Dis Child Fetal Neonatal Ed* 95:F132-F136, 2010.

Sabin FR: The origin and development of the lymphatic system, *Johns Hopkins Hosp Rep* 17:347-440, 1916.

Schulte-Merker S, Sabine A, Petrova TV: Lymphatic vascular morphogenesis in development, physiology, and disease, *J Cell Biol* 193:607-618, 2011.

Tavian M, Péault B: Embryonic development of the human hematopoietic system, *Int J Dev Biol* 49:243-250, 2005.

Tille J-C, Pepper MS: Heredity vascular anomalies: new insights into their pathogenesis, *Arterioscler Thromb Vasc Biol* 24:1578-1590, 2004.

Tomanek RJ: Formation of the coronary vasculature during development, *Angiogenesis* 8:273-284, 2005.

Tzahor E, Evans SM: Pharyngeal mesoderm development during embryogenesis: implications for both heart and head myogenesis, *Cardiovasc Res* 91:196-202, 2011.

van Weerd JH and others: Epigenetic factors and cardiac development, *Cardiovasc Res* 91:203-211, 2011.

van Wijk B, van den Hoff M: Epicardium and myocardium originate from a common cardiogenic progenitor pool, *Trends Cardiovasc Med* 20:1-7, 2010.

Vincentz JW, Barnes RM, Firulli AB: Hand factors as regulators of cardiac morphogenesis and implications for congenital heart defects, *Birth Def Res A Clin Mol Teratol* 91:485-494, 2011.

Wang J, Greene SB, Martin JF: BMP signaling in congenital heart disease: new developments and future directions, *Birth Def Res A Clin Mol Teratol* 91:441-448, 2011.

Weichert J, Hartge DR, Axt-Fliedner R: The fetal ductus arteriosus and its abnormalities: a review, *Congenit Heart Dis* 5:398-408, 2010.

Weinstein BM: Vessels and nerves: marching to the same tune, *Cell* 120:299-302, 2005.

Wilting J and others: Dual origin of avian lymphatics, *Dev Biol* 292:165-173, 2006.

Yamashita JS: Differentiation of arterial, venous, and lymphatic endothelial cells from vascular precursors, *Trends Cardiovasc Med* 17:59-63, 2007.

Fetal Period and Birth

After the eighth week of pregnancy, the period of organogenesis (embryonic period) is largely completed, and the fetal period begins. By the end of the embryonic period, almost all the organs are present in a grossly recognizable form. The external contours of the embryo show a very large head in proportion to the rest of the body and greater development of the cranial than of the caudal part of the body (**Figs. 18.1** and **18.2**).

The fetal period has often been considered a time of growth and physiological maturation of organ systems, and it has not received much attention in traditional embryology courses. Advances in imaging and other diagnostic techniques, however, have provided considerable access to the fetus. Determining the fetus' pattern of growth and state of well-being with remarkable accuracy is now possible. Improved surgical techniques and the realization that surgical wounds in the fetus heal without scarring have led to the field of fetal surgery.

This chapter emphasizes the functional development of the fetus and the adaptations that ensure a smooth transition to independent living after the fetus has passed through the birth canal and the umbilical cord is cut. Techniques that are used to monitor the functional state of the fetus are also described in Clinical Correlation 18.1, later in the chapter.

Growth and Form of the Fetus

Despite the intense developmental activity that occurs during the embryonic period (3 to 8 weeks), the absolute growth of the embryo in length and mass is not great (**Fig. 18.3**). The fetal period (9 weeks to birth), however, is characterized by rapid growth. The change in proportions of the various regions of the body during the prenatal and postnatal growth periods is as striking as the absolute growth of the embryo. The early dominance of the head is reduced as development of the trunk becomes a major factor in the growth of the early fetus. Even later, a relatively greater growth of the limbs changes the proportions of various regions of the body. During the early fetal period, the entire body is hairless and very thin because of the absence of subcutaneous fat (**Fig. 18.4**). By midpregnancy, the contours of the head and face approach those of the neonate, and the abdomen begins to fill out. Beginning at around week 27, the deposition of subcutaneous fat causes the body to round out. (Some major

developmental landmarks during the fetal period are summarized in the table on pp. xii and xiii.)

Fetal Physiology

Circulation

The circulation of the human embryo can be first studied at about 5 weeks by means of ultrasound. At that time, the heart beats at a rate of approximately 100 beats/minute. This probably represents an inherent atrial rhythm. The pulse rate increases to about 160 beats/minute by 8 weeks and then decreases to 150 beats/minute by 15 weeks, with a further slight decline near term. The pulse rate in utero is remarkably constant, and embryos exhibiting **bradycardia** (slow pulse rate) often die before term. Near term, the pulse rate varies to some extent if conditions in the uterus change or if the embryo is stressed. This variation is related to the functional establishment of the autonomic innervation of the heart (**Fig. 18.5**).

The heart of the fetus has gross physiological properties quite different from those of the postnatal heart. The myocardial force, the velocity of shortening, and the extent of shortening all are less in the fetal heart. Some gross functional characteristics of the fetal heart are related to the presence of fetal isoforms of contractile proteins in the cardiac myocytes. In fetal heart cells, the β-myosin heavy chain isoform predominates. This is advantageous because a lower oxygen requirement and less adenosine triphosphate are needed to develop the same amount of force as the α-myosin isoform in the adult heart.

The **stroke volume** (blood expelled with one heartbeat) of the early (18 to 19 weeks' gestation) fetus is very small (<1 mL), but it increases rapidly with continued growth of the fetus. In a term human fetus, the combined ventricular output is about 450 mL/kg/minute. The right ventricle of a human fetus has a greater stroke volume than the left ventricle. This is correlated with an 8% greater diameter of the pulmonary artery than of the fetal aorta.

Quantitative studies have shown a good correlation between blood flow and functional needs of various regions of the embryo. Approximately 40% of the combined cardiac output goes to the head and upper body and supplies the relatively

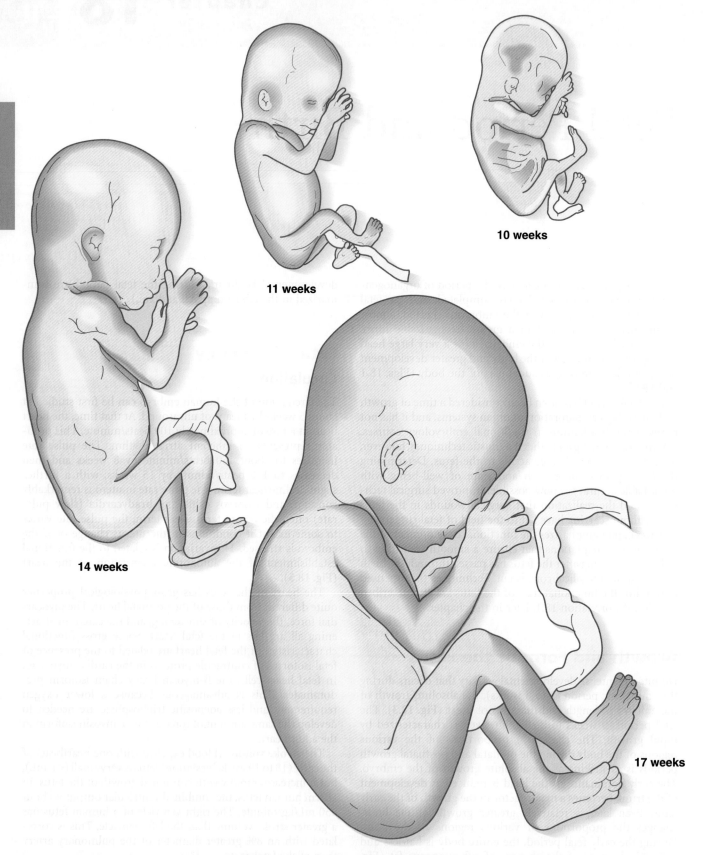

Fig. 18.1 **Drawings of fetuses from 8 to 25 weeks' fertilization age.** The fetuses from 8 through 17 weeks are drawn actual size. The fetuses from 20 and 25 weeks are drawn two-thirds actual size.

10 weeks

11 weeks

14 weeks

17 weeks

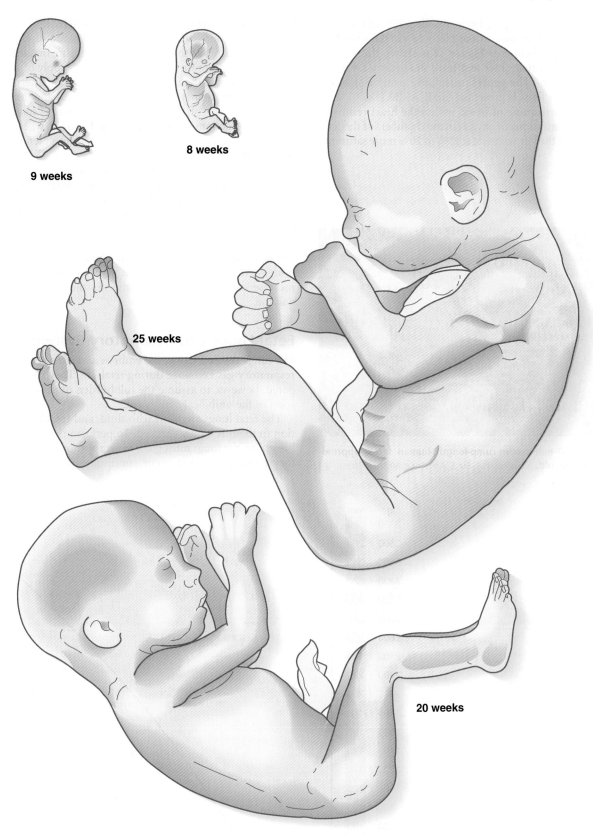

9 weeks

8 weeks

25 weeks

20 weeks

Fig. 18.1, cont'd

great needs of the developing brain. Another 30% of the combined cardiac output goes to the placenta via the umbilical arteries for replenishment. **Figure 18.6** shows the relative amounts of blood that enter and leave the heart via various vascular channels. (The general qualitative pattern of blood flow in a human fetus is presented in Fig. 17.30.)

Differential streaming of blood within the heart results in different concentrations of oxygen in the chambers of the fetal heart. Blood in the left ventricle is 15% to 20% more saturated with oxygen than blood in the right ventricle. This increased oxygen saturation and the high volume of blood supplying the head via branches of the ascending aorta ensure that the developing heart and brain receive an adequate supply of oxygen.

A key factor in the maintenance of the fetal pattern of circulation is the patency of the ductus arteriosus and the ductus venosus. Patency of the fetal ductus venosus is maintained through the actions of **prostaglandins** E_2 and I_2, whereas only prostaglandin E_2 is involved in maintaining patency of the ductus arteriosus.

Myocardial cells of the developing atrium gradually produce and store granules containing **atrial natriuretic peptide**, a hormone that has pronounced vasodilatory, natriuretic, and diuretic properties. This hormone is released after the atrial walls are stretched, normally a sign of increased blood volume. It has been detected in atrial cardiomyocytes as early as 8 to 9 weeks' gestation. After intrauterine blood transfusions during midgestation or later, blood levels of atrial natriuretic peptide increase significantly in response to the increased blood volume.

Fetal Lungs and Respiratory System

The lungs develop late in the embryo and are not involved in respiratory gas exchange during fetal life. They must be prepared, however, to assume the full burden of gas exchange as soon as the umbilical cord is cut.

The fetal lungs are filled with fluid, and the blood circulation to them is highly reduced. To perform normal postnatal breathing, the lungs must grow to an appropriate size, respiratory movements must occur continuously, and the air sacs

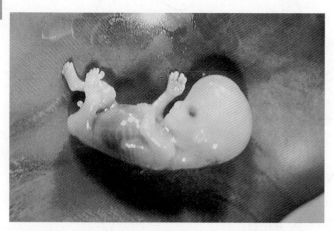

Fig. 18.2 **A 37-mm crown-rump-length human fetus, approximately 9 weeks old.** *(Courtesy of A. Burdi, Ann Arbor, Mich.)*

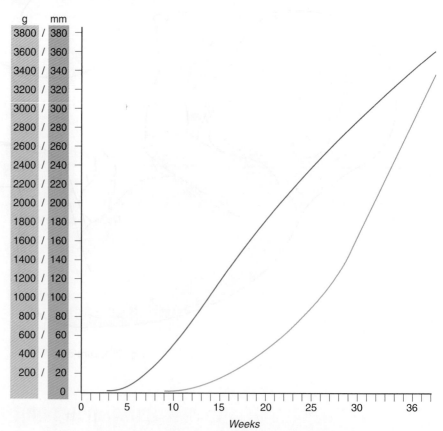

Fig. 18.3 **Growth in crown-rump length *(green)* and weight *(orange)* of the human fetus.** *(Data from Patten BM: Human embryology, New York, 1968, McGraw-Hill.)*

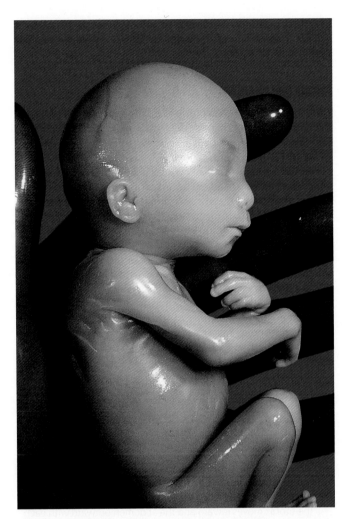

Fig. 18.4 **A 3¾-month-old human fetus (130-mm crown-rump length).** *(EH 902 from the Patten Embryological Collection at the University of Michigan. Courtesy of A. Burdi, Ann Arbor, Mich.)*

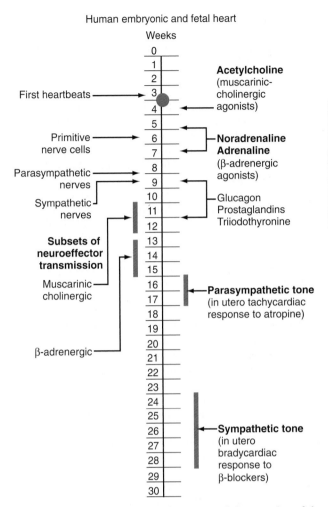

Fig. 18.5 **Sequence of events in the autonomic innervation of the heart.** *(Based on Papp JG: Basic Res Cardiol 83:2-9, 1988.)*

(**alveoli**) must become appropriately configured for air exchange.

Normal growth of the fetal lungs depends on their containing an adequate amount of fluid. During the last trimester of pregnancy, fluid constitutes 90% to 95% of the total weight of the lung. The fluid filling the fetal lungs differs in composition from amniotic fluid, and it has been shown to be secreted by the pulmonary epithelial cells. Secretion begins with a net movement of chloride ions into the lumina of the pulmonary passages. Water movement follows the chloride ions. Studies have shown a relationship between total fluid volume in the lungs and fetal breathing movements, with dilation and constriction of the larynx serving a valvelike function. In vitro studies have shown that proliferation of lung epithelial cells is stimulated by mechanical stretching. In vivo, the internal pressure of the lung fluid serves as the stretching agent. A reduced volume of lung fluid is associated with **pulmonary hypoplasia**.

Ultrasound analysis has shown that the fetus begins to make gross breathing movements as early as 10 weeks. These movements are periodic rather than continuous, and they have two forms. One type of movement is rapid and irregular, with

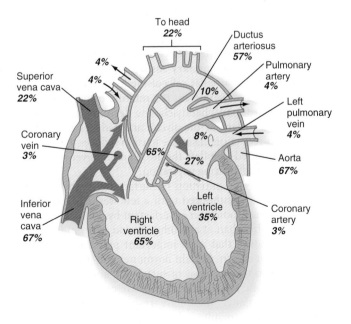

Fig. 18.6 **Percentages of blood entering and leaving the fetal heart via various channels.** *(From Teitel DF: Physiologic development of the cardiovascular system in the fetus. In Polin R, Fox W, eds: Fetal and neonatal physiology, vol 1, Philadelphia, 1992, Saunders, pp 609-619.)*

varying rate and amplitude. The other form is represented by isolated, slow movements, almost like gasps. The former type is more prominent and is associated with conditions of rapid eye movement (REM) sleep. Periods of rapid breathing (often for about 10 minutes) alternate with periods of **apnea** (cessation of breathing).

Breathing movements in the adult are controlled by two centers located in the medulla. One of these centers controls inspiration, and the other controls expiration. In rodent fetuses, the center that controls expiration becomes functional in the midfetal period, slightly before functional maturation of the inspiratory center. Breathing movements are known to be responsive to maternal factors, many of which remain to be identified. The amount of breathing (minutes of breathing per hour) is highest in the evening and lowest in the early morning. The fetal breathing rate increases after the mother has eaten. This increase is related to the concentration of glucose in the maternal blood. Maternal smoking causes a rapid decrease in the rate of fetal breathing for up to 1 hour and is linked to impaired lung development.

Fetal breathing movements are essential for postnatal survival. One function of fetal breathing is to condition the respiratory muscles so that they can perform regular postnatal contractions. Another important function is to stimulate the growth of the embryonic lungs. If intrauterine breathing movements are suppressed, lung growth is retarded. This is a result of a reduction in the production of **platelet-derived growth factor, insulinlike growth factor**, and **thyroid transcription factor-1**, which stimulate cell proliferation and reduce apoptosis in the peripheral parts of the fetal lungs.

An important developmental adaptation of the fetal respiratory system is growth of the upper airway. Although a newborn is about 4% the weight of an adult, the diameter of its trachea is one third that of the adult trachea. Other components of the airway are similarly proportioned. If the trachea were narrower, the physical resistance to airflow would be so great that movement of air would be almost impossible. Even with these adaptations, the resistance of a neonate's airway is five to six times greater than that of an adult.

A functionally important aspect of fetal lung development is the secretion of **pulmonary surfactant** by the newly differentiating type II alveolar cells of the lung, starting around 24 weeks' gestation. Surfactant is a mixture of phospholipids (about two thirds phosphatidylcholine [lecithin]) and protein that lines the surface of the alveoli and reduces the surface tension. This reduction in surface tension reduces the inspiratory force required to inflate the alveoli and prevents the collapse of the alveoli during expiration.

Despite the early initiation of surfactant synthesis, large amounts are not synthesized until a few weeks before birth. At this time, the production of surfactant by the type II alveolar cells is higher than at any other period in an individual's life, an adaptation that is an important preparation for the newborn's first breath. Certain hormones and growth factors are involved in the synthesis of surfactant, and the effects of thyroid hormone and glucocorticoids are particularly strong.

Premature infants often have **respiratory distress syndrome**, which is manifested by rapid, labored breathing shortly after birth. This condition is related to a deficiency in pulmonary surfactant and can be ameliorated by the administration of glucocorticoids, which stimulate the production of surfactant by the alveolar epithelium. The risk of respiratory distress syndrome in infants born at 29 weeks is greater than 60% and decreases to 20% at 34 weeks and less than 5% at 37 weeks.

Fetal Movements and Sensations

Ultrasonography has revolutionized the analysis of fetal movements and behavior because the fetus can be examined virtually undisturbed (except for an increase in vascular activity induced by the ultrasound) for extended periods. Earlier studies of fetal movements were principally concerned with the development of reflex responses, and the information was obtained largely by the analysis of newly aborted fetuses (see Chapter 11). Although valuable information on maturation of reflex arcs was obtained in this manner, many of the movements elicited were not those normally made by the fetus in utero.

The undisturbed embryo does not show any indication of movement until about $7\frac{1}{2}$ weeks. The first spontaneous movements consist of slow flexion and extension of the vertebral column, with the limbs being passively displaced. Within a short time, a large repertoire of fetal movements evolves. After study by numerous investigators, a classification of fetal movements has been suggested (**Box 18.1**). The first fetal movements are followed in a few days by startle and general

---

Box 18.1 Major Types of Fetal Movements

Anteflexion of the head: Normally slow, forward bending of the head

Fetal breathing movements: Paradoxical movements in which the thorax moves inward and the abdomen outward with each contraction of the diaphragm

General movements: Slow gross movements involving the whole body lasting several seconds to a minute

Hand-face contact: Contact that occurs any time the moving hand touches the face or mouth

Hiccups: Repetitive phasic contractions of the diaphragm (an episode may last several minutes)

Isolated arm or leg movements: Movements of extremities that occur without movement of the trunk

Lateral rotation of the head: Movement that involves isolated turning of the head from side to side

Opening of mouth: Isolated movement that may be accompanied by protrusion of the tongue

Retroflexion of the head: Slow to jerky backward bending of the head

Startle movements: Quick (1-second), generalized movements that always start in the limbs and may spread to the trunk and neck

Stretch: Complex movement that involves overextension of the spine, retroflexion of the head, and elevation of the arms

Sucking: Burst of rhythmical jaw movements that is sometimes followed by swallowing (with this movement, the fetus may be drinking amniotic fluid)

Yawn: Movement in which the mouth is slowly opened and rapidly closed after a few seconds

Based on Prechtl HFR: In Hill A, Volpe J, eds: Fetal neurology, New York, 1989, Raven Press, pp 1-16.

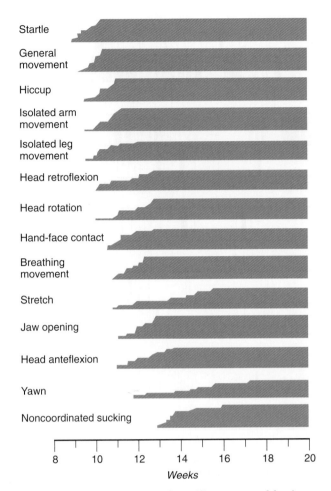

Fig. 18.7   Time of appearance of specific patterns of fetal motor movements. *Each corner on the jagged edge at the left of the broad lines represents one fetus. (From Prechtl HFR: Fetal behavior. In Hill A, Volpe J, eds:* Fetal neurology, *New York, 1989, Raven Press, pp 1-16.)*

movements. Shortly thereafter, isolated limb movements are added (**Fig. 18.7**). Movements associated with the head and jaw appear later. Toward the end of the fourth month, the fetus begins a pattern of periods of activity, followed by times of inactivity. Many women first become aware of fetal movements at this time. Between the fourth and fifth months, the fetus becomes capable of gripping firmly onto a glass rod. Although weak protorespiratory movements are possible, they cannot be sustained.

Continuous ultrasound monitoring for extended periods reveals patterns involving many types of movements (**Fig. 18.8**). At different weeks of pregnancy, some movements are predominant, whereas others are in decline or are just beginning to take shape. Analysis of anencephalic fetuses has shown that although many movements occur, they are poorly regulated. These movements start abruptly, are maintained at the same force, and then stop abruptly. These abnormal patterns of movements are considered evidence for strong supraspinal modulation of movement in the fetus.

Human fetal activities, as reflected in breathing or general activity level, show distinct diurnal rhythms beginning at about 20 to 22 weeks' gestation. There is a strong negative correlation between maternal plasma glucocorticoid levels and fetal activity. Fetal activity is highest in the early evening, when maternal blood glucocorticoid levels are lowest, and lowest in the early morning, when the concentration of maternal hormone peaks. Studies of women who have been given additional glucocorticoids or inhibitors have shown increased fetal activity when maternal corticoid levels are low. Usually, when the overall fetal activity is low, the fetus is in a state of REM sleep, but definitions of sleep and wakefulness in the fetus need further clarification.

Several sensory systems also begin to function during the fetal period. Near-term fetuses are responsive to 2000-Hz stimuli when in a state of wakefulness, but they are unresponsive during periods of sleep. Loud vibroacoustic stimuli applied to the maternal abdomen produce a fetal response consisting of an eye blink, a startle reaction, and an increase in heart rate. Although the fetus is constantly in the dark, the **pupillary light reflex** can usually be elicited by 30 weeks.

The issue of fetal pain has remained controversial with respect to fetal surgery and abortion techniques. Key is whether fetal withdrawal reactions to noxious stimuli really represent an adultlike sensation of pain. Because of the late development of thalamocortical fibers, which are required for awareness of noxious stimuli, and the presence of other endogenous inhibitors, many fetal physiologists believe that functional perception of pain in utero is unlikely to exist before 29 or 30 weeks of gestation.

Fetal Digestive Tract

The fetal digestive tract is not functional in the standard sense because the fetus obtains its nutrition from the maternal blood via the placenta. The digestive tract must be prepared, however, to assume the full responsibility for nutritional intake after birth. When the basic digestive tube and glands have formed in the early embryo, the remainder of the intrauterine period is devoted to cellular differentiation of the epithelia of the gut and preparation of the numerous cells involved for their specific roles in the digestive process. Beneath the epithelium, the walls of the digestive tube must become capable of propelling ingested food and liquid. Analysis of development of the fetal digestive tract has concentrated on (1) the biochemical adaptations of the epithelium of the various regions for digestive function and (2) the development of motility of the digestive tube.

The development and differentiation of epithelia or specific regional characteristics of the gut lining typically follow gradients along the length of the segment of the gut specifically involved. In the esophagus and stomach, differentiation of the mucosal epithelium is well under way starting around 4 months. Although **parietal cells** (hydrochloric acid–producing cells) and **chief cells** (pepsinogen-producing cells) are first seen at 11 and 12 weeks, there is little evidence of their secretions during fetal life. The contents of the stomach have a nearly neutral pH until after birth, but then gastric acid production increases greatly within a few hours.

In the small intestine, villi begin to form in the upper duodenum at the end of the second month, and crypts appear 1 to 2 weeks later. The formation of villi and crypts spreads along the length of the intestine in a spatiotemporal gradient. By approximately 16 weeks of gestation, villi have formed along the entire length of the intestine, and crypts appear in the lower ileum by 19 weeks. Villi even form in the colon

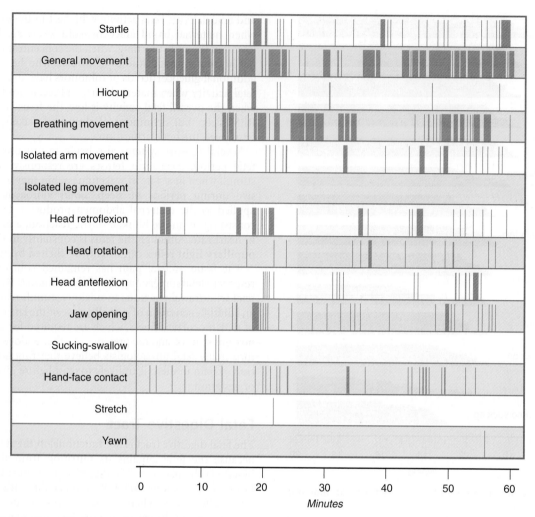

Fig. 18.8   Actogram record of types of movements of a 14-week fetus taken over 60 minutes. *(Adapted from Prechtl HFR: Fetal behavior. In Hill A, Volpe J, eds: Fetal neurology, New York, 1989, Raven Press.)*

during the third and fourth months, but they then regress and are gone by the seventh or eighth month.

Individual epithelial cell types, including **Brunner's glands**, which protect the duodenal lining from gastric acid, appear in the small intestine early in the second trimester. Although the presence of most enzymes or proenzymes characteristic of the intestinal lining can be shown histochemically during the midfetal period, the amounts of these substances are generally quite small. Activity of some of the enzymes secreted by the exocrine pancreatic tissue can also be shown between 16 and 22 weeks' gestation. **Meconium**, a greenish mixture of desquamated intestinal cells, swallowed lanugo hair, and various secretions, begins to fill the lower ileum and colon late in the fourth month (**Fig. 18.9**).

Differentiation of the neuromuscular complex of the digestive tract also follows a gradient, with the circular layer of smooth muscle forming in the esophagus at 6 weeks. **Myenteric plexuses** (parasympathetic neurons) take shape after the inner circular muscle layer is present, but before the formation of the outer longitudinal layer of muscle a couple weeks later in any given region. Starting in the esophagus at 6 weeks, the final formation of myenteric plexuses throughout the length

of the digestive tract is complete at 12 weeks. The first spontaneous rhythmical activity in the small intestine is seen in the seventh week, at approximately the time of formation of the inner circular muscular layer. Recognizable peristaltic movements do not begin until the fourth month, however. Fetuses older than 34 weeks are able to pass meconium in utero.

Another intrauterine preparation for feeding is the development of swallowing and the sucking reflex. Swallowing is first detected at 10 to 11 weeks of gestation, and then its incidence gradually increases. The function of fetal swallowing is unclear, but by term, fetuses swallow 200 to 750 mL or more of amniotic fluid per day. The amniotic fluid contains protein, and much of this is absorbed through the gut by a process of intracellular digestion, occurring by the uptake of macromolecules by fetal enterocytes. According to some estimates, 15% to 20% of total body protein deposition is derived from protein found in amniotic fluid. The swallowed amniotic fluid may contain growth factors that facilitate the differentiation of epithelial cells in the digestive tract. To a certain extent, taste seems to regulate fetal swallowing. Taste buds are seemingly mature by 12 weeks, and the amount of swallowing increases if saccharin is introduced into the amniotic fluid. Conversely,

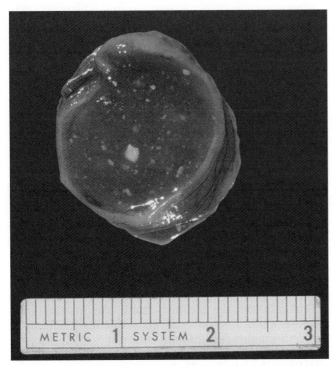

Fig. 18.9 Meconium ileus (accumulation of fetal meconium *[the greenish material]*) in the fetal small intestine. *(Photo 2536 from the Arey-DaPeña Pediatric Pathology Photographic Collection, Human Developmental Anatomy Center, National Museum of Health and Medicine, Armed Forces Institute of Pathology, Washington, D.C.)*

swallowing is reduced if noxious chemicals are added. Fetal swallowing is followed by gastric peristalsis and gastric emptying, which begin at 12 weeks. The gastric emptying cycles mature throughout the fetal period and are important in maintaining the overall balance of amniotic fluid.

Coordinated sucking movements do not appear until late in fetal development. Although noncoordinated precursors of sucking movements occur by 18 weeks, it is not until 32 to 36 weeks that the fetus undertakes short bursts of sucking. These short bursts are not associated with effective swallowing movements. Ineffective sucking is the main reason that premature infants of this age must be fed through a nasogastric tube. Mature sucking capability appears after 36 weeks.

Fetal Kidney Function

Although the placenta performs most excretory functions characteristic of the kidney during prenatal life, the developing kidneys also function by producing urine. As early as 5 weeks of gestation, the mesonephric kidneys produce small amounts of very dilute urine, but the mesonephros degenerates late in the third month, after the metanephric kidneys have taken shape. Tubules of the metanephric kidneys begin to function between 9 and 12 weeks of gestation, and resorptive functions involving the loop of Henle occur by 14 weeks, even though new nephrons continue to form until birth. The urine produced by the fetal kidney is hypotonic to plasma throughout most of pregnancy. This is a reflection of immature resorptive mechanisms, which are manifested morphologically by short loops of Henle. As the neural lobe of the hypophysis produces antidiuretic hormone beginning at 11

weeks, another mechanism for the concentration of urine begins to be established.

Intrauterine renal function is unnecessary for fetal life because embryos with bilateral renal agenesis survive in utero. Bilateral renal agenesis, however, is commonly associated with oligohydramnios (see Chapter 7), thus indicating that the overall balance of amniotic fluid requires a certain amount of fetal renal function.

Endocrine Function in the Fetus

The development of prenatal endocrine function occurs in several phases. Most endocrine glands (e.g., thyroid, pancreatic islets, adrenals, gonads) form early in the second month as the result of epithelial-mesenchymal interactions. As these glands differentiate late in the second month or early in the third, they develop the intrinsic capacity to synthesize their specific hormonal products. In most cases, the amount of hormone secreted is initially very small; increased secretion often depends on the stimulation of the gland by a higher-order hormone produced in another gland.

The anterior pituitary gland develops similarly to many other endocrine glands. Its hormonal products generally stimulate more peripheral endocrine glands, such as the thyroid, adrenals, and gonads, to produce or release their specific hormonal products. Pituitary hormones can be shown immunocytochemically within individual pituitary epithelial cells as early as 8 weeks (adrenocorticotropic hormone [ACTH]) or 10 weeks (luteinizing hormone and follicle-stimulating hormone). Most pituitary hormones are typically not present in the blood in detectable quantities, however, until a couple of months after they can be shown in the cells that produce them. An exception is growth hormone, which can be detected in plasma by 10 weeks of gestation.

While the anterior pituitary is developing its intrinsic synthetic capacities, the hypothalamus also takes shape and develops its capacity to produce the various releasing and inhibitory factors that modulate the function of the pituitary gland. Regardless of its intrinsic capacities, the hypothalamus is limited in its influence on the embryonic pituitary gland until about 12 weeks of gestation, when the neurovascular links between the hypothalamus and pituitary become established.

At each level in the control hierarchy, a generally low intrinsic level of hormone production can be stimulated by the actions of hormones produced by the next higher-order gland. The amount of thyroid hormone released is considerably increased when thyroid-stimulating hormone, released by the anterior pituitary, acts on the thyroid gland. The release of this hormone by the pituitary is regulated by thyrotropin-releasing hormone, which is produced in the hypothalamus. Regardless of the nature of the upstream stimulation of the thyroid, the forms of thyroid hormone released by the fetal thyroid are largely biologically inactive because of enzymatic modifications or through sulfation. Late in gestation, thyroid hormone accelerates the development of brown fat in the fetus. Brown fat, much of which is stored in depots in the upper back, maintains body temperature in the neonate through a process of nonshivering thermogenesis. Studies on anencephalic fetuses have shown that the anterior pituitary can produce and release most of its hormones in the absence of hypothalamic input, although plasma concentrations of some are reduced.

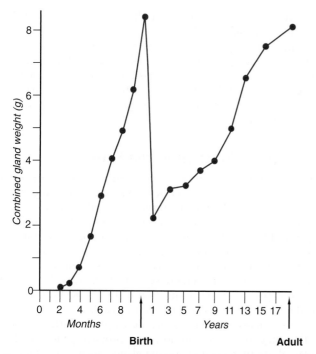

Fig. 18.10 **Graph of the weights of the human adrenal glands during prenatal and postnatal development.** After birth, the weight of the gland decreases dramatically with the reorganization of the cortex of the gland. *(Based on data from Neville AM, O'Hare MJ:* The human adrenal cortex, *Berlin, 1982, Springer.)*

Among the fetal endocrine glands, the adrenal remains the most enigmatic. By 6 to 8 weeks of development, the inner cortex enlarges greatly to form a distinct fetal zone, which later in pregnancy occupies about 80% of the gland. By the end of pregnancy, the adrenal glands weigh 4 g each, the same mass as that of the adult glands (**Fig. 18.10**). The fetal adrenal cortex produces 100 to 200 mg of steroids each day, an amount several times higher than that of the adult adrenal glands. The main hormonal products of the fetal adrenal are Δ^5-3β-hydroxysteroids, such as dehydroepiandrosterone, which are inactive alone, but are converted to biologically active steroids (e.g., estrogens, especially estrone) by the placenta and liver. The fetal adrenal cortex depends on the presence of pituitary ACTH; in its absence, the fetal adrenal cortex is small. If exogenous ACTH is administered, the fetal adrenal cortex persists after birth.

There is a parallel between the presence of the fetal adrenal cortex and functions of the embryonic testis. In the testis, a population of Leydig cells produces testosterone, which is necessary for the morphogenesis of many components of the male reproductive system. Then these cells regress until after birth, when a new population of Leydig cells takes over the production of testosterone to fulfill the needs of postnatal life.

Despite the prominence of the fetal adrenal cortex, its specific functions during pregnancy are still unclear. The large fetal zone of the adrenal cortex produces a steroid precursor for estrogen biosynthesis by the placenta. Fetal adrenal hormones influence maturation of the lungs (as prolactin has also been postulated to do), liver, and epithelium of the digestive tract (**Box 18.2**). In sheep, products of the adrenal cortex prepare the fetus for independent postnatal life and influence

---

Box 18.2 **Maturation Processes That Are Affected by Fetal Cortisol**

1. Shift of the main hematopoietic site to bone marrow
2. Storage of glycogen by the liver
3. Final digestive enzymatic differentiation of intestinal absorptive cells
4. Stimulation of surfactant synthesis in the lung
5. Involvement in parturition

the initiation of parturition, but the situation in primates is considerably less clear. Shortly after birth, the fetal adrenal cortex rapidly involutes (see Fig. 18.10). Within 1 month after birth, the weight of each gland is reduced by 50%, and the volume of the fetal cortex decreases from 70% of total adrenal volume to approximately 3%. By 1 year of age, each gland weighs only 1 g. The mass of the adrenal glands does not return to that of the late fetus until adulthood.

Fetal endocrinology is complicated by the presence of the placenta, which can synthesize and release many hormones, convert hormones released from other glands to active forms, and potentially exchange other hormones with the maternal circulation. By 6 to 7 weeks, hormone production (e.g., progesterone) by the placenta is enough to maintain pregnancy even if the ovaries are removed.

One of the earliest placental hormones produced is human **chorionic gonadotropin** (**HCG**) (see Chapter 7). One later function of HCG is to stimulate steroidogenesis by the placenta. The synthesis of HCG by the syncytiotrophoblast of the placenta is regulated by the production of **gonadotropin-releasing hormone** by cells of the cytotrophoblast. Synthesis of this hormone by the placenta supplants its normal production by the hypothalamus and is probably an adaptation that allows earlier and more local control of HCG than could be accomplished by the hypothalamus.

Clinical Correlation 18.1 discusses the clinical study and manipulation of the fetus.

Parturition

Parturition, the process of childbirth, occurs approximately 38 weeks after fertilization (**Fig. 18.16**). The process of childbirth consists of three distinct stages of labor. The first, the **stage of dilation**, begins with the onset of regular, hard contractions of the uterus and ends with complete dilation of the cervix. Although the contractions of the uterine smooth muscle may seem to be the dominant process in the first stage of labor, the most important components are the effacement and dilation of the cervix. During the entire pregnancy, the cervix functions to retain the fetus in the uterus. For childbirth to proceed, the cervix must change consistency from a firm, almost tubular structure to one that is soft, distensible, and not canal-like. This change involves a reconfiguration and removal of much of the cervical collagen. Although many of the factors underlying the reconfiguration of the cervix during the first stage of labor remain undefined, considerable evidence exists for an important role of **prostaglandin F$_{2\alpha}$** in the process. Although there is great variation, the average length of the first stage of labor is approximately 12 hours.

Text continued on p. 467

CLINICAL CORRELATION 18.1
Clinical Study and Manipulation of the Fetus

New imaging and diagnostic techniques have revolutionized the study of living fetuses. Many congenital malformations can be diagnosed in utero with accuracy. On the basis of this information, the surgeon can treat some congenital malformations through fetal surgery much more efficiently than by traditional surgery on infants or older children.

Fetal Diagnostic Procedures

Imaging Techniques

Because of its safety, cost, and ability to look at the fetus in real time, **ultrasonography** is currently the most widely used obstetrical imaging technique (**Figs. 18.11** and **18.12**). It is useful for the simple diagnosis of structural anomalies and can be used in real time to guide fetal invasive procedures, such as chorionic villus sampling and intrauterine transfusions. The major uses of ultrasonography are summarized in **Box 18.3**.

Conventional **radiographs** continue to be used in certain circumstances, but because of the potential for radiation damage to

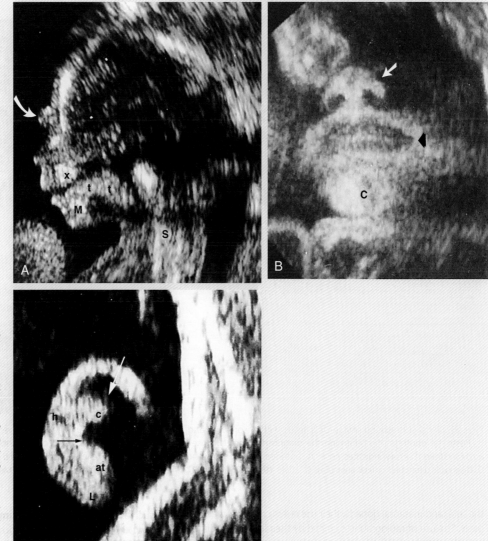

Fig. 18.11 **Ultrasound images of the normal fetal head.** A, Lateral profile showing the nose, mandible (M), maxilla (x), tongue (t), and medial aspect of the bony orbit *(curved arrow)*. S, Sternocleidomastoid muscle. B, Frontal view of the face showing the nose *(white arrow)*, chin (C), and corner of the mouth *(black arrow)*. C, Ear. *Black arrow* indicates antihelix; *white arrow* indicates scaphoid fossa. at, antitragus; c, antihelical crus; h, helix; L, earlobe. (A and B, *From Bowerman RA:* Atlas of normal fetal ultrasonographic anatomy, *ed 2, St. Louis, 1992, Mosby;* C, *from Nyberg DA and others:* Diagnostic ultrasound of fetal anomalies, *St. Louis, 1990, Mosby.)*

Continued

CLINICAL CORRELATION 18.1
Clinical Study and Manipulation of the Fetus—cont'd

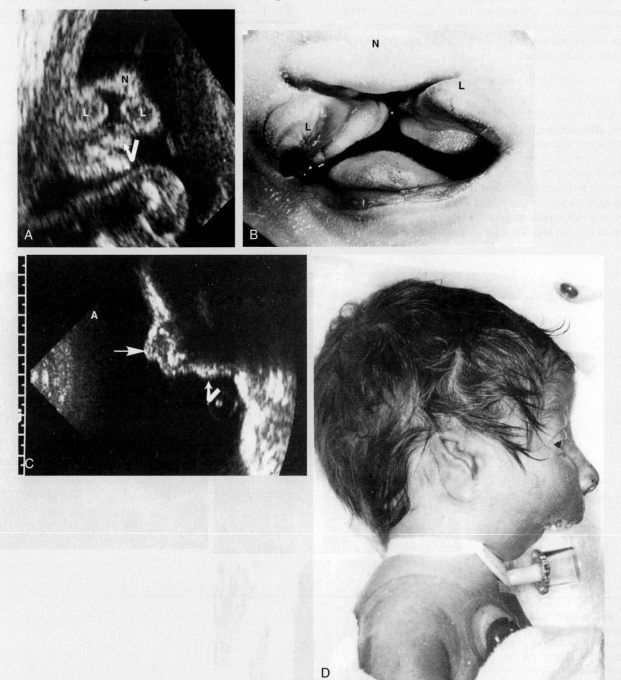

Fig. 18.12 **A,** Ultrasound image of a fetus with trisomy 13 and a midline cleft lip (L) and palate. *Curved arrow* indicates tongue. N, nose. **B,** Postnatal photograph confirming the diagnosis. L, lip; N, nose. **C,** Ultrasound image of facial profile showing marked micrognathia *(curved arrow)*. *Straight arrow* indicates nose. A, anterior. **D,** Postnatal photograph confirming the diagnosis. *(A and B, From Nyberg D, Mahony B, Pretorius D: Diagnostic ultrasound of fetal anomalies,* St. Louis, 1990, Mosby; **C and D,** *from Benson CB and others:* J Ultrasound Med *7:163-167, 1988.)*

the fetal and maternal gonads, their use is less common than previously. The use of radiographs is limited by their inability to discriminate the details of soft tissues, including cartilaginous components of the skeleton. By injecting radiopaque substances into the amniotic cavity (**amniography, fetography**), clinicians can obtain outlines of the fetus and amniotic cavity. Other imaging techniques,

such as **magnetic resonance imaging, computed tomography**, and **xeroradiography**, produce useful images of the fetus, but their use is limited because of factors such as cost and availability (**Figs. 18.13** and **18.14**).

Fetoscopy is the direct visualization of the fetus through a tube inserted into the amniotic cavity. This technique is accomplished

CLINICAL CORRELATION 18.1
Clinical Study and Manipulation of the Fetus—cont'd

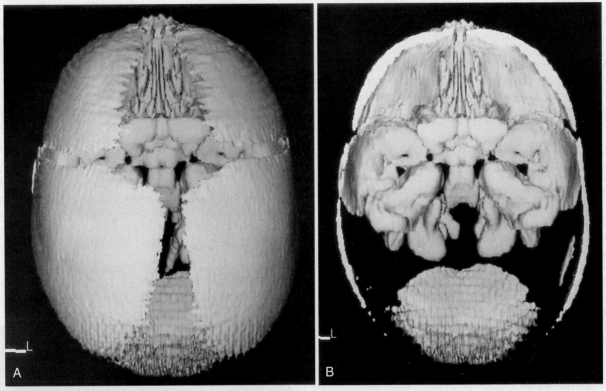

Fig. 18.13 High-resolution computed tomographic reconstructions of the skull of an 18-week fetus. *A,* Focus on superficial bones of the skull. *B,* Deeper bones from the same skull. *(Courtesy of R.A. Levy, H. Maher, and A.R. Burdi, Ann Arbor, Mich.)*

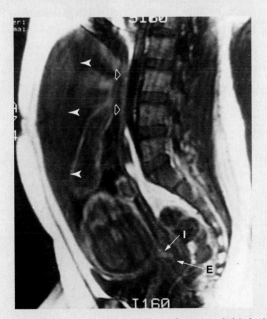

Fig. 18.14 Magnetic resonance image of a normal third trimester fetus inside the uterus. The head of the fetus is near the point of the *arrow* from I (internal cervical os). *Closed arrowheads* indicate placenta; *open arrowheads* indicate uterine wall. E, external cervical os. *(From Friedman AC and others:* Clinical pelvic imaging, *St. Louis, 1990, Mosby.)*

principally through the use of fiberoptic technology. Because of the risk of spontaneous abortion and infection, this technique is not normally used for purely diagnostic purposes, but rather as an aid to intrauterine sampling procedures. Its use has been largely supplanted by other techniques that rely on ultrasound guidance.

Sampling Techniques

The classic sampling technique is **amniocentesis**, which involves the insertion of a needle into the amniotic sac and removal of a small amount of amniotic fluid for analysis. Amniocentesis is normally not done before 13 weeks of gestation because of the relatively small amount of amniotic fluid.

Amniocentesis was originally used for detecting chromosomal anomalies (e.g., Down syndrome) in fetal cells found in the amniotic fluid and for the determination of levels of **α-fetoprotein**, a marker for closure defects of the neural tube and certain other malformations. Analysis of the fetal cells in amniotic fluid is also the basis for determining the gender of embryos. This is typically accomplished by the use of a fluorescent dye that intensely stains the Y chromosome. At present, various analytic procedures on amniotic fluid and cells cultured from the fluid are used to detect many enzymatic and biochemical defects in embryos and to monitor the condition of the fetus.

Another widely used diagnostic technique is **chorionic villus sampling**. In this technique, ultrasonography is used as a guide

Continued

CLINICAL CORRELATION 18.1
Clinical Study and Manipulation of the Fetus—cont'd

to insert a biopsy needle into the placenta, from which a small sample of the villi is removed for diagnostic purposes. This technique is typically used at earlier periods of pregnancy (6 to 9 weeks) than amniocentesis.

With increasing sophistication of fetal imaging techniques, especially ultrasonography, direct sampling of fetal tissues is possible. Ultrasonography-guided sampling of fetal blood, mainly from umbilical vessels, is now common for the diagnosis of hereditary and pathological conditions, such as immunodeficiencies, coagulation defects, hemoglobin abnormalities, and fetal infections. It is also possible to obtain biopsy specimens of fetal skin and even the fetal liver for organ-specific abnormalities.

Therapeutic Manipulations on the Fetus

Some conditions are better treated in the fetal period than after birth (**Box 18.4**). In some cases involving blockage, severe structural damage to the fetus can be prevented. In other cases, the buildup of toxic waste products can be reduced. The recognition that fetal surgery produces essentially scarless results has stimulated some surgeons to consider corrective surgery in utero rather than waiting until after birth.

Fetal shunts can be applied to correct specific conditions in which major permanent damage would result before the time of birth. One such situation is a shunt into the urinary bladder to relieve the pressure and subsequent kidney damage caused by anatomical obstructions of the lower urinary tract. **Figure 18.15** shows the consequence of nontreatment of a persistent cloacal plate, which results in **megacystitis** (enlarged bladder). Fetal shunts have also been used in attempts to relieve the cerebrospinal pressures that result in hydrocephaly (see Fig. 11.38), but the results of these procedures have been equivocal.

Fetal blood transfusions are used for the treatment of fetal anemia and severe erythroblastosis fetalis (see Chapter 7). Earlier, the blood was introduced intraperitoneally. With the increasing sophistication of umbilical cord blood sampling techniques, direct intravascular transfusions are now possible.

Open fetal surgery can now be performed because of the diagnostic procedures that allow an accurate assessment of the condition of the fetus. This is still a new and highly experimental procedure, and its application has been confined to cases of fetal anomalies that would cause grave damage to the fetus if left uncorrected before birth. Currently, the principal indications for open fetal surgery are blockage of the urinary tract, severe diaphragmatic hernia, and some cases of hydrocephalus. Open fetal surgery entails a risk to the mother as well, and the advisability of such a procedure must be carefully considered. With future improvements in procedures, correcting other malformations, such as cleft lip and palate or limb deformities in utero, may be possible.

Box 18.4 Fetal Conditions Treatable in Utero

Surgical Treatment

1. Urinary obstruction (urethral valves)
2. Diaphragmatic hernia
3. Sacrococcygeal teratoma
4. Chylothorax
5. Cystic adenomatoid malformation of lung
6. Twin-twin transfusion syndrome
7. Hydrocephalus caused by aqueductal stenosis
8. Complete heart block
9. Spina bifida

Medical Treatment

1. Erythroblastosis fetalis
2. Adrenal hyperplasia
3. Hyperthyroidism and hypothyroidism
4. Fetal dysrhythmias
5. Diabetes
6. Agammaglobulinemia
7. Vitamin-responsive metabolic disorders
8. Idiopathic thrombocytic purpura

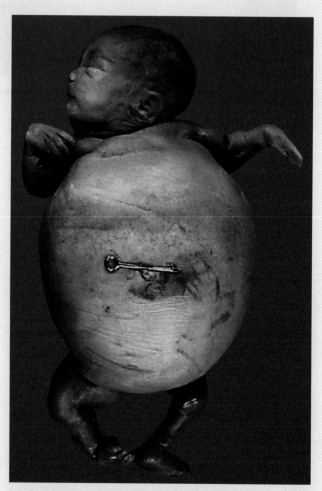

Fig. 18.15 **Fetus with great abdominal distention as a result of megacystis (large bladder) caused by a cloacal plate.** *(Courtesy of M. Barr, Ann Arbor, Mich.)*

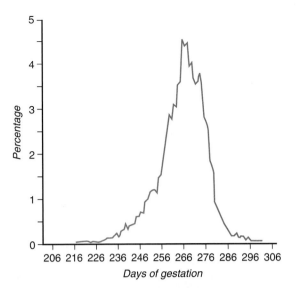

Fig. 18.16 **Graph showing the distribution in days of normal pregnancy in 1336 spontaneous, full-term deliveries.** *(Adapted from Wigglesworth J, Singer D:* Textbook of fetal and perinatal pathology, *London, 1991, Blackwell Scientific.)*

The second stage of labor (**stage of expulsion**) begins with complete dilation of the cervix and ends with the passage of the infant from the birth canal. During this stage, which typically lasts 30 to 60 minutes, depending on the number of previous deliveries of the mother, the infant still depends on a functioning umbilical circulation for survival.

The third stage of labor (**placental stage**) represents the period between delivery of the infant and expulsion of the placenta. Typically, the umbilical cord is cut within minutes of delivery, and the infant must then quickly adapt to independent living. During the next 15 to 30 minutes, continued contractions of the uterus separate the placenta from the maternal decidua, and the intact placenta is delivered. After delivery of the placenta, major hemorrhage from the spiral uterine arteries is prevented naturally by continued contraction of the myometrium. In actual clinical practice, the third stage is commonly abbreviated by the intramuscular injection of synthetic oxytocin and external manipulation of the uterus to reduce the amount of uterine blood loss.

The mechanisms underlying the initiation and progression of parturition in humans remain remarkably poorly understood, even though considerable progress has been made in uncovering the stimuli for parturition in certain domestic animals. In sheep, parturition is initiated by a sharp increase in the cortisol concentration in the fetal blood. As a result, placental enzyme activity changes, resulting in the conversion of placental progesterone to estrogen synthesis. This increase in estrogen stimulates the formation and release of prostaglandin $F_{2\alpha}$.

In humans, there is less dependence on activity of the pituitary-adrenal cortical axis for the initiation of parturition. More recent research on primate embryos has suggested that **corticotropin-releasing hormone** (**CRH**), which is normally released by the hypothalamus, is produced in significant amounts by the placenta, starting at about 12 weeks of pregnancy (**Fig. 18.17**). Some of the placental CRH stimulates the fetal adenohypophysis to release **ACTH**. ACTH stimulates the adrenal cortex to produce **cortisol**, which is necessary for many maturation processes in the fetus (see Box 18.2). Much of the CRH acts directly on the fetal adrenal cortex, however, and stimulates it to produce dehydroepiandrosterone sulfate, which the placenta uses directly as a substrate for the synthesis of estrogen.

High levels of **estrogen**, accompanied by lowered levels of progesterone, during late pregnancy tip the scales in favor of parturition, but the exact nature of the trigger for parturition is obscure. Progesterone, as its name implies, acts to maintain pregnancy; increasing amounts of estrogen prepare the female reproductive tissues for parturition. Estrogen stimulates the production of **connexins**, which form junctions that electrically connect uterine smooth muscles to one another. It also stimulates the uterine smooth muscle cells to produce receptors for oxytocin, a major stimulus for the contraction of uterine smooth muscle during labor. Estrogen, through promoting the action of prostaglandins, stimulates the degradation of collagen fibers in the cervix to make them flexible enough to expand to accommodate the fetus during childbirth.

The production of CRH by the placenta helps to explain why a fetus with pituitary or adrenal hypoplasia, or even anencephaly, is typically delivered within the normal time frame. Spontaneous labor occurs in cases of pituitary or adrenal hypoplasia of the fetus or even in anencephaly, but the timing of parturition typically has a considerably wider range than normal. As in sheep, the local release of prostaglandins E_2 and $F_{2\alpha}$ may be important in the initiation of labor in humans. In rare cases of human twins implanted in different horns of a double uterus, one member of the pair may not be born until several days or even weeks after the first delivery.

Adaptations to Postnatal Life

When the umbilical cord is clamped after birth, the neonate is suddenly thrust into a totally independent existence. The respiratory and cardiovascular systems must almost instantaneously assume a type and level of function quite different from those during the fetal period. Within hours or days of birth, the digestive system, immune system, and sense organs must also adapt to a much more complex environment.

Circulatory Changes at Birth

Two major events drive the functional adaptations of the circulatory system immediately at birth. The first is the cutting of the umbilical cord, and the second comprises the changes in the lungs after the first breaths of the newborn. These events stimulate a series of sweeping changes that not only alter the circulatory balance, but also result in major structural alterations in the circulatory system of the infant.

Cutting the umbilical cord results in an immediate cessation of blood entering the body via the umbilical vein. The major blood flow through the ductus venosus is eliminated, and the amount of blood that enters the right atrium via the inferior vena cava is greatly reduced. A consequence of this activity is a reduction of the stream of blood that was directly shunted from the right to the left atrium via the foramen ovale during fetal life.

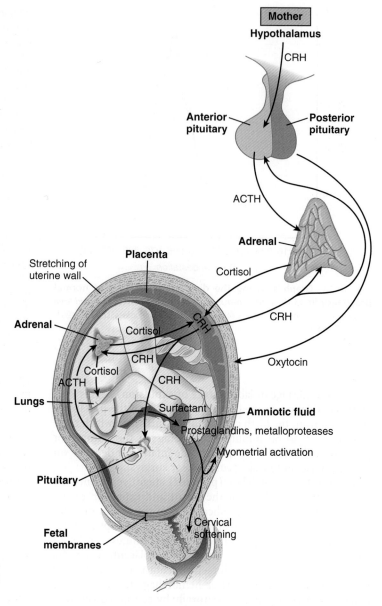

Fig. 18.17 **Factors involved in the initiation of parturition.**
ACTH, adrenocorticotropic hormone; CRH, corticotropin-releasing hormone.

After just a few breaths, the pulmonary circulatory bed expands and can accommodate much greater blood flow than during the fetal period. Consequences of this change are a reduced flow of blood through the ductus arteriosus and a correspondingly greater return of blood into the left atrium via the pulmonary veins. Within minutes after birth, the ductus arteriosus undergoes a reflex closure. This shunt, which in prenatal life is actively kept open in great part through the actions of **prostaglandin E₂**, rapidly constricts after the oxygen concentration in the blood increases. The mechanism for constriction seems to involve the action of cytochrome P450, but the way it is translated into contraction of the smooth musculature of the ductus is unclear. Shortly after birth, platelets form a plug that seals the lumen of the constricted ductus arteriosus. The principal tissue involved in closure of the ductus is smooth muscle; the shunt also experiences a breakdown of elastic fibers and a thickening of the inner intimal layer. Although initial closure of the ductus arteriosus is based on a reflex mechanism, over the next few weeks it is followed by a phase of anatomical closure, during which cell death and proliferation of connective tissue combine to reduce the ductus into a fibrous cord.

Because of closure of the ductus arteriosus, increased pulmonary venous flow, and loss of 25% to 50% of the peripheral vasculature (placental circulation) when the umbilical cord is cut, the blood pressure in the left atrium becomes slightly increased over that in the right atrium. This increase leads to physiological closure of the interatrial shunt, with the result that all the blood entering the right atrium empties into the right ventricle (**Fig. 18.18**). Structural closure of the valve at the foramen ovale is prolonged, occurring over several months after birth. Before complete structural obliteration of the interatrial valve, it possesses the property of "probe patency," which allows a catheter inserted into the right atrium to pass freely through the foramen ovale into the left atrium. As structural fusion of the valve to the interatrial septum progresses, the property of probe patency is gradually reduced and ultimately disappears. In approximately 20% of individuals,

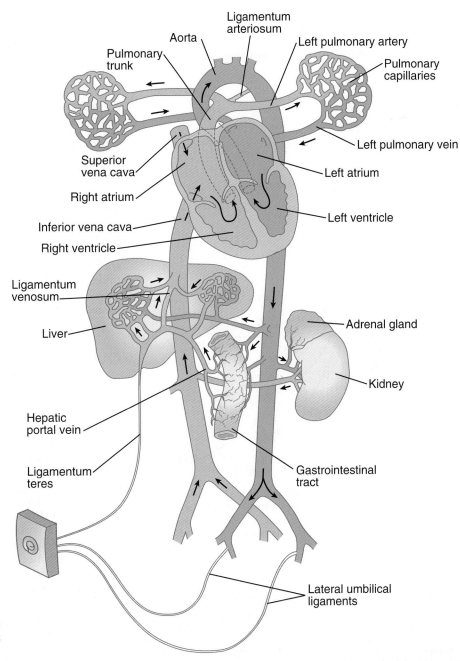

Fig. 18.18 **Postnatal circulation showing the location of remnants of embryonic vessels.**

structural closure of the interatrial valve is not completed, thus leading to the normally asymptomatic condition of **probe patent foramen ovale**.

Although the ductus venosus also loses its patency after birth, its closure is more prolonged than that of the ductus arteriosus. The tissue of the wall of the ductus venosus is not as responsive to increased oxygen saturation of the blood as that of the ductus arteriosus.

After the postnatal pattern of the circulation is fully established, obliterated vessels or shunts that were important circulatory channels in the fetus either are replaced by connective tissue strands, forming ligaments, or are represented by relatively smaller vessels (see Fig. 18.18; **Fig. 18.19**). These changes are summarized in **Table 18.1**. In early postnatal life, the umbilical vein can still be used for exchange transfusions (in

cases of hemolytic disease resulting from erythroblastosis fetalis) before its lumen becomes obliterated.

Lung Breathing in the Perinatal Period

Immediately after birth, the infant must begin to breathe regularly and effectively with the lungs to survive. The initial breaths are difficult because the lungs are filled with fluid and the alveoli are collapsed at birth. On a purely mechanical basis, air breathing is facilitated by a proportionally large diameter of the trachea and major airways. The large diameter reduces resistance to airflow, which would be insurmountable if these passageways were proportionally as small as the lungs.

Just before birth, increased levels of **arginine vasopressin** and **adrenaline** suppress the secretion of fetal lung fluid and

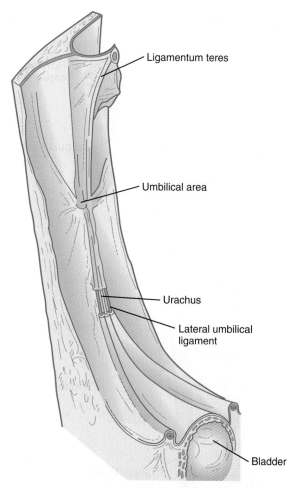

Fig. 18.19 Posterior view of the umbilical region of the abdominal wall showing the two obliterated umbilical arteries flanking the urachus leading from the bladder to the umbilicus. The single strand at the top is the ligamentum teres (remnant of the umbilical vein) leading from the umbilicus to the liver.

Table 18.1 Postnatal Derivatives of Prenatal Circulatory Shunts or Vessels

Prenatal Structure	Postnatal Derivative
Ductus arteriosus	Ligamentum arteriosum
Ductus venosus	Ligamentum venosum
Interatrial shunt	Interatrial septum
Umbilical vein	Ligamentum teres
Umbilical arteries	Distal segments, lateral umbilical ligaments; proximal segments, superior vesical arteries

stimulate its resorption by the pulmonary epithelial cells. At birth, the lungs contain about 50 mL of alveolar fluid, which must be removed for adequate air breathing. Approximately half of that volume enters the lymphatic system. Of the remainder, perhaps half may be expelled during birth. The remainder enters the bloodstream.

The alveolar sacs in the lungs begin to inflate on the first inspiration. The pulmonary surfactant, which was secreted in increasing amounts during the last few weeks of a term pregnancy, reduces the surface tension that would otherwise be present at the air-fluid interface on the alveolar surfaces and facilitates inflation of the lungs. With the rush of air into the lungs, the pulmonary vasculature opens and allows a greatly increased flow of blood through the lungs. This increased flow results in an increased oxygen saturation of the blood; the color of the caucasian newborn changes from a dusky purple to pink.

Breathing movements in the fetus are intermittent and irregular even after birth. Many factors can affect the frequency of breathing, but the factors responsible for the transition from intermittent to regular breathing remain poorly understood. Factors such as cold, touch, chemical stimuli, sleep patterns, and signals emanating from the carotid and aortic bodies have been implicated. During periods of wakefulness, breathing of the neonate soon stabilizes, but for several weeks after birth, short periods of apnea (5 to 10 seconds) are common during REM sleep.

Overview

The story of prenatal development is complex but fascinating. Many generalizations can be extracted from the study of embryology, but one dominant theme is that of an overall coordination of a large number of very complex integrative processes that range from the translation of information encoded in structural genes, such as the homeobox-containing genes, to the influence of physical factors, such as pressure and tension, on the form and function of the developing embryo.

Sometimes, things go wrong. Studies of spontaneous abortions show that nature has provided a screening mechanism that eliminates many of the embryos least capable of normal development or independent survival. A simple base substitution in the DNA of an embryo can produce a defect that may be highly localized or have far-reaching consequences on the development of a variety of systems.

With ever greater insight into the molecular and cellular mechanisms underlying normal and abnormal development, and with increasingly sophisticated technology, biomedical scientists and physicians can manipulate the embryo in ways that were unimaginable not long ago. It is an exciting era that is rapidly increasing in technological complexity and uncertain in many social and ethical aspects. It also has an economic impact that is difficult to predict.

Clinical Vignette

A routine ultrasound examination of a woman in her eighth month of pregnancy shows a swelling of the abdomen of the fetus. A repeat examination 1 week later shows a further progression of the swelling. The woman is referred to an academic medical center and is told that without intrauterine surgery there is a very good chance that the fetus would undergo irreversible pathological changes before birth. She is informed that the condition affecting her fetus is one in which fetal surgery has produced a good record of success.

From this history, what is the likely type of condition, and what would be the nature of the surgical procedure?

Summary

- The fetal period is characterized by intense growth in length and mass of the embryo. With time, the trunk grows relatively faster than the head, and later the limbs show the greatest growth. The early fetus is thin because of the absence of subcutaneous fat. By midpregnancy, subcutaneous fat is deposited.
- At 5 weeks' gestation, the heart beats at 100 beats/minute; the heart rate increases to 160 beats/minute by 8 weeks and then declines slightly during the remainder of pregnancy. Some different physiological properties of the fetal heart can be explained by the presence of fetal isozymes in the cardiac muscle. The patency of the ductus arteriosus in the fetus is actively maintained through the actions of prostaglandin E_2.
- The fetal lungs are filled with fluid, but they must be prepared for full respiratory function within moments after birth. The fetus begins to make anticipatory breathing movements as early as 11 weeks. Fetal breathing is affected by maternal physiological conditions, such as eating and smoking. Disproportionate growth in diameter of the upper airway is important in allowing a newborn to take the first breath. The secretion of pulmonary surfactant begins at about 24 weeks, but large amounts are not synthesized until just a few weeks before birth. Premature infants with a deficiency of pulmonary surfactant often have respiratory distress syndrome.
- Fetal movements begin at about $7\frac{1}{2}$ weeks and increase in complexity thereafter. The maturation of fetal movements mirrors the structural and functional maturation of the nervous system. Diurnal rhythms in fetal activity appear at 20 to 22 weeks. The fetus has alternating periods of sleep and wakefulness. Near term, the fetus responds to vibroacoustic stimuli, and by 30 weeks, the pupillary light reflex can be elicited.
- The fetal digestive tract is nonfunctional in the usual sense, but maturation of enzyme systems for digestion and absorption occurs. Spontaneous rhythmical movements of the small intestine begin by 7 weeks' gestation. Meconium begins to fill the lower intestinal tract by midpregnancy. By term, the fetus typically swallows more than half a liter of amniotic fluid per day.
- Fetal kidneys produce small amounts of dilute urine. Fetal endocrine glands produce small amounts of hormones that can be histochemically shown in glandular tissue early in the fetal period, but several months often pass before the same hormones can be measured in the blood. The fetal adrenal cortex is very large and produces 100 to 200 mg of steroids per day. The exact functions of the fetal adrenal gland are poorly understood, but fetal cortisol seems to prepare certain organ systems for the transition to independent life after birth. The placenta continues to produce a variety of hormones throughout most of pregnancy.
- Many new diagnostic techniques have considerably improved access to the fetus. Among the imaging techniques, ultrasonography has emerged as the most widely used in obstetrics. Through sampling techniques such as amniocentesis and chorionic villus sampling, fluids or cells of the embryo and fetus can be removed for analysis. These techniques allow certain manipulations on the fetus (e.g., fetal blood transfusions, fetal surgery for certain anomalies).
- Parturition occurs in three stages of labor. The first is the stage of dilation, which culminates with effacement of the cervix. The second stage culminates with expulsion of the infant. The third stage represents the period between delivery of the infant and expulsion of the placenta. The mechanisms underlying the initiation of parturition in the human remain poorly understood.
- After birth and cutting of the umbilical cord, a newborn must quickly adapt to an independent existence in terms of breathing and cardiac function. After the first breaths and severing of the umbilical cord, the pulmonary circulation opens. In response to increased flow into the left atrium, the interatrial shunt undergoes physiological closure, and the ductus arteriosus undergoes a reflex closure. Closure of the ductus venosus in the liver is more prolonged.

Review Questions

1. The ligamentum teres is the postnatal remains of the:
A. Ductus arteriosus
B. Ductus venosus
C. Umbilical vein
D. Umbilical artery
E. Urachus

2. Insufficient production of which of the following is a major cause of poor viability of infants born 24 to 26 weeks after conception?
A. Pulmonary surfactant
B. α-Fetoprotein
C. Meconium
D. Lanugo
E. Urine

3. Meconium is formed in the fetal:
A. Liver
B. Ileum
C. Lungs
D. Amniotic fluid
E. Kidneys

4. Which organ is much larger in the fetus than it is shortly after birth?
A. Kidneys
B. Heart
C. Liver
D. Urinary bladder
E. Adrenal gland

5. Fetal movements usually can first be detected by ultrasound at how many weeks?
A. 6
B. 8
C. 10
D. 12
E. 14

6. What blood vessel is commonly used for exchange transfusions in newborns?

A. Umbilical artery
B. Jugular vein
C. Femoral artery
D. Umbilical vein
E. None of the above

7. A premature infant develops labored breathing and dies within a few days. What is the likely cause?

8. A rare condition that can persist into adulthood is caput medusae ("Medusa's head"), in which a dark vascular ring with irregular radiations appears around the umbilicus with straining of the abdomen. What is an embryological basis for this condition?

9. A pregnant woman typically first feels fetal movements about 15 weeks into pregnancy. The movements become more noticeable during succeeding weeks, but are commonly reduced during the last couple of weeks before parturition. What is the explanation for this?

10. List some medical advances that have allowed fetal surgery to become a reality.

References

Avery ME, Wang N-S, Taeusch HW: The lung of the newborn infant, *Sci Am* 228:75-85, 1973.

Barclay AE, Franklin KJ, Prichard MML: *The foetal circulation and cardiovascular system, and the changes that they undergo at birth*, Oxford, 1944, Blackwell Scientific.

Barcroft J: *Researches on prenatal life*, vol 1, Oxford, 1946, Blackwell Scientific.

Barron DH: The changes in the fetal circulation at birth, *Physiol Rev* 24:277-295, 1944.

Beeshay VE, Carr BR, Rainey WE: The human fetal adrenal gland, corticotropin-releasing hormone, and parturition, *Semin Reprod Med* 25:14-20, 2007.

Bowerman RA: *Atlas of normal fetal ultrasonographic anatomy*, ed 2, St. Louis, 1992, Mosby.

Busnel MC, Granier-Deferre C, LeCanuet JP: Fetal audition, *Ann N Y Acad Sci* 662:118-134, 1992.

Chaoui R: Coronary arteries in fetal life: physiology, malformations and the "heart-sparing effect," *Acta Paediatr Suppl* 446:6-12, 2004.

Coceani F, Olley PM: The control of cardiovascular shunts in the fetal and perinatal period, *Can J Physiol Pharmacol* 66:1129-1134, 1988.

Cook AC, Yates RW, Anderson RH: Normal and abnormal fetal cardiac anatomy, *Prenat Diagn* 24:1032-1048, 2004.

Dawes GS: The development of fetal behavioural patterns, *Can J Physiol Pharmacol* 66:541-548, 1988.

Duenholter JH, Pritchard JA: Fetal respiration, *Am J Obstet Gynecol* 129:326-338, 1977.

Fortin G, Thoby-Brisson M: Embryonic emergence of the respiratory rhythm generator, *Respir Physiol Neurobiol* 168:86-91, 2009.

Fuse Y: Development of the hypothalamic-pituitary-thyroid axis in humans, *Reprod Fertil Dev* 8:1-21, 1996.

Gnanalingham MG and others: Developmental regulation of the lung in preparation for life after birth: hormonal and nutritional manipulation of local glucocorticoid action and uncoupling protein-2, *J Endocrinol* 188:375-386, 2006.

Grenache DG, Gronowski AM: Fetal lung maturity, *Clin Biochem* 39:1-10, 2006.

Groenman F, Unger S, Post M: The molecular basis for abnormal human lung development, *Biol Neonate* 87:164-177, 2005.

Harding R, Bocking AD, eds: *Fetal growth and development*, Cambridge, 2001, Cambridge University Press.

Hooker D: *The prenatal origin of behavior*, Lawrence, Kan, 1952, University of Kansas Press.

Hooper SB, Harding R: Fetal lung liquid: a major determinant of the growth and functional development of the fetal lung, *Clin Exp Pharmacol Physiol* 22:235-247, 1995.

Inanlou MR, Baguma-Nibasheka M, Kablar B: The role of fetal breathing-like movements in lung organogenesis, *Histol Histopathol* 20:1261-1266, 2005.

Jones CT, Nathanielsz PW, eds: *The physiological development of the fetus and newborn*, London, 1985, Academic Press.

Kline RM: Whose blood is it, anyway? *Sci Am* 284:42-49, 2001.

Lagercrantz H, Slotkin TA: The "stress" of being born, *Sci Am* 254:100-107, 1986.

Larsen T and others: Normal fetal growth evaluated by longitudinal ultrasound examinations, *Early Hum Dev* 24:37-45, 1990.

Lee AJ and others: Fetal pain: a systematic multidisciplinary review of the evidence, *JAMA* 294:947-954, 2005.

Liu M and others: Stimulation of fetal rat lung cell proliferation in vitro by mechanical stretch, *Am J Physiol* 263:L376-L383, 1992.

Malendowicz LK: 100th anniversary of the discovery of the human adrenal fetal zone by Stella Starkel and Leslaw Wegrzyowski: how far have we come? *Folia Histochem Cytobiol* 48:491-506, 2010.

Manning FA: Fetal breathing movements, *Postgrad Med* 61:116-122, 1977.

Maritz GS, Morley CJ, Harding R: Early developmental origins of impaired lung structure and function, *Early Hum Dev* 81:763-771, 2005.

McArdle HJ, Ashworth CJ: Micronutrients in fetal growth and development, *Br Med Bull* 55:499-510, 1999.

Mellor DJ and others: The importance of "awareness" for understanding fetal pain, *Brain Res Rev* 49:455-471, 2005.

Naeye RL: *Disorders of the placenta, fetus, and neonate: diagnosis and clinical significance*, St. Louis, 1992, Mosby.

Nyberg DA, Mahony BS, Pretorius DH: *Diagnostic ultrasound of fetal anomalies: text and atlas*, St. Louis, 1990, Mosby.

Polin RA, Fox WW, eds: *Fetal and neonatal physiology*, vols 1 and 2, Philadelphia, 1992, Saunders.

Prechtl HFR: Fetal behavior. In Hill A, Volpe J, eds: *Fetal neurology*, New York, 1989, Raven Press, pp 1-16.

Rigatto H: Control of breathing in fetal life and onset and control of breathing in the neonate. In Polin, R, Fox W, eds: *Fetal and neonatal physiology*, vol 1, Philadelphia, 1992, Saunders, pp 790-801.

Sase M and others: Gastric emptying cycles in the human fetus, *Am J Obstet Gynecol* 193:1000-1004, 2005.

Smith BR: Visualizing human embryos, *Sci Am* 280:76-81, 1999.

Smith R: Parturition, *N Engl J Med* 356:271-283, 2007.

Smith R: The timing of birth, *Sci Am* 280:68-75, 1999.

St. John Sutton M, Gill T, Plappert T: Functional anatomic development in the fetal heart. In Polin R, Fox W, eds: *Fetal and neonatal physiology*, vol 1, Philadelphia, 1992, Saunders, pp 598-609.

Teitel DF: Physiologic development of the cardiovascular system in the fetus. In Polin R, Fox W, eds: *Fetal and neonatal physiology*, vol 1, Philadelphia, 1992, Saunders, pp 609-619.

Vanhatalo S, van Nieuwenhuizen O: Fetal pain? *Brain Dev* 22:145-150, 2000.

Warburton D and others: Lung organogenesis, *Curr Top Dev Biol* 90:73-158, 2010.

Wetzel GT, Klitzner TS: Developmental cardiac electrophysiology: recent advances in cellular physiology, *Cardiovasc Res* 31:E52-E60, 1996.

Winter JSD: Fetal and neonatal adrenocortical physiology. In Polin R, Fox W, eds: *Fetal and neonatal physiology*, vol 2, Philadelphia, 1992, Saunders, pp 1829-1841.

Wright C, Sibley CP, Baker PN: The role of fetal magnetic resonance imaging, *Arch Dis Child Fetal Neonatal Ed* 95:F137-F141, 2010.

Answers to Clinical Vignettes and Review Questions

Chapter 1

Clinical Vignette

C. Although most tissues of the body are affected to some extent, the heart is not a primary target tissue of ovarian steroid hormones.

Review Questions

1. D
2. E
3. B
4. A mediastinal teratoma, which is likely to have arisen from an aberrant primordial germ cell that became lodged in the connective tissue near the heart.
5. In the female, meiosis begins during embryonic life; in the male, meiosis begins at puberty.
6. At prophase (diplotene stage) of the first meiotic division and at metaphase of the second meiotic division.
7. Chromosomal abnormalities, such as polyploidy or trisomies of individual chromosomes.
8. Spermatogenesis is the entire process of sperm formation from a spermatogonium. It includes the two meiotic divisions and the period of spermiogenesis. Spermiogenesis, or sperm metamorphosis, is the process of transformation of a postmeiotic spermatid, which looks like an ordinary cell, to a highly specialized spermatozoon.
9. Estrogens, secreted by the ovary, support the preovulatory proliferative phase. From the time of ovulation, progesterone is secreted in large amounts by the corpus luteum and is responsible for the secretory phase, which prepares the endometrium for implantation of an embryo.
10. Follicle-stimulating hormone (FSH) produced by the anterior lobe of the pituitary gland and testosterone produced by the Leydig cells of the testis.

Chapter 2

Clinical Vignette

1. Before the plane crash, the issue is who is the "real" mother. After the crash, the issue is who gets the money—the surrogate mother who claims that she is the real mother or the aunt who claims a blood affinity. Although these present as legal issues that would likely be decided in a court, the concept of what is meant by surrogacy also involves psychological and religious issues.
2. This is a very important issue that has not been resolved. If the parents had been of a religion that strongly supports the rights of embryos to life, should the remaining embryos also be implanted into someone, and, if so, into whom? In a case where a considerable inheritance is involved, the financial implications could cloud the issue. If there were no parental money, who would undergo the risk and the expense to prevent the frozen embryos from being simply thrown out? In many cases of in vitro fertilization and embryo transfer, the question of what to do with the "extra" frozen embryos when the first transfer is successful is a real one. Many frozen embryos are being stored at various sites around the world.

Review Questions

1. E
2. D
3. The sharp surge of luteinizing hormone produced by the anterior lobe of the pituitary gland.
4. Capacitation is a poorly understood interaction between a spermatozoon and female reproductive tissues that increases the ability of the sperm to fertilize an egg. In some mammals, capacitation is obligatory, but in humans the importance of capacitation is less well established.
5. Fertilization usually occurs in the upper third of the uterine tube.
6. The ZP_3 protein acts as a specific sperm receptor through its O-linked oligosaccharides; much of its polypeptide backbone must be exposed to stimulate the acrosomal reaction.
7. Polyspermy is the fertilization of an egg by more than one spermatozoon. It is prevented through the fast electrical block on the plasma membrane of the egg and by the later zona reaction, by which products released from the cortical granules act to inactivate the sperm receptors in the zona pellucida.
8. She had probably taken clomiphene for the stimulation of ovulation. Natural septuplets are almost never seen.
9. The introduction of more than one embryo into the tube of the woman is commonly done because the chance that any single implanted embryo will survive to the time of birth is quite small. The reasons for this are poorly understood. Extra embryos are frozen because if a

pregnancy does not result from the first implantation, the frozen embryos can be implanted without the inconvenience and expense of obtaining new eggs from the mother and fertilizing them in vitro.

10. In cases of incompatibility between the sperm and egg, poor sperm motility, or deficient sperm receptors in the zona, introducing the sperm directly into or near the egg can bypass a weak point in the reproductive sequence of events.

Chapter 3

Clinical Vignette

She had an ectopic pregnancy in her right uterine tube. With the rapidly increasing size of the embryo and its extraembryonic structures, her right uterine tube had ruptured.

Review Questions

1. D
2. E
3. A
4. C
5. The embryonic body proper arises from the inner cell mass.
6. Trophoblastic tissues.
7. They allow the trophoblast of the embryo to adhere to the uterine epithelium.
8. Cells derived from the cytotrophoblast fuse to form the syncytiotrophoblast.
9. In addition to the standard causes of lower abdominal pain, such as appendicitis, the physician should consider ectopic pregnancy (tubal variety) as a result of stretching and possible rupture of the uterine tube containing the implanted embryo.

Chapter 4

Review Questions

1. A homeobox is a highly conserved region consisting of 180 nucleotides that is found in many morphogenetically active genes. Homeobox gene products act as transcription factors.
2. B
3. In contrast to many receptors, the receptors for retinoic acid (α, β, and γ) are located in the nucleus.
4. A
5. D
6. D
7. E

Chapter 5

Clinical Vignette

C. Because of the respiratory problems associated with his situs inversus, this man probably has a mutation of a dynein gene. Commonly, such individuals also have immotile spermatozoa, a condition that would lead to infertility.

Review Questions

1. D
2. A
3. B
4. B

5. C
6. The epiblast.
7. The primitive node acts as the organizer of the embryo. Through it pass the cells that become the notochord. The notochord induces the formation of the nervous system. The primitive node is also the site of synthesis of morphogenetically active molecules, such as retinoic acid. If a primitive node is transplanted to another embryo, it stimulates the formation of another embryonic axis.
8. Hyaluronic acid and fibronectin.
9. Vg1 and activin.
10. Cell adhesion molecules are lost in a migratory phase. When the migratory cells settle down, they may re-express cell adhesion molecules.

Chapter 6

Review Questions

1. B
2. C
3. E
4. C
5. A
6. A change in cell shape at the median hinge point and pressures of the lateral ectoderm acting to push up the lateral walls of the neural plate.
7. Neuromeres provide the fundamental organization of parts of the brain in which they are present. Certain homeobox genes are expressed in a definite sequence along the neuromeres.
8. The somites. Axial muscles form from cells derived from the medial halves of the somites, and limb muscles arise from cellular precursors located in the lateral halves of the somites.
9. In blood islands that arise from mesoderm of the wall of the yolk sac.

Chapter 7

Clinical Vignette

D. α-Fetoprotein, which is produced principally by the fetal liver, is found in many tissues of the body, but normally, only small amounts are excreted into the amniotic fluid. With open neural tube defects, large quantities of α-fetoprotein escape through the opening and enter the amniotic fluid.

Review Questions

1. A
2. E
3. B
4. D
5. E
6. C
7. Because the placental villi (specializations of the chorion) are directly bathed in maternal blood.
8. This depends on the age of the embryo. In an early fetus, the molecule may have to pass through the following layers: syncytiotrophoblast, cytotrophoblast, basal lamina underlying cytotrophoblast, villous mesenchyme, basal lamina of a fetal capillary, and endothelium of the fetal capillary. In a mature placenta, the same molecule may pass from the maternal to the fetal circulation by traversing as few layers as syncytiotrophoblast, a fused

basal lamina of trophoblast and capillary endothelium, and the endothelium of a fetal capillary.

9. Human chorionic gonadotropin. This is the first distinctive embryonic hormone to be produced by the trophoblastic tissues. Early pregnancy tests involved injecting small amounts of urine of a woman into female African clawed toads *(Xenopus laevis)*. If the woman was pregnant, the chorionic gonadotropin contained in the urine stimulated the frogs to lay eggs the next day. Contemporary pregnancy tests, which can be done using kits bought over the counter, give almost instantaneous results.

10. Many substances that enter a woman's blood are now known to cross the placental barrier, including alcohol, many drugs (both prescribed and illicit), steroid hormones, and other low-molecular-weight substances. Generally, molecules with molecular weights less than 5000 daltons should be assumed to cross the placental barrier with little difficulty.

Chapter 8

Clinical Vignette

Although this woman's history suggests many risk factors, none of her children's problems could be definitely attributed to any specific cause. Nevertheless, there is a good likelihood that the spina bifida in the first child and the anencephaly of the third child could be related to overall poor nutrition and a specific deficiency in folic acid because poor nutrition is common in persons with alcoholism. The small stature of the middle child could possibly result from the mother's heavy smoking. On the one hand, the behavior problem of the middle child could be a consequence of the mother's cocaine use, smoking, or alcohol consumption. On the other hand, there could be no relation between any of the mother's risk factors and a prenatal influence on the child's later behavior. An important point is that despite many well-known risk factors, it is very difficult, if not impossible, to assign a given congenital anomaly to a specific cause. Realistically, one can only speak in terms of probabilities.

Review Questions

1. E
2. C
3. E
4. B
5. A
6. B
7. The conditions that result in cleft palate occur during the second month of pregnancy. By the fourth month, the palate is normally completely established. It is almost certain that this malformation had already been established by the time of the accident.
8. Although there may be a connection between the drug and the birth defect, proving a connection between an individual case and any drug, especially a new one, is very difficult. The woman's genetic background, other drugs that she may have taken during the same period, her history of illnesses during early pregnancy, and her nutritional status should be investigated. Even in the best of circumstances, the probability that a specific malformation is caused by a particular factor can only be estimated in many cases.

9. A common cause of such malformations is an insufficiency of amniotic fluid (oligohydramnios), which can place exposed parts of fetuses under excessive mechanical pressure from the uterine wall and lead to deformations of this type.

10. Dysplasia of ectodermal derivatives is a likely cause.

Chapter 9

Clinical Vignette

1. B
2. A

Review Questions

1. D
2. E
3. D
4. B
5. C
6. B
7. The dermis. Recombination experiments have clearly shown that the dermis confers regional morphogenetic information on the epidermis and instructs it to form, for example, cranial hair or abdominal hair.
8. They may be supernumerary nipples located along the caudal ends of the embryonic milk lines.
9. In the early embryo, brain tissue induces the formation of the surrounding membranous skeletal elements. If a significant region of the brain is missing, the inductive interaction does not occur.
10. In experiments involving the use of the quail nuclear marker, quail somites were grafted in place of the original somites in chick embryos. The muscles in the developing limbs all contained quail and not chick nuclei.

Chapter 10

Clinical Vignette

At a descriptive level, the mirror image asymmetry of the duplicated foot and digits is a classic example of Bateson's rule of symmetry in duplicated structures (see p. 50). The best explanation for this malformation would be the presence of a duplicated zone of polarizing activity (ZPA) in the anterior margin of the affected limb. Through the actions of sonic hedgehog, secreted by the duplicated ZPA, a secondary gradient of morphogenetic activity could have instructed the anterior mesoderm of the leg bud to form an additional set of posterior structures. This rare defect in humans parallels almost exactly the formation of supernumerary wing structures produced by ZPA transplant experiments on chickens (see Fig. 10.12).

Review Questions

1. B
2. D
3. A
4. C
5. E
6. D
7. A tear of the amnion during the chorionic villus sampling procedure could have resulted in an amniotic band wrapping around the digits and strangulating their blood supply, thereby causing the tips to degenerate and fall off.

8. This defect is unlikely to be related to the amniocentesis procedure because the morphology of the digits is well established by the time such a procedure is undertaken (usually around 15 to 16 weeks). The most likely cause is a genetic mutation.
9. Muscle-forming cells arise from the somites.
10. The immediate cause is likely the absence of programmed cell death in the interdigital mesoderm. The cause of the disturbance in cell death is currently not understood.

Chapter 11

Clinical Vignette

The most immediate problems are surgery to deal with (1) the open spinal cord and (2) the developing hydrocephalus. The surgery for the rachischisis first must be directed toward closing the open lesion to prevent infection and to prevent the leakage of cerebrospinal fluid. Later, surgery will likely be necessary for problems associated with traction on the spinal cord and spinal nerves as the child grows. The hydrocephalus is typically treated by implantation of a shunt to lead excess cerebrospinal fluid from the ventricular system of the brain. Patency of the shunt must be maintained.

In addition to surgery, this infant will face many problems associated with impaired function of the lower spinal nerves. Impaired urinary bladder function is common among such infants, as is impaired mobility of the lower extremities. Infection is a constant threat because of the problem with containment and circulation of the cerebrospinal fluid in the spinal cord lesion. Children with various forms of spina bifida typically require intensive physical therapy for a variety of problems. Clogging of the shunt leading from the ventricular system is a recurring threat. In this patient, the thinning of the walls of the brain indicates that the brain tissue itself has been compromised. Some element of mental retardation and the associated educational and socialization problems represent other components of the problem.

Even relatively simple cases of spina bifida pose many chronic problems. The total annual medical, rehabilitation, and education costs of treatment for one individual are often significant. In addition to these are the stresses in the family, which is faced with a continuing need to care for the affected individual. There is a significantly higher than normal divorce rate among parents of children with chronic problems resulting from congenital malformations. A strong support network is important for successfully dealing with children with spina bifida conditions.

Review Questions

1. E
2. A
3. D
4. B
5. D
6. C
7. C
8. B
9. Congenital megacolon (Hirschsprung's disease), in which a segment of large intestine develops without parasympathetic ganglia. Intestinal contents cannot actively move through such an aganglionic segment.
10. The nerves would be hypoplastic (much smaller than normal), and the spinal cord would be thinner than normal in the area from which the nerves supplying the affected limb arise. The likely cause is excessive neuronal cell death because of the absence of an end organ for many of the axons that normally supply the limb.

Chapter 12

Clinical Vignette

With the diagnosis of immunodeficiency along with the infant's outflow tract defect of the heart, the pediatrician's differential diagnosis included DiGeorge's syndrome. This was confirmed when the blood levels of parathyroid hormone were found to be low. The cause of the infant's problem probably goes back to the fourth week of pregnancy or possibly earlier, when the cranial neural crest supplying the outflow tract of the heart and pharynx was migrating or preparing to migrate into the affected regions.

Review Questions

1. E
2. B
3. C
4. D
5. A
6. B
7. C
8. D
9. Along the length of the spinal cord, migrating neural crest cells are funneled into the anterior sclerotomal region of the somites and are excluded from the posterior half. This results in the formation of a pair of ganglia for each vertebral segment and space between ganglia in the craniocaudal direction.
10. Cranial crest cells can form skeletal elements; trunk crest cells cannot. Migrating cranial neural crest cells have more morphogenetic information encoded in them than trunk crest cells do. (For example, craniocaudal levels are specified in cranial crest, whereas they are not fixed in trunk crest cells.) Cranial crest cells form large amounts of dermis and other connective tissues, whereas trunk crest cells do not.

Chapter 13

Clinical Vignette

1. The common embryological denominator is a deficit in the neural crest associated with the first pharyngeal arch. The first arch gives rise to the lower jaw, much of the middle ear complex, and a significant part of the external ear.
2. Because of the statistical association between abnormalities of the external ear and kidney defects, the physician wants to be sure that there are no underlying abnormalities of the urinary system.

Review Questions

1. D
2. B

3. A
4. D
5. C
6. Coloboma of the iris is caused by failure of the choroid fissure to close during the sixth week of pregnancy. Because the area of the defect remains open when the rest of the iris constricts in bright light, excessive unwanted light can enter the eye through the defect.
7. Some of the secretions of the lacrimal glands enter the nasolacrimal ducts, which carry the lacrimal fluid into the nasal cavity.
8. Hyaluronic acid. Migration of neural crest cells into the developing cornea occurs during a period when large amounts of hyaluronic acid have been secreted into the primary corneal stroma.
9. During the fetal period, the middle ear cavity is filled with a loose connective tissue that dampens the action of the middle ear ossicles. After birth, the connective tissue is resorbed.
10. Similar to the lower jaw, much of the external ear arises from tissue of the first arch bordering the first pharyngeal cleft.

Chapter 14

Clinical Vignette

The woman had a hormone-secreting thyroid adenoma that gave her the symptoms of hyperthyroidism. Because the radioactivity was concentrated at the base of her tongue, its location suggested that the tumor formed within a remnant of thyroid tissue left behind at the beginning of the pathway of thyroid tissue migration from its site of origin at the midline base of the future tongue.

Review Questions

1. B
2. B
3. C
4. E
5. A
6. One option is simply acne. Another more significant possibility is a branchial cyst. Branchial cysts are typically located along the anterior border of the sternocleidomastoid muscle. One possible reason for its late manifestation is that the same conditions that resulted in the boy's acne caused a simultaneous reaction in the epidermis lining the cyst.
7. First, all epithelium lining the cyst must be removed, or the remnants could reform into a new cyst, and the symptoms could recur. The surgeon also must determine that the cyst is isolated and not connected to the pharynx via a sinus, which would result from an accompanying persistence of the corresponding pharyngeal pouch.
8. Some of the secretions of the lacrimal glands enter the nasolacrimal ducts, which carry the lacrimal fluid into the nasal cavity.
9. By 10 weeks, all the processes of fusion of facial primordia have already been completed. The cause of the defects could almost certainly be attributed to something that influenced the embryo long before the time when the anticonvulsant therapy was initiated, probably before the seventh week of pregnancy.
10. These defects could be a manifestation of fetal alcohol syndrome. They could represent a mild form of holoprosencephaly, which in this case would relate to defective formation of the forebrain (prosencephalon). The defects in olfaction and in the structure of the upper lip could be secondary effects of a primary defect in early formation of the prosencephalon.

Chapter 15

Clinical Vignette

The girl had Meckel's diverticulum that contained ectopic endometrial tissue. When she had menstrual periods, the reaction of the ectopic endometrial tissue gave her upper abdominal cramps. After surgery, her symptoms disappeared.

Review Questions

1. E
2. A
3. C
4. C
5. B
6. A
7. Esophageal atresia or a tracheoesophageal fistula. In the former, the milk fills the blind esophageal pouch and then spills into the trachea via the laryngeal opening. In the latter, milk may pass directly from the esophagus into the trachea depending on the type of fistula.
8. Congenital pyloric stenosis. Projectile vomiting is a common symptom of this condition, and palpation of the knotted pyloric opening of the stomach confirmed the diagnosis.
9. The most likely diagnosis is a vitelline duct fistula connecting the midgut with the umbilicus. This allows some contents of the small intestine to escape through the umbilicus. Another possibility is a urachal fistula (see Chapter 16), which connects the urinary bladder to the umbilicus through a persistent allantoic duct. In this case, however, the escaping fluid would be urine and would likely not be accompanied by mucus.
10. Imperforate anus. When examining a newborn, clinicians must ensure that there is an anal opening.

Chapter 16

Clinical Vignette

1. To be diagnosed as a male by the sex chromatin test, none of the cells that were examined should have exhibited Barr bodies (condensed X chromosomes).
2. The most likely cause for her female phenotype is a lack of testosterone receptors (androgen insensitivity syndrome).
3. Her gonads would be male (internal testes). Because in the embryo her testes produced müllerian inhibitory substance, her paramesonephric ducts regressed. She formed no uterus, uterine tubes, or upper vagina. This is also the reason for her amenorrhea, although amenorrhea is common in female athletes who train intensively. Although her testes produce abundant testosterone, the lack of receptors resulted in the lack of differentiation of male genital duct structures or external genitalia that would normally be dependent on testosterone.

Review Questions

1. A
2. D
3. B
4. A
5. A
6. D
7. B
8. The most likely cause is a urachal fistula connecting the urinary bladder to the umbilicus and allowing the leakage of urine. This is caused by the persistence of the lumen in the distal part of the allantois.
9. Bilateral renal agenesis. The first clue was the mother's low weight gain, which could have been the result of oligohydramnios (although this is not the only cause for low weight gain during pregnancy). The infant's appearance showed many of the characteristics of Potter's syndrome, which is caused by intrauterine pressures on the fetus when the amount of amniotic fluid is very low.
10. In normal development, the caudal ends of the paramesonephric ducts swing toward the midline and fuse. In this patient's case, the point of fusion probably occurred more caudally than normal. This condition is compatible with a normal pregnancy and delivery, although in some cases, pain or problems with delivery can occur.

Chapter 17

Clinical Vignette

The physician suspected high coarctation of the aorta with accompanying patent ductus arteriosus. In high and low coarctation of the aorta, the pulse in the lower part of the body is commonly reduced. The cyanosis of the feet resulted from the spillage of venous blood into the systemic circulation through the patent ductus arteriosus.

Review Questions

1. A
2. D
3. C
4. D
5. E
6. A
7. C
8. The physician suspected that the boy had a double aortic arch or a right aortic arch, either of which can cause difficulty in swallowing (dysphagia) during childhood, especially when growth spurts are occurring. Another possibility for an embryologically based dysphagia is esophageal stenosis.
9. To survive, an individual with this defect would have to have a means for draining blood entering the left atrium and a means of transporting blood into the left ventricle or the systemic circulation. One combination that would compensate is an atrial septal defect, which allows incoming blood to escape from the left atrium, and an associated ventricular septal defect, which allows blood from the right ventricle to pass into the left ventricle. A patent ductus arteriosus added to the first two compensatory defects could also help to balance the circulation by adding blood to the systemic side of the circulation. This would not be very helpful in the physiological sense, however, because the added blood would be unoxygenated blood entering the aorta from the pulmonary artery.

10. Persistence of the caudal segment of the left supracardinal vein. Normally, the caudal segment of the right supracardinal vein persists and becomes incorporated into the inferior vena cava, and the corresponding segment of the left supracardinal vein disappears (see Fig. 17.12). A wide variety of anomalous patterns of the abdominal veins exist.

Chapter 18

Clinical Vignette

The fetus likely had an obstruction of the lower urinary tract (perhaps urethral valves) leading to megacystis (see Fig. 18.15). This condition is treated by inserting a shunt into the bladder, thus allowing the urine to escape into the amniotic fluid. If the shunt functions during the remainder of pregnancy, the actual problem with the urinary outlet can be treated surgically after birth. The placement of shunts into the fetal bladder is one of the more successful types of intrauterine surgery practiced to date.

Review Questions

1. C
2. A
3. B
4. E
5. B
6. D
7. Hyaline membrane disease of the newborn. If the infant was born prematurely, the fetal lungs had not produced sufficient pulmonary surfactant to support normal breathing.
8. This condition is the result of persistence of a patent ductus venosus and umbilical vein. When the individual strains, venous blood fills these vessels and small venous branches radiating from the umbilical area. This is reminiscent of the head of Medusa, with snakes taking the place of hairs.
9. One simple explanation is that with continued growth of the fetus, the extremities become so tightly packed in the uterus that there is little room for movement.
10. More refined imaging techniques, such as ultrasonography, that allow more accurate diagnosis in utero of congenital malformations; recognition of the exceptional healing powers of the fetus; improved intrauterine and extrauterine surgical techniques, some assisted by ultrasound visualization, which allows direct surgery on the fetus; and increased ability to forestall or prevent premature labor after surgery on the uterus.

Index

Page numbers followed by "f" indicate figures, "t" indicate tables, and "b" indicate boxes.